AF796697
WITHDRAWN

BIOTECHNOLOGY IN AGRICULTURE SERIES

General Editor: Gabrielle J. Persley, International Service for National Agricultural Research (The Hague, Netherlands), and Project Manager, World Bank/ISNAR/ACIAR/AIDAB Biotechnology Study.

For a number of years, biotechnology has held out the prospect for major advances in agricultural production, but only recently have the results of this new revolution started to reach application in the field. The potential for further rapid developments is however immense.

The aim of this new book series is to review advances and current knowledge in key areas of biotechnology as applied to crop and animal production, forestry and food science. Some titles will focus on individual crop species or groups of species, others on specific goals such as plant protection or animal health, with yet others addressing particular methodologies such as tissue culture, transformation or immunoassay. In some cases, relevant molecular and cell biology and genetics will also be covered. Issues of relevance to both industrialized and developing countries will be addressed, and social, economic and legal implications will also be considered. Most titles will be written for research workers in the biological sciences and agriculture, but some will also be useful as textbooks for senior-level students in these disciplines.

Editorial Advisory Board:
P.J. Brumby, formerly of the World Bank, Washington DC, USA.
E.P. Cunningham, FAO, Rome, Italy.
P. Day, Rutgers University, New Jersey, USA.
J.H. Dodds, International Potato Center (CIP), Peru.
J.J. Doyle, International Laboratory for Research on Animal Diseases, Nairobi, Kenya.
S.L. Krugman, United States Department of Agriculture, Forest Service.
W.J. Peacock, CSIRO, Division of Plant Industry, Australia.

Titles Available:

1: Beyond Mendel's Garden: Biotechnology in the Service of World Agriculture *G.J. Persley*
2: Agricultural Biotechnology: Opportunities for International Development *Edited by G.J. Persley*
3: The Molecular and Cellular Biology of the Potato *Edited by M.E. Vayda and W.D. Park*
4: Advanced Methods in Plant Breeding and Biotechnology *Edited by D.R. Murray*
5: Barley: Genetics, Biochemistry, Molecular Biology and Biotechnology *Edited by P.R. Shewry*
6: Rice Biotechnology *Edited by G.S. Khush and G.H. Toenniessen*
7: Plant Genetic Manipulation for Crop Protection *Edited by A. Gatehouse, V. Hilder, and D. Boulter*

Titles in Preparation

Biotechnology of Perennial Fruit Crops *Edited by F. Hammerschlag and R. Litz*

Non-Food Uses of Agricultural Raw Materials: Economics, Biotechnology and Politics *C. Spelman*

Barley:

Genetics, Biochemistry, Molecular Biology and Biotechnology

Edited by

Peter R. Shewry

Department of Agricultural Sciences,
University of Bristol,
Long Ashton Research Station,
Long Ashton,
Bristol BS18 9AF,
UK

C·A·B International

QK
495
.G74
B322
1992

C·A·B International
Wallingford
Oxon OX10 8DE
UK

Tel: Wallingford (0491) 32111
Telex: 847964 (COMAGG G)
Telecom Gold/Dialcom: 84: CAU001
Fax: (0491) 33508

© C·A·B International 1992. All rights reserved. No part of this publication may be reproduced in any form or by any means, electronically, mechanically, by photocopying, recording or otherwise, without the prior permission of the copyright owners.

A CIP catalogue record for this book is available from the British Library

ISBN 0 85198 725 7
ISSN 0960 202X

Printed and bound in the UK by The Alden Press Ltd, Oxford

Contents

Contributors ix

Preface xiii

Section I : Origin, Evolution and Wild Relatives

1. The Wild Species of *Hordeum*: Relationships and Potential Use for Improvement of Cultivated Barley 3
Roland von Bothmer

2. Origin, Evolution, Population Genetics and Resources for Breeding of Wild Barley, *Hordeum spontaneum*, in the Fertile Crescent 19
Eviatar Nevo

3. Intergeneric Hybrids with *Hordeum* 45
George Fedak

Section II : Genetics

4. Cloned and Mapped Genes: Current Status 73
Penny von Wettstein-Knowles

5. Progress in the Production of Wheat–Barley Addition and Recombination Lines and their use in Mapping the Barley Genome 99
K.W. Shepherd & A.K.M.R. Islam

6. DNA Marker Techniques for Genetic Analysis in Barley 115
D.A. Laurie, J.W. Snape & M.D. Gale

7. Nuclear Genome Structure and Organization 133
E. Ananiev

Section III : Analysis of Metabolism and Development

8. Molecular Analysis of Barley Chloroplast Proteins 153
Jens Sigurd Okkels

9. Analysis of Barley Metabolism using Mutant Genes 181
P.J. Lea, R.D. Blackwell & R.A. Azevedo

10. Molecular Analysis of Nitrate Metabolism 209
Andris Kleinhofs & Robert L. Warner

11. Genetics of Barley Development: Mutant Phenotypes and Molecular Aspects 231
Gerd Bossinger, Wolfgang Rohde, Udda Lundqvist & Francesco Salamini

12. The *Slender* Mutation of Barley 265
Christopher J. Pollock, Helen J. Ougham & John L. Stoddart

13. Biochemical and Molecular Studies of Stress Tolerance in Barley 277
A.M. Stanca, V. Terzi & L. Cattivelli

Section IV : Seed Development, Composition, Germination and Utilization

14. Grain Structure and Composition 291
C.M. Duffus & M.P. Cochrane

15. The Control of Protein Synthesis in Developing Barley Seeds 319
Martin Kreis & Peter R. Shewry

16. Trypsin/α-Amylase Inhibitors and Thionins: Possible Defence Proteins from Barley 335
F. García-Olmedo, G. Salcedo, R. Sanchez-Monge, C. Hernandez-Lucas, M.J. Carmona, J.J. Lopez-Fando, J.A. Fernandez, L. Gomez, J. Royo, F. García-Maroto, A. Castagnaro & P. Carbonero

17. Nutritional Aspects of Barley Seed Structure and Composition 351
C.W. Newman & R.K. Newman

18. Barley Germination: Biochemical Changes and Hormonal Control 369
Dennis E. Briggs

19. Gibberellin Responses in Barley 403
Peter M. Chandler

20. Cell Wall Metabolism in Barley 413
Geoffrey B. Fincher

Section V : Pathogen Resistance

21. Sources and Genetics of Resistance to Fungal Pathogens 441
J. Helms Jørgensen

22. Biochemical and Molecular Analyses of the Response of Barley to Infection by Powdery Mildew 459
Tomas Bryngelsson & David B. Collinge

23. The Molecular Analysis of Barley Resistance to Powdery Mildew 481
K.J. Scott

24. Strategies for Cloning Disease Resistance Genes 497
Shauna Somerville

Section VI : Biotechnology

25. Haploid Production: Approaches and Use in Plant Breeding 519
R.A. Pickering & P. Devaux

26. Regeneration, Stability and Transformation of Barley 549
Angela Karp & Paul A. Lazzeri

27. The Case of High-lysine Barley Breeding 573
L. Munck

Index 603

Contributors

Chapter 1

Roland von Bothmer
The Swedish University of Agricultural Sciences, Department of Crop Genetics and Breeding, S-268 00 Svalöv, Sweden

Chapter 2

Eviatar Nevo
Institute of Evolution, University of Haifa, Mount Carmel, Haifa 31999, Israel

Chapter 3

George Fedak
Plant Research Centre, Agriculture Canada, Building 50, CEF, Ottawa, Ontario K1A 0C6, Canada

Chapter 4

Penny von Wettstein-Knowles
Carlsberg Laboratory, Department of Physiology, Gamle Carlsberg Vej 10, DK-2500 Copenhagen Valby, Denmark

Chapter 5

K.W. Shepherd and A.K.M.R. Islam
Department of Agronomy, Waite Agricultural Research Institute, Glen Osmond, South Australia 5064, Australia

Chapter 6

D.A. Laurie, J.W. Snape & M.D. Gale
AFRC Institute of Plant Science Research, Cambridge Laboratory, Colney Lane, Norwich NR4 7UJ, UK

Chapter 7

E.V. Ananiev
N.I. Vavilov Institute of General Genetics, USSR Academy of Sciences, Gubkin St. 3, 117809 Moscow B-333, USSR

Chapter 8

Jens Sigurd Okkels
Plant Biochemistry Laboratory, Department of Plant Biology, Royal Veterinary and Agricultural University, 40 Thorvaldsensvej, DK-1871 Frederiksberg C, Denmark

Chapter 9

P.J. Lea, R.D. Blackwell & R.A. Azevedo
Division of Biological Science, University of Lancaster, Institute of Environmental & Biological Science, Lancaster LA1 4YQ, UK

Chapter 10

Andris Kleinhofs
Department of Crop and Soil Sciences and Program in Genetics & Cell Biology, Washington State University, Pullman, Washington 99164-6420, USA

Robert L. Warner
Department of Crop and Soil Sciences, Washington State University, Pullman, Washington 99164-6420, USA

Chapter 11

Gerd Bossinger, Wolfgang Rohde & Francesco Salamini
Max-Planck-Institut für Züchtungsforschung, Erwin-Baur Institut, D-5000 Köln 30, Germany

Udda Lundqvist
Svalöf AB, 268 00 Svalöf, Sweden

Chapter 12

Christopher J. Pollock, Helen J. Ougham & John L. Stoddart
AFRC Institute for Grassland and Environmental Research, Welsh Plant Breeding Station, Plas Gogerddan, Aberystwyth, Dyfed SY23 3EB, UK

Chapter 13

A.M. Stanca, V. Terzi & L. Cattivelli
Experimental Institute for Cereal Research, Via S. Protaso 308, I-29017 Fiorenzuola d'Arda (PC), Italy

Chapter 14

C.M. Duffus
Crop Sciences Division, The Scottish Agricultural College, West Mains Road, Edinburgh EH9 3JG, UK

M.P. Cochrane
International Centre for Brewing and Distilling, Heriot-Watt University, Riccarton, Edinburgh EH14 4AS, UK

Chapter 15

Martin Kreis
Université de Paris-XI, Institut de Recherches sur les Plantes, Biologie du Developpement des Plantes, Bâtiment 430, F-91405 Orsay Cedex, France

Peter R. Shewry
Department of Agricultural Sciences, University of Bristol, AFRC Institute of Arable Crops Research, Long Ashton Research Station, Bristol BS18 9AF, UK

Chapter 16

F. García-Olmedo, G. Salcedo, R. Sanchez-Monge, C. Hernandez-Lucas, M.J. Carmona, J.J. Lopez-Fando, J.A. Fernandez, L. Gomez, J. Royo, F. García-Maroto, A. Castagnaro & P. Carbonero
Cátedra de Bioquímica y Biología Molecular, Escuela Tecnica Superior de Ingenieros Agronomos, E-28040 Madrid, Spain

Chapter 17

C.W. & R.K. Newman
Montana Agricultural Experiment Station, Bozeman, Montana 59717-0290, USA

Chapter 18

Dennis E. Briggs
Birmingham Malting and Brewing Group, University of Birmingham, Department of Biochemistry, PO Box 363, Birmingham B15 2TT, UK

Chapter 19

Peter M. Chandler
CSIRO Division of Plant Industry, GPO Box 1600, Canberra, ACT 2601, Australia

Chapter 20

Geoffrey B. Fincher
Department of Biochemistry, La Trobe University, Bundoora, Victoria 3083, Australia

Chapter 21

J. Helms Jørgensen
Plant Biology Section, Risø National Laboratory, Postboks 49, DK-4000 Roskilde, Denmark

Chapter 22

Tomas Bryngelsson
The Swedish University of Agricultural Sciences, Department of Crop Genetics and Breeding, S-268 00 Svalöv, Sweden

David B. Collinge
Section for Plant Pathology, Department of Plant Biology, The Royal Veterinary and Agricultural University, Thorvaldsensvej 40, 1871 Fredericksberg C, Copenhagen, Denmark

Chapter 23

K.J. Scott
Department of Biochemistry, University of Queensland, St Lucia, Queensland 4072, Australia

Chapter 24

Shauna Somerville
DOE-Plant Research Laboratory, Michigan State University, East Lansing, Michigan 48824, USA

Chapter 25

R.A. Pickering
Crop Research Division, DSIR, Private Bag, Christchurch, New Zealand

P. Devaux
Florimond Desprez, BP 41, Cappelle-en-Pévèle, 59242 Templeuve, France

Chapter 26

Angela Karp
AFRC Institute of Arable Crops Research, Long Ashton Research Station, Bristol BS18 9AF, UK

Paul A. Lazzeri
Institut für Allgemeine Botanik und Botanischer Garten, Universitat Hamburg, Ohnhorstrasse 18, D-2000 Hamburg 52, Germany

Chapter 27

L. Munck
Department of Biotechnology, Carlsberg Research Laboratory, Gamle Carlsberg Vej 10, DK-2500 Valby, Copenhagen, Denmark

Preface

Barley is not only the fourth most important cereal crop in the world, but also an excellent experimental system for biochemists, physiologists, geneticists and molecular biologists. The reasons for this should become clear to readers of the individual chapters, but important features are the ease of growth under laboratory conditions, and its inbred diploid nature which facilitates the isolation of mutants and construction of genetic maps. There are, of course, also drawbacks, most notably its genomic complexity and intransigence to genetic engineering. However, it is certainly less problematical in these respects than wheat (and probably also rye), and is therefore a good model species for the Triticeae.

The present book provides an up-to-date view of many diverse aspects of barley research, through a mixture of detailed review articles and shorter accounts of specific topics in which exciting advances are being made. It is not intended to be exhaustive, but rather to provide interesting and stimulating reading for young researchers as well as experienced barley workers. I hope the reader will share my view that barley is indeed the system of choice for many studies.

I am particularly indebted to Dr Penny von Wettstein-Knowles for advice on designations of barley genes and to Mrs Pat Baldwin for secretarial assistance.

Peter R. Shewry

I

Origin, Evolution and Wild Relatives

1

The Wild Species of *Hordeum*: Relationships and Potential Use for Improvement of Cultivated Barley

Roland von Bothmer

Plant breeding uses selected germplasm for incorporation of new variation or of defined agronomic traits like disease resistance into commercial varieties. New sources of genes are becoming increasingly important to reduce the input of agrochemicals in agriculture and to manipulate qualitative traits. In this respect wild relatives of crops have received increased attention as gene donors in recent years. Earlier difficulties in gene transfer using conventional methods can now be overcome by the development of new techniques, which also make more distantly related species accessible for gene transfer.

The present chapter aims to give a comprehensive overview of the genus *Hordeum*, and the relationships of the species, which include cultivated barley, to summarize our current knowledge of agronomic traits and the applications of *in vitro* techniques to wild species. For more detailed information on the individual species the reader is referred to Bothmer & Jacobsen (1985), and Bothmer *et al.* (1991).

The genus *Hordeum*

Taxonomy

Cultivated barley, *H. vulgare*, belongs to the tribe Triticeae in the grass family, Poaceae. The Triticeae is a temperate plant group and is distributed in most areas in the world but with a main centre in Central and Southwestern Asia. The common temperate annual cereal crops, wheat (*Triticum aestivum*), rye (*Secale cereale*) and triticale, are also placed in this tribe together with several very important perennial forage grasses in many temperate areas of the world, for example, Russian wild rye (*Psathyrostachys fragilis*), the crested wheatgrasses (*Agropyron cristatum* s.l.), Altai wild rye (*Leymus angustus*) and intermediate wheatgrass (*Thinopyrum intermedium*). Apart from the economically important cereals and forages, the tribe Triticeae also contains about 350 wild species and due to the great possibilities for interspecific and intergeneric hybridization

Table 1.1. The species/taxa of *Hordeum*. The basic chromosome number is x = 7. A, annual; P, Perennial. *H. *arizonicum* is sometimes biennial or even with a shorter life cycle

Species/taxon	Ploidy level	Distribution	Life form
H. vulgare L. subsp. *vulgare*	2x(4x)	Cultivated	A
subsp. *spontaneum* (C. Koch) Thell.	2x	SW Asia	A
H. bulbosum L.	2x,4x	Mediterranean	P
H. murinum L. subsp. *murinum*	4x	Mediterranean	A
subsp. *glaucum* (Steud.) Tzvelev	2x	Mediterranean	A
subsp. *leporinum* (Link) Arc.	4x,6x	Mediterranean	A
H. marinum Huds. subsp. *marinum*	2x	Mediterranean	A
subsp. *gussoneanum* (Parl.) Thell.	2x,4x	Mediterranean	A
H. muticum Presl	2x	South America	P
H. cordobense Bothmer *et al.*	2x	South America	P
H. chilense Roem. & Schult.	2x	South America	P
H. stenostachys Godr.	2x	South America	P
H. erectifolium Bothmer *et al.*	2x	South America	P
H. flexuosum Steud.	2x	South America	P
H. euclaston Steud.	2x	South America	A
H. intercedens Nevski	2x	W USA	A
H. pusillum Nutt.	2x	North America	A
H. jubatum L.	4x	North America–East Asia	P
H. comosum Presl	2x	South America	P
H. pubiflorum Hook. f.	2x	South America	P
H. lechleri (Steud.) Schenck	6x	South America	P
H. procerum Nevski	6x	South America	P
H. arizonicum Covas & Stebbins	6x	SW USA	A/P*
H. secalinum Schreb.	4x	Europe–(North Africa)	P
H. bogdanii Wil.	2x	Central Asia	P
H. roshevitzii Bowden	2x	Central Asia	P
H. brevisubulatum (Trin.) Link subsp. *brevisubulatum*	2x,4x	E Asia	P
subsp. *violaceum* (Boiss. & Hohen.) Tzvelev	2x,4x	SW Asia	P
subsp. *turkestanicum* (Nevski) Tzvelev	6x	Central Asia	P
subsp. *nevskianum* (Bowden) Tzvelev	4x	Central Asia	P
subsp. *iranicum* Bothmer	6x	W Asia	P
H. brachyantherum Nevski	4x,6x	W North America–E Asia	P
H. californicum Covas & Stebbins	2x	W USA	P
H. depressum (Scribn. & Sm.) Rydb.	4x	W USA	A
H. guatemalense Bothmer *et al.*	4x	Central America	P
H. capense Thunb.	4x	South Africa	P
H. parodii Covas	6x	South America	P
H. tetraploidum Covas	4x	South America	P
H. fuegianum Bothmer *et al.*	4x	South America	P
H. patagonicum (Haum.) Covas subsp. *patagonicum*	2x	South America	P
subsp. *mustersii* (Nicora) Bothmer *et al.*	2x	South America	P
subsp. *santacrucense* (Parodi & Nicora) Bothmer *et al.*	2x	South America	P
subsp. setifolium (Parodi & Nicora) Bothmer *et al.*	2x	South America	P
subsp. *magellanicum* (Parodi & Nicora) Bothmer *et al.*	2x	South America	P

all of these are of putative interest as gene donors in breeding. Before the wild taxa can be utilized more generally for breeding and for genetic engineering more basic research on their taxonomy, distribution, variation patterns, species relationships, and cytogenetic behaviour needs to be carried out. Some of the genera in Triticeae are fairly well known, especially the species of *Aegilops* and *Triticum*. The knowledge of *Hordeum* is also rapidly increasing.

As most scientists involved in studies of the Triticeae have noticed, conflicting opinions on generic delimitations have caused major disagreements between taxonomists over the last couple of decades and so far no universal acceptance of generic boundaries has been achieved.

The generic circumscription of *Hordeum* has been rather stable over the years and most scientists agree on the species content of the genus. However, there has been a suggestion to separate cultivated barley, *H. vulgare*, together with *H. bulbosum* into a genus of its own (*Hordeum* in a narrow sense). All other species would then be referred into a separate genus, *Critesion* (cf. Dewey, 1984; Löve, 1984). This view does not reflect the actual relationships and it has not been widely accepted. With the traditional view of the genus (which is retained in this presentation) *Hordeum* comprises about 30 species (45 taxa/cytotypes, see Table 1.1).

The morphology of *Hordeum* is very specialized in comparison with most other genera in the Triticeae. The unifying morphological characters for the *Hordeum* species are the one-flowered spikelets which are borne three together at each rachis (main axis of the spike) node (so-called 'triplets'). The central spikelet is sessile (except in *H. murinum*) and both male and female fertile. The two lateral spikelets, which are stalked except in cultivated barley, where they are sessile, may be fertile (as in six-rowed barley, Fig. 1.1.) or sterile (as in two-rowed barley, Fig. 1.1). In many of the wild species the lateral spikelets are often rudimentary and thus sterile. The glumes, which in most other Triticeae species are broad and cover the base (or even more or less the whole) spikelet, are reduced and situated on the dorsal side of each spikelet. In most species the glumes are setaceous (thread-like), but in some species, e.g. *H. vulgare*, *H. bulbosum*, and *H. pusillum*, the glumes are flattened at the base. Most species of *Hordeum* have long awns on the lemmas of the spikelets as well as on the glumes.

Fig. 1.1. A. Two-rowed barley. B. Six-rowed barley. The arrows indicate the sterile and fertile lateral spikelets in A and B, respectively.

The taxonomic delimitations between the species are mainly clear-cut. However, due to lack of morphological features the species may sometimes be difficult to distinguish for a non-specialist. Four sections have been established based on morphology (see Bothmer *et al.*, 1991).

Life forms and reproductive patterns

Unlike most genera in the Triticeae, *Hordeum* comprises both annual and perennial species. Annual species like *H. vulgare*, *H. marinum* and *H. murinum* have developed different life forms responding to the geographical area of distribution. For example, rapid life cycles (summer annuals) developed in many parts of the dry Eastern Mediterranean. The progenitor form of cultivated barley, *H. vulgare* subsp. *spontaneum*, may have winter and summer annual forms depending on the local climate, but the different life forms may also be present even within the same population, thus showing large genetic diversity (Giles & Lefkovitch, 1984). The annuals are mainly inbreeders, although some forms of subsp. *spontaneum* and *H. murinum* may have rather open flowers and larger anthers and are thus adapted to a higher rate of outbreeding (Brown *et al.*, 1978). Cultivated barley is estimated to have around 1% outcrossing per generation (Allard, 1988). Some annuals are more specialized self-fertilizers (e.g. *H. marinum* and *H. pusillum*) but none of them is obligate inbreeding.

The perennials are very variable in their breeding behaviour. Among them are specialized outbreeders with a two-locus self-incompatibility system, namely *H. bulbosum* and *H. brevisubulatum* (Lundqvist, 1962; Bothmer, 1979; Dewey, 1979). Some species like *H. tetraploidum* from South America and *H. secalinum* from Europe have large, protruding anthers, large feather-like stigmas and very open flowers at anthesis and are mainly out-breeding although they have no self-incompatibility system (Bothmer & Jacobsen, 1980; Bothmer *et al.*, 1986a). The other perennial species have smaller floral parts, are not so open-flowering and have a higher rate of self-pollination.

The seed-dispersal differs among species. *H. vulgare*, *H. bulbosum*, and *H. murinum* have larger seeds than the other species. They have special, tough, straight hairs (bristles) on the glumes, awns and rachis internodes which make them adhere well to the furs of larger animals, the feathers of birds, and to the clothing of humans, and the seeds get dispersed in this way. The other extreme is illustrated by, for example, *H. jubatum* from North America, *H. lechleri* and *H. comosum* from South America. These species have very light seeds, with long, setaceous awns and glumes which bend out at a right angle at seed maturity. These species are thus wind-dispersed and the light seed with its flying apparatus may 'sail' on upwinds for large distances.

Cultivated barley, *H. vulgare* subsp. *vulgare*, and the Asiatic perennial species *H. bogdanii*, have a tough rachis. The single seed, enclosed in the lemma and palea, is broken off at the base at seed maturity. All other species have a brittle rachis, i.e. at seed maturity the whole rachis disarticulates at the nodes (shattering). The dispersal unit in this case is not a single seed but a triplet of spikelets together with one rachis internode. However, the genetic background for brittle versus tough rachis is very simple (two genes; see Takahashi, 1972; Konishi, 1988) and a very strong selection for the mutant forms (or recombinants) with tough rachis was made during the early phase of the domestication process of barley.

Another important feature for the domestication, which also affects the reproductive and dispersal patterns, is the selection for non-dormancy of seeds resulting in high and even germination which is of great importance for an efficient agriculture. In the wild form, subsp. *spontaneum*, the seed germination is highly irregular whereas in cultivated barley more than 90% of all seeds germinate within 4 days of imbibition.

Ecology

The *Hordeum* species occur in a wide array of habitats. The majority of the wild, perennial species grow in rather moist environments like grassy meadows and pastures, along streams, lake shores, marshes, etc. The annuals are mostly restricted to open habitats and ephemeral areas, such as roadsides, with comparatively low competition from other species. Some annuals are adapted to very special habitats; for example, *H. intercedens* in California is restricted to vernal pools, sets seeds and dies off very early in the season when the pools dry up. Some species, like *H. bogdanii* in Central Asia, some forms of *H. patagonicum* in southern South America, and *H. depressum*, endemic to western USA, grow mainly in saline environments, at lakes, streams, etc. *H. patagonicum* subsp. *magellanicum* occurring in Tierra del Fuego and adjacent mainland grows mainly on the sandy shores at the Atlantic coast. Many species have adapted to extreme environments. For example, *H. comosum* has a very well developed root system and grows on the extremely dry steppe sides on the Andes. Tolerance to cold environments in many species is prevalent. In Central Asia barley is cultivated at very high altitudes; up to 4400 m has been observed by the author in Tibet. The perennial species *H. brevisubulatum* subsp. *nevskianum* grows up to an altitude of 5000 m in the Himalayas and *H. comosum* is recorded up to 4000 m in the Andes.

Distribution

The genus has centres of diversity in central and southwestern Asia, western North America, southern South America, and in the Mediterranean. The highest number of species occur in these areas. Three species have particularly developed as noxious weeds in many parts of the world, namely the two annual, originally Mediterranean species, *H. murinum* and *H. marinum*, the latter in somewhat more humid environments, and the perennial North American species, *H. jubatum*.

Most species have distinct, continuous distribution areas within a continent and larger disjunctions are rare. However, *H. brachyantherum* and *H. jubatum* occur in western North America and extend their range to Kamtchatka, via the Aleutian islands, and the former species also has a small, disjunct area in easternmost Canada. *H. brevisubulatum* extends from western Turkey to eastern China. One interesting biogeographic connection is between the single, endemic South African species *H. capense* and its closest relative, the European (North African) *H. secalinum* (Bothmer & Jacobsen, 1980).

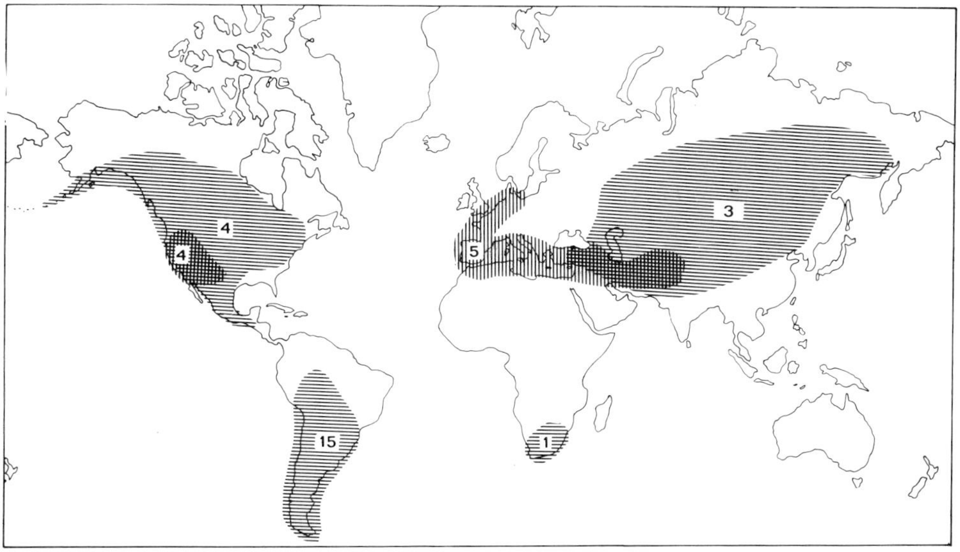

Fig. 1.2. Major distribution area of the genus *Hordeum*. The figures indicate number of species in each area. Only native distributions are shown.

The relationships among species

During recent years a number of studies of relationships among the *Hordeum* species have been published, using different methodological approaches. The general patterns seem to agree fairly well among investigations.

In larger crossing programmes Rajhathy *et al.* (1964) and Bothmer & Jacobsen (1986) produced a number of the possible hybrid combinations in the genus. The intraspecific hybrids were generally fully fertile, for example within *H. marinum* and *H. brevisubulatum*, *H. patagonicum* (Landström *et al.*, 1984; Bothmer *et al.*, 1986b, 1989a). All interspecific combinations were seed and pollen sterile except the tetraploid combination *H. jubatum* × *H. brachyantherum*, and the hexaploid combination *H. lechleri* × *H. procerum* which showed occasional seed set.

Meiotic investigations of *Hordeum* hybrids reveal a rather clear pattern regarding the genomic system in the genus. The hybrid of *H. vulgare* and *H. bulbosum* has high meiotic pairing and the two species are considered to share the common basic genome, called I (Kasha and Sadasivaiah, 1971; Lange, 1971; Bothmer *et al.*, 1983, 1986c). This genome is not related to any other genomes in the genus. *H. bulbosum* occurs in two cytotypes, a diploid in the western part of the Mediterranean and a tetraploid in the eastern part (Katznelson & Zohary, 1967; Jørgensen, 1982). The 4*x* cytotype is an autotetraploid as shown by the high frequency of multivalents (thus with the genome composition II).

Of the two annual Mediterranean species, *H. marinum* and *H. murinum*, the former is diploid and tetraploid, while the latter occurs in diploid, tetraploid and hexaploid cytotypes. The two species are distantly related to each other and to the I genome species (*H. vulgare* and *H. bulbosum*) as well as to the other *Hordeum* species (Bothmer *et al.*, 1986c, 1988, 1989a, b). The genomes of these two annuals thus seem not to be closely related to the other genomes in *Hordeum* and have been primarily designated as 'X' (*H. marinum*) and 'Y' (*H. murinum*) by Bothmer *et al.* (1986c, 1987). In contrast with studies of karyotypes and C-bands (Linde-Laursen & Bothmer, 1989), and isoenzyme variation (Jaaska & Jaaska, 1986; Jørgensen, 1986) which show differentiation within the genus (i.e. an alloploid origin), the cytogenetic studies show a close relationship between the genomes of the polyploid cytotypes within *H. marinum* and *H. murinum*, respectively (Pohler *et al.*, 1986; Bothmer *et al.*, 1988, 1989a, b). The two species would thus be more of an autoploid nature.

The remaining, diploid *Hordeum* species (see Table 1.1) are all closely related, sharing the same basic genome (H), which is somewhat differentiated between American and Asiatic taxa (Bothmer *et al.*, 1986c). Nearly all the polyploids are segmental alloploids in which the genomes are closely related. The H genome and differentiations of this constitute the basis (Bothmer *et al.*, 1988, 1989b, c). *H. brevisubulatum* is a polyploid complex where the 4*x* and 6*x* cytotypes are of a true autoploid nature (Bothmer, 1979; Dewey, 1979; Bothmer *et al.*, 1988, 1989b, c).

The general pattern of the relationships as revealed by the genome studies is mainly confirmed by studies of the isoenzyme variation (Jørgensen, 1986), chloroplast DNA (Doebley *et al.*, in press), C-banding patterns (see Linde-Laursen *et al.*, 1980, 1989, 1990), and repeated sequences of nuclear DNA (Gupta *et al.*, 1989).

There have been several theories concerning the evolution and domestication of cultivated barley. The current view is that the earliest agriculture including barley started about 10 000 years ago in the area of Southwestern Asia called 'The Fertile Crescent' (Harlan, 1968; Zohary, 1973). The immediate progenitor of the crop, i.e. the two-rowed form *H. vulgare* subsp. *spontaneum*, is conspecific to cultivated barley (*H. vulgare* subsp. *vulgare*) and is still growing in this area in natural habitats as well as in ruderate ground (Giles & Bothmer, 1985; Nevo, this volume).

Disease resistance and agricultural traits

Further information on the agricultural traits of wild barleys is required to facilitate their utilization in conventional plant breeding or in biotechnology. Disease resistant has so far attracted most interest but other characters may also prove to be important.

Powdery mildew

Powdery mildew, caused by the biotrophic fungus *Erysiphe graminis* f. sp. *hordei*, is one of the most important diseases of barley on a world-wide basis and it is the most severe disease in northern temperate areas. In Sweden the average loss in yield due to the attack of powdery mildew is estimated to be 5–10% annually (Sundell, 1977). It is of great importance to find new sources of resistance for breeding since the pathogen usually easily overcomes single-gene resistances. Scientists and breeders have therefore looked to wild species of *Hordeum* for new sources of resistance. A substantial amount of research and breeding has been carried out on host–parasite interactions, the genetics of resistance and virulence and the utilization in breeding of cultivated barley as well as its wild progenitor *H. vulgare* subsp. *spontaneum* (see Segal *et al.*, 1978; Moseman *et al.*, 1983; Nevo, this volume). There is a vast knowledge of the genetic basis for resistance and some genes for resistance to powdery mildew have been transferred into breeding lines of barley (see Lehmann & Bothmer, 1988).

Powdery mildew is also the most studied pathogen in wild species of *Hordeum*, but no successful transfers to cultivated barley have been reported despite many attempts.

Hordeum bulbosum has long been known to contain several interesting agronomic characters, especially resistance to powdery mildew (Kuckuck, 1934; Hardison, 1944; Eshed & Wahl, 1970, 1975; Kasha & Sadasivaiah, 1971). Jones & Pickering (1978) studied the variation in resistance to powdery mildew in *H. bulbosum*. Isolates of mildew from two susceptible genotypes of *H. bulbosum* were found to be avirulent to more than 30 genotypes of cultivated barley. Some of the most susceptible barley lines (e.g. cvs Proctor and Golden Promise) were resistant to the *H. bulbosum* forms of mildew. One genotype of *H. bulbosum* was resistant to five isolates of powdery mildew of *H. vulgare* with most of the common virulence factors. The conclusions of the study were that the mildews from *H. bulbosum* and *H. vulgare*, respectively, are widely different rather than closely related.

Six tetraploid strains of *H. bulbosum* also showed large variation between individual plants in their resistance/susceptibility among plants against three Japanese races of powdery mildew. This reflects heterogeneity in populations of *H. bulbosum* due to the outbreeding habit of the species (Prasad *et al.*, 1983).

Giles & Barrett (1983) investigated 25 accessions of *H. murinum* subsp. *murinum* from nine countries. In a greenhouse experiment the accessions were exposed to three isolates of *Erysiphe graminis* f. sp. *hordei* from the UK, containing a total of five virulence factors. The majority of the reactions were of resistant type and only two populations showed a susceptible reaction. When exposed to powdery mildew of agricultural origin, *H. murinum* showed high levels of resistance, but there was variation between populations both in type of resistance and timing of its expression during plant development.

Giles & Barrett (1983) also performed a second experiment where they inoculated 20 individuals from each of four *H. murinum* populations with two different races of *Erysiphe graminis* originating from a bulk collection from Israel. There was great variation in resistance reactions, mainly between populations. When *H. murinum* is exposed to powdery mildew of wild origin it exhibits higher levels of susceptibility than when exposed to isolates of cultivated origin. Isolates from cultivated barley have probably been exclusively selected for performance on the crop whereas the Israeli isolates may have been exposed to a wider range of host species and hence may have a wider host-range adaptation (see Wahl *et al.*, 1978).

In an extensive study of resistance of wild species of *Hordeum* to *Erysiphe graminis* f. sp. *hordei*, Gustafsson & Claesson (1988) screened 155 populations representing 28 *Hordeum* species. They used four cultures of the powdery mildew, the isolates being chosen because they exhibited virulence to most known resistance genes. Apart from the two completely susceptible cultivars of cultivated barley (cvs Tellus and Vada) and a single susceptible accession of *H. marinum*, all populations of the wild species showed immune or resistant reactions.

Gustafsson & Claesson (1988) also made a study of the infection process where they compared reactions in susceptible barley with those in more or less resistant or immune wild species. The frequency of conidial germination varied significantly between the species, from 35% in *H. brachyantherum* up to 79% in *H. jubatum*. Appressoria were almost always formed when conidia had germinated, and most of them developed infection hyphae, which could penetrate the epidermis. There was a significant difference between the susceptible cultivars of *H. vulgare*, in which up to 85% of the germinated spores formed well-developed primary haustoria. The variation in the wild species was 2–39%. Differences between compatible and incompatible reactions were most pronounced for penetration of secondary hyphae and formation of secondary haustoria. Very few abnormal haustoria were formed in *H. vulgare* subsp. *spontaneum*, *H. bulbosum* and *H. procerum*. All the other wild species had no secondary haustoria. The susceptible barley varieties had 20 and 60% formation of haustoria, respectively (Fig. 1.3). The authors concluded that the resistance reactions in the wild species are due to a combination of different characters, for example, variation in the waxy layer of leaves, variation in pubescence of leaves, and variation in biochemical factors.

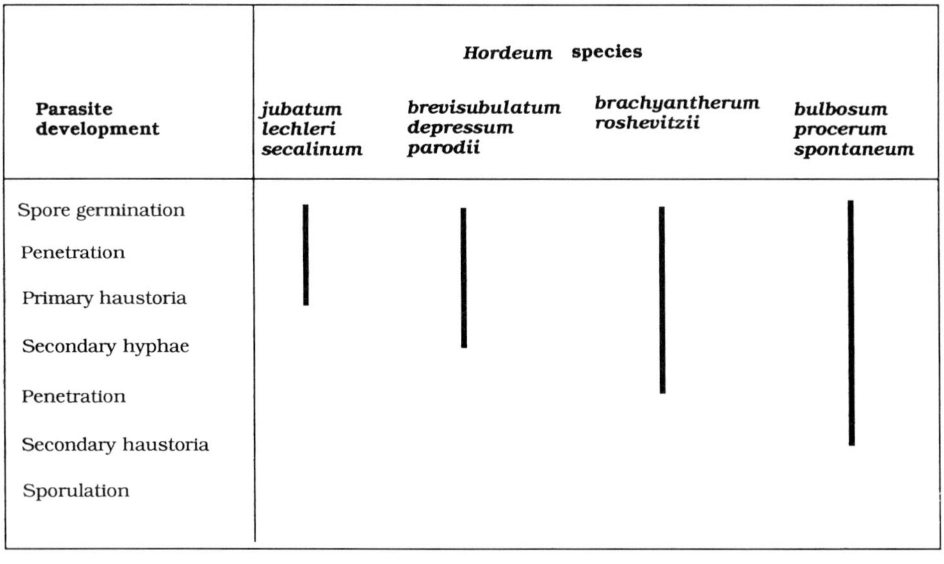

Fig. 1.3. Development of the fungus *Erysiphe graminis* f. sp. *hordei* (powdery mildew) on different *Hordeum* species (after Gustafsson & Claesson, 1988).

Other pathogens

Typhula blight is caused by two species of a parasitic fungus, namely *Typhula incarnata* and *T. ischicariensis* (Mielke, 1978). Both are severe pathogens on winter barley and the disease is one of the major obstacles to increased cultivation of winter barley in more northern areas. It is thus a component of winter hardiness, which apart from cold tolerance also includes resistance to several different pathogens. Mielke (1980) studied the resistance against *T. incarnata* in eight *Hordeum* species, including *H. vulgare* cv. Dura of German origin as a standard, during two successive years. *H. bogdanii*, *H. jubatum*, and *H. brevisubulatum* subsp. *violaceum* and subsp. *turkestanicum* were especially free of infection after 2 years. *H. vulgare* subsp. *spontaneum* had a similar degree of infection as the *vulgare*-line 'Dura'. In a larger study P. Lundin (cited by Bothmer & Hagberg, 1983, and unpublished) screened 28 *Hordeum* species, with at least 2–3 populations of each, for resistance to Typhula blight. There was particularly good resistance in some of the American species. The most prominent resistance reactions were found in some populations of *H. lechleri* and *H. pubiflorum* from Argentina.

Loose smut is a world-wide disease on barley and wheat caused by the two embryo-infecting fungi *Ustilago nuda* and *U. tritici*. The disease can also be frequently found attacking wild species of *Hordeum* under natural conditions in almost all parts of the world (author, personal observations). Nielsen (1978, 1987) investigated resistance to the disease in 19 *Hordeum* species, many of them with 2–3 accessions. He found good resistance to both species of *Ustilago* in *H. murinum*, *H. chilense*, *H. marinum*, and *H. depressum*, and several other species were resistant to one of the pathogens. Nielsen (1987) concluded that outside *Hordeum* only species of *Agropyron* (*Elymus*) are susceptible to both fungal pathogens. During the evolution of *Hordeum* many of the species retained susceptibility to both the ancestral *U. tritici* and to the co-evolving *U. nuda*.

Different species of aphids are regular and severe parasites of spring barley in many areas and screening for sources of resistance is urgently required. Annual losses have been estimated at 10% of the total yield in Scandinavia (Sundell, 1977). Some studies have also been performed on wild *Hordeum* species. Weibull (1987) studied resistance reactions against *Rhophalosiphum padii* in 27 accessions of 10 perennial *Hordeum* taxa and 6 interspecific F_1 hybrids between wild species and cultivated barley. Susceptibility was expressed as the total number of aphids per plant and per original female aphid after 16 days of exposure. The mean number of nymphs produced per female was only on average about 14% of that in the control barley cv. Tellus (206 nymphs per plant), with *H. bogdanii* being the most resistant species (4 nymphs per plant) and one accession of *H. lechleri* the most susceptible (58 nymphs per plant). The most resistant species were all diploids (*H. bogdanii*, *H. californicum*, and *H. comosum*) with the interspecific hybrids being more or less intermediate between the parents.

A current study at ICARDA in Syria showed that one accession of *H. murinum* subsp. *murinum* had moderate tolerance and *H. secalinum* excellent tolerance to greenbug infections (Dr R.H. Miller, ICARDA, personal communication).

Resistances to other pathogens which have been studied so far include barley yellow dwarf virus (BYDV) (Schooler, 1980; Schooler & Franckowiak, 1981) and yellow rust in *H. vulgare* subsp. *spontaneum* (Anikster *et al.*, 1975). Xu & Snape (1989) screened *H. bulbosum* lines and found resistance to both yellow and brown rusts and to powdery mildew. The pathogenicity of 182 single spore isolates of *Rhyncosporium secalis* from *H. leporinum* (= *H. murinum* subsp. *leporinum*) on 15 barley varieties was tested by Brown (1990): 45% of the isolates were pathenogenic on 5 or more varieties.

Quality and other traits

Since barley is the major grain used for production of malt the malting quality is very important. Lines of the Japanese barley cultivar Satsukiyo containing the cytoplasm of *H. bulbosum* were produced through the chromosome elimination system. These alloplasmic lines generally had inferior malting quality, but the kernels of 2.5–2.8 mm in length had similar quality to the original line (Aida *et al.*, 1981).

Other potentially useful traits present in wild species may be cytoplasmic male sterility (Schooler, 1967), leaf pubescence (as a mechanical resistance character), outcrossing habit, protein content, lysine content and salt tolerance (see Schooler, 1974; Orton, 1979). Gorham *et al.* screened 13 species of wild *Hordeum* for their accumulation of Na at low external salt concentrations and found that all except *H. secalinum* differed from *H. vulgare* (Dr J. Gorham, personal communication; Gorham *et al.*, 1990).

Application of *in vitro* techniques to wild *Hordeum* species

There are significant difficulties for interspecific gene transfer through sexual hybridization with barley (Bothmer *et al.*, 1983; Fedak, 1985 and this volume; Bothmer & Linde-Laursen, 1989; Bothmer & Jacobsen, 1990). One way to overcome these strict sterility barriers would be to pass through a cycle of *in vitro* cultivation which may induce various kinds of somaclonal variants. In callus of primary sterile interspecific hybrids there may be possibilities for induction of translocations or partial elimination of alien chromosomes with a subsequent establishment of substitution or addition lines.

Rather few investigations of *in vitro* techniques have, however, been carried out on wild species of *Hordeum*. Studies have mainly been made to induce somaclonal variation after cycles of *in vitro* culture and successive regeneration of plants, either in the individual species or in interspecific or intergeneric hybrids.

In a number of articles Orton (1979, 1980a, b, c) studied the effects of tissue cultures on *H. vulgare*, *H. jubatum*, and the interspecific hybrid between the two species. He found that the individual species as well as their hybrid exhibited similar patterns of media-induced callus growth and regeneration. Immature ovaries generated the most vigorous and easily regenerated calluses compared with several other explanted tissues (Orton, 1979). Immature embryos have also been found by others to be the most suitable tissue for callus induction and plant regeneration (Orton, 1980a, for *H. vulgare* × *H. jubatum*; Breiman, 1985, for *H. vulgare* subsp. *spontaneum* and *H. bulbosum*; Jørgensen & Bothmer, 1988, for *H. vulgare* × *H. marinum*; Pickering, 1989, for *H. vulgare* × *H. bulbosum*). Immature spikes have also been used with good results (Jørgensen *et al.*, 1986, for several interspecific hybrids; Pershina *et al.*, 1989, for *H. marinum* × *Triticum aestivum*).

The regeneration capacity of *H. bulbosum* and *H. vulgare* subsp. *spontaneum* was investigated by Breiman (1985). No significant differences were observed between the two species with respect to their ability to form callus or to regeneration capacity and differentiation in shoots and roots. A low percentage of albinos were observed among the regenerated plants, but most of the mature plants were morphologically similar to the donor plants and were fertile.

The regeneration capacity of a callus is usually lost after a certain time. In Breiman's study (1985) the regeneration capacity was lost after 6 months and in the study by Jørgensen & Andersen (1989) no plants regenerated after 9 months as callus.

Pershina *et al.* (1989) used young inflorescences to produce calluses of the hybrid *H. geniculatum* (= *H. marinum*) × *Triticum aestivum*. Among the regenerants were plants with different levels of female fertility, and seeds were produced after backcrossing to common wheat. Some regenerants were treated with colchicine to induce chromosome doubling, producing partially self-fertile plants.

The regeneration of callus and suspension cultures of immature embryos of *H. jubatum*, *H. marinum*, and *H. vulgare* cv. Borwina was studied by Omelianchuk, Novosibirsk, USSR (personal communication). All species performed well in callus cultures, with 80–90% callus formation and regeneration. However, only *H. marinum* regenerated in suspension cultures, forming dividing cells and some embryoids which developed into mature but sterile plants.

The induction of callus and regenerative capacity are highly influenced by environmental factors, for example media composition as shown by Orton (1979) (*H. vulgare* × *H. jubatum*) and Breiman (1985) (*H. vulgare* subsp. *spontaneum* and *H. bulbosum*), and genotype as shown by Pickering (1989) (*H. vulgare* × *H. bulbosum*). Jørgensen *et al.* (1986) found that the capacity for plant regeneration is highly dependent on the wild parental species. Out of interspecific hybrids with cultivated barley and 10 wild species the authors found that combinations including *H. jubatum*, *H. roshevitzii*, *H. lechleri* and *H. procerum* regenerated particularly high numbers of plants from calluses.

Jørgensen & Andersen (1989) investigated primary hybrids between cultivated barley and seven wild species (*H. brachyantherum*, *H. jubatum*, *H. roshevitzii*, *H. tetraploidum*, *H. brevisubulatum*, *H. lechleri*, and *H. procerum*), one triple hybrid ((*H. lechleri* × *H. procerum*) × *H. vulgare*) and one first backcross to cultivated barley of *H. lechleri* × *H. vulgare*. The phase of callus culture induced a wide array of morphological and cytological variation as observed in the 1315 regenerated plants. Some regenerants looked more like *H. vulgare* than the primary hybrid while others were more like the wild species parent. There was also a significant increase in the proportion of variant plants (karyotypic and morphological variants) after about 9 months in callus culture. The numerical chromosome changes comprised aneuploids with loss of individual chromosomes from either cultivated barley (chromosomes 5 or 7) or from the wild species parent. Seven amphiploids were regenerated from combinations with *H. jubatum* and *H. lechleri*. Structural rearrangements were found: deletions, inversions (in *H. roshevitzii* × *H. vulgare*), duplications and one possible case of translocation (*H. lechleri* × *H. vulgare*). The regenerated plants were pollen and seed sterile like the primary hybrids, except for a few plants which were probably doubled haploids (of *H. lechleri*) arising from elimination of the barley chromosomes. Many regenerated plants were aneusomatic with normal cells together with aberrant ones. Of the cytologically studied material more than 50% of the plants had abnormal chromosome constitutions.

Haploids of barley were obtained by Orton (1980b) after a phase of callus culture of *H. vulgare* × *H. jubatum*; 5–10% of all regenerants were haploids resembling the *H. vulgare* parent. Some of the regenerants also exhibited activity of specific *H. jubatum* esterases, indicating that chromosomal or subchromosomal substitutions or additions may have occurred *in vitro*. An interesting feature was that all *H. vulgare*-like regenerants (thus more or less barley haploids) could be maintained as perennial stocks, like the *H. jubatum* parent and in contrast to the annual *H. vulgare*.

A high frequency of chromosomal variation was also observed by Orton (1980a) in the hybrid of *H. vulgare* × *H. jubatum*. He found a high frequency of spontaneous polyploidy, aneuploidy and chromosomal rearrangements in both callus and suspension cultures of the pure species and of the hybrid. The chromosomal aberrations were highly dependent on the differentiation stage, age and history of the culture. The genome interactions of the parental species were not responsible for the chromosomal variability. By applying colchicine in a low concentration Orton & Steidl (1980) tried to induce the regeneration of chromosome doubled, fertile amphidiploids. Around 20% of the regenerated plants were aneuploids with a mean of $2n = 37$, and about 40% were stable and fertile amphidiploids of *H. vulgare* × *H. jubatum*.

Pickering (1989) studied regeneration from callus and somaclonal variation in several interspecific hybrid combinations including *H. vulgare* × *H. bulbosum*. There was large genotypic variation in regeneration capacity among the barley cultivars as well as among the *H. bulbosum* genotypes. He found 0–33% of spontaneous doubling and all (except one amphidiploid plant) had $2n = 14$, thus being doubled haploids of cultivated barley. Eleven regenerants from *vulgare–bulbosum* (VB) callus had haploid and hybrid sectors. Fertility was restored after colchicine treatment. Apart from several cases of polyploidy and albinism, Pickering (1989) identified three variants, one of which showed maternal inheritance for yellow leaf striping, while the other two were controlled by single recessive genes. One of these resembled a chlorina mutant and the other possessed an increased resistance to scald (*Rhyncosporium secalis*).

In vitro cultures may be ideal systems for physiological studies of stress factors of agricultural importance. They may also be used as rapid and comparatively cheap methods for screening and selection. Orton (1980c) used callus cultures for studies of salt tolerance in *H. vulgare* and *H. jubatum*. Growth and ion accummulation were measured under different salt concentrations. *H. jubatum* showed a considerably higher tolerance to high salt concentrations in comparison with cultivated barley. The tolerance is expressed at the tissue level and *H. jubatum* cultures were able to adjust to the salt stress up to 0.86 M, while the cultivated barley cultures did not recover after treatment with 0.17 M NaCl.

Current use and potentialities in the wild species of *Hordeum*

The wild species of *Hordeum* have so far played only a minor role in barley improvement. *H. vulgare* subsp. *spontaneum* belongs to the primary gene pool of barley, has no biological isolation barriers for crosses with the cultivated form and hybrids between them are formed in nature under sympatric conditions (Harlan, 1976; Asfaw & Bothmer, 1990). There are thus good prospects for gene transfer and subsp. *spontaneum* has great potential for immediate use in barley breeding for agronomic characters such as disease resistance, and drought and salt tolerance (Segal *et al.*, 1987; Lehmann & Bothmer, 1988).

Of the other wild *Hordeum* species greatest interest has been shown in *H. bulbosum*, especially for utilization of the selective chromosome elimination system. This has been widely used in breeding for the production of homozygous barley lines ('doubled haploids'), the so-called, bulbosum method (see Pickering & Devaux, this volume). Interest has also been focused on *H. bulbosum* for transfer of desirable genes. This species is the sole representative in the secondary gene pool of barley, and it is more compatible in crosses with cultivated barley than the remaining wild species. For this reason *H. bulbosum* has been investigated in more detail than the other *Hordeum* species. However, despite the high crossability and high meiotic pairing among the parental genomes, no fertile gametes have so far been reported in the primary hybrid between the two species at the diploid level (Pickering, 1988).

Another wild species of *Hordeum* which has been of considerable interest in recent years is the South American diploid perennial *H. chilense*. It is readily crossable with different species of *Triticum* and it was reported by Martin (1988) to have a great agricultural potential. Both hexaploid (e.g. with *T. turgidum* and *T. dicoccoides*) and octaploid (with *T. aestivum* cv. Chinese Spring) fertile amphiploids have been produced, which all have higher protein content than the wheat parents and small but not shrivelled kernels (see Martin & Chapman, 1977; Chapman & Miller, 1978; Martin & Cubero, 1981; Thomas & Pickering, 1985). Additions of *H. chilense* chromosomes to *T. turgidum* convar.

durum have been reported by Fernandez & Jouve (1988) who found 50 monosomic or polysomic addition forms. Addition and substitution lines in *T. aestivum* cv. Chinese Spring were also reported by Miller *et al.* (1981, 1985).

Conclusions

The general pattern for interspecific hybridization shows strong incompatibility barriers, usually as F_1 sterility, between cultivated barley and the wild species. So far, no breakthrough of significance for breeding has been reported using conventional sexual crossing techniques (see Bothmer *et al.*, 1983, 1989; Fedak, 1985 and this volume; Bothmer & Jacobsen, 1990). A better understanding of the incompatibility mechanisms and genetic variation for the characters of interest as well as the development of new procedures for gene transfer are necessary for the future utilization of wild *Hordeum* species.

Acknowledgements

Prof. Peter Shewry kindly gave valuable comments on the manuscript. I am most grateful to several scientists the unpublished results of whom I was allowed to cite in this paper.

References

Aida, Y., Muramatsu, M., Fukui, T. & Tsubuko, S. (1981) Malting quality of a barley cultivar with *Hordeum bulbosum* cytoplasm. *Bulletin of Brewing Sciences* 27, 17–21.

Allard, R.W. (1988) Genetic changes associated with the evolution of adaptedness in cultivated plants and their wild progenitors. *Journal of Heredity* 79, 225–38.

Anikster, Y., Moseman, J.G. & Wahl, I. (1975) Parasite specialization of *Puccinia hordei* Otth. and sources of resistance in *Hordeum spontaneum* C. Koch. *Barley Genetics 3* (Proc. 3rd Int. Barley Genet. Symp., Garching, 1975), 412–17.

Asfaw, Z. & Bothmer, R. von (1990) Hybridization between landrace varieties of Ethiopian barley (*Hordeum vulgare* subsp. *vulgare*) and the progenitor of barley (*H. vulgare.* subsp. *spontaneum*). *Hereditas* 112, 57–64.

Bothmer, R. von (1979) Revision of the Asiatic taxa of *Hordeum* sect. *Stenostachys. Botanisk Tidsskrift* 74, 117–47.

Bothmer, R. von & Hagberg, A. (1983) Prebreeding and wide hybridization in barley. *Genetika (Beograd)* 15, 185–97.

Bothmer, R. von & Jacobsen, N. (1980) A taxonomic revision of *Hordeum secalinum* and *H. capense. Botanisk Tidsskrift* 74, 223–35.

Bothmer, R. von & Jacobsen, N. (1985) Origin, taxonomy and related species. In: Rasmusson, D.C. (ed.), *Barley*. American Society of Agronomists, pp. 19–56.

Bothmer, R. von & Jacobsen, N. (1986) Interspecific crosses in *Hordeum* (Poaceae). *Plant Systematics and Evolution* 153, 49–64.

Bothmer, R. von & Jacobsen, N. (1990) Interspecific hybrids within the genus *Hordeum*. In: Gupta, P.K. & Tsuchyia, T. (eds), *Chromosome Engineering in Plants: Genetics, Breeding, Evolution. Part A*. Elsevier. Amsterdam, Oxford, New York, Tokyo, pp. 411–31.

Bothmer, R. von & Linde-Laursen, I. (1989) Backcrosses to cultivated barley (*Hordeum vulgare* L.) and partial elimination of alien chromosomes. *Hereditas* 111, 145–7.

Bothmer, R. von, Flink, J., Jacobsen, N., Kotimäki, M. & Landström, T. (1983) Interspecific hybridization with cultivated barley (*Hordeum vulgare* L.). *Hereditas* 99, 219–44.

Bothmer, R. von, Jacobsen, N. & Jørgensen, R.B. (1986a) Taxonomy variation and relationships in the *Hordeum parodii* group. *Nordic Journal of Botany* 6, 399–410.

Bothmer, R. von, Giles, B.E. & Jacobsen, N. (1986b) Crosses and genome relationships in the *Hordeum patagonicum* group. *Genetica* 71, 75–80.

Bothmer, R. von, Flink, J. & Landström, T. (1986c) Meiosis in interspecific *Hordeum* hybrids. I. Diploid combinations. *Canadian Journal of Genetics and Cytology* 28, 525–35.

Bothmer, R. von, Flink, J. & Landström, T. (1987) Meiosis in interspecific *Hordeum* hybrids. II. Triploid hybrids. *Evolutionary Trends in Plants* 1, 41–50.

Bothmer, R. von, Flink, J. & Landström, T. (1988) Meiosis in interspecific *Hordeum* hybrids. IV. Tetraploid ($4x \times 4x$) hybrids. *Genome* 30, 479–85.

Bothmer, R. von, Flink, J., Jacobsen, N. & Jørgensen, R.B. (1989a) Variation and differentiation in *Hordeum marinum*. *Nordic Journal of Botany* 9, 1–10.

Bothmer, R. von, Flink, J., Landström, T. & Thomas, H.M. (1989b) Meiosis in interspecific *Hordeum* hybrids. V. Pentaploid hybrids. *Hereditas* 110, 217–26.

Bothmer, R. von, Flink, J. & Landström, T. (1989c) Meiosis in interspecific *Hordeum* hybrids. VI. Hexaploid hybrids. *Evolutionary Trends in Plants* 3, 53–8.

Bothmer, R. von, Jacobsen, N., Baden, C., Linde-Laursen, I. & Jørgensen, R.B. (1991) An ecogeographical study of the genus *Hordeum*. *Systematic and Ecogeographic Studies on Crop Genepools* 7. International Board for Plant Genetic Resources, Rome.

Breiman, A. (1985) Plant regeneration from *Hordeum spontaneum* and *Hordeum bulbosum* immature embryo derived calli. *Plant Cell Reports* 4, 70–3.

Brown, A.H.D., Nevo, E. & Zohary, D. (1978) Outcrossing rates and hetereozygosity in natural populations of *Hordeum spontaneum* in Israel. *Heredity* 41, 49–62.

Brown, J.S. (1990) Pathogenic variation among isolates of *Rhynchosporium secalis* from barley grass growing in South Eastern Australia. *Euphytica* 50, 81–9.

Chapman, V. & Miller, T.E. (1978) The amphiploid of *Hordeum chilense* × *Triticum aestivum*. *Cereal Research Communication* 6, 351–2.

Dewey, D.R. (1979) The *Hordeum violaceum* complex of Iran. *American Journal of Botany* 66, 166–72.

Dewey, D.R. (1984) The genome system of classification as a guide to hybridization with the perennial *Triticeae*. In: Gustafsson, J.P. (ed.), *Gene Manipulation in Plant Improvement*. Plenum, New York, pp. 209–70.

Doebley, J., Bothmer, R. von & Larson, S. Chloroplast DNA variation and phylogeny in the genus *Hordeum* (Poaceae). *American Journal of Botany* (in press).

Eshed, N. & Wahl, I. (1970) Host ranges and interrelations of *Erysiphe graminis hordei*, *E. graminis tritici* and *E. graminis avenae*. *Phytopathology* 60, 628–34.

Eshed, N. & Wahl, I. (1975) Role of wild grasses in epidemics of powdery mildew on small grains in Israel. *Phytopathology* 65, 57–62.

Fedak, G. (1985) Wide crosses in *Hordeum*. In: Rasmusson, D.C. (ed.), *Barley*. ASA no. 26, Agronomy, pp. 156–89.

Fernandez, J.A. & Jouve, N. (1988) The addition of *Hordeum chilense* chromosomes to *Triticum turgidum* conv. *durum*. Biochemical, karyological and morphological characterization. *Euphytica* 37, 247–59.

Giles, B.E. & Barrett, J. (1983) *Erysiphe graminis* resistance in the *Hordeum murinum* complex. *Barley Genetics Newsletter* 13, 78–82.

Giles, B.E. & Bothmer, R. von (1985) The progenitor of barley (*Hordeum vulgare* ssp. *spontaneum*) – its importance as a gene source. *Sveriges Utsädesförenings Tidskrift* 95, 53–61.

Giles, B.E. & Lefkovitch, L.P. (1984) Differential germination in *Hordeum spontaneum* from Iran and Morocco. *Zeitschrift für Pflanzenzüchtung* 92, 234–8.

Gorham, J., Bristol, A., Young, E.M., Wyn Jones, R.G. & Kashour, G. (1990) Salt tolerance in the *Triticeae*: K/Na discrimination in barley. *Journal of Experimental Botany* 41, 1095–1101.

Gupta, P.K., Fedak, G., Molnar, S.J. & Wheatcroft, R. (1989) Distribution of a *Secale cereale* DNA repeat sequence among 25 *Hordeum* species. *Genome* 383–8.

Gustafsson, M. & Claesson, L. (1988) Resistance to powdery mildew in wild species of barley. *Hereditas* 108, 231–7.

Hardison, J.R. (1944) Specialization of pathogenicity in *Erysiphe graminis* on wild and cultivated grasses. *Phytopathology* 34, 1–20.

Harlan, J.R. (1968) On the origin of barley. In: *Barley: Origin, Botany, Culture, Winterhardiness, Genetics, Utilization, Pests*. USDA Agricultural Handbook No. 338, pp. 9–31.

Harlan, J. (1976) Barley *Hordeum vulgare* (Gramineae-Triticinae). In: Simmonds, N.W. (ed.), *Evolution of Crop Plants*. Longman, Essex, pp. 93–8.

Jaaska, V. & Jaaska, V. (1986) Isoenzyme variation in the barley genus *Hordeum*. I. Alcohol dehydrogenase and superoxide dismutase. *Biochemie und Physiologie der Pflanzen* 181, 301–20.

Jones, I.T. & Pickering, R.A. (1978) The mildew resistance of *Hordeum bulbosum* and its transference into *H. vulgare* genotypes. *Annals of Applied Biology* 88, 295–8.

Jørgensen, R.B. (1982) Biosystematics of *Hordeum bulbosum* L. *Nordic Journal of Botany* 2, 421–34.

Jørgensen, R.B. (1986) Relationships in the barley genus (*Hordeum*): An electrophoretic examination of proteins. *Hereditas* 104, 273–91.

Jørgensen, R.B. & Andersen, B. (1989) Karyotype analysis of regenerated plants from callus cultures of interspecific hybrids of cultivated barley (*Hordeum vulgare* L.). *Theoretical and Applied Genetics* 77, 343–51.

Jørgensen, R.B. & Bothmer, R. von (1988) Haploids of *Hordeum vulgare* and *H. marinum* from crosses between the two species. *Hereditas* 108, 207–12.

Jørgensen, R.B., Jensen, C.J., Andersen, B. & Bothmer, R. von (1986) High capacity of plant regeneration from callus of interspecific hybrids with cultivated barley (*Hordeum vulgare* L.). *Plant Cell, Tissue and Organ Culture* 6, 199–207.

Kasha, K.J. & Sadasivaiah, R.S. (1971) Genome relationships between *Hordeum vulgare* L. and *H. bulbosum* L. *Chromosoma* 35, 264–87.

Katznelson, J. & Zohary, D. (1967) Diploid and tetraploid *Hordeum bulbosum. Israel Journal of Botany* 16, 57–62.

Konishi, T. (1988) Genetic differentiation and geographical distribution of barley. *Crop Genetics Resources of East Asia* 22, 237–43.

Kuckuck, H. (1934) Artkreuzungen bei Gerste. *Der Züchter* 6, 270–3.

Landström, T., Bothmer, R. von & Dewey, D.R. (1984) Genomic relationships in the *Hordeum brevisubulatum* complex. *Canadian Journal of Genetics and Cytology* 26, 569–77.

Lange, W. (1971) Crosses between *Hordeum vulgare* and *H. bulbosum*. 1. Production, morphology and meiosis of hybrids, haploids and dihaploids. *Euphytica* 20, 14–29.

Lehmann, L. & Bothmer, R. von (1988) *Hordeum spontaneum* and landraces as a gene resource for barley breeding. In: Jorn, M. & Slootmaker, C.A.J. (eds), *Cereal Breeding Related to Integrated Cereal Production*. pp. 190–4.

Linde-Laursen, I. & Bothmer, R. von (1989) Giemsa C-banded karyotypes of *Hordeum marinum* and *H. murinum. Genome* 32, 629–39.

Linde-Laursen, I., Bothmer, R. von & Jacobsen, N. (1980) Giemsa C-banding in asiatic taxa of *Hordeum* section *Stenostachys* with notes on chromosome morphology. *Hereditas* 93, 235–54.

Linde-Laursen, I., Bothmer, R. von & Jacobsen, N. (1989) Giemsa C-banded karyotypes of South American *Hordeum* (Poaceae). I. Diploid taxa. *Hereditas* 110, 289–305.

Linde-Laursen, I., Bothmer, R. von & Jacobsen, N. (1990) Giemsa C-banded karyotypes of South and Central American *Hordeum* (Poaceae). II. 6 polyploid taxa. *Hereditas* 112, 93–107.

Löve, A. (1984) Conspectus of the *Triticeae. Feddes Repertorium* 95, 425–521.

Lundqvist, A. (1962) Self-incompatibility in diploid *Hordeum bulbosum* L. *Hereditas* 48, 138–52.

Martin, A. (1988) Tritordeum: the first ten years. *Rachis* 7, 12–15.

Martin, A. & Chapman, V. (1977) A hybrid between *Hordeum chilense* and *Triticum aestivum. Cereal Research Communication* 5, 365–8.

Martin, A. & Cubero, J.I. (1981) The use of *Hordeum chilense* in cereal breeding. *Cereal Research Communication* 9, 317–23.

Mielke, H. (1978) Untersuchungen zur Schadwirkung von *Typhula incarnata* Lasch ex Fr. an Wintergerstensorten. *Journal of Agronomy and Crop Science* 147, 161–79.

Mielke, H. (1980) Untersuchungen über die Anfälligkeit verschiedener Wildgerstenarten gegenüber *Typhula incarnata. Nachrichtenblatt Deutches Pflanzenschutzdienst (Braunschweig)* 32, 39.

Miller, T.E., Reader, S.M. & Chapman, V. (1981) The addition of *Hordeum chilense* chromosome to wheat. In: *Induced variability in Plant Breeding*, (Int. Symp. Eucarpia, Pudoc, Wageningen), pp. 79–81.

Miller, T.E., Reader, S.M. & Ainsworth, C.C. (1985) A chromosome of *Hordeum chilense* homoeologous to group 7 of wheat. *Canadian Journal of Genetics and Cytology* 27, 101–4.

Moseman, J.G., Nevo, E. & Zohary, D. (1983) Resistance of *Hordeum spontaneum* collected in Israel to infection with *Erysiphe graminiis hordei. Crop Science* 23, 1115–19.

Nielsen, J. (1978) Host range of the smut species *Ustilago nuda* and *Ustilago tritici* in the tribe Triticeae. *Canadian Journal of Botany* 56, 901–15.

Nielsen, J. (1987) Reaction of *Hordeum* species to the smut fungi *Ustilago nuda* and *U. tritici. Canadian Journal of Botany* 65, 2024–27.

Orton, T.J. (1979) A quantitative analysis of growth and regeneration from tissue cultures of *Hordeum vulgare, H. jubatum* and their interspecific hybrid. *Environmental and Experimental Botany* 19, 319–35.

Orton, T.J. (1980a) Chromosomal variability in tissue cultures and regenerated plants of *Hordeum. Theoretical and Applied Genetics* 56, 101–12.

Orton, T.J. (1980b) Haploid barley regenerated from callus cultures of *Hordeum vulgare* × *H. jubatum*. *Journal of Heredity* 71, 280–2.

Orton, T.J. (1980c) Comparison of salt tolerance between *Hordeum vulgare* and *H. jubatum* in whole plants and callus cultures. *Zeitschrift für Pflanzenphysiology* 98, 105–18.

Orton, T.J. & Steidl, R.P. (1980) Cytogenetic analysis of plants regenerated from colchicine-treated callus cultures of an interspecific *Hordeum* hybrid. *Theoretical and Applied Genetics* 57, 89–95.

Pershina, L., Numerova, O.M., Belova, L.J. & Derjatkina, E.P. (1989) Cultivation of callus tissues for multiplication of barley × wheat hybrids. *Izvestija Sibirskogo Otdelenija Academii NAUK SSSR* 1, 27–32 (in Russian).

Pickering R.A. (1988) The attempted transfer of disease resistance from *Hordeum bulbosum* L. to *H. vulgare*. *Barley Genetics Newsletter* 18, 5–8.

Pickering, R.A. (1989) Plant regeneration and variants from calli derived from immature embryos of diploid barley (*Hordeum vulgare* L.) and *H. vulgare* L. × *H. bulbosum*. L. crosses. *Theoretical and Applied Genetics* 78, 105–12.

Pohler, W., Clauss, E. & Staat, B. (1986) Confirmation of genetic diploidization in *Hordeum geniculatum* 4x by meiotic investigation of hybrid plants. *Archiv für Züchtungsforschung* 16, 91–4.

Prasad, G., Yasuda, S. & Konishi, T. (1983) Reaction of *Hordeum bulbosum* L. to Japanese races of powdery mildew (*Erysiphe graminis* f. sp. *hordei*). *Barley Genetics Newsletter* 13, 60–2.

Rajhathy, T., Morrison, J.W. & Symko, S. (1964) Interspecific and intergeneric hybrids in *Hordeum*. *Barley Genetics* 1, 195–212.

Schooler, A.B. (1967) A form of male sterility in barley hybrids. *Journal of Heredity* 58, 207–11.

Schooler, A.B. (1974) Progress report on interspecific hybrids of *Hordeum*. *Barley Newsletter* 18, 49–51.

Schooler, A.B. (1980) Intergeneric and interspecific barley hybrids show tolerance to barley yellow dwarf virus. *North Dakota Farm Research* 38, 19–21.

Schooler, A.B. & Franckowiak, J.D. (1981) Registration of North Dakota 497 and 586 barley germplasm. *Crop Science* 21, 352.

Segal, A., Dörr, K.H., Fischbeck, G., Zohary, D. & Wahl, I. (1987) Genotypic composition and mildew resistance in a natural population of wild barley, *Hordeum spontaneum*. *Plant Breeding* 99, 118–27.

Sundell, B. (1977) Pests in agriculture. Part 1: economic evaluation of crop losses. *Report 109, Dept. of Economics and Statistics, The Agricultural College of Sweden*. IV, 119 pp.

Takahashi, R. (1972) Non-brittle rachis 1, bt(-Bt 2) and non-brittle rachis 2, bt 2-(Bt). *Barley Genetics Newsletter* 2, 181–2.

Thomas, H.M. & Pickering, R.A. (1985) Comparisons of the hybrids *Hordeum chilense* × *H. vulgare*, *H. chilense* × *H. bulbosum*, *H. chilense* × *Secale cereale* and the amphidiploid of *H. chilense* × *H. vulgare*. *Theoretical and Applied Genetics* 69, 519–22.

Wahl, I., Eshed, N., Segal, A. & Sobel, Z. (1978) Significance of wild relatives of small grains and other wild grasses in cereal powdery mildews. In: Spencer, D.M. (ed.), *The Powdery Mildews*. Academic Press, London, pp. 83–100.

Weibull, J. (1987) Screening for resistance against *Rhopalosiphum padi* (L.). 2. *Hordeum* species and interspecific hybrids. *Euphytica* 36, 571–6.

Xu, J. & Snape, J.W. (1989) The resistance of *Hordeum bulbosum* and its hybrids with *H. vulgare* from common fungal pathogens. *Euphytica* 41, 273–6.

Zohary, D. (1960) Studies of the origin of cultivated barley. *Bulletin of Research Council Israel, D* 9, 21–42.

Zohary, D. (1973) The origin of cultivated cereals and pulses in the near east. *Chromosomes Today* 4, 307–20.

2

Origin, Evolution, Population Genetics and Resources for Breeding of Wild Barley, *Hordeum spontaneum*, in the Fertile Crescent

Eviatar Nevo

Cultivated barley: origins and domestication

Cultivated barley, *Hordeum vulgare* L., is one of the main cereals of the belt of Mediterranean agriculture, a founder crop of Old World Neolithic food production and one of the earliest crops domesticated (Harlan & Zohary, 1966; Harlan, 1968; Zohary & Hopf, 1988). It is an important crop, ranking fifth in world crop production. In order of importance barley is used for: animal feed; brewing malts; and human food. Barley is a short season, early maturing grain with a high yield potential and may be found on the fringes of agriculture, in widely varying environments, including extremes of latitude and altitude where other crops are not adapted (Harlan, 1976a). It extends far into the Arctic, reaching the upper limit of cultivation in high mountains; it grows in desert oases and desert fringes, where it is more salt-and drought-tolerant than other cereals. Barley is a cool season crop; it can tolerate high temperatures if the humidity is low, but is not suited to warm humid climates. It is rarely grown in the tropics except in cool highlands, as in Mexico, the Andes and East Africa. Major production areas are Europe, the Mediterranean fringe of North Africa, Ethiopia, the Near East, Russia, China, India, Canada, the United States and Australia.

Barley was domesticated about 10 000 years ago in the Near East Fertile Crescent from brittle, two-rowed forms, morphologically identical with present-day *Hordeum spontaneum* (Zohary & Hopf, 1988). *Hordeum spontaneum* C. Koch is the wild ancestor of the cultivated barley (Harlan & Zohary, 1966; Zohary, 1969). It is considered systematically the wild race or subspecies of the cultivated crop, and is therefore designated *H. vulgare* L. subsp.

spontaneum (C. Koch) Korn (Zohary & Hopf, 1988; p. 56). I will treat it here traditionally, however, as *Hordeum spontaneum*, for the following reasons: (i) its ancestry and pre-domesticated genetic polymorphisms (Brown *et al.*, 1978a); (ii) its distinct morphogenetic identity (Baum, 1983; Chakrabarti & Subrahmanyan, 1986; Jaradat, 1989a, b); and (iii) its biological uniqueness in nature. These reasons may justify retaining its traditional biological species status. *H. spontaneum* has differentiated adaptively in Israel into four eco-geographical races from robust mesic plants to slender xeric plants (Snow & Brody, 1984). Barley has been extensively studied genetically at the isozyme level and the chromosomal localization of genes (Brown, 1983) has been facilitated by euplasmic wheat–barley chromosome addition lines (Islam *et al.*, 1981).

Origin, cytotaxonomy and evolution of wild barley

Wild barley, *H. spontaneum*, is an annual, brittle, two-rowed diploid ($2n = 14$) (Takahashi, 1955; Anonymous, 1968; Nilan, 1964, 1971). It is a predominantly self-pollinated, and colonizing species (Brown & Marshall, 1981) and the only wild *Hordeum* species that is cross-compatible and fully interfertile with the cultivated barley. Hybrids of *H. vulgare* × *H. spontaneum* show normal chromosome pairing and segregation in meiosis. There is also great morphological similarity between *spontaneum* and cultivated two-rowed *distichum* varieties. They differ mainly in their modes of seed dispersal. *H. spontaneum* ears are brittle and at maturity disarticulate into individual arrow-like triplets. These are highly specialized devices which ensure the survival of the plant in nature. Under cultivation, this specialization was replaced by non-brittle mutants which were immediately selected by man for reaping, threshing and sowing (Zohary & Hopf, 1988).

The genetic affinity between cultivated barley and wild *H. spontaneum* is also indicated by the spontaneous hybridization that occurs sporadically when wild and cultivated forms grow together. Some of these hybrids, combining brittle ears and fertile lateral spikelets, i.e. six-rowed plants, were erroneously regarded in the past as genuine wild-types designated *Hordeum agriocrithon* Aberg. Hybrid plants are fully fertile. In *H. spontaneum* the spikes fragment at maturity and the grains fall, while in the domesticated races, the spike is tough and the grains persist. The difference is controlled by either one or two tightly linked 'brittle' (*Bt* and Bt_1) genes. The wild-types are two-rowed, i.e. of the three spikelets at each node of the ear the two lateral ones are female sterile and only the central one develops to form a grain. Six-rowed races, in which all three spikelets produce grains, appeared about 8000 years ago under domestication (Helbaek, 1966). Two major genes are involved, both with multiple allelic series, but a single recessive mutation (*vv*) is sufficient to cause a two-rowed barley to become six-rowed. Additional studies by Gustafsson & Lundquist (1980), however, have shown that more than two genes are involved.

Wild barley has a two-rowed brittle rachis and a hulled (covered) grain (genetic constitution *Bt V N*). In cultivated naked barley, which also appeared 8000 years ago (Helbaek, 1966), the husks do not adhere to the grain, which falls free on threshing. The naked grain is much more acceptable as human food, and such cultivars are preferred where barley is a major part of the human diet. The character is controlled by a single recessive gene (*n*). Cultivated barley has been crossed with several other species of *Hordeum* (Bothmer *et al.*, 1983), but the hybrids, except those with *H. spontaneum*, are highly sterile or anomalous. No other species is known to be involved in the evolution of cultivated barley, except *Hordeum spontaneum* (Harlan, 1976a).

The genetic basis of cultivated barley: consequences of narrowing and restoration

Until the late nineteenth century, all barleys existed as highly heterogeneous 'landraces', mixtures of inbred lines and hybrid segregates, the products of a low level of random crossing in earlier generations. Many landraces still exist today, displaying intermediate (Brown & Munday, 1982) or high (Jana & Pietrzak, 1988) levels of genetic diversity; but over the past 100 years the landraces have been almost wholly displaced in advanced agricultures by pureline cultivars. These were initially selections from landraces, but later they derived from successive cycles of crosses between established purelines, sometimes of diverse geographical origins. The result has been a marked narrowing of the genetic base in many, probably all, advanced agricultures, partially concealed by a great diversity of cultivar names (Harlan, 1976a). While barley yield increased in advanced agricultures (e.g. Avery, 1985), genetic homogeneity also increased due to modern purebreeding practices (Frankel & Bennett, 1970; Frankel & Hawkes, 1975; Harlan, 1975, 1976b; Nevo *et al.*, 1979a, b; Frankel & Soulé, 1981; Plucknett *et al.*, 1983, 1987).

The loss of genetic diversity of some of the world's crops has accelerated in recent decades with many crops becoming increasingly susceptible to diseases, pests and environmental stresses (Plucknett *et al.*, 1983, 1987). A global network of gene banks has been established to provide plant breeders with the genetic resources for developing more resistant and tolerant crops that will maintain stable and high yields (Plucknett *et al.*, 1987; Brown *et al.*, 1989a, b). Currently, dynamic *in situ* conservation of wild relatives of cultivated plants is being actively discussed as an optimal conservation strategy (for example, the International Workshop held in Israel, 2–3 April (Ministry of Science and Technology, Israel 1990)). However, not only the conservation of diverse germplasms is needed. To achieve a more efficient and comprehensive utilization of the conserved wild gene pool, it is essential to be able to predict, screen, and evaluate promising genetic resources in natural populations and genotypes, thereby optimizing breeding. The best hope for future crop improvement seems to be in exploiting the rich gene pools of wild relatives from their centre of origin (Vavilov, 1926, 1951; Zohary, 1970, Feldman & Sears, 1981; Nevo, 1983, 1986, 1987, 1988a; Brown *et al.*, 1989b; *Biological Journal of the Linnean Society*, 1990).

The centre of origin and diversity of many wild progenitors of cultivated plants is the Near East Fertile Crescent, particularly Israel (Zohary, 1983; Nevo, 1986). Here, wild relatives of many cultivated plants have long evolved across a wide ecological range. This has generated in these wild ancestors a wide range of adaptive diversity to multiple diseases, pests and ecological stresses. Israel appears to be unique in the Near East Fertile Crescent in its physical and biotic heterogeneity which generated highly adaptive genetic diversity in the ancestors of many cultivated plants (Nevo, 1986). These wild genetic resources are as yet mostly untapped and are crucial for crop improvement by traditional plant breeding or modern biotechnology (e.g. Plucknett *et al.*, 1987; Schell, 1987; Gasser & Fraley, 1989).

Multidisciplinary research programme on wild barley

A multidisciplinary research programme on wild barley, *H. spontaneum*, was started at the Institute of Evolution, University of Haifa, Israel, in 1976. The programme involved macrogeographic population genetics and ecology studies in the Near East Fertile Crescent in three countries, Israel (Brown *et al.*, 1977, 1978a, b, 1980; Nevo *et al.*, 1979a, b), Iran (Nevo *et al.*, 1986a), and Turkey (Nevo *et al.*, 1986b), giving a broad view of the Fertile Crescent (Nevo *et al.*, 1986c). We also conducted a series of microgeographical population genetics and ecology studies in Israel on allozyme

(Nevo *et al.*, 1981, 1986d) and hordein (Nevo *et al.*, 1983) polymorphisms. In addition, we conducted studies on agronomic traits (Nevo *et al.*, 1984a), disease resistance (Moseman *et al.*, 1983, 1990; Nevo *et al.*, 1984b), protein content (Nevo *et al.*, 1985) and drought and salt tolerance (unpublished). A methodology was developed for assessing wild genetic resources based on ecological factors and allozyme markers, enabling us to screen for single and multiple trait élite genotypes in nature (Nevo, 1987).

Israel is central and unique, due to its physical and biotic diversity, as a natural laboratory for genetic resources for crop improvement (Nevo, 1986). The evidence relevant to barley improvement derives from studies of population genetics and ecology, climatology, phytopathology and breeding practices. The following questions have been explored in our analysis of wild barley in Israel, paralleling our analysis of wild emmer wheat (Nevo, 1983, 1988a).

1. What is the genetic structure and geographical differentiation of populations?
2. What are the ecological–genetic correlates and mechanisms generating population genetics structure at the macro- and microgeographical levels?
3. To what extent are the levels and patterns of genetic diversity and differentiation determined by natural selection as opposed to migration and random factors?
4. What is the importance of multilocus organization in population genetics dynamics?
5. What are the phenotypic and genotypic diversities of economically significant quantitative agronomic traits within and between populations?
6. What are the geographical patterns of disease resistance to pathogens, and can they be predicted, either singly or in combination, for a better screening of populations and genotypes for breeding?
7. What are the patterns and ecogeographical correlates of seed proteins?

A generalized overview of present answers to these theoretical and applied problems follows.

Population genetics and ecology of wild barley

Ecology

H. spontaneum is widely distributed over the eastern Mediterranean basin and in western Asiatic countries (Fig. 2.1), penetrating into Turkmenia, Afghanistan, western China and India, across a wide range of climates and soils. It occupies both primary habitats, and secondary, segetal, man-made habitats. In the Fertile Crescent, central populations of wild *H. spontaneum* are often continuously and massively distributed. *H. spontaneum* constitutes an important annual component of open herbaceous formations, and it is particularly common in the summer-dry deciduous oak park–forest belt, east, north, and west of the Syrian desert and the Euphrates basin, and on the slopes facing the Jordan Rift Valley (Zohary, 1973). From here, *H. spontaneum* spills over the drier steppes and semideserts. In the Near Eastern countries, wild barley also occupies a whole array of secondary habitats, such as opened-up Mediterranean maquis, abandoned fields and roadsides. It infests cereal cultivation and fruit tree plantations (Harlan & Zohary, 1966).

Peripheral populations of *H. spontaneum* become increasingly sporadic, semi-isolated and isolated. This is true in the Aegean region, southern Israel, Mediterranean shore of Egypt, Cyrenaica and Morocco, and further east in northeast Iran, Soviet central Asia, Afghanistan and India. In these peripheral regions *H. spontaneum* rarely builds large stands and seems to be largely restricted in most localities to segetal habitats, ruins, or to sites which have been drastically disturbed by man's activity. In general, wild barley is not tolerant to extreme cold temperatures, and it is only occasionally found above 1500 m. It is almost completely absent from the elevated continental plateaus of Turkey and Iran. On the other hand, it is more drought resistant than the wild wheats and penetrates relatively deeply into the warm steppes and deserts (Zohary & Hopf, 1988).

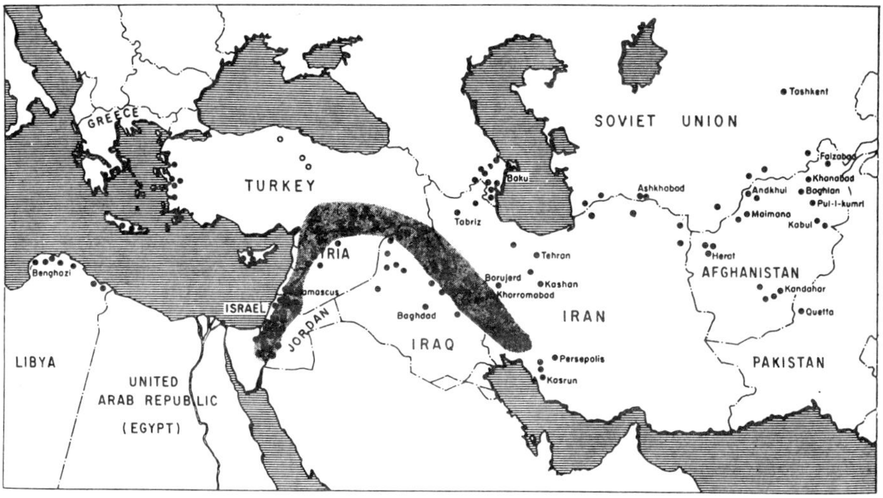

Fig. 2.1. Distribution of wild *Hordeum spontaneum* and of early archaeological sites for cultivated barley, *H. vulgare* (after Harlan & Zohary, 1966).

Population genetics: macrogeographic differentiation

We analysed enzyme polymorphisms encoded by 27 shared loci in 2125 individuals representing 52 populations of wild barley in Israel (Nevo *et al.*, 1979a, b), Iran (Nevo *et al.*, 1986a), Turkey (Nevo, 1986b) and the Near East Fertile Crescent (Nevo *et al.*, 1986c, where all isozyme loci are specified). Our main population genetics results from across the Fertile Crescent are outlined next.

Patterns of genetic variation

H. spontaneum in the Fertile Crescent is very variable genetically (Table 2.1). Overall, 127 alleles at 27 loci were described: 103, 63 and 62 for Israel, Turkey and Iran, respectively. The number of alleles increases with the average range of annual rainfall in the three countries: 100, 19 and 21% of the total range for Israel, Turkey and Iran, respectively; see explanation in Tables 1 and 2 of Nevo *et al.* (1986c) for climatic variation and allele frequencies at 27 loci in the three countries. The mean proportion of polymorphic loci per population (1% criterion of polymorphism) was P-1% = 0.312 (range, 0.0–0.41), and gene diversity was He = 0.103 (range, 0.0–0.182), displaying high levels and wide ranges of genetic diversity in comparison with mean genetic estimates in nature (Nevo *et al.*, 1984c; Nevo, 1988b).

Genetic differentiation

Wild barley populations in the Near East differ considerably in their allelic content (Nevo *et al.*, 1986c). Out of 127 alleles at 27 shared loci, 65 alleles (51%) were unique, that is they occurred only in one country. Out of these 65 unique alleles, the distribution between Israel, Turkey and Iran was 50, 7 and 8, and their percentage in each of the countries was 48.5, 11.1 and 12.9% respectively. Most of the unique alleles (51 out of 65, or 78.5%) were not widespread but localized or sporadic. Remarkably, 40 out of the 65 unique alleles (61.5%) were common (frequency >10%). Thus, most importantly, a major population genetics pattern of wild barley, like that of wild emmer wheat though somewhat less dramatic, is the 'island' structure. A substantial number of the alleles of *H. spontaneum* are both unique and common locally. This 'island'

Table 2.1. Comparison of overall genetic indices (A, *p*-1%, *p*-5%, *H*, H_e, *F*), mean and ranges, based on 27 shared gene loci in: A, 28 Israeli, 11 Turkish and 13 Iranian populations; and B, three regions (Israel, Turkey and Iran), of wild barley, *H. spontaneum* (from Nevo *et al.*, 1986c).

	Sample size (*N*)	No. of populations	Mean no. of alleles per locus (*A*)	Mean proportion of loci: Polymorphic per population (*P*) 1%	Mean proportion of loci: Polymorphic per population (*P*) 5%	Mean proportion of loci: Heterozygous per individual (*H*) Mean	Mean proportion of loci: Heterozygous per individual (*H*) S.E.	Wright's fixation index (*F*)	Gene diversity (H_e)
A									
Israel	1179	28	1.509	0.312	0.247	0.003	0.001	0.962	0.103
range			1.000–1.778	0–0.407	0–0.407	0–0.020	0–0.007	0.790–1.000	0–0.182
Turkey	437	11	1.428	0.306	0.253	0	0	0.994	0.099
range			1.111–1.593	0.111–0.370	0.037–0.333	0–0.004	0–0.002	0.954–1.000	0.022–0.156
Iran	509	13	1.507	0.356	0.328	0	0	0.997	0.130
range			1.296–1.778	0.259–0.481	0.185–0.481	0–0.003	0–0.002	0.979–1.000	0.057–0.194
Total	2125	52							
Mean			1.492	0.320	0.271	0	0	0.984	0.108
range			1.000–1.778	0–0.481	0–0.481	0–0.020	0–0.007	0.790–1.000	0–0.194
B									
Israel	1179		3.815	0.667	0.444	0.003	0.001		0.198
Turkey	437		2.333	0.444	0.444	0	0		0.180
Iran	509		2.296	0.519	0.481	0	0		0.184
Mean	2125		2.815	0.543	0.457	0.002	0.001		0.187

population genetics structure has theoretical and practical implications.

The average relative genetic differentiation (*Gst*) (Nei, 1973) was 54% within populations, 39% among populations within countries, and 8% between the three countries. This pattern is also substantiated by the average genetic distances, *D* (Nei, 1972), between and within countries. Mean *D* decreased within countries in the following order: Israel ($D = 0.119$), Turkey ($D = 0.110$), and Iran ($D = 0.065$). The genetic distances also indicate sharp genetic differentiation over short geographic distances. The genetic distance is larger between Israel and either Turkey or Iran ($D = 0.134$ and $D = 0.129$, respectively) than between Turkey and Iran ($D = 0.111$). Discriminant analysis by allele frequencies successfully clustered wild barley of the three countries (96% correct classification) (Fig. 2.2).

Environmental correlates of allozyme diversity

A substantial portion of allozyme diversity in the wild gene pool is significantly correlated with the environment and is predictable ecologically, chiefly by a combination of climatic factors of rainfall, humidity and temperature variables. For example, about a third of the variance of *P*-1% is significantly explained by a three-variable combination of mean annual rainfall, humidity, and August temperature ($R^2 = 0.31$, $P < 0.001$), (Nevo *et al.*, 1986c, p. 370). It appears that rainfall is the best predictor of the genetic indices. Out of 29 alleles representing 11 polymorphic loci, the geographic variation in the frequency of 20 (69%) alleles was explained by a combination of climatic and geographical variables ($R^2 = 17$–46%), primarily with factors of water availability and temperature (Table 8 in Nevo *et al.*, 1986c). This is essentially the pattern found earlier, both qualitatively and quantitatively, for each country (Israel, Turkey, Iran). Notably, geographic variation in *H. spontaneum* from Israel was also described for cytoplasmic male sterility genes and in nuclear genes for restoration (Ahokas, 1980).

Multilocus structure of natural populations

The association of alleles among different loci (linkage or gametic phase disequilibrium) has been primarily studied between two-locus associations (e.g. Clegg *et al.*,

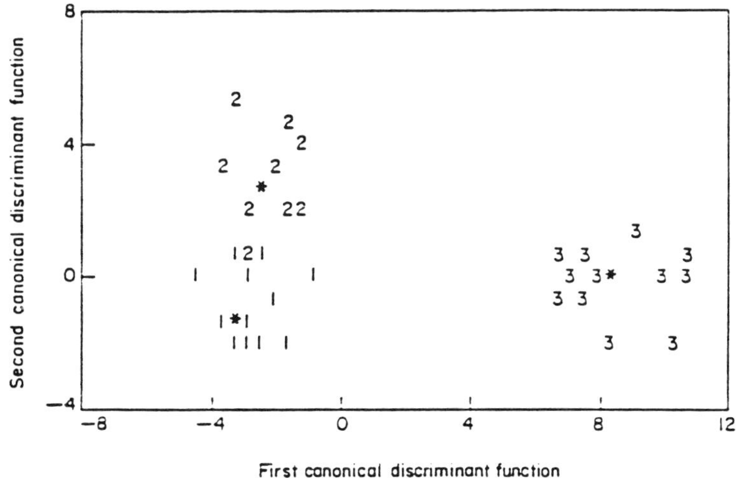

Fig. 2.2. Discriminant analysis of populations of wild barley, *H. spontaneum* from Israel (1), Turkey (2), and Iran (3). *Centroid (from Nevo *et al.*, 1986c).

1972; Karlin, 1975; Hedrick *et al.*, 1978; Weir, 1979; Brown, 1984). We have reported the widespread occurrence of co-adapted combinations of alleles at the four esterase loci in 11 populations of wild barley from Israel (Brown *et al.*, 1977). However, multilocus organization can effectively be summarized over *many* loci rather than by two-locus disequilibria (Brown *et al.*, 1980). The variance of the number of heterozygous loci in two randomly chosen gametes affords a useful measure of such association. Generally, linkage disequilibrium tends to increase the variance above the value expected under complete independence. When data on several loci per individual are available, the observed variance may be tested for agreement with that expected under the hypothesis of complete interlocus independence, using the sampling theory of this model, and generating an index of multilocus organization.

Multilocus organization, based on allozyme data from 26 polymorphic populations of wild barley, was shown to be geographically widespread (Brown *et al.*, 1980). On average, the variance was 80% higher than expected under random association. Most generations of barley composite crosses showed less multilocus structure of four esterase loci than the wild populations. Additional multilocus allozyme analysis indicated a repetitive pattern of the multilocus associations in the composite crosses in contrast to the regionally localized and geographically variable pattern in wild *H. spontaneum* (Brown & Feldman, 1981). While multilocus associations could also be generated by historic and non-selective factors, we interpreted these associations in wild barley and in wild emmer (Nevo & Beiles, 1989) as the result of natural selection.

Population genetics: microgeographic differentiation

Microsite allozyme polymorphisms

Microgeographic allozymic differentiation can take place over remarkably short distances, despite considerable gene flow, and is therefore adaptive at the morphological, physiological (Bradshaw, 1972) and allozymic levels (Nevo *et al.*, 1977, 1981, 1983, 1986d, 1988a, b, in press). Microgeographic allozyme differentiation is partially related to environmental factors (Brown, 1979). We have conducted two microsite allozyme studies in *H. spontaneum* (Nevo *et al.*, 1981, 1986d) and one on the storage protein, hordein (Nevo *et al.*, 1983). Our microsite allozyme studies involved first two soil types, basalt and terra-rossa, along 100 m transects at Tabigha (Nevo *et al.*, 1981); and secondly, six climatic microniches, mosaically distributed, in an open oak forest in the Lower Galilee mountains (Nevo *et al.*, 1986d) involving the following microniches: (i) sun–soil; (ii) sun–rock; (iii) shade–soil; (iv) shade–rock, and the contact zones; (v) soil periphery of the sun–rock microniche; and (vi) soil periphery of the shade–rock microniche. In both microsite allozyme studies, involving 22 and 25 enzyme loci, respectively, we obtained similar results. Allozyme polymorphisms in wild barley are patterned and not random even in a microsite. They are at least partly adaptive and differentiate predominantly by microniche ecological selection, rather than by stochastic processes and/or neutrality of allozyme variants.

Similar results of microsite genetic differentiation were obtained in wild emmer wheat, *T. dicoccoides* by: (i) climatic selection (Nevo *et al.*, 1988a); (ii) edaphic, topographical and temporal selections (Nevo *et al.*, 1988b); and (iii) topographical selection (Nevo *et al.*, in press). The underlying major evolutionary force causing genetic differentiation in overall genic diversity, allele frequencies, gametic phase disequilibria, and multilocus organization of wild barley, as well as wild emmer wheat, appears to be the aridity stress.

Microsite hordein polymorphisms

The hordein storage protein consists of two major groups of polypeptides governed by co-dominant alleles at two loci, *Hor1* and *Hor2*, which are highly polymorphic in cultivated barley, but particularly in *H. spontaneum* (Doll & Brown, 1979). Recently, a third locus, *Hor3*, was described, located with *Hor1* and *Hor2* on chromosome 5 (Blake *et al.*, 1982; Shewry *et al.*, 1983).

Microgeographic edaphic differentiation was also demonstrated in hordein polymorphisms of wild barley (Nevo *et al.*, 1983) along the same 100 m soil transect which displayed allozymic differentiation in an earlier study (Nevo *et al.*, 1981). Genetic diversity in *Hor1* and *Hor2* was analysed electrophoretically in seeds from 123 plants of wild barley, *H. spontaneum*. The test was conducted in two topographically different 100 m transects in Israel, each equally divided into basalt and terra-rossa soil types. Altogether 15 *Hor1* and 16 *Hor2* phenotypes were detected; 7 phenotypes of *Hor1* and 5 of *Hor2* were common. Significant differentiation of both *Hor1* and *Hor2* phenotypes and their associations was found to be correlated with soil type and topography. Likewise, significant correlations were found between hordein phenotypes and allozyme types detected in a previous study (Nevo *et al.*, 1981). Our results suggest that at least part of the hordein polymorphisms in wild barley are adaptive and selected by soil and topographic differences over very short distances. Many different allozymic and hordein homozygous genotypes ('clones') occur differentially on the basalt and terra-rossa soil types, suggesting adaptive 'clonal' microsite distribution (Beiles & Nevo, in preparation).

Genetic diversity in wild and in cultivated barley

We compared total allozymic diversity of 19 shared loci of wild barley in Iran, Turkey and Israel with that of barley landraces from Iran and a collection of European landraces, also including two accessions from Egypt and one from India, and two composite crosses (Tables 2.2 and 2.3). Allozymic diversity was lowest in the composite crosses, intermediate in the landraces, and highest in the wild germplasm in the following order: Iran < Turkey < Israel. These results appear to hold even if the world collection of cultivated barley, which presumably represents the allelic diversity maintained in the domestic gene pool, is compared. When this comparison is made for four esterase loci, the differences between the wild and cultivated gene pools in number of alleles are: *Est1* (8,7); *Est2* (17,12); *Est4* (10,6); *Est5* (6,6) (Nevo *et al.*, 1986c; for cultivars see Kahler & Allard, 1981). These results emphatically demonstrate the potential role of the wild progenitor as well as that of the landraces (Brown & Munday, 1982) as valuable genetic resources for plant breeding. Higher diversity was also obtained for α-amylases than in cultivated barley (Brown & Jacobsen, 1982).

Chloroplast DNA (cpDNA) diversity in 11 lines of *H. spontaneum* was significantly higher than in 9 lines of *H. vulgare*, suggesting that the level of cytoplasmic diversity was markedly restricted during barley domestication (Clegg *et al.*, 1984). cpDNA diversity was found within and among

Table 2.2. Comparison of genetic indices among *Hordeum spontaneum*, and landraces and composite crosses of *H. vulgare*, based on 19 shared loci (from Nevo *et al.*, 1986c).

	N	*A*	*P*	H_e
H. vulgare				
Composite cross XXI (F_{17})	50	1.58	0.42	0.099
Composite cross XXXIV (F_4)	50	1.68	0.37	0.139
Landraces				
'European'*	178	2.26	0.47	0.146
Iranian	290	2.21	0.74	0.164
H. spontaneum				
Iran	509	2.58	0.68	0.218
Turkey	437	2.63	0.74	0.226
Israel	1179	4.11	1.00	0.228

* Twenty-nine accessions from the collection of Landbrugets Plantekulturs Sortssamling, Denmark, which include European landraces, plus one Indian and two from Egypt.

Table 2.3. Comparison of numbers of alleles and total diversity (H_t) among *Hordeum spontaneum* and *H. vulgare* (from Nevo *et al.*, 1986c).

	H. vulgare				*H. spontaneum*							
	European*		Iranian		Near East		Iran		Turkey		Israel	
Locus	Allele	H_t	Allele	H_t	Allele	H_t	Allele	H_t	Allele	H_t	Allele	H_t
Est-2	6	0.28	3	0.38	17	0.83	7	0.61	6	0.72	15	0.86
Acph-3	4	0.47	4	0.57	8	0.73	4	0.48	5	0.63	4	0.66
Est-4	4	0.58	3	0.45	10	0.70	4	0.62	4	0.63	7	0.70
Est-5	4	0.40	2	0.01	6	0.59	4	0.57	5	0.64	6	0.53
Nadh-1	2	0.01	3	0.50	4	0.47	3	0.29	2	0.48	4	0.54
Adh-1	1	0	2	0.04	3	0.29	2	0.17	3	0.52	3	0.24
Est-1	3	0.48	3	0.44	8	0.25	4	0.44	3	0.18	6	0.13
Aat-1	1	0	1	0	3	0.17	2	0.30	2	0.28	2	0.03
Pgi	1	0	3	0.19	5	0.16	4	0.22	4	0.17	3	0.13
Pgm	1	0	2	0.01	3	0.10	2	0.004	2	0.02	3	0.18
Adh-2	1	0	1	0	5	0.10	3	0.11	2	0.04	3	0.11
Cat	1	0	2	0.15	3	0.09	2	0.34	2	0.01	2	0.01
Mdh-2	1	0	1	0	7	0.05	2	0.004	3	0.02	4	0.08
Gdh	1	0	2	0.05	3	0.04	1	0	1	0	3	0.07
Aat-2	2	0.04	2	0.02	4	0.02	1	0	1	0	4	0.03
6Pgd-2	4	0.46	3	0.21	3	0.02	1	0	1	0	3	0.02
Acph-1	1	0	1	0	2	0.01	1	0	1	0	2	0.01
6Pgd-1	6	0.07	1	0	3	0	1	0	1	0	3	0.01
Mdh-1	1	0	2	0.11	2	0	1	0	2	0.002	2	0.01

*Twenty-nine accessions from the collection of Landbrugets Plantekulturs Sortssamling, Denmark which include European landraces, plus one Indian and two from Egypt.

populations of *H. spontaneum* from Israel and Iran, displaying ecogeographic correlates, and diversity was higher in *H. spontaneum* than in *H. vulgare* (Neale *et al.*, 1988). By contrast, however, landraces of *H. vulgare* proved more polymorphic in cpDNA and in mitochondrial DNA (mtDNA) diversity than *H. spontaneum* (Holwerda *et al.*, 1986). More genotypes and populations need to be studied to make a meaningful comparison.

Outcrossing in natural populations of *H. spontaneum*

The mating system is crucial to understanding the evolution of plant populations. We used our allozyme data to estimate outcrossing rates across the range of wild barley in Israel (Brown *et al.*, 1978a). The extent of cross-pollination was estimated quantitatively in 26 populations of wild barley (*H. spontaneum*) in Israel, using allozyme variation at 22 polymorphic loci. Individual population estimates varied from 0 to 9.6% outcrossing. The overall average was a very low rate, 1.6%, with a 95% confidence range of 0.8–2.7%. Outcrossing varied with ecology and was significantly higher in populations from mesic (2.1%), than from xeric regions (0.4%). The 0.037 multilocus heterozygosity encountered at 22 loci displayed a very high degree of multiple disequilibrium. In general, these results indicate that extensive genetic polymorphisms in the progenitor probably predate the domestication of the crop, and are maintained independently of introgression, presumably by natural selection in accordance with environmental heterogeneity.

Phenotypic and genotypic quantitative morphological and physiological variation within and between populations of *H. spontaneum*

Cultivated barley, even some composite crosses, harbour substantially less genetic diversity as compared with the wild gene pool (Nevo *et al.*, 1979a, b), or some tested landraces (Brown & Munday, 1982). Opposite results have been reported on landraces from the eastern Mediterranean countries Jordan, Syria, Turkey and Greece by Jana & Pietrzak (1988). This discrepancy may be derived from a lower number and a somewhat different set of tested loci. The reduction in genetic diversity of purebred lines makes cultivated barley increasingly vulnerable to new pathogens, pests and adverse climates, primarily hot ones. We therefore tested variation in quantitative traits of agronomic importance of *H. spontaneum* as potential genetic resources for breeding.

Variations in morphological and phenological characteristics

Populations of wild barley, *H. spontaneum*, in Israel, originating from diverse habitats, and tested earlier for allozymes (Nevo *et al.*, 1979a, b) and disease resistance polymorphisms (Moseman *et al.*, 1983; Nevo *et al.*, 1984b), were compared and contrasted for performance in agronomically important quantitative phenotypic traits (Nevo *et al.*, 1984a). The traits compared involved ten variables of germination, earliness, biomass and yield. The field experiments were conducted in 1980–2 in two relatively standardized and contrasting environments: mesic (Mount Carmel, Haifa) and xeric (Avedat farm and Sede Boqer, both in the northern Negev desert). The experimental design involved 26 population quadrangles at Avedat farm (1980), and rows of randomized plants of 11 populations in both Haifa and Sede Boqer (1981 and 1982).

The results indicated that the characters studied are partly genetically determined (Fig. 2.3). Striking genetic variation was found between and within populations in each site, whereas remarkable environmental variation, including genetic × environmental interaction, was found between the mesic and the xeric sites, as well as between populations and years. We concluded that wild barley in Israel varies not only in

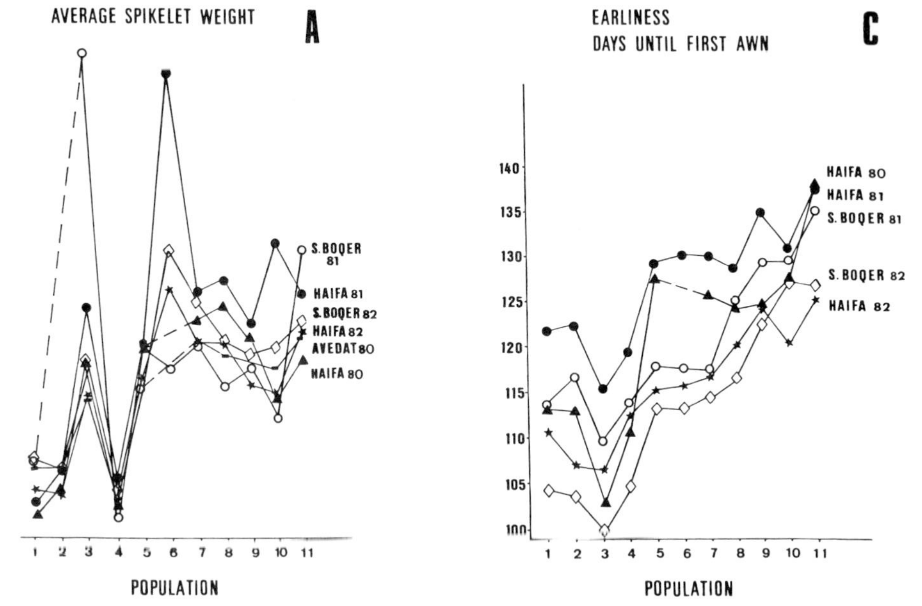

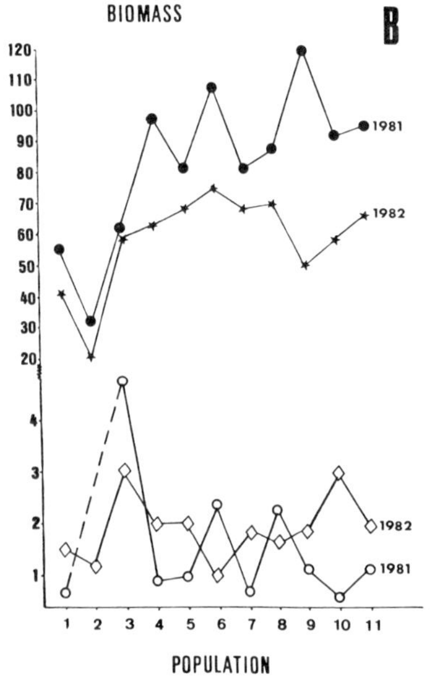

Fig. 2.3. Variation in population means of three characters of agronomic importance: average spikelet weight (A), biomass (B), and earliness (C) in 11 populations of *H. spontaneum* in Israel grown in two different environments: mesic (Haifa) and xeric (desert of Sede Boqer). The 11 populations are ordered from low to high rainfall. (See additional examples of variation in nine agronomic characters in Nevo *et al.*, 1984a.)

allozyme polymorphisms and disease resistances, but also in quantitative traits of agronomic importance. Similar results were obtained by others (Vega & Frey, 1981; Snow & Brody, 1984; Giles & Lefkovitch, 1985; Jaradat, 1989a). These traits are economically significant and should be conserved and utilized in barley improvement.

These agronomically important traits, which differ remarkably both within and between populations, are also geographically structured, and originated by natural selection as adaptations to macro- and microgeographic environments. The variance in earliness, for example, is highly and significantly explained ($R^2 = 0.70$, $P < 0.001$) by a three-variable combination involving evaporation, humidity and the number of rainfall days (see other examples in Table 5 in Nevo *et al.*, 1984a). Outstanding earliness was found, as expected, in populations from steppe (Beth Shean, Mehola) and desert (Wadi Qilt), warm and dry habitats. Likewise, the yield in both the mesic and xeric test sites was higher in populations derived from optimal rather than marginal habitats for wild barley.

Advanced cultivars have been purebred throughout the last hundred years for high performance under intensive high-input cultivation, which tends to even out environmental, as well as genetic differences of site and season. Modern varieties, in contrast to either their progenitors or to landraces, are purebred cultivars. They are deliberately selected for performance over a range of environments, thus reducing the need for specific local adaptation. Hence, they are selected for a high degree of genetic homogeneity. Consequently, the genetic bases of cultivated crops have been narrowed and placed under serious risks and their future evolution placed in jeopardy (Frankel & Soule, 1981; Plucknett *et al.*, 1987). Our results substantiated earlier claims that wild barley in Israel displays remarkable morphological and physiological variation both within and between populations in each of the agronomic characters studied (Table 2, Fig. 2 in Nevo *et al.*, 1984a). Geographic variation in the Near East was also described in another study of reproductive structures of *H. spontaneum*, but its biological correlates require further investigation (Giles & Bengston, 1988).

The interest in the progenitors of crop species should not reside only in their wealth of resistance genes to pathogens acquired over long periods of host–pathogen co-evolution. *H. spontaneum* harbours a wealth of additional genetic traits of economic significance, such as earliness, high yield and high biomass. Apart from the direct importance of these traits in breeding, they can also improve breeding indirectly. When parental genotypes of the progenitor are used as donors of alien genes for disease resistance to a cultivar, it is desirable that they also combine the genetic capacity for high yield and biomass to avoid yield depression in the resulting crosses.

Variation in physiological performances: drought and salt tolerance

Natural populations of *H. spontaneum* involve abundant genetic variation in resistance to physiological stresses, such as drought and salinity. This is highlighted by the tremendous variation both between and within populations observed in our extreme desert station at Sede Boqer (Table 2 and Fig. 2 in Nevo *et al.*, 1984a). Recently, we tested populations and genotypes of wild barley *H. spontaneum* and wild emmer wheat, *T. dicoccoides*, for salt tolerance under controlled conditions, using 150, 250 and 350 mM NaCl (equivalent to 25, 40 and 60% sea water) with non-saline controls (Nevo, Krugman & Beiles, unpublished). We found high tolerances for salt within and between populations of *H. spontaneum* and *T. dicoccoides*, and identified genotypes of wild barley from coastal Mediterranean populations (e.g. Caesarea), the Jordan Valley (Mehola) and the northern Negev desert (Makhtesh Gadol) that spiked in 350 mM NaCl. Natural populations of wild barley extend southwards to the nothern Negev, yet they only grow there in dry water courses thereby effectively utilizing the extra run-off water. However, in our Sede Boqer experiments, under a water regime of 130 mm and, under highly saline soil conditions, some populations and superior genotypes have proved relatively outstanding in both growth and yield, despite these extremely harsh ecological conditions. Thus, the genetic resources of the progenitor display striking morphological and physiological performances, which are of potential economic importance for barley improvement.

Variation in vegetative parameters related to the nitrogen economy

The nitrogen contents and dry weights of leaf and 'stem' (stem plus sheath) at anthesis, and the final grain size and grain protein content were measured in 33 *H. spontaneum* and two *H. vulgare* genotypes (Corke *et al.*, 1988). *H. spontaneum* was generally higher in nitrogen content of leaves and stems, but lower in dry weight at anthesis. Consistent with previous reports, the *H. spontaneum* genotypes were considerably higher in grain protein (see next section) than the cultivars. There was wide variation between and within populations of *H. spontaneum*, suggesting that lines combining high vegetative nitrogen content, dry weight and grain protein content can be selected for breeding.

Variation in protein resources

The initial protein studies of *H. spontaneum* were conducted by Ladizinsky (1975) and Ahokas (1977, 1982). These studies indicated many significant differences in protein content between wild barley populations from different biogeographical regions. We have recently determined the protein content of *H. spontaneum* in Israel, and tested for associations with ecological and allozyme markers in an attempt to derive predictive guidelines for conservation and utilization in breeding programmes (Nevo *et al.*, 1985). The study involved 195 genotypes of wild barley from 25 populations, 15 central and 10 marginal, tested earlier for allozymic variation (Nevo *et al.*, 1979a, b). The results indicated that protein content varies within but particularly between populations. Notably, the ten marginal populations exhibited high protein contents (15.06%, S.E. ± 0.281), but low kernel weights (21.4 mg, S.E. ± 5.8), as compared with the 15 central populations which displayed lower protein contents (13.92%, S.E. ± 0.138), but high kernel weights (36.0 mg, S.E. ± 0.914). Thus, we demonstrated negative correlation between mean population protein percentage and mean kernel weight ($r_s = -0.56$; $P = 0.004$, for $n = 25$ populations). Notably 96% of our populations were correctly classified in a discriminant analysis to either marginal or central populations. Three-variable combinations of climatic factors explain 40% of the variability in protein content among populations. Likewise, three-variable combinations of allozyme allele frequencies explain a significant degree of spatial variance in protein content ($R^2 = 0.63$). The genetic determination estimate *within* populations was 57% and the overall genetic determination was 80%. Thus, protein content is amenable to cultivar improvement. Later, 21.5% grain protein content was described in *H. spontaneum* versus 13% in a standard Israeli barley cultivar (Friedman & Atsmon, 1988). We concluded that natural populations of wild barley in Israel contain large amounts of yet untapped genes for protein content. These could be effectively screened and utilized for producing high-protein cultivars of barley by following ecological and allozymic markers as predictive guidelines in screening natural populations of wild barley.

Disease resistance polymorphisms in *Hordeum spontaneum*

Host–pathogen co-evolution in the centre of diversity

Crop diseases prevent a rapid increase in food production. Crop improvement should, therefore, aim to minimize the risks of both epidemic and endemic diseases in an attempt to increase production efficiency, without affecting the agronomic properties of existing varieties (Horsfall & Cowling, 1977). In general, natural plant ecosystems are not devastated, but develop efficient defence against diseases, due to the long-lasting host–pathogen co-evolution. In a centre of diversity, this evolutionary interaction results in a balanced genetic polymorphism with the highest genetic diversity for virulence in the pathogen, and for resistance in the host (Dinnor, 1974; Anikster & Wahl, 1979).

Israel is one of the epicentres where active co-evolution of several crop ancestors and their pathogens occurs, including wild barley, wild wheat and their rust and powdery mildew pathogens. Following the theory and genetics of host–parasite co-evolution, the highest and broadest genetic immunity against devastation in the progenitors of some crops would be expected in the Fertile Crescent in general, and in Israel in particular. Israel is ideal for the exploitation of resistance genes which could provide useful material for crop improvement, for both the short-term, and more importantly, the long-term solution to disease problems in agriculture (Segal *et al.*, 1980, 1982). This includes the clarification of the controversy related to specific and general resistance (Anikster & Wahl, 1979).

Resistance to powdery mildew (*Erysiphe graminis hordei*)

The world-wide importance of yield losses in barley caused by the powdery mildew disease of barley (*Erysiphe graminis hordei*) is well known (e.g. Brooks, 1970), and was discussed as an objective in breeding cereal varieties (Slootmaker, 1974). Many sources of barley resistant to *E. graminis hordei* have been identified (Moseman *et al.*, 1983), and used extensively in developing resistant cultivars in many countries. Plant diseases in natural populations of wild barley were recently reviewed by Dinoor & Eshed (1990a, b). Many *H. spontaneum* accessions obtained from plants growing in Israel have been found to be highly resistant to *E. graminis hordei* (Moseman *et al.*, 1980). Moseman & Craddock (1976) reported that 75% of 277 accessions of *H. spontaneum* collected in Israel were resistant to two cultures of *E. graminis hordei*. Fischbeck *et al.* (1976a) reported that about 50% of 2700 unselected accessions of *H. spontaneum* were attacked only slightly or not at all by European field isolates of the pathogen. They found accessions resistant to the pathogen in all 69 collection sites in Israel with natural stands (see also Segal *et al.*, 1980, 1982). Baenziger *et al.* (1981) have developed a composite cross population with recessive male-sterile genes consisting of 19 *H. spontaneum* accessions which are resistant to many cultures of *E. graminis hordei*.

The percentage of *H. spontaneum* accessions resistant to *E. graminis* depends on where the plants are collected in Israel. Wahl *et al.* (1978) found that *H. spontaneum* plants with resistant types were prevalent in the Golan Heights, the Galilee and in the Samarian and Judean highlands. Fischbeck *et al.* (1976b) showed that there are few plants resistant to *E. graminis hordei* in regions of Israel with low rainfall. The number and relationships of resistance genes in cultivated barley and in *H. spontaneum* have been determined by using cultures of *E. graminis hordei* collected in Israel and other cultures with specific virulences. Moseman *et al.* (1981) showed that the number of identifiable genes conditioning the resistance of *H. spontaneum* accessions was determined by the virulence of cultures of *E. graminis hordei*. They used culture CR3 and culture 64–54 isolated from a plant in Israel. Moseman & Jørgensen (1973) used culture 64–54 and other cultures with specific virulences to identify and differentiate between resistance genes within the complex *Ml-a* locus on chromosome 5 of barley. A total of 56 mildew resistance genes participate in the inheritance of mildew resistance of *H. spontaneum* derived lines (Jahoor & Fischbeck 1987a). Thus, *H. spontaneum* in Israel forms a large gene pool for mildew resistance which is not yet widely used in barley breeding (Jahoor & Fischbeck, 1987b).

In the work conducted at the Institute of Evolution we found geographic variation in resistance of *H. spontaneum* against powdery mildew (Moseman *et al.*, 1983). Eighty-one resistant accessions and 180 moderately resistant accessions were identified against cultures of powdery mildew. Resistance was prevalent in northern and central Israel whereas sensitivity was distinct in the Negev and on Mt Hermon, where the pathogen is seldom observed. The resistance against powdery mildew is predictable ecologically and allozymically (Nevo *et al.*, 1984b).

Resistance to leaf rust (*Puccinia hordei*)

The reactions to infection with two cultures of *Puccinia hordei* were determined for 292 *H. spontaneum* accessions, collected at 16 sites which encompassed the entire ecological range of *H. spontaneum* in Israel (Moseman *et al.*, 1990). Culture Tel-Aviv was from Israel and culture 57.19 was from the United States. Of the accessions tested, 52% were resistant to culture Tel-Aviv, 67% were resistant to culture 57.19, and 43% were resistant to both cultures. The average infection type (IT) of accessions within sites ranged from 2.7 to 7.5 on a 0–9 rating scale. The results showed that the presence of *Ornithogalum* species, the alternate host of *P. hordei*, may increase the percentage of *H. spontaneum* accessions resistant to *P. hordei*. More accessions were resistant at sites where average annual humidity at 14:00 hours was higher, the annual evaporation was lower, and where the glumes were shorter. Kernel weight and annual rainfall were not correlated with resistance. A lower percentage of *H. spontaneum* accessions was resistant to *P. hordei* culture Tel-Aviv from Israel, than to culture 57.19 from the United States. Earlier, a lower percentage of *H. spontaneum* accessions was also found to be resistant to a culture of *Erysiphe graminis hordei* from Israel than to cultures from other countries. Previous studies have also shown that cultures of *P. hordei* and *E. graminis hordei* from Israel have many genes for virulence on barley, and that *H. spontaneum* accessions from Israel have many genes for resistance to these two pathogens. Previous results (e.g. Manisterski *et al.*, 1986; Treeful & Wilcoxson, 1986) and the results discussed here confirm that resistance to leaf rust is common in wild barley. The evidence supports the hypothesis of co-evolution of resistant host genes and virulent pathogen genes where hosts and pathogens have co-existed for a long evolutionary period.

We compared and contrasted disease resistance polymorphisms to infection with *Erysiphe graminis hordei* and *Puccinia hordei* in the Fertile Crescent (Jana & Nevo, in press). Random samples of 330 genotypes of wild barley plants collected from 14 sites in Iran, Turkey and Syria were compared and contrasted with 258 genotypes from 15 sites in Israel, studied earlier for allozyme diversity by Nevo *et al.* (1979a, b), and later for disease resistance for their reactions to infection with three cultures of *Erysiphe graminis hordei* (Moseman *et al.*, 1983) and two cultures of *Puccinia hordei* (Moseman *et al.*, 1990). One culture of each of the two pathogens was isolated in Israel, and the other cultures were isolated either in Japan or the United States. Of a total of 330 wild barley plants collected from 14 sites in Iran, Turkey and Syria, only 18.8% were resistant to the Israeli culture, and 14.8% were resistant to a composite of the Japanese and American cultures of *E. graminis hordei*. Of 105 plants collected from six sites in Iran and Turkey, none was found to be resistant to the Israeli culture and only 34.3% were resistant to the American culture of *P. hordei*. These frequencies were significantly lower than the corresponding frequencies of resistant wild barley plants from Israel. Phenotypic diversity for reactions to different cultures of the two pathogens was also higher in Israel. Considerable variations were observed both within and among sites for reactions to infection with different cultures of each of the two pathogens. These results illustrate the relative importance of wild barley accessions from Israel as a source of new genes for resistance to *E. graminis hordei* and *P. hordei*. Israel also appears to be one of the most desirable locations in west Asia for the *in situ* conservation of genetic diversity in wild barley in a dynamically evolving state (Dinoor & Eshed, 1990b; Ministry of Science and Technology, Israel, 1990).

Chromosomal localization and isolation of resistance genes

The utilization of major genes to detect and map genes controlling quantitative traits was described by Thoday (1961), and elaborated by Thompson & Thoday (1979). It was pioneered by Tanksley *et al.* (1982), using isozyme markers as major genes in analysing quantitative traits in tomato. The chromosomal localization of genes for isozymes has been summarized by Brown (1983).

As described earlier, we have used isozyme marker genes to identify élite genotypes of wild cereals (Nevo, 1987). However, the weakness of this predictive technique is in establishing correlations between the isozyme markers and the traits in question, while indisputable confirmation calls for tests by crossings. The segregation analysis of several generations can unequivocally determine the association or linkage relationships between isozyme loci and genes controlling the quantitative trait under study. We are currently analysing time of flowering (earliness) through breeding programmes in *H. spontaneum*. This trait displays wide geographic variation and has agronomic importance in enhancing cultivated germplasm.

Aside from its theoretical aspects, chromosomal localization and linkage analysis of isozyme loci (McIntosh, 1983; Brown 1983; Brown & Clegg, 1983; Brown *et al.*, 1989c; Tanksley & Orton, 1983; Hart & Gale, 1990) may permit direct genetic analysis of quantitative traits, and substantiate correlative results. It may also facilitate identification of these traits in segregants in breeding programmes, and of traits determined mainly by major genes, such as those for herbicide resistance in wild emmer wheat (Snape *et al.*, 1991a, b). Chromosomal localization of genes can be determined by analysis of addition lines (Islam *et al.*, 1981). As more enzymes and other DNA markers (Beckman & Soller, 1986) are located on the chromosomes, it will become easier and faster to analyse quantitative and single major gene traits of agronomic importance. A detailed review of isozymes in plant genetics and breeding appears in Tanksley & Orton (1983); of DNA markers in Beckman & Soller (1986); and of genetic engineering in Gasser & Fraley (1989). Chromosomal localization of the complex powdery resistant locus *Ml-a* (Moseman & Jorgensen, 1973), as well as the location of *Hor1, 2, 3* (Blake *et al.*, 1982; Shewry *et al.*, 1983), on chromosome 5 of barley, was described earlier. Chromosomal localization of salt-tolerant genes in wheat and barley was described recently by Forster *et al.* (1990). Genes with positive salt tolerance effects were located to chromosomes 4 (4H) and 5 (1H) of *H. vulgare*. The transfer of these genes from *H. spontaneum* to *H. vulgare* should be straightforward.

Genetically engineered plants for crop improvement

Through recombinant DNA technology, molecular biology can now probe much deeper into the genome, giving detailed information on the structure and function of genes. It also provides plant breeders with numerous potential genetic markers (reviewed in Beckman & Soller, 1986) for isolating genes of interest from genomic libraries. Dramatic progress has been made in the development of gene transfer systems for higher plants (Schell, 1987; Gasser & Fraley, 1989). The ability to introduce foreign genes into plant cells and tissues and to regenerate viable, fertile plants, has facilitated an explosive expansion of our understanding of plant biology and provided an unparalleled opportunity to modify and improve crop plants. Genetic engineering of plants offers significant potential for crop improvement. Transgenic plants have already been produced in many species, including herbaceous dicots (e.g. tomato, potato, tobacco), monocots (e.g. asparagus, rice, corn, orchard grass, rye), and woody dicots (e.g. poplar, walnut, apple). In particular, virus-, herbicide- and insect-resistant plants have been produced, and other genes conferring traits of agronomic importance will be transferred in the future.

The wild gene pool of barley and future barley breeding

The wild gene pool of *H. spontaneum* in the entire Near East Fertile Crescent is very rich in genetic variation, notably in important agronomic characters, such as earliness, biomass and yield (Nevo *et al.*, 1984a; Ceccarelli & Grando, 1987; Giles & Lefkovitch, 1985), proportions of lysine and other amino acids (Ahokas, 1977, 1982; Corke & Atsmon, 1988), nitrogen content (Corke *et al.*, 1988), protein content (Nevo *et al.*, 1985; Friedman & Atsmon, 1988), carbohydrate composition (Henry & Brown, 1987; Ahokas & Nashali, 1990), hordein patterns (Doll & Brown, 1979; Nevo *et al.*, 1983) and disease resistances (Moseman *et al.*, 1980; Lehmann & Bothmer (in Jorna & Slootmaker, 1988); Dinoor & Eshed 1990a, b), including a high proportion of resistant genotypes against powdery mildew (Fischbeck *et al.*, 1976a, b; Moseman *et al.*, 1983; Jahoor & Fischbeck 1987a, b; Jana & Nevo, in press) or leaf rust (Moseman *et al.*, 1990; Jana & Nevo, in press); and grain milling energy variation (Ellis, R.P., Forster, B.P. & Nevo, E., in preparation). Furthermore, we now have preliminary evidence that this vast biochemical, morphological and physiological diversity is distributed non-randomly in nature. This information is of paramount importance for optimizing sampling strategies, and for outlining guidelines for *in situ* conservation sites in nature.

The association of ecological (primarily climatic and edaphic factors) and allozyme markers as a predictive method to screen for genotypes resistant against powdery mildew was explored by Nevo *et al.* (1984b). *H. spontaneum* populations in Israel, which grow in the species centre of diversity, contain large amounts of unexploited disease-resistant polymorphisms significantly exceeding in resistance genotypes from other regions of the Fertile Crescent (Iran, Turkey, Syria) (Jana & Nevo, in press). This may also be true for other important agronomic characters discussed above. These traits could be effectively screened and utilized for producing resistant barley varieties by using ecological factors and allozymic markers as guidelines. Transfer of alien genes could proceed by classical plant breeding and/or by modern biotechnology. The use of isozyme-marked segments from *H. spontaneum* in barley breeding was described by Brown *et al.* (1988). Chromosomal localization of major genes and their transgenic utilization in barley and improvement of other crops, may become of paramount importance in the future.

The studies of population genetics structure reported here, as well as the availability of predictive screening methods based on ecology and allozymes, support a basic change in sampling strategies for *in situ* conservation in nature and in gene banks, respectively, and utilization in breeding. The number of sample sites based on the largest ecological spectrum available, should increase. This will increase the probability of collecting and/or preserving *in situ* the locally common alleles rather than concentrating on capturing the rare conspicuous morphological variant by intensive or biased sampling (see critical discussions in Marshall & Brown, 1975; Brown & Munday, 1982; Brown, 1983). Furthermore, if allozymes and restriction fragment length polymorphisms (RFLPs) are indeed good markers for screening disease-resistant genotypes, or genotypes with important agronomic traits such as high yield or earliness, then they could be used on an extensive scale because electrophoresis and RFLP analysis can be conducted quickly and relatively cheaply on a massive scale. By using allozyme markers (Tanksley & Orton, 1983) and RFLPs (Beckman & Soller, 1986) genes of agronomic interest could be chromosomally located, isolated from genomic libraries, and used to engineer improved cultivated barley (Gasser & Fraley, 1989).

The utilization of wild relatives in barley breeding was reviewed by Moseman *et al.* (1980), Grando *et al.* (1985), Brown &

Marshall (1986), and by several papers in Jorna & Slootmaker (1988). The potential of *H. spontaneum* for malting barley was described by Ahokas & Nashali (1990). High-yielding genotypes can be bred from *H. spontaneum* × *H. vulgare* crosses (Vega & Frey, 1981; Brown *et al.*, 1988). The wild germplasm of barley across the Fertile Crescent exhibits vast intra- and inter-population variation in a range of agronomic and quality traits (see earlier), as well as ecological tolerances to extremes of drought, salt and temperature. This vast potential store of genetic resources is as yet largely unexploited. Exploration, sampling, evaluation, documentation *in situ* and *ex situ* conservation, and utilization of these wild resources are of a supreme importance for future breeding. They should be actively and wisely explored and exploited to improve the genetically impoverished cultivated germplasm of barley.

Conclusions and prospects

The demands of an increasing and stabilized food production are extremely urgent in view of the population explosion (11–14 milliard people expected at the end of the twenty-first century). This dictates a concerted effort to advance crop improvement, not only agrotechnically, but, most importantly, genetically. Yet a major problem preventing effective crop improvement, and causing growing concern, lies in the drastic erosion of the genetic resources of the world's crops as compared with those of their ancestors. This process has been accelerated during domestication, particularly over the last century, due to modern pureline breeding practices which have increased genetic homogeneity in the cultivars. Thus, many important food crops have become increasingly susceptible to diseases, pests and environmental stresses.

The present interdisciplinary review of wild barley, *H. spontaneum*, the progenitor of cultivated barley, supports the idea that the greatest hope for future crop improvement lies in rationally and effectively exploring and exploiting the rich gene pool of the plant's wild relative. The centre of origin and diversity of wild barley is the Near East Fertile Crescent. Particularly in Israel, with its extraordinary biotic and genetic diversity, wild barley developed over a long evolutionary history, both within and between populations, a wide range of adaptive diversity to multiple diseases, pests and ecological stress. Most importantly, this variation is neither random nor neutral. By contrast, it displays at all levels adaptive genetic diversity for biochemical, morphological, physiological, and immunological characteristics which contribute to its ability to adapt to diverse climatic and edaphic conditions. The long-lasting co-evolution with parasites and with the ecologically heterogeneous nature of the Fertile Crescent, particularly in Israel, caused the development of numerous adaptive traits. These involve single variable genes, as well as multilocus co-adapted structures, locally and regionally, for both short- and long-term survival. Genetic resources of *H. spontaneum* harbour disease resistances, physiological adaptations for environmental stresses, high protein and nitrogen contents, nutritional quality, and diverse agronomic traits of economic importance. The last group includes growth rates, ripening and yield, in accordance with specific habitats, as well as grain size, grain milling energy variation and malting quality. A sound prediction would be that wild barley will display parallel patterns in other characters already found in wild emmer wheat, such as herbicide resistance (Snape *et al.*, 1991a, b) and photosynthetic yield (Carver & Nevo, 1990; Nevo *et al.*, 1991).

The genetic resources of wild barley, wheat and oats, and most probably of other progenitors of many crops found in Israel and elsewhere across their entire ranges, exhibit non-random, geographic distribution. This distribution is correlated with, and predicted by, climatic and soil factors as well as by allozyme and RFLP markers. We have suggested a predictive methodology, based on ecology and genetic markers to optimize sampling strategies, and maximize the screening and evaluation of élite populations and genotypes appropriate for breeding.

Chromosomal localization, crossing and recombinant DNA techniques are available, and pioneering success has been achieved in transferring alien genes from wild emmer and wild barley to their respective cultivars. The diverse genetic resources of these wild cereals are far from being fully exploited. Our predictive method of screening for optimal genotypes may also be extended to other progenitors, to help in maximizing the conservation and utilization of the genetic resources of wild germplasm. Thus, the genetic resources from the wild gene pool could substantially contribute to crop improvement by means of traditional plant breeding, complemented by the new biotechnology of gene transfer to overcome barriers to crossing due to incompatibility.

Acknowledgements

I am grateful for comments on the manuscript to A. Beiles, A.H.D. Brown, R.V. Bothmer, L. Lehman, P. Shewry and D. Zohary. This research was supported by grants from the Wolfson Foundation and the Israel Discount Bank Chair of Evolutionary Biology; the 'Ancell-Teicher Research Foundation for Genetics and Molecular Evolution' established by Florence and Theodore Baumritter of New York, and by the Humana Inc., Kentucky.

References

Ahokas, H. (1977) Lysine and tryptophan content of some high-protein stocks and species of *Hordeum*. *Hereditas* 86, 143–5.

Ahokas, H. (1980) Cytoplasmic male sterility in barley. 7 Nuclear genes for restoration. *Applied Genetics* 57, 193–202.

Ahokas, H. (1982) Variation of kernel protein and lysine in the wild progenitor of barley. *Hereditas* 96, 29–37.

Ahokas, H. & Nashali, L (1990) Variation of α amylase, β amylase, β glucanase, pullanase, proteinase and chitinase activity in germinated samples of the wild progenitor of barley. *Journal of the Institute of Brewing* 96, 27–32.

Anikster, Y. & Wahl, I. (1979) Coevolution of the rust fungi on *Gramineae* and *Liliaceae* and their host. *Annual Review of Phytopathology* 17, 367–403.

Anonymous (1968) Barley: origin, botany, culture, winter-hardiness, genetics, utlization, pests. *USDA Agriculture Handbook* 338, p. 127.

Avery, D. (1985) U.S. Farm dilemma: The global bad news is wrong. *Science* 230, 408–12.

Baenziger, P.S., Moseman, J.G. & Kilpatrick, R.A. (1981) Registration of barley composite crosses XXXVII-A, -B and -C. *Crop Sciences*, 21, 351–2.

Baum, B.R. (1983) Morphometric relationships in *Hordeum-vulgare* Triticae-Poaceae. 2. *Hordeum agriocrithon, Hordeum-distichum, Hordeum-lagunchliforme, Hordeum spontaneum* and *Hordeum vulgare*. *Canadian Journal of Botany* 61, 2023–31.

Beckman, J.S. & Soller, M. (1986) Restriction fragment length polymorphism in plant genetic improvement. In: Miflin, B.J. (ed.), *Oxford Surveys of Plant Molecular and Cell Biology*. Oxford University Press, Oxford, pp. 196–250.

Biological Journal of the Linnean Society (1990) Bicentenary Joint Meeting with the Institute of Archeology, University College, London, 39, 1–93.

Blake, T.K., Ullrich, S.E. & Nilan, R.A. (1982) Mapping of the *Hor 3* locus encoding 'D' hordeins in barley. *Theoretical and Applied Genetics* 63, 367–71.

Bothmer, R. von, Flink, J., Jacobsen, N., Kotimaki, M. & Landström, T. (1983) Interspecific hybridization with cultivated barley (*Hordeum vulgare* L.). *Hereditas* 99, 219–44.

Bradshaw, A.D. (1972) Some evolutionary consequences of being a plant. *Evolutionary Biology* 5, 25–47.

Brooks, D.A. (1970) Powdery mildew of barley and its control. *Outlook Agriculture* 6, 122–7.

Brown, A.H.D. (1979) Enzyme polymorphism in plant populations. *Theoretical Population Biology* 15, 1–42.

Brown, A.H.D. (1983) Barley. In: Tanksley, S.D. & Orton, T.J. (eds), *Isozymes in Plant Genetics and Breeding*. Elsevier, Amsterdam, pp. 57–74.

Brown, A.H.D. (1984) Multilocus organization of plant populations. In: Wohrmann, K. &

Loeschke, V. (eds), *Population Biology and Evolution*. Springer Verlag, Berlin.
Brown, A.H.D. & Clegg, M.T. (1983) Isozyme assessment of plant genetic resources. In: Rattazzi, M.C., Scandalios, G.J. & Whitt, A.R. (eds), *Isozymes: Current Topics in Biological and Medical Research*, Vol. 11, *Medical and Other Applications*. Alan R. Liss, New York, pp. 285–95.
Brown, A.H.D. & Feldman, M.W. (1981) Population structure of multilocus associations. *Proceedings of the National Academy of Sciences (USA)* 78, 5913–16.
Brown, A.H.D. & Jacobsen, J. (1982) Genetic basis and natural variation of α-amlyase isozymes in barley. *Genetical Research (Cambridge)* 40, 315–24.
Brown, A.H.D. & Marshall, D.R, (1981) Evolutionary changes accompanying colonization in plants. In: Scudder, G.G.E. & Reveal, J.L. (eds), *Evolution Today* (Proceedings of the Second International Congress of Systematic and Evolutionary Biology), pp. 351–63.
Brown, A.H.D. & Marshall, D.R. (1986) Wild species as genetic resources for plant breeding. In: Williams, T.A. & Wratt, G.S. (eds), *Plant Breeding Symposium DSIR*. Christchurch, Agronomy Society of New Zealand, pp. 9–13.
Brown, A.H.D. & Munday, J. (1982) Population genetic structure and optimal sampling of land races of barley from Iran. *Genetica* 58, 85–96.
Brown, A.H.D., Nevo, E. & Zohary, D. (1977) Association of alleles at esterase loci in wild barley, *Hordeum spontaneum*. *Nature* 268, 430–2.
Brown, A.H.D., Nevo, E., Zohary, D. & Dagan, O. (1978b) Genetic variation in natural populations of wild barley *Hordeum spontaneum*. *Genetica* 49, 97–108.
Brown, A.H.D., Zohary, D. & Nevo, E. (1978a) Outcrossing rates and heterozygosity in natural populations of *Hordeum spontaneum* Koch in Israel. *Heredity* 41, 49–62.
Brown, A.H.D., Feldman, M.W. & Nevo, E. (1980) Multilocus structures of natural populations of *Hordeum spontaneum*. *Genetics* 96, 523–36.
Brown, A.H.D., Munday, J. & Oram, R.N. (1988) Use of isozyme-marked segments from wild barley *Hordeum spontaneum* in barley breeding. *Plant Breeding* 100, 80–8.
Brown, A.H.D., Clegg, M.T., Kahler, A.L. & Weir, B.S. (eds) (1989a) *Plant Population Genetics, Breeding, and Genetic Resources*. Sinauer, Sunderland, MA, USA.
Brown, A.H.D., Frankel, O.H., Marshall, D.R. & Williams, J.T. (eds) (1989b) *The Use of Plant Genetic Resources*. Cambridge University Press, New York.
Brown, A.H.D., Lawrence, G.J., Jenkin, M., Douglass, J. & Gregory E. (1989c) Linkage drag in backcross breeding in barley. *Journal of Heredity* 80, 234–9.
Carver, B.F. & Nevo, E. (1990) Genetic diversity of photosynthetic characters in native populations of *Triticum dicoccoides*. *Photosynthesis Research* 25, 119–28.
Ceccarelli, S. & Grando, S. (1987) Diversity for morphological and agronomic characters in *Hordeum vulgare ssp. spontaneum* C. Koch. *Genetica Agraria* 41, 131–42.
Chakrabarti, T. & Subrahmanyan, N.C. (1986) Analysis of DNA from related and diverse species of barley. *Plant Science* 42, 183–90.
Clegg, M.T., Allard, R.W. & Kahler, A.L. (1972) Is the gene the unit of selection? Evidence from two experimental plant populations. *Proceedings of the National Academy of Sciences (USA)* 69, 2474–8.
Clegg, M.T., Brown, A.H.D. & Whitefeld, P.R. (1984) Chloroplast DNA diversity in wild and cultivated barley: implications for genetic conservation. *Genetical Research (Cambridge)* 43, 339–43.
Corke, H. & Atsmon, D. (1988) Amino acid resources of the wild relative of barley *Hordeum spontaneum* in Israel. *Plant Breeding* 101, 89–94.
Corke, H., Nevo, E. & Atsmon, D. (1988) Variation in vegetative parameters related to the nitrogen economy of wild barley, *Hordeum spontaneum*, in Israel. *Euphytica* 39, 227–32.
Dinnor, A. (1974) Role of wild and cultivated plants in the epidemiology of plant diseases in Israel. *Annual Review of Phytopathology* 12, 413–36.
Dinnor, A. & Eshed, N. (1990a) The analysis of host and pathogen populations in natural ecosystems. In: Wolfe, M.S. & Caten, C.E. (eds), *Populations of Plant Pathogens: Their Dynamics and Genetics*. Blackwell Scientific, Oxford, pp. 75–88.
Dinnor, A. & Eshed, N. (1990b) Plant diseases in natural populations of wild barley (*Hordeum spontaneum*). In: Burdon, J.J. & Leather, S.R. (eds), *Pests, Pathogens and Plant Communities*. Blackwell Scientific, Oxford, pp. 169–86.

Doll, H. & Brown, A H.D. (1979) Hordein variation in wild *Hordeum spontaneum* and cultivated *Hordeum vulgare* barley. *Canadian Journal of Genetics and Cytology* 21, 391–404.

Feldman, M. & Sears, E.R. (1981) The wild gene resources of wheat. *Scientific American* 244, 102–12.

Fischbeck, G., Schwarzbach, E., Sobek, F. & Wahl, I. (1976a) Mehltauresistenz aus Israelischen populationen der zwizeiligen wildgerste (*Hordeum spontaneum*). *Annual Pflanzenzucht* 76, 163–6.

Fischbeck, G., Schwarzbach, E., Sobek, F. & Wahl, I. (1976b) Types of protection against barley powdery mildew in Germany and Israel selected from *Hordeum spontaneum*. In: Guel, H. (ed.), *Barley Genetics III*. Verlag Karl Thieming, Munchen, pp. 412–17.

Forster, B.P., Law, C.N., Nevo, E. & Powell, W. (1990) Genetic strategies for improving the salt tolerance of wheat and barley. *Proceedings of the International Symposium on Molecular and Genetic Approaches to Plant Stress*. New Dehli, India, pp. T31.1–3.

Frankel, O.H. & Bennett, E. (1970) *Genetic Resources in Plants – their Exploration and Conservation*. Blackwell, Oxford.

Frankel, O.H. & Hawkes, J.G. (eds) (1975) *Crop Genetic Resources for Today and Tomorrow*. Int. Biol. Progr. 2. Cambridge University Press, Cambridge.

Frankel, O.H. & Soulé, M.E. (1981) *Conservation and Evolution*. Cambridge University Press, Cambridge.

Friedman, M. & Atsmon, D. (1988) Comparison of grain composition and nutritional quality in wild barley, *Hordeum spontaneum* and in a standard cultivar. *Journal of Agricultural Food Chemistry* 36, 1167–72.

Gasser, C.S. & Fraley, R.T. (1989) Genetically engineering plants for crop improvement. *Science* 244, 1293–9.

Giles, B.E. & Bengtson, B.O. (1988) Variation in anther size in wild barley *Hordeum vulgare* ssp. *spontaneum*. *Hereditas* 108, 199–206.

Giles, B.W. & Lefkovitch, L.P. (1985) Agronomic differences in *Hordeum spontaneum* from Iran and Morocco. *Zeitschrift für Pflanzensüchtung* 94, 25–40.

Grando, S., Falistocco, E. & Ceccarelli, S. (1985) Use of wild relatives in barley breeding. *Genetica Agraria* 39, 65–76.

Gustafsson, A. & Lundquist, I. (1980) Hexastichon and intermedium mutants in barley. *Hereditas* 92, 229–36.

Harlan, J.R. (1968) On the origin of barley. *USDA Agriculture Handbook* 338, 9–31.

Harlan, J.R. (1975) Our vanishing genetic resources. *Science* 188, 618–21.

Harlan, J.R. (1976a) Barley. In: Simmonds, N.W. (ed.), *Evolution of Crop Plants*. Longman, London, pp. 93–8.

Harlan, J.R. (1976b) Genetic resources in wild relatives of crops. *Crop Sciences* 16, 329–33.

Harlan, J.R. & Zohary, D. (1966) Distribution of wild wheats and barley. *Science* 153, 1074–80.

Hart, G. & Gale, M.D. (1990) Biochemical molecular loci of hexaploid wheat (*Triticum aestivum* $2n = 42$, Genomes AABBDD). In O'Brien, S.J. (ed.), *Genetic Maps 1990. A Compilation of Linkage and Restriction Maps of Genetically Studied Organisms*. Vol. 5, 28–38.

Hedrick, P.W., Jain, S. & Holden, L. (1978) Multilocus systems in evolution. *Evolutionary Biology* 11, 101–82.

Helbaek, H. (1966) Commentary on the phylogenesis of *Triticum* and *Hordeum*. *Economic Botany* 20, 350–60.

Henry, R.J. & Brown, A.H.D. (1987) Variation in the carbohydrate composition of wild barley (*Hordeum spontaneum*) grain. *Plant Breeding* 98, 97–103.

Holwerda, B.C., Jana, S. & Crosby, W.L. (1986) Chloroplast and mitochondrial DNA variation in *Hordeum vulgare* and *Hordeum spontaneum*. *Genetics* 114, 1201–92.

Horsfall, J.G. & Cowling, E.B. (eds) (1977) *Plant Diseases. An Advanced Treatise*, Vol 1. *How Disease is managed*. Academic, New York.

Islam, A.K.M.R., Shepherd, K.W. & Sparrow, D.H.B. (1981) Isolation and characterization of euplasmic wheat–barley chromosome addition lines. *Heredity* 46, 161–74.

Jahoor, A. & Fischbeck, G. (1987a) Genetical studies of resistance of powdery mildew in barley lines derived from *Hordeum spontaneum* collected from Israel. *Plant Breeding* 99, 265–73.

Jahoor, A. & Fischbeck, G. (1987b) Sources of resistance to powdery mildew in barley lines derived from *Hordeum spontaneum* collected in Israel. *Plant Breeding* 99, 274–81.

Jana, S. & Pietrzak, L.N. (1988) Comparative assessment of genetic diversity in wild and primitive cultivated barley in a center of diversity. *Genetics* 119, 981–90.

Jana, S. & Nevo, E. Variation in response to infection with *Erysiphe graminis hordei* and

Puccinia hordei in some wild barley populations in a centre of diversity. *Canadian Journal of Botany* (in press).

Jaradat, A.A. (1989a) Diversity within and between populations of two sympatrically distributed *Hordeum*-spp. in Jordan. *Theoretical and Applied Genetics* 78, 653–6.

Jaradat, A.A. (1989b) Ecotypes and genetic divergence among sympatrically distributed populations of *Hordeum-vulgare* and *Hordeum-spontaneum* from the xeric region of Jordan. *Theoretical and Applied Genetics* 78, 857–62.

Jorna, M. & Slootmaker, L.A.J. (eds) (1988) *Cereal Breeding Related to Integrated Cereal Production* (Proceedings Conference of the Cereal Section of Eucarpia (European Association for Research on Plant Breeding) Wageningen, Netherlands, February 24–26, 1988). Center for Agricultural Publishing and Documentation (Pudoc), Wageningen, The Netherlands.

Kahler, A.L. & Allard, R.W. (1981) Worldwide patterns of genetic variation among 4 esterase loci in barley *Hordeum vulgare. Theoretical and Applied Genetics* 59, 101–11.

Karlin, S. (1975) General two-locus selection models. Some objectives, results and interpretations. *Theoretical Population Biology* 7, 364–98.

Ladizinsky, G. (1975) Collection of wild cereals in the Upper Jordan Valley. *Economic Botany* 29, 264–7.

Manisterski, J., Treeful, L.M., Tomerlin, J.R., Anikster, Y., Moseman, J.G., Wahl, I. & Wilcoxson, R.D. (1986) Resistance of wild barley *Hordeum spontaneum* accessions from Israel to leaf rust *Puccinia-hordei*, collected in the USA and Israel. *Crop Science* 26, 727–30.

Marshall, D.R. & Brown, A.H.D. (1975) Optimum sampling strategies in genetic conservation. In: Frankel, O.H. & Jawkes, J.G. (eds), *Crop Genetic Resources for Today and Tomorrow*. Cambridge University Press, Cambridge, pp. 58–80.

McIntosh, R.A. (1983) A catalogue of gene symbols for wheat (1983 edition). In: Sakamoto, S. (ed.), *Proceedings of the Sixth International Wheat Genetics Symposium*. Alan R. Liss, New York, pp. 1197–254.

Ministry of Science and Technology, Israel (1990) *International Workshop on dynamic* in situ *conservation of wild relatives of major cultivated plants*. Scientific Programs and Abstracts. National Council Research Development, Israel.

Moseman, J.G. & Craddock, J.C. (1976) Genetic basis for barley germ-plasm conservation. In: Gaul, H. (ed.), *Barley Genetics III*. Verlag Karl Thieming, Munich, pp. 51–7.

Moseman, J.G. & Jorgensen, J.H. (1973) Differentiation of resistance genes at the *Ml-a* locus in six pairs of isogenic barley lines. *Euphytica* 22, 189–96.

Moseman, J.G., Baenziger, P.F. & Kilpatrick, R.A. (1980) *Hordeum spontaneum* – an overlooked source of disease resistance. *Europe and Mediterranean Cereal Rust Foundation*, pp. 91–3.

Moseman, J.G., Baenziger, P.F. & Kilpatrick, R.A. (1981) Genes conditioning resistance of *Hordeum spontaneum* to *Erysiphe graminis* f. sp. *hordei*. *Crop Science* 21, 229–32.

Moseman, J.G., Nevo, E. & Zohary, D. (1983) Resistance of *Hordeum spontaneum* collected in Israel to infection with *Erysiphe graminis hordei*. *Crop Science* 23, 1115–19.

Moseman, J.G., Nevo, E. & El-Morshidy, M.A. (1990) Reactions of *Hordeum spontaneum* to infection with two cultures of *Puccinia hordei* from Israel and United States. *Euphytica* 49, 169–75.

Neale, D.B., Saghai-Maroof, M.A., Allard, R.W., Zhang, Q. & Jorgensen, R.A. (1988) Chloroplast DNA diversity in populations of wild and cultivated barley. *Genetics* 120, 1105–10.

Nei, M. (1972) Genetic distance between populations. *American Naturalist* 106, 283–92.

Nei, M. (1973) Analysis of gene diversity in subdivided populations. *Proceedings of the National Academy of Sciences (USA)* 70, 3321–3.

Nevo, E. (1983) Genetic resources of wild emmer wheat: structure, evolution and application in breeding. In: Sakamoto, S. (ed.), *Proceedings of the 6th International Wheat Genetic Symposium*. Alan R. Liss, New York, pp. 421–31.

Nevo, E. (1986) Genetic resources of wild cereals and crop improvement: Israel, a natural laboratory. *Israel Journal of Botany* 35, 255–78.

Nevo, E. (1987) Plant genetic resources: prediction by isozyme markers and ecology. In: Rattazzi, M.C., Scandalios, J.G. & Whitt, G.S. (eds), *Isozymes: Current Topics in Biological and Medical Research*. Vol. 16, *Agriculture, Physiology, and Medicine*. Alan R. Liss, New York, pp. 247–67.

Nevo, E. (1988a) Genetic resources of wild emmer wheat revisited: genetic evolution, conservation and utilization. *Proceedings of the 7th International Wheat Genetics Symposium*. Cambridge, UK, pp. 121–6.

Nevo, E. (1988b) Genetic diversity in nature: patterns and theory. *Evolutionary Biology* 23, 217–46.

Nevo, E. & Beiles, A. (1989) Genetic diversity of wild emmer wheat in Israel and Turkey: structure, evolution, and application in breeding. *Theoretical and Applied Genetics* 77, 421–55.

Nevo, E., Shimony, T. & Libni, M. (1977) Thermal selection of allozyme polymorphisms in barnacles. *Nature* 267, 699–701.

Nevo, E., Brown, A.H.D. & Zohary, D. (1979a) Genetic diversity in the wild progenitor of barley in Israel. *Experientia* 35, 1027–9.

Nevo, E., Brown, A.H.D. & Zohary, D. (1979b) Genetic diversity and environmental associations of wild barley *Hordeum spontaneum* in Israel. *Evolution* 33, 815–33.

Nevo, E., Brown, A.H.D., Zohary, D., Storch, N. & Beiles, A. (1981) Microgeographic edaphic differentiation of allozyme polymorphism of wild barley. *Plant Systematics and Evolution* 138, 287–92.

Nevo, E., Beiles, A., Storch, N., Doll, H. & Anderson, B. (1983) Microgeographic edaphic differentiation in hordein polymorphisms of wild wheat. *Theoretical and Applied Genetics* 64, 123–32.

Nevo, E., Beiles, A., Gutterman, Y., Storch, N. & Kaplan, D. (1984a) Genetic resources of wild cereals in Israel and vicinity: II. Phenotypic variation within and between populations of wild wheat, *Hordeum spontaneum. Euphytica* 33, 737–56.

Nevo, E., Moseman, J.G., Beiles, A. & Zohary, D. (1984b) Correlation of ecological factors and allozymic variations with resistance to *Erysiphe graminis hordei* in *Hordeum spontaneum* in Israel: patterns and application. *Plant Systematics and Evolution* 145, 79–96.

Nevo, E., Beiles, A. & Ben-Shlomo, R. (1984c) The evolutionary significance of genetic diversity: ecological, demographic and life history correlates. *Lecture Notes Biomathematics* 53, 13–213.

Nevo, E., Atsmon, D. & Beiles, A. (1985) Protein resources in wild barley *Hordeum spontaneum* in Israel. Predictive method by ecology and allozyme markers. *Plant Systematic Evolution* 150, 205–22.

Nevo, E., Beiles, A., Kaplan, D., Storch, N. & Zohary, D. (1986a) Genetic diversity and environmental associations of wild barley, *Hordeum spontaneum* (Poaceae), in Iran. *Plant Systematics and Evolution* 153, 141–64.

Nevo, E., Zohary, D., Beiles, A., Kaplan, D. & Storch, N. (1986b) Genetic diversity and environmental associations of wild barley, *Hordeum spontaneum* in Turkey. *Genetica* 68, 201–13.

Nevo, E., Beiles, A. & Zohary, D. (1986c) Genetic resources of wild barley in the Near East: structure, evolution and application in breeding. *Biological Journal of the Linnean Society* 27, 355–80.

Nevo, E., Beiles, A., Kaplan, D., Golenberg, E.M., Olsvig-Whittaker, L.S. & Naveh, Z. (1986d) Natural selection of allozyme polymorphisms: a microsite test revealing ecological genetic differentiation. *Evolution* 40, 13–20.

Nevo, E., Beiles, A. & Krugman, T. (1988a) Natural selection of allozyme polymorphisms: a microgeographical differentiation by edaphic, topographical, and temporal factors in wild emmer wheat (*Triticum dicoccoides*). *Theoretical and Applied Genetics* 75, 529–38.

Nevo, E., Beiles, A. & Krugman, T. (1988b) Natural selection of allozyme polymorphisms: a microgeographic climatic differentiation in wild emmer wheat (*Triticum dicoccoides*). *Theoretical and Applied Genetics* 76, 737–52.

Nevo, E., Noy-Meir, E., Beiles, A., Krugman, T. & Agami, M. Natural selection of allozyme polymorphisms: a microgeographic spatial and temporal ecological differentiation in wild emmer wheat (*Triticum dicoccoides*). *Israel Journal of Botany* (in press).

Nevo, E., Carver, B.F. & Beiles, A. (1991) Photosynthetic performance in wild emmer wheat *Triticum dicoccoides*: ecological and genetic predictability. *Theoretical and Applied Genetics* 81, 445–60.

Nilan, R.A. (1964) The cytology and genetics of barley, 1957–1962. *Research Studies*, Washington State University 32, 278 pp.

Nilan, R.A. (ed.) (1971) *Barley Genetics II.* (Proceedings 2nd International Barley Genetic Symposium, Washington), 622 pp.

Plucknett, D.L., Smith, N.J.H., Williams, J.T. & Anishetty, N.M. (1983) Crop germplasm conservation and developing countries. *Science* 220, 163–9.

Plucknett, D.L., Smith, N.J.H., Williams, J.T. & Anishetty, N.M. (1987) *Gene Banks and the World's Food.* Princeton University Press, Princeton, NJ, USA.

Schell, J.S.T. (1987) Transgenic plants as tools to study the molecular organization of plant genes. *Science* 237, 1176–83.

Segal, A., Manisterki, J., Fischbeck, J. & Wahl, I. (1980) How plant populations defend themselves in natural ecosystems. In: Horsfall, J.G. & Cowling, E.B. (eds), *Plant Disease*, Vol. V. Academic, New York, pp. 75–102.

Segal, A., Manisterski, J., Browning, J.A. Fischbeck, J. & Wahl, I. (1982) Balance in indigenous plant populations. In: Heybroek, H.M., Stephan, B.R. & Van Weissenberg, K. (eds), *Resistance to Disease and Pests in Forestries*. Center for Agriculture Publishing and Documentation (Pudoc), Wageningen, pp. 361–70.

Shewry, P.R., Finch, R.A., Parmar, S., Franklin, J. & Miflin, B.J. (1983) Chromosomal location of *Hor 3*, a new locus governing storage proteins in barley. *Heredity* 50, 179–89.

Slootmaker, L.A.J. (1974) Aims and objectives in breeding cereal varieties. *Outlook Agriculture* 8, 133–40.

Snape, J.W., Leckie, D., Parker, B.B. & Nevo, E. (1991a) The genetical analysis and exploitation of differential responses to herbicides in crop species. In: *Proceedings of the 11th Herbicide Resistant Symposium*, Long Ashton, Bristol, UK. Butterworths, Oxford, pp. 305–19.

Snape, J.W., Nevo, E., Parker, B.B., Leckie, D. & Morgunov, A. (1991b) Herbicide response polymorphisms in wild populations of emmer wheat *Heredity*, 66, 251–7.

Snow, L. & Brody, T. (1984) Genetic variation of *Hordeum spontaneum* in Israel, ecogeographical races detected by trait measurements. *Plant Systematics and Evolution* 145, 15–28.

Takahashi, R. (1955) The origin and evolution of cultivated barley. *Advances in Genetics* 7, 227–66.

Tanksley, S.D. & Orton, T.J. (eds) (1983) *Isozyme in Plant Genetics and Breeding*. Part A, 516 pp; Part B, 472 pp Elsevier, Amsterdam.

Tanksley, D., Medina-Filho, H. & Rick, C.M. (1982) Use of naturally occurring enzyme variation to detect and map genes controlling quantitative traits in an interspecific backcross of tomato. *Heredity* 49, 11–25.

Thoday, J.M. (1961) Location of polygenes. *Nature* 191, 368–70.

Thompson, J.N. Jr. & Thoday, J.M. (eds) (1979) *Quantitative Genetic Variation*. Academic, New York.

Treeful, L.M. & Wilcoxson, R.D. (1986) Specific and slow rusting of *Hordeum spontaneum* infected with three races of *Puccinia hordei*. *Phytopathology* 76, 1005.

Vavilov, N.I. (1926) *Studies on the Origin of Cultivated Plants*. Institut Botanique Applique et d'Amelioration des Plantes, Leningrad.

Vavilov, N.I. (1951) The origin, variation, immunity and breeding of cultivated plants (translated from the Russian by K. Starr Chester). *Chronica Botanica* 13, i–xviii, 1–364.

Vega, U. & Frey, K.J. (1981) Transgressive segregation in interspecific and intraspecific crosses of barley *Hordeum*. *Euphytica* 29, 585–94.

Wahl, I., Eshed, N., Segal, A. & Sobel, Z. (1978) Significance of wild relatives of small grains and other wild grasses in cereal powdery mildews. In: Spencer, D.M. (ed.), *Powdery Mildews*. Academic, London, pp. 83–100.

Weir, B.S. (1979) Inferences about linkage disequilibrium. *Biometrics* 35, 235–54.

Zohary, D. (1969) The progenitors of wheat and barley in relation to domestication and agriculture dispersal in the Old World. In: Ucko, P.J. & Dimbleby, G.W. (eds), *The Domestication and Exploitation of Plants and Animals*. Duckworth, London, pp. 47–66.

Zohary, D. (1970) Centers of diversity and centers of origin. In: Frankel, O.H. & Bennett, E. (eds), *Genetic Resources in Plants – their Exploration and Conservation*. Blackwell, Oxford, pp. 33–42.

Zohary, D. (1973) *Geobotanical foundations to the Middle East*, Vols 1 and 2. G. Fischer, Stuttgart and Swets and Zeitlinger, Amsterdam.

Zohary, D. (1983) Wild genetic resources of crops in Israel. *Israel Journal of Botany* 21, 97–122.

Zohary, D. & Hopf, M. (1988) *Domestication of Plants in the Old World*. Clarendon, Oxford.

3

Intergeneric Hybrids with *Hordeum*

George Fedak

Cultivated barley, *Hordeum vulgare* L., is one of the major world cereal crops ranking fourth in world acreage and production behind wheat, corn and rice. In 1988, barley occupied over 72 million hectares and produced 169 million metric tonnes of grain.

Hordeum vulgare is one of the species in the genus *Hordeum* within the tribe Triticeae that consists of 24 other species at three ploidy levels (Jacobsen & Bothmer, 1981). The interspecific aspect of wide crosses in barley has been quite extensive and will be dealt with in a separate chapter.

The presumed close phytogenetic relationships between genera within the tribe Triticeae has been a stimulus to plant geneticists to attempt to obtain hybrids between species of the various genera. The elucidation of crossability genes, identification of crossable genotypes and impact of improved tissue culture techniques on the embryo rescue process have served to improve success rates in wide crosses in the major cereal crops and their wild relatives. Approximately 325 species of temperate climate grasses are represented in the tribe, and thus an unlimited resource of genetic variability exists not only for agronomic traits but for genetic parameters such as additional crossability genes, mechanisms for altering meiotic pairing controls, mechanisms affecting chromosome elimination and mechanisms for molecular manipulations. Thus the interests in intergeneric hybridization have a practical and applied basis.

Barley–wheat hybrids

Plant geneticists have been interested in hybridizing barley with wheat for more than 100 years but without any success, until 1973 when Anton Kruse demonstrated that diploid *Hordeum vulgare* L. could be crossed as a maternal parent with diploid (*Triticum monococcum*), tetraploid (*T. turgidum*, L.) and hexaploid (*T. aestivum* L.) wheat (Kruse, 1973). This initial success stimulated considerable additional activity so that numerous hybrid combinations have now been reported. A recent compilation shows that 32 hybrid combinations have been produced between diploid barley cultivars as maternal parents with hexaploid wheat in 17 reports (Table 3.1). The highest crossability, expressed as number of plants produced as a proportion of florets pollinated was 3.7% in a Betzes × Chinese

Table 3.1. Hybrids between cultivars of *H. vulgare* and *T. aestivum*.

Cultivar combinations	References
Bonus × Koga	Kruse (1973)
Emir × Cato	Kruse (1973)
Lofa × Starke	Kruse (1973)
Manker × Tobari	Thomas *et al.* (1977)
Manker × Chinese Spring (CS)	Mujeeb-Kazi (1979)
Manker × Bonza	Mujeeb-Kazi (1979)
Manker × Pavon	Mujeeb-Kazi (1979)
Manker × Zaragoza 75	Mujeeb-Kazi (1979)
Celaya × CS	Mujeeb-Kazi (1979)
Celaya × Nyu Bay	Mujeeb-Kazi (1979)
Celaya × Zaragoza 75	Mujeeb-Kazi (1979)
Dickson-Hiproly × Zaragoza 75	Mujeeb-Kazi (1979)
Betzes × CS	Islam *et al.* (1975)
Betzes × CS	Cauderon *et al.* (1978)
Betzes × CS	Fedak (1980)
Ketch × CS	Islam *et al.* (1975)
Clipper × CS	Islam *et al.* (1975)
Prior × CS	Islam *et al.* (1975)
Ketch × Gabo	Islam *et al.* (1975)
Clipper × Gabo	Islam *et al.* (1975)
Ketch × Falcon	Islam *et al.* (1975)
Betzes × Koga	Fedak (1982)
Bonus × Koga	Fedak (1982)
Bonus × CS	Fedak (1982)
Emir × CS	Fedak (1982)
Sofia × Saratovskaya 29	Shumny *et al.* (1981)
Galina × Saratovskaya 29	Shumny *et al.* (1981)
PF 51811 × CS	Chu *et al.* (1984)
PF 5801 × CS	Chu *et al.* (1984)
Emir × CS	Wojciechowska (1985)
Boashan × CS	Jiang (1986)
Jumbo × CS	Wojciechowska (1986)
Emir × CS	Wojciechowska (1986)
Carina × WW T. aestivum 427, L-86	Surikov & Kissel (1987)
Martonvasari 50 × CS	Molnar-Lang *et al.* (1991)

Spring (CS) combination (Islam *et al.*, 1975) and the same in a Jumbo × CS combination (Wojciechowska, 1985a). The lowest reported crossability was 0.13% in a Manker × Tobari combination (Thomas *et al.*, 1977). The wheat cultivar used most frequently was CS followed by Koga while the most frequently used barley cultivars were Manker, Emir and Betzes. By crossing the CS–Hope series of substitution lines on to Betzes barley it was shown that crossability genes in CS were located on chromosomes 5A, 5B and 5D (Fedak & Jui, 1982). It has not been shown as yet whether the loci controlling crossability of wheat with barley are the same as those controlling crossability of wheat with rye or wheat with *H. bulbosum* though the same chromosomes are involved.

With the exception of one report of an alloplasmic haploid ($2n = 21$) obtained from a Betzes × CS cross (Fedak, 1980), the chromosome numbers of all other hybrids were $2n = 28$ as expected and the chromosomes were somatically stable. In contrast, considerable variation in chromosome number was observed between meiocytes at metaphase I (MI) of meiosis in the hybrids. For example, ranges in chromosome number of 22–42 (Wojciechowska, 1985a) and 28–63 (Fedak, 1980) were reported in some studies, and pollen mother cells (PMCs) with 56 chromosomes in another combination (Surikov *et al.*, 1986).

Meiotic chromosome pairing in F_1 hybrids was generally low with some variation between individual hybrids, e.g. a range in chiasma frequency of 0.38–3.27 was reported for the hybrids Emir × CS and Bonus × Koga respectively (Fedak, 1982), Mujeeb-Kazi & Rodriquez (1982) reported a range of 2.47–2.63, Wojciechowska (1985a) a range of 1.17–1.98, Cauderon *et al.* (1978) a level of 1.16, and Jiang (1986) a level of 0.45. The differences observed in chromosome pairing may have been caused by genomic interactions between meiotic pairing control mechanisms or simply by differences in autosyndesis and thus provide no indication of homology between parental genomes. Other common observations at MI in the hybrids were secondary associations, isochromosomes and precocious separations of chromosomes at anaphase. Some of these abnormalities were attributed to the formation of multipolar spindles at premeiotic mitosis (Fedak, 1979a).

All barley–wheat hybrid combinations were wheat-like in morphology, completely self-sterile and no one has been successful thus far in obtaining a fertile amphiploid from this combination. The hybrids were female-fertile, however, and when pollinated by wheat gave rise to progenies with chromosome numbers of 49, 48 or 47 (Islam *et al.*, 1975; Fedak, 1980). Apomictic barley–wheat backcross progenies with chromosome numbers similar to F_1 hybrids ($2n = 28$ and $2n = 21$) were recovered in both *T. aestivum* and *T. turgidum* backcross progenies (Mujeeb-Kazi, 1981). The 49 chromosome backcross-1 progenies were the result of the fertilization by wheat of a 28 chromosome egg cell that arose through a restitution nucleus in the F_1 hybrid plant (Islam & Shepherd, 1980). The backcross (BC) progeny were also female-fertile so that with further backcrossing, BC_2 progeny with chromosome numbers of 44–52 were obtained (Fedak, unpublished). Pistilloidy, accompanied by male and female sterility which was attributed to an incompatible interaction between the chromosomes of the wheat parent and the cytoplasm of barley, occurred at this stage in reverse proportion to the number of barley chromosomes present. Thus the attempts to produce addition lines of barley in wheat by backcrossing of wheat on to barley × wheat hybrids have all resulted in failure because of the problems of pistilloidy even though some laboratories were able to advance the backcrossing up to BC_5 (Pershina *et al.*, 1986) to produce progenies with partial fertility and variable chromosome numbers. A number of laboratories have now succeeded in regenerating barley–wheat hybrid plants from callus induced on immature inflorescence explants. The crossability of regenerated plants was higher than that of the original hybrids even though the meiotic stability of regenerants was lower (Molnar-Lang *et al.*, 1990; Fedak, unpublished), however there are no reports of prolonged backcrossing attempts with such materials.

Wheat–barley hybrids

It has been much more difficult to obtain hybrids with wheat as the maternal parent than it was with the reciprocal cross described above. With wheat as the maternal parent the crossability varied from 0.04% in a CS × Ketch hybrid (Islam *et al.*, 1978) to 1.7% in a Fukuhokomugi × Luther combination (Jauhar, in press). In the earliest attempts it was necessary to use immature endosperms of one of the parents as nurse cultures to rescue the tiny undifferentiated hybrid embryos. Thus far, progeny have been reported from ten hybrid combinations (Table 3.2). Many of these combinations yielded, in addition to hybrids with 28

Table 3.2. Hybrids between cultivars of *Triticum aestivum* and *Hordeum vulgare*.

Cultivar combinations	References
CS × Betzes	Islam *et al.* (1978)
	Fedak (1980)
CS × Ketch	Islam *et al.* (1981)
T. aestivum × *H. vulgare*	Kruse (1976)
T. durum × *H. vulgare*	Kruse (1976)
CS × *H. vulgare*	Korolyuk & Simonenko (1984)
CS (*ph1b*) × Tuleen 346	Sethi *et al.* (1986)
CS × Gatchinsky	Surikov *et al.* (1986)
Lines 403, 405, 407 × Bemir-1, Cyclone,	Surikov & Kissel (1987)
421, 427, 430 Urfa	Surikov & Kissel (1987)
CS × CI 3208–4, Atlas	McGuire & Qualset (1990)
Fukuhokomugi × Luther	Jauhar (in press)

chromosomes, progeny with chromosome numbers of 21 (Islam *et al.*, 1978; Fedak, 1980; Jauhar, in press), 31 (Sethi *et al.*, 1986), 36 (Islam *et al.*, 1978) and 56 (Kruse, 1976). Some combinations of wheat and barley hybrids gave no 28 chromosome progeny but only 21 and 22 chromosome plants (Finch & Bennett, 1982) or 23 or 26 chromosome plants (Fedak, unpublished). It was suggested by Shepherd & Islam (1981) that abnormal disjunction at mitosis in young embryos may result in the complete elimination of barley chromosomes or in duplication or deficiency of chromosomes, particularly those of barley.

In addition to the variable chromosome numbers observed between individual hybrids, chromosomal mixoploidy was observed within cells of individual plants giving chromosome numbers of 23–27 in root meristems (Sethi *et al.*, 1986) and 28–56 in meiocytes (Fedak, 1980). Chromosome pairing in 28 chromosome hybrids with CS as the maternal parent gave average chiasma frequencies of 0.72 (Islam *et al.*, 1978) to 1.21 (Fedak, 1980) but this increased to 6.46 when the CS *ph1b* mutant was used as the maternal parent (Sethi *et al.*, 1986). The observed pairing, however, was restricted to homoeologous associations with no evidence for intergenomic pairing.

Chiasma frequencies of 5.18 and 6.28 with up to ten bivalents in some cells and heteromorphic bivalents, trivalents and quadrivalents in others were detected in hybrids of Fukuhokomugi (Fukuho) × Luther (Jauhar, in press). The obvious intergenomic pairing was attributed to the effect of the Luther genotype on the *Ph1* system of Fukuho. Islam & Shepherd (1988) were able to induce low levels of pairing of barley telocentrics with homoeologous arms of wheat chromosomes by using either wheat *ph1b* mutants or the *Ph* suppressing activity of *Aegilops speltoides*. In this way recombinant chromosomes were produced for barley chromosome mapping studies.

Wheat and barley hybrid plants have all been wheat-like in morphology and self-sterile. They were female-fertile, however, and through repeated backcrossing the addition lines of Betzes barley chromosomes in CS have been produced (Islam & Shepherd, 1978). The disomic addition lines are meiotically stable and quite fertile with the exception of barley chromosome 5 which was sterile and could not be maintained. However, for restriction fragment length polymorphism (RFLP) studies for example, the DNA has been extracted from the monosomic addition of that chromosome (Islam, personal communication). A translocation line involving Betzes 5S with wheat 2A is also available (Shepherd & Islam, 1987). Based on a combination of isozyme similarity and morphological compensation, the homoeology between barley and wheat chromosomes was determined (Islam & Shepherd, 1981). Barley chromosomes 1–7 (numbered according to Tsuchiya, 1983) were homoeologous to wheat chromosomes 7, 2, 3, 4, 1, 6 and 5 respectively. Similarly Benito *et al.* (1985) by means of isozyme analysis were able to show that Betzes chromosomes 3, 4, 5 and 6 were homoeologous to rye chromosomes 3, 4, 1 and 6 respectively. The 12 ditelosomic additions of Betzes have been produced from the corresponding monosomic additions (Islam, 1983). A number of substitutions of barley for wheat chromosomes have also been produced. These include the substitution of barley chromosomes 3, 6 and 7 for all the wheat chromosomes of homoeologous groups 3, 6 and 5 respectively plus barley chromosome 1 for 7B of wheat (Shepherd & Islam, 1987). Substitution lines of barley chromosome 4 for 4A of wheat had very low fertility which was partially restored by substituting a barley chromosome 4 with a duplication that served as a restorer.

Hybrids between cultivars of *H. vulgare* with diploid and tetraploid wheat strains–species

Intergeneric hybrids have been obtained between barley cultivars and diploid and tetraploid wheats (Table 3.3). A total of 16 combinations were made with tetraploid wheats: 6 with *T. timopheevi* (Cauderon, 1978; Shumny *et al.*, 1981; Surikov *et al.*, 1986), 5 with *T. dicoccum* (Kruse, 1976; Jiang, 1986), 4 with *T. turgidum* (Thomas *et al.*, 1977; Mujeeb-Kazi, 1979; Jiang, 1986) and one with *T. persicum* (Wojciechowska, 1986). A single hybrid has been reported between a diploid wheat species and a barley cultivar, between Bonus barley and *T. monococcum* (Kruse, 1973). The crossability between the above combinations was low; in the order of less than 1%. All hybrids involving tetraploid wheats were cytologically stable with chromosome numbers of 21. A negligible level of chromosome pairing was observed at MI of meiosis with bivalent frequencies of less than one per cell (Mujeeb-Kazi, 1979). All hybrids were wheat-like in morphology and self-sterile, but in one case BC progeny with 35 chromosomes were obtained (Mujeeb-Kazi, 1979).

Other *Hordeum–Triticum* hybrid combinations

An additional 29 hybrids between *Hordeum* and *Triticum* have been obtained from crosses between species of the two genera or crosses between species and cultivars (Table 3.4). The results obtained from these combinations are as diverse as the parents them-

Table 3.3. Hybrids of *Hordeum vulgare* cultivars with diploid and tetraploid wheat.

Cultivar–species combinations	References
Bomi × *T. monococcum*	Kruse (1973)
Bomi Emir, Lofa Bonus × *T. dicoccum*	Kruse (1976)
Apizaco × Cocorit 71 (*T. turgidum*)	Thomas *et al.* (1977)
Ketch × *T. timopheevi*	Cauderon *et al.* (1978)
Manker, Celaya × Cocorit 71	Mujeeb-Kazi (1979)
Sofia B-139 × *T. timopheevi* Omsky, Moskovsky	Shumny *et al.* (1981)
H. vulgare × *T. durum, T. dicoccum*	Jiang (1986)
L-1215 × *T. timopheevi*	Surikov *et al.* (1986)
Emir × *T. persicum*	Wojciechowska (1986)

Table 3.4. Intergeneric hybrids involving wild species of *Hordeum* and *Triticum*.

Combinations	References
(a) Combinations giving hybrid progeny	
Ae. crassa 6x × *H. vulgare* cv. Bomi	Fedak & Nakamura (1981)
H. chilense 2x × *Ae. squarrosa* 2x	Martin (1983)
H. geniculatum 2x × *T. aestivum* 6x cv. CS	Clauss (1983), Pershina *et al.* (1988)
T. turgidum 4x × *H. bulbosum* 2x	Blanco *et al.* (1986)
H. jubatum 4x × *T. aestivum* 6x cv. Fukuhokomugi	Comeau *et al.* (1988)
H. pubiflorum 2x × *T. aestivum* 6x cv. CS	Fedak (unpublished)
H. marinum 2x × *T. aestivum* 6x cv. CS	Jiang & Lui (1987)
H. depressum 4x × *T. aestivum* 6x cv. CS	Jiang & Lui (1987)
H. chilense 4x × *T. aestivum* 6x cv. CS	Pohler & Schrader (1990)
(b) Combinations giving amphiploids	
T. timopheevi 4x × *H. bogdanii* 2x	Kimber & Sallee (1978)
H. chilense 2x × *T. aestivum* 6x cv. (CS)	Chapman & Miller (1978)
H. chilense 2x × *T. turgidum* 4x	Martin & Sanchez-Monge Laguna (1982)
H. californicum 2x × *T. aestivum* 6x cv. (CS)	Fedak (1987)
T. aestivum 6x cv. CS × *H. bulbosum* 4x	Rong & Wang (1989)
H. chilense 2x × *T. aestivum* 6x subsp. *sphaerococcum*	Martin *et al.* (1987)
H. chilense 2x × *T. turgidum* 4x subsp. *dicoccoides*	Padilla & Martin (1987)
H. chilense 2x × *T. turgidum* 4x subsp. *georgicum*	Padilla & Martin (1987)
H. chilense 2x × *T. turgidum* 4x subsp. *polonicum*	Padilla & Martin (1987)
H. chilense 2x × *T. monococcum* 2x	A. Martin (pers. comm.)
(c) Combinations giving subviable progeny	
Ae. squarrosa 2x × *H. bulbosum* 2x	Shigenobu & Sakamoto (1977)
H. pusillum 2x × *T. aestivum* 6x cv. (CS)	Finch & Bennett (1980)
H. vulgare 2x × *Ae. squarrosa* 2x	Jiang (1986)
Ae. squarrosa 2x × *H. bulbosum* 2x	Fedak (unpublished)
H. stenostachys 2x × *T. aestivum* 6x cv. (CS)	Fedak (unpublished)
(d) Combinations giving haploid progeny	
T. aestivum 6x cv. CS × *H. bulbosum* 4x, 2x	Barclay (1975), Miller & Chapman (1976)
Ae. crassa 6x × *H. bulbosum* 4x	Shigenobu & Sakamoto (1977)
Ae. ventricosa 4x × *H. bulbosum* 4x	Fedak (1983)
H. vulgare 2x cv. Elrose × *Dasypyrum villosum* 2x	Fedak (unpublished)

selves, ranging from subviable hybrids and haploids to viable hybrids some of which yielded amphiploids and others which were backcrossed to produce addition lines.

Hybrid progeny

Hybrids were obtained from ten combinations of crosses between cultivars/species of *Hordeum* and *Triticum* (Table 3.4a). Vigorous but sterile hybrid plants were obtained from crossing barley cultivar Bomi on to *Ae. crassa 6x* (Fedak & Nakamura, 1981). Average meiotic configuration in the hybrids was no different than in polyhaploids of the maternal parent indicating no homology between parental genomes. These hybrids responded well to callusing and regeneration in tissue culture systems and were used to study the effects of callusing on meiotic stability in regenerated plants (Fedak & Grainger, 1986).

Several types of progeny have been

obtained from crossing *Hordeum bulbosum* 2*x* or 4*x* on to the various wheat species/cultivars. Most frequently a high frequency of wheat haploids is obtained following the elimination of *bulbosum* chromosomes during mitosis in the hybrid embryo (Barclay, 1975; Miller & Chapman, 1976). The tetraploid form of *H. bulbosum* induces a higher frequency of haploid embryos than the diploid form. The tetraploid form of *H. bulbosum* has been effective in inducing haploidy in any cultivars having the right combination of crossability genes (*Kr, kr*) and in those cultivars the haploid technique can be an effective breeding tool (Falk & Kasha, 1981).

In contrast to the above findings, intergeneric hybrids have been reported between CS and *H. bulbosum* 4*x* (Wang *et al.*, 1982) and *T. turgidum* and *H. bulbosum* 2*x* (Blanco *et al.*, 1986). The genetic control of *H. bulbosum* chromosome elimination by CS has not been determined but it is obviously complex and must be influenced by the interaction of both parental genotypes and therefore the two types of progeny have been reported from crosses of this species onto wheat. *H. bulbosum*-induced haploidy has also been reported in *Ae. crassa* (Shigenobu-Kishimoto & Sakamoto, 1977) and in *Ae. ventricosa* (Fedak, 1983), while haploids have been induced in *H. vulgare* (2*x*) by *Dasypyrum villosum* (2*x*) (Fedak, unpublished).

In the production of the intergeneric hybrid the crossability of *H. bulbosum* 4*x* on to CS was 0.65%, the F_1 hybrids were intermediate in morphology between the parents, had chromosome numbers of 24–30 in the PMCs, were completely sterile and did not respond to colchicine treatments (Wang *et al.*, 1982). The F_1 hybrid was female-fertile however and was backcrossed with CS monosomic 5B and by self-pollinating the BC_1F_3, six disomic addition lines with chromosome numbers of $2n = 44$ were obtained. One euploid segregant with $2n = 42$ carrying resistance to wheat yellow mosaic virus was obtained from one of the addition lines (Wang *et al.*, 1987). In the BC_1F_5 progeny, plants with chromosome numbers of $2n = 56$ were isolated which were octoploids, or amphiploids of the original hybrid. The amphiploids were highly fertile with approximately 80% of progeny being normal (Rong & Wang, 1989).

The crossability of *H. bulbosum* 2*x* on to *T. turgidum* cv. Capelli was 0.16%, the plants were wheat-like in morphology, with a stable chromosome number of 21 in somatic and meiotic cells (Blanco *et al.*, 1986). The average meiotic configuration was $19.07^{I} + 0.92^{II} + 0.03^{III}$. The plants were self-sterile but formed restitution nuclei and upon backcrossing gave progenies with chromosome numbers of 30 to 35.

The *H. chilense* × *Ae. squarossa* hybrids were totally sterile with an average bivalent frequency of 0.66% at meiosis. The hybrid resembled the paternal parent but had a non-brittle rachis whereas both parents had brittle counterparts (Martin, 1983).

Following regeneration through callus culture of the immature inflorescence of the hybrid *H. geniculatum* × CS, BC_1 seeds were obtained and subsequent progeny appeared to be self-fertile (Pershina *et al.*, 1988). The hybrid produced from crossing the wheat cultivar Fukuhokomugi on to *H. jubatum* was intermediate in morphology between the two parents and not totally wheat-like as with other barley–wheat hybrids (Comeau *et al.*, 1988). Some of the maternal parent features included a brittle rachis; long lemma awns and a waxy bloom on the foliage. Total meiotic pairing was about equal to that of a typical wheat polyhaploid indicating no homology between parental genomes. The hybrid was highly sterile, recalcitrant to backcrossing and the anthers contained only very few meiocytes.

The hybrid between *H. marinum* 2*x* and CS wheat resembled the wheat parent whereas that between *H. depressum* 4*x* and CS had some morphological traits from the maternal parent such as reduced plant height, narrow leaves and thin stems (Jiang & Liu, 1987). The mean bivalent frequency in the two hybrids was 0.11 and 0.04, respectively, and both were sterile.

Chinese Spring wheat was crossed on to an induced tetraploid form of *H. chilense* (Pohler & Schrader, 1990). The hybrid plants resembled the wheat parent but were perennial like *H. chilense*. Pairing among homologous chromosomes of *H. chilense* was nearly complete but negligible pairing was seen among wheat chromosomes indicating that the extra genome of *H. chilense* had no effect on wheat chromosome pairing.

In a general way it has been shown that in hybrids between diploid *Hordeum* cultivars/species and hexaploid wheat cultivars/species the plant morphology of the latter predominates. In the case of the *H. chilense* (2*x*) × *T. turgidum* (4*x*) hybrid, some characteristics of the *Hordeum* parent were obvious (Martin & Sanchez-Monge Laguna, 1982) as they were in the CS × *H. bulbosum* 4*x* hybrid (Wang *et al.*, 1982) and in the *H. jubatum* (4*x*) × *T. aestivum* (6*x*) hybrid (Comeau *et al.*, 1988). In the *H. chilense* (2*x*) × *Ae. squarossa* 2*x* hybrid a sort of complementation occurred for certain traits, e.g. the hybrid had a tough rachis whereas both parents were brittle (Martin, 1983). The differences in hybrid plant morphology expression may be explained on the basis of genome ratios. In the case of a 1:3 ratio of *Hordeum*:*Triticum* genomes, the phenotype of the *Triticum* parent predominates; in the case of 1:2 or 2:3 ratios, *Hordeum* characteristics are displayed, whereas in a 1:1 ratio a complementation between parental genomes may take place.

Amphiploid progeny

Chromosome doubling to produce amphiploids has been achieved with nine *Hordeum–Triticum* hybrids (Table 3.4b). A stable, highly fertile amphiploid with a high transmission rate of balanced gametes was induced in the *T. timopheevi* × *H. bogdanii* hybrid by Kimber & Sallee (1978). This amphiploid has been successfully hybridized with other amphiploids (Balyan & Fedak, 1990).

The average meiotic chromosome configuration in meiocytes of the *H. chilense* (2*x*) × *T. aestivum* (6*x*) amphiploid was $1.90^{I} + 27.05^{II}$ and the progeny of this amphiploid consisted of 55 and 56 chromosome individuals (Chapman & Miller, 1978). Following repeated backcrossing a series of aneuploid stocks were produced (Miller *et al.*, 1981). The latter report lists four disomic additions (chromosomes 1, 4, 5, 7), one monosomic addition (chromosome 2), one translocation line, three ditelo additions and six substitution lines.

Average meiotic configuration in the first *H. chilense* × *T. turgidum* amphiploid produced was 2.68^{I} and 19.65^{II} (Martin & Sanchez-Monge Laguna, 1982). The line had high fertility with a high transmission rate of balanced gametes. The seven monosomic addition lines of *H. chilense* in a *T. turgidum* background have been produced (Fernandez & Jouve, 1988). Using isozyme analysis the homoeology of *H. chilense* chromosomes with those of wheat have been established as follows: Hch 3, 4 and 6 correspond to wheat chromosomes 3, 4, 6 (Fernandez & Jouve, 1987a) while Hch 5 corresponds to 1W, 1R and 5H (Fernandez & Jouve, 1987b). An additional four lines of *Tritordeum* have subsequently been produced using *H. chilense* as the maternal parent and various subspecies of *T. turgidum* (4*x*) (Padilla & Martin, 1987) to produce the hexaploid *Tritordeum*, or *T. aestivum* to produce the octoploid form (Martin *et al.*, 1987). The aneuploid frequency in these lines, that had been selfed for 2–10 generations, varied from 3.9 to 31.7% with a mean of 17.1%. Compared with the original *Triticales*, the *Tritordeum* lines are much more fertile and stable and with a further widening of the genetic base offer the potential of a new crop species. It is interesting to note that hybrids between various octoploid *Tritordeum* strains were sterile whereas hybrids between hexaploid and octoploid forms were fertile and thus produced fertile secondary *Tritordeums* (Martin *et al.*, 1987).

In contrast to the regular meiotic behaviour described for some of the amphiploids above, meiosis in the *H. californicum* × CS was somewhat less regular. The average meiotic chromosome pairing in the latter was reported to be $.81^{I} + 24.9^{II} + 0.2^{III} + .05^{IV}$ (Fedak, 1987). Despite the high univalency at meiosis the transmission of balanced gametes must be quite high since the chromosome number in a sample of 30 seeds was 2n = 56 which implies a perfect transmission rate. This aspect, however, requires additional study. Through repeated backcrossing on to the amphiploid, a series of monosomic 43 chromosome addition lines was produced. However, disomic addition lines were not found in the selfed progeny presumably due to a failure of transmission of the monosome through the pollen. Twenty-two chromosome and $2n = 21 +$ telo haploid embryos were recovered in the progeny of monosomic addition lines pollinated by 4*x H. bulbosum*. On chromosome doubling a series of disomic and ditelo additions have been produced that are now being identified by a combination of banding and isozyme techniques (Fedak, unpublished).

One additional amphiploid, between *H. chilense* and *T. monococcum* is reported to have been made but is apparently sterile (A. Martin, personal communication).

Subviable progeny

An additional five hybrid combinations between species/cultivars of *Hordeum* and *Triticum* are shown in Table 3.4c. In all cases the hybrid plantlets grew to a rosette stage but perished soon after. In the case of the *H. stenostachys* × CS hybrid (Fedak, unpublished) the hybrid plant remained at a rosette stage for a prolonged period of time and headed only after being treated with an aqueous solution of GA_3. Subviable hybrids indicate that fertilization has taken place and for future crossing attempts the substitution of alternate accessions of parental strains may lead to the production of viable hybrids.

Trigeneric hybrids involving *Hordeum*, *Triticum* and *Secale*

One method of producing trigeneric hybrids is by crossing cultivars/species of *Secale* on to *Hordeum–Triticum* hybrids (Kimber & Sallee, 1979; Fedak & Armstrong, 1980; Clauss, 1980). Such hybrids produced viable seed directly without requiring embryo rescue. Some of the first hybrids were vegetatively vigorous, but totally sterile with no intergenomic chromosome pairing at meiosis. More recently Clauss (1983) obtained seed set when pollinating a colchicine-treated (barley and wheat) × rye hybrid with octoploid *Triticale*.

Another method of producing trigeneric hybrids is to cross barley or a species of *Hordeum* with *Triticale* or by intercrossing *Triticale* and *Tritordeum*. Barley and *Triticale* hybrids have been reported by Clauss (1980), Balyan & Fedak (1989) and Fedak (unpublished). Generally these hybrids were sterile and exhibited no homoeologous chromosome pairing at meiosis. Schrader *et al.* (1988) were able to backcross 6*x Triticale* on to a colchicine-treated (barley × 6*x Triticale*) hybrid to isolate 43 and 44 chromosome plants from such manipulation. The trigeneric hybrids produced by crossing three strains of 6*x Triticale* on to 6*x H. parodii* were totally sterile (Gupta & Fedak, 1985a) however, selfed seeds were obtained from a colchicine-treated *H. chilense* (2*x*) × *Triticale* (6*x*) hybrid (Schrader & Pohler, 1985). Trigeneric hybrids have also been produced by intercrossing various combinations of *Triticale* and *Tritordeum*. No embryo rescue was required to produce the *Triticale* (6*x*) × *Tritordeum* (6*x*) hybrid (Fernandez-Escobar & Martin, 1985; Fernandez & Jouve, 1985). The chromosome pairing was reduced even among the wheat chromosomes of the hybrid, it was self-sterile but good seed set was obtained following pollination with wheat. No embryo rescue was required to produce hybrids between *Triticale* 6*x* and *Tritordeum* 8*x* but the hybrids were sterile and no pairing between homoeologous chromosomes was observed at meiosis (Fernandez-Escobar & Martin, 1988). In contrast to the above, the hybrids between *Triticale* (8*x*) and *Tritordeum* (8*x*) were self-fertile and chromosome numbers in F_2 progeny ranged between 42 and 52. No homoeologous pairing between *Hordeum* and *Secale* chromosomes was observed in the hybrid (Fernandez-Escobar & Martin, 1989). Trigeneric hybrids are not overly difficult to produce and a total of approximately 30 combinations have been successful in the reports cited above. One of the objectives of such hybrids was to obtain recombinations between *Hordeum* and *Secale* chromosomes and it is not clear that this has been achieved as yet.

Hordeum–Secale hybrids

Interest in hybridizing barley with rye dates back to the 1940s when the first crosses were attempted (Quincke, 1940). The interest stems from the fact that the two genera share a close phytogenetic affinity (Sakamoto, 1973), both contain diploid species for ease of cytogenetic manipulation and finally from a plant improvement perspective, reciprocal exchanges of traits would be beneficial for both crops. The subject of *Hordecale* has been reviewed in some detail (Fedak, 1986) so this will be a more brief overview. A summary of results obtained from intercrossing barley cultivars with cultivars and species of rye is presented in Table 3.5. Such hybridizations are characterized by high seed set, a relatively high yield of embryos from certain combinations but also a very high, at times total seedling necrosis from some combinations. The unusual types of progeny obtained include vegetative clumps, haploids from most combinations, hybrids with incomplete genomic numbers of chromosomes and finally hybrids. The hybrids between barley cultivars and *Secale* cultivars that survived seedling necrosis were only reasonably vigorous and were self-sterile. The nucleolar organizer (NOR) chromosomes of the barley genome suppressed those of the rye parent. Backcross progenies were obtained in one case by pollinating

Table 3.5. Hybrids of *Hordeum vulgare* with *Secale* cultivars and species.

H. vulgare parent	Type of progeny	References
Pallas × Petkus	Hybrid and haploids	Kruse (1967)
H. distichum × *S. africanum*	Vegetative	Tozu (1975)
Pallas × Petkus	Hybrid and haploids	Kruse (1976)
Fort, Minn 66-151, OB128-1 × Prolific	Barley haploid	Fedak (1977a, b)
Betzes × Sangaste	Seedling necrosis	Fedak (1977b)
Betzes × Beagle	Haploid	Fedak (unpublished)
Sundance × Petkus Somro	6 barley chromosomes–1 rye	Cooper *et al.* (1978)
Bomi × Petkus	Hybrid	Fedak (1979b)
Ceresia 4*x* × Rogo	Hybrid–5 barley chromosomes	Pickering & Thomas (1979)
Amsel 4*x* × Rogo	Callus, variable chromosome numbers	Pickering & Thomas (1979)
Vada × Rheidol	Hybrids–amphiploids	Thomas & Pickering (1979)
Sultan × *S. africanum*	Subviable	Finch & Bennett (1980)
Betzes × *S. vavilovii*	Triploid hybrid	Fedak & Nakamura (1981)
Galina, B-319, Sofia × *S. cereale* 2*x*, 4*x*	Hybrid	Shumny *et al.* (1981)
Emir, Menuet, Aramir, Trumpf × (*H. jubatum* 4*x* × *S. cereale* 4*x*	Hybrids	Wojciechowska (1984)
Dornburger × *S. montanum*	Amphiploid	Pohler & Clauss (1984)
Aramir, Atlas 4*x*, Emir × Strzekecinskie 2*x*	Hybrids, haploids	Wojciechowska (1987)
Elgina × *S. kuprijanovii*	Hybrid–amphiploid, backcross	Pohler & Clauss (1987)

an (Elgina × *S. kuprijanovii*) hybrid with *S. cereale* to produce a triploid hybrid that was perennial in growth habit but sterile (Pohler & Clauss, 1982). Amphiploids between barley and rye (*S. cereale*) (Thomas & Pickering, 1979) and barley × *S. montanum* (Pohler & Clauss, 1984) were likewise sterile because of instability at meiosis, especially among the rye chromosomes. The instability was attributed to allocycly of DNA replication times between barley and rye chromosomes. Because the chromosome replication time of rye chromosomes was longer, they would be delayed in their orientation on the metaphase plate, delayed in anaphase I separation, and hence excluded from daughter nuclei and thus cause fertility problems. The triploid hybrid, consisting of 14 barley and 7 *S. vavilovii* chromosomes was also sterile because of chromosome instability at meiosis, probably caused by the allocycly phenomenon (Fedak & Nakamura, 1982). *Secale kuprijanovii* and *S. montanum* were successfully crossed on to the amphiploid obtained from the (*H. vulgare* and *S. kuprijanovii*) hybrid (Pohler & Clauss, 1987). In this case the triploid hybrid was meiotically more stable than either the diploid hybrid or amphiploid, possibly because of the genome ratio, however, the frequency of *Secale* chromosome bivalents was reduced relative to the amphiploid. Wojciechowska (1984) had some success in obtaining hybrids by pollinating barley with pollen from the fertile hybrid (*H. jubatum* × *S. cereale* 4*x*), however, not when alloplasmic rye, derived from BC_3 generations of the latter parent was used as the pollen parent.

Another common feature of virtually all of the hybrids/amphiploids listed in Table 3.5 was the lack of allosyndetic pairing between chromosomes of parental genomes thus precluding any intergenomic gene transfer. Thus from the results presented here the prospects of producing either a fertile amphiploid between barley and rye or recombinant backcross progenies have not materialized. However, considering the vast array of cultivars within both species, a compatible combination of cultivars may yet be found. Forster & Dale (1983) have shown that both barley and rye cultivars can have marked differences in embryo growth rates in terms of mean number of cells at day 5 and mean cell doubling time. Selection of cultivar combinations with closer parameters may improve success rates in such wide crosses.

Most authors have reported a higher success rate in obtaining hybrids between species of *Hordeum* and *Secale* than between cultivars of the two genera and the number of examples listed in Table 3.6 provides support for this. Certainly the first species hybrids predate any hybrids between cultivars (e.g. Brink *et al.*, 1944 vs Kruse, 1967). Thus, the wild species would offer some potential as bridges between *Hordeum* and *Secale*. In addition some success has been reported in cytogenetic manipulations of hybrids obtained from the species crosses.

The hybrids obtained from *H. jubatum* 4*x* × *S. cereale* 2*x* and 4*x* provide good examples of possible cytogenetic manipulations. For example, Wojciechowska (1985b) has backcrossed rye on to this hybrid for up to five generations to produce an alloplasmic rye which has been used as a parent for crosses to barley. Staat & Pohler (1984) obtained a fertile amphiploid from the same combination which was backcrossed to rye. In their studies they noted pairing irregularities at meiosis, especially between the rye chromosomes in the amphiploid.

Hordeum maritimum 2*x* was crossed with both 2*x* and 4*x S. cereale* (Pohler & Clauss, 1982). The hybrid resembled rye phenotypically but the *Hordeum* NORs predominated. Although there was only an average of 0.42 bivalents per PMC in the hybrid, 31% were of the allosyndetic type, and it was noted that the rye chromosomes with heterochromatic telomeres were less frequently associated in allosyndetic configurations. In backcross progenies with rye, however, there were fewer allosyndetic configurations and relatively high rye chromosome asynapsis.

Hordeum geniculatun 4*x* has been hybridized with both 2*x* and 4*x* rye (Pershina *et al.*, 1985; Staat *et al.*, 1985).

Table 3.6. Hybrids between other *Hordeum* species and *Secale* cultivars and species.

Parental combination		References
H. chilense 2x × *S. cereale* cv. UC90		Finch & Bennett (1980)
H. chilense 2x × *S. cereale* cv. Petkus Somro		Finch & Bennett (1980)
H. chilense 2x × *S. africanum*		Finch & Bennett (1980)
H. chilense 2x × *S. montanum*	vegetative	Finch & Bennett (1980)
H. chilense 2x × *S. vavilovii*	vegetative	Finch & Bennett (1980)
H. jubatum 4x × *S. africanum*		Finch & Bennett (1980)
H. jubatum 4x × *S. cereale* cv. Antelope, Sangaste		Wagenaar (1959)
H. jubatum 4x × *S. cereale* 2x, 4x		Shumny *et al.* (1978)
H. jubatum 4x × *S. cereale* cv. Imperial		Brink *et al.* (1944)
H. depressum 4x × *S. cereale*		Morrison & Rajhathy (1959)
(*H. depressum* 4x *H. compressum* 2x) × cv. Svalov 4x		Schooler (1967)
H. parodii 6x × *S. cereale*, cv. Prolific		Fedak & Armstrong (1981)
H. lechleri 6x × *S. cereale*, cv. Gazelle		Fedak & Armstrong (1981)
H. jubatum 4x × *S. kuprijanovii*		Shumny *et al.* (1978)
H. californicum 2x × *S. montanum*		Wang (pers. comm.)
H. jubatum 4x × *S. cereale* 2x		Pohler & Clauss (1981)
		Staat & Pohler (1984)
H. maritimum 2x × *S. cereale* 2x, 4x		Pohler & Clauss (1982)
H. geniculatum 4x × *S. cereale* 2x, 4x		Staat *et al.* (1985)
		Pershina *et al* (1985)
H. chilense 2x × *S. cereale* 2x		Pohler & Schrader (1986)
		Martin *et al.* (1988)
		Thomas & Pickering (1985)
H. californicum 2x × *S. vavilovii*		Gupta & Fedak (1987a, b)
H. californicum 2x × *S. anatolicum*		Gupta & Fedak (1987a, b)
H. parodii 2x × *S. anatolicum*		Gupta & Fedak (1987a, b)
H. parodii 6x × *S. cereale*		Gupta & Fedak (1985b)
H. lechleri 6x × *S. cereale*		Gupta & Fedak (1985b)
H. jubatum 4x × *S. cereale*		Gupta & Fedak (1985b)
H. jubatum 4x × *S. anatolicum*		Gupta & Fedak (1985b)
H. jubatum 4x × *S. vavilovii*		Gupta & Fedak (1985b)
H. bogdanii 2x × *S. cereale*		Gupta & Fedak (1985b)
H. brevisubulatum 2x × *S. dighoricum*		Gupta & Fedak (1985b)
H. jubatum 4x × *S. cereale* 4x (Bialotetra)		Wojciechowska (1984)
H. jubatum 4x × *S. cereale* 2x cv. Prolific		Wojciechowska (1984)
H. pubiflorum 2x × *S. africanum*		Fedak (1985c)

The progeny obtained with diploid rye were sterile, did not respond to colchicine and would not backcross to rye (Pershina *et al.*, 1985). The hybrid with $4(x)$ rye, however, was self-fertile and was backcrossed to rye giving fertile progeny. Some apomictic-type plants were observed among the selfed progeny. It was interesting to note that F_1 hybrids originating from both $2x$ and $4x$ rye had nearly identical frequencies of *Hordeum* bivalents, indicating that *H. geniculatum* has a strong meiotic pairing control system that is not affected by the *Secale* genome unlike many other *Hordeum* species.

Another fertile amphiploid has been obtained from a hybrid between *H. pubiflorum* and *S. africanum* (Fedak, 1985c). This amphiploid was about 40% fertile producing shrivelled but viable seeds that were stable in terms of chromosome number.

A high frequency of allosyndetic paired configurations was reported in hybrids between *H. chilense* and *S. cereale* (Thomas

& Pickering, 1985; Pohler & Schrader, 1986) which in addition to the *H. maritimum* × *S. cereale* hybrid are the only combinations to show this phenomenon. Allosyndetic pairing has not been reported for any other combinations of *Hordeum–Secale* hybrids. The amphiploids between *H. chilense* and *S. cereale* were sterile (Pohler & Schrader, 1986; Martin *et al.*, 1988) and the allosyndetic pairing was reduced in this combination plus in the triploid hybrid that was obtained from backcrosses to rye.

The nucleolar dominance of *Hordeum* over *Secale* has been a universal phenomenon in such hybrids. However, Pohler (1990) produced addition lines of both *H. chilense* NOR chromosomes in a rye background and noticed that one suppressed rye NORs while the other did not.

The lack of allosyndetic pairing between *Hordeum* and *Secale* chromosomes in hybrids was a general phenomenon with the exception of the few examples noted above. In addition many reports indicate marked effects of *Secale* genomes on meiotic pairing control systems of *Hordeum* polyploids and consequent alterations in patterns of homoeologous chromosome pairing. There appears to be a pattern to these systems as described by Gupta & Fedak (1985b).

Hordeum–Elymus hybrids

To date 13 hybrids have been produced between different combinations of barley cultivars and *Elymus* species plus another two combinations using *Hordeum spontaneum* as the *Hordeum* parent (Table 3.7). The majority of the hybrids involved diploid barley cultivars with tetraploid *Elymus* species. The hybrid between *H. spontaneum* and *E. arenarius* died at the seedling stage (Bakhteyev & Darevskaya, 1975). The *H. vulgare* × *E. arenarius* cross produced not only a haploid barley plant (Pissarev *et al.*, 1945) but also one monoploid of *E. arenarius* plus intergeneric hybrids (Ahokas, 1970) and only *E. arenarius* monoploid from the reciprocal cross (Ahokas, 1973). The general features of viable hybrids included excellent vegetative vigour, plant morphology combining both parental traits, self-sterility, recalcitrance to chromosome doubling and backcrossing. There was no indication of intergeneric pairing in any such hybrids indicating no apparent homology between the SH genomes of *Elymus* and the I genome of *Hordeum vulgare*. Some specific examples are as follows. Hybrids of *H. vulgare* cv. Betzes (2*x*) with *E. caninus*, and *E. lanceolatus* 4*x* were

Table 3.7. Hybrids between *Hordeum vulgare* and species of *Elymus*.

Parental combination	References
H. vulgare × *Elymus*	Korablin (1937)
H. vulgare × *E. arenarius* 6x	Pissarev *et al.* (1945)
H. vulgare × *E. arenarius* 6x	Ahokas (1970)
E. arenarius 6x × *H. vulgare*	Ahokas (1973)
H. distichum × *E. giganteus* 4x	Bakhteyev & Darevskaya (1975)
H. spontaneum × *E. arenarius* 6x	Bakhteyev & Darevskaya (1975)
H. vulgare cv. Dicktoo × *E. mollis* 4x	Schooler & Anderson (1980)
E. canadensis 4x × *H. vulgare* (cv. Manker)	Mujeeb-Kazi & Rodriquez (1982)
H. vulgare cv. Betzes × *E. caninus* 4x	Fedak (1985a)
H. vulgare cv. Betzes × *E. lavceolatus* 4x	Fedak (1985a)
H. vulgare cv. Betzes × *E. patagonicus* 6x	Mujeeb-Kazi (1985)
H. vulgare cv. Betzes × *E. canadensis* 4x	Joppa (pers. comm.)
H. vulgare cv. Klages × *E. trachycaulus* 4x	Aung (1991)
E. tsukushiensis 6x × *H. spontaneum*	Lu & Bothmer (1990)
H. vulgare cv. Betzes × *E. angustiglumis* 4x	Fedak (unpublished)
H. vulgare cv. Elrose × *E. sibiricus* 4x	Fedak (unpublished)
H. vulgare cv. Bomi × *E. ugamicus* 4x	Fedak (unpublished)
H. vulgare cv. Betzes × *E. canadensis* 4x	Fedak (unpublished)

obtained at frequencies of 1.4 and 6.1% of pollinated florets respectively (Fedak, 1985a). The floral structure of the hybrid spikes combined traits from both parents and thus had three spikelets at each rachis node, each with 4–8 florets. No chromosome instability was observed in the hybrid between Betzes and *E. caninus* and the average chromosome configuration at MI was $20.40^{I} + 0.30^{II}$. The above characteristics typify most of the other triploid hybrids in Table 3.7. One of the exceptions was the hybrid between Betzes and *E. lanceolatus* in which about 40% of meiocytes were somewhat abnormal in that some bivalents resembled C-pairs type of structures and chromosome numbers per meiocyte varied from 2 to 30 with a mean of 9.7. The unusual bivalent structures and their behaviour may have been caused by deficiencies in the structure or function of the spindle apparatus.

Mujeeb-Kazi (1985) produced a tetraploid (4*x*) hybrid between diploid *H. vulgare* and *E. patagonicus* (6*x*) that was perennial with a codominant phenotype. The average chromosome association per meiocyte was $19.6^{I} + 2.6^{II} + 0.8^{III} + 0.41^{IV}$ with an average chiasma frequency of 4.77. The observed chromosome synapsis was attributed to autosyndetic pairing between the H_1 and H_2 genome chromosomes of the paternal parent.

The autotetraploid (4*x*) form of the *H. vulgare* cultivar Dicktoo was crossed with *E. mollis* (4*x*) to give a partially fertile hybrid that resembled the barley parent. Meiotic pairing in the F_1 showed 6–8 bivalents and 3–4 trivalents. Following backcrossing to diploid barley and selection in the progeny, two Barley Yellow Dwarf Virus tolerant lines were obtained (Schooler & Anderson, 1980).

The hybrid between Klages and *E. trachycaulus* had a very low chiasma frequency (0.3) and has proved to be resistant to the Russian Wheat Aphid (Aung, 1991).

Some species of *Hordeum* and some of *Elymus* are sufficiently close in their phylogeny that they intercross naturally. *Hordeum jubatum* and *H. brachyantherum* intercross in nature with a total of seven species of *Elymus* in areas where the two genera co-exist to produce the intergeneric hybrid described as X *Elymordeum*. A total of 11 such hybrids were listed in a recent review (Fedak, 1984) which represents the work of Stebbins *et al.* (1946), Mitchell & Hodgson (1968) among others. The most significant habitats include the James Bay region of northern North America, the US West Coast and the Great Plains. The natural hybrids have phenotypes intermediate between those of the parents and are completely sterile. In many cases synthetic hybrids have been made between the same combinations and they share the same traits.

Extensive cytogenetic investigations have been conducted between *Hordeum* species and those of the genus *Elymus* as shown in Table 3.8 where 44 hybrid combinations are listed. The main objective was to determine the extent of genome homology between species in the two genera. For example, the earlier work (Mitchell & Hodgson, 1968) on meiotic chromosome pairing in hybrids had shown that *H. jubatum* had one genome in common with *E. sibiricus*. Similarly, Dewey (1971b) reported that *H. bogdanii* 2*x* and *E. canadensis* 4*x* shared a common genome and proposed that *E. glaucus* and *E. virginicus* also had the *H. bogdanii*, genome. He further proposed that the basic genome formula of *E. canadensis* be designated SSHH with the S genome probably derived from *Agropyron spicatum* (*Pseudoroegneria spicata*) and H genome from *Hordeum bogdanii*, although the latter may not be the only source of the H genome.

The present review will regard the basic genome of diploid *Hordeum* species as H, although as shown elsewhere in this monograph, the genome formulas designations for *Hordeum* species have been somewhat refined so that the genome in *H. vulgare*, *H. spontaneum* and *H. bulbosum* is designated as I, *H. murinum* as Y, that of *H. marinum* as X with H in the remaining 20 or so species (Bothmer & Jacobsen, 1989). The existence of the H genome in *Elymus* species has given rise to the genomic constitution classification scheme (Dewey, 1984) and some species formerly classified as *Agropyron* are now classified as *Elymus*. Dewey's classification scheme has been used in this review with the exception that *Hordeum* has been retained in favour of *Critesion*.

Based on a high frequency of bivalent formation at meiosis in hybrids, genome homologies were established in numerous additional combinations of *Hordeum* and *Elymus* species. A common genome was detected in hybrids of *E. caninus* 4*x* with *H. californicum* 2*x*, *H. violaceum* 2*x* × *H. parodii* 6*x* (Gupta & Fedak, 1985a), also in combinations of *H. brachyantherum* 4*x* × *E. scribneri* (Dewey, 1968), *E. canadensis* 4*x* × *H. californicum* 2*x* (Wang & Hsiao, 1986), *H. procerum* 6*x* × *E. virginicus* 4*x* (Gupta & Fedak, 1986) and *H. procerum* 6*x* × *E. caninus* 4*x* (Gupta & Fedak, 1987a), *H. procerum* 6*x* × *E. trachycaulus* 4*x* (Gupta *et al.*, 1988), *H. jubatum* 4*x* × *E. canadensis* 4*x* (Balyan & Fedak, submitted), *H. lechleri* 6*x* × *E. canadensis* 4*x*, *H. arizonicum* 6*x* × *E. canadensis* 4*x* (Gupta *et al.*, submitted) and *H. parodii*

Table 3.8. Hybrids between species of *Hordeum* and *Elymus*

Parental combination	References
E. trachycaulus 4x × *H. jubatum* 4x	Boyle & Holmgren (1955)
H. brachyantherum 4x × *E. condensatus* 4x	Stebbins (cited by Bowden, 1958)
H. brachyantherum 4x × *E. glaucus* 4x	Stebbins *et al.* (1946)
E. trachycaulus 4x × *H. brachyantherum* 4x	Stebbins *et al.* (1946)
H. nodosum 4x × *E. arenarius* 6x	Tsitsin (1956)
E. virginicus 4x × *H. jubatum* 4x	Bowden (1958)
H. brachyantherum 4x × *E. scribneri* 4x	Dewey (1968)
E. latiglume 4x × *H. brachyantherum* 4x	Dewey (1968)
E. canadensis 4x × *H. bogdanii* 2x	Dewey (1971a)
H. californicum 2x × *E. caninus* 4x	Gupta & Fedak (1985a)
H. violaceum 2x × *E. caninus* 4x	Gupta & Fedak (1985a)
H. parodii 6x × *E. caninus* 4x	Gupta & Fedak (1985a)
E. canadensis 4x × *H. californicum* 2x	Wang & Hsiao (1986)
H. procerum 6x × *E. virginicus* 4x	Gupta & Fedak (1986b)
H. procerum 6x × *E. caninus* 4x	Gupta & Fedak (1987a)
E. semicostatus 4x × *H. bogdanii* 2x	Lu & Bothmer (1990)
E. semicostatus 4x × *H. roshevitzii* 2x	Lu & Bothmer (1990)
E. parviglume 4x × *H. bogdanii* 2x	Lu & Bothmer (1990)
E. parviglume 4x × *H. guatamalense* 4x	Lu & Bothmer (1990)
E. tsukushiensis 6x × *H. bogdanii* 2x	Lu & Bothmer (1990)
E. tsukushiensis 6x × *H. guatamalense* 4x	Lu & Bothmer (1990)
H. procerum 4x × *E. trachycaulus* 4x	Gupta *et al.* (1988)
H. parodii 6x × *E. scabriglumis* 6x	Fedak (unpublished)
E. lanceolatus 4x × *H. bulbosum* 2x	Wang (pers. comm.)
E. ciliaris 4x × *H. bulbosum* 2x	Bothmer (pers. comm.)
H. brevisubulatum 4x × *E. dahuricus* 6x	Bothmer (pers. comm.)
H. arizonicum 6x × *E. dentatus* 6x	Bothmer (pers. comm.)
E. pseudonutans 4x × *H. spontaneum* 2x	Lu *et al.* (1990)
E. semicostatus 4x × *H. bulbosum* 2x	Lu *et al.* (1990)
8 *Hordeum* sp. × *E. gayanus* 6x	Jacobsen & Bothmer (1981)
E. jubatum 4x × *E. canadensis* 4x	Balyan & Fedak (1991)
H. parodii 6x × *E. canadensis* 4x	Balyan & Fedak (1991)
H. parodii 6x × *E. innovatus* 4x	Gupta *et al.* (1991)
H. lechleri 6x × *E. caninus* 4x	Gupta *et al.* (1991)
H. arizonicum 6x × *E. canadensis* 4x	Gupta *et al.* (1991)
H. parodii 6x × *E. scabriglumis* 6x	Lee *et al.* (1991)

6x × *E. scabriglumis* 6x (Lee *et al.*, submitted). However, the genome of *H. bulbosum* is not as closely related to that of *E. canadensis* 4x as those of the other *Hordeum* species (Wang & Hsiao, 1986).

Of the six additional *Elymus–Hordeum* hybrids produced by Lu & Bothmer (1990) and listed in Table 3.8, genome homologies were only detected in the combinations involving *E. semicostatus* 4x × *H. bogdanii* 2x and *E. tsukushiensis* × *H. guatemalense* 4x. Dihaploids of *E. pseudonutans* 4x and *E. semicostatus* 4x were produced by pollination with *H. spontaneum* 2x and *H. bulbosum* 2x respectively (Lu *et al.*, 1990). In addition to the 36 *Hordeum–Elymus* combinations listed in Table 3.8, *Elymus gayanus* 6x has been crossed onto eight *Hordeum* species (Jacobson & Bothmer, 1981).

Other intergeneric combinations involving *Hordeum* species

Psathyrostachys fragilis 2x has been crossed onto several cultivars of barley (Bothmer *et al.*, 1984) and progeny obtained at the rate of 0.08% pollinated florets. The progeny were mixoploid plants with some tillers being haploid with only the seven chromosomes of barley being retained, whereas other tillers exhibited a phenotype and chromosome numbers intermediate between the parents. Thus, in this intergeneric hybrid the chromosomes of barley are not eliminated similar to the case in some barley × rye hybrids (Fedak, 1977a, b) and reciprocal hybrids between *E. arenarius* 6x and *H. vulgare* (Ahokas, 1973). In interspecific hybrids within the genus *Hordeum* there are many examples of chromosome elimination and in all cases but the barley × *H. bulbosum* combination, the chromosomes of barley are eliminated. With regard to chromosome elimination in other intergeneric hybrids, *Dasypyrum villosum* induces haploidy in barley (Fedak, unpublished) while in crosses of *Taeniatherum caput-medusae* 2x × *H. bulbosum* 2x, the chromosomes of the latter species are eliminated (Frederiksen, 1989).

Species of *Psathyrostachys* have been crossed onto numerous other *Hordeum* species. For example Baden *et al.* (1989) report on the production of 23 new hybrid combinations between *Psathyrostachys fragilis* and numerous *Hordeum* species. Based on the study of meiosis in eight of these combinations they concluded that the H genome of *Psathyrostachys* has no homology with either the I, X or H genomes of *Hordeum*. Bothmer (personal communication) also reports the hybridization of *Psathyrostachys juncea* 2x with *H. arizonicum*

Table 3.9. Other intergeneric hybrids involving species of *Hordeum*.

Parental combination	References
S. leptorini 4x × *H. bogdanii* 2x	Dewey (1971b)
H. brachyantherum 4x × *Ps. spicata* 2x	Dewey (1971b)
H. arizonicum 6x × *Ps. spicata* 2x	Dewey (1971b)
Ps. spicata 2x × *H. violaceum* 2x	Wang *et al.* (1986b)
H. brachyantherum 4x × *Ps. cognata* 4x	Bothmer (pers. comm.)
Thinopyrum repens 6x × *H. secalinum* 4x	Cauderon & Saigne (1961)
Th. repens 6x × *H. bulbosum* 4x	Shigenabu-Kishimoto & Sakamoto (1987)
Th. elongatum 2x × *H. violaceum* 2x	Wang (1986b)
Agropyron tsukushiense 6x × *H. bulbosum* 4x	Shigenabu & Sakamoto (1981)
H. vulgare 2x × *Th. intermedium* 6x	Fedak (1985b)
H. roshevitzii 4x × *Dasypyrum villosum* 2x	Bothmer (pers. comm.)
H. chilense 2x × *Dasypyrum villosum* 2x	Martin (pers. comm.)
T. caput-medusae 2x × *H. bulbosum* 2x	Frederiksen (1989)
H. violaceu 2x × *Taeniatherum caput-medusae* 2x	Frederiksen & Bothmer (1989)
H. vulgare 2x × *H. patula* 4x	Fedak (unpublished)

6x, *H. jubatum* 4x, *H. patagonicum* 2x, *H. brevisubulatum* 4x and *H. violaceum* 2x and *Psathyrostachys fragilis* 2x with *H. jubatum* 4x and *H. marinum* 4x. In addition *Psathyrostachys haushanica* 2x has been crossed with *H. vulgare* (C. Yen, personal communication) and *Psathyrostachys juncea* 2x with *H. violaceum* 2x (Wang, 1986a). In the latter case a low c value of 0.06 indicated that only homoeology exists between parental genomes.

Hordelymus europus 4x (H.T. genomes) was crossed on to nine *Hordeum* species and meiosis reported for four combinations involving *H. pusillum* 2x, *H. jubatum* 4x, *H. brachyantherum* 4x and *H. depressum* 4x (Bothmer & Jacobsen, 1989). There was an intermediate level of pairing in the hybrid involving *H. depressum* 4x, indicating some homoeology in one genome but the other *Hordeum* species had no genomes in common with *Hordelymus europus*.

Hybrids between the cultivar Betzes and *Hystrix patula* 4x were obtained at a frequency of 3.0% (Fedak, unpublished). The hybrid plants were vegetatively very vigorous, the spike morphology was a combination of traits from both parents but the florets were largely pistilloid and hence sterile.

Hybrids between Betzes barley and *Thinopyrum intermedium* 6x were obtained at a frequency of 1.7% of pollinated florets. The hybrid spikes resembled the paternal parent in phenotype having single spikelets per rachis node but 4–5 florets per spikelet (Fedak, 1985b). At meiosis the chromosome numbers between PMCs varied from 14 to 42 and only 50% of PMCs had the expected chromosome number of 28. The low level of chromosome pairing in normal meiocytes indicated no homology between parental genomes.

It has been well established that the diploid and tetraploid forms of *H. bulbosum* induce haploidy in barley and wheat cultivars plus other species by the selective elimination of their own chromosomes soon after egg cell fertilization. A few exceptions to this generalization have been reported, namely the production of hybrids between *H. bulbosum* and both bread wheat (Wang *et al.*, 1982) and durum wheat (Blanco *et al.*, 1986) as described earlier in this review. Hybrids were also produced when *H. bulbosum* 4x was used to pollinate *Agropyron tsukushiense* 6x (*Elymus tsukushiensis*) (Shigenobu & Sakamoto, 1981) and *Elytrigia repens* (Shigenobu-Kishimoto & Sakamoto, 1987). In the former case the hybrid had the expected somatic chromosome number of $2n = 35$ but chromosome numbers among PMCs varied from 29 to 35 and the average bivalent frequency was 5.5. In the latter hybrid, one chromosome was lost during the vegetative phase leaving $2n = 34$. The hybrid resembled the maternal parent in morphology. PMCs with chromosome numbers of 30 and 31 predominated and the average bivalent frequency was 3.1. Aung & Pickering (personal communication) were successful in inducing haploidy in both *E. trachycaulus* 4x and *E. patagonicus* 6x and thus determined that the genomic structure of the latter was an autoallohexaploid. A hybrid between *Elytrigia repens* 6x × *H. secalinum* 4x was reported by Cauderon & Saigne (1961).

Several species of *Hordeum* have been used as parents in intergeneric hybridization in an effort to elucidate genome relationships between the various genera. For example, Wang (1986b) crossed *Pseudoroegneria spicata* (2x – S genome) and *Thinopyrum elongatum* (2x – J^e genome) with *H. violaceum* (2x – H^v genome). On the basis of comparative levels of chromosome synapsis at MI he concluded that the S haplome is more closely related to H^v than to J^e and that J^e is further diverged from H^v. Further, based on the meiotic pairing in a hybrid between *Pseudoroegneria juncea* (2x – N genome) and *H. violaceum*, he concluded that the relationship between the P and S genomes is closer than between H^v and N genomes. A similar hybrid with genome formula of HHS was produced by Bothmer (personal communication) by crossing *H. brachyantherum* 4x × *P. cognata* 2x. Additional hybrids related to the above were produced by Dewey (1971b) with *H. brachy-*

antherum 4*x* × *H. arizonicum* 6*x* being crossed by *Pseudoroegneria spicata* 2*x*. No genome homologies were detected in either case.

Taeniatherum caput-medusae 2*x* based on morphological similarity had at one time been classified as a species of *Hordeum*, *Elymus* or *Hordelymus*. However, based on negligible chromosome synapsis at meiosis in hybrids with *H. jubatum* 4*x* (Schooler *et al.*, 1972) and *H. violaceum* 2*x* (Frederiksen & Bothmer, 1989) plus other Triticeae it was concluded that *Taeniatherum* is a distinct genus within the Triticeae.

Other examples of intergeneric hybrids involving *Hordeum* include the crossing of *Dasypyrum villosum* on to *H. chilense* 2*x* (Martin, personal communication) and onto *H. roshevitzii* 4*x* (Bothmer, personal communication) and the crossing of *H. bulbosum* 4*x* onto *Leymus angustus* 8*x* (Wang, personal communication). In the last case a hybrid with complex meiotic chromosome synapsis patterns was produced.

Agronomic potential of wild species and intergeneric hybrids

Table 3.10 gives a list of agronomic and value added traits that have either been detected in intergeneric hybrids or were sought from the wild species with the objective of enhancing the crop plant. In many cases the traits in question were not available in the primary gene pool of the crop and

Table 3.10. Agronomic potential of wild species and intergeneric hybrids.

Species or hybrid	Trait(s)	Reference
ND 497 (CI 15859)	Leaf rust resistance	Schooler & Franckowiak (1981)
H. vulgare 4x × *H. bulbosum* × *E. mollis* 4x	Barley Yellow Dwarf Virus tolerance net blotch resistance spot blotch resistance	Schooler (1980)
H. jubatum 4x × *S. cereale*	Winter hardiness	Wagenaar (1959)
H. arizonicum 6x × *P. spicata* 2x	Forage grass Perennial Post-harvest dormancy	Dewey (1971b) Ahokas (1970)
Betzes barley × *Th. intermedium* 6x	*Helminthosporium* resistance	Fedak (unpublished)
CS wheat × CI 3208-4	BYDV resistance	McGuire & Qualset (1990) Fedal (unpublished)
Klages barley × *E. trachycaulus* 4x	Russian wheat aphid resistance	Aung (1991)
Fukuho wheat × Luther barley	Karnal bunt resistance	Jauhar (1990)
CS wheat × *H. bulbosum* 4x	Wheat yellow mosaic virus resistance Mildew resistance Winter hardiness Protein content	Rong & Wang (1989)
H. californicum 2x × CS wheat	Unique storage proteins Possible *Fusarium* resistance	Fedak & Lukow (unpublished) Fedak (unpublished)
H. chilense 2x × *T. turgidum* 4x	Unique storage proteins	Payne *et al.* (1987)
H. vulgare 2x × *T. turgidum* 4x	Apomixis	Mujeeb-Kazi (1981)
H. vulgare 2x × *R. scabrus* 6x	Apomixis	Carman (pers. comm.)
H. chilense 2x × *T. turgidum* 4x (*Tritordeum*)	Potential new crop	Cubero *et al.* (1986)

thus necessitated the more laborious and time-consuming process of intergeneric hybridization. To complete the successful transfer of a trait into a crop plant from a species in a related genus requires an elaborate process of growth regulator applications during crossing followed by embryo rescue on a nurse culture or artificial medium. Invariably intergeneric hybrids are sterile and very few respond to colchicine doubling, and backcrossing to one of the parents can be more difficult than producing the original hybrid. Many intergeneric hybrids, particularly with barley, have never been successfully backcrossed and in such cases this is the limiting step in the whole process. Backcrossing if successful is repeated until single chromosome addition lines are produced. At this point, recombination between alien chromosome and crop plant chromosomes must be induced since strong meiotic pairing control mechanisms generally inhibit synapsis between homoeologous chromosomes.

There are several examples of successful intergeneric transfers of useful traits into wheat from rye and *Thinopyrum* species, however, the transfers from *Leymus* species into wheat have proved to be very difficult for some of the reasons cited above, and also chromosome instability at many stages of the process. In barley there are few examples of successful germplasm enhancement by interspecific or intergeneric hybridization. Some of the few examples include the transfer of barley yellow dwarf virus (BYDV) into barley from *E. mollis* (Schooler, 1980) and leaf rust resistance from another complex hybrid (Schooler & Franckowiak, 1981).

The cultivars Massey (Fedak *et al.*, 1980) and Leger (Fejer *et al.*, 1984) have a hybrid between barley and *H. leporinum* (Hamilton *et al.*, 1955) as one of the parents. Although it is difficult to unequivocally document the traits derived from the wild species, the mildew and scald tolerance of Massey and the lodging resistance, straw elasticity and seed dormancy of Leger are attributed to the wild parent.

A list of other unique traits that appear in intergeneric hybrids is given in Table 3.10. For example, the resistance to leaf spotting caused by *Helminthosporium sativum* is much better in a hybrid between Betzes and *Th. intermedium* than in the best barley cultivar Dickson (Fedak, unpublished). There is no resistance to BYDV in any wheat cultivar or *Aegilops* species but some barley strains such as CI3208-4 carry the *Yd2* gene for resistance on chromosome 3 and for this reason that barley strain has been crossed on to CS wheat and appropriate addition lines produced (McGuire & Qualset, 1990).

Luther barley carries resistance to carnal bunt and has been crossed to Fukuho wheat to transfer the resistance (Jauhar, in press). Resistance to Russian wheat aphid has been expressed in a hybrid between Klages barley and *E. trachycaulus* (Aung, in press). The amphiploid between CS wheat and *H. bulbosum* has a seed protein content in excess of 20%, improved winterhardiness and resistance to mildew and wheat yellow mosaic virus (Rong & Wang, 1989). The last has been transferred into a euploid wheat derivative obtained from one particular *H. bulbosum* addition line (Wang *et al.*, 1987). The *H. californicum* × CS amphiploid has glutenin bands not present in wheat (Fedak & Lukow, unpublished) and preliminary results indicate that one of the addition lines carries resistance to *Fusarium graminearum* (Fedak, unpublished). Payne *et al.* (1987) analysed the addtion lines of *H. chilense* in *T. turgidum* and showed that the majority of prolamin genes are located on *H. chilense* chromosome 1 with smaller numbers on chromosomes 5 and 7.

Apomixis has been detected in backcross progenies of an *H. vulgare* × *T. turgidum* hybrid (Mujeeb-Kazi, 1981) and an attempt was made to transfer this trait from *Recticetus scabrum* into barley by making the appropriate hybrids (Carman, personal communication).

Amphiploids have been produced from a number of combinations between *H. chilense* 2x and tetraploid wheat cultivars. These amphiploids are quite stable, fertile and vigorous and offer the potential of a new crop species not unlike *Triticale* (Cubero *et al.*, 1986).

Even though the barley crop has not benefited greatly thus far in terms of introgression of genes through intergeneric hybridization the potential is substantial and already in progress. As shown in Table 3.10, many hybrids carrying useful traits have already been made and it will only be a matter of time before the traits are introgressed into the crop plant, be it wheat or barley.

Acknowledgements

I wish to thank T. Aung, R. von Bothmer, J. Carman, E. Clauss, A.K.M.R. Islam, P. Jauhar, L. Jappa, A. Martin, R. Pickering, W. Pohler and R. Wang for sharing their unpublished information in order to make this review complete and up to date.

References

Ahokas, H. (1970) Some artificial intergeneric hybrids in the Triticeae. *Annales Botanica Fennici* 7, 182–92.

Akohas, H. (1973) Evidence of cytoplasmic evolution from matings between barley and *Elymus*. *Barley Genetics Newsletter* 3, 5–7.

Aung, T. (1991) An intergeneric hybrid between *Hordeum vulgare* and *Elymus trachycaulus* with resistance to Russian Wheat Aphid *Genome* 34, in press.

Baden, C., Bothmer, R. von, Flink, J. & Jacobsen, N. (1989) Intergeneric hybridization between *Psathyrostachys* and *Hordeum*. *Nordic Journal of Botany* 9, 333–42.

Bakhteyev, F. Kh. & Darevskaya, E.M. (1975) The intergeneric hybrid *Hordeum* and *Elymus*. *Barley Genetics Newsletter* 5, 1.

Balyan, H.S. & Fedak, G. (1989) Meiotic study of hybrids between barley (*Hordeum vulgare* L.) and Triticale (× *Triticosecale* Wittnock). *Journal of Heredity* 80, 460–3.

Balyan, H.S. & Fedak, G. (1990) Hybrids of an amphiploid (*Triticum timopheevi* × *Hordeum bogdanii*) with cultivars of triticale (× *Triticosecale* Wittmack). *Cytologia* 55, 65–9.

Balyan, H.S. & Fedak, G. A study on genome relationships in *Hordeum parodii*, *Hordeum jubatum*, *Elymus trachycaulus* and *Elymus canadensis* (submitted).

Barclay, I.R. (1975) High frequencies of haploid production in wheat (*Triticum aestivum*) by chromosome elimination. *Nature (London)* 256, 410–11.

Benito, C., Fegueiras, A.M., Gonzalez-Jaen, M.T. & Salinas, J. (1985) Biochemical evidence of homoeology between wheat and barley chromosomes. *Zeitschrift fur Pflanzenzuchtung* 94, 307–20.

Blanco, A., Fracchiolla, G.V. & Greco, B. (1986) Intergeneric wheat × barley hybrid. *Journal of Heredity* 77, 98–100.

Bothmer, R. von & Jacobsen, N. (1989) Intergeneric hybridization between *Hordeum* and *Hordelymus poaceae*. *Nordic Journal of Botany* 9, 113–17.

Bothmer, R. von, Jacobsen, N., Jorgensen, R.B. & Linde Laursen, I. (1984) Haploid barley from the intergeneric *Hordeum vulgare* × *Psathyrostachys fragilis*. *Euphytica* 33, 363–7.

Bowden, W.M. (1958) Natural and artificial × *Elymodeum* hybrids. *Canadian Journal of Botany* 36, 101–23.

Boyle, W.S. & Holmgren, A.N. (1955) A cytogenetic study of natural and controlled hybrids between *Agropyron trachycaulum* and *Hordeum jubatum*. *Genetics* 40, 539–45.

Brink, R.A., Cooper, D.C. & Ausherman, L.E. (1944) A hybrid between *Hordeum jubatum* and *Secale cereale*. *Journal of Heredity* 35, 66–75.

Cauderon, Y. & Saigne, B. (1961) New interspecific and intergeneric hybrids involving *Agropyron*. *Wheat Information Service* 12, 13–14.

Cauderon, Y., Tempe, J. & Gay, G. (1978) Creation et analyze cytogenetique d'un nouvel hybride: *Hordeum vulgare* spp. *distichon* × *Triticum timopheevi*. *Comptes Rendus de L'Academie des Sciences (Paris)* 286, 1687–90.

Chapman, V. & Miller, T.E. (1978) The amphiploids of *Hordeum chilense* × *Triticum aestivum*. *Cereal Research Communications* 6, 351–2.

Chu, C.C, Sun, C.S., Chen, X, Zhang, U.X. & Du, Z.H. (1984) Somatic embryogenesis and plant regeneration in callus from inflorescences of *Hordeum vulgare* × *Triticum aestivum* hybrids. *Theoretical and Applied Genetics* 68, 375–9.

Clauss, E. (1980) Trigeneric hybrids between barley, wheat and rye. *Cereal Research Communications* 8, 341–7.

Clauss, E. (1983) Bastarde aus *Hordeum geniculatum*. All und *Triticum aestivum* L. *Archives Zuchtungsforschung* 13, 413–18.

Comeau, A., Fedak, G., St. Pierre, C.A. & Cazeault, R. (1988) Intergeneric hybrids between *Hordeum jubatum* (4*x*) and *Triticum aestivum* (6*x*). *Genome* 30, 245–9.

Cooper, K.V., Dale, J.E., Dyer, A.F., Lyne, R.L. & Walker, J.T. (1978) Hybrid plants from barley and rye crosses. *Plant Science Letters* 12, 293–8.

Cubero, J.I., Martin, A., Millan, T., Gomez-Cabera, A. & de Haro, A. (1986) Tritordeum: a new alloploid of potential importance as a protein source crop. *Crop Science* 26, 1186–90.

Dewey, D.R. (1968) Synthetic hybrids among *Hordeum brachyantherum*, *Agropyron scribneri* and *Agropyron latiglume*. *Bulletin of the Torrey Botanical Club* 95, 454–64.

Dewey, D.R. (1971a) Synthetic hybrids of *Hordeum bogdanii* with *Elymus canadensis* and *Sitanion hystrix*. *American Journal of Botany* 58, 902–8.

Dewey, D.R. (1971b) Genome relations among *Agropyron spicatum*, *A. scribneri*, *Hordeum brachyantherum* and *H. arizonicum*. *Bulletin of the Torrey Botanical Club* 98, 200–6.

Dewey, D.R. (1984) The genomic system of classification as a guide to intergeneric hybridization with the perennial Triticeae. In: Gustafson, J.P. (ed.), *Gene Manipulation in Plant Improvement*, (16th Stadler Genetics Symposium), pp. 209–79.

Falk, D.E. & Kasha, K.J. (1981) Comparison of the crossability of rye (*Secale cereale*) and *Hordeum bulbosum* onto wheat (*Triticum aestivum*). *Canadian Journal of Genetics and Cytology* 23, 81–8.

Fedak, G. (1977a) Haploids from barley × rye crosses. *Canadian Journal of Genetics and Cytology* 19, 15–19.

Fedak, G. (1977b) Barley monoploids and hybrids from barley × rye crosses. In: Sanchez-Monge, E. & Garcia Olmedo, F. (eds), *Interspecific Hybridization in Plant Breeding*, (Proceedings of 8th Congress Eucarpia, Madrid), pp. 269–73.

Fedak, G. (1979a) Source of meiotic abnormalities in barley–wheat hybrids. *Wheat Information Service* 50, 10–11.

Fedak, G. (1979b) Cytogenetics of a barley × rye hybrid. *Canadian Journal of Genetics and Cytology* 21, 543–8.

Fedak, G. (1980) Production, morphology and meiosis of reciprocal barley wheat hybrids. *Canadian Journal of Genetics and Cytology* 22, 117–23.

Fedak, G. (1982) Effect of cultivar combination on meiosis in barley–wheat hybrids. *Canadian Journal of Genetics and Cytology* 24, 575–82.

Fedak, G. (1983) Haploids in *Triticum ventricosum* via intergeneric hybridization with *Hordeum bulbosum*. *Canadian Journal of Genetics and Cytology* 25, 104–6.

Fedak, G. (1984) Wide crosses in *Hordeum*. In: Rasmussen, D.C. (ed.), *Barley Agronomy Monograph*, No. 26, ASA-CSSA-SSSA, 677 South Segoe Road, Madison, WI 53711, USA, pp. 155–86.

Fedak, G. (1985a) Intergeneric hybrids of *Horeum vulgare* with *Agropyron caninum* and *A. dasystachyum*. *Canadian Journal of Genetics and Cytology* 27, 387–92.

Fedak, G. (1985b) Intergeneric hybrids between *Hordeum vulgare* and *Agropyron intermedium* var. *trichophorum*. *Zeitschrift fur Pflanzenzuchtung* 95, 45–9.

Fedak, G. (1985c) Cytogenetics of a hybrid and amphiploid between *Hordeum pubiflorum* and *Secale africanum*. *Canadian Journal of Genetics and Cytology* 27, 1–5.

Fedak, G. (1986) Hordecale (*Hordeum vulgare* L. × *Secale cereale* L.). In: Bajaj, Y.P.S. (ed.), *Biotechnology in Agriculture and Forestry*, Vol. 2. Springer-Verlag, Berlin, pp. 544–55.

Fedak, G. (1987) Intergeneric hybrids in *Hordeum*. In: *Barley Genetics V* (Proceedings of 5th International Barley Genetics Symposium, Okayama, Japan), pp. 301–10.

Fedak, G. & Armstrong, K.C. (1980) Production of trigeneric (barley × wheat) × rye hybrids. *Theoretical and Applied Genetics* 56, 221–4.

Fedak, G. & Armstrong, K.C. (1981) Hybrids of *Hordeum parodii* and *H. lechleri* with *Secale cereale*. In: Whitehouse, R.N.H. (ed.), *Barley Genetics IV* (Proceedings of 4th International Barley Genetics Symposium), pp. 740–5.

Fedak, G. & Grainger, J. (1986) Chromosome instability in somaclones of a *Triticum crassum* × *Hordeum vulgare* hybrid. *Canadian Journal of Genetics and Cytology* 28, 430–2.

Fedak, G. & Jui, P.Y. (1982) Chromosomes of Chinese Spring wheat carrying genes for crossability with Betzes barley. *Canadian Journal of Genetics and Cytology* 24, 227–33.

Fedak, G. & Nakamura, C. (1981) Intergeneric hybrids between *Triticum crassum* and

Hordeum vulgare. *Theoretical and Applied Genetics* 60, 349–52.

Fedak, G. & Nakamura, C. (1982) Chromosomal instability in a hybrid between *Hordeum vulgare* and *Secale vavilovii*. *Canadian Journal of Genetics and Cytology* 24, 207–12.

Fedak, G., Loiselle, R. & Fejer, S.O. (1980) Massey Barley. *Canadian Journal of Plant Science* 60, 277–8.

Fejer, S.O., Fedak, G. & Ho, K.M. (1984) Leger Barley. *Canadian Journal of Plant Science* 64, 195–6.

Fernandez, J.A. & Jouve, N. (1985) Meiotic pairing in hybrids of 6*x* triticale and the amphiploid *Hordeum chilense* × *triticum turgidum* conv. durum. *Journal of Heredity* 76, 63–4.

Fernandez, J.A. & Jouve, N. (1987a) Chromosomal location of structural genes controlling isozymes in *Hordeum chilense*. 3. Esterases, glutamate oxaloacetate transaminase and phosphoglucomutase. *Theoretical and Applied Genetics* 73, 690–8.

Fernandez, J.A. & Jouve, N. (1987b) Chromosomal location of structural genes controlling isozymes in *Hordeum chilense*. 1. 6-phosphogluconate dehydrogenase and malate dehydrogenase. *Theoretical and Applied Genetics* 73, 433–9.

Fernandez, J.A. & Jouve, N. (1988) The addition of *Hordeum chilense* chromosomes to *Triticum turgidum* conv. durum. Biochemical, karyological and morphological characteristics. *Euphytica* 37, 247–59.

Fernandez-Escobar, J. & Martin, A. (1985) Morphology, cytology and fertility of a trigeneric hybrid from triticale × tritordeum. *Zeitschrift fur Pflanzenzuchtung* 95, 311–18.

Fernandez-Escobar, J. & Martin, A. (1988) A hybrid between hexaploid triticale and octoploid tritordeum. *Cereal Research Communications* 16, 45–51.

Fernandez-Escobar, J. & Martin, A. (1989) A self-fertile trigeneric hybrid in the Triticeae involving *Triticum*, *Hordeum*, and *Secale*. *Euphytica* 42, 291–6.

Finch, R.A. & Bennett, M.D. (1980) Mitotic and meiotic chromosome behaviour in new hybrids of *Hordeum* with *Triticum* and *Secale*. *Heredity* 44, 201–9.

Finch, R.A. & Bennett, M.D. (1982) Preferential survival of wheat haploids over hybrids in a wheat × barley cross. *Heredity* 48, 293–8.

Forster, B.P. & Dale, P.E. (1983) Effects of parental embryo and endosperm mitotic cycle times on development of hybrids between barley and rye. *Annals of Botany (London)* 52, 613–20.

Frederiksen, S. (1989) Chromosome elimination in a hybrid between *Taenietherum* and *Hordeum bulbosum*. *Hereditas* 110, 121–3.

Frederiksen, S. & Bothmer, R. von (1989) Intergeneric hybridization between *Taenietherum* and different genera of *Triticeae poaceae*. *Nordic Journal of Botany* 9, 229–40.

Gupta, P.K. & Fedak, G. (1985a) New hybrids of *Hordeum parodii* with *Hordeum vulgare*, *H. bogdanii*, *Agropyron caninum* and X*Triticosecale*. *Theoretical and Applied Genetics* 70, 265–70.

Gupta, P.K. & Fedak, G. (1985b) Meiosis in seven intergeneric hybrids between *Hordeum* and *Secale*. *Zeitschrift·fur Pflanzenzuchtung* 95, 262–73.

Gupta, P.K. & Fedak, G. (1986) Meiotic pairing in new hybrids of *Hordeum procerum* (6*x*) with *H. parodii* (6*x*) and *Elymus virginicus* (4*x*). *Canadian Journal of Genetics and Cytology* 28, 416–19.

Gupta, P.K. & Fedak, G. (1987a) Meiosis in a hybrid between *Hordeum procerum* (6*x*) and *Agropyron caninum* (4*x*). *Canadian Journal of Genetics and Cytology* 28, 416–19.

Gupta, P.K. & Fedak, G. (1987b) Preferential intragenomic chromosome pairing in two diploid *Hordeum–Secale* species. *Genome* 29, 594–7.

Gupta, P.K., Balyan, H.S. & Fedak, G. (1988) A study on genomic relationships of *Agropyron trachycaulum* with *Elymus scabriglumis*, *E. innovatus* and *Hordeum procerum*. *Genome* 30, 525–30.

Gupta, P.K., Balyan, H.S. & Fedak, G. Variation in meiotic pairing in four pentaploid hybrids and its implications on genomic relationships among seven species from *Agropyron*, *Elymus* and *Hordeum* (submitted).

Hamilton, D.G., Symko, S. & Morrison, J.W. (1955) An anomalous cross between *Hordeum leporinum* and *Hordeum vulgare*. *Canadian Journal of Agricultural Science* 35, 287–93.

Islam, A.K.M.R. (1983) Ditelosomic additions of barley chromosomes to wheat. In: *Proceedings of the 6th International Wheat Genetics Symposium*, Kyoto, Japan, pp. 233–8.

Islam, A.K.M.R., Shepherd, K.W. & Sparrow, D.H.B. (1975) Addition of individual barley chromosomes to wheat. In: *Barley Genetics III*, (Proceedings of 3rd International Barley Genetics Symposium, Garching, W. Germany), pp. 260–70.

Islam, A.K.M.R., Shepherd, K.W. & Sparrow, D.H.B. (1978) Production and characterization of wheat–barley addition lines. In: *Proceedings of Fifth International Wheat Genetics Symposium*, New Delhi, India, pp. 365–71.

Islam, A.K.M.R. & Shepherd, K.W. (1980) Meiotic restitution in wheat–barley hybrids. *Chromosoma* (Berl.) 70, 363–72.

Islam, A.K.M.R. & Shepherd, K.W. (1981) Wheat–barley addition lines: their use in genetic and evolutionary studies of barley. In: *Barley Genetics IV*, (Proceedings of 4th International Barley Genetics Symposium, Edinburgh), pp. 729–39.

Islam, A.K.M.R. & Shepherd, K.W. (1988) Induced pairing between wheat and barley chromosomes. In: *Proceedings of Seventh International Wheat Genetics Symposium*, Cambridge, England, pp. 309–14.

Jacobsen, N. & Bothmer, R. von (1981) Interspecific hybridization in the genus *Hordeum*. In: *Barley Genetics IV* (Proceedings of 4th International Barley Genetics Symposium, Edinburgh), pp. 710–15.

Jauhar, P.P. High pairing breadwheat and barley hybrids. In: *Proceedings of Second International Symposium on Chromosome Manipulations in Plants*, Columbia, MO, USA (in press).

Jiang, J. (1986) Morphology and cytogenetics of intergeneric hybrids between *Hordeum vulgare* L. and *Triticum* L. and its BC_1 derivatives. MSc Thesis, Narying Agricultural University.

Jiang, J. & Lui, D. (1987) New *Hordeum–Triticum* hybrids. *Cereal Research Communications* 15, 95–9.

Kimber, A. & Sallee, P.J. (1978) An amphiploid of *Triticum timopheevi* × *Hordeum bodganii*. In: *Cytogenetics and Crop Improvement Symposium*, Varanasi, India, pp. 98–136.

Kimber, G. & Sallee, P.J. (1979) A trigeneric hybrid in the Triticeae. *Cereal Research Communications* 7, 5–9.

Korablin, I.I. (1937) New varieties of barley bred at Omsk. *Selektsia i Semenovodstvo* 11, 57–9.

Koroylyuk, S.S & Simonenko, V.K. (1984) Hybridization of wheat with barley. In: *Ecological Genetics of Plants and Animals, International Conference*, Kishinev, p. 180.

Kruse, A. (1967) Intergeneric hybrids between *Hordeum vulgare* L. spp. *disticum* (V. Pallas 2n + 14) In: *Yearbook: Royal Veterinary and Agricultural College*, Copenhagen, pp. 82–92.

Kruse, A. (1973) *Hordeum* × *Triticum* hybrids. *Hereditas* 73, 157–61.

Kruse, A. (1976) Reciprocal hybrids between the genera *Hordeum*, *Secale*, *Triticum*. *Hereditas* 84, 244.

Lee, Y.H., Wang, B. & Fedak, G. Cytogenetics of three *Elymus* × *Hordeum* (Critesion) hybrids (submitted).

Lu, B.R. & Bothmer, R. von (1990) Intergeneric hybridization between Hordeum and Asiatic Elymus. *Hereditas* 112, 109–16.

Lu, B.R., Salomon, B. & Bothmer, R. von (1990) Cytogenetic studies of progeny from the intergeneric crosses *Elymus* × *Hordeum* and *Elymus* × *Secale*. *Genome* 33, 425–32.

Martin, A. (1983) Cytology and morphology of the hybrid *Hordeum chilense* × *Aegilops squarrosa*. *Journal of Heredity* 74, 487.

Martin, A. & Sanchez-Monge Laguna, E. (1982) Cytology and morphology of the amphiploid *Hordeum chilense* × *Triticum turgidum* conv. durum. *Euphytica* 31, 261–7.

Martin, A., Padilla, J.A. & Fernandez-Escobar, J. (1987) The amphiploid *Hordeum chilense* × *Triticum aestivum* spp. *sphaerococcum* — variability in octoploid triticale. *Zeitschrift fur Pflanzenzuchtung* 99, 336–9.

Martin, A., Millan, T. & Fernandez-Escobar, J. (1988) Morfologia y cytologia del hybrids of amfiploide *Hordeum chilense* × *Secale cereale*. *Anales de la Estacion Experimental De Aula Dei* 19, 135–42.

McGuire, P.E. & Qualset, C.O. (1990) Transfer of the *Yd2* barley yellow dwarf virus gene from barley to wheat. In: Burnett, P.A. (ed.), *World Perspectives of Barley Yellow Dwarf Virus*. CIMMYT, Mexico, pp. 476–81.

Miller, T.E. & Chapman, V. (1976) Aneuhaploids in bread wheat. *Genetical Research* 28, 37–45.

Miller, T.E., Reader, S.M. & Chapman, V. (1981) The addition of *Hordeum chilense* chromosomes to wheat. Induced variability in plant breeding. *Int. Symp., Eucarpia, Mutation and polyploid section*, Pudoc, Wageningen, pp. 79–81.

Molnar-Lang, M., Galiba, G., Kovacs, G. & Sutka, J. (1990) Changes in the fertility and meiotic behaviour of barley (*Hordeum vulgare* L.) × wheat (*Triticum aestivum* L.) hybrids regenerated from tissue cultures. *Genome* 34, 261–6.

Morrison, J.W. & Rajhathy, T. (1959) Cytogenetic studies in the genus *Hordeum*. III. Pairing in some interspecific and intergeneric hybrids. *Canadian Journal of Genetics and Cytology* 1, 65–7.

Mitchell, W.W. & Hodgson, H.J. (1968) Hybridization within Triticeae of Alaska: A new X *Elymordeum* and comments. *Rhodora* 70, 466–73.

Mujeeb-Kazi, A. (1979) Wide crosses. In: *CIMMYT Report on Wheat Improvement*. International Maize and Wheat Improvement Centre, Mexico City, pp. 121–7.

Mujeeb-Kazi, A. (1981) Apomictic progeny derived from intergeneric *Hordeum–Triticum* hybrids. *Journal of Heredity* 72, 282–5.

Mujeeb-Kazi, A. (1985) Cytogenetics of a *Hordeum vulgare* × *Elymus patagonicus* hybrid ($n = 4x = 28$). *Theoretical and Applied Genetics* 69, 475–9.

Mujeeb-Kazi, A. & Rodriquez, R. (1982) Cytogenetics of hybrids of *Elymus canadensis* × *Hordeum vulgare*. *Journal of Heredity* 73, 77–9.

Padilla, J.A. & Martin, A. (1987) Cytology, fertility and morphology of amphiploids *Hordeum chilense* × tetraploid wheats (*Tritordeum*). *Plant Breeding* 99, 295–302.

Payne, P.J., Holt, L.M., Reader, S.M. & Miller, T.E. (1987) Chromosomal location of genes coding for endosperm proteins of *Hordeum chilense* determined by two-dimensional electrophoresis of wheat-*H. chilense* addition lines. *Biochemical Genetics* 25, 53–65.

Pershina, L.A., Shumny, V.K., Belova, L.I. & Numerova, O.M. (1985) *Hordeum geniculatum* All × *Secale cereale* L. hybrids and their backcross progenies with rye. *Cereal Research Communications* 13, 141–7.

Pershina, L.A., Shumny, V.K., Numerova, O.M. & Belova, L.I. (1986) Progeny of barley × wheat hybrids *H. vulgare* × *T. aestivum* L. from backcrosses to common wheat. *Cereal Research Communications* 14, 371–8.

Pershina, L.A., Numerova, O.M., Belova, L.I., Devyatkina, E.P. & Shumny, V.K. (1988) Fertility in barley and wheat hybrids *H. geniculatum* All × *T. aestivum* L., their regenerants and hybrid progeny of backcrosses to *T. aestivum* L. *Cereal Research Communications* 16, 157–63.

Pickering, R.A. & Thomas, H.M. (1979) Crosses between tetraploid barley and diploid rye. *Plant Science Letters* 16, 291–6.

Pissarev, V.E., Vinogradova, H.M. & Poddubnaya-Arnoldi, B.A. (1945) A haploid barley plant produced by remote hybridization. *Doklady Akademia Nauk SSSR* 49, 372.

Pohler, W. (1990) Different ability of the two SAT-chromosomes of *Hordeum chilense* to suppress rye nucleolar activity in chromosome additions to *S. cereale*. *Biologische Zentralblatt* 109, 79–80.

Pohler, W. & Clauss, E. (1981) Meiotirche chromosomenpaarung bei amphidiploiden *Hordeum jubatum–Secale cereale* – lostarden und das Wesen des diploidahnlicken paarungsverhaltens bei *H. jubatum* L. (2n = 28). *Archives Zuchtungsforschung (Berlin)* 11, 311–21.

Pohler, W. & Clauss, E. (1982) The behaviour of the parental chromosomes in the meiosis of hybrids between *Hordeum maritimum* and *Secale cereale*. *Tagerische Berlin Akademica Landivirtschaftern Wissen. D.D.R.* 198, 199–206.

Pohler, W. & Clauss, E. (1984) Differences in chromosomal stability, chromosome pairing and rhythm of nuclear division between the parental genomes in an amphidiploid hybrid from the cross *Hordeum vulgare* L. × *Secale montanum*. Goss. *Archives Zuchtungsforschung* 14(3), 215–24.

Pohler, W. & Clauss, E. (1987) High somatic stability but serious meiotic irregularities in triploid *Hordeum* and *Secale* hybrids (3x = HSS). *Archives Zuchtungsforschung (Berlin)* 17(4), 259–63.

Pohler, W. & Schrader, O. (1986) Meiotic chromosome pairing in hybrids and a maternal haploid from the cross *Hordeum chilense* × *Secale cereale*. *Archives Zuchtungsforschung (Berlin)* 16, 211–22.

Pohler, W. & Schrader, O. (1990) Nearly total absence of homoeologous pairing in a hybrid between tetraploid *Hordeum chilense* and *Triticum aestivum*. *Plant Breeding* 104, 255–7.

Quincke, F.L. (1940) Interspecific and intergeneric crosses with *Hordeum*. *Canadian Journal of Research* 18, 372–3.

Rong, J. & Wang, L. (1989) Selection of wheat barley octoploid. *Hereditas (Beijing)* 11, 1–4.

Sakamoto, S. (1973) Patterns of phylogenetic differentiation in the tribe Triticeae. *Seiken Ziho* 24, 11–31.

Schooler (1967) Intergeneric hybrids between allohexaploid (*Hordeum depressum* × *Hordeum compressum*) and autotetraploid Svalov rye (*Secale cereale*). *Botanical Gazette* 128, 113–16.

Schooler, A.B. (1980) Intergeneric and interspecific barley hybrids show tolerance to barley yellow dwarf virus. *North Dakota Farm Research* 38, 19–21.

Schooler, A.B. & Anderson, M.K. (1980) Behaviour of the intergeneric hybrids between *Hordeum vulgare* (4*x*) and an *Elymus mollis* type. *Cytologia* 45, 157–62.

Schooler, A.B. & Franckowiak, J.D. (1981) Registration of North Dakota 497 and 586 barley germplasm. *Crop Science* 21, 352.

Schooler, A.B., Bell, A.R. & Nalewaja, J.D. (1972) Crosses between *Hordeum jubatum* × *Taeniatherum asperum* (Sim) Nevsk: to determine genomic relationships. *Crop Science* 12, 542–4.

Schrader, O. & Pohler, W. (1985) Seed set from a colchicine treated trigeneric hybrid of the cross *Hordeum chilense* 2*x* × triticale 6*x*. *Cereal Research Communications* 13(1), 63–9.

Schrader, O., Schmidt, J-C. & Zerneke, F. (1988) Selbstungsansatz nach Ruckbreuzung gerstenplasmetischer Dreigattungsbastarde der kombination *Hordeum vulgare* × 6*x* Triticale. *Archives Zuchtungsforschung (Berlin)* 18, 269–74.

Sethi, G.S., Finch, R.A. & Miller, T.E. (1986) A bread wheat (*Triticum aestivum*) × cultivated barley (*Hordeum vulgare*) hybrid with homoeologous chromosome pairing. *Canadian Journal of Genetics and Cytology* 28, 777–82.

Shepherd, K.W. & Islam, A.K.M.R. (1981) Wheat–barley hybrids. The first eighty years. *Wheat Science – Today and Tomorrow*, pp. 107–8.

Shepherd, K.W. & Islam, A.K.M.R. (1987) Cytogenetic manipulation of barley chromosomes in a wheat background. In: *Barley Genetics V* (Proceedings of 5th International Barley Genetics Symposium, Okayama, Japan), pp. 375–87.

Shigenobu, T. & Sakamoto, S. (1977) Production of a polyploid plant of *Aegilops crassa* (6*x*) pollinated by *Hordeum bulbosum*. *Japanese Journal of Genetics* 52, 397–401.

Shigenobu, T. & Sakamoto, S. (1981) Intergeneric hybridization between *Agropyron tsukushiense* and *Hordeum bulbosum*. *Japanese Journal of Genetics* 56, 505–17.

Shigenobu-Kishimoto, T. & Sakamoto, S. (1987) Cytogenetic studies on intergeneric hybrids between two *Agropyron* species and *Hordeum bulbosum* (4*x*) In: *Barley Genetics V* (Proceedings of 5th International Barley Genetics Symposium, Okayama, Japan), pp. 351–8.

Shumny, V.K., Pershina, L.A. & Belova, L.I. (1981) Production of barley × rye and barley × wheat hybrids. *Cereal Research Communications* 9, 265–72.

Shumny, V.K., Pershina, L.A. & Shchapova, A.T. (1978) Use of different *Hordeum* L. species in remote hybridization. I. Production of intergeneric barley hybrids. *Tsitologia i Genetika* 8, 110–16.

Staat, B. & Pohler, W. (1984) Meiotische chromosomenpaaring bei *Hordeum jubatum–Secale cereale* – bastarden und ruckkreuzenpprodukten mit kulturroggen. *Archives Zuchtungsforschung (Berlin)* 14, 57–68.

Staat, B., Pohler, W. & Clauss, E. (1985) Unexpectedly strong homoeologous pairing in hybrids of *Hordeum geniculatum* All. with *Secale cereale* L. as compared with the wild barley parent. *Archives Zuchtungsforschung (Berlin)* 15(1), 19–25.

Stebbins, G.L., Valencia, J.I. & Valencia, R.M. (1946) Artificial and natural hybrids in the Gramineae, Tribe Hordeae, II. Agropyron *Elymus* and *Hordeum*. *American Journal of Botany* 33, 570–86.

Surikov, I.M. & Kissel, N.I. (1987) Crossing of barley with soft wheat. *Cytology and Genetics* (Russian) 21, 221–5.

Surikov, I.M., Kissel, N.I. & Orlova, I.N. (1986) Cytogenetical characteristics of hybrids between barley (*Hordeum vulgare* L.) and wheat (*Triticum aestivum* L.) and *T. timopheevi* Zhuk. *Genetika* 22, 290–5.

Thomas, H.M. & Pickering, R.A. (1979) Barley × rye crosses. The morphology and cytology of the hybrids and the amphiploid. *Zeitschrift fur Pflanzenzuchtung* 82, 192–200.

Thomas, H.M. & Pickering, R.A. (1985) Comparisons of the hybrids *Hordeum chilense* × *H. vulgare*, *H. chilense* × *H. bulbosum*, *H. chilense* × *S. cereale* and the amphiploid of *H. chilense* × *H. vulgare*. *Theoretical and Applied Genetics* 69, 519–22.

Thomas, J.B., Mujeeb-Kazi, A., Rodriquez, R. & Bates, L.W. (1977) Barley × wheat hybrids. *Cereal Research Communications* 5, 181–91.

Tozu (1975) On the breeding method of *Hordeum–Secale* hybrid. In: Gaul, N. (ed.), *Barley Genetics III*, (Proceedings of 3rd International Barley Genetics Symposium, Garching, Germany), p. 306.

Tsitsin, N.V. (1956) Distant hybridization in

plants. Summary of work on distant hybridization in plants. *Plant Breeding Abstracts* 26, 283.

Tsuchiya, T. (1983) Linkage maps of barley. *Barley Genetics Newsletter* 13, 101–6.

Wagenaar, E.B. (1959) Intergeneric hybrids between *Hordeum jubatum* and *Secale cereale*. *Journal of Heredity* 50, 195–202.

Wang, R. R-C. (1986a) Diploid perennial intergeneric hybrids in the tribe Triticeae. I. *Agropyron cristatum* × *Pseudoroegneria libanotica* and *Critesion violaceum* × *Psathyrostachys juncea*. *Crop Science* 26, 75–8.

Wang, R. R-C. (1986b) Diploid perennial intergeneric hybrids in the tribe Triticeae. II. Hybrids of *Thinopyrum elongatum* and *Pseudoroegneria spicata* and *Critesion violaceum*. *Biologische Zentralblatt* 105, 362–8.

Wang, R. R-C. & Hsiao, C. (1986) Differentiation of H genomes in the genus Critesion: evidence from synthetic hybrids involving *Elymus* and *Critesion* and one natural hybrid of *C. violaceum* and *C. bogdanii*. *Canadian Journal of Genetics and Cytology* 28, 947–53.

Wang, L.Q., Zhu, H.R., Liang, S.Q., Zheng, Y.R., Guan, Q.L. & Yuan, M.G. (1982) A preliminary study on intergeneric crosses between the wheat variety Chinese Spring (6*x*) and *Hordeum bulbosum* (4*x*). *Acta Agronomica Sinica* 8, 95–101.

Wang, L.Q., Zhu, H.R., Guan, Q.L. & Rong, J.Q. (1987) Production of *T. aestivum* (6*x*)–*H. bulbosum* (4*x*) alien disomic addition lines and introgression of resistant gene (WYMV) from *H. bulbosum* to common wheat. In: *Barley Genetics V*, (Proceedings of 5th International Barley Genetics Symposium, Okayama, Japan), pp. 359–68.

Wojciechowska, B. (1984) Hybrids between *Hordeum vulgare* L. and *Hordeum jubatum* × 4*x* *Secale cereale*. *Genetica Polonica* 25, 247–53.

Wojciechowska, B. (1985a) Hybrids between *Hordeum vulgare* L. and *Triticum aestivum* L. *Genetica Polonica* 26, 457–62.

Wojciechowska, B. (1985b) Hybrids between barley and alloplasmic rye. *Genetica Polonica* 26, 317–26.

Wojciechowska, B. (1986) Hybrid plants from crosses between *Hordeum vulgare* L. and *Triticum persicum*. *Genetica Polonica* 27, 19–23.

Wojciechowska, B. (1987) Barley haploids and hybrids regenerated from callus tissue of *Hordeum vulgare* L. × *Secale cereale* L. embryos. *Genetica Polonica* 28, 217–23.

II

Genetics

4

Cloned and Mapped Genes: Current Status

Penny von Wettstein-Knowles

This chapter summarizes selected aspects of the currently available knowledge of the barley genome. A list of markers assigned to each of the seven chromosomes is presented. Included are maps of each chromosome, a map of the chloroplast genome and one of the first RFLP (restriction fragment length polymorphism) maps. With respect to the chromosome maps I have used the term centiMorgan (cM) to distinguish results which have been subjected to the Kosambi function. Since both the list of markers and maps are updates of the summary found in reference 177, which also covers markers not localized to chromosomes, only references not found therein are included here. Finally, three tables tabulate the barley genes and sequences which have been cloned to date. In those cases where a cloned gene of known function cannot be assigned to a classically defined marker with the same function, the symbol is followed by a small c or g and a number denoting the member of a cDNA or genomic DNA family (but not a locus).

Markers on chromosome 1

Acl3 acyl carrier protein III [187], *Adk* adenylate kinase [8], *alb-m*(a_c2) albino seedling, *Amp3* [98] *AmpA* aminopeptidase, *Amy2* α-amylase, *ant-1* anthocyaninless, *ant-22* proanthocyanidin, catechin and anthocyaninless [81], *ari-d(lk8) ari-i(br1)* breviaristatum, bl_2 bl_5 white aleurone, *Bz1* homology to maize *bz1* [192], *cer-a(gs3) cer-f* eceriferum, *Chi1* 28 kDa chitinase, *Chi2* 33 kDa chitinase [183], *clo-fc*(f_c) *clo-f8(f8)* chlorina seedling, *Iat1* (*Cma1*) subunit of tetrameric inhibitor of heterologous α-amylases; *Itr 2* (*CMc1*) BTI (CMc) inhibitor of trypsin, *des1 des4 des5* desynapsis, *e2* fine awned glume, *Enp1* endopeptidase, *ert-a ert-d ert-m* erectoides, *Est3 Est5* esterase, *f4 f5* chlorina seedling, *fs3* fragile stem, *Ga* gametophytic competition (certation), *gl5* glossy leaf, *Grb(Rsg1)* reaction to *Schizaphis graminum*, *kl* short awned, *L L2 L7 La* lax (long) spike, *lb3* long weak basal rachis internode, $lb3^m$ moderately long weak basal rachis internode, *Lk2* long awn, *lys2* high lysine, *mo1* multiovary, *msg10 msg14 msg22 msg23*, *n* naked caryopsis, *op* opposite spikelets, *Pbg* pubescence on outer glume, *Plm* anthocyanin rich, *Prx4* peroxidase, *R4* rough awn, *Rs* red stem and leaf sheath, *sb* subnodal bract, *seg1 seg2 seg4 seg5 seg8* shrunken endosperm genetic, *sex6* shrunken endosperm xenia, *smn* seminaked caryopsis, *T(Rpg1)* reaction to *Puccinia*

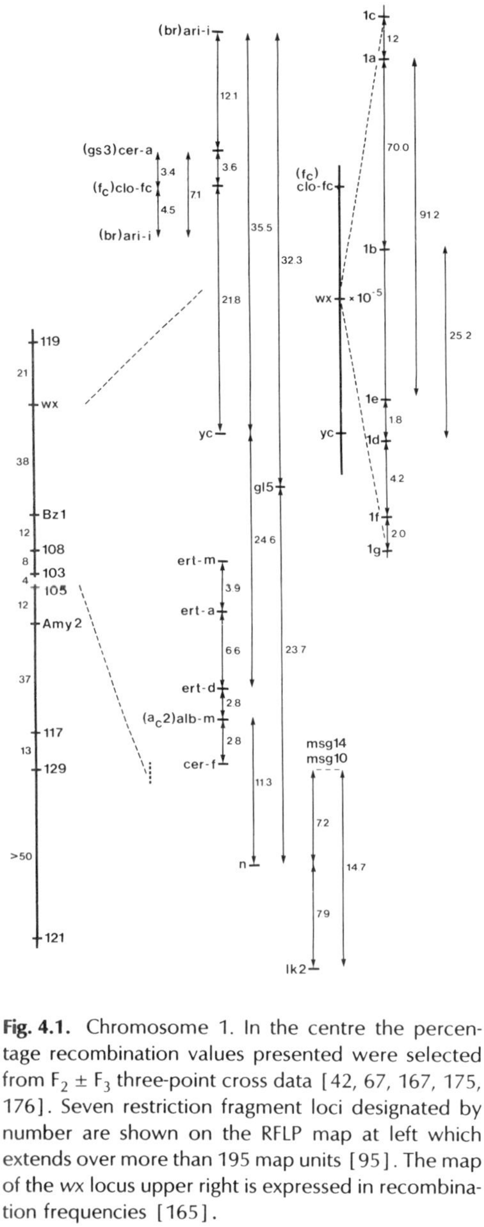

Fig. 4.1. Chromosome 1. In the centre the percentage recombination values presented were selected from $F_2 \pm F_3$ three-point cross data [42, 67, 167, 175, 176]. Seven restriction fragment loci designated by number are shown on the RFLP map at left which extends over more than 195 map units [95]. The map of the *wx* locus upper right is expressed in recombination frequencies [165].

graminis, *u4* unbranched style, *Un(Run1)* reaction to *Ustilago nuda*, *vbn* low number vascular bundles, *wnd* winding peduncles semidwarf, *wx* high amylopectin endosperm, X_a *xan,,i* xantha seedling, y_c virescent seedling, *Ym2(Rym2)* reaction to barley yellow mosaic virus (BYMV).

Markers on chromosome 2

a a2 a4 a5 albino seedling, *abr1* accordian basal rachis internode, *alb-c(y* y_x*) alb,,p* albino seedling, *ant-2(pr)* anthocyaninless, *Bmy2* β-amylase [108], *cer-g cer-n cer-s(gs5) cer-v* eceriferum, *Cig6* cold-induced gene [38], *D* sterile dwarf, *da* dehiscent awn, *des7* desynapsis, *e(log lep-2)* elongated outer glume awn, *f(lg)* chlorina seedling, *flo-c* extra floret, *fol-a* angustifolium, *G* teeth on lemma, *gai* gibberellin insensitive, *Gln2* chloroplastic glutamine synthetase [21], *glo-c glo-d* globe-shaped grain, *gp* grandpa, *Gpd1* glucose-6-phosphate dehydrogenase, *gs9* glossy sheath and ear, *H* high culm, *Ha2(Rha2)* reaction to *Heterodera avena*, *hl* reaction to *Cochliobolus sativus*, *Hr* hairy rachis, *Idh2* isocitrate dehydrogenase, In^e inhibitor of awns in *e/e* genotypes, *Int-d* intermedium spike [126], *Isa1* inhibitor of subtilisin and α-amylase, *L5* lax (long) spike, *li(aur-a)* ligule and auricle less, *lr* awnless lateral floret, *m2* many-noded dwarf, *msg2 msg3 msg27 msg37* [54] male sterile, *mss* midseason stripe, *mt mt,,e* mottled leaves, *mul* multiflorous, *Ndh4* NADH dehydrogenase, *or* orange seedling, *P* purple lemma, palea and pericarp, P_c purple veined lemma, *Pa1(Rph1) Pa2 (Rph2)* [155] reaction to *Puccinia hordei*, *Pau* purple auricle, *Pbq* basic protein Q, *Pn* purple node, *Prx2 Prx?* [9] peroxidase, *Pt3(Rpt3)* reaction to *Pyrenophora teres*, *ra* short rachilla, *rin(lin)* rachis (lesser) internode number, *Rrn5S* ribosomal 5S RNA, shortest repeat [100], *rt* rat tail spike, *sk* subjacent hood, *sld2* short slender culm, *Sod1* superoxide dismutase [101], *Ssu(RbcS)* ribulose bisphosphate carboxylase small subunit [21], *tr* triple awned lemma, *v(hex-v)* six row spike, *wst4* white striped leaf, *yst3* yellow striped leaf.

Markers on chromosome 3

a_c a_n albino seedling, *Aat3* aspartate aminotransferase, *al* albino lemma, *alb-ze* albino seedling, *als* absent lower laterals, *amo1* high amylose, *ant-17* flavanone-3-hydroxy-

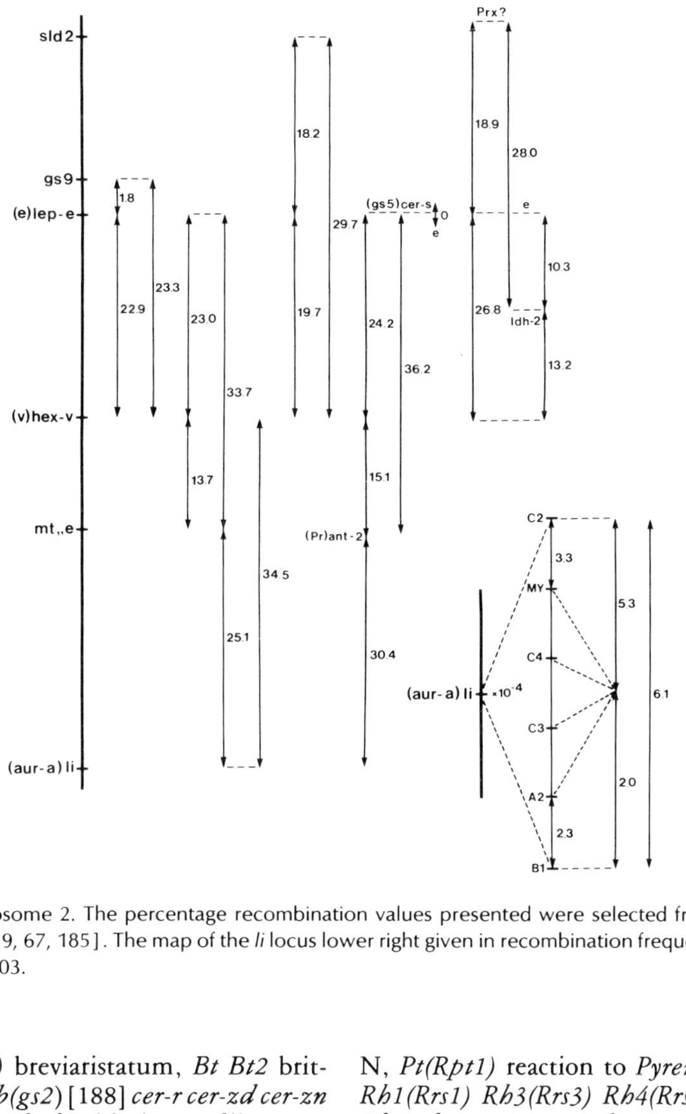

Fig. 4.2. Chromosome 2. The percentage recombination values presented were selected from $F_2 \pm F_3$ multi-point cross data [9, 67, 185]. The map of the *li* locus lower right given in recombination frequencies is modified from reference 103.

lase, *ari-a(lk7)* breviaristatum, *Bt Bt2* brittle rachis, *cer-b(gs2)* [188] *cer-r cer-zd cer-zn* eceriferum, *clo-f2(f2)* chlorina seedling, *cu2* curly, *des2* desynapsis, *dw1* light green dwarf, *ert-c ert-ii* erectoides, *Est1 Est2 Est4 Est10* esterase, *Ga2* gametophytic competition (certation) [105], *Gr* growth factors, *gra* granum, *int-e* intermedium spike, *Itr1 (Itc1 CMe1)* BTI-CMe (TI-1) inhibitor of trypsin, L_c lax (long) spike, *lnt* low number of tillers, *lzd* dwarf, *m3* many-noded dwarf, *Mal1* malic enzyme [124], *Mdh2* malate dehydrogenase, *msg5* male sterile, *nec4 nec5* $Nec\text{-}B_1$ necrotic leaf, *Pa7(Rph7)* reaction to *Puccinia hordei*, *Pbn1* basic protein N, *Pt(Rpt1)* reaction to *Pyrenophora teres*, *Rh1(Rrs1) Rh3(Rrs3) Rh4(Rrs4)* reaction to *Rhynchosporium secalis*, *sca* short crooked awn, *sdw-b* semidwarf bracteatum, *seg3 seg6* shrunken endosperm genetic, *sld* dwarf, *st* striped, *Tpi1* cytoplasmic triosephosphate isomerase, *uz* 'uzu' semi-brachytic, v_4 'six row' (intermedium spike), $vir\text{-}l(x_c)$ viridis seedling, *wst wst,,j* white striped leaf, x_s *xan,,h* xantha seedling, *Yd2(Ryd2)* reaction to barley yellow dwarf virus (BYDV), *yst yst2* yellow striped leaf, *zb* zebra striped (tigrina) leaf.

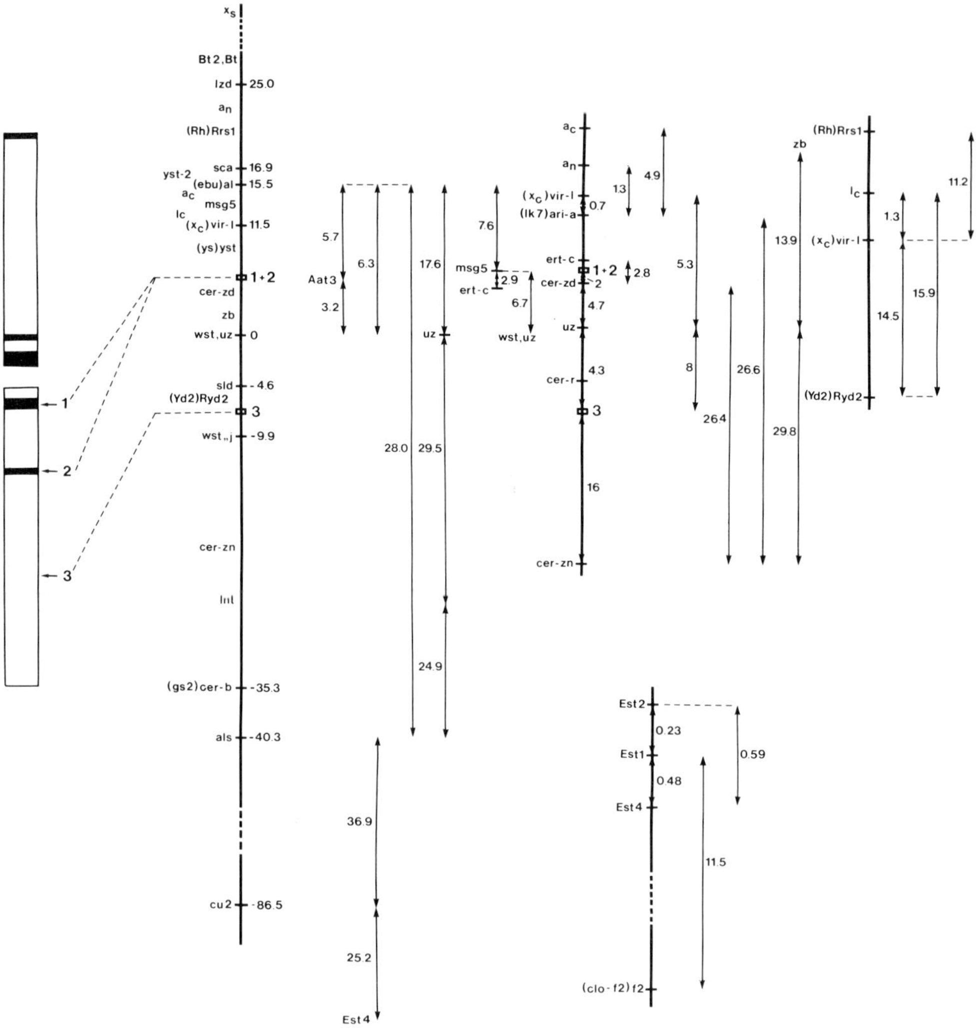

Fig. 4.3. Chromosome 3. At the far left is a cytological map showing the relationship among the centromere and the C-bands [121]. Adjacent is a genetic map modified from references 104, 106 and 184. The presented sequence and distances in cM were derived using a maximum likelihood estimation procedure and assigning *uz* and *wst* to the zero position. The SD ranges from 1.2 to 26.4% of the distances specified between genes. *Rh(Rrs1)* has been used to symbolize the one or more genes – *Rh(Rrs1) Rh3(Rrs3) Rh4(Rrs4)* – clustered at the position shown (upper right). Most other values given are percentage recombinations selected from $F_2 \pm F_3$ multipoint cross data among genes and/or the numbered C-bands [12, 43, 90, 106, 122, 133, 135, 168, 177]. The 17.6 value from *uz* to *al* as well as some of those in the insets with *msg5* and *vir-l(x$_c$)* are from two-point crosses.

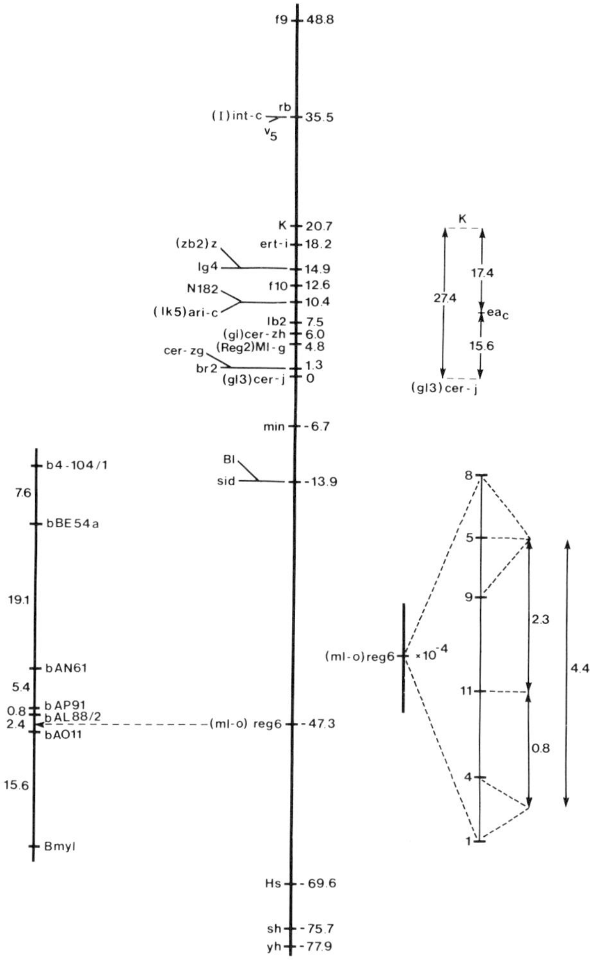

Fig. 4.4. Chromosome 4. The map in the centre, modified from reference 83, presents gene sequence and distances in cM derived using a maximum likelihood map estimation procedure and assigning *cer-j(gl3)* to the zero position. For markers from *yh* to *ml-o(reg6)* the SD ranges from 2.3 to 3.1 cM, and for markers *Bl* to *int-c(I)* from 0.7 to 1.3 cM except for *ari-c(lk5)* and *cer-zg* with SDs of 2.1 and 1.8, respectively. The SD of 6.9 for *rb* precludes its being assigned a numerical position on the map. Grouped together at the average position on the map are all genes located within 0.8 cM of one another. The percentage recombination values from $F_2 + F_3$ three-point cross data locates ea_c between *K* and *cer-j(gl3)* [193]. The RFLP map of 50.8 map units on the left includes the *ml-o(reg6)* locus at ←, the *Bmy1* locus and six restriction fragment loci [72]. The map of the *ml-o(reg6)* locus on the right is expressed in recombination frequencies [87].

Markers on chromosome 4

Acp2 acid phosphatase, *Adh1 Adh2* alcohol dehydrogenase, *Amp2* aminopeptidase [98], *ari-c(lk5)* breviaristatum, *bl* white aleurone xenia, bl_3 white aleurone, bl_4 pink aleurone, *Bmy1* β-amylase, *br2* brachytic dwarf, *Cat1* catalase, *cer-cqu(gs1) cer-j(gl3) cer-zg cer-zh(gl)* eceriferum, *Ch-e* chlorotic hybrids, *Ea2* ea_c early maturity, *ert-i* erectoides, *f9 f10* chlorina seedling, *flo-a* extra floret, *glo-a* globe-shaped grain, *H4* high (tall), H_n hair on lemma nerves, *Hs* hairy leaf sheath, *I*/I^b*(int-c)*/*i* infertile intermedium / fertile intermedium / deficiens spike, I_l intermedium spike, *Iat2 (cMb1) Iat3 (CMd1)* BTAI-CMb and BTAI-CMd subunits of tetrameric inhibitor of heterologous α-amylases, *K*/K^e/*k* hooded/elevated hooded/awned lemma, *lb2* long weak basal rachis internode, *lg2 lg4 lg5* light green seedling, *Lpx1* lipoxygenase, *Mas* mugineic acid synthesis from 2′deoxymugineic acid [128], *min* semiminute (dwarf), *N182 Ml-(CP)* reaction to *Erysiphe graminis*, *msg24 msg25* male sterile, *Ndh1* NADH dehydrogenase, *ovl* ovaryless, *Paz1* protein Z4, *Pgm1* phosphoglucomutase, *Pht* pH3 soil

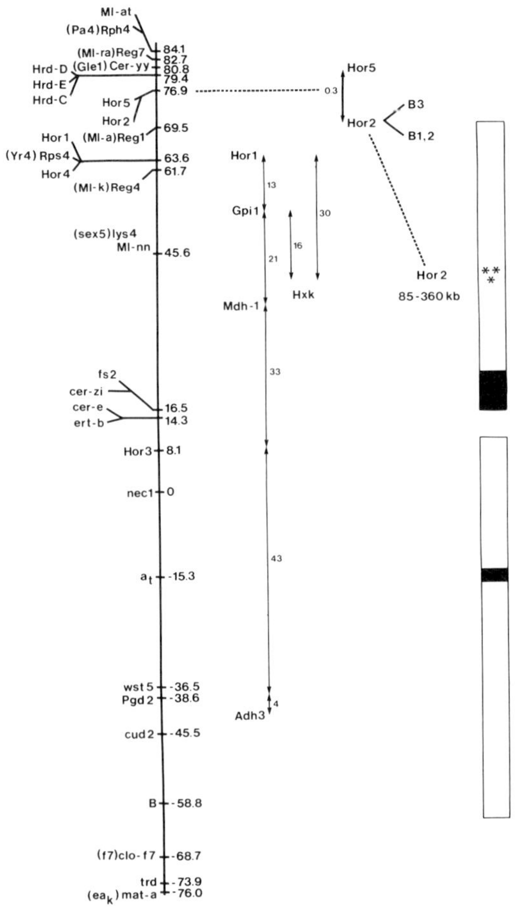

Fig. 4.5. Chromosome 5. The map is modified from reference 85. The presented sequence and distances in cM were derived using a maximum likelihood map estimation procedure and assigning *nec1* to the zero position. For markers from *Ml-at* to *cer-e* the SDs are ≥ 3.0 cM, while from *ert-b* to ea_k the SDs range from 1.5 to 2.9 cM. The SD of 8.6 for *Lys4(sex5)* precludes its being assigned a position on the map. Grouped together at their average position on the map are all genes located within 0.9 cM of one another. An expansion of the *Hor2* locus includes its approximate size and the orientation of the different B-hordein subgroups therein [173, 179]. The map of the isozyme loci in the centre is in percentage recombination [19]. At the right is a cytological map showing the relationship among the centromere, C-bands [121] and the position of hybridization (stars) [31] to the *Hor2* locus as revealed using a subclone of pHBR 184 (Table 3A).

tolerance, *rb* ribbon grass, *Reg2(Ml-g) reg6(ml-o)* reaction to *Erysiphe graminis*, *rh6(rrs6) Rh9 (Rrs9)* reaction to *Rynchosporium secalis*, *Sh* winter habit of growth, *sid* single elongated internode dwarf, v_5 'six row' (intermedium spike), *yh* yellow head (spike), *zb2*(zb_cz) zebra striped leaf.

Markers on chromosome 5

a_t albino seedling, *Adh3* alcohol dehydrogenase [19], *B* black lemma and pericarp, *cer-e cer-zi Cer-yy(Gle1)* [188] eceriferum, *clh* curly leaf dwarf, *clo-f7(f7)* chlorina seedling, *cm2* cream seedling, *cud2* curly dwarf, *des6* desynapsis, ea_k*(mat-a)* early maturity, *ert-b* erectoides, *f3* chlorina seedling, *fs2* fragile stem, *Gdh1* glutamate dehydrogenase, *glo-e* globe-shaped grain, *Glu*A ≈ 30 kDa glutelin [112], *Glu*B ≈ 55 kDa glutelin [112], *GluE* ≈ 33.5 kDa glutelin [113], *GluF* ≈ 34.5 kDa glutelin [113], *Gpi* glucose phosphate isomerase, *hl2* reaction to *Cochliobolus sativus*, *Hor1* C hordeins, *Hor2* B hordeins, *Hor3* D hordeins, *Hor4 (HrdG)* [174] *Hor5(HrdF)* [172] *HrdC HrdD HrdE* hordeins, *Hxk* hexokinase [19], *Ica1 Ica2* inhibitor of chymotrypsin and *Aspergillus* protease, *int-a* intermedium spike, *Lys4(sex5)* high lysine (shrunken endosperm xenia), *Mdh1* malate dehydrogenase, *Ml-a(Reg1) Ml-at Ml-ga* [86] *Ml-i* [79] *Ml-k(Reg4) Ml-n Ml-nn Ml-ra(Reg7)*

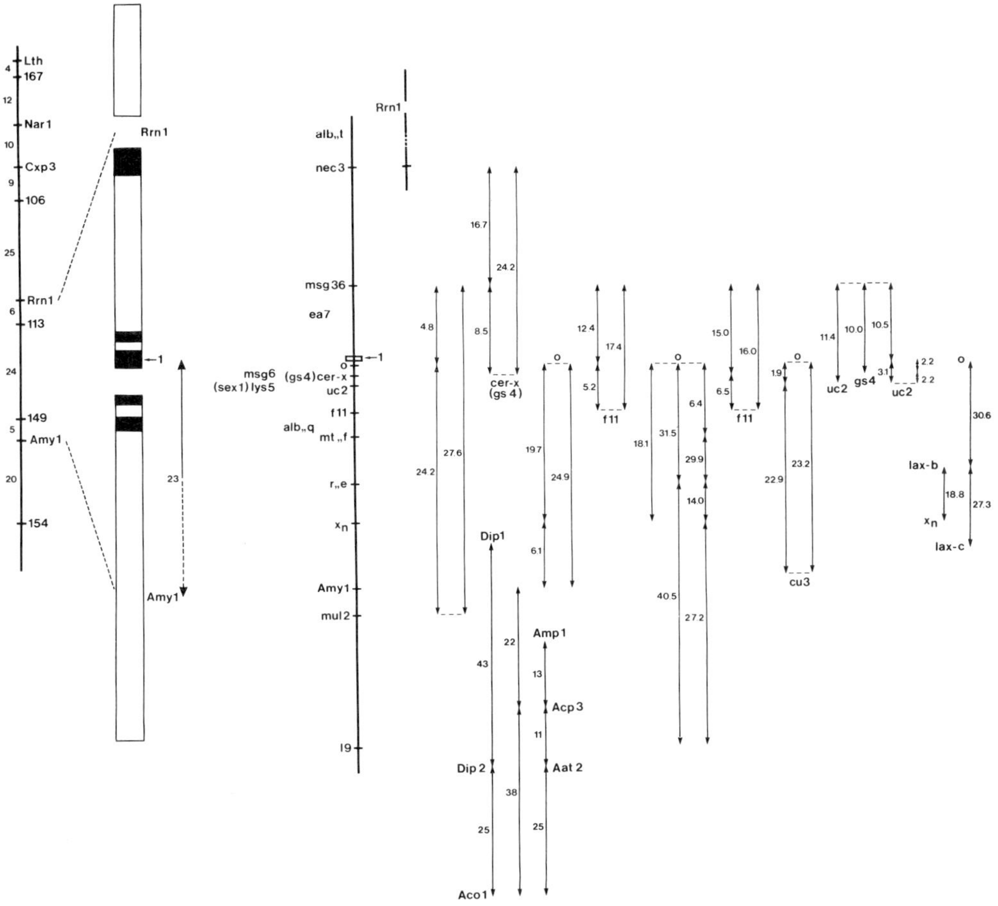

Fig. 4.6 Chromosome 6. The RFLP map at the left extending over 115 map units includes five restriction fragment loci designated by number and five gene loci [96, 181]. Two of the latter have been localized [5, 120, 122] on the adjacent cytological map in relationship to the centromere, nucleolus organizer region and C-bands [119]; that is, *Rrn1* at the nucleolus organizer region and *Amy1* in the long arm. One possible sequence of 12 genes and the C band S1 (←1) are shown on the genetic map in the centre. If the order of *msg6* and *lys5(sex)1* should turn out to be inverted then *alb,,q* and *alb,,t* must be interchanged. Approximate locations of three additional loci are indicated. The unknown relationship of *nec3* to *Rrn1* is indicated by a broken line. The percentage recombination values presented were selected from $F_2 \pm F_3$ multipoint cross data [19, 44, 46, 47, 66, 68, 92, 93, 102, 114, 120, 131, 132].

[37] reaction to *Erysiphe graminis*, *msg msg20* male sterile, *nec1* necrotic leaf, *Pa4 (Rph4)* reaction to *Puccinia hordei*, *Pgd2* phosphogluconate dehydrogenase, *Prx1* peroxidase, *Pt2(Rpt2)* reaction to *Pyrenophora teres*, *Re* red ear, *rvl* revoluted leaf, *Sh3* spring habit of growth, *trd* third outer glume, *v3* 'six row' (intermedium spike), *wst5* white striped leaf, *xan,,g* xantha seedling, *Yr4(Rps4)* reaction to *Puccinia striiformis*, *zb3* zebra striped leaf.

Markers on chromosome 6

Aat2 aspartate aminotransferase, *Aco1* aconitate hydratase, *Acp3* acid phosphatase, *ag* agropyroides, *alb,,q alb,,t* albino *seedling*, *Amp1* aminopeptidase, *Amyl* α-amylase, *ant13* regulatory gene; proanthocyanidin, catechin and anthocyaninless, *cat2* catalase, *cer-x(gs4)* [187] eceriferum, *cu cu3* curly, *Cxp3* carboxypeptidase III [96, 181], *Dhn3* 16.2 kDa dehydrin [32], *Dhn4* 22.6 kDa dehydrin [32], *Dip1 Dip2* [66] dipeptidase, *ea7* early maturity, *f11* chlorina seedling, *L9 lax-c* lax (long) spike, *lax-b* lax (long) spike; shrunken endosperm xenia, *Lth* leaf thionin [14], *lys5(sex1) lys6* high lysine (shrunken endosperm xenia), *msg6 msg34 msg36* [53, 147] *msg39* [54] male sterile, *mt,,f* mottled leaf, *mul2* multiflorous, *nar1* NADH-nitrate reductase apoprotein [96], *nec2 nec3* necrotic leaf, *o* orange lemma base and node, *Pa11 (Rph11)* reaction to *Puccinia hordei* [47], *Pgk2* cytoplasmic phosphoglycerate kinase [25], *r,,e* smooth awn, *Reg3(Ml-h)* reaction to *Erysiphe graminis*, *Rrn1* cytoplasmic ribosomal RNA repeat short intergenic spacer [5, 166], *uc2* uniculm, x_n xantha seedling.

Markers on chromosome 7

a_b albino seedling, *Acl1* acyl carrier protein I [187], *ari-e(lk9)* breviaristatum, *bb* broad leaf, *bir* branching inflorescent rachilla, *cer-i cer-t cer-w cer-zj cer-zp* eceriferum, *cl* curly lateral, *Clt* coloured leaf tip, *cm* cream seedling, *com1* compositum, *cs* corn stalk, *cud* curly dwarf, *ddt* resistance to DDT, *dex1* defective endosperm xenia, *Dhn1* 14.2 kDa dehydrin [32], *Dhn2* 14.4 kDa dehydrin [32], *dsk* dusky, *Dt* dented leaf margin, *Ea5* early maturity, *ert-g ert-n* erectoides, *Est9* esterase, *f6(yv)* chlorina seedling, *flo-b* extra floret, *fs* fragile stem, *glo-b glo-f* globe-shaped grain, *Gr2* high yield (growth factor), *H2* high (tall), *hl3* reaction to *Cochliobolus sativum*, *Hr2* hairy rachis, *Inc* partial incompatibility with *H. bulbosum*, *Kw1* kernel weight [84], *L3 lax-a* lax (long) spike; five anthers, *lb* long basal rachis internode, *let* pollen lethal, *Lfb* leafy bracht, *lys lys3(sex3)* high lysine (shrunken endosperm xenia), *Mdh3* malate dehydrogenase [125], *msg7* [52] *msg8* [52] *msg16 msg19 msg29* [52] male sterile, *mt2* mottled leaf, *nar2*

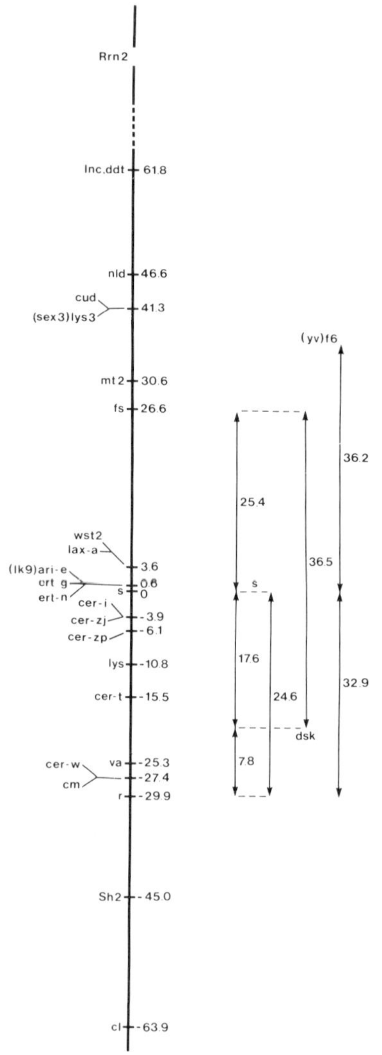

Fig. 4.7. Chromosome 7. The map on the left is modified from reference 82. The presented sequence and distances in cM were derived using a maximum likelihood map estimation procedure and assigning *s* to the zero position. For markers from *wst2* to *ert-n* the SD ranges from 0.7 to 1.5 cM and for *cer-i* to *r* from 0.7 to 1.3 cM. Grouped together at their average position on the map are all genes located within 1.1 cM of one another. All other specified values are percentage recombinations selected from $F_2 + F_3$ multipoint cross data [45, 67]. The unknown relationship of *ddt* to *Rrn2* is indicated with a broken line. The location of *Rrn2* at the nucleolus organizer region is from reference 5 and of *Inc* from 141.

deficient in molybdenum enzymes nitrate reductase, xanthine dehydrogenase and aldehyde oxidase, *nld* narrow-leafed dwarf, *Paz2* protein Z7, *R/r* rough/smooth awn, *R2* rough awn, *Rrn2* cytoplasmic ribosomal RNA repeat long intergenic spacer [5, 166], *s* short rachilla hair, *Sdh1* shikimate dehydrogenase, *Sh2* spring habit of growth, *stb* starch binding in endosperm [70], *tg* third inner glume, *tig-d* tigrina seedling [91], *tig-o* tigrina to zonata seedling, *Tpi2* chloroplastic triosephosphate isomerase [142], *u* unbranched style, v_2 'six row' (intermedium spike), *va* va_3 variegated, *wst2* white striped leaf.

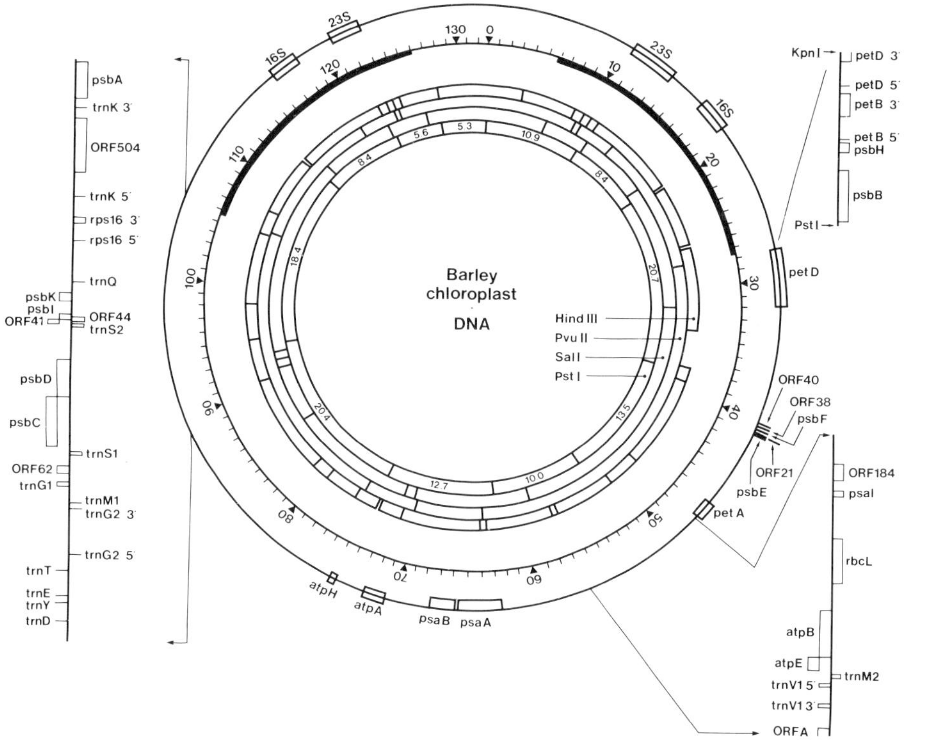

Fig. 4.8. Chloroplast DNA. The physical map of barley cpDNA as derived from digests with *Hin*dIII, *Pvu*II, *Sal*I and *Pst*I [130, 144] is shown in the four innermost circles. The zero position on the kb circle is located midway between the two inverted repeats (thick lines). Reference 35 contains restriction maps of barley cpDNA after *Hpa*I, *Pst*I, *Pvu*II, *Sal*GI, *Sst*I, *Xho*I and *Bam*HI digests. The boxes and/or lines on the outermost circle or enlargements thereof represent mapped genes. Sequenced genes on the inside of the circle are transcribed in a clockwise direction from DNA strand B and those on the outside of the circle in an anticlockwise direction from the complementary DNA strand A. Open-ended boxes indicate that gene or ORF (open reading frame) was not completely sequenced. Boxes bisected by the circle represent genes localized within that region by heterologous hybridization whose direction of transcription is unknown. Boxes or lines displaced from the circle indicate overlapping reading frames. Exons of split genes are labelled 5′ and 3′. See Table 4.1.

Table 4.1. Cloned chloroplast genes and sequences with their position on the physical map in kb and if known the strand from which they are transcribed (Fig. 4.8).

Symbol	Character	Plasmid		Strand	Position	Refs
		Name	kb			
atp A	α-subunit of	*Hin*dIII pHvC21	7.4		72	138
	CF_1-ATP synthase					
atp B	β-subunit of	*Pst*I pHvC209	10	B	54	144
	CF_1-ATP synthase	*Sma/Hind* pHvu-SH13	3.7	B	54	195
atp E	ε-subunit of	*Pst*I pHvC209	10	B	55	144
	CF_1-ATP synthase	*Sma/Hind* pHvu-SH13	3.7	B	55	195
atp H	subunit III of	*Hin*dIII pHvC174	1.7		75	138
	CF_0-ATP synthase					
ORF A	open reading frame of	*Sma/Hind* pHvu-SH13	3.7	B	57	195
	unknown length					
ORF 21	open reading frame for	*Bam*HI pHvC46-4.2	4.2	A	41	110
	21 amino acids	*Pst*I	13.5	A	41	23
ORF 38	open reading frame for	*Pst*I	13.5	A	41	23
	38 amino acids					
ORF 40	open reading frame for	*Pst*I	13.5	A	41	23
	40 amino acids					
ORF 41	open reading frame for	*Eco*RI K	6.25	A	98	171
	41 amino acids					
ORF 44	open reading frame for	*Eco*RI K	6.25	B	98	171
	44 amino acids					
ORF 62	open reading frame for	*Hin*dIII pHvC35	4.3	A	94	139
	62 amino acids					
ORF 184	open reading frame for	*Pst*I pHvC186	13.5	A	49	169
	184 amino acids					
ORF 504	open reading frame for	*Eco*RI	2.19	B	103	15
	504 amino acids	*Eco*RI K	6.25	B	103	171
pet A	cytochrome *f*	*Pst*I pHvC186			47	138
pet B	cytochrome b_6					
	3′ exon	*Pst/Kpn*	5.2	A	32	153
	5′ exon	*Pst/Kpn*	5.2	A	33	153
pet D	subunit 4 of cytochrome	*Hin*dIII pHvC79	8.4		30	138
	b_6/f complex					
	3′exon	*Pst/Kpn*	5.2	A	30	153
	5′exon	*Pst/Kpn*	5.2	A	31	153
psa A	photosystem I P700	*Eco*RI pBE432-3	7.6	B	64	10
	reaction centre					
	apoprotein A					
psa B	photosystem I P700	*Eco*RI pBE432-3	7.6	B	67	10
	reaction centre					
	apoprotein B					
psa I	4 kDa photosystem I	*Pst*I pHvC186	13.5	A	50	169
	protein					
psb A	photosystem II reaction	*Eco*RI	2.19	B	105	15
	centre P680 protein D1	*Hind/Eco*R	1.6	B	105	130
		*Sma/Eco*R	1.43	B	105	40
psb B	photosystem II	*Pst/Kpn*	5.2	A	34	153
	chlorophyll					
	a-protein CP47					

Table 4.1. *cont.*

Symbol	Character	Plasmid		Strand	Position	Refs
		Name	kb			
psb C	photosystem II	–	–	A	95	154
	chlorophyll *a*-protein CP43	*Bam*HI pBB531-4	3.4	A	95	10, 130
psb D	photosystem II reaction	*Bam*HI pBB531-4	3.4	A	96	10, 130
	centre P 680 protein D2	*Eco*R/*Sac*	1.41	A	96	39
psb E	photosystem II	*Bam*HI pHvC46-4.2	4.2	A	41	110, 111
	cytochrome *b*559 α-subunit	*Pst*I	13.5	A	41	23
psb F	photosystem II	*Bam*HI pHvC46-4.2	4.2	A	41	110, 111
	cytochrome *b*559 β-subunit	*Pst*I	13.5	A	41	23
psb H	photosystem II 10 kDa phosphoprotein	*Pst/Kpn*	5.2	A	33	153
psb I	photosystem II 36 amino acid protein	*Eco*RI K	6.25	A	98	171
psb K	photosystem II 61 amino acid protein	*Eco*RI K	6.25	A	99	171
rbc L	large subunit of ribulose bisphosphate carboxylase	*Pst*I pHvC186	13.5	A	52	144, 145
rps 16	ribosomal protein S16					
	3′ exon	*Eco*RI K	6.25	B	101	171
	5′ exon	*Eco*RI K	6.25	B	100	171
16S	chloroplast 16S ribosomal RNA	*Pst*I pHvC205	8.4		17 + 117	144
23S	chloroplast 23S ribosomal RNA	*Pst*I pHvC203	5.6		12 + 122	144
trn D	t-RNA-aspartate (GUC)	*Hind/Eco*R pHvC208-1.6	1.6	A	90	11
trn E	tRNA-glutamate (UUC)	*Hind/Eco*R pHvC208-1.6	1.6	A	90	11
trn G1	tRNA-glycine (GCC)	*Hind*III pHvC35	4.3	A	94	139
trn G2	tRNA-glycine (UCC)					
	3′ exon	*Hind*III pHvC35	4.3	B	93	139
	5′ exon	*Hind*III pHvC35	4.3	B	92	139
trn K	tRNA-lysine (UUU)					
	3′ exon	*Eco*RI	2.19	B	104	15
		*Eco*RI K	6.25	B	104	171
	5′ exon	*Eco*RI K	6.25	B	101	171
trn M1	tRNA-formyl methionine (CAU)	*Hind*III pHvC35	4.3	B	93	139
trn M2	tRNA-methionine (CAU)	*Sma/Hind* pHvu-SH13	3.7	A	55	195
trn Q	tRNA-glutamine (UUG)	*Eco*RI K	6.25	A	99	171
trn S1	tRNA-serine (UGA)	*Hind*III pHvC35	4.3	B	94	139
trn S2	tRNA-serine (GCU)	*Eco*RI K	6.25	B	98	171
trn T	tRNA-threonine (GCU)	*Hind*III pHvC35	4.3	A	91	139
trn V1	tRNA-valine (UAC)					
	3′ exon	*Sma/Hind* pHvu-SH13	3.7	B	56	195
	5′ exon	*Sma/Hind* pHvu-SH13	3.7	B	56	195
trn Y	tRNA-tyrosine (GUA)	*Hind/Eco*R pHvC208-1.6	1.6	A	90	11

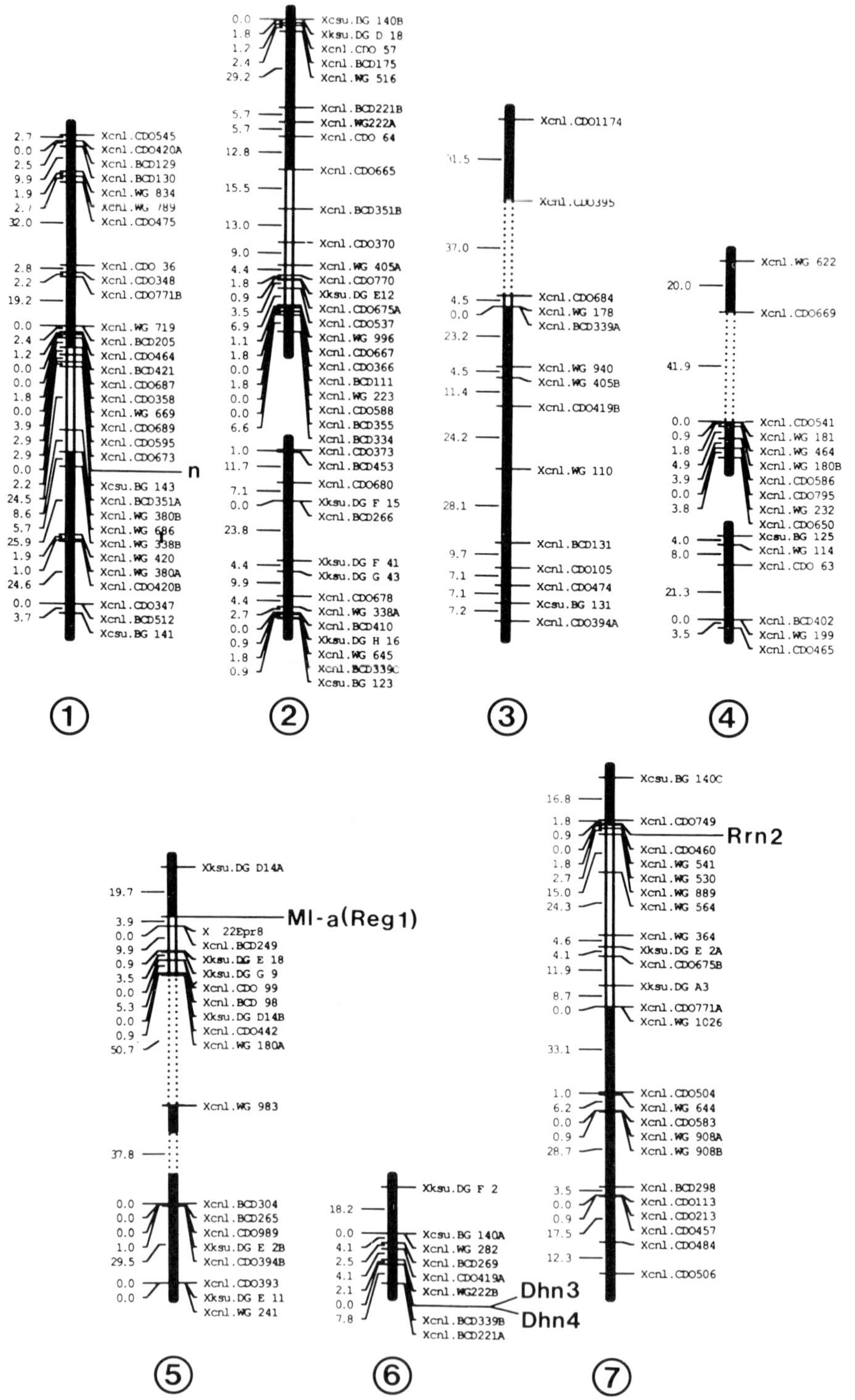

Fig. 4.9. A restriction fragment length polymorphism (RFLP) map. A map of the barley genome constructed using doubled haploids derived from cultured anthers from the F_1 of cv. Proctor x cv. Nudinka defines 125 loci, including the genes *n*, *Ml-a(Reg1)*, *Dhn3*, *Dhn4* and *Rrn2* plus 150 RFLP markers. Distances between loci (left side of the chromosomes) are in cM; values greater than 33 are highlighted by broken lines indicating their uncertainty. Markers definitively assigned to the short (upper) and long (lower) arms are mapped on the black portions of the chromosomes. The orientation of the lower segments of chromosomes 2 and 4 is unknown. This figure is modified from reference 71a.

Table 4.2. Cloned barley genes – cDNAs.

Symbol	Character	Plasmid name	cDNA bp	Refs
–	maize a1 analogue	pHva1	1175	163
–	developing aleurone specific	B11E	485	80
Acl 1	acyl carrier protein I	pACP11	768	65
Acl 3	acyl carrier protein III	pACP1	520	65
Adh 1	alcohol dehydrogenase	–	1654	60
Ale	aleurain (thiol protease)	–	1403	159
Amy 1	α-amylase (high pI, type B, type 2)	103	≈630	77
		pHV19	1119	24
		pM/C	1479	94, 158
Amy 2	α-amylase (low pI, type A, type 1)	E	1588	160
Ant 18	dihydroquercitin reductase	pHva 14	1258	109, 163
Bmy 1	β-amylase	pcβC51	1754	107
Cab c1	photosystem II light-harvesting chlorophyll a/b protein	pHvLF2	≈610	59
Cam	calmodulin	pBcam-1	489	123
		pBcam-3	1086	123
Cep B	cysteine endopeptidase B (EP-B)	pHV14	≈1000	24
		pHVEP1	1427	99
		pHVEP4	1450	99
		pHVEP7	1462	99
Chi 1	28 kDa endochitinase (basic protein C)	cCHI26	1001	118
Chi 2	33 kDa endochitinase	10	556	183
Chs	chalcone synthase	pcchs11	1477	163
Cig 1	≈22 kDa cold-induced gene	pV60	≈550	22
Cig 2	potential cold-induced gene	pT59	652	22
Cig 3	potential cold-induced gene	pAO86	770	22
Cig 4	potential cold-induced gene	pAO29	≈300	22
Cig 5	potential cold-induced gene	pAF93	1154	22
Cig 6	cold-induced gene	BLT14	514	38
Cxp 1	carboxypeptidase I	λc1	1424	36
Dhn 1	14.2 kDa dehydrin	B8	683	33
Dhn 2	14.4 kDa dehydrin	B9	683	33
Dhn 3	16.2 kDa dehydrin	B17	812	33
Dhn 4	22.6 kDa dehydrin	B18	1049	33
Eli 1	≈20 kDa early light-induced chloroplast membrane protein	HV 58	860	63
Eli 2	≈14 kDa early light-induced chloroplast	HV 60	719	63, 62
	membrane protein	HV 90	685	63
Glb 2	(1→3, 1→4)-β-D-glucan	pA5	874	48
	4-glucanohydrolase isoenzyme II	pBA170	287	78
		pYBAG17	1088	78
Gld	glyceraldehyde-3-phosphate dehydrogenase	pc gld-1	1147	29
Glg 2	(1→3)-β-glucan 3-glucanohydrolase	λ3	1250	73
	isoenzyme II	cBGL32	1234	118
		pRP2	≈414	34
Gln 2	chloroplast glutamine synthetase	pGS8	1182	6
		pcHvGS14	1096	55
		pcHvGS6	1560	55
Grp 1	glycine-rich protein	cgrphv27	939	164
Gsa	glutamate 1-semialdehyde aminotransferase	–	1620	61

Table 4.2. *cont.*

Symbol	Character	Plasmid name	cDNA bp	Refs
H 3	histone H3	pc h31-1	522	30
Hor 1	C hordein	pc P387	475	50
		pc hor1-3	362	148
Hor 1 + 2	C + B1 hordeins	pc-919	252	148
Hor 2	B1 hordein	pB11	884	50
		pc hor2-1	254	149
		pc hor2-2	153	76, 149
		pc hor2-3	257	76, 149
		pc hor2-4	720	76, 149
		pc hor2-5	506	17
		pc hor2-7	390	17
		pc hor2-8	398	17
		pc hor2-9	≈350	16, 180
	B3 hordein	pB7	977	50
Hor 3	D hordein	pHvE-c133	–	50
		pHvE-c135	–	51
		pHvE-c155	≈900	51
		pc-918	≈400	16, 180
Hth 1	α-hordothionin	pTH1	569	143
Hth 2	β-hordothionin	pTH2	489	71
		pTH3	471	71
Iad 1	dimeric inhibitor of heterologous α-amylases	pUP-44	661	116
Iah 1	homology to inhibitors of heterologous α-amylases	pUP-13	506	140
Iah 2	homology to inhibitors of heterologous α-amylases	pUP-23	672	115
Iat 1	CMd subunit of tetrameric inhibitor of heterologous α-amylases	pCMd	579	64
		pUP-38	496	140
Ica 1	inhibitor of chymotrypsin and *Aspergillus* protease (CI-1, SP-1)			
	A subfamily	pCI1-21	447	191
		pCI1-27	≈435	191
		pCI1-28	454	191
		pCI1-39	448	191
		pCI1-4	448	191
		pCI1-6	≈435	191
	B subfamily	pCI1-7	435	191
		pCI1-14	≈448	191
Ica 2	inhibitor of chymotrypsin and *Aspergillus* protease (CI-2, SP-II)			
	A subfamily	pcIC 38	451	190
	B subfamily	pcIH 161	≈225	190
Isa	inhibitor of subtilisin and α-amylase (BASI, ASI)	BASI cDNA	792	117
Itr1	inhibitor of trypsin and (BTI-CMe, TI-1)	λ CMe	583	157
Ldh 1	lactate dehydrogenase polypeptide A	LDHA	1297	74
Ldh 2	lactate dehydrogenase polypeptide B	LDHB	1166	74
Lea	22 kDa late embryogenesis abundant protein	pHV A1	998	75

Table 4.2. *cont.*

Symbol	Character	Plasmid name	cDNA bp	Refs
Lec	root lectin	BLc3	972	119
Lth	leaf thionins			
	group 1	pHvDA6	ns	13
		pHvDD3	488	13
	group 2	pGK1348	608	56
		pGK1908	304	56
		pGK1940	445	56
		pGK2872	612	56
		pHvDB4	635	13
		pHvDC4	656	13
		pHvDF2	451	13
		pHvDG3	543	13
Ltp	lipid transfer protein (probable α-amylase/protease inhibitor)	Alf	650	129
Myb 1	myb-related DNA-binding protein, meristematic tissue	c-myb.Hv1	1280	127
Myb 2	myb-related DNA-binding protein, meristematic tissue	c-myb.Hv5	1100	163
Myb 6	myb-related DNA-binding protein, meristematic tissue	c-myb.Hv33	1100	127
Nar 1	NADH-nitrate reductase	bNRp10	≈ 1100	26
Paz 1	protein Z_4 (antigen 1a)	pc paz1-1	≈ 360	150
		pc paz1-2	325	69
		pc paz1-3	409	69
Pc	plastocyanin	pKG1409	616	134
		pKG3541	545	134
		pKG4438	522	134
		pKG755	203	134
		pKG951	193	134
Pcr	NADPH-protochlorophyllide oxidoreductase	pHvA4	1222	170
		pHvA7	1480	170
		pHvDF1	524	4, 170, 182
Pr 1	Psaknon-related protein produced upon fungal infection	pRP1	862	34, 89
Pr 3	Psaknon-related protein produced upon fungal infection	pRP3	≈ 357	34
Pr 4	Psaknon-related protein produced upon fungal infection	pRP4	≈ 280	34
Pr 5	Psaknon-related protein produced upon fungal infection	pRP5	≈ 270	34
Pr 6	Psaknon-related protein produced upon fungal infection	pRP6	≈ 160	34
Prx c1	37 kDa endosperm basic peroxidase (BP1)	pc13A9	1177	151
		pcR7	620	152
Prx c2	leaf peroxidase	pcD1311	≈ 1400	151
Psa E	16 kDa photosystem I protein	3.1	610	136
Psa H	10.2 kDa photosystem I protein	9.5-1	613	137
Rca	rubisco activase protein	9.5-2	280	137
	Al larger spliced message	λ Brca-19	≈ 1400	194
Rip	ribosome-inactivating protein	cRIP30	1031	118
Rip 2	ribosome-inactivating protein II	cRIP30A	479	118
Ub 1	ubiquitin	pKG3730	702	57
Wx	starch synthase	pcwx27	2397	162

Table 4.3. Cloned barley genes and repetitive sequences – genomic.

Symbol	Characteristics (copy number/N)	Plasmid name	Plasmid length (bp)	Refs
A. Genes				
Adh 1	alcohol dehydrogenase	pM915.1	≈ 1.5 kb, partly	186
Adh 2	alcohol dehydrogenase	pM91.2	−287 to +2033	186
Adh 3	alcohol dehydrogenase	pPr7.1	−99 to +2171	186
Ale	aleurain (thiol protease)	λ aleurain	−1457 to +3919	189
Amy 1	α-amylase (high pI, type B, type 2)	λAmy 46	−463 to −1	94
		gKAmy 141	−301 to +1690	97
		λAmy 6-4	−1175 to +1735	94
		gRAmy 152	−705 to +1920	146
		gRAmy 56	−504 to +1699	146
Amy 2	α-amylase (low pI, type A, type 1)	λamy32b	−688 to +2029	161, 189
		gKAmy 155	−324 to +2013	97
Bz 1	75% homology to the maize *Bz* gene product	pBz.Hv102-1	−476 to +1850	192
Cab 2	≈ 29.5 kDa photosystem II light-harvesting chlorophyll a/b protein	pCab-2	−180 to +849	28
Grp 1	glycine-rich protein	ggrphv2	1877	164
Hor 5	γ-hordein	λhor δ-1	−378 to +1236	20
Hor 1	C hordein	λhor 1-14	−428 to +1637	41
Hor 2	B hordein	λhor 2-4	−550 to +1229	17
		λyag9	−539 to + 988	27
		pBHR 184	−563 to +2337	49
Hth 1	α-hordothionin	pTHG1	−371 to +1444	156
		λBTH6	not published	14
Paz 1	protein Z_4 (antigen 1a)	λPaz 1-2	−1079 to +2054	18
Paz Ψ	protein Z_4 pseudogene	Paz Ψ	1554	18
Prx g1	seed basic peroxidase	W7422	≈ 7.8 kb, partly	151
Rrn	cytoplasmic ribosomal RNA repeat (≈ 1200 tandem copies/N): 18S, 5. 8S, 25S long intergenic spacer	pHV 103	not	58
		pHV 132	not	58
		pHV 144	not	58
		pHV 158	not	58
		pHV 256	not	58
		pHV 279	not	58
		pHV 29	not	58
		pHV 294	not	58
		pHV 309	not	58
		pHV361	not	58
Rrn 2	cytoplasmic ribosomal RNA repeat (≈ 600 tandem copies/N): 18S, 5.8S, 25S short intergenic spacer	pHV 107	not	58
		pHV 287	not	58
		pHV 99	not	58
Rrn 5S	ribosomal RNA: 5S, shortest repeat	–	301	100
Wx	starch synthase	–	−839 to +3061	162
B. Repetitive sequences				
ARS	functions as an autonomous replicating sequence in yeast	pHv 10	≈ 7.8 kb	2
		pHv 13-7	≈ 7 kb	2
		pHv 26-20	≈ 20 kb	2
		pHv 67-2	≈ 2 kb	2
		pHv 67-2.9	≈ 2.9 kb	2

Table 4.3. *cont.*

Symbol	Characteristics (copy number/N)	Plasmid name	Plasmid length (bp)	Refs
D	*Eco*RI-generated, dispersed, three variants ($>5 \times 10^3$)	D1	≈ 1.85 kb	178
		D2	≈ 2.3 kb	178
		D3	≈ 3.4 kb	178
H	*Eco*RI-generated, dispersed, heterogeneous ($>5 \times 10^4$)	Hv006	≈ 8 kb	3
		Hv1029	≈ 14.5 kb	1
		Hv19	≈ 12 kb	1
R	118 bp *Alu*I-generated relic DNA, tandem, homogeneous (7×10^5)	HvT 01	208	7
		HvT 02	236	7
		HvT 06	148	7
S	*Eco*RI-generated relic DNA, dispersed, heterogeneous (3×10^5)	MHV11	106	88
		MHV16	101	88
		MHV47	> 108	88
		MHV56	36	88

References

1. Ananiev, E.V., Ryzik, M.V., Sonina, N.V., Golovkin, M.V. & Shipkova, N.I. (1988) Mobile genetic elements of cereal genomes. *Biochemie und Physiologie der Pflanzen* 183, 129–41.
2. Ananiev, E.V., Shchipkova, N.I., Khvyrleva, T.D. & Danilevskaya, Q.N. (1989) Isolation of fragments of the barley genome DNA, supporting autonomous replication of plasmids in *Saccharomyces* yeast. *Molecular Genetics, Microbiology and Virology*. 2, 46–9.
3. Ananiev, E.V., Bochkanov, S.S., Ryzhik, M.V., Soniva, N.V., Chernyshev, A.I., Shchipkova, N.I. & Yakovleva, E.Y. (1986) Characteristics of cloned repeated DNA sequences in the barley genome. *Genetika* 22, 920–8.
4. Apel, K., Gollmer, I. & Batschauer, A. (1983) The light-dependent control of chloroplast development in barley (*Hordeum vulgare* L.). *Journal of Cellular Biochemistry* 23, 181–9.
5. Appels, R., Gerlach, W.L., Dennis, E.S., Swift, H. & Peacock, W.J. (1980) Molecular and chromosomal organization of DNA sequences coding for the ribosomal RNAs in cereals. *Chromosoma (Berlin)* 78, 293–311.
6. Baima, S., Haegi, A., Strøman, P. & Casadoro, G. (1989) Characterization of a cDNA clone for barley leaf glutamine synthetase. *Carlsberg Research Communications* 54, 1–9.
7. Belostotsky, D.A. & Ananiev, E.V. (1990) Characterization of relic DNA from barley genome. *Theoretical and Applied Genetics* 80, 374–80.
8. Benito, C., Gallego, F.J., Frade, J.M., Zaragoza, C. & Figueiras, A.M. (1990). Chromosomal location of adenylate kinase isozymes in Triticeae species. *Theoretical and Applied Genetics* 79, 157–60.
9. Benito, M.C., Sanchez, M., Shin, J.S. & Blake, T. (1988) A map of barley chromosome 2 using isozymic and morphological markers. *Biochemical Genetics* 26, 387–94.
10. Berends, T., Gamble, P.E. & Mullet, J.E. (1987) Characterization of the barley chloroplast transcription units containing psaA-psaB and psbD-psbC. *Nucleic Acids Research* 15, 5217–40.
11. Berry-Lowe, S. (1987) The chloroplast glutamate tRNA gene required for δ-aminolevulinate synthesis. *Carlsberg Research Communications* 52, 197–210.
12. Bockelman, H.E. & Eslick, R.F. (1977) Absence of crossing-over of *ert-c* in crosses involving T2–3a and additional information on the linkage of *uz* and *msg5*. *Barley Genetics Newsletter* 7, 15–16.
13. Bohlmann, H. & Apel, K. (1987) Isolation and characterization of cDNAs coding for leaf-specific thionins closely related to the

endosperm-specific hordothionin of barley (*Hordeum vulgare* L.). *Molecular and General Genetics* 207, 446–54.

14. Bohlmann, H., Clausen, S., Behnke, S., Giese, H., Hiller, C., Reimann-Philipp, U., Schrader, G., Barkholt, V. & Apel, K. (1988) Leaf-specific thionins of barley – a novel class of cell wall proteins toxic to plant-pathogenic fungi and possibly involved in the defence mechanism of plants. *EMBO Journal* 7, 1559–65.
15. Boyer, S.K. & Mullet, J.E. (1988) Sequence and transcript mapping of barley chloroplast *psbA* gene. *Nucleic Acids Research* 16, 8184.
16. Brandt, A. (personal communication).
17. Brandt, A., Montembault, A., Cameron-Mills, V. & Rasmussen, S.K. (1985) Primary structure of a B1 hordein gene from barley. *Carlsberg Research Communications* 50, 333–45.
18. Brandt, A., Svendsen, I. & Hejgaard, J. (1990) A plant serpin gene: structure, organization and expression of the gene encoding barley protein Z_4. *European Journal of Biochemistry* 194, 499–505.
19. Brown, A.H.D., Lawrence, G.J., Jenkin, M., Douglass, J. & Gregory, E. (1989) Linkage drag in backcross breeding in barley. *Journal of Heredity* 80, 234–9.
20. Cameron-Mills, V. & Brandt, A. (1988) A γ-hordein gene. *Plant Molecular Biology* 11, 449–61.
21. Cannell, M., Mulcahy, M., Karp, A. & Shewry, P.R. (1989) Chromosomal assignment of characterized cDNA's and genes using RFLP analysis of ditelosomic addition lines. *Barley Genetics Newsletter* 19, 9–11.
22. Cattivelli, L. & Bartels, D. (1990) Molecular cloning and characterization of cold-regulated genes in barley. *Plant Physiology* 93, 1504–10.
23. Chakhmakhcheva, O.G., Andreeva, A.V., Buryakova, A.A., Reverdatto, S.V. & Efimov, V.A. (1989) Nucleotide sequence of the barley chloroplast *psbE, psbF* genes and flanking regions. *Nucleic Acids Research* 17, 2858.
24. Chandler, P.M., Zwar, J.A., Jacobsen, J.V., Higgins, T.J. & Inglis, A.S. (1984) The effects of gibberellic acid and abscisic acid on α-amylase mRNA levels in barley aleurone layers studies using an α-amylase cDNA clone. *Plant Molecular Biology* 3, 407–18.
25. Chao, S., Raines, C.A., Longstaff, M., Sharp, P.S., Gale, M.D. & Dyer, T.A. (1989) Chromosomal location and copy number in wheat and some of its close relatives of genes for enzymes involved in photosynthesis. *Molecular and General Genetics* 218, 423–30.
26. Cheng, C.-L., Dewdney, J., Kleinhofs, A. & Goodman, H.M. (1986) Cloning and nitrate induction of nitrate reductase mRNA. *Proceedings of the National Academy of Sciences (USA)* 83, 6825–8.
27. Chernyshev, A.I., Davletova, S.K., Bashkirov, V.I., Shakhmanov, N.B., Mekhedov, S.L. & Ananiev, E.V. (1989) Nucleotide sequence of the B1 hordein gene of barley (Hordeum vulgare L.). *Genetika* 25, 1349–55.
28. Chitnis, P.R., Morishige, D.T., Nechushtai, R. & Thornber, J.P. (1988) Assembly of the light-harvesting chlorophyll *a*/*b* proteins in barley etiochloroplasts involves processing of the precursor on thylakoids. *Plant Molecular Biology*, 11, 95–107.
29. Chojecki, J. (1986) Identification and characterisation of a cDNA clone for cytosolic glyceraldehyde-3-phosphate dehydrogenase in barley. *Carlsberg Research Communications* 51, 203–10.
30. Chojecki, J. (1986) Identification and characterisation of a cDNA clone for histone H3 in barley. *Carlsberg Research Communications* 51, 211–17.
31. Clark, M., Karp, A. & Archer, S. (1989) Physical mapping of the B-hordein loci on barley chromosome 5 by *in situ* hybridization. *Genome* 32, 925–9.
32. Close, T.J. & Chandler, P.M. (1990) Cereal dehydrins: serology, gene mapping and potential functional roles. *Australian Journal of Plant Physiology* 17, 333–44.
33. Close, T.J., Kortt, A.A. & Chandler, P.M. (1989) A cDNA-based comparison of dehydration-induced proteins (dehydrins) in barley and corn. *Plant Molecular Biology* 13, 95–108.
34. Davidson, A.D., Manners, J.M., Simpson, R.S. & Scott, K.J. (1987) cDNA cloning of mRNAs induced in resistant barley during infection by *Erysiphe graminis* f. sp. *Hordei*. *Plant Molecular Biology* 8, 77–85.
35. Day, A. & Ellis, T.H. (1985) Deleted forms of plastid DNA in albino plants from cereal anther culture. *Current Genetics* 9, 671–8.
36. Doan, N.P. & Fincher, G.B. (1988) The A- and B-chains of carboxypeptidase I from

germinated barley originate from a single precursor polypeptide. *Journal of Biological Chemistry* 263, 11106–10.

37. Doll, H. & Jensen, H.P. (1986) Localization of powdery mildew resistance gene *Ml-ra* on barley chromosome 5. *Hereditas* 105, 61–5.
38. Dunn, A.M., Hughes, M.A., Pearce, R.S. & Jack, P.L. (in press) Molecular characterization of a barley gene induced by cold treatment. *Journal of Experimental Botany* 41,
39. Efimov, V.A., Andreeva, A.V., Reverdatto, S.V. & Chakhmakhcheva, O.G. (1988) Nucleotide sequence of the barley chloroplast *psb*D gene for the D2 protein of photosystem II. *Nucleic Acids Research* 16, 5686.
40. Efimov, V.A., Andreeva, A.V., Reverdatto, S.V., Jung, R. & Chakhmakhcheva, O.G. (1988) Nucleotide sequence of the barley chloroplast psbA gene for the Q_B protein of photosystem II. *Nucleic Acids Research* 16, 5685.
41. Entwistle, J. (1988) Primary structure of a C-hordein gene from barley. *Carlsberg Research Communications* 53, 247–58.
42. Eslick, R.F., Hockett, E.A. & Kushnak, G.D. (1972) Recombination values of four genes on chromosome 1. *Barley Genetics Newsletter* 2, 123–4.
43. Eslick, R.F. & McProud, W.L. (1974) Positioning of genetic male sterile 5 (*msg5*) on chromosome 3. *Barley Genetics Newsletter* 4, 16–23.
44. Falk, D.E. & Kasha, K.J. (1979) The map location of a pale green mutant on chromosome 6. *Barley Genetics Newsletter* 9, 17–18.
45. Falk, D.E. & Kasha, K.J. (1983) Linkage of the *f6* chlorina mutant on chromosome 7. *Barley Genetics Newsletter* 13, 50–1.
46. Falk, D.E., Swartz, M.J. & Kasha, K.J. (1980) Linkage data with genes near the centromere of barley chromosome 6. *Barley Genetics Newsletter* 10, 13–16.
47. Feuerstein, U., Brown, A.H.D. & Burdon, J.J. (1990) Linkage of rust resistance genes from wild barley (*Hordeum spontaneum*) with isozyme markers. *Plant Breeding* 104, 318–24.
48. Fincher, G.B., Lock, P.A., Morgan, M.M., Lingelbach, K., Wettenhall, R.E., Mereer, J.F., Brandt, A. & Thomsen, K.K. (1986) Primary structure of the (1 → 3,1 → 4)-β-D-glucan 4-glucanohydrolase from barley aleurone. *Proceedings of the National Academy of Sciences (USA)* 83, 2081–5.
49. Forde, B.G., Heyworth, A., Pywell, J. & Kreis, M. (1985) Nucleotide sequence of a B1 hordein gene and the identification of possible upstream regulatory elements in endosperm storage protein genes from barley, wheat and maize. *Nucleic Acids Research* 13, 7327–39.
50. Forde, B.G., Kreis, M., Williamson, M.S., Fry, R.P., Pywell, J., Shewry, P.R., Bunce, N. & Miflin, B.J. (1985) Short tandem repeats shared by B- and C-hordein cDNAs suggest a common evolutionary origin for two groups of cereal storage protein genes. *EMBO Journal* 4, 9–15.
51. Forde, J., Forde, B.G., Fry, R.P., Kreis, M., Shewry, P.R. & Miflin, B.J. (1983) Identification of barley and wheat cDNA clones related to the high-M_r polypetides of wheat gluten. *FEBS Letters* 162, 360–6.
52. Franckowiak, J.D. (1987) Multiple recessive marker stocks for mapping male-sterile genes in barley. *Barley Genetics V* (Proceedings of the 5th International Barley Genetics Symposium) Sanyo Press, Nakasange, pp. 209–12.
53. Franckowiak, J.D. & Hockett, E.A. (1987) Allelism tests for the genetic male sterile *msg, bk*. *Barley Genetics Newsletter* 17, 77–8.
54. Franckowiak, J.D. & Hockett, E.A. (1988) Identification of three new loci which control male sterility of barley. *Barley Genetics Newsletter* 18, 11–13.
55. Freeman, J., Marquez, A.J., Wallsgrove, R.M., Saarelainen, R. & Forde, B.G. (1990) Molecular analysis of barley mutants deficient in chloroplast glutamine synthetase. *Plant Molecular Biology* 14, 297–311.
56. Gausing, K. (1987) Thionin genes specifically expressed in barley leaves. *Planta* 171, 241–6.
57. Gausing, K. & Barkardottir, R. (1986) Structure and expression of ubiquitin genes in higher plants. *European Journal of Biochemistry* 158, 57–62.
58. Gerlach, W.L. & Bedrook, J.R. (1979) Cloning and characterization of ribosomal RNA genes from wheat and barley. *Nucleic Acids Research* 7, 1869–85.
59. Gollmer, I. & Apel, K. (1983) The phytochrome-controlled accumulation of mRNA sequences encoding the light-harvesting chlorophyll *a*/*b* protein of barley (*Hordeum vulgare* L.). *European Journal of*

Biochemistry 133, 309–13.

60. Good, A.G., Pelcher, L.E. & Crosby, W.L. (1988) Nucleotide sequence of a complete barley alcohol dehydrogenase 1 cDNA. *Nucleic Acids Research* 16, 7182.

61. Grimm, B. (1990) Primary structure of a key enzyme in plant tetrapyrrole synthesis: Glutamate 1-semialdehyde aminotransferase. *Proceedings of the National Academy of Sciences (USA)* 87, 4169–73.

62. Grimm, B. & Kloppstech, K. (1987) The early light-inducible proteins of Barley. Characterization of two families of 2-h-specific nuclear-coded chloroplast proteins *European Journal of Biochemistry* 167, 493–9.

63. Grimm, B., Kruse, E. & Kloppstech, K. (1989) Transiently expressed early light-inducible thylakoid proteins share transmembrane domains with light-harvesting chlorophyll binding proteins. *Plant Molecular Biology* 13, 583–93.

64. Halford, N.G., Morris, N.A., Urwin, P., Williamson, M.S., Kasarda, D.D., Lew, E., Kreis, M. & Shewry, P. (1988) Molecular cloning of the barley seed protein CMd. *Biochimica et Biophysica Acta* 950, 435–40.

65. Hansen, L. (1987) Three cDNA clones for barley leaf acyl carrier proteins I and III. *Carlsberg Research Communications* 52, 381–92.

66. Hanson, A.D. & Brown, A.H.D. (1984) Three alcohol dehydrogenase genes in wild and cultivated barley: characterization of the products of variant alleles. *Biochemical Genetics* 22, 495–515.

67. Hayashi, J., Konishi, T., Moriya, I. & Takahashi, R. (1984) Inheritance and linkage studies in barley VI. Ten mutant genes located on chromosomes 1 to 7, except 3. *Berichte des Ohara Instituts für Landwirtschaftliche Biologie, Okayama Universitat* 18, 227–50.

68. Hayashi, J., Takahashi, R., Moriya, I. & Nobuoka, T. (1978) Linkage studies of two mutant genes on chromosomes 1 and 6. *Barley Genetics Newsletter* 8, 48–50.

69. Hejgaard, J., Rasmussen, S.K., Brandt, A. & Svendsen, I. (1985) Sequence homology between barley endosperm protein Z and protease inhibitors of the α_1-antitrypsin family. *FEBS Letters* 180, 89–94.

70. Helm, J.H., Metzer, R.J. & Kronstad, W.E. (1974) Inheritance of high lysine in Hiproly barley and its association with the Hiproly endosperm gene. *Crop Science* 14, 637–40.

71. Hernández-Lucas, C., Royo, J., Paz-Ares, J., Ponz, F., García-Olmedo, F. & Carbonero, P. (1986) Polyadenylation site heterogeneity in mRNA encoding the precursor of the barley toxin β-hordothionin. *FEBS Letters* 200, 103–6.

71a. Heun, M., Kennedy, A.E., Anderson, J.A., Lapitan, N.L., Sorrells, M.E. & Tanksley, S.D. (in press) Construction of a restriction fragment length polymorphism map for barley (*Hordeum vulgare*) *Genome* 34.

72. Hinze, K., Thompson, R.D., Ritter, E., Salamini, F. & Schulze-Lefert, P. (1991) Restriction fragment length polymorphism-mediated targeting of the *ml-o* resistance locus in barley (*Hordeum vulgare*). *Proceedings of the National Academy of Sciences (USA)* 88, 3691–5.

73. Høj, P.B., Hartman, D.J., Morrice, N.A., Doan D. & Fincher, G.B. (1989) Purification of $(1 \rightarrow 3)$-β-glucan endohydrolase isoenzyme II from germinated barley and determination of its primary structure from a cDNA clone. *Plant Molecular Biology* 13, 31–42.

74. Hondred, D. & Hanson, A.D. (1990) Hypoxically inducible barley lactate dehydrogenase: cDNA cloning and molecular analysis. *Proceedings of the National Academy of Sciences (USA)* 87, 7300–4.

75. Hong, B., Uknes, S.J. & Ho, T.D. (1988) Cloning and characterization of a cDNA encoding a mRNA rapidly-induced by ABA in barley aleurone layers. *Plant Molecular Biology* 11, 495–506.

76. Hopp, H.E., Rasmussen, S.K. & Brandt, A. (1983) Organization and transcription of B1 hordein genes in high lysine mutants of barley. *Carlsberg Research Communications* 48, 201–16.

77. Huang, J.-K., Swegle, M., Dandekar, A.M. & Muthukrishnan, S. (1984) Expression and regulation of α-amylase gene family in barley aleurones. *Journal of Molecular and Applied Genetics* 2, 579–88.

78. Jackson, E.A., Ballance, G.M. & Thomsen, K.K. (1986) Construction of a yeast vector directing the synthesis and release of barley $(1 \rightarrow 3,1 \rightarrow 4)$-$\beta$-glucanase. *Carlsberg Research Communications* 51, 445–58.

79. Jahoor, A. & Fischbeck, G. (1987) Genetical studies of resistance of powdery mildew in barley lines derived from

Hordeum spontaneum collected from Israel. *Plant Breeding* 99, 265–73.

80. Jakobsen, K., Klemsdal, S., Aalen, R., Bosnes, M., Alexander, D. & Olsen, O.-A. (1989) Barley aleurone cell development: molecular cloning of aleurone-specific cDNAs from immature grains. *Plant Molecular Biology* 12, 285–93.

81. Jende-Strid, J. (1991) Gene-enzyme relations in the pathway of flavonoid biosynthesis in barley. *Theoretical and Applied Genetics* 81, 668–74.

82. Jensen, J. (1981) Construction of a barley chromosome 7 linkage map. *Barley Genetics IV* (Proceedings of the 4th International Barley Genetics Symposium) Edinburgh Univ. Press, Edinburgh, pp. 927–39.

83. Jensen, J. (1987) Linkage map of barley chromosome 4. *Barley Genetics V* (Proceedings of the 5th International Barley Genetics Symposium) Sanyo Press, Nakasange, pp. 189–99.

84. Jensen, J. (1989) Estimation of recombination parameters between a quantitative trait locus (QTL) and two marker gene loci. *Theoretical and Applied Genetics* 78, 613–18.

85. Jensen, J. (1989) Coordinator's report: chromosome 5. *Barley Genetics Newsletter* 19, 74–7.

86. Jorgensen, J.H. (1988) Coordinator's report: disease and pest resistance genes. *Barley Genetics Newsletter* 18, 65–9.

87. Jørgensen, J.H. & Jensen, H.P. (1979) Inter-allelic recombination in the *ml-o* locus in barley. *Barley Genetics Newsletter* 9, 37–9.

88. Junghans, H. & Metzlaff, M. (1988) Genome specific, highly repeated sequences of *Hordeum vulgare*: cloning, sequencing and squash dot test. *Theoretical and Applied Genetics* 76, 728–32.

89. Jutidamrongphan, W., Mackinnon, G., Manners, J.M. & Scott, K.J. (1989) Sequence of a near-full length cDNA clone for a mRNA of barley induced by fungal infection. *Nucleic Acids Research* 17, 9478.

90. Kahler, A.L. & Allard, R.W. (1970) Genetics of isozyme variants in barley. I. Esterases. *Crop Science* 10, 444–8.

91. Kahn, A., Avivi-Bleiser, N. & von Wettstein, D. (1976) Genetic regulation of chlorophyll synthesis analyzed with double mutants in barley. In: Bücher Th., Neuport, W., Sebald, W. & Werner, S. (eds), *Genetics and Biogenesis of Chloroplasts and Mitochondria*. Elsevier/North Holland Biomedical Press, Amsterdam, pp. 119–31.

92. Kasha, K.J. (1982) Coordinator's report: chromosome 6. *Barley Genetics Newsletter* 12, 90–2.

93. Kasha, K.J., Falk, D.E. & Ho-Tsai, A. (1978) Linkage data with genes on chromosome 6. *Barley Genetics Newsletter* 8, 61–5.

94. Khursheed, B. & Rogers, J.C. (1988) Barley α-amylase genes. Quantitative comparison of steady-state mRNA levels from individual members of the two different families expressed in aleurone cells. *Journal of Biological Chemistry* 263, 18953–60.

95. Kilian, A. & Gale, M.D. (personal communication).

96. Kleinhofs, A., Chao, S. & Sharp, P.J. (1988) Mapping of nitrate reductase genes in barley and wheat. In: Miller, T.E. & Koebner, R.M. (eds), *Proceedings of the 7th International Wheat Genetics Symposium*. Bath Press, Bath, pp. 541–6.

97. Knox, C., Sonthayanon, B., Chandra, G.R. & Muthukrishnan, S. (1987) Structure and organization of two divergent α-amylase genes from barley. *Plant Molecular Biology* 9, 3–17.

98. Koebner, R.M. & Martin, P.K. (1989) Chromosomal control of the aminopeptidases of wheat and its close relatives. *Theoretical and Applied Genetics* 78, 657–64.

99. Koehler, S.M. & Ho, T.D. (1990) Hormonal regulation, processing, and secretion of cysteine proteinases in barley aleurone layers. *Plant Cell* 2, 769–83.

100. Kolchinsky, A., Kanazin, V., Yakovleva, E., Gazumyan, A., Kole, C. & Ananiev, E. (1990) *5S*-RNA genes of barley are located on the second chromosome. *Theoretical and Applied Genetics* 8, 333–6.

101. König, S. (1988) Biochemical markers for barley chromosomes (*Hordeum vulgare* L.). *Biochemie und Physiologie der Pflanzen* 183, 345–9.

102. Konishi, T. (1978) New linkage data on chromosome 6 of barley. *Barley Genetics Newsletter* 8, 71–2.

103. Konishi, T. (1981) Reverse mutation and interallelic recombination at the liguleless locus of barley. *Barley Genetics IV* (Proceedings of the 4th International Barley Genetics Symposium) Edinburgh

University Press, Edinburgh, pp. 838–45.
104. Konishi, T., Hayashi, J., Moriya, I. & Takahashi, R. (1984) Inheritance and linkage studies in barley VII. Location of six new mutant genes on chromosome 3. *Berichte des Ohara Instituts für Landwirtschaftliche Biologie, Okayama Universitat*, 18, 251–64.
105. Konishi, T. & Matsuura, S. (1988) A new certation gene *Ga-2* on the long arm of chromosome 3 in barley. *Barley Genetics Newsletter* 18, 33–7.
106. Konishi, T. & Matsuura, S. (1989) Chromosomal localization of two esterase isozyme loci on chromosome 3 in barley. *Barley Genetics Newsletter* 19, 46–8.
107. Kreis, M., Williamson, M., Buxton, B., Pywell, J., Hejgaard, J. & Svendsen, I. (1987) Primary structure and differential expression of β-amylase in normal and mutant barleys. *European Journal of Biochemistry* 169, 517–25.
108. Kreis, M., Williamson, M.S., Shewry, P.R., Sharp, P. & Gale, M. (1988) Identification of a second locus encoding β-amylase on chromosome 2 of barley. *Genetical Research (Camb.)* 51, 13–16.
109. Kristiansen, K.N. (personal communication).
110. Krupinska, K. (1988) Characterization and in vitro expression of the cytochrome *b*-559 genes of barley. II. In vitro transcription and translation. *Carlsberg Research Communications* 53, 233–46.
111. Krupinska, K. & Berry-Lowe, S. (1988) Characterization and in vitro expression of the cytochrome *b*-559 genes of barley I. Localization and sequence of the genes. *Carlsberg Research Communications* 53, 43–55.
112. Ladogina, M.P., Netsvetaev, V.P., Pomortsev, A.A., Ovchinnikov, A.N. & Sozinov, A.A. (1987) Two new glutelin-coding loci *Glu A* and *Glu B* on chromosome 5 of barley. *Barley Genetics Newsletter* 17, 28–31.
113. Ladogina, M.P., Pomortsev, A.A., Netsvetaev, V.P. & Sozinov, A.A. (1989) Identification of three loci controlling low-molecular-weight subunits of glutenin in the barley plant. *Genetika* 25, 1818–26.
114. Larsson, H.E. (1985) Linkage studies with genetic markers and some *laxatum* barley mutants. *Hereditas* 103, 269–79.
115. Lázaro, A., Rodriguez-Palenzuela, P., Maraña, C., Carbonero, P. & Garcia-Olmedo, F. (1988) Signal peptide homology between the sweet protein thaumatin II and unrelated cereal α-amylase/trypsin inhibitors. *FEBS Letters* 239, 147–50.
116. Lázaro, A., Sanchez-Monge, R., Salcedo, G., Paz-Ares, J., Carbonero, P. & García-Olmedo, F. (1988) A dimeric inhibitor of insect α-amylase from barley. Cloning of the cDNA and identification of the protein. *European Journal of Biochemistry* 172, 129–34.
117. Leah, R. & Mundy, J. (1989) The bifunctional α-amylase/subtilisin inhibitor of barley: nucleotide sequence and patterns of seed-specific expression. *Plant Molecular Biology* 12, 673–82.
118. Leah, R. Tommerup, H., Svendsen, I. & Mundy, J. (1991) Biochemical and molecular characterization of three anti-fungal proteins from barley seed. *Journal of Biological Chemistry* 266, 1564–73.
119. Lerner, D.R. & Raikhel, N.V. (1989) Cloning and characterization of root-specific barley lectin. *Plant Physiology* 91, 124–9.
120. Linde-Laursen, I. (1979) Giemsa C-banding of barley chromosomes. III. Segregation and linkage of C-bands on chromosomes 3, 6, and 7. *Hereditas* 91, 73–7.
121. Linde-Laursen, I. (1981) Giemsa banding patterns of the chromosomes of cultivated and wild barleys. *Barley Genetics IV* (Proceedings of the 4th International Barley Genetics Symposium) Edinburgh University Press, Edinburgh, pp. 786–95.
122. Linde-Laursen, I. (1982) Linkage map of the long arm of barley chromosome 3 using C-bands and marker genes. *Heredity* 49, 27–35.
123. Ling, V. & Zielinski, R.E. (1989) Cloning of cDNA sequences encoding the calcium-binding protein, calmodulin, from barley (*Hordeum vulgare* L.) *Plant Physiology* 90, 714–19.
124. Liu, C.J. & Gale, M.D. (1988) Three new marker systems, iodine binding factor (*Ibf-1*), malic enzyme (*Mal-1*) and malate dehydrogenase (*Mdh-3*) in wheat and related species. In: Miller, T.E. & Koebner, R.M. (eds), *Proceedings of the 7th International Wheat Genetics Symposium*. Bath Press, Bath, pp. 550–60.
125. Liu, C.J. & Gale, M.D. (1989) The chromosomal location of a third set of malate dehydrogenase loci, *Mdh-3*, in wheat, barley and related species. *Theoretical and*

Applied Genetics 78, 349–52.
126. Lundqvist, U. & Lundqvist, A. (1988) Induced *intermedium* mutants in barley: origin, morphology and inheritance. *Hereditas* 108, 13–26.
127. Marocco, A., Wissenbach, M., Becker, D., Paz-Ares, J., Saedler, H., Salamini, F. & Rohde, W. (1989) Multiple genes are transcribed in *Hordeum vulgare* and *Zea mays* that carry the DNA binding domain of the *myb* oncoproteins. *Molecular and General Genetics* 216, 183–7.
128. Mori, S. & Nishizawa, N. (1989) Identification of barley chromosome no. 4, possible encoder of genes of mugineic acid synthesis from 2′-deoxymugineic acid using wheat-barley addition lines. *Plant Cell Physiology* 30, 1057–61.
129. Mundy, J. & Rogers, J.C. (1986) Selective expression of a probable amylase/protease inhibitor in barley aleurone cells: Comparison to the barley amylase/subtilisin inhibitor. *Planta* 169, 51–63.
130. Neumann, E.M. (1988) Primary structure of barley genes encoding quinone and chlorophyll *a* binding proteins of photosystem II. *Carlsberg Research Communications* 53, 259–75.
131. Nielsen, G. & Frydenberg, O. (1974) Linkage between the loci Amy1 (α-amylase), o (orange lemma) and x_n (xantha seedling). *Barley Genetics Newsletter* 4, 53–4.
132. Nielsen, G. & Hejgaard, J. (1987) Mapping of isozyme and protein loci in barley. In: Scandalios, J.G. (ed.) *Isozymes: Current Topics in Biological and Medical Research*, Vol. 15, *Genetics, Development and Evolution*. Alan R. Liss, New York, pp. 77–95.
133. Nielsen, G. & Hvid, S. (1976) Location of the esterase locus *Est1* on barley chromosome 3. *Barley Genetics Newsletter* 6, 45–6.
134. Nielsen, P.S. & Gausing, K. (1987) The precursor of barley plastocyanin. Sequence of cDNA clones and gene expression in different tissues. *FEBS Letters* 225, 159–62.
135. Nonaka, S. (1974) Further information of the *rnt* locus on chromosome 3. *Barley Genetics Newsletter* 4, 56–8.
136. Okkels, J.S., Jepsen, L.B., Hønberg, L.S., Lehmbeck, J., Scheller, H.V., Brandt, P., Høyer-Hansen, G., Stummann, B., Henningsen, K.W., von Wettstein, D. & Møller, B.L. (1988) A cDNA clone encoding a 10.8 kDa photosystem I polypeptide of barley. *FEBS Letters* 237, 108–12.
137. Okkels, J.S., Scheller, H.V., Jepsen, L.B. & Møller, B.L. (1989) A cDNA clone encoding the precursor for a 10.2 kDa photosystem I polypeptide of barley. *FEBS Letters* 250, 575–9.
138. Oliver, R.P. (1984) Location of the genes for cytochrome f, subunit IV of the b_6/f complex, the α-subunit of CF_1 ATP-synthase and subunit III of the CF_0 ATP-synthase on the barley chloroplast genome. *Carlsberg Research Communications* 49, 555–7.
139. Oliver, R.P. & Poulsen, C. (1984) Structure of a heavily transcribed region of barley chloroplast DNA. Transfer RNA genes for serine (UGA), glycine (GCC, UCC), formyl-methionine and threonine (GGU). *Carlsberg Research Communications* 49, 647–73.
140. Paz-Ares, J., Ponz, F., Rodríguez-Palenzuela, P., Lázaro, A., Hernández-Lucas, C., García-Olmedo, F. & Carbonero, P. (1986) Characterization of cDNA clones of the family of trypsin/α-amylase inhibitors (CM-proteins) in barley (*Hordeum vulgare L.*). *Theoretical and Applied Genetics* 71, 842–6.
141. Pickering, R.A. (1983) The location of a gene for incompatability between *Hordeum vulgare* L. and *H. bulbosum* L. *Heredity* 51, 455–9.
142. Pietro, M.E. & Hart, G.E. (1985) The genetic control of triosephosphate isomerase of hexaploid wheat and other Triticeae species. *Genetical Research (Camb.)* 45, 127–42.
143. Ponz, F., Paz-Ares, J., Hernández-Lucas, C., García-Olmedo, F. & Carbonero, P. (1986) Cloning and nucleotide sequence of a cDNA encoding the precursor of the barley toxin α-hordothionin. *European Journal of Biochemistry* 156, 131–5.
144. Poulsen, C. (1983) The barley chloroplast genome: physical structure and transcriptional activity in vivo. *Carlsberg Research Communications* 48, 57–80.
145. Poulsen, C. (1984) Two mRNA species differing by 258 nucleotides at the 5′ end are formed from the barley chloroplast rbcL gene. *Carlsberg Research Communications* 49, 89–104.
146. Rahmatullah, R.J., Huang, J.-K., Clark, K.L., Reeck, G.R., Chandra, G.R. & Muthukrishnan, S. (1989) Nucleotide and predicted amino acid sequences of two

different genes for high-pI α-amylases from barley. *Plant Molecular Biology* 12, 119–21.

147. Ramage, R.T. & Paluska, M. (1975) Mapping chromosome 6. *Barley Genetics Newsletter* 5, 49–51.

148. Rasmussen, S.K. & Brandt, A. (1986) Nucleotide sequences of cDNA clones for C-hordein polypeptides. *Carlsberg Research Communications* 51, 371–9.

149. Rasmussen, S.K., Hopp, H.E. & Brandt, A. (1983) Nucleotide sequences of cDNA clones for B1 hordein polypeptides. *Carlsberg Research Communications* 48, 187–99.

150. Rasmussen, S.K., Hopp, H.E., Brandt, A., Svendsen, I. & Hejgaard, J. (1984) A cDNA clone for protein Z, a major barley endosperm albumin. *Carlsberg Research Communications* 49, 385–90.

151. Rasmussen, S.K., Johansson, A., Rasmussen, H.N. & Theilade, B. (in press) Molecular analysis and cloning of barley peroxidase genes. *Molecular and physiological aspects of plant peroxidases* (Proceedings of the 2nd International Symposium, Lublin), 1990.

152. Rasmussen, S.K., Welinder, K.G. & Hejgaard, J. (1991) cDNA cloning, characterization and expression of an endosperm specific barley peroxidase. *Plant Molecular Biology* 16, 317–27.

153. Reverdatto, S.V., Andreeva, A.V., Buryakova, A.A., Chakhmakhcheva, O.G. & Efimov, V.A. (1989) Nucleotide sequence of the 5.2 kbp barley chloroplast DNA fragment containing *psbB-psbH-petB-petD* gene cluster. *Nucleic Acids Research* 17, 2859–60.

154. Reverdatto, S.V., Andreeva, A.V., Buryakova, A.A., Chakhmakhcheva, O.G. & Efimov, V.A. (1989) Nucleotide sequence of the barley chloroplast *psbC* gene. *Nucleic Acids Research* 17, 3996.

155. Roane, C.W. & Starling, T.M. (1989) Linkage studies with genes conditioning leaf rust reaction in barley. *Barley Newsletter* 33, 190–2.

156. Rodríguez-Palenzuela, P., Pintor-Toro, J.-A., Carbonero, P., & García-Olmedo, F. (1988) Nucleotide sequence and endosperm-specific expression of the structural gene for the toxin α-hordothionin in barley (*Hordeum vulgare* L.). *Gene* 70, 271–81.

157. Rodríguez-Palenzuela, P., Royo, J., Gómez, L., Sanchez-Monge, R., Salcedo, G., Molina-Cano, J., Garcia-Olmedo, F., & Carbonero, P. (1989) The gene for trypsin inhibitor CMe is regulated in *trans* by the *lys 3a* locus in the endosperm of barley (*Hordeum vulgare* L.). *Molecular and General Genetics* 219, 474–9.

158. Rogers, J.C. (1985) Two barley α-amylase gene families are regulated differently in aleurone cells. *Journal of Biological Chemistry* 260, 3731–8.

159. Rogers, J.C., Dean, D. & Heck, G.R. (1985) Aleurain: A barley thiol protease closely related to mammalian cathepsin H. *Proceedings of the National Academy of Sciences (USA)* 82, 6512–16.

160. Rogers, J.C. & Milliman, C. (1983) Isolation and sequence analysis of a barley α-amylase cDNA clone. *Journal of Biological Chemistry* 258, 8169–74.

161. Rogers, J.C. & Milliman, C. (1984) Coordinate increase in major transcripts from the high pI α-amylase multigene family in barley aleurone cells stimulated with gibberellic acid. *Journal of Biological Chemistry* 259, 12234–40.

162. Rohde, W., Becker, D. & Salamini, F. (1988) Structural analysis of the *waxy* locus from *Hordeum vulgare*. *Nucleic Acids Research* 16, 7185–6.

163. Rohde, W., Marocco, A., Wissenbach, M., Barzen, E., Kristiansen, K. & Salamini, F. (1989) Anthocyanin biosynthesis in barley: Characterization of structural and putative regulatory genes. In: Styles, D.E., Gavazzi, G. & Racchi, M.L. (eds), *The Genetics of Flavonoids*. Edizioni Unicopli, Milano, pp. 79–95.

164. Rohde, W., Rosch, K., Kröger, K. & Salamini, F. (1990) Nucleotide sequence of a *Hordeum vulgare* gene encoding a glycine-rich protein with homology to vertebrate cytokeratins. *Plant Molecular Biology* 14, 1057–9.

165. Rosichan, J., Nilan, R.A., Arenaz, P. & Kleinhofs, A. (1979) Intragenic recombination at the *waxy* locus in *Hordeum vulgare*. *Barley Genetics Newsletter* 9, 79–85.

166. Saghai-Maroof, M.A., Soliman, K.M., Jorgensen, R.A. & Allard, R.W. (1984) Ribosomal DNA spacer-length polymorphisms in barley: Mendelian inheritance, chromosomal location, and population dynamics. *Proceedings of the National Academy of Sciences (USA)* 81, 8014–18.

167. Säll, T. (1990) Genetic control of recom-

bination in barley II. Variation in linkage between marker genes. *Hereditas* 112, 171–8.

168. Schaller, C.W., Qualset, C.O. & Rutger, J.N. (1964) Inheritance and linkage of the Yd2 gene conditioning resistance to the barley yellow dwarf virus disease in barley. *Crop Science* 4, 544–8.

169. Scheller, H.V., Okkels, J.S., Høj, P.B., Svendsen, I., Roepstorff, P. & Møller, B.L. (1989) The primary structure of a 4.0-kDa photosystem I polypeptide encoded by the chloroplast *psaI* gene. *Journal of Biological Chemistry* 264, 18402–6.

170. Schulz, R., Steinmüller, K., Klaas, M., Forreiter, C., Rasmussen, S.K. & Hiller, C. (1989) Nucleotide sequence of a cDNA coding for the NADPH-protochlorophyllide oxidoreductase (PCR) of barley (*Hordeum vulgare* L.) and its expression in *Escherichia coli. Molecular and General Genetics* 217, 355–61.

171. Sexton, T.B., Jones, J.T. & Mullet, J.E. (1990) Sequence and transcriptional analysis of the barley ctDNA region upstream of *psb*D-*psb*C encoding *trn*K(UUU), *rps*16, *trn*Q(UUG), *psb*K, *psb*I, and *trn*S(GCU). *Current Genetics* 17, 445–54.

172. Shewry, P.R. & Parmar, S. (1987) The HrdF (Hor5) locus encodes γ-type hordeins. *Barley Genetics Newsletter* 17, 32–4.

173. Shewry, P.R., Parmar, S., Franklin, J. & Burgess, S.R. (1990) Analysis of a rare recombination event within the multigenic *Hor 2* locus of barley (*Hordeum vulgare* L.). *Genetical Research (Camb.)* 55, 171–6.

174. Shewry, P.R., Parmar, S., White, R.P. & Franklin, J. (1987) Mapping and biochemical analysis of the *Hor 4 (Hrd G)* locus encoding B hordein-like polypeptides. *Barley Genetics Newsletter* 17, 87–9.

175. Søgaard, B. (1974) The localization of *eceriferum* loci in barley III. Three point tests of genes on chromosome 1 in barley. *Hereditas* 76, 41–8.

176. Søgaard, B. (1977) The localization of eceriferum loci in barley V. Three point tests of genes on chromosome 1 and 3 in barley. *Carlsberg Research Communications* 42, 67–75.

177. Søgaard, B. & von Wettstein, P. (1987) Barley: genes and chromosomes. *Carlsberg Research Communications* 52, 123–96.

178. Sonina, N.V., Lushnikova, A.A., Tihonov, A.P. & Ananiev, E.V. (1989) Dialect-I, species specific repeated DNA sequences from barley, *Hordeum vulgare. Theoretical and Applied Genetics* 78, 589–93.

179. Sørensen, M.B. (1989) Mapping of the *Hor2* locus in barley by pulsed field gel electrophoresis. *Carlsberg Research Communications* 54, 109–20.

180. Sørensen, M.B., Cameron-Mills, V. & Brandt, A. (1989) Transcriptional and post-transcriptional regulation of gene expression in developing barley endosperm. *Molecular and General Genetics* 217, 195–201.

181. Sørensen, S.B., Svendsen, I. & Breddam, K. (1989) Primary structure of carboxypeptidase III from malted barley. *Carlsberg Research Communications* 54, 193–202.

182. Steinmüller, K., Batschauer, A., Mösinger, E., Schäfer, E., Rasmussen, S.K. & Apel, K. (1986) The light-induced greening of barley. In: van Vloten-Doting, L., Groot, G.S., & Hall, T.C. (eds), *Molecular Form and Function of the Plant Genome.* Nato ASI Series A, Plenum Press, NY, 83, 277–90.

183. Swegle, M., Huang, J.-K., Lee, G. & Muthukrishnan, S. (1989) Identification of an endochitinase cDNA clone from barley aleurone cells. *Plant Molecular Biology* 12, 403–12.

184. Takahashi, R. (1983) Coordinator's report: chromosome 3. *Barley Genetics Newsletter* 13, 89–93.

185. Takahashi, R., Hayashi J., Konishi, T. & Moriya, I. (1972) Inheritance and linkage studies in barley. V. Locating of seven new mutant genes. *Berichte des Ohara Institut für Landwirtschaftliche Biologie, Okayama Universitat*, 15, 147–68.

186. Trick, M., Dennis, E.S., Edwards, K.J. & Peacock, W.J. (1988) Molecular analysis of the alcohol dehydrogenase gene family of barley. *Plant Molecular Biology* 11, 147–60.

187. von Wettstein-Knowles, P. (1989) Facets of the barley genome. *Vorträge Pflanzenzüchtung* 16, 107–24.

188. von Wettstein-Knowles, P. (in press) New alleles of *Cer-yy* and *cer-b*. *Barley Genetics Newsletter* 20,

189. Whittier, R.F., Dean, D.A. & Rogers, J.C. (1987) Nucleotide sequence analysis of alpha-amylase and thiol protease genes that are hormonally regulated in barley aleurone cells. *Nucleic Acids Research* 15, 2515–35.

190. Williamson, M., Forde, J., Buxton, B. &

Kreis, M. (1987) Nucleotide sequence of barley chymotrypsin inhibitor (CI-2) and its expression in normal and high-lysine barley. *European Journal of Biochemistry* 165, 99–106.

191. Williamson, M., Forde, J. & Kreis, M. (1988) Molecular cloning of two isoinhibitor forms of chymotrypsin inhibitor 1 (CI-1) from barley endosperm and their expression in normal and mutant barleys. *Plant Molecular Biology* 10, 512–35.

192. Wise, R.P., Rohde, W. & Salamini, F. (1990) Nucleotide sequence of the *Bronze-1* homologus gene from *Hordeum vulgare*. *Plant Molecular Biology* 14, 277–9.

193. Yasuda, S. & Hayashi, J. (1981) Effects of two kinds of very early genes on yield and its components in barley. *Nogaku Kenkyu* 59, 113–24.

194. Zielinski, R.E., Werneke, J.M. & Jenkins, M.E. (1989) Coordinate expression of rubisco activase and rubisco during barley leaf cell development. *Plant Physiology* 90, 516–21.

195. Zurawski, G. & Clegg, M.T. (1984) The barley chloroplast DNA *atpBE*, *trnM2* and *trnV1* loci. *Nucleic Acids Research* 12, 2549–59.

5

Progress in the Production of Wheat–Barley Addition and Recombination Lines and their use in Mapping the Barley Genome

K.W. Shepherd & A.K.M.R. Islam

Introduction

A comprehensive review of the literature on the production of wheat × barley hybrids, the isolation of wheat–barley addition lines and their utilization has been published recently (Islam & Shepherd, 1990). The current review is designed to provide an update of developments in this general area since the previous literature survey was completed in 1988.

Some new plant materials of interest were developed during this period. Recently, Koba *et al.* (1991) described how particular wheat × barley hybrids can be produced with a much higher success rate than achieved by earlier workers. This work provides optimism that new sets of barley chromosome addition lines might be produced in the future with less difficulties than encountered previously. Also, there have been some recent innovations in solving the problem posed for gene mapping studies stemming from our inability to maintain chromosome 5 of Betzes barley as a fertile addition line. Other new materials of value for gene mapping studies, and some novel applications in barley cytogenetics, are the wheat–barley recombinant chromosomes currently being isolated and characterized in our laboratory (Islam & Shepherd, in press).

However, the main progress achieved over the last three years has come from the utilization of the existing Betzes addition lines to determine the chromosomal location of genes, in barley, controlling an ever increasing range of physiological, biochemical and molecular characters. The value of the addition lines to barley researchers is attested to by the fact that since 1981, seed of these lines have been distributed on request to more than 100 scientists working in 24 different countries around the world. A major new application for these lines has been their use as a set of reference genotypes for allocating molecular markers to particular chromosomes, or chromosome arms, in the initial stages of constructing a high density linkage map of the barley genome, being carried out in several laboratories around the world. The first two genetic maps of the barley genome based partly (Shin *et al.*, 1990) and mainly (Heun *et al.*, 1991) on molecular markers have now been published.

Another recent development of more practical interest is the attempt being made to transfer the *Yd2* gene for resistance to barley yellow dwarf virus (BYDV) from barley into wheat (McGuire & Qualset, 1990).

The progress, problems and implications of recent work in each of these various areas are discussed below.

Production of wheat × barley hybrids

Following the pioneering work of Kruse (1973), there have been numerous reports of successful hybridization between hexaploid wheat and diploid barley, but the majority of these crosses were made using barley as the ovule parent. The problem of pistillody which invariably accompanied barley × wheat crosses (Islam *et al.*, 1975), made it necessary later to attempt the more difficult reciprocal cross using wheat as the ovule parent. The details of these latter programmes (Fedak, 1980; Islam *et al.*, 1981; Sethi *et al.*, 1986) have been covered extensively in the earlier review (Islam & Shepherd, 1990). The most serious problems encountered were the very low yield of hybrid plants and their abnormal cytology. The progeny usually included wheat haploids, hypoploids and hyperploids with some of the latter having as many as 36 chromosomes. Frequently these plants also exhibited chromosome number mosaicism among different root tips from a given plant and also in pollen mother cells (PMC). Although Fedak (1980) reported several F_1 hybrids with the expected 28 chromosomes in his programme, only one of these true hybrids was obtained in each of the other two studies.

There are two additional programmes to be considered in the present review. One was reported very recently (Koba *et al.*, 1991) and the other had been published in a Conference Proceedings (McGuire, 1984) but was overlooked in the earlier review.

McGuire & Qualset at the University of California, Davis, were interested in incorporating the *Yd2* gene on chromosome 3 from BYDV resistant barley cultivars into susceptible wheat cultivars. The wheat parent chosen was Chinese Spring and the barley parents all possessed *Yd2* and included two Ethiopian introductions and two Californian cultivars. The hybridization and embryo-culturing procedures were similar to those used earlier at the Waite Institute (Islam *et al.*, 1981) and their success rate was similar to ours with Betzes in terms of numbers of embryos cultured per floret pollinated (0.38% vs. 0.56%, respectively). However, they had less success (35%) in germinating these embryos into green plants than we did in our programme (100%). The chromosome constitution of the progeny plants was examined by C-banding and they were found to have variable somatic chromosome numbers just as observed earlier by Islam *et al.* (1981) and Sethi *et al.* (1986) when wheat was used as the ovule parent. Moreover, only one of the 12 progeny plants obtained was a genuine F_1 hybrid carrying the full complement of 21 wheat and seven barley chromosomes and the other plants included nine wheat haploids and three 'partial' hybrids carrying only two to four barley chromosomes. As described below, they succeeded in developing some new disomic addition lines from these hybrid plants.

The recent report of Koba *et al.* (1991) is of special interest because they obtained a very high rate of success, in terms of embryos produced per floret pollinated, when some Japanese wheat varieties were pollinated with Betzes and some Japanese barley cultivars. The overall success rate was 82 embryos produced from a total of 3845 floret pollinations (2.13%), but the best result was obtained with the cross Norin 12 × Betzes (8.25%). They ascribed this high rate of success to the genotypes of the parental cultivars, the favourable environmental conditions supplied for plant growth and crossing and the procedure used for embryo rescue. Unlike earlier workers who regularly applied gibberellic acid to the stigma following pollination to promote growth of the embryo, Koba *et al.* followed the technique of Inagaki (1986) and injected the first wheat internode with a dilute 2, 4-D solution at the time of emasculation. They claimed that this promoted both ovary and embryo development following pollination resulting in the improved rate of embryo rescue. However, they did not report any control crosses lacking the 2, 4-D treatment for comparison. Although the progeny had variable chromosome numbers and included haploid (13.3%) and aneuploid (40%) individuals, a high frequency of 28-chromosome euploids was obtained (46.7%). In particular, when barley cultivar Harunanijou was used to pollinate three different wheat cultivars, all of the 11 progeny obtained were euploid. Thus, this barley cultivar may be particularly suitable for the production of new addition lines and it will be of some interest to find if the adverse effect of Betzes chromosome 5 on meiosis when added to wheat (Shepherd & Islam, 1987) is absent when chromosome 5 is derived from cultivar Harunanijou.

Isolation of new addition lines

Many attempts have been made to produce euplasmic wheat–barley addition lines from wheat × barley hybrids and the main problems that have been encountered were covered extensively in the recent review (Islam & Shepherd, 1990) and will not be repeated here. The most successful programme has involved Chinese Spring × Betzes crosses. The origin and characteristics of the 6 disomic addition lines having pairs of complete chromosomes of Betzes barley added to Chinese Spring wheat (Islam *et al.*, 1981) and the 12 corresponding ditelosomic additions (Islam, 1983) have also been described in detail in that review. As indicated therein, when Betzes chromosome 5 or its long arm is added to the genome of Chinese Spring, a suite of meiotic upsets ensues, resulting in complete male sterility and an inability to maintain this line as a stable addition. This problem has been partially overcome by the recent isolation of a fertile ditelosomic addition line carrying the short arm of barley chromosome 5 (Islam & Shepherd, 1990). Prior to the isolation of this line, we had identified a spontaneously occurring translocation chromosome which carries the short arm of barley 5 joined on to part of chromosome 2A of wheat and we assumed that this had arisen from centric fusion following misdivision of the respective centromeres. In summary, the maintainable Betzes addition lines available represent the addition of all the barley genome to Chinese Spring wheat except for the long arm of chromosome 5.

As far as we are aware, the only other study which has been successful in producing disomic wheat–barley addition lines is the work of McGuire & Qualset referred to earlier and further described in a recent publication (McGuire & Qualset, 1990). Monosomic addition lines carrying barley chromosome 3 from Atlas 68 and CI 3208-4 and bearing the *Yd2* gene were isolated in the BC_2 generation from the F_1 plants which possessed barley chromosome 3. Initial attempts at isolating the disomic addition by screening 450 BC_2F_2 progeny from the monosomics were unsuccessful, presumably because of very low transmission of the complete chromosome 3 through the pollen. Nevertheless, some monotelodisomic additions involving both the long and short arms of barley 3 as the telosome, were isolated among these progeny and on selfing them it was possible to isolate two chromosome 3 disomic addition lines, one deriving its chromosome from Atlas 68 and the other from CI 3208-4. Despite the difficulty in isolating these lines as disomics, they appear to be relatively stable as additions since with the Atlas 68 line, 16 of 17 progeny were disomic whereas with the other line, 66 of 77 progeny were disomic. The main interest in these lines is that they carry the barley *Yd2* gene. Their level of resistance to BYDV is now being evaluated and as described later attempts are being made to transfer the barley segment carrying the *Yd2* gene into a susceptible wheat cultivar.

In the course of this work some ditelosomic addition lines carrying the short arm of chromosome 3 were also isolated. However, these workers did not report any attempt to isolate addition lines involving the other barley chromosomes present in their F_1 hybrids.

Production of substitution lines

The isolation of addition lines allowed barley chromosomes to be substituted for particular wheat chromosomes using standard procedures. The isolation and characteristics of lines having chromosomes 1, 3 and 6 of barley substituted for the homoeologous group 7, 3 and 6 chromosomes of wheat were covered in our last review (Islam & Shepherd, 1990) and in a recent manuscript (Islam & Shepherd, submitted). These chromosomes showed specific substituting ability, in terms of plant vigour and fertility, as predicted earlier from their similar gene synteny relationships (Hart *et al.*, 1980; Powling *et al.*, 1981; Nielsen & Hejgaard, 1987). Thus, the earlier conclusion of Islam & Shepherd (1981) that barley chromosomes 1, 3 and 6 should be renamed 7H, 3H and 6H, respectively, to conform with the wheat homoeologous groups was confirmed. Substitutions of barley chromosomes 2, 4 and 7 have also been made for at least one chromosome from wheat homoeologous groups 2, 4 and 5, respectively, and compensation for plant vigour was good in all cases but there were problems with reduced fertility in the group 4 substitutions (Islam & Shepherd, unpublished). Confirmation of the chromosome constitution of all of these putative substitution lines has been greatly facilitated by the release of the set of 14 cDNA probes (Sharp *et al.*, 1989) which allow each chromosome arm of the A, B, D and H genomes in the Triticeae to be distinguished.

In summary, data from the substitution lines, where available, support the earlier conclusion from similarity of gene synteny that barley chromosomes 1, 2, 3, 4, 5, 6, and 7 should be redesignated 7H, 2H, 3H, 4H, 1H, 6H and 5H, respectively, to reflect their relationship with the wheat groups.

Isolation of wheat–barley recombinant chromosomes

The availability of lines having individual barley chromosomes added to the wheat genome, and subsequently substituted for related wheat chromosomes, has made it possible to begin a new phase of cytogenetic manipulation of barley chromosomes in wheat, namely the production of wheat–barley recombinant chromosomes. These recombinant chromosomes are expected to have several uses in barley genome mapping and cytogenetic studies and our progress in isolating such lines is outlined below.

The first requirement is to be able to induce meiotic pairing and recombination between the barley chromosomes and their related wheat chromosomes since normally these chromosomes do not pair because of the activity of the *Ph* gene on chromosome arm 5BL of wheat which restricts pairing to homologous chromosomes (Riley & Chapman, 1958; Sears & Okamoto, 1958). Initially we investigated the effect of suppressing the activity of the *Ph* locus by hybridizing several ditelosomic addition lines with a high-pairing form of *Triticum speltoides* (Riley *et al.*, 1968). The barley telocentrics (7Hβ, 2HS, 4HS, 4HL, 6HS and 5HL) were observed to pair in heteromorphic bivalent configurations at low frequency (1.2–4.5%) in PMCs from these 29 chromosome (28′ + t′) hybrids, but because of the presence of the *speltoides* genome it was not clear how much of this pairing involved wheat and barley chromosomes (Islam & Shepherd, 1988). A more critical test of wheat–barley chromosome pairing was devised using Sears' (1977) *ph1b* mutant to allow homoeologous pairing to occur without interference from the *speltoides* genome. In this test, several barley ditelosomic substitution lines were crossed with appropriate aneuploid wheat lines following the procedure described earlier (Shepherd & Islam, 1987) and lines were isolated containing a single dose of three critical chromosomes (triple monosomics). These three univalent chromosomes were the barley telocentric, a related wheat chromosome and chromosome 5B carrying the *ph1b* allele. The level of pairing between barley telocentrics and related wheat univalents involving combinations of barley telocentrics 6HS, 6HL and 3HL with wheat chromosomes of groups 6 and 3, respectively, was recorded in PMCS of the appropriate triple monosomics (Islam & Shepherd, 1988, in press). The level of pairing involving telocentrics 6HS and 3HL was generally very low ($\leq 1\%$) but barley 6HL showed a higher level of pairing (2.6%). However, all of these values are likely to be underestimates of the actual pairing frequency because of the likely occurrence of desynapsis as discussed earlier (Islam & Shepherd, 1988).

The induced pairing between barley and wheat chromosomes in the triple monosomics was expected to generate the required wheat–barley recombinant chromosomes. However, given the low frequency of pairing an efficient selection procedure was required to allow large numbers of progeny to be screened. We have chosen to screen for two or more biochemical characters (e.g. isozymes) controlled by genes on the barley telocentric under study which can be easily scored after electrophoretic separation in gels and which are recognizably different from the equivalent markers in the wheat parent. Progeny from the triple monosomics are then scored for these markers to search for possible wheat–barley recombinants showing dissociation of the parental marker combinations. So far, we have analysed progeny involving barley chromosome arms 6HL and 3HL using the isozyme markers glutamic oxaloacetic transaminase, aconitase hydratase and dipeptidase on 6HL and esterase and malate dehydrogenase on 3HL (Islam & Shepherd, in press). Putative recombinants were obtained in low frequency with both 6HL (1.4%) and 3HL (1.1%) barley telocentrics. All except one of these recombinants have been confirmed in progeny tests and they are now being further characterized to try and define more precisely the structure of the recombinant chromosomes. One of the recombinants obtained from each telocentric appeared to involve a double crossover event and this was surprising given that only six recombinants had been detected in each group. With the 3HL telocentric, progeny from two different stocks were screened for recombinants; one was the triple monosomic line carrying a single dose of 3HL, 3A, 5B and *ph1b* and the other stock differed by having two doses of 5B and *ph1b*. The latter stock yielded a much higher frequency of recombinants than the former (5/337 vs. 1/222) and although this difference was not statistically significant, it was consistent with the earlier finding of Wang *et al.* (1977) who observed a fourfold increase in pairing between a wheat chromosome and a chromosome of *Agropyron intermedium* when the dosage of *ph1b* was increased from 1 to 2. Clearly we need further information on this possible effect of *ph1b* dosage on the yield of recombinants to plan more efficient strategies for isolating recombinants between the other barley and wheat chromosomes.

Another question to be considered in future work is whether it will be more efficient to select for recombinants by including complete barley chromosomes in the triple monosomic stock, and selecting for dissociation of two biochemical markers located distally on opposite arms of the barley chromosome. This procedure will allow selection for recombinants over nearly the whole length of the barley chromosome in the one population. However, there are two disadvantages in that double crossovers would not be detected and telocentrics resulting from misdivision of the centromere of the barley chromosome would be selected as putative recombinants. The telocentrics could be detected and eliminated by cytological screening of the suspected recombinants, but double crossovers could only be detected if additional marker genes were included. Lines possessing an entire barley chromosome and a homoeologous wheat univalent in a homozygous *ph1b* background have now been constructed to find whether screening for two distal markers can result in an increased yield of recombinants and therefore a more efficient system for detecting these recombinants.

Utilization of addition lines

The value and scope of using wheat–barley addition lines for assigning genes in barley to particular barley chromosomes or arms has been discussed in detail in earlier reviews (Islam & Shepherd, 1981, 1990; Shepherd & Islam, 1987). The main requirements are that the barley character is expressed in a wheat background and that it is recognizably different from the wheat phenotype. The characters most amenable to such analysis have proved to be biochemical characters such as seed proteins and isozymes and, more recently, molecular markers. The current situation with respect to these two types of markers is summarized below.

Mapping of biochemical markers

A comprehensive listing of the biochemical markers which had been mapped up until 1988 by addition line analysis was given recently (Islam & Shepherd, 1990, Table 3). It included a total of at least 50 different loci of which 39 were isozymes and 11 were seed proteins of various types. Some more of these markers can now be added to the list, including a few which were overlooked previously and others which have recently been allocated to chromosomes. Thus Hart & Tuleen (1983) had assigned an alcohol dehydrogenase (*Adh3*) and a lipoxygenase (*Lpx1*) locus to chromosomes 6H and 4H, respectively, by isozyme analysis of addition lines. Also Nielsen & Hejgaard (1987) had determined by immunoblotting that some basic proteins (C, N, Q) of unknown function produced in the barley endosperm during grain filling were controlled by loci *Pbc1*, *Pbn1* and *Pbq1* on chromosomes 7H, 3H and 2H, respectively. Recent reports of new gene locations include a malate dehydrogenase locus (*Mdh3*) on chromosome 5H (Liu & Gale, 1989), a leaf esterase (*Est7*) on chromosome arm 2HL (Liu & Gale, 1990), and a gene controlling the conversion of deoxymugineic acid to mugineic acid located to chromosome 4H of barley (Mori & Nishizawa, 1989). Soliman & Allard (1989) recently investigated the chromosomal control of nine different isozymes in barley. However, there are some inconsistencies in their results and use of nomenclature, compared to the earlier studies. For example, the locus designated *Acp1* and located on chromosome 6H seems to correspond to the *Acp3* of Brown & Munday (1982), *Est4* on chromosome 7H to *Est3, 5* described earlier (Hart *et al.*, 1980; Nielsen & Johansen, 1986), *Prx2* on chromosome 7H to the earlier *Prx4* (Nielsen & Johansen, 1986) and *Mdh3* inferred to be on chromosome 1H corresponds with *Mdh1* of Powling *et al.* (1981). Also there is an inconsistency between their findings and those of Powling *et al.* (1981) in that they have located the locus designated *G6pd2* on chromosome 7H whereas Powling *et al.* assigned this locus to chromosome 2H. Soliman & Allard also assigned some new isozyme loci to barley chromosomes including the locus for leucine amino peptidase (*Lap1*) to chromosome 7H and a malate dehydrogenase (*Mdh1*) to chromosome 4H. Their results with the other isozymes investigated agreed with previous allocations.

Genes controlling several other enzymes and proteins in barley have been assigned to barley chromosomes by addition line analysis during the period under review but by using the very different method based on restriction fragment length polymorphism (RFLP) analysis with DNA clones and these analyses will be described in detail in the next section.

A continuing problem in allocating genes controlling biochemical markers to barley chromosomes has been the lack of an addition line for barley chromosome 1H. A common procedure has been to infer that the gene controlling a discrete barley character difference from wheat is located on chromosome 1H if it cannot be assigned to any of the other 6 chromosomes in addition line analysis, but there are obvious dangers inherent in relying on this negative evidence for gene assignment. The isolation of the supposed 2AS.1HS translocation line (Shepherd & Islam, 1987) and more recently the 1HS telocentric addition line has overcome this problem for the short arm. A solution for the long arm has also now been devised (Islam & Shepherd, unpublished). This utilizes a double monosomic addition line which carries both chromosomes 1H and 6H of barley and although meiosis in PMCs is very abnormal and it is self-sterile, this line will set a few seed if pollinated with normal wheat. Seeds carrying one dose of complete chromosome 1H alone can then be selected among their progeny and grown into plants which are vegetatively vigorous even though completely sterile. Thus leaf material can be extracted and analysed for those isozymes and proteins which are expressed in the vegetative tissue and suspected to be controlled by chromosome 1H. Recently, Koebner (1990) made a comparative study of the location of genes controlling proteins which inhibit the microbial protease subtilisin in barley, rye and wheat. He drew attention to an apparent anomaly whereby the genes controlling this protein are located on the short arms of group 1 chromosomes in wheat and rye but have been assigned to the long arm of barley 5 (1HL), based on the failure to detect the enzyme inhibitor in any of the 6 addition lines or the supposed 2AS.1HS translocation line (Hejgaard *et al.*, 1984). He suggested as alternative possibilities that the barley inhibitor may not be expressed in a wheat background or else the translocation may not be centric and includes only a portion of chromosome 1HS. The availability of these new chromosome 1H stocks will allow at least this last possibility to be tested.

Mapping of molecular markers

The addition lines were first used to assign molecular markers to barley chromosomes in 1984 and since then there has been an accelerating use of them in mapping studies with molecular markers. Lists of genes (clones) which had been mapped utilizing DNA clones were given in earlier reviews but, because of the rapid developments in this area over the last few years, a summary of all the mapping data is included herein. The basic principle of this approach is to use the DNA clone (genomic or cDNA) as a probe in Southern blots and hybridization to identify at least one restriction fragment length difference (RFLD) between wheat and barley and then to identify which addition line carries the distinctive barley fragment or fragments.

The potential of this molecular approach to supply markers for genome mapping is almost unbounded because of the virtually unlimited number of clones that can be produced from a plant allowing probes to be mapped to most parts of its genome. The types of clones used as probes can be divided into three broad categories including genes of known function, random low copy number genomic clones and cDNAs with unknown function (uncharacterized probes) and repetitive sequence DNA of various kinds. Recently a new approach using polymerase chain reaction (PCR) has been developed to produce additional molecular markers.

DNA clones with known function

The clones of known function which have been mapped to barley chromosomes so far utilizing addition lines are listed in Table 5.1. They include barley enzymes, some of which (e.g.α-amylases) had been mapped earlier by analysis of isozyme variation, a range of barley proteins including some from the seed (e.g. subtilisin inhibitors) and others from leaf (e.g. thionins) and genes for ribosomal RNA species. This approach has also allowed mapping of genes controlling characters which were not previously amenable to analysis because the gene products had not been identified. Included here are the physiological responses of barley to environmental extremes, such as acute water

Table 5.1. Assignment of genes of known function to barley chromosomes by identification of RFLDs (Restriction fragment length differences) in wheat–barley addition lines.

Probe	Gene product	Restriction enzyme	Barley DNA fragment size (Kb)	Barley chromosome/arm	Reference
pA5	β-glucanase	*Hin*dIII	4.0	1HL*	3
P7	chl PGK	*Hin*dIII	?	1H*	7
pcβC51	β-amylase	*Bam*HI	10.5	2HS	4
pW9	Rubisco (SSU)	*Hin*dIII	several (5–8)	2H	7
pKG4626	Rubisco (SSU)	*Eco*RI	three (2.5–5.2)	2HS	8
GS6A	Glutamine synthetase	*Hin*dIII	6 + 0.8	2HL	8
pB22A-2	BASI	*Hin*dIII	2–13	2HL	8
BLT14	LTR	*Hin*dIII	1–11	2H	9
5S DNA	5S RNA	*Mva*I	0.3 + 0.9	2H	10
Sorco	PEPC	*Hin*dIII	~ 12	3H(?)	7
pcβC51	β-amylase	*Bam*HI	13.8	4HL	4
dhn-1, 2	Barley dehydrins	n.p.	n.p.	5H	11
pTA71	Ribosomal RNA	*Sst*I	6	6H	1
pHv103	α-amylase 1	*Hin*dIII	several (6.2–18.8)	6H	2
pHv19	α-amylase 1	*Eco*RI	3.5	6HL	8
DB4	Leaf thionins	*Hin*dIII	several	6H	5
pKG1348	Leaf thionins	*Eco*RI	several (4.1–16.5)	6HS	8
P20	cyt PGK	*Hin*dIII	2.3	6H	7
dhn-3, 4	Barley dehydrins	*Bam*HI	3.6	6H	11
pTA71	Ribosomal RNA	*Sst*I	5	7H	1
pHvE	α-amylase 2	*Hin*dIII	three (4–14)	7H	2
P1	Sucrose synthase	*Eco*RI	1.9	7H	6
pA5	β-glucanase	*Hin*dIII	1.8	7H	3
pA5	β-glucanase	*Eco*RI	2.0	7Hα	8
pW9	Rubisco (SSU)	*Hin*dIII	~ 9.2	7H	7
Sorco	PEPC	*Hin*dIII	~ 9	7H	7

* Inferred location based on failure to detect barley restriction fragment in the other 6 lines
n.p. = not published
BASI = Bifunctional α-amylase/subtilisin inhibitor
chl PGK = chloroplast phosphoglycerate kinase
cyt PGK = cytosolic phosphoglycerate kinase
LTR = low temperature response polypeptide
PEPC = phosphoenol pyruvate carboxylase
Rubisco (SSU) = Ribulose-1, 5-bisphosphate carboxylase small subunit

References
1 Saghai-Maroof *et al.* (1984)
2 Muthukrishnan *et al.* (1984)
3 Loi *et al.* (1988)
4 Kreis *et al.* (1988)
5 Bohlmann *et al.* (1988)
6 Maraña *et al.* (1988)
7 Chao *et al.* (1989a)
8 Cannell *et al.* (1989)
9 Dunn *et al.* (1990)
10 Kolchinsky *et al.* (1990)
11 Close & Chandler (1990)

stress (Close & Chandler, 1990) and exposure of seedlings to very low temperatures (Dunn *et al.*, 1990). The potential of adding new markers to this list is enormous as illustrated by the example in Table 5.1 showing how this approach has been successfully applied to the enzymes involved in the photosynthetic carbon reduction (Calvin) cycle (Chao *et al.*, 1989a). In fact, a further 19 clones of genes controlling a diverse range of identified functions in barley are currently being mapped to chromosome arms using this approach (P.R. Shewry, pers. comm.).

Random low copy number DNA clones

It was recognized earlier (Shepherd & Islam, 1987) that the greatest potential value of the addition lines would be in assigning random cDNA and low copy genomic DNA clones to particular barley chromosomes whenever RFLDs could be detected between the barley and wheat parents. This potential is now being realised through the large efforts currently being made by co-operators in North America (the North American Barley Genome Mapping Project) and several other laboratories in other countries, to produce a high density genetic map of the barley genome based on molecular markers.

The role of the barley addition lines in this work is to allow the DNA clones which are to be used in mapping to be sorted into groups relating to the individual barley chromosomes, or preferably the chromosome arms, before analysing the mapping population. For example, both Kam-Morgan *et al.* (1989) and Chao *et al.* (1989b) who were planning to use cDNA clones to map the genome of *Aegilops squarrosa* and group 7 chromosomes of wheat, respectively, first used the wheat–barley addition lines as a reference set to assign the cDNA sequences to homoeologous groups before using them on the mapping populations. However, it should be noted that the validity of this approach depends on a high degree of conservation of gene synteny between wheat and barley chromosomes.

The most comprehensive genetic map of barley available based on restriction fragment length polymorphisms (RFLPs) is that recently published by Heun *et al.* (1991). Out of a total of 603 DNA clones of diverse origin, they identified 137 clones which gave RFLPs between two barley cultivars, Proctor and Nudinka. A mapping population of 113 doubled haploids was produced from this cross and probed with these clones (155 RFLPs identified) and scored for two other characters. The best fit linkage map was determined with the aid of the computer-driven mapping programme Mapmaker (Lander *et al.*, 1987). To assist in the orientation of the linkage groups determined from the mapping population the same clones were used to probe the six available Betzes addition lines and allocations to chromosome 5 were based on obtaining negative results with all of the other lines. A feature of their map was the marked clustering of markers into groups with long interstitial gaps devoid of markers along some of the chromosomes. Thus the addition line data were required to join the separate groups together to form the seven linkage groups.

DNA extracts from ditelosomic lines of wheat and five of the ditelosomic barley additions were also tested with these same clones in an attempt to locate the centromere positions on the genetic map. However, the centromeres could not be precisely defined because of the paucity of markers in these regions. It should be noted that use of the wheat telosomics for this purpose depends on strict conservation of gene synteny between wheat and barley chromosomes, hence it would have been more appropriate to use all of the available barley telosomic additions for such analyses. Nevertheless, there is still the assumption with these lines that the barley parents used to produce the mapping population have the same gene sequence as Betzes, that is they do not differ by one or more chromosome inversions or translocations. Finally it should be noted that the problem Heun *et al.* had in allocating RFLDs to chromosome 5 of barley could be overcome in the future when we isolate plants having this chromosome as a monosomic addition. Then it will be possible to isolate large amounts of genomic DNA from these individuals and this could be put into long-term storage to provide a regular source of addition line DNA which could be probed with DNA clones in the same way as the other addition lines.

Repetitive DNA sequence clones

Since approximately 75% of the barley genome consists of repetitive DNA sequences (Flavell *et al.*, 1977), many of the random genomic clones isolated will consist of this form of DNA. Although these cannot be used as normal genetic markers, there is the possibility of selecting species specific or even chromosome specific probes from among them. For example, Sonina *et al.* (1989) recently isolated a sequence of moderately repeated DNA from barley (Dialect-1) which hybridizes with genomic DNA from barley but not from wheat. Furthermore, when Dialect-1 was used as the probe in *in situ* hybridization of barley chromosomes and Southern blot hybridizations of the addition lines, it was shown that this clone is dispersed throughout the barley genome. Thus it could be used as a molecular marker in *in situ* hybridization to identify barley chromosome segments in a wheat background and it may prove very useful in characterizing the wheat–barley recombinant chromosomes.

A family of highly-repeated DNA sequences was recently isolated by Belostotsky & Ananiev (1990) from high-molecular weight 'relic' DNA which remained undigested after treatment with the *Alu*I restriction enzyme. One of these families (HvRT) has been studied in detail and was shown to hybridize specifically with DNA from some *Hordeum* species, including barley, but not with wheat. Furthermore, analysis of addition lines showed that the probe was specific for certain barley chromosomes and they suggested that it may have potential value as a chromosome specific marker in cytogenetic studies.

Genetic markers generated by PCR

A potent new source of genetic variation for use in genome mapping comes from application of PCR to amplify DNA sequences of different length in the two parents. Shin *et al.* (1990) showed how given relevant sequence information and the existence of a small insertion of or deletion of DNA in the target zone, selection of suitable primers could be used with PCR to generate a polymorphism in the mapping population. A different approach for generating markers with PCR has recently been described by Weining & Langridge (1991), requiring less sequence information and hence having wider application in cereal mapping. They showed that PCR of DNA extracts from wheat and barley using a primer for the highly conserved intron-exon splice junction of plant genes combined with a random primer of about 20 bases, frequently gave rise to differences in the pattern of DNA fragments amplified, and these differences can be used as genetic markers. Furthermore, they used the addition lines to map the different length DNA fragments detected in barley to specific barley chromosome(s). Another modification of PCR is the RAPD (Random Amplified Polymorphic DNA) procedure of Williams *et al.* (1990) which uses a single primer with arbitrary nucleotide sequence and, although they did not test it on cereal DNA extracts, it yielded a high degree of polymorphism with the other plant species tested. The above authors emphasize that methods based on PCR have several advantages over normal RFLP analysis in respect to increased speed and reduced cost and complexity of the analyses. Thus it is likely that many more markers of this type will be used in barley genome mapping in the future and the addition lines will play the same role in locating them to chromosomes as they do presently in mapping RFLPs.

Utilization of wheat–barley recombinant lines

The first wheat–barley recombinant chromosomes are currently being isolated so it is too early to provide any examples of their use in gene mapping or exploring the organization of the barley genome. However, at this stage it is useful to discuss their potential value in barley genetics and cytogenetics.

After their isolation it is necessary first to characterize the recombinant chromosomes as fully as possible in terms of determining how much barley chromatin has been transferred to the wheat chromosome and thereby inferring where the point(s) of crossing over took place. The recent progress in developing a high density linkage map of barley based on molecular probes will greatly facilitate their characterization, because it will now be possible to use molecular probes to determine whether wheat markers have been replaced by barley markers and, if so, which ones. It is also planned to utilize *in situ* hybridization with barley specific probes to obtain information on the physical size of the barley segment present in the recombinant chromosomes. The dispersed moderately repeated DNA sequences such as Dialect-1 (Sonina *et al.*, 1989) which are specific to barley chromatin, could be used for such *in situ* hybridization. However, recently it has been shown that segments of chromatin from an alien species transferred into wheat can be detected by *in situ* hybridization using biotin-labelled genomic DNA from the alien species (Le *et al.*, 1989; Schwarzacher *et al.*, 1989), in competition with unlabelled wheat genomic DNA. It appears that this procedure might provide a simpler approach for characterizing the recombinants cytologically than using genome specific probes. Ultimately, these studies would lead to the development of a physical map of the barley genome to compare with the genetic map.

When a large number of these wheat–barley recombinants have been isolated and characterized for each barley chromosome arm, it is envisaged they will have several uses in barley genetics and cytogenetics and perhaps in breeding programmes also. First, there is the intrinsic interest in the phenomenon of pairing and recombination between chromosomes of wheat and barley. When the frequency and position of chiasmata giving rise to the recombinants have been determined, this will provide important information on genome organization and evolution in these two important cereals.

The second and principal immediate use of the recombinant lines will be in barley gene mapping studies with molecular markers. Although the detailed map of the barley genome will ultimately depend on the analysis of mapping populations as used recently by Heun *et al.* (1991), the recombinant lines are expected to provide valuable supplementary information for these studies. For example, when sufficiently numerous they will allow the gene order of new markers to be decided by a form of deletion analysis as discussed earlier (Shepherd & Islam, 1987). A problem that has been encountered in analysing barley mapping populations is the relatively low frequency of RFLPs among barley cultivars (Graner *et al.*, 1990; Heun *et al.*, 1991) but because of the much greater genetic distance between wheat and barley, RFLDs are expected to be much more frequent. For example, many more than 130 of the 603 probes tested by Heun *et al.* (1991) giving RFLPs with Proctor and Nudinka are likely to give RFLDs between Chinese Spring and Betzes and hence be available for gene order mapping with recombinant lines. Furthermore, these recombinant lines could be multiplied as stable, homozygous stocks and distributed to other laboratories as reference sets just like the addition lines earlier.

Another situation where the recombinants may prove valuable is to allow probes to be selected for specific regions of the barley genome of particular interest to cytogeneticists or barley breeders. Thus if the barley chromosome segment of interest has been incorporated into a Chinese Spring wheat background by recombination, homozygous recombinant lines can be multiplied and probed with a large collection of cDNA clones and the Southern patterns compared with that of the Chinese Spring parent. Probes giving RFLDs between these lines must map within the chromosome segment transferred by recombination and if they are single copy clones they will be specific for this segment. In principle, this strategy could allow any chromosome region carrying genes of interest to barley breeders to be saturated with RFLP markers.

The recombinant lines could also have some practical application in facilitating the transfer of useful barley characters into wheat. Thus any desirable gene identified in Betzes barley could be transferred directly into wheat by inducing wheat–barley recombination as described above. However, if the desirable gene is present in a different barley cultivar or species such as *H. spontaneum*, as is more likely, the Betzes recombinant lines could also be used to facilitate its transfer. In this case it would be necessary first to produce a new addition line having the barley chromosome bearing the desirable gene, and then to hybridize the addition line with existing Betzes recombinant lines involving the requisite barley arm. The desirable gene could then be transferred from the intact barley chromosome into the recombinant Betzes chromosome by homologous pairing and recombination thus bypassing the need to reisolate new wheat–barley recombinants. In fact, attempts are currently being made to utilize the existing 3HL recombinants isolated in our laboratory (Islam & Shepherd, in press) to facilitate the transfer of the *Yd2* gene conferring BYDV resistance from the chromosome 3H addition lines produced by McGuire & Qualset (1990). Holloway & Arundel (1988), using high resolution two-dimensional electrophoresis, have detected a polypeptide difference between two near isogenic lines of California Mariout which differ in reaction to BYDV and, if the gene responsible for this polypeptide proves to be closely linked with *Yd2*, this polypeptide will be a valuable marker to assist in the transfer of *Yd2* into wheat.

Overview

Six of the seven possible disomic wheat–barley addition lines carrying complete barley chromosomes and 12 of the 14 possible telosomic additions have now been available for 13 and 8 years, respectively. In that time they have been used extensively, particularly the former group, as a primary resource for gene mapping studies in barley. Initially mapping was restricted to genes controlling biochemical characters but now the main source of markers are those revealed by molecular methods. The number of these molecular markers made available is expected to expand exponentially in the future as more researchers turn to these procedures, and new methods such as PCR are adopted. Thus there is likely to be an increasing demand in the future for the Betzes additions and any other newly developed addition lines. The recent report of Koba *et al.* (1991) describing their high rate of success in producing some new wheat × barley hybrids with the full complement of wheat and barley chromosomes provides a good prospect for the development of some new addition lines.

Mapping genes to their chromosome arms has lagged behind the use of the complete additions but Islam, Hart & Shepherd (unpublished) and Shewry (pers. comm.) are currently using the ditelosomic additions to map a wide range of enzymes and cDNA clones of known function, respectively, to chromosome arms. The telosomic additions will be particularly valuable for defining the centromere position as high density mapping of the barley genome progresses. However, the early experience of Heun *et al.* (1991) indicates that these regions of barley chromosomes are poorly marked with molecular markers, at least with the probes used so far.

The initial problems which arose because of the unavailability of the 1H addition are now being overcome by some alternative strategies. Initially, a supposed 2AS.1HS translocation line was used to allocate genes to 1HS but recently Koebner (1990) has questioned whether the whole arm of 1HS is present in this line. However, the recent isolation of a 1HS telosomic addition has circumvented this potential problem. Furthermore, the problem with mapping genes on to 1HL will be overcome, at least for molecular markers, when we have isolated the monosomic 1H addition and extracted a bulk supply of genomic DNA from it.

The first wheat–barley recombinants have now been isolated but since these have not yet been fully characterized it is too early to be certain about their future value in barley genetics and breeding. However, we believe they have potential value in several areas of research and intend to make a large effort to develop in the long term many recombinants for each of the barley chromosome arms, except possibly 1HL. However, one immediate problem is to find suitable isozyme markers for some of the arms to allow large numbers of progeny to be screened efficiently.

As outlined earlier we believe they will be useful in barley genome mapping studies to complement the information coming from the analysis of conventional mapping populations as used by Heun *et al.* In particular, when sufficient recombinants have been produced they could be made homozygous and used as reference stocks for determining the gene order and approximate location of any new molecular markers as they arise, with the added benefit of a higher chance of detecting RFLDs between wheat and barley than between barley varieties. Furthermore, these lines will allow a physical map of the barley chromosomes to be generated for comparison with the genetic map now being produced and also there is the prospect of using them to detect markers for specific regions of the barley chromosomes. One major problem that needs to be addressed in all of these projections is the likely occurrence of structural variation between wheat and barley chromosomes. For example if these chromosomes differ by any inversions of significant length, the recovery of recombinants will be restricted and the gene order will be reversed.

The final potential use for the recombinant lines is as vehicles to facilitate the transfer of desirable genes from barley into wheat. The prospects of achieving this transfer are already under investigation with the *Yd2* gene but there are other sources of resistance in barley that may be useful in wheat, e.g. resistance to the cereal cyst nematode. However, the risk of increasing the genetic vulnerability of both wheat and barley to new virulent forms of the disease or pest organisms must be carefully considered before embarking on such programmes.

References

Belostotsky, D.A. & Ananiev, E.V. (1990) Characterization of relic DNA from barley genomes. *Theoretical and Applied Genetics* 80, 374–80.

Bohlmann, H., Clausen, S., Behnke, S., Giese, H., Hiller, C., Reimann-Philipp, U., Schrader, G., Barkholt, V. & Apel, K. (1988) Leaf-specific thionins of barley – a novel class of cell wall proteins toxic to plant-pathogenic fungi and possibly involved in the defence mechanism of plants. *EMBO Journal* 7, 1559–65.

Brown, A.H.D. & Munday, J. (1982) Population genetic structure of land races of barley from Iran. *Genetica* 58, 85–96.

Cannell, M., Mulcahy, M., Karp, A. & Shewry, P.R. (1989) Chromosomal assignment of characterized cDNAs and genes using RFLP analysis of ditelosomic addition lines. *Barley Genetics Newsletter* 19, 9–11.

Chao, S., Raines, C.A., Longstaff, M., Sharp, P.J., Gale, M.D. & Dyer, T.A. (1989a) Chromosomal location and copy number in wheat and some of its close relatives of genes for enzymes involved in photosynthesis. *Molecular and General Genetics* 218, 423–30.

Chao, S., Sharp, P.J., Worland, A.J., Warham, E.J., Koebner, R.M.D. & Gale, M.D. (1989b) RFLP-based genetic maps of wheat homoeologous group 7 chromosomes. *Theortical and Applied Genetics* 78, 495–504.

Close, T.J. & Chandler, P.M. (1990) Cereal dehydrins: serology, gene mapping and potential functional roles. *Australian Journal of Plant Physiology* 17, 333–44.

Dunn, M.A., Hughes, M.A., Pearce, R.S. & Jack, P.L. (1990) Molecular characterization of a barley gene induced by cold treatment. *Journal of Experimental Botany* 41, 1405–13.

Fedak, G. (1980) Production, morphology and meiosis of reciprocal barley–wheat hybrids. *Canadian Journal of Genetics and Cytology* 22, 117–23.

Flavell, R.B., Rimpau, J. & Smith, D.B. (1977) Repeated sequence DNA relationships in four cereal genomes. *Chromosoma* 63, 205–22.

Graner, A., Siedler, H., Jahoor, A., Herrmann, R.G. & Wenzel, G. (1990) Assessment of the degree and type of restriction fragment length polymorphism in barley (*Hordeum vulgare*). *Theoretical and Applied Genetics* 80, 826–32.

Hart, G.E. & Tuleen, N.A. (1983) Introduction and characterization of alien genetic material. In: Tanksley, S.D. & Orton, T.J. (eds), *Isozymes in Plant Genetics and Breeding, Part A*. Elsevier, Amsterdam, pp. 339–62.

Hart, G.E., Islam, A.K.M.R. & Shepherd, K.W. (1980) Use of isozymes as chromosome markers in the isolation and characterization of wheat–barley chromosome addition lines. *Genetical Research* 36, 311–25.

Hejgaard, J., Bjorn, S.E. & Nielsen, G. (1984) Localization to chromosomes of structural genes for the major protease inhibitors of barley grains. *Theoretical and Applied Genetics* 68, 127–30.

Heun, M., Kennedy, A.E., Anderson, J.A., Lapitan, N.L.V., Sorrells, M.E. & Tanksley, S.D. (1991) Construction of a restriction fragment length polymorphism map of barley (*Hordeum vulgare*). *Genome* 34, 437–47.

Holloway, P.J. & Arundel, P.H. (1988) High-resolution two-dimensional electrophoresis of plant proteins. *Analytical Biochemistry* 172, 8–15.

Inagaki, M. (1986) Crossability of Japanese wheat cultivars with *Hordeum bulbosum* L. *Japanese Journal of Breeding* 36, 363–70.

Islam, A.K.M.R. (1983) Ditelosomic additions of barley chromosomes to wheat. In: Sakamoto, S. (ed.), *Proceedings of the 6th International Wheat Genetics Symposium*. Maruzen, Kyoto, pp. 233–8.

Islam, A.K.M.R. & Shepherd, K.W. (1981) Wheat–barley addition lines: their use in genetic and evolutionary studies of barley. In: Asher, M.J.C., Ellis, R.P., Hayter, A.M. & Whitehouse, R.N.H. (eds), *Barley Genetics IV* (Proceedings of the 4th International Barley Genetics Symposium), Edinburgh University Press, Edinburgh, pp. 729–39.

Islam, A.K.M.R. & Shepherd, K.W. (1988) Induced pairing between wheat and barley chromosomes. In: Miller, T.E. & Koebner, R.M.D. (eds), *Proceedings of the 7th International Wheat Genetics Symposium*. The Institute of Plant Science Research, Cambridge, pp. 309–14.

Islam, A.K.M.R. & Shepherd, K.W. (1990) Incorporation of barley chromosomes into wheat. In: Bajaj, Y.P.S. (ed.), *Biotechnology in Agriculture and Forestry; Wheat*. Vol. 13. Springer-Verlag, Berlin, Heidelberg, pp. 128–51.

Islam, A.K.M.R. & Shepherd, K.W. Production of wheat–barley recombinant chromosomes through induced homoeologous pairing. 1. Isolation of recombinants involving barley arms 3HL and 6HL. *Theoretical and Applied Genetics* (in press).

Islam, A.K.M.R. & Shepherd, K.W. Substituting ability of individual barley chromosomes for wheat chromosomes. 1. Substitutions involving barley chromosomes 1, 3 and 6. (submitted).

Islam, A.K.M.R., Shepherd, K.W. & Sparrow, D.H.B. (1975) Addition of individual barley chromosomes to wheat. In: Gaul, H. (ed.), *Barley Genetics III: Proceedings of the 3rd International Barley Genetics Symposium*, Verlag Karl Thiemig, Munchen, pp. 260–70.

Islam, A.K.M.R., Shepherd, K.W. & Sparrow, D.H.B. (1981) Isolation and characterization of euplasmic wheat–barley chromosome addition lines. *Heredity* 46, 161–74.

Kam-Morgan, L.N.W., Gill, B.S. & Muthukrishnan, S. (1989) DNA restriction fragment length polymorphisms: a strategy for genetic mapping of D genome of wheat. *Genome* 32, 724–73.

Koba, T., Handa, T. & Shimada, T. (1991) Efficient production of wheat–barley hybrids and preferential elimination of barley chromosomes. *Theoretical and Applied Genetics* 81, 285–92.

Koebner, R.M.D. (1990) Subtilisin inhibitor – a polymorphic protein produced by a gene on the short arms of wheat homoeologous group 1 chromosomes. *Journal of Genetics and Breeding* 44, 49–52.

Kolchinsky, A., Kanazin, V., Yakovleva, E., Gazumyan, A., Kole, C. & Ananiev, E. (1990) 5S-RNA genes of barley are located on the second chromosome. *Theoretical and Applied Genetics* 80, 333–6.

Kreis, M., Williamson, M., Shewry, P.R., Sharp, P.J. & Gale, M. (1988) Identification of a second locus encoding β-amylase on chromosome 2 of barley. *Genetical Research* 51, 13–16.

Kruse, A. (1973) *Hordeum* × *Triticum* hybrids. *Hereditas* 73, 157–61.

Lander, E.S., Green, P., Abrahamson, J., Barlow, A., Daly, M.J., Lincoln, S.E. & Newburg, L. (1987) MAPMAKER: An interactive computer package for constructing primary genetic linkage maps of experimental and natural populations. *Genomics* 1, 174–81.

Le, H.T., Armstrong, K.C. & Miki, B. (1989) Detection of rye DNA in wheat–rye hybrids and wheat translocation stocks using total genomic DNA as a probe. *Plant Molecular Biology Reporter* 7, 150–8.

Liu, C.J. & Gale, M.D. (1989) The chromosomal location of a third set of malate dehydrogenase loci, *Mdh-3*, in wheat, barley and related species. *Theoretical and Applied Genetics* 78, 349–52.

Liu, C.J. & Gale, M.D. (1990) *Est-7*, a set of genes controlling green tissue esterases in wheat and related species. *Theoretical and Applied Genetics* 79, 781–4.

Loi, L., Ahluwalia, B. & Fincher, G.B. (1988) Chromosomal location of genes encoding barley (1–3, 1–4)-β-glucan 4-glucanohydrolases. *Plant Physiology* 87, 300–2.

Maraña, C., Garcia-Olmedo, F. & Carbonero, P. (1988) Equivalent locations of sucrose synthase genes in chromosomes 7D of wheat, 7Ag of *Agropyron elongatum*, and 7H of barley. *FEBS Letters* 234, 417–20.

McGuire, P.E. (1984) Status of an attempt to transfer the barley yellow dwarf virus resistance gene Yd_2 of barley to hexaploid wheat. *Proceedings of the Workshop on Barley Yellow Dwarf.* CIMMYT, Mexico, pp. 113–19.

McGuire, P.E. & Qualset, C.O. (1990) Transfer of the Yd_2 barley yellow dwarf virus resistance gene from barley to wheat. In: Burnett, P.A. (ed.), *World Perspective on Barley Yellow Dwarf.* CIMMYT, Mexico, pp. 476–81.

Mori, S. & Nishizawa, N. (1989) Identification of barley chromosome No 4, possible encoder of genes of mugineic acid synthesis from 2'-Deoxymugineic acid using wheat–barley addition lines. *Plant Cell Physiology* 30, 1057–61.

Muthukrishnan, S., Gill, B.S., Swegle, M. & Ram Chandra, G. (1984) Structural genes for α-Amylases are located on barley chromosomes 1 and 6. *Journal of Biological Chemistry* 259, 13637–9.

Nielsen, G. & Hejgaard, J. (1987) Mapping of isozyme and protein loci in barley. In: Scandalios, J.G. (ed.), *Isozymes: Current Topics in Biological and Medical Research. Volume 15: Genetics, Development and Evolution.* Alan R. Liss, New York, pp. 77–95.

Nielsen, G. & Johansen, H.B. (1986) Proposal for the identification of barley varieties based on the genotypes for 2 hordein and 39 isozyme loci of 47 reference varieties. *Euphytica* 35, 717–28.

Powling, A., Islam, A.K.M.R. & Shepherd, K.W. (1981) Isozymes in wheat–barley hybrid derivative lines. *Biochemical Genetics* 19, 237–54.

Riley, R. & Chapman, V. (1958) Genetic control of the cytologically diploid behaviour of hexaploid wheat. *Nature* 182, 713–15.

Riley, R., Chapman, V. & Johnson, R. (1968) The incorporation of alien disease resistance in wheat by genetic interference with the regulation of meiotic chromosome synapsis. *Genetical Research* 12, 199–219.

Saghai-Maroof, M.A., Soliman, K.M., Jorgensen, R.A. & Allard, R.W. (1984) Ribosomal DNA spacer-length polymorphisms in barley: Mendelian inheritance, chromosomal location, and population dynamics. *Proceedings of the National Academy of Science (USA)* 81, 8014–18.

Schwarzacher, T., Leitch, A.R., Bennett, M.D. & Heslop-Harrison, J.S. (1989) *In situ* localization of parental genomes in a wide hybrid. *Annals of Botany* 64, 315–24.

Sears, E.R. (1977) An induced mutant with homoeologous pairing in common wheat. *Canadian Journal of Genetics and Cytology* 19, 585–93.

Sears, E.R. & Okamoto, M. (1958) Intergenomic chromosome relationships in hexaploid wheat. In: *Proceedings of the 10th International Congress of Genetics.* Montreal pp. 258–9.

Sethi, G.S., Finch, R.A. & Miller, T.E. (1986) A bread wheat (*Triticum aestivum*) × cultivated barley (*Hordeum vulgare*) hybrid with homoeologous chromosome pairing. *Canadian Journal of Genetics and Cytology* 28, 777–82.

Sharp, P.J., Chao, S., Desai, S. & Gale, M.D. (1989) The isolation, characterization and application in the Triticeae of a set of wheat RFLP probes identifying each homoeologous chromosome arm. *Theoretical and Applied Genetics* 78, 342–8.

Shepherd, K.W. & Islam, A.K.M.R. (1987) Cytogenetic manipulation of barley chromosomes in a wheat background. In: Yasuda, S. & Konishi, T. (eds), *Barley Genetics V: Proceedings of the 5th International Barley Genetics Symposium*, Maruzen, Okayama, Japan, pp. 375–87.

Shin, J.S., Chao, S., Corpuz, L. & Blake, T. (1990) A partial map of the barley genome incorporating restriction fragment length polymorphism, polymerase chain reaction, isozyme, and morphological marker loci. *Genome* 33, 803–10.

Soliman, K.M. & Allard, R.W. (1989) Chromosome locations of additional barley enzyme loci identified using wheat–barley addition lines. *Plant Breeding* 102, 177–81.

Sonina, N.V., Lushnikova, A.A., Tihonov, A.P. & Ananiev, E.V. (1989) Dialect – I, species-specific repeated DNA sequence from barley, *Hordeum vulgare. Theoretical and Applied Genetics* 78, 589–93.

Wang, R.C., Liang, G.H. & Heyne, E.G. (1977) Effectiveness of *ph* gene in homoeologous chromosome pairing in Agrotricum. *Theoretical and Applied Genetics* 51, 139–42.

Weining, S. & Langridge, P. (1991) Identification and mapping of polymorphisms in cereals based on the polymerase chain reaction. *Theoretical and Applied Genetics* 82, 209–16.

Williams, J.G.K., Kubelik, A.R., Livak, K.J., Rafalski, J.A. & Tingey, S.V. (1990) DNA polymorphisms amplified by arbitrary primers are useful as genetic markers. *Nucleic Acids Research* 18, 6531–5.

6

DNA Marker Techniques for Genetic Analysis in Barley

D.A. Laurie, J.W. Snape & M.D. Gale

The rationale for developing comprehensive genetic maps of barley

The development of comprehensive genetic maps of crop species has long been a major goal in plant science since such maps lead to a greater understanding of the relationship between chromosome structure, gene function and the genetical architecture of plant characteristics. They will also have important impacts on crop improvement because plant breeders will have the opportunity for the directed manipulation of specific genes by exploiting 'marker-mediated selection'. Properly used, this should increase the efficiency of breeding programmes by reducing the time and cost involved in the production of superior varieties.

Marker-mediated selection can be used in two ways. Firstly, where the marker allele is itself desirable and can be selected directly. Two examples of this are the avoidance of haze in beer by using proanthocyanidin-free barley mutants (von Wettstein *et al.*, 1977, 1985) and the improvement of bread-making quality in wheat by the selection of particular high-molecular-weight glutenin subunits (Payne, 1987). Secondly, where a marker, which may itself be phenotypically neutral, is tightly linked to an allele with an important agronomic effect. This second strategy is of greatest use where the target allele is difficult or expensive to detect in single plants or requires large-scale tests. Two examples which are being used in breeding programmes are the linkage of an endopeptidase isozyme to eyespot resistance in wheat (Worland *et al.*, 1988) and the linkage of esterase isozymes to barley yellow mosaic virus resistance in barley (Konishi *et al.*, 1989).

Such work requires an understanding of the genetical control of the trait of interest and the availability of suitable markers in breeding material. Until recently, the opportunities for this type of approach in barley were severely limited because its genetic map was composed of a relatively small number of morphological, disease resistance, isozyme and seed storage protein markers (Søgaard & von Wettstein-

Knowles, 1987; Shahla & Tsuchiya, 1990). Consequently most crosses, particularly those between commercial varieties, show few polymorphisms for these types of markers. However, recent developments in the use of DNA polymorphisms allow much greater numbers of loci to be identified. This should enable the accurate location and controlled manipulation of many more agronomically important genes, including those affecting characters which exert important influences on quantitative traits such as yield and quality.

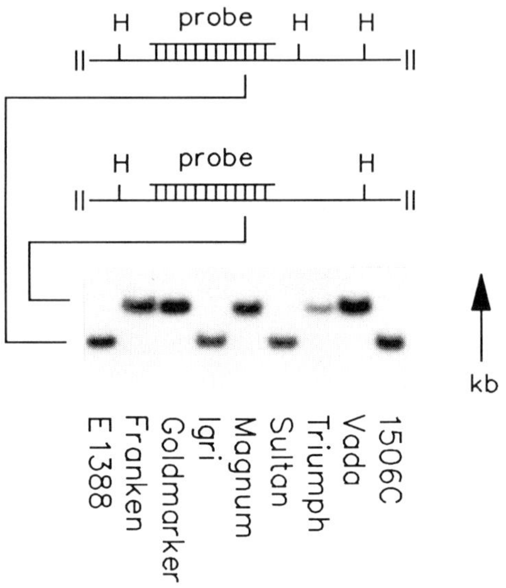

Fig. 6.1. Autoradiograph of a Southern blot of DNA from nine barley varieties digested with *Hin*dIII and probed with PSR489, a cDNA isolated from wheat. H indicates *Hin*dIII cleavage sites.

Methods for identifying DNA polymorphisms

In the late 1960s the discovery of bacterial restriction endonucleases which cleave DNA at defined sequences, usually 4–8 base pairs in length, led to the development of techniques for isolating and handling defined pieces of DNA. One major application of this technology, largely developed in the field of human genetics, was the assembly of comprehensive genetic maps using restriction fragment length polymorphism (RFLP) (Botstein *et al.*, 1980; Donis-Keller *et al.*, 1987).

This technique exploits variations in DNA sequence that create or abolish sites for restriction endonuclease cleavage, or which involve addition or deletion of DNA segments between sites. In addition, some restriction endonucleases are unable to cut DNA if the recognition sequence contains one or more methylated cytosine residues. Consequently, such enzymes may detect RFLP if there is variation in methylation patterns. RFLP can be visualized because the DNA from different genotypes is cleaved into fragments of different size which are separated electrophoretically and identified by hybridization with an appropriately labelled homologous probe sequence (Figs 6.1 and 6.2) or by direct amplification using the polymerase chain reaction (PCR) (Fig. 6.2).

RFLP analysis uses DNA sequences which are cloned from, or homologous to, DNA from the organism of interest. However, to recognize a specific piece of DNA, a probe is needed that discriminates this particular sequence from the remainder of the genome. Probes hybridizing to single- or low-copy DNA sequences are the easiest to analyse, but the isolation of appropriate probes in species such as barley is not straightforward, since at least 80% of the genome consists of repeated DNA sequences (Flavell *et al.*, 1977; Rimpau *et al.*, 1980). Single- or low-copy number probes can be obtained by the construction of cDNA libraries from cellular mRNA, but the most common method that has been used in plants is to select probes from genomic libraries made with the methylation-sensitive restriction enzyme *Pst* I. Work in vertebrates suggests that the use of methylation-sensitive restriction enzymes such as *Not* I, *Pst* I, or *Hpa* II generate libraries enriched in hypomethylated 'CpG island' regions which have a high probability of association with genes that are transcriptionally active in several tissues (so-called 'house-keeping' genes) (Bird, 1987). It is not yet known whether

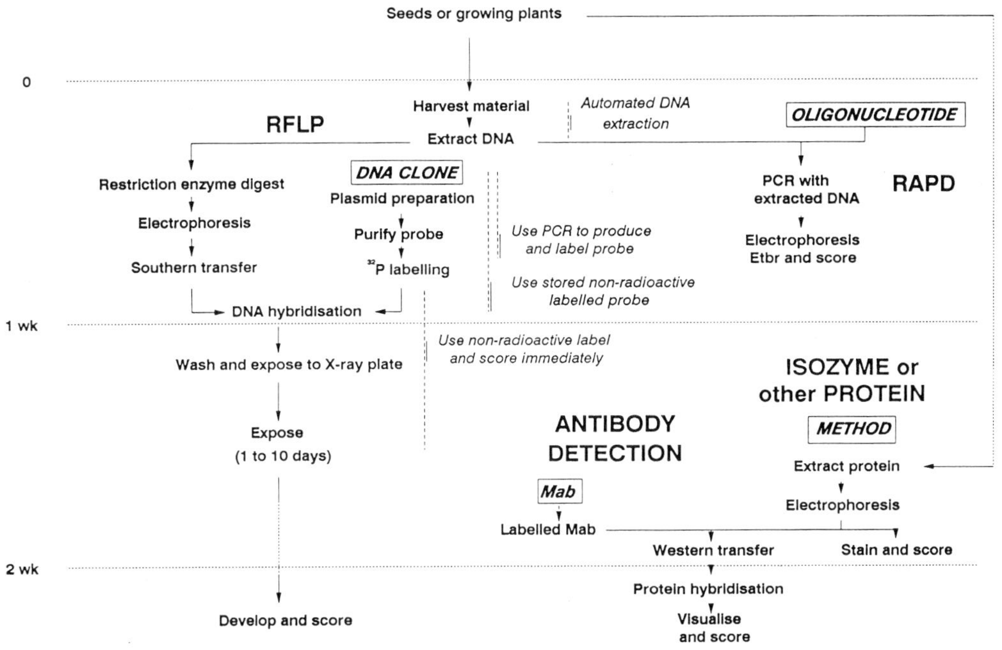

Fig. 6.2. The time scale and methods of restriction fragment length polymorphism (RFLP) analysis in comparison to random amplified polymorphic DNA (RAPD), and biochemical marker analyses. The various steps have been placed on an approximate time scale, although reductions in the time needed will be obtained with automation, increased use of polymerase chain reaction (PCR) and the use of non-radioactive detection methods (italics). The current methods of RFLP analysis in cereals are given in Sharp *et al.* (1988) and Chao *et al.* (1989a), and those for RAPD analysis by Williams *et al.* (1990). More general information can be obtained from laboratory manuals such as Sambrook *et al.* (1989).

this association also holds for plants but unmethylated sites have been found in association with the *a1*, *Sh1* and *Adh1* genes in maize, though not with the zein genes (Antequera & Bird, 1988) and the α-amylase-1 gene cluster in wheat (Cheung & Gale 1989).

An alternative approach is to analyse RAPDs (random amplified polymorphic DNA markers). In this method short synthetic oligonucleotides (about 10 bp), which can be of arbitrary or selected sequence, are used as primers for PCRs with genomic DNA as a template (Williams *et al.*, 1990). The products of the PCR are separated by electrophoresis and can be visualized directly using ethidium bromide. Polymorphism is revealed when DNA is amplified from only one parent or where DNA fragments of differing size are generated. The speed of this technique and the ease with which oligonucleotides can be made is expected to lead to its rapid and widespread adoption. The relative time scales and procedures of several methods of DNA and conventional marker analysis are illustrated in Fig. 6.2.

The advantages of DNA markers

RFLPs and RAPDs have major advantages over conventional markers. Firstly, conventional markers rely on differences between gene products which arise from alterations to a gene's coding sequence or from gross alterations in the level of gene expression. RFLPs and RAPDs, in contrast, directly identify variation in the DNA. Because coding sequences and their accompanying control regions account for only a small part of the genome of organisms such as barley, the number of potential sites of DNA polymorphism is much greater than the number of conventional markers. Furthermore, sequence variation even within coding regions may generate RFLP or RAPD without generating any other observable phenotypic variation.

Secondly, most RFLPs, like most proteins, are co-dominant, enabling both alleles to be distinguished in heterozygous individuals. This is a major advantage when one considers the transfer of a linked recessive gene into a potential crop variety. If RFLP-based marker-mediated selection can be used then the need for the production and testing of homozygous progeny at each generation is avoided. This advantage does not apply to those RAPD markers where DNA is amplified only from one parent. Thirdly, large numbers of DNA polymorphisms can be recognized using a single technique, enabling routine analytical procedures to be established. Fourthly, DNA markers offer entry points into the genome. Thus, in principle, any gene can be isolated using 'chromosome walking' techniques (see the section on the integration of physical and genetic maps). This is of great importance since advances in understanding the action of agronomically important genes will increasingly depend on their isolation and characterization at the DNA level.

Strategies for genetic analysis using DNA markers

In barley, as in other species, two important aspects of analysis are the selection of appropriate populations for mapping and the methodology of linkage analysis. The type of population used to produce the genetic map will depend largely on the extent to which the progenies are also to be evaluated for agronomic characters. If only a genetic map is required then the parents for the cross can be selected simply on the basis of a high frequency of polymorphism. For barley, F_2 populations from crosses with its wild relative *Hordeum spontaneum* would be suitable provided that there were no structural chromosomal differences between the parents. These 'species' are completely cross-compatible and mapping populations are therefore easily developed.

The preferred materials for basic mapping purposes are F_2 or backcross populations. Both are technically easy to develop, but F_2 populations require less crossing. When complete or almost complete genotypic classification of plants is possible, as is the case with most RFLP loci, F_2 plants provide twice as much information as backcross plants since two recombinant gametes are assessed per plant rather than one. Thus, estimating recombination frequencies to any given level of accuracy requires fewer individuals.

In practice it is usually convenient to derive mapping populations from varieties previously used to characterize morphological or agronomic characters or in which defined genetic stocks are available. For example, a primary population of the Cambridge Laboratory barley mapping project is the cross between the Canadian variety Betzes (the donor parent of the wheat–barley addition series, Islam *et al.*, 1981) and Golden Promise, a UK commercial semidwarf variety. Similarly, crosses between feed- and malting-quality barley varieties are being used in North American mapping projects. Many such crosses would be suitable for mapping purposes but it is important that those used to establish recombination frequencies (rather than to establish simply which chromosome a particular gene is on) should avoid using stocks with aneuploidy (such as trisomics) or structural chromosomal differences because of the risk of intra- and interchromosomal effects on recombination.

The application of RFLP to the study of agronomically important traits requires additional types of mapping populations. In particular, genes controlling quantitative traits (often referred to as quantitative trait loci or QTL), or single genes which are difficult to detect in individual plants, can be mapped more efficiently using populations of homozygous or near-homozygous lines grown and evaluated in replicated trials. As detailed genetic maps become available, the most efficient strategy to locate QTL in a particular cross will probably be to start with a selected subset of markers regularly spaced along the map. This should enable the approximate position of QTL to be determined without the need to analyse very large numbers of probes. More detailed analysis can then proceed using additional probes from the regions of interest.

Barley is well suited to the analysis of quantitative traits because of the ease with which homozygous material can be developed. Doubled haploid lines, each of which represents a single recombinant gamete, are particularly useful and can readily be produced by the *H. bulbosum* system or by anther or microspore culture (Dunwell, 1986). Homozygous material such as doubled haploid and single-seed descent populations also have the advantage that they can be retested indefinitely, since individual lines maintain their genetic integrity over selfed generations. Thus new markers can easily be fitted into the existing genetic map.

The use of doubled haploid lines highlights the need for genetic mapping to consider the possibility of variation in the frequency or distribution of recombination between male and female meiosis (as, for example, was reported in Shewry *et al.*, 1978). Significant differences in the amount of recombination would have important implications for the choice of doubled haploid production methods. For most uses, including efficient backcross breeding and the production of near-isogenic lines, it would be preferable to have the higher level of recombination.

Methods of genetic map construction

Data from segregating material are assembled into a genetic map which shows the linear sequence of loci on the chromosome and the observed frequency of recombination between them. Since barley is readily crossed of selfed, and homozygous lines are the parental material for crosses, genetic mapping procedures are straightforward, and well established. The book by Mather (1938) and the seminal paper of Allard (1956), for example, set out the methods for, first, the *detection* and, secondly, the *estimation* of linkage between pairs of loci in F_2 and backcross populations.

Linkage is detected when the proportion of progeny inheriting recombinant combinations of any given pair of markers (Θ) falls significantly below 0.5 (or 50%). Chi-square tests are commonly used to analyse the segregation ratios at individual loci and to determine if given pairs of loci segregate independently. If linkage is established then an estimate of the recombination frequency (Θ) can be obtained from maximum likelihood estimation equations. The type of equation used depends on the type of population and the segregation frequencies at the individual loci. The commonest situation when two co-dominant RFLP loci are involved is that the nine observable phenotypes (AABB, AaBB, AABb, AAbb, AaBb, aaBB, Aabb, aaBb, aabb) will segregate in a 1:2:2:1:4:1:2:2:1 ratio in the absence of linkage. This gives almost complete genotypic classification since only the coupling, AB.ab, and repulsion, Ab.aB, heterozygotes cannot be separated. If this ratio fails because of the presence of linkage, then a maximum likelihood solution for p is obtained by substituting observed numbers of genotypes in equation (1):

$$(1)\quad (\text{Nos. AABB} + \text{aabb})\left[\frac{2}{p}\right] + (\text{Nos. AABb} + \text{AaBB} + \text{Aabb} + \text{aaBb})\left[\frac{(1-2p)}{p(1-p)}\right] + (\text{Nos. AAbb} + \text{aaBB})\left[\frac{2}{(p-1)}\right] + (\text{Nos. AaBb})\left[\frac{2(2p-1)}{(1-2p+2p^2)}\right] = 0$$

and solving for p. Since estimates of p must take a value between 0 and 0.5 a simple numerical solution is obtained. Obviously when many loci are involved the calculations become extremely laborious. However, computer programs are now available to carry out the estimation procedure for F_2 and backcross data. The most comprehensive and efficient appears to be MAPMAKER, developed by Eric Lander and colleagues at the Whitehead Institute, Cambridge, Massachusetts (Lander *et al.*, 1987; Lander & Botstein, 1989). This calculates two-point recombination frequencies between all possible loci, but, more importantly, offers the facility for estimating three and higher point linkage distances and their associated likelihoods. In this way gene order and the best fit genetic map can be obtained for individual linkage groups.

The detection and estimation of linkage with doubled haploid and single-seed descent (SSD) populations has been described by Snape (1988). With doubled haploid populations the number of genotypes is reduced (to 2^n instead of the 3^n genotypes of a F_2) because heterozygotes are not present. Consequently, genotypic classification is easier and the estimation procedure is identical to a backcross with complete classification. With single-seed descent lines the estimation equation varies with the generation assessed. Nevertheless, an approximate solution can be obtained by considering only the homozygous loci. For example, for advanced SSD generations ($>F_5$) an approximate estimate of p can be obtained by ignoring the relatively rare heterozygotes and using equation (2) which is the maximum likelihood solution at complete homozygosity (F_∞) (Snape, 1988).

$$(2) \qquad p = \frac{\text{Nos(AAbb + aaBB)}}{\text{Nos(AABB + aabb)}}$$

An exact solution requires a computer program such as that described by Snape (1988) which calculates exact maximum likelihood recombination frequencies for any single-seed descent generation using a general optimizing routine (Ross, 1980).

Integration of quantitative trait loci into genetic maps

One of the promising uses of detailed RFLP maps is the identification and location of loci affecting quantitative (or polygenic) characters. Such quantitative trait loci (QTL) are not qualitatively different from 'major' genes, they are just more difficult to detect, and their analysis therefore requires a more biometrical approach. Methods for locating QTL derive largely from work on *Drosophila* by Thoday and colleagues in the 1960s (Thoday, 1961), and from work in cereals by Law and colleagues (Law, 1966, 1967; Law *et al.*, 1987). In these methods a QTL is detected by associating marker allele variation with phenotypic variation for the character under study. This requires statistical comparison of the character means of the various allelic groups for each marker in turn. Determining the chromosomal location of the QTL involves estimating the recombination frequencies between the QTL and adjacent flanking markers. Several methods of doing this are now available (Weller, 1987; Jensen, 1989; Lander & Botstein, 1989; Luo & Kearsey, 1990) all of which use maximum likelihood techniques to fit distributions to the observed phenotypic variation between and within the marker allele classes. The best fit obtained provides an estimate of the location of the QTL relative to the markers and an estimate of its average genotypic effect.

These methods contain assumptions (such as no epistasis) that are restrictive, and further refinement will therefore be needed to deal with more complex genetical architectures. Nevertheless, the statistical packages currently available (such as MAPMAKER/QTL (Lander & Botstein, 1989)) appear to be able to establish QTL location for F_2 and backcross populations.

The relationship between recombination fraction and genetic distance

The above discussions deal with estimates of Θ, but genetic maps can also be presented in Morgans. It is important to distinguish between these two scales. Θ is the proportion of progeny with recombinant combinations of a given pair of markers and in the absence of chromatid interference this will not statistically exceed 0.5, irrespective of the number of crossovers that occur between these loci. In contrast, the genetic distance in Morgans (more often shown as centiMorgans (cM)), is the sum of all crossovers and is therefore equal to half the mean chiasma frequency between the two loci. Thus Θ equals the genetic distance in Morgans only for intervals where no more than one crossover occurs. If two or more crossovers occur between a pair of loci then the relationship between Θ and genetic distance (cM) becomes much less clear since the number of crossovers can rarely be ascertained directly. Programs such as MAPMAKER therefore estimate genetic distance from recombination fractions using mapping functions such as those of Haldane (1919) or Kosambi (1944).

However, estimates of map distance derived in this way must be treated with caution since it is not clear to what extent mapping functions reflect biological reality. In particular, mapping functions do not allow for chiasma interference or the effects of gene conversion. Fortunately, these difficulties are unlikely to be a serious problem since the application of genetic maps for marker-mediated selection or chromosome walking must use very close linkages where the occurrence of more than one crossover is unlikely. In these cases the differences between the values of Θ and the genetic distance in cM are expected to be negligible.

When considering values of Θ or cM distances it must also be remembered that the amount of recombination between any given pair of loci is not a fixed character and it is therefore to be expected that values from separate investigations will vary. The frequency of recombination can be affected by parental genotype (Säll, 1990) and by environmental effects such as temperature during meiosis (Powell & Nilan, 1963).

Methods for determining the chromosomal location of DNA polymorphisms

Figures 6.1 to 6.4 illustrate the approach used in this laboratory to map single-copy and low-copy probes. Firstly, probes are tested on a selection of diverse varieties used in our genetic analysis programmes (Fig. 6.1). Probes which fail to show variation with four enzymes (*Eco*RI, *Eco*RV, *Dra*I, and *Hin*dIII) used in this primary screen are stored and are not immediately analysed further.

When polymorphism is identified the probe is located to a particular chromosome using defined aneuploid stocks. The most widely used are wheat nullisomic/tetrasomic lines in the variety Chinese Spring. These lack a particular pair of chromosomes which is compensated for by an additional homoeologous pair from one of the other genomes (for example, one such line is nullisomic for chromosome 5B and tetrasomic for chromosome 5D).

Alien chromosome addition lines,

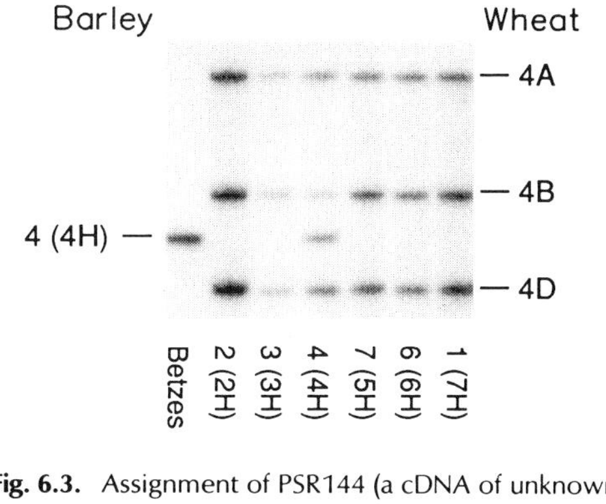

Fig. 6.3. Assignment of PSR144 (a cDNA of unknown function isolated from wheat) to chromosome 4 (4H) of barley using *Hin*dIII digests of Chinese Spring/Betzes single chromosome addition lines.

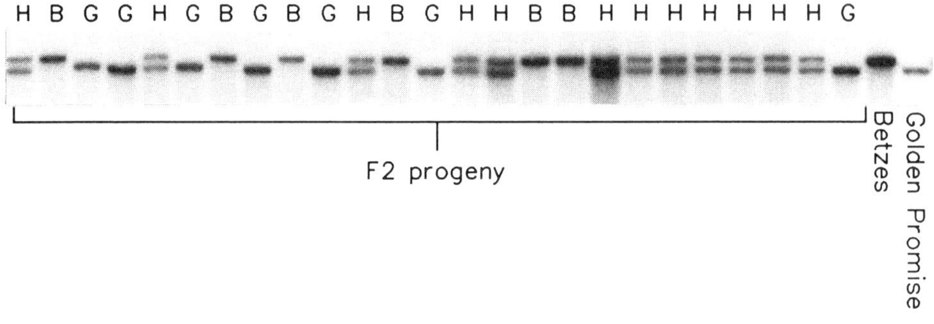

Fig. 6.4. Segregation of restriction fragment length polymorphisms in 25 F_2 progeny of a cross between Betzes and Golden Promise. The blot shows *Dra*I digests probed with PSR74 (a cDNA of unknown function isolated from wheat) (B, Betzes homozygotes; G, Golden Promise homozygotes; H, heterozygotes).

usually in the wheat variety Chinese Spring, are then used to confirm the chromosome location in other species. An example is the set of barley cv. Betzes addition lines, which have a full complement of wheat chromosomes plus a pair of barley chromosomes or chromosome arms. Six addition lines are available but the barley chromosome 5 (1H)[1] addition line is male-sterile and cannot be maintained as a disomic addition line (Islam *et al.*, 1981)[2]. Telosomic addition lines are also available which are extremely valuable for rapidly determining the arm locations of particular probes. For example, Kreis *et al.* (1988) used Betzes telosomic addition lines to identify β-amylase sequences on chromosome arms 2S(2HS) and 4(4HL) of barley. Figure 6.3 shows the assignment of a cDNA to barley chromosome 4(4H) using the Betzes addition lines.

The use of alien addition lines exploits a useful and widespread feature of cDNA clones, namely an evolutionary conservation of coding sequences which enables probes to identify equivalent sequences in related species. A cDNA isolated from barley will usually cross-hybridize well to wheat and rye and often to more distantly related species. For example, genes encoding enzymes controlling anthocyanin biosynthesis in barley have been isolated using probes from maize (Rohde *et al.*, 1987) and sequences encoding some enzymes involved in photosynthesis in wheat have been mapped to specific chromosomes using probes from maize or sorghum (Chao *et al.*, 1989b).

After polymorphic probes have been assigned to chromosomes they are used to analyse DNA isolated from individual plants of segregating populations (Fig. 6.4 shows an example of an RFLP segregating in an F_2 population derived from a cross between the barley varieties Betzes and Golden Promise). Individual plants are categorized for each polymorphic probe and data on the cosegregation of all loci is then used to determine the linkage relationships of the RFLPs. This analysis should, of course, include any conventional markers which are also segregating in the progeny in order to produce complete integrated maps (e.g. Shin *et al.*, 1990).

[1] In order to compare barley with other Triticeae species we prefer to describe barley chromosomes by homoeology. Thus 1H=chromosome 5, 2H=2, 3H=3, 4H=4, 5H=7, 6H=6 and 7H=1.
[2] DNA can be obtained from monosomic or disomic addition plants.

The frequency of RFLP in barley

Table 6.1 shows results from 45 probes isolated from wheat (13 cDNA and 32 genomic) screened with three restriction enzymes in 13 diverse barleys, and illustrates typical frequencies of RFLP. Polymorphisms in this sample are common, particularly with the cDNAs, so that overall some 50–65% of probes show RFLP between two or more of the 13 varieties with each enzyme. When data from the three enzymes are combined 88% of loci tested with the cDNA probes and 65.8% of the loci tested with genomic DNA probes showed polymorphism.

In several cases, however, RFLP is detected only between *H. spontaneum* and the 12 barley varieties so that exclusion of the *spontaneum* data reduces the frequencies of RFLP to 40 and 50% respectively. An alternative estimate of RFLP frequency is pairwise comparisons (i.e. the percentage of possible crosses between the 13 varieties that would give RFLP for a given probe). These values are naturally lower, typically being about half the frequency of overall polymorphism. Table 6.1 also shows the percentage of probes which are polymorphic in particular crosses used in our mapping programme. About 40% of probes are informative when the cross includes *H. spontaneum* and about 20% are informative in other crosses.

In the present sample, cDNA probes appeared to be more polymorphic than genomic probes even though the opposite has been observed when these probes are tested on wheat (unpublished data). Further work is needed to determine if this is a genuine difference in behaviour, or if it is a

Table 6.1. Frequencies of polymorphism obtained with three restriction enzymes in 12 barley varieties and one *H. spontaneum* accession. The varieties used in the screen were Betzes, Golden Promise, Captain, *H. spontaneum*, E1388, Franken, Goldmarker, Igri, Magnum, Sultan, Triumph, Vada and 1506C.

		*Eco*RI		*Eco*RV		*Hind*III	
13 cDNA probes	**Number of loci scored**	23	%	24	%	19	%
no. polymorphic	overall	19	82.6	19	79.2	10	52.6
	pairwise comparisons		33.6		37.6		22.2
no. polymorphic excluding *H. spontaneum*	overall	16	69.6	16	66.7	9	47.4
	pairwise comparisons		29.3		31.8		19.7
32 genomic DNA probes[1]	**Number of loci scored**	34		37		31	
no. polymorphic	overall	19	55.9	14	37.8	16	51.6
	pairwise comparisons		21.6		15.2		20.1
no. polymorphic excluding *H. spontaneum*	overall	13	38.2	9	24.3	11	35.5
	pairwise comparisons		17.2		12.2		16.7
	Total number of loci scored	57		61		50	
no. polymorphic	overall	38	66.7	33	54.1	26	52.0
	pairwise comparisons		26.4		24.1		20.9
no. polymorphic excluding *H. spontaneum*	overall	29	50.9	25	41.0	20	40.0
	pairwise comparisons		22.1		20.0		17.8
Polymorphisms detected in crosses	**Total**						
Captain × *H. spontaneum*	26 (41.9%)[2]	16	32.7	15	31.3	11	20.4
Betzes × Golden Promise	14 (22.6%)	6	12.2	6	12.5	7	13.0
Igri × Triumph	12 (20.0%)	6	12.2	6	12.5	6	11.1
Magnum × Goldmarker	12 (20.0%)	6	12.2	6	12.5	5	9.3

[1] All cDNA clones hybridized to both wheat and barley but a further 13 genomic DNA clones did not hybridize to barley or gave a signal which was too weak to allow mapping.

[2] From a total sample of 62 loci. The number tested with *Eco*RI, *Eco*RV and *Hind*III was 49, 48 and 54 respectively. The total number of polymorphic probes is less than the sum of the column totals because most probes are polymorphic with more than one enzyme.

property of wheat probes applied to barley, or if it is simply a sampling phenomenon. The frequencies for the cDNA probes are similar to those obtained by Graner *et al.* (1990) using genomic single-copy sequences derived from barley. In their experiment the overall percentage of informative pairwise comparisons was 30% for *Eco*RI, 37% for *Eco*RV and 13% for *Hin*dIII.

The genetic map of barley chromosome 1 (7H)

Figure 6.5 shows a map of barley chromosome 1 (7H) derived from 120 progeny from Captain × *H. spontaneum* crosses and compiled using the MAPMAKER program. Comparisons of the genetic maps of wheat, rye and barley show that apart from a number of gross chromosome rearrangements (such as the position of the nucleolus organizer regions) the order of loci is very similar, reflecting a general evolutionary conservation of linkage group structure. This is illustrated by Figure 6.5 which shows that the chromosome 1 (7H) map of barley shares a common order of loci with the consensus map of the group 7 chromosomes of wheat.

The integration of physical maps and genetic maps

The efficient use of genetic maps requires an understanding of the non-linear relationship between physical and genetic distance. This non-linearity arises from localization of recombination events and results in regions of the genome where tight linkage between loci does not reflect close physical proximity of the corresponding genes. This discrepancy is important for two reasons. Firstly, regions with little recombination may present problems for marker-mediated selection because of the increased likelihood of co-transference of undesired alleles. Secondly, tightly linked loci in regions where recombination is rare may be physically too far apart to be analysable by current chromosome walking methods.

Several workers have analysed the relationship between physical and genetic distance in barley. Results indicate that recombination is uncommon in regions close to the centromere, but most of these studies involved translocated or inverted chromosomes (e.g. Künzel, 1982). Thus, there is no guarantee that they accurately reflect the behaviour of karyotypically normal material. However, strong supporting evidence is provided by studies on the segregation of C-band polymorphisms. For example, Linde-Laursen (1982) showed that recombination was largely restricted to the distal half of chromosome arm 3L (3HL). Similar results were obtained for the short arms of wheat chromosomes 1B and 6B by Snape *et al.* (1985) and Dvořák & Chen (1984) respectively. The effect was particularly striking in 6B where no recombination was detected between the centromere and the NOR even though this region accounted for 68% of the physical length of the arm. Such findings are expected since cells at metaphase I in barley and wheat show bivalents which are predominantly in the form of open rings, indicating that chiasmata are preferentially formed in distal regions of the arm. These conclusions are also supported circumstantially by RFLP analysis which shows tight linkage of loci around the centromere. For instance, a recent analysis of rye

Fig. 6.5. The relationship between the genetic map of chromosome 1 (7H) of barley and the consensus map of the homoeologous group 7 chromosomes of wheat. The points of correspondence are all identified by cDNA probes isolated from wheat that hybridize equally well to wheat and barley and for which RFLP has been detected in mapping populations of both species. The barley map is described in Kilian *et al.* (in press) and the wheat map in Chao *et al.* (1989a) with alterations and additions from Liu *et al.* (1989, 1990, in press). X preceding a symbol denotes a DNA locus. The known-function probes were all kindly supplied by the authors: *XWx*, waxy

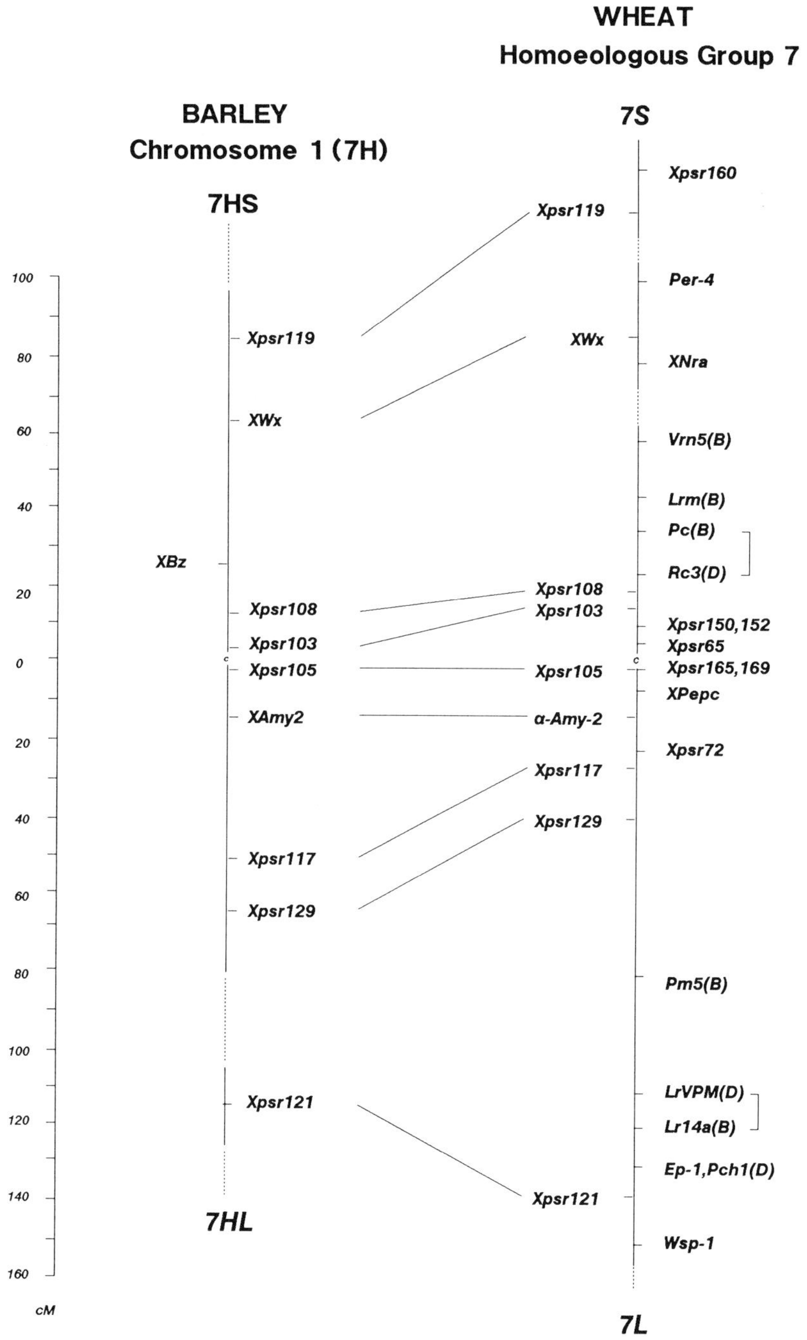

(Rohde *et al.*, 1988); *XBz*, bronze (W. Rohde, personal communication); *XNra*, nitrate reductase (Cheng *et al.*, 1986); *XAmy2*, low pI α-amylase (Lazarus *et al.*, 1985); and *XPepc*, PEP-carboxylase (P. Westhoff, personal communication). Other loci are: *Per-4*, peroxidase (Liu *et al.*, 1990); *Wsp-1*, water-soluble protein (Liu *et al.*, 1989); *Ep-1*, endopeptidase (Koebner *et al.*, 1988); *Vrn5*, vernalization response; *Lr*, leaf rust resistance; *Pc*, purple culms; *Rc*, red coleoptile; *Pm5*, powdery mildew resistance; *Pch-1*, eyespot resistance. In instances where the wheat locus is known in only one genome (A, B or D) the location is indicated in brackets.

Table 6.2. Barley chromosome arm sizes in megabases (10^6 bp).

Chromosome	Long arm	Short arm	Satellite
1H(5)	374	270	
2H(2)	425	386	
3H(3)	447	350	
4H(4)	412	367	
5H(7)	481	214	61
6H(6)	383	268	89
7H(1)	445	380	

Based on the 1C DNA content of 5.55 pg measured in cv. Sultan (Bennett & Smith, 1976), the relative chromosome sizes in cv. Sultan (Heslop-Harrison & Bennett, 1983) and 1 pg = 0.965×10^9 bp (Lewin, 1980). Barley chromosome numbers are shown in brackets.

In comparison with other organisms popular with geneticists the total barley genome (5356 mb) is 1.6 times bigger than humans[1] (3281 mb), 2.2 times bigger than maize[2] (*Zea mays*) (2413 mb for a line with a 1C DNA content of 2.5 pg), 28 times bigger than *Arabidopsis thaliana*[3] (193 mb), 31 times bigger than *Drosophila melanogaster*[1] (174 mb) and 67 times bigger than *Caenorhabditis elegans*[1] (80 mb). DNA contents are from [1]Lewin (1980), [2]Laurie & Bennett (1985), [3]Bennett & Smith (1976).

showed that the rRNA genes mapped only 2 cM from the centromere despite the fact that cytological studies show the NOR to be located at a distal position on the arm (Wang *et al.*, 1991). It is likely that other non-NOR bearing arms behave in the same way, but this requires further study.

These results highlight the need for integrating physical and genetic maps, but this is not a simple matter in an organism such as barley which has a large genome (5.36×10^9 bp, Table 6.2), of which perhaps no more than 0.5% is coding sequences (as in wheat, Flavell *et al.*, 1987). Of the remainder, at least 75% is made up of sequences repeated thousands to millions of times (Flavell *et al.*, 1977; Rimpau *et al.*, 1980). Some highly repeated sequences are concentrated in blocks of heterochromatin which can be visualized cytologically as C-bands (Linde-Laursen, 1981; Søgaard & von Wettstein-Knowles, 1987), but most are dispersed. Integration of physical and genetic data therefore requires detailed information on how unique and repeated sequences are organized within the genome. Interestingly, recent work suggests that much of the genome of wheat and barley consists of blocks, made up of diverse repeated elements, which are flanked by clusters of restriction sites for the methylation-sensitive enzymes *Not*I and *Mlu*I (Moore *et al.*, in press a, b, c). It appears that a proportion of these clusters of endonuclease sites are associated with genes, and these may therefore serve as genomic landmarks (Cheung & Gale, 1989; Moore *et al.*, 1991c).

Physical mapping using pulsed-field gel electrophoresis (PFGE)

Methods for physical mapping in cereals are, for technical reasons, currently restricted to relatively small regions of the genome. The most useful approach presently available is the one described by Cheung & Gale (1989) where PFGE was used to determine the full structure of the *α-Amy-B1* locus on wheat chromosome 6B. The six structural genes were located within a region spanning 4.3 megabases (10^6 bp). This approach was also used by Sørensen (1989) to analyse the *Hor2* locus (encoding the B-hordein seed storage proteins) on chromosome 5 (1H) of barley. In this case, two fragments were identified which hybridized with a cDNA clone specific for the B hordein genes (of which there are expected to be at least 13 copies per haploid genome). The fragments were of 200 and 160 kb in size, indicating that all the B hordein genes were located within a total of 360 kb of DNA divided into two subloci. However, the two fragments were separated by an unknown length of DNA, so the full structure of the *Hor2* locus could not be determined. It was also not determined whether the two fragments corresponded to separate subclasses of the B hordein proteins (class III B hordein genes can be separated from classes I and II by rare recombination events, Shewry *et al.*, 1990). Analysis of lambda clones showed that none contained more than one copy of a B hordein gene, indicating that the coding sequences were dispersed within the overall 360 kb structures. It should be possible to clone the fragments in yeast artificial chromosomes (YACs), thus enabling a more detailed analysis of gene order and polarity.

Other approaches to physical mapping

The assembly of cloned sequences into physical maps, or 'contigs', is an important future goal, at least for those regions which contain coding sequences, since this information will be needed for chromosome walking (Coulson *et al.*, 1988; Garza *et al.*, 1989). The ease with which clones can be assembled into contigs depends largely on the extent to which repeated DNA sequences, which make up the greater part of the genome, can be ordered (see Evans & Lewis, 1989; Stallings *et al.*, 1990). To put the problem of contig assembly and chromosome walking into perspective, Table 6.2 gives the estimated sizes of each barley arm in megabases. This shows that a hypothetical contig of 100 YACs each containing 500 kb and overlapping by 20% would contain only about 15% of the smallest barley chromosome arm (5S (1HS)).

The most useful approach for integrating physical and genetic maps over the genome as a whole would be to use *in situ* hybridization to locate the physical positions of RFLP probes. Recent advances in the use of non-radioactive detection systems offer the most promise for this approach because they offer far greater resolution than can be obtained from radiolabelled probes. To date, most work in plants has located blocks of repeated genes, such as the 18S–28S rRNA genes (Flavell *et al.*, 1986, 1987), the 5S rRNA genes (Mukai *et al.*, 1990) or the B hordein genes of the *Hor2* locus (Clark & Karp, 1989). Successful hybridization of the single-copy waxy gene has also been reported using maize pachytene chromosomes (Shen *et al.*, 1987). However, it should soon be possible to locate single-copy sequences by *in situ* hybridization to somatic chromosomes. Work on human chromosomes has already shown that the location of single-copy sequences can be determined in this way. For example, Lichter *et al.* (1990) and Fan *et al.* (1990) respectively mapped 50 and 23 probes to specific G-band regions of chromosome 11. Moreover, it should be possible to use different fluorescent labels on probes to label simultaneously at least three probes (Nederlof *et al.*, 1989) to rapidly confirm gene order, as well as gene location, on the chromosome. By using interphase nuclei as well as chromosome spreads it should be feasible to order sequences separated by as little as 100 kb (Lawrence *et al.*, 1990).

Applications of RFLP markers in plant genetics and plant breeding

High-density genetic maps based on RFLPs are certain to have major impacts on our understanding of plant genetics, and in particular on the genetical control of quantitative characters. Investigations designed to correlate RFLP variation with quantitative traits have already produced interesting results in crop species such as tomato (Paterson *et al.*, 1988, 1990). Similar approaches will soon be feasible in barley, where previous work has been limited by the paucity of conventional markers. For instance, Jensen (1989) used doubled haploid lines to map a locus designated *Kw1* (which accounted for 25% of the variance in kernel weight), in relation to two chromosome 7(5H) markers, DDT resistance (*ddt*) and short rachilla hairs (*s*). However, linkages were weak, with values for Θ of 0.42 for *ddt-Kw1* and 0.26 for *s-Kw1*. RFLP analysis should enable such genes to be mapped with much greater precision.

RFLP markers will also be extremely important as starting points for gene isolation by chromosome walking. One implication of this is that the increasing interest in the genetical control of development and differentiation in plants should focus renewed attention on many of the morphological mutants of barley. These genes, many of which have previously been used as conventional markers, are likely to become targets for analysis at the DNA level.

Success in gene isolation is crucially dependent on the development of transformation technologies which allow the activity of cloned genes to be assayed. Once transgenic plants can be produced, RFLP analysis will find an additional role in determining whether the position of integration within the chromosome affects the level of gene expression. Physical and genetic mapping data will also be important if transposable elements can be introduced into the barley genome since evidence from maize indicates that elements such as *Ac* tend to move to closely linked sites (Greenblatt, 1984; Dooner & Belachew, 1989). Thus, transposon tagging strategies will probably be most efficient if the starting position of the element is close to the target gene.

Other important applications of RFLPs include: (i) defining breakpoints in translocations, which would be valuable in studies relating physical and genetic maps; (ii) analysis of somaclonal variation, including the selection of mutants at particular loci of interest; (iii) analysis of the amount of DNA from the non-recurrent parent that is left in backcross populations, including the development of near-isogenic lines (an example of such an analysis in tomato has been described by Young & Tanksley (1989)); (iv) rapid development of alien addition lines (Sharp *et al.*, 1989).

The above discussion focuses on applications of RFLP analysis in fundamental research. But it is also hoped that this technology will rapidly find use in plant breeding programmes since tags for genes desired by plant breeders should, in theory, enable selections to be carried out more efficiently. 'Fingerprinting' RFLPs (Dallas, 1988) are also expected to find application in maintaining varietal purity and in preventing varietal piracy. All major cereal breeders in the UK are investigating the use of RFLP techniques, but it is not yet clear to what extent RFLP analysis will be adopted as a routine method in the cost/benefit world of commercial companies.

Technical advances in RFLP analysis

RFLP analysis has already developed to a considerable level of sophistication but there are several areas in which technical advances are expected to increase its usefulness still further. These include:

1. the development of rapid cost-effective non-radioactive hybridization and detection systems (Ishii *et al.*, 1990; Kreike *et al.*, 1990; Pollard-Knight, 1990)
2. single-copy *in situ* hybridization and improved chromosome banding techniques for physical mapping, particularly in extended prometaphase chromosomes
3. the application of flow-cytometry techniques developed in human genetics for the production of chromosome-specific or arm-specific libraries (Van Dilla *et al.*, 1989). In addition, chromosomes can be sorted into discrete regions on filters which can then be used to determine the chromosome locations of probes (Lebo, 1989)
4. the development of routine systems for the detection of polymorphism using short nucleotide sequences (RAPD) (Williams *et al.*, 1990)
5. exploitation of variation in methylation patterns for the detection of RFLP (Silva & White, 1988; Flavell & O'Dell, 1990)
6. the development of 'designer endonucleases' that recognize large (at least 15–20 bp) restriction sites. These very rare cutting enzymes, and synthetic oligonucleotides tipped with EDTA.Fe that cleave at specific sequences of comparable size (Strobel & Dervan, 1990), are expected to be of considerable benefit to long-range mapping in organisms such as barley which have large genomes. Eventually it may even be possible specifically to cut out a segment of DNA between two endonuclease sites defined by DNA polymorphism.

These developments are expected to

increase significantly the power of DNA marker analysis, to improve gene isolation techniques and to increase the usefulness of DNA marker analysis as a tool for plant breeding. However, it is also important to stress that progress will be most rapid if the information generated in different laboratories can be combined. It is to be hoped that a set of common markers can be adopted that will serve as landmarks on the genetic map.

References

Allard, R.W. (1956) Formulas and tables to facilitate the calculation of recombination values in heredity. *Hilgardia* 24, 235–78.

Antequera, F. & Bird, A.P. (1988) Unmethylated CpG islands associated with genes in higher plant DNA. *EMBO Journal* 7, 2295–9.

Bennett, M.D. & Smith, J.B. (1976) Nuclear DNA amounts in angiosperms. *Philosophical Transactions of the Royal Society of London* B 274, 227–74.

Bird, A.P. (1987) CpG islands as gene markers in the vertebrate nucleus. *Trends in Genetics* 3, 342–7.

Botstein, D., White, R.L., Skolnick, M. & Davis, R.W. (1980) Construction of a genetic linkage map in man using restriction fragment length polymorphisms. *American Journal of Human Genetics* 32, 314–31.

Chao, S., Sharp, P.J., Worland, A.J., Warham, E.J., Koebner, R.M.D. & Gale, M.D. (1989a) RFLP-based genetic maps of wheat homoeologous group 7 chromosomes. *Theoretical and Applied Genetics* 78, 495–504.

Chao, S., Raines, C.A., Longstaff, M., Sharp, P.J., Gale, M.D. & Dyer, T.A. (1989b) Chromosomal location and copy number in wheat and some of its close relatives of genes for enzymes involved in photosynthesis. *Molecular and General Genetics* 218, 423–30.

Cheng, C.L, Dewdney, J., Kleinhofs, A. & Goodman, H.M. (1986) Cloning and nitrate induction of nitrate reductase mRNA. *Proceedings of the National Academy of Sciences (USA)* 83, 6825–8.

Cheung, W.Y. & Gale, M.D. (1989) Long-range physical mapping of the α-amylase (*α-Amy-1*) loci on homoeologous group 6 chromosomes of wheat. *IPSR Cambridge Laboratory Annual Report*, 1989, pp. 4–7.

Clark, M. & Karp, A. (1989) Physical mapping of the B-hordein loci on barley chromosome 5 by *in situ* hybridization. *Genome* 32, 925–9.

Coulson, A., Waterston, R., Kiff, J., Sulston, J. & Kohara, Y. (1988) Genome linking with yeast artificial chromosomes. *Nature* 335, 184–6.

Dallas, J.F. (1988) Detection of DNA 'fingerprints' of cultivated rice by hybridization with a human minisatellite DNA probe. *Proceedings of the National Academy of Sciences (USA)* 85, 6831–5.

Donis-Keller, H. & 32 others (1987) A genetic linkage map of the human genome. *Cell* 51, 319–37.

Dooner, H.K. & Belachew, A. (1989) Transposition pattern of the maize element *Ac* from the *bz-m2(Ac)* allele. *Genetics* 122, 447–57.

Dunwell, J.M. (1986) Barley. In: Evans, D.A., Sharp, W.R. & Ammirato, P.V. (eds), *Handbook of Plant Cell Culture*, Vol. 4, *Techniques and Applications*. Macmillan, New York, pp. 339–69.

Dvořák, J. & Chen, K.-C. (1984) Distribution of nonstructural variation between wheat cultivars along chromosome arm 6Bp: evidence from the linkage map and physical map of the arm. *Genetics* 106, 325–33.

Evans, G.A. & Lewis, K.A. (1989) Physical mapping of complex genomes by cosmid multiplex analysis. *Proceedings of the National Academy of Sciences (USA)* 86, 5030–4.

Fan, Y.-S., Davis, L.M. & Shows, T.B. (1990) Mapping small DNA sequences by fluorescence *in situ* hybridization directly on banded metaphase chromosomes. *Proceedings of the National Academy of Sciences (USA)* 87, 6223–7.

Flavell, R.B. & O'Dell, M. (1990) Variation and inheritance of cytosine methylation patterns in wheat at the high molecular weight glutenin and ribosomal RNA gene loci. *Development* (1990 supplement) 15–20.

Flavell, R.B., Rimpau, J. & Smith, D.B. (1977) Repeated sequence DNA relationships in four cereal genomes. *Chromosoma* 63, 205–22.

Flavell, R.B., O'Dell, M., Sharp, P.J., Nevo, E. & Beiles, A. (1986) Variation in the intergenic spacer of ribosomal DNA of wild wheat, *Triticum dicoccoides* in Israel. *Molecular Biology and Evolution* 3, 547–58.

Flavell, R.B., Bennett, M.D., Seal, A.G. & Hutchinson, J. (1987) Chromosome structure and organization. In: Lupton F.G.H. (ed.),

Wheat Breeding. Its Scientific Basis. Chapman & Hall, London, pp. 211–68.

Garza, D., Ajioka, J.W., Burke, D.T. & Hartl, D.L. (1989) Mapping the *Drosophila* genome with yeast artificial chromosomes. *Science* 246, 641–6.

Graner, A., Siedler, H., Jahoor, A., Herrmann, R.G. & Wenzel, G. (1990) Assessment of the degree and the type of restriction fragment length polymorphism in barley (*Hordeum vulgare*). *Theoretical and Applied Genetics* 80, 826–32.

Greenblatt, I.M. (1984) A chromosome replication pattern deduced from pericarp phenotypes resulting from movements of the transposable element, Modulator, in maize. *Genetics* 108, 471–85.

Haldane, J.B.S. (1919) The combination of linkage values, and the calculation of distances between the loci of linked factors. *Journal of Genetics* 8, 299–309.

Heslop-Harrison, J.S. & Bennett, M.D. (1983) Prediction and analysis of spatial order in haploid chromosome complements. *Proceedings of the Royal Society of London B* 218, 211–23.

Ishii, T., Panaud, O., Brar, D.S. & Khush, G.S. (1990) Use of non-radioactive digoxigenin-labelled DNA probes for RFLP analysis in rice. *Plant Molecular Biology Reporter* 8, 167–71.

Islam, A.K.M.R., Shepherd, K.W. & Sparrow, D.H.B. (1981) Isolation and characterization of euplasmic wheat–barley chromosome addition lines. *Heredity* 46, 161–74.

Jensen, J. (1989) Estimation of recombination parameters between a quantitative trait locus (QTL) and two marker gene loci. *Theoretical and Applied Genetics* 78, 613–18.

Kilian, A., Kleinhofs, A. & Gale, M.D. Colinearity of genetic maps of barley chromosome 1 (7H) and wheat group 7 chromosomes. *Theoretical and Applied Genetics* (submitted).

Koebner, R.N.D., Miller, T.E., Snape, J.W. & Law, C.N. (1988) Wheat endopeptidase: genetic control, polymorphism, intrachromosomal gene location, and alien variation. *Genome* 30, 186–92.

Konishi, T., Kawada, N., Yoshida, H. & Sohtome, K. (1989) Linkage relationship between two loci for the barley yellow mosaic resistance of Mokusekko 3 and esterase isozymes in barley (*Hordeum vulgare* L.). *Japanese Journal of Breeding* 39, 423–30.

Kosambi, D.D. (1944) The estimation of map distances from recombination values. *Annals of Eugenics* 12, 172–5.

Kreike, C.M., de Koning, J.R.K. & Krens, F.A. (1990) Non-radioactive detection of single-copy DNA – DNA hybrids. *Plant Molecular Biology Reporter* 8, 172–9.

Kreis, M., Williamson, M.S., Shewry, P.R., Sharp, P.J. & Gale, M.D. (1988) Identification of a second locus encoding β-amylase on chromosome 2 of barley. *Genetical Research (Cambridge)* 51, 13–16.

Künzel, G. (1982) Differences between genetic and physical centromere distances in the case of two genes for male sterility in barley. *Theoretical and Applied Genetics* 64, 25–9.

Lander, E.S. & Botstein, D. (1989) Mapping Mendelian factors underlying quantitative traits using RFLP linkage maps. *Genetics* 121, 185–99.

Lander, E.S., Green, P., Abrahamson, J., Barlow, A., Daly, M.J., Lincoln, S.E. & Newburg, L. (1987) MAPMAKER: an interactive computer package for constructing primary genetic linkage maps of experimental and natural populations. *Genomics* 1, 174–81.

Laurie, D.A. & Bennett, M.D. (1985) Nuclear DNA content in the genera *Zea* and *Sorghum*. Intergeneric, interspecific and intraspecific variation. *Heredity* 55, 307–13.

Law, C.N. (1966) The location of genetic factors affecting a quantitative character in wheat. *Genetics* 53, 487–98.

Law, C.N. (1967) The location of genetic factors controlling a number of quantitative characters in wheat. *Genetics* 56, 445–61.

Law, C.N., Snape, J.W. & Worland, A.J. (1987) Aneuploidy in wheat and its uses in genetic analysis. In: Lupton, F.G.H. (ed.), *Wheat Breeding. Its Scientific Basis*. Chapman & Hall, London, pp. 71–108.

Lawrence, J.B., Singer, R.H. & McNeil, J.A. (1990) Interphase and metaphase resolution of different distances within the human dystrophin gene. *Science* 249, 928–32.

Lazarus, C.M., Baulcombe, D.C. & Martienssen, R.A. (1985) α-Amylase genes of wheat are two multigene families which are differentially expressed. *Journal of Plant Molecular Biology* 5, 13–24.

Lebo, R.V. (1989) Gene mapping strategies and bivariate flow cytogenetics. In: Gray, J.W. (ed.), *Flow Cytogenetics*. Academic, London, pp. 225–42.

Lewin, B. (1980) *Gene Expression 2*, 2nd edn. John Wiley, New York.

Lichter, P., Tang, C-J.T., Call, K., Hermanson,

G., Evans, G.A., Housman, D. & Ward, D.C. (1990) High-resolution mapping of human chromosome 11 by *in situ* hybridization with cosmid clones. *Science* 247, 64–9.

Linde-Laursen, I. (1981) Giemsa banding patterns of the chromosomes of cultivated and wild barleys. In: *Barley Genetics IV* (Proceedings of the Fourth International Barley Genetics Symposium). Edinburgh University Press, Edinburgh, pp. 786–95.

Linde-Laursen, I. (1982) Linkage map of the long arm of barley chromosome 3 using C-bands and marker genes. *Heredity* 49, 27–35.

Liu, C.J., Chao, S. & Gale, M.D. (1989) *Wsp-1*, a set of genes controlling water-soluble proteins in wheat and related species. *Genetical Research (Camb.)* 54, 173–81.

Liu, C.J., Chao, S. & Gale, M.D. (1990) The genetical control of tissue-specific peroxidases *Per-1, Per-2, Per-3, Per-4* and *Per-5* in wheat. *Theoretical and Applied Genetics* 79, 305–13.

Liu, C.J., Atkinson, M.D., Chinoy, C.N., Devos, K.M. & Gale, M.D. Non-homologous translocations between group 4,5 and 7 chromosomes in wheat and rye. *Theoretical and Applied Genetics* (in press).

Luo, Z.W. & Kearsey, M.J. (1990) Maximum likelihood estimation of linkage between a marker gene and a quantitative locus. *Heredity* 63, 401–8.

Mather, K. (1938) *The Measurement of Linkage in Heredity*. Methuen, London.

Moore, G., Lucas, H., Batty, N. & Flavell, R.B. (1991a) A family of retrotransposons and associated genomic variation in wheat. *Genomics* 10, 461–8.

Moore, G., Cheung, W.Y., Schwarzacher, T. & Flavell, R.B. (1991b) BIS 1, a major component of the cereal genome and a tool for studying genomic organization. *Genomics* 10, 469–76.

Moore, G., Cheung, W.Y., Stancombe, P., Money, T., Foote, T. & Flavell, R.B. The distribution of methylation free islands reveals key features of cereal genome organization. (submitted).

Mukai, Y, Endo, T.R. & Gill, B.S. (1990) Physical mapping of the 5S rRNA multigene family in common wheat. *Journal of Heredity* 81, 290–5.

Nederlof, P.M., Robinson, D., Abuknesha, R., Wiegant, J., Hopman, A.H.N., Tanke, H.J. & Raap, A.K. (1989) Three-color fluorescence *in situ* hybridization for the simultaneous detection of multiple nucleic acid sequences. *Cytometry* 10, 20–7.

Paterson, A.H., Lander, E.S., Hewitt, J.D., Peterson, S., Lincoln, S.E. & Tanksley, S.D. (1988) Resolution of quantitative traits into Mendelian factors by using a complete linkage map of restriction fragment length polymorphisms. *Nature* 335, 721–6.

Paterson, A.H., DeVerna, J.W., Lanini, B. & Tanksley, S.D. (1990) Fine mapping of quantitative trait loci using selected overlapping recombinant chromosomes, in an interspecies cross of tomato. *Genetics* 124, 735–42.

Payne, P.I. (1987) Genetics of wheat storage proteins and the effect of allelic variation on bread-making quality. *Annual Review of Plant Physiology* 38, 141–53.

Pollard-Knight, D. (1990) Current methods in nonradioactive nucleic acid labelling and detection. *Technique* 2, 113–32.

Powell, J.B. & Nilan, R.A. (1963) Influence of temperature on crossing over in an inversion heterozygote in barley. *Crop Science* 3, 11–13.

Rimpau, J., Smith, D.B. & Flavell, R.B. (1980) Sequence organisation in barley and oat chromosomes revealed by interspecies DNA/DNA hybridisation. *Heredity* 44, 131–49.

Rohde, W., Barzen, E., Marocco, A., Schwarz-Sommer, Z., Saedler, H. & Salamini, F. (1987) Isolation of genes that could serve as traps for transposable elements in *Hordeum vulgare*. In: *Barley Genetics V*. pp. 533–41.

Rohde, W., Becker, D. & Salamini, F. (1988) Structural analysis of the *waxy* locus from *Hordeum vulgare*. *Nucleic Acids Research* 16, 7185–6.

Ross, G.J.S. (1980) *Maximum Likelihood Programme*. Rothamstead Experiment Station, Harpenden, UK.

Säll, T. (1990) Genetic control of recombination in barley. II. Variation in linkage between marker genes. *Hereditas* 112, 171–8.

Sambrook, J., Fritsch, E.F. & Maniatis, T. (1989) *Molecular Cloning: A Laboratory Manual*, 2nd edn. Cold Spring Harbor, New York.

Shahla, A. & Tsuchiya, T. (1990) Genetic analysis in six telotrisomic lines in barley (*Hordeum vulgare* L.). *Journal of Heredity* 81, 127–30.

Sharp, P.J., Desai, S. & Gale, M.D. (1988) Isozyme variation and RFLPs at the β-amylase loci in wheat. *Theoretical and Applied Genetics* 76, 691–9.

Sharp, P.J., Chao, S., Desai, S. & Gale, M.D. (1989) The isolation, characterization and application in the Triticeae of a set of wheat RFLP probes identifying each homoeologous chromosome arm. *Theoretical and Applied Genetics* 78, 342–8.

Shen, D.L., Wang, Z.F. & Wu, M. (1987) Gene mapping on maize pachytene chromosomes by *in situ* hybridization. *Chromosoma* 95, 311–14.

Shewry, P.R., Pratt, H.M., Finch, R.A. & Miflin, B.J. (1987) Genetic analysis of hordein polypeptides from single seeds of barley. *Heredity* 40, 463–6.

Shewry, P.R., Parmar, S., Franklin, J. & Burgess, S.R. (1990) Analysis of a rare recombination event within the multigenic *Hor 2* locus of barley (*Hordeum vulgare* L.). *Genetical Research (Camb.)* 55, 171–6.

Shin, J.S., Chao, S., Corpuz, L. & Blake, T. (1990) A partial map of the barley genome incorporating restriction fragment length polymorphism, polymerase chain reaction, isozyme and morphological marker loci. *Genome* 33, 803–10.

Silva, A.J. & White, R. (1988) Inheritance of allelic blueprints for methylation patterns. *Cell* 54, 145–52.

Snape, J.W. (1988) The detection and estimation of linkage using doubled haploid or single seed descent populations. *Theoretical and Applied Genetics* 76, 125–8.

Snape, J.W., Flavell, R.B., O'Dell, M., Hughes, W.G. & Payne, P.I. (1985) Intrachromosomal mapping of the nucleolar organiser region relative to three marker loci on chromosome 1B of wheat (*Triticum aestivum*). *Theoretical and Applied Genetics* 69, 263–70.

Søgaard, B. & von Wettstein-Knowles, P. (1987) Barley: genes and chromosomes. *Carlsberg Research Communications* 52, 123–96.

Sørensen, M.B. (1989) Mapping of the *Hor2* locus in barley by pulsed field gel electrophoresis. *Carlsberg Research Communications* 54, 109–20.

Stallings, R.L., Torney, D.C., Hildebrand, C.E., Longmire, J.L., Deaven, L.L., Jett, J.H., Doggett, N.A. & Moyzis, R.K. (1990) Physical mapping of human chromosomes by repetitive sequence fingerprinting. *Proceedings of the National Academy of Sciences (USA)* 87, 6218–22.

Strobel, S.A. & Dervan, P.B. (1990) Site-specific cleavage of a yeast chromosome by oligonucleotide-directed triple-helix formation. *Science* 249, 73–5.

Thoday, J.M. (1961) Location of polygenes. *Nature* 191, 368–70.

Van Dilla & 35 others (1989) The National Laboratory gene library project. In Gray, J.W. (ed.), *Flow Cytogenetics*. Academic, London, pp. 257–74.

Wang, M.L., Atkinson, M.D., Chinoy, C.N., Devos, K.M., Harcourt, R.L., Liu, C.J., Rogers, W.J. & Gale, M.D. (1991) RFLP-based genetic map of rye (*Secale cereale*) chromosome 1R. *Theoretical and Applied Genetics* 82, 174–8.

Weller, J.I. (1987) Mapping and analysis of quantitative trait loci in *Lycopersicon* (tomato) with the aid of genetic markers using approximate maximum likelihood methods. *Heredity* 59, 413–21.

Wettstein, D. von, Jende-Strid, B., Ahrenst-Larsen, B. & Sørensen, J.A. (1977) Biochemical mutant in barley renders chemical stabilization of beer superfluous. *Carlsberg Research Communications* 42, 341–51.

Wettstein, D. von, Nilan, R.A., Ahrenst-Larsen, B., Erdal, K., Ingversen, J., Jende-Strid, B., Nyegaard Kristiansen, K., Larsen, J., Outtrup, H. & Ullrich, S.E. (1985) Proanthocyanidin-free barley for brewing: progress in breeding for high yield and research tool in polyphenol chemistry. *MBAA Technical Quarterly* 22, 41–52.

Williams, J.G.K., Kubelik, A.R., Livak, K.J., Rafalski, J.A. & Tingey, S.V. (1990) DNA polymorphisms amplified by arbitrary primers are useful as genetic markers. *Nucleic Acids Research* 18, 6531–5.

Worland, A.J., Law, C.N., Hollins, T.W., Koebner, R.M.D. & Giura, A. (1988) Location of a gene for resistance to eyespot (*Pseudocercosporella herpotrichoides*) on chromosome 7D of bread wheat. *Plant Breeding* 101, 43–51.

Young, N.D. & Tanksley, S.D. (1989) RFLP analysis of the size of chromosomal segments retained around the *Tm-2* locus of tomato during backcross breeding. *Theoretical and Applied Genetics* 77, 353–9.

7

Nuclear Genome Structure and Organization

E. Ananiev

Introduction

The genome of barley (*Hordeum vulgare*) has been estimated to contain around 5.5 picograms (pg) of DNA per haploid nucleus ($2n = 14$), which is equivalent to approximately 5.3×10^9 bp (Bennett & Smith, 1976). Relative chromosome length at metaphase was found not to be a very reliable parameter to identify individual chromosomes, as each accounts for about 14% of the total length (Fukui *et al.*, 1987). This means that each chromosome contains a DNA molecule about 7.5×10^8 bp in size.

Base composition in barley DNA was determined by paper chromatography, the proportion of G + C bases comprising 41% according to Chakrabarti & Subrahmanyam (1985) and 43.1% according to Vanyushin *et al.* (1971). Calculations of the G + C content based on buoyant density measurements in a CsCl gradient (1.702 g/cm^3) and on the melting point (88°C) gave slightly different values, 42.6% and 45.6% respectively (Melzer & Kleinhofs, 1987). Various analyses of plant DNA have indicated a high level of methylation of cytosine residues relative to animal DNA, and in barley 23% of the total cytosine residues as well as 0.2% of the adenine residues are methylated (Vanyushin *et al.*, 1971). Owing to this some restriction enzymes, for example *Pst*I, *Sal*I and *Hpa*II, which are sensitive to methylation of C residues, cleave cereal DNA very poorly (Gruenbaum *et al.*, 1981).

In barley, as in the other cereals, the genome consists of a complex mixture of unique and repeated nucleotide sequences (Flavell, 1980). The proportion of a genome that is classified as repeated or single-copy depends upon the method of determination. The approach based on studying the renaturation kinetics of DNA fragmented into small pieces indicated that at least 70–80% of the barley genome consists of repeated nucleotide sequences (Rimpau *et al.*, 1980; Sivolap *et al.*, 1983). However, this proportion depends strongly upon the stringency of the renaturation conditions and should be taken only as a rough estimate because individual members of families of repeated sequences are frequently not identical. Also, it is not easy to distinguish single-copy nucleotide sequences from small families of sequences consisting of three to four related members.

The repeated sequences of the barley genome can be subdivided into several classes. According to Sivolap *et al.* (1983), about 6% of the genome exists as inverted repeats ranging from 300 to 3000 bp in size, with the majority in the 400–700 bp range. According to Rimpau *et al.* (1980), 10–20% of barley genome is tandemly arranged repeated sequences while 50–60% is repeated sequences which are interspersed among one another or among unique nucleotide sequences.

Barley DNA does not usually form a distinct satellite DNA fraction in a neutral CsCl gradient, but several small satellites can be revealed in a Ag^+/Cs_2SO_4 gradient. One such satellite has been isolated by Dennis *et al.* (1980). It consists of about 2×10^7 copies of a 12 bp sequence, accounting for about 4% of the total genome. This satellite is made up of the simple sequence (GAA)*m* (CAC)*n* which appears to form long uninterrupted tracks which are mainly situated in pericentromeric regions of all seven chromosomes.

Restriction enzymes provide another means of separating and classifying various families of DNA sequences. The parameter used with this technique is the spacing of specific restriction enzyme sites throughout the genome. Even in agarose gels it is possible to count about 30–50 distinct bands in restriction enzyme digests of total DNA, some of which may comprise up to 1% of the barley genome (Fig. 7.1a). For example, the heavy bands in the 9–10 kb region of *Eco*RI digest are 18–26S rRNA repeats, while the heavy bands in 4.8 kb and 2.3 kb regions of *Bam*HI digests belong to the H-family (Ananiev *et al.*, 1988). The appearance of these bands on agarose gels means that they are families of repeated DNA which are similar in size as well as in nucleotide sequences.

The part of the high-molecular-weight nuclear DNA which can be separated from the bulk of restriction fragments obtained after digestion with any restriction enzyme has been named 'relic' DNA. It is especially easy to see such 'relic' DNA after digestion with *Alu*I or *Sau*3A, which cleave most of barley genomic DNA into small fragments (around 500 bp) (Junghans & Metzlaff, 1988; Belostotsky & Ananiev, 1990).

At the same time it was shown that it is possible to observe sharp and rather complex patterns with many bands of restriction fragments in the range of 50–500 bp after polyacrylamide gel electrophoresis of barley DNA digests (Fig. 7.1b). To identify DNA fragments in this range they were end-labelled using T4-DNA polymerase and separated on a sequencing polyacrylamide gel which provides high resolution of DNA fragments with an accuracy of one nucleotide. Such 'direct dactyloscopy' demonstrated that approximately 40% of the total DNA ends obtained after treatment with an enzyme such as *Eco*RI belong to fragments shorter than 500 bp. It is possible to see dozens of such bands on the autoradiograph (Fig. 7.1b) (Kolchinsky *et al.*, 1990a).

The patterns of restriction fragments are very similar for different varieties of barley. However, different species of *Hordeum* produce specific sets of bands and this method can be used for their identification. The differences between barley and wheat are even greater.

The proportion of functional nucleotide sequences which are involved in the developmental processes of the plant from the single zygotic cell up to the male and female gametes has not been determined accurately for flowering plants. However, Kiper *et al.* (1980) presented data based on RNA-excess DNA/RNA hybridization experiments showing that 20 000–30 000 different genes are expressed in etiolated seedlings of barley. These results and studies of other species such as tobacco indicate that plant organs may share a common set of mRNA sequences, consisting of approximately 8000 diverse messages, which encode proteins required by all cells. Altogether at least 70 000 diverse structural genes are expressed collectively in the mature plant and are probably required to programme and maintain the sporophytic phase of the life cycle (Kamalay & Goldberg, 1984). Assuming an

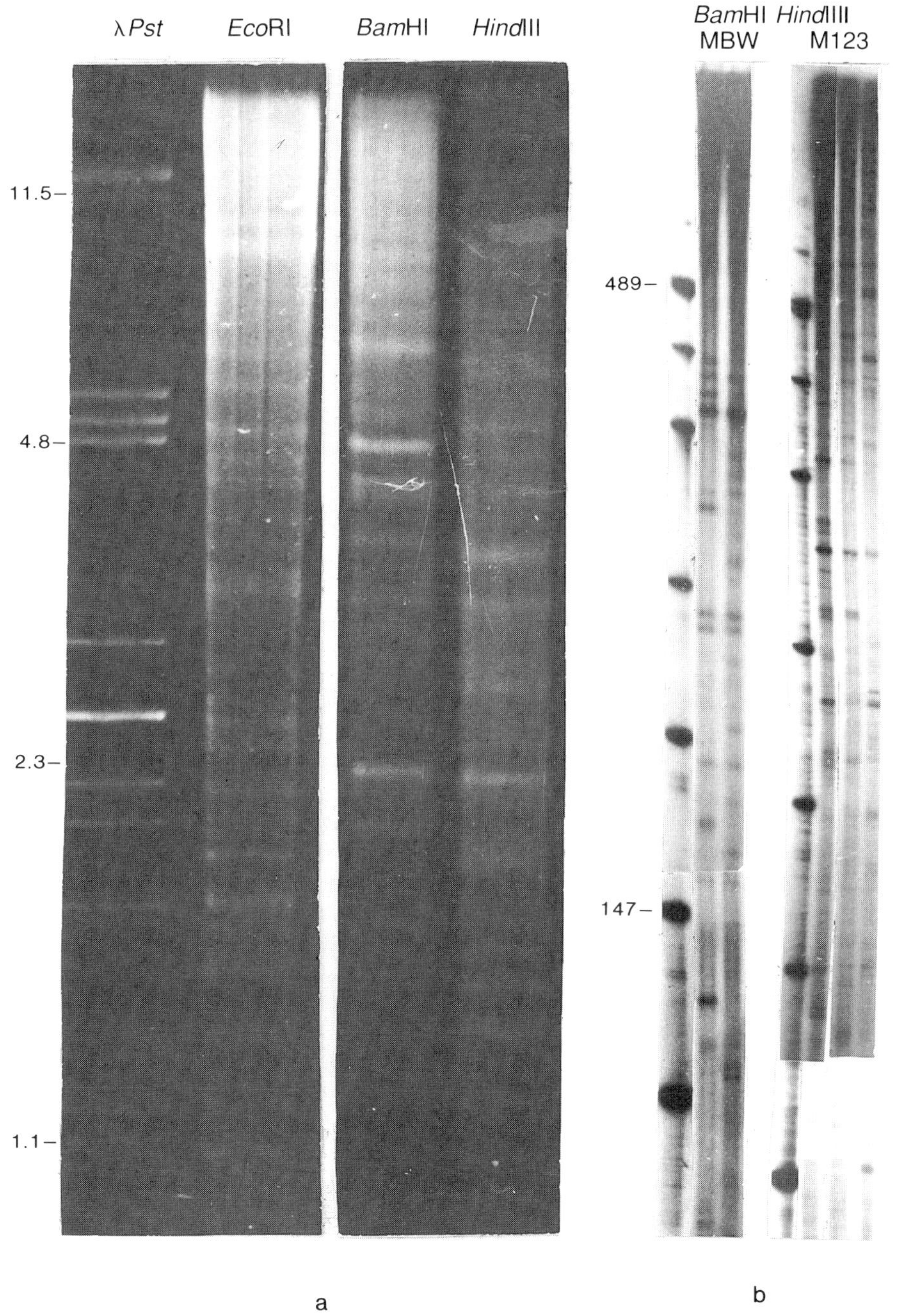

Fig. 7.1. Gel electrophoresis of restriction fragments of barley DNA. (a) Agarose gel with *Pst*I-digested λ-phage DNA as molecular size markers and digests of barley DNA with *Eco*RI, *Bam*HI and *Hin*dIII restriction enzymes. The most prominent bands in the *Eco*RI digests corresponding to the 8.8 and 9.8 kb repeating units of rDNA and in the *Bam*HI digests (2.3 and 4.5 kb) to the H-family repeating unit. (b) Polyacrylamide sequencing gel, M, molecular size markers (489 and 147 bp); B, *Hordeum vulgare*; W, *Triticum aestivum*; 1, *H. leporinum*; 2, *H. jubatum*; 3, *H. murinum*. There are many sharp bands on the sequencing gel which are characteristic for each species of plant.

average mRNA length of 1300 bp, the 30 000–70 000 genes would require about 9×10^7 bp to 21×10^7 bp, or 2–5% of the nuclear barley genome. Recent information on the molecular organization of structural genes is reviewed by von Wettstein-Knowles (1989, this volume) while RFLP mapping of barley chromosomes is reviewed by Blake (1987) and Laurie *et al.* in this volume.

Barley chromosomes in different tissues

There have been several reports of barley karyotypes and idiograms based on the relative length, area density volume (Fukui *et al.*, 1987) and C-banding (Noda & Kasha, 1978; Linde-Laursen, 1981) of metaphase chromosomes from root tips. C-banding allows one to distinguish each chromosome of the barley genome, which consists of meta- or submetacentric chromosomes of approximately similar lengths. Metaphase chromosomes from other tissues have not been studied so thoroughly, but can provide additional information on the structural characteristics of individual chromosomes. For example, the prometaphase chromosomes of triploid endosperm cells reveal quite obvious segments or even chromomeres in heterochromatic pericentromeric regions, as well as in subtelomeric euchromatic regions (Fig. 7.2AB). Electron microscopy revealed loop-like chromatin structures attached to the longitudinal axis of the chromatid (Fig. 7.2C, D) (Pushkina & Barsky, 1989).

Endopolyploidization of chromosomal sets was found in several tissues of barley and especially in the antipodal cells in the embryonic sac (Ivanovskaya, 1968). In the early stages of development of antipodal cells the nuclei contain diffuse chromatin with visible chromomeric structure and well-developed nucleoli (Fig. 7.2E, F) (Pushkina *et al.*, 1989). Between 1 and 4 days after pollination the chromosomes of the antipodal cell begin to shorten and seven separate chromosomes with a specific structure are visible in the nuclei (Fig. 7.2K). They look like bundles of chromatin fibres condensed and attached together in the pericentromeric regions. Telomeric regions (Fig. 7.2G) of the polytenic chromosomes are dispersed in most parts of the nuclei. In some antipodal cells it is possible to observe highly condensed polytenic chromosomes which resemble metaphase chromosomes, some of which may be separated on many condensed chromatids. Cytological preparations of barley polytenic chromosomes were used for *in situ* hybridization of different types of repeated DNA sequences (Fig. 7.2K) (Ananiev *et al.*, 1986a).

The morphology of polytenic chromosomes and the level of polyteny vary between different cultivars. It would be very useful to induce homologous conjugation between sister chromosomes to obtain specific patterns of bands, as in polytenic chromosomes of *D. melanogaster*. At present it is possible to identify only chromosomes 6 and 7 due to the presence of nucleoli in the antipodal cells.

Sequences selected for autonomous replicating activity

The replication of the barley nuclear genome, like that of all large eukaryotic genomes, presents a formidable problem of

Fig. 7.2. Morphology of chromosomes from triploid endosperm cells (A, B, C, D) and antipodal cells (E, F, G, K). (A) Light microscopy of prophase chromosomes. (B) *In situ* hybridization with the 18S–26S rDNA. (C, D) Electron microscopy of early (C) and late (D) prophase chromosomes. Chromatin loops are seen attached to the longitudinal axis of the chromosomes. (E) Endopolyploid nucleus of antipodal cell at early stage containing diffuse chromatin and several chromatin clumps corresponding to the centromeric heterochromatin. (F) Advanced stage of contraction of polytenic chromosomes. (G) A single polytenic chromosome; centromeric regions are compact but telomeric regions are dispersed. Separate chromomeres are seen. (K) *In situ* hybridization of the 18S–26S rDNA with polytenic chromosomes at the stage of terminal condensation. Two clusters of label are seen on chromosomes 6 and 7. Scale 10 μm. Platinum shadowed.

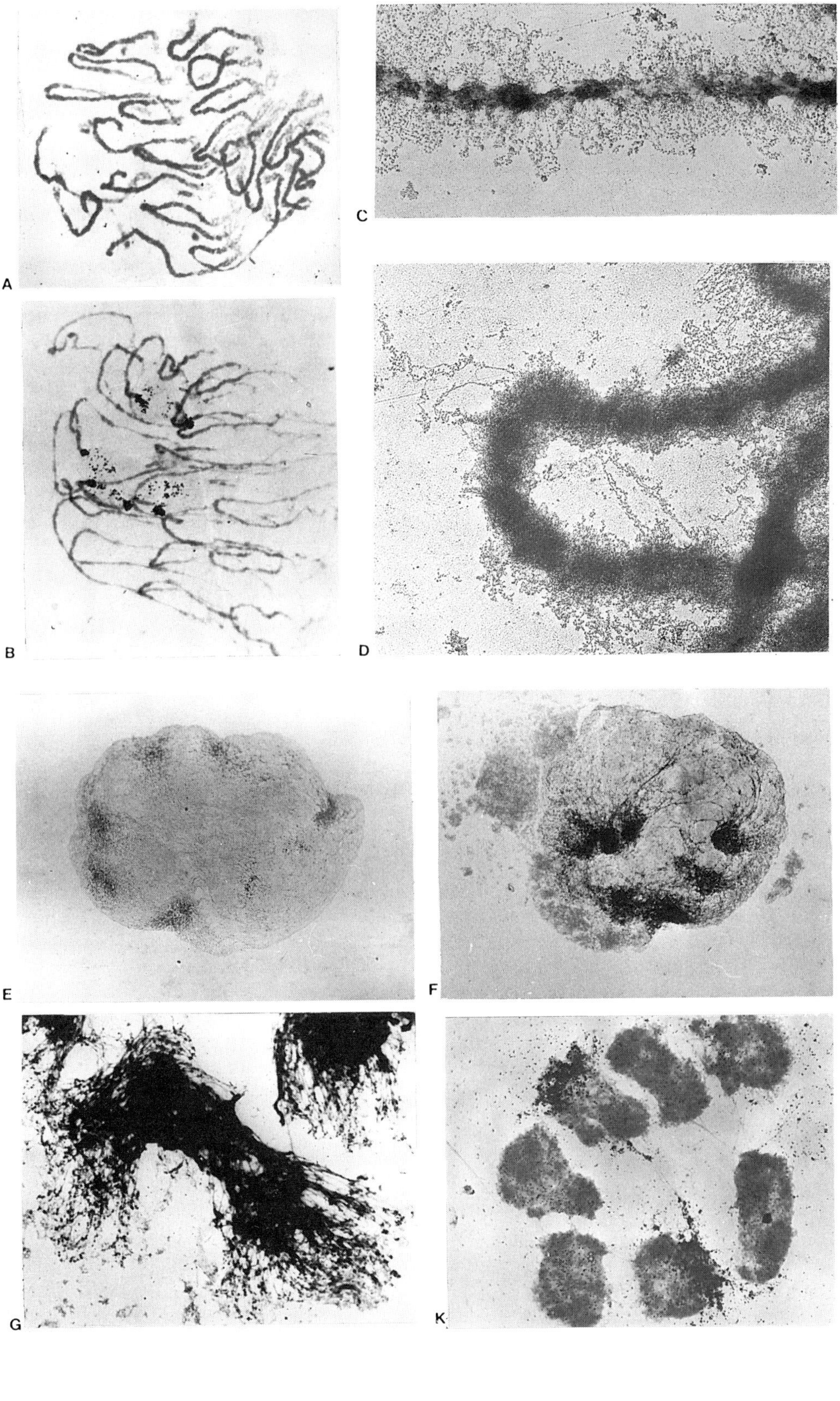
A
B
C
D
E
F
G
K

organization to ensure that all sequences are replicated co-ordinately during the S phase from thousands of dispersed sites of initiation of replication (Hand, 1978).

The replication of chromosomes has been studied in very few plant species. It begins at a very large number of sites, thereby ensuring a somewhat synchronous replication of the whole genome during part of the cell cycle. The number of sites used may vary in different tissues, and sites appear to be grouped into families which are activated at approximately the same time (Hand, 1978; Vant Hof *et al.*, 1978). Barley nuclear DNA sequences have been isolated which may play a role in the initiation of chromosomal replication (Ananiev *et al.*, 1989). These sequences were selected for their ability to promote autonomous replication of chimeric plasmids in yeast. Autonomously replicating sequences (ARS) are defined as DNA sequences which confer the ability to maintain all collinear DNA in an autonomous state in yeast (Stinchcomb *et al.*, 1979). *Bam*HI fragments of barley genomic DNA were cloned into the YIp5 plasmid of *S. cerevisiae* containing the URA3 gene as a selectable marker. Only 5 of 69 clones supported replication of the plasmid in yeast. The average size of the cloned barley DNA fragments was around 5 kb. This means that functional ARSs are about 50–70 kb apart. The frequency of ARSs in the barley genome corresponds broadly to the number of origins calculated for genomic DNA of rye on the basis of fibre autoradiography (Francis & Bennett, 1982). However, in maize one copy of an ARS-related sequence was found in each 12–15 kb fragment of genomic DNA (Berlani *et al.*, 1988). All five barley ARS fragments are moderately repeated nucleotide sequences and are shown by *in situ* hybridization to be dispersed throughout all chromosomes. But the question remains open whether these sequences function as chromosomal origins of replication in the organism from which they were isolated.

Dispersed repeated nucleotide sequences

To isolate individual copies of repeated nucleotide sequences a partial library of genomic *Eco*RI DNA fragments was constructed in plasmid pBR325 (Ananiev *et al.*, 1986b). 104 clones were isolated with inserts of between 1 and 14 kb. Dot hybridization of the cloned DNA with ^{32}P-labelled barley genomic DNA revealed about 20 strong signals which probably corresponded to abundant families of repeated DNA sequences. All other clones gave lower levels of hybridization, 20 clones containing fragments of chloroplast DNA. Southern blot hybridization of the labelled cloned DNA fragments with barley genomic DNA digested with *Eco*RI or *Bam*HI revealed several patterns of signals on the autoradiograms. Most of the clones (Fig. 7.3A,B) gave complicated hybridization patterns with many bands in combination with a strong smear in the region of high-molecular-weight DNA fragments. A limited number of discrete bands was found for only one recombinant clone which contains an *Eco*RI repeat of 18S–26S rDNA and represents a family of tandemly repeated DNA sequences. *In situ* hybridization with metaphase or polytenic chromosomes showed that most clones contained dispersed repeated nucleotide sequences. This indicates that almost all cloned DNA fragments are members of highly heterogeneous families of repeated DNA sequences.

It is well known that the dispersed repetitive DNA sequences of the *Drosophila* genome are species-specific and many correspond to mobile genetic elements (Young, 1979). To determine which of the barley clones were species-specific, labelled DNA of each clone was hybridized with total genomic DNA of barley, wheat, rye and maize (Ananiev *et al.*, 1986c). Three clones showed strong hybridization with barley DNA, but gave very weak signals with wheat, rye and maize DNAs (Fig. 7.3C). Of the clones 35% did not hybridize or gave only very weak signals with maize genomic

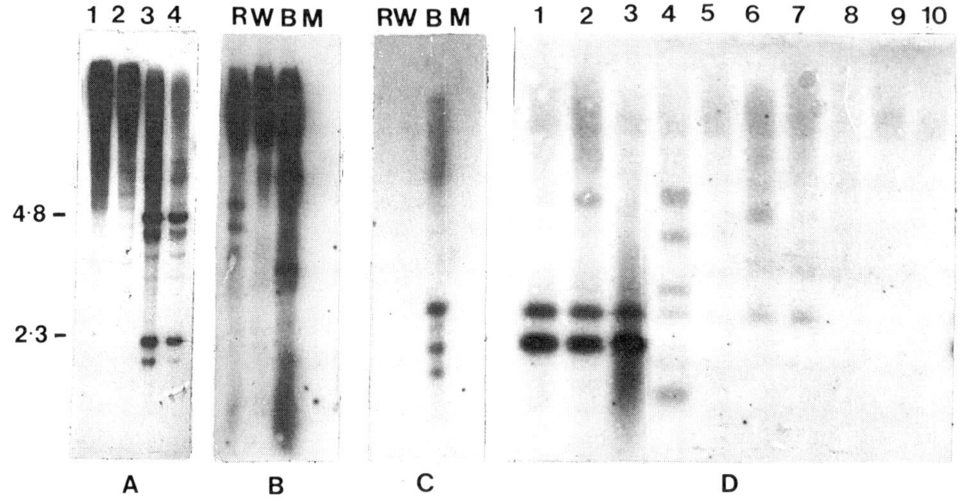

Fig. 7.3. Blot hybridization of cloned DNA fragments with barley genomic DNA. (A) Blot hybridization of H-element with *Eco*RI (1 ,2) and *Bam*HI (3, 4) digests of barley DNA. 1, 3, spring cv. Donetsky 4; 2, 4, winter cv. Oksamyt. Two prominent bands are seen in the *Bam*HI digest, corresponding to the main 4.8 and 2.3 kb bands. (B) Blot hybridization of the H.v. 029 cloned barley DNA fragment with *Eco*RI digests of rye (R), wheat (W), barley (B) and maize (M) DNAs. No cross-hybridization is seen with maize DNA. (C) Blot hybridization of the H.v. 028 cloned barley DNA fragment with genomic DNAs from the same set of species. No cross-hybridization is seen with rye, wheat and maize DNAs. (D) Blot hybridization of Dialect-1 element with *Eco*RI DNA digests of: 1, *Hordeum vulgare*; 2, *H. agricrithon*; 3, *H. spontaneum*; 4, *H. bulbosum*; 5, *H. geniculatum*; 6, *H. leporinum*; 7, *H. murinum*; 8, *H. marinum*; 9, *H. violaceum*; 10, *H. jubatum*. Two main bands are seen in tracks 1, 2 and 3. All other species have less copies of D-1.

DNA (Fig. 7.3B) but cross-hybridized with wheat and rye DNA. Two clones were studied more thoroughly. One of them (H.v. 006) gave a very low level of hybridization with maize genomic DNA. The *Hin*dIII subfragment of H.v. 006 hybridized with the two major bands of 4.8 and 2.3 kb (Fig. 7.3A), visible after staining of a *Bam*HI digest of fractionated genomic DNA with ethidium bromide (Fig. 7.1A).

These two bands are probably the internal parts of the same large repeated element. This suggestion is supported by the fact that the *Hin*dIII DNA fragment may be cut in two parts by *Bam*HI and one of the resulting subfragments hybridizes with the 4.8 kb genomic band while the other one hybridizes with the 2.3 kb band. The DNA sequences homologous to the *Hind*III fragment were called the H-family. However, several additional faint bands corresponding to DNA fragments of different sizes on the genomic blot also hybridized with the *Hin*dIII fragment. This means that H-family consists of a population of copies heterogeneous in size and structure. There are about 50 000 copies of these elements per barley genome as shown by titration of the copy number using blotting hybridization of *Hin*dIII digested genomic DNA and calibrated amounts of *Hin*dIII DNA fragment with the same DNA fragment. Thus it comprises about 3% of barley genomic DNA (Ananiev *et al.*, 1988).

Several clones homologous to the H-element were isolated from a λ-phage genomic DNA library. All clones were heterogeneous in size and had different restriction maps (Ananiev *et al.*, 1988). However, one clone presumably contained a 10 kb full size copy of the H-element. Cross-hybridization of its subfragments revealed that the H-element is flanked in this clone by long direct terminal repeats of about

1 kb. Partial sequencing of these terminal repeats showed that they have 90% homology. Other clones of this family are characterized by a common 1.5 kb *Hin*dIII fragment but have heterogeneous flanks. Two cDNA clones homologous to the terminal repeats of the H-element were isolated from a cDNA library constructed using the poly(A)+ RNA from barley endosperm. Thus H-elements comprise a significant part of the barley genome, are dispersed along the chromosomes and are unstable during the evolution; i.e. they resemble retrovirus-like mobile genetic elements.

The second clone chosen for analysis contained a 1.85 kb insert. This DNA fragment gave specific hybridization only with barley DNA and not with DNA from rye and wheat (Fig. 7.3C). This element was named Dialect-1 (D-1) (Sonina *et al.*, 1989). There are about 5000 copies per barley genome and according to *in situ* hybridization they are dispersed along the chromosomes. Southern blot hybridization of a D-1 element with barley genomic DNA digested with a number of restriction enzymes revealed distinct bands. Besides one or two major bands there were a number of weak bands corresponding to DNA fragments of different sizes. This means that the family consists of a heterogeneous population of copies, and that a full size copy of the D-1 element may be rather large.

Blotting hybridization experiments performed with DNA of different species of barley such as *Hordeum agriocrithon, H. spontaneum, H. bulbosum, H. geniculatum, H. leporinum, H. murinum, H. violaceum* and *H. jubatum* showed drastic variation in copy number as well as in the structure of the element. This indicates that D-1 is evolutionarily unstable and it is therefore a good candidate for a mobile genetic element (Fig. 7.3D).

In spite of huge amounts of repetitive nucleotide sequences in the barley genome it is rather stable and it is difficult to find genes which are unstable and could lead to the isolation of mobile genetic elements. However, taking into consideration the fact that repeated DNA sequences are most likely a result of amplification of different types of mobile genetic elements during evolution, the isolation and characterization of anonymous dispersed repeated DNA sequences may be a way to isolate potential mobile genetic elements.

Organization of the 5S rRNA gene cluster

As in other cereals, 5S rDNA of barley comprises a multigene family, with several thousand copies of 5S rDNA coding sequences per nucleus. They exist as tandemly repeated sequences. However, these conclusions are based on the general similarities of 5S rRNA genes in barley and other cereals and on the results of Southern blot hybridization experiments rather than on precise determinations. The first attempts to investigate 5S rRNA genes in barley were undertaken by Appels *et al.* (1980). They showed considerable heterogeneity in 5S rDNA repeats and localized the barley 5S rRNA genes in the telomeric region of a single non-nucleolar chromosome.

The repeating unit of 5S rDNA was recently isolated (Chvyrleva *et al.*, 1988) by cloning a *Bam*HI digest of barley nuclear DNA (cv. Donetsky 4). The results of sequence analysis show that the repeating unit is 301 bp long (Fig. 7.4) and the 5S rRNA gene-coding segment comprises 120 bp. The coding sequence differs in one position from the sequences reported for rye and wheat 5S rRNAs. At the same time the spacer region of barley 5S rRNA differs dramatically from those of rye and wheat, especially in the central part. The non-coding spacer of the 5S rDNA insert has three characteristic regions. The 3′ and 5′ flanking regions which are adjacent to the structural gene have 60–70% homology with corresponding regions of repeating units of wheat and rye. The middle segment is about 50 bp long and has no homology with corresponding segments in wheat and rye 5S rDNAs. The 3′ flanking region contains an AT tract of 17 bp with blocks of 5T residues which is probably important in ter-

GGATGCGATCATACCAGCACTAAATCACCG
GATCCCATCAGAACTCCGAAGTTAAGCGTG

CTTGGGTGAGAGTAGTACTAGGATGGGTGA
CCTCCTGGGAAGTCCTCGTGTTGCATTCCC

CTTTTTAAATATATTTTTGCGCCACGTGAC
AAGGATGACGCGGGAGCGTGATCTATATGA

CCTCATTTACTTATTTTTGACGATTACTGT
GTGACTTTTCCCACCGCGCTTGACACCCAA

CGACTAGTAGTAGTAGTAGTAGTAGGGGCA
AGCATAAGGAACAAATAGATAGTTGCATGTC

Fig. 7.4. The nucleotide sequence of the short repeating unit of barley 5S rDNA. The coding part is underlined and the TAG repeats with a dotted line.

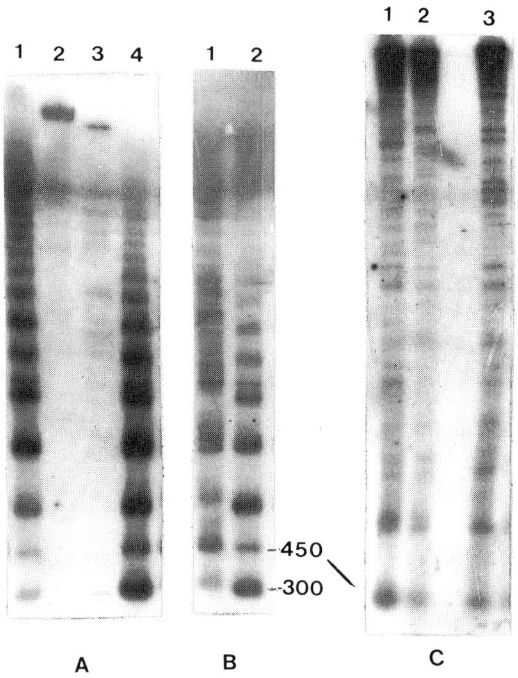

Fig. 7.5. Blot hybridization of 5S rDNA with genomic DNA of barley. (A) *Msp*I (1), *Hpa*II (2), *Hae*III (3) and *Bam*HI (4) digests of barley DNA (cv. Donetsky 4). (B, C) Restriction fragment length polymorphism of barley 5S rDNA. (B) *Bam*HI DNA digests of cv. Oksamyt (1) and Nutans 970 (2). (C) *Alu*I digests of DNA from cvs Violaceum (1), Oderbruker (2) and Donetsky 4 (3).

mination of transcription. Another characteristic tract consisting of 7 TAG triplets is present in the 5′ flanking region and is probably involved in creating length heterogeneity of the short 5S rDNA repeat unit (see later).

Thus it appears that the differences which distinguish these segments of spacers have probably arisen by base changes as well as by complex structural rearrangements during evolution.

The genomic organization of 5S rRNA genes was studied using Southern blot hybridization and polymerase chain reaction (PCR) techniques. A complex ladder of labelled bands was observed on blots after digestion of nuclear barley DNA with restriction enzymes having a recognition site within the 5S rRNA structural gene. The ladder consists of two regularly spaced groups of bands, one of which starts from the monomeric unit corresponding to the cloned short repeat of 5S rRNA gene while the other one starts from the long repeat of 5S RNA, having a size about 450 bp (Fig. 7.5). This means that there are at least two clusters of tandemly arranged 5S RNA genes. The *Bam*HI recognition site partially overlaps with a *Msp*I site. *Msp*I digests of genomic DNA showed ladders having the same periodicity but with different distribution of sensitivity to the restriction enzyme sites within the clusters of 5S rRNA genes.

The ladders most probably resulted from inability of the restriction enzymes to cut some recognition sites with methylated C rather than from base changes. The isoschizomer of *Msp*I is *Hpa*II, which is sensitive to methylation of the internal C in its recognition site and is incapable of cutting total chromosomal DNA as well as the 5S rRNA gene cluster (Fig. 7.5). *Mva*I and *Eco*RII cut the same recognition site (Kolchinsky *et al.*, 1990b) in the coding segment of the 5S rDNA repeat but *Mva*I is insensitive to methylation of C. Nevertheless it produces a ladder with a complex pattern of bands. This means that base substitutions occur even in the coding sequence of 5S rRNA genes. *Alu*I has no recognition sites in the small repeat of the 5S rDNA unit but produces a ladder with the 450 bp monomer which corresponds to the long 5S rDNA repeat. At the same time high-molecular-weight DNA fragments are seen on the blots which contain long stretches of clustered 5S rRNA short repeats. These results indicate that the short and long repeating units are not interspersed but form separate clusters. Comparison of restriction fragment length in different stocks of barley revealed polymorphic variants of 5S rDNA after cutting the genomic DNA with different restriction enzymes. Restriction fragment length polymorphism (RFLP) is evident especially among DNA fragments between 2 and 20 kb in size.

To assign the 5S rRNA genes to a specific barley chromosome the wheat–barley chromosome additional lines obtained by Islam *et al.* (1981) were used as proposed by Appels & Moran (1984). The DNA from the addition lines and from donor plants (wheat cv. Chinese Spring and barley cv. Betzes) were digested with *Mva*I and analysed by Southern blot hybridization after polyacrylamide gel electrophoresis (Kolchinsky *et al.*, 1990b). The result of this experiment is shown in Fig. 7.6A. The restriction enzyme *Mva*I cuts the 5S RNA genes of barley and most of the 5S genes of wheat once per repeat and is not sensitive to methylation. DNAs from different barley cultivars have two major repeats, 300 and 450 bp long (Chvyrleva *et al.*, 1988). Their relative amounts in the cultivars studied varied significantly.

The wheat cv. Chinese Spring gives two sets of bands corresponding to unit repeats of about 410 and 500 bp (Gerlach & Dyer, 1980). This fine heterogeneity of wheat 5S rDNA within the 410 and 500 bp repetitive groups could not be shown earlier by agarose gel electrophoresis.

The picture clearly shows that only addition line 2 contains the 301 and 900 bp bands corresponding to barley 5S genes. The 450 bp band of barley migrates close to wheat bands and it is probably masked.

It has not been possible to produce an addition line containing chromosome 5 of barley. Hence, the possibility cannot be excluded that there is another cluster of

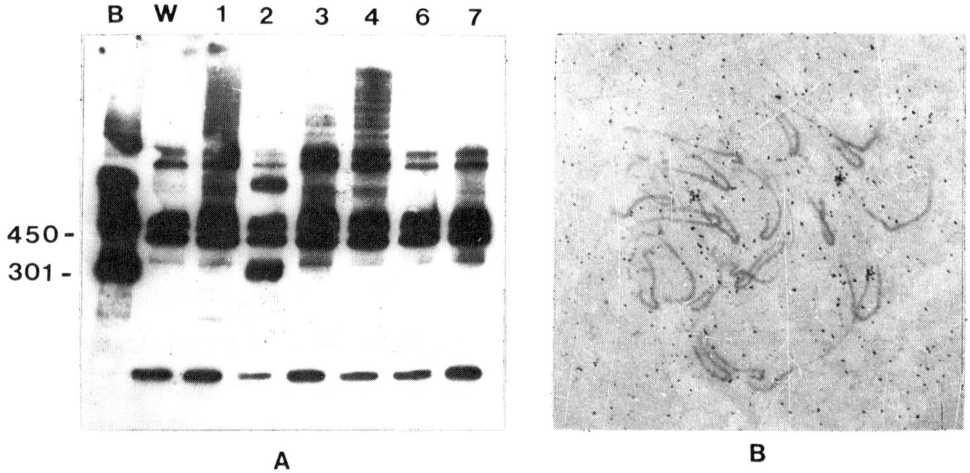

Fig. 7.6. Blot hybridization of 5S rDNA with MvaI digests of DNA from wheat–barley addition lines after polyacrylamide gel electrophoresis: B, barley cv. Betzes; W, wheat cv. Chinese Spring. 1, 2, 3, 4, 6, 7 addition lines. The sizes of some bands are indicated in base pairs. (B) *In situ* hybridization of 5S rDNA with triploid endosperm prophase chromosomes. Three clusters of label are seen on subtelomeric regions of three chromosomes.

5S genes in this chromosome. Moreover, Appels *et al.* (1980) suggested that additional site(s) of 5S genes in cereals exist. To test this possibility, we carried out *in situ* hybridization of the 5S gene with triploid nuclei from barley endosperm (Fig. 7.6B) and haploid nuclei with polytenic chromosomes of antipodal cells from the embryonic sac. The preparations never showed more than one clear-cut cluster of silver grains per haploid genome. Unfortunately, the cytological identification of chromosomes in these preparations is not possible. Nevertheless the blot-hybridization with DNA of wheat–barley addition lines and *in situ* hybridization with chromosomes allow us to conclude that there is only one cluster of 5S rRNA genes in the barley genome, which is present on chromosome 2.

Recently a new technique was introduced to amplify the required segment of the genome with the aid of the polymerase chain reaction (PCR) directed by oligonucleotide primers. This technique has been successfully used to study gene structure as well as genome mapping (Saiki *et al.*, 1985; Mullis & Faloona, 1987; Ochman *et al.*, 1988). The PCR technique was used to reveal the heterogeneity of the non-transcribed spacer (NTS) of the 5S rRNA repeat (Kolchinsky *et al.*, in press). The 'left' primer is a direct sequence of 5S gene on the 5′ end (bases 101–120, see Fig. 7.4) the 'right' primer is a reversed sequence of bases 1–20 complementary to the 3′ end of the 5S structural gene. Thus, theoretically these two primers will amplify the non-transcribed spacers of any higher plant. It should be noted that they will not anneal with chloroplast 5S RNA genes.

Figure 7.7 shows the products of PCR with the 5S primers stained with ethidium bromide. Amplified 5S genes of wheat, barley and maize produce characteristic patterns consistent with available data on the organization of 5S genes in these species (Appels *et al.*, 1980; Mascia *et al.*, 1981; Vakhitov & Gimalov, 1988). Owing to multiplicity of the 5S genes the reaction is very effective; 10 ng of total genomic DNA and 15 cycles of amplification are quite sufficient. To improve the resolution of the technique we used PCR with ^{32}P-end-labelled primers with subsequent electrophoresis in non-denaturating (Fig. 7.7A) or sequencing (Fig. 7.7B) polyacrylamide gels. The short non-transcribed spacer of the 5S repeat of barley produced an unusual ladder of 15–20 bands. This pattern is unique among all the species studied. This short 5S gene repeat contains an unusual stretch consisting of 7 TAG triplets (see earlier). A first explanation of this result is that the resulting picture is an artefact of the PCR technique because the simple repeat in the non-transcribed spacer may cause the 'slippage' of DNA chains during the amplification process (Tautz, 1989).

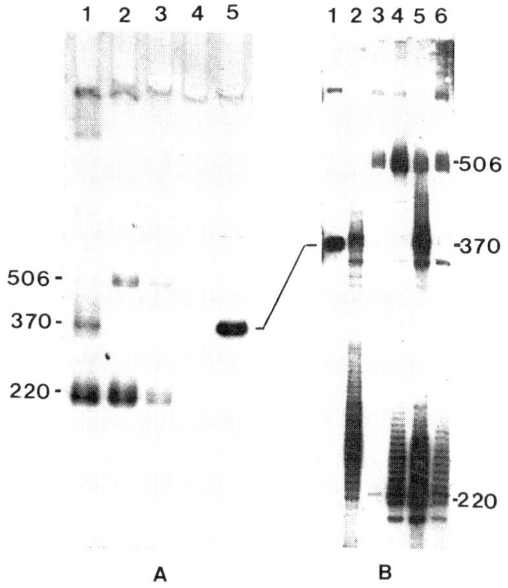

Fig. 7.7. Separation of polymerase chain reaction (PCR) products obtained from the non-transcribed spacer of 5S rDNA of different barley cultivars on normal (A) and sequencing (B) polyacrylamide gels. (A) 1, *Hordeum vulgare* cv. Djao; 2, cv. Hiproly; 3, cv. Nutans 970; 4, cv. Odessky-9; 5, *Hordeum bulbosum*. (B) 1, *Hordeum bulbosum*; 2, *H. vulgare* cv. Odessky-9; 3, cv. Nutans 970; 4, cv Hiproly; 5, cv. Djao. Different products are present in the different barley cultivars. The fraction of PCR products corresponding to the short non-transcribed spacer of rDNA may be resolved as a set of DNA fragments which form a ladder with 3 bp steps on the sequencing gel (tracks 2, 3, 4, 5).

A second explanation is that there is heterogeneity within the cluster of short 5S repeats which provided variable numbers of TAG triplets in spacer regions of different subgroups of repeating units. Additional experiments using Southern blot hybridization of 5S rDNA with genomic DNA after separation of *Bam*HI fragments on a sequencing gel and analysis of the PCR products obtained from plasmid DNA carrying 5S rDNA provided support for the second explanation.

Organization of the 18S–26S rRNA gene cluster

Barley DNA enriched for ribosomal RNA genes was isolated from an actinomycin D–CsCl gradient and used to clone the ribosomal repeating unit. Restriction mapping revealed two size classes of ribosomal gene repeats (8.9 and 9.9 kb) which differed only in the length of their spacer regions. The transcribed regions of these two size classes appear identical (Gerlach & Bedbrook, 1979; Appels *et al.*, 1980).

It was shown that the barley non-transcribed spacer of 18S–26S DNA repeating units is species-specific and may be used for identification of alien genetic material in hybrids (Appels & Moran, 1984). Southern blot hybridization of barley ribosomal DNA spacer sequences with DNA of wheat–barley chromosome addition lines revealed that barley chromosomes 6 and 7 carry ribosomal RNA genes and demonstrate that the two loci are distinguishable by length of the *Taq*I spacer fragment. The long rDNA repeat cluster localized on chromosome 6 and the short repeat cluster on chromosome 7 were designated as *Rrn1* and *Rrn2* respectively. The total number of rDNA repeats has been estimated to be about 2000 copies per haploid genome distributed between the two clusters as 1200 copies on chromosome 6 and 600 copies on chromosome 7 (Appels *et al.*, 1980). Saghai-Maroof *et al.* (1984) found 15 different rDNA spacer length variants among cultivated and wild (*H. spontaneum*) barleys. The variants form a ladder with 115 nucleotide step increases in *Sst*I fragments on Southern blots probed with the cloned wheat rDNA repeat pTA71 (Gerlach & Bedbrook, 1979). The 15 spacer length variants result in 17 different phenotypes with 2–4 variants per phenotype. Chvyrleva *et al.* (1987) found several polymorphic variants of 18S–26S rDNA repeats in 20 different barley varieties. Most of the varieties had a short repeat which was 8.8 kb long while some other varieties had short repeats which were 9.0 and 9.4 kb long. Only one variety (cv. Nutans) had a long repeat unit (9.5 kb) which differed from all others (9.8 kb). Some varieties had additional minor bands of 18S–26S rDNA repeats.

Cloned rDNA repeats have three *Bam*HI recognition sites. When present in tandem array in chromosomal DNA they are cut into three fragments of 1.8, 3.8 and 4.2 kb for the long rDNA repeat and an additional fragment of 3.2 kb corresponding to the short repeat. However, besides these fragments there are six additional bands from 5 to 9.8 kb in size (Fig. 7.8, track 4) which appeared as a result of the inability of *Bam*HI to cut some of the recognition sites. Gerlach & Bedbrook (1979) suggested that these additional bands could not be mutational variants which had lost *Bam*HI recognition sites or spacer length variants as neither they nor other investigators had detected such examples in cloned barley rDNA. They suggested that methylation of C residues in some of *Bam*HI recognition sites is the reason for the additional bands. Data supporting this hypothesis were obtained by Chvyrleva *et al.* (1986). They studied the structure of rDNA in barley callus cells after *in vitro* cultivation for different lengths of time. It was found that DNA isolated from the callus had the same pattern of restriction fragments as in cloned rDNA (Fig. 7.8, tracks 2 and 3). The conclusion was that rDNA became undermethylated in callus cells after long-term cultivation. This experiment was repeated with other restriction enzymes which are sensitive to methylation of C residues. No visible changes were observed in agarose gel

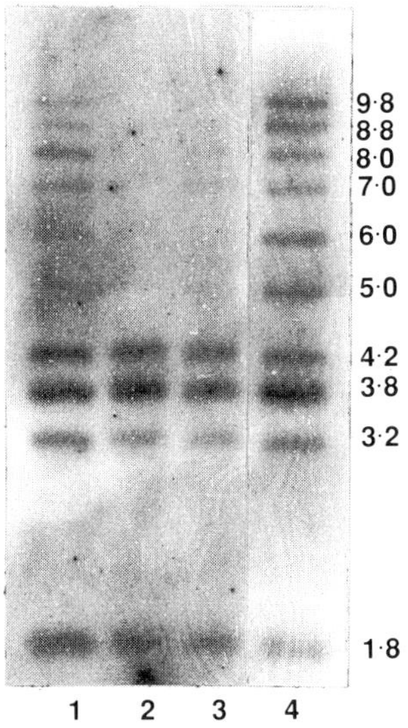

Fig. 7.8. Blot hybridization of 18S–26S rDNA with *Bam*HI digests of DNA from calluses (1, 2, 3) and seedlings (4) of barley (cv. M-121). 1 – 1–2-month-old calluses which are capable of producing regenerants. 2, 3 – 6–12-month-old calluses which cannot be regenerated. Numbers show the size of DNA fragments in kb.

after treating the callus DNA with *Pst*I, *Sal*I or *Hpa*II but Southern blot hybridization showed the appearance of some characteristic *Hpa*II fragments corresponding to rDNA. However, most of the rDNA repeats were not cleaved by this enzyme. Thus only some *Hpa*II recognition sites were undermethylated. To determine whether C residues in other nucleotide sequences were also undermethylated the same experiment was also performed with the H.v 006 repeat, which belongs to the H-family of dispersed repeated nucleotide sequences comprising 50 000 copies per haploid genome (see earlier). No fragmentation of H repeats was observed after treatment of callus DNA with the same restriction enzymes. Thus the level of undermethylation of C residues varies considerably in different families of nucleotide sequences in callus cells and may be related to their function in the nucleus.

The results of these experiments show that the level of methylation of C residues in barley genomic DNA may be changed during *in vitro* cultivation of callus cells and that this process may directly or indirectly induce somaclonal variation (Larkin & Scowcroft, 1981) by changing the specificity of regulatory elements or by activating silent mobile genetic elements which can induce different type of mutations (Peschke *et al.*, 1985).

The stable patterns of restriction fragment variants in plant genomic DNA revealed by some enzymes suggest very stable inheritance of patterns of methylated C residues through many generations.

'Relic' DNA

Bedbrook *et al.* (1980) described 'relic' DNA, which consists of highly repeated DNA lacking recognition sites for the enzyme used. The high-molecular-weight 'relic' DNA fraction can be easily observed after electrophoretic separation of total barley DNA digested with different restriction enzymes (Junghans & Metzlaff, 1988; Belostotsky & Ananiev, in press). The fraction consists of a mixture of long restriction fragments which are most probably built up from tandemly arranged monomeric units. *Alu*I-generated relic DNA was chosen for analysis (Belostotsky & Ananiev, 1990) because this particular fraction can be efficiently separated from the bulk of digested DNA.

It was found that *Alu*I-relic DNA cannot be completely converted into low-molecular-weight fragments by digestion of either total barley DNA or of isolated *Alu*I-relic DNA with restriction enzymes such as *Eco*RI, *Bam*HI, *Bgl*II, *Hin*dIII, *Alu*I, *Sau*3A, *Msp*I, *Bsp*RI, *Taq*I and their pairwise combinations. However, when different restriction enzyme digests of barley DNA were blot-hybridized with *Alu*I-relic DNA as a probe, a set of fragments of low molecular weight was observed as well as the band of relic DNA. At least two different ladders consisting of regularly spaced bands were revealed (e.g. in *Eco*RI and *Sau*3A digests) against rather complex background hybridization. This type of pattern is typical of tandemly organized repetitive sequence families (Horz & Zachau, 1977).

*Alu*I relic DNA was cloned in appropriate vectors after digestion with *Taq*I or DNase I. After cross-hybridization the clones were separated into three groups of related nucleotide sequences. When a member of the first group, namely the HvRII insert, was hybridized to a Southern blot of different restriction digests of barley DNA, it revealed only the band of relic DNA in all digests. The most probable explanation of this hybridization pattern is that the HvRII family consists of fragments of barley cryptic satellite DNA with simple sequence such as (GAA)*m*(GAC)*n*. This was confirmed by partial sequencing of the HvR05 clone insert, which is also a member of the HvRII family. This cryptic satellite DNA has been previously investigated in detail (Dennis *et al.*, 1980).

Three additional clones which were selected turned out to be members of another family which was named HvRT family. Nucleotide sequence analysis of these clones demonstrated that the HvRT family consists of tandemly repeated units 118 bp long. Computer comparison of these sequences revealed some variation between the monomers due to point mutations, insertions and deletions. However, their average homology to the consensus is high (94%). Similar divergence values were observed for tandemly repeated sequence families of *Arabidopsis thaliana* (Martinez-Zapater *et al.*, 1986), *Allium cepa* (Barnes *et al.*, 1985), *Brassica oleracea* (Benslimane *et al.*, 1986) and *Zea mays* (Dennis & Peacock, 1984). A search for nucleotide homology with known repeated DNA sequences of different origin, including recently published plant satellite DNA sequences and those contained in GENBANK (release 54), did not show any significant homologies.

Southern blot hybridization of the HvRT probe with different restriction enzyme digests of barley DNA revealed typical ladder patterns with *Hin*fI, *Alu*I, *Sau*3A, *Bsp*RI and *Taq*I digests, indicating tandem arrangement of the HvRT family. In particular, *Bsp*RI totally cleaved this family to mono-, di-, tri- and tetramers (Type A pattern according to Horz & Zachau, 1977), while four other enzymes produced relic DNA and ladders starting from the dimer. *Bam*HI produced a series of larger oligomers corresponding to 5,6,7,9 and 11 multimers. Finally, *Eco*RI, *Hin*dIII and *Bgl*II sites are rare in HvRT clusters, since they produce no ladders. Estimation of the copy number of the HvRT monomer in the barley genome was performed by quantitative dot hybridization and found to be approximately 7×10^5. Thus the HvRT family comprises about 1.7% of barley DNA, corresponding to a total length of HvRT clusters of about 90 000 kbp.

Blot hybridization of the HvRT probe with total barley DNA digested with different restriction enzymes and their methylation-sensitive isoschizomers allowed the following conclusions to be drawn. First, comparison of *Msp*I with *Hpa*II patterns, as well as *Mva*I with *Eco*RII and *Atu*BI patterns, shows that cytosine residues are highly methylated in CG, CAG and CTG sites in HvRT clusters. At the same time, external cytosines in CCG triplets are methylated to a lesser extent, since *Msp*I produces a characteristic ladder rather than a high-molecular-weight (relic) band. Comparison of *Mbo*I and *Sau*3A patterns shows that all bands larger than pentamers in the *Sau*3A pattern are due to methylation of the 3′-cytosine in the GATC sites. Moreover, *Sau*3A patterns of the two barley cultivars (Donetsky 4 & Odessky 9) differ in the presence of 'amplified' bands of different size (Fig. 7.9).

The distribution of the HvRT family in the genus *Hordeum* was studied by Southern hybridization to *Bam*HI-digested DNAs of several barley species under different stringency conditions (3 × SSC and 30, 40 or 50% formamide at 48°C). HvRT homologous families were detected in the genomes of *H. violaceum, H. murinum, H. jubatum* and *H. leporinum*, but not in the genomes of *H. marinum, H. geniculatum* and wheat. This finding is in accordance with recent phylogenetic studies

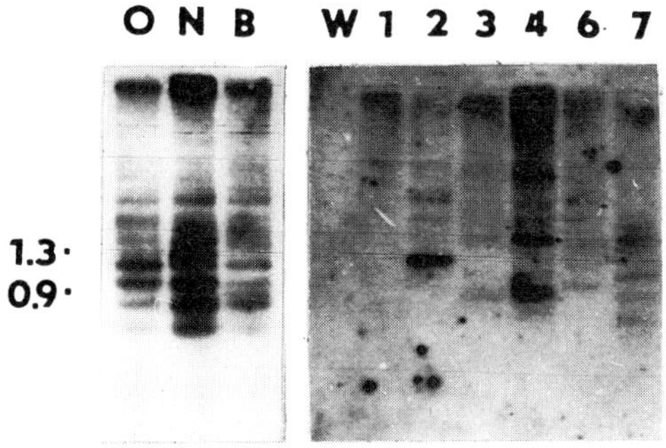

Fig. 7.9. Organization of HvRT sequences in individual barley chromosomes. Southern blot hybridization of *Bam*HI digested DNA from barley, wheat and wheat–barley chromosome addition lines (AD); O, barley cv. Odessky-9; N, barley cv. Nutans 45; B, barley cv. Betzes; W, wheat cv. Chinese spring; 1, 2, 3, 4, 6, 7 addition lines.

(Jørgensen, 1986). In addition, restriction fragment length polymorphism can be observed between species, which is more pronounced in the high-molecular-weight range. Variation in the hybridization stringency did not result in qualitative changes in hybridization patterns (not shown). This suggests that the homology of the HvRT-related families from different barley species is relatively high. High copy number together with apparent species specificity indicate that amplification of the HvRT ancestor has occurred relatively recently in evolution, after the divergence between *Hordeum* and *Triticum* and separation of the *marinum* group of *Hordeum*.

To investigate the localization of HvRT family sequences on barley chromosomes, we used Southern blot hybridization of *Bam*HI digested DNAs from wheat/barley addition lines (Islam *et al.*, 1981). Barley chromosomes 2,3,6 and 7 appeared to possess chromosome-specific hybridization patterns, chromosomes 3 and 4 showing similar patterns (Fig. 7.9). Each individual band was also present in the digest of barley cv. Betzes, which was used as a source of barley chromosomes for the production of the addition lines. The conclusion from this experiment is that barley chromosomes 2,3,6 and 7 bear chromosome-specific HvRT clusters (Belostosky & Ananiev, 1990).

To estimate the upper size limit of HvRT clusters, pulsed-field gel electrophoresis (PFGE) (Schwartz & Cantor, 1984) was employed. This technique allowed the fraction appearing as a high-molecular-weight relic DNA upon conventional gel electrophoresis to be resolved into a number of bands. In particular, the largest *Hin*dIII fragment was approximately 810 kp in length.

It can be inferred from the results of restriction analysis and molecular cloning of the *Alu*I relic DNA that the fraction is heterogeneous and contains cryptic satellite DNA, the HvRT family of tandemly arranged repeated sequences, another family with a similar organization pattern but with longer periodicity and, as has been shown earlier, the family of tandemly arranged genes coding for short unit of 5S RNA (Chvyrleva *et al.*, 1988).

Junghaus & Metzlaf (1988) isolated four barley-specific nucleotide sequences from *Eco*RI relic DNA. The repeats differ from each other in their length, sequence and redundancy. Southern blot hybridization showed that the repeats belong to different families which differ from HvRT and HvRII families (Belostotsky *et al.*, 1990). The possibility of using these clones as probes for simple and rapid genome analysis was demonstrated in squash dot experiments.

Conclusions

The systematic accumulation of data about different families of nucleotide sequences is only just beginning and it is not clear how much time and effort will be required to obtain knowledge of their structure and function in the barley genome. At present it is only possible to conclude that most of the repeated sequences are dispersed throughout the genome and interspersed with other repeats, only a few being organized in tandem arrays in specific regions of chromosomes. The factors which are responsible for the structural organization of the barley genome are not known, but the mechanisms are probably the same as in the other eukaryotic organisms: duplications of nucleotide sequences, and activity of mobile genetic elements which remains to be discovered. Theoretically the dispersed repeated nucleotide sequences might serve as sources of genetic polymorphisms, but efficient ways to use them have not yet been found. In addition there is no information about interactions of this 'silent' part of barley genome and the structural genes which are embedded in it.

References

Ananiev, E.V., Yakovleva, E.Yu., Chernyshov, A.I. & Chvyrleva, Ts.D. (1986a) Usage of plant polytenic chromosomes for localization of cloned genes. *Doklady Akademii Nauk SSSR* (Russian) 286, 982–4.

Ananiev, E.V., Bochkanov, S.S., Ryzik, M.V., Sonina, N.V., Chernyshov, A.I., Shchipkova, N.I. & Yakovleva, E.Yu. (1986b) Characterization of the cloned repeated DNA sequences from the barley genome. *Genetika* (Russian) 22, 920–8.

Ananiev, E.V., Bochkanov, S.S., Ryzik, M.V., Chernyshov, A.I., Sonina, N.V. & Shchipkova, N.I. (1986c) Cross-hybridization of cloned repeated sequences of barley genome with DNA from wheat, rye and maize. *Genetika* (Russian) 22, 2021–4.

Ananiev, E.V., Ryzik, M.V., Sonina, N.V., Golovkin, M.V. & Shchipkova, N.I. (1988) Mobile genetic elements of cereal genomes. *Biochemie und Physiologie der Pflanzen* 182, 129–41.

Ananiev, E.V., Shchipkova, N.I., Chvyrleva, Ts.D. & Danilevskaya, O.N. (1989) Barley DNA sequences selected for autonomous replicating activity in yeast. *Molekulyarnaya Genetika, Mikrobiologiya i Viroligiya* (Russian) 2, 46–9.

Appels, R. & Moran, L.B. (1984) Molecular analysis of alien chromatin introduced into wheat. In: *Gene Manipulation in Plant Improvement* (16th Stadler Genetic Symposium), New York, pp. 529–58.

Appels, R., Gerlach, E.S., Dennis, H., Swift, H. & Peacock, W.I. (1980) Molecular and chromosomal organization of DNA sequences coding for the ribosomal RNAs in cereals. *Chromosoma* 78, 293–311.

Barnes, S.M., James, A.M. & Jamieson, G. (1985) The organization, nucleotide sequence and chromosomal distribution of a satellite DNA from *Allium cepa. Chromosoma* 92, 185–92.

Bedbrook, J.R., Jones, J., O'Dell, M., Thompson, R.D. & Flavell, R.B. (1980) A molecular description of telomeric heterochromatin in Secale species. *Cell* 19, 545–60.

Belostotsky, D.A. & Ananiev, E.V. (1990) Characterization of relic DNA from barley genome. *Theoretical and Applied Genetics* 80, 374–80.

Bennett, M.D. & Smith, L.B. (1976) Nuclear DNA amounts in angiosperms. *Philosophical Transactions of the Royal Society (London) B. Biological Sciences* 274, 227–74.

Benslimane, A.A., Dron, M., Hartmann, C. & Rode, A. (1986) Small tandemly repeated DNA sequences of higher plants likely originate from a tRNA gene ancestor. *Nucleic Acids Research* 14, 8111–19.

Berlani, R.E., Davis, R.W. & Walbot, V. (1988) Genomic organization of two families of highly repeated nuclear DNA sequences of maize selected for autonomous replicating activity in yeast. *Plant Molecular Biology* 11, 161–72.

Blake, T.K. (1987) Strategies for restriction fragment length polymorphism analysis in barley. In: *Barley Genetics V* (Okayama, Japan), pp. 503–8.

Chakrabarti, T. & Subrahmanyam, G. (1985) Analysis of DNA from related and diverse species of barley. *Plant Science* 42, 183–90.

Chvyrleva, Ts.D., Ryzik, M.V., Ananiev, E.V. & Gaponenko, A.K. (1986) Undermethylation of rDNA repeats in callus tissues of barley cultured in vitro. *Doklady Akademii Nauk SSSR* (Russian) 290, 1249–52.

Chvyrleva, Ts.D., Chernyshov, A.I., Bochkanov, S.S., Yakovleva, E.Yu. & Ananiev E.V. (1987) Length polymorphism and organization of 18S–26S rRNA repeats in barley. *Genetika* (Russian) 23, 717–24.

Chvyrleva, Ts.D., Gazumyan, A.K. & Ananiev, E.V. (1988) Organization of the primary nucleotide sequence of the 5S rRNA genes in barley (Hordeum vulgare). *Genetika* (Russian) 24, 1830–40.

Dennis, E.S. & Peacock, W.J. (1984) Knob heterochromatin homology in maize and its relatives. *Journal of Molecular Evolution* 20, 341–50.

Dennis, E.S., Gerlach, W.L. & Peacock, W.J. (1980) Identical polypirimidine–polypurine satellite DNAs in wheat and barley. *Heredity* 44, 349–66.

Flavell, R. (1980) The molecular characterization and organization of plant chromosomal DNA sequences. *Annual Review of Plant Physiology* 31, 569–96.

Francis, D. & Bennett, M.D. (1982) Replicon size and mean rate of DNA synthesis in rye (Secale cereale L. cv Petkins Spring). *Chromosoma* 86, 115–21.

Fukui, K., Kakeda, K. & Matsuno, T. (1987) Karyotypes of barley chromosomes constructed by several parameters. In: Yasuda, S. & Konishi, T. (eds), *Barley Genetics V* (Okayama, Japan), pp. 415–22.

Gerlach, W.L. & Bedbrook, J.R. (1979) Cloning and characterization of ribosomal RNA genes from wheat and barley. *Nucleic Acids Research* 7, 1869–85.

Gerlach, V.L. & Dyer, T.A. (1980) Sequence organization of the repeat units in the nucleus of wheat which contain 5S rRNA genes. *Nucleic Acids Research* 8, 4851–65.

Gruenbaum, J., Naveh-Many, T., Cedar, H. & Razin, A. (1981) Sequence specificity of methylation in higher plant DNA. *Nature* 292, 860–2.

Hand, R. (1978) Eukaryotic DNA. Organization of the genome for replication. *Cell* 15, 137–325.

Horz, W. & Zachau, H.G. (1977) Characterization of distinct segments in mouse satellite DNA by restriction nucleases. *European Journal of Biochemistry* 73, 383–92.

Islam, A.K.M.R., Shepherd, R. & Sparrow, D.H.B. (1981) Isolation and characterization of euplasmic wheat–barley chromosome addition lines. *Heredity* 46, 161–74.

Ivanovskaya, E.V. (1968) Endopolyploidization in tissues of cereals. *Izvestiya Akademii Nauk SSSR* (Russian) 4, 507–16.

Jorgensen, R.B. (1986) Relationships in the barley genus (Hordeum): an electrophoretic examination of proteins. *Hereditas* 104, 273–91.

Junghans, H. & Metzlaff, M. (1988) Genome specific, highly repeated sequences of Hordeum vulgare: cloning, sequencing and squash dot test. *Theoretical and Applied Genetics* 76, 728–32.

Kamalay, J.C. & Goldberg, R.B. (1984) Organ-specific nuclear RNAs in tobacco. *Proceedings of the National Academy of Sciences (USA)* 81, 2801–5.

Kiper, M., Bolte, M. & Herzfeld, F. (1980) Reiteration frequency of genes coding for abundant and rare messenger RNA in greened barley seedlings. *Heredity* 45, 233–43.

Kolchinsky, A.M., Trifonova, I.M., Gazumyan, A.K. & Ananiev, E.V. (1990a) The use of end-labelled digests of genomic DNA for 'direct dactiloscopy'. *Abstracts of the Ist International Conference on DNA Fingerprinting*, Bern, 21.

Kolchinsky, A.M., Kanazin, V.I., Yakovleva, E.Yu., Gazumyan, A.K., Cole C. & Ananiev, E.V. (1990b) 5S-RNA genes of barley are located on the second chromosome. *Theoretical and Applied Genetics* 80, 333–6.

Kolchinsky, A.M., Kolesnikova, M.A. & Ananiev, E.V. Analysis of 5S rRNA gene cluster by PCR technique (in preparation).

Larkin, P.I. & Scowcroft, W.R. (1981) Somaclonal variation — a novel source of variability from cell culture for plant improvement. *Theoretical and Applied Genetics* 60, 197–206.

Linde-Laursen, I. (1981) Giemsa banding patterns of the chromosomes of cultivated and wild barleys. In: *Barley Genetics* (Proc. 4th International Barley Genetics Symposium), Edinburgh University Press, Edinburgh, pp. 786–95.

Martinez-Zapater, J.M., Estelle, M.A. & Somerville, C.R. (1986) A highly repeated DNA sequence in *Arabidopsis thaliana*. *Molecular and General Genetics* 204, 417–23.

Mascia, P.N., Rubenstein, I., Phillips, R.L., Wang, A.S. & Lu Zhen Xiang (1981) Localization of the 5S rRNA genes and evidence for diversity in the 5S rDNA region of maize. *Gene* 15, 7–20.

Melzer, I.M. & Kleinhofs, A. (1987) Molecular genetics of barley. In: Yasuda, S. & Konishi, T. (eds), *Barley Genetics V* (Okayama, Japan), pp. 481–91.

Mullis, K.B. & Faloona, K.B. (1987) Specific synthesis of DNA in vitro via a polymerase catalysed chain reaction. *Methods in Enzymology* 155, 335–50.

Noda, K. & Kasha, K.I. (1978) A proposed barley caryotype revision based on C-banding chromosome identification. *Crop Science* 18, 925–30.

Ochman, H., Gerber, A.S. & Hartl, D.L. (1988) Genetic application of an inverse polymerase chain reaction. *Genetics* 120, 621–3.

Peschke, V.M., Phillips, R.L. & Gengenbach, B.G. (1985) Discovery of Ac activity among progeny of regenerated maize plants. *Maize*

Genetic Cooperation Newsletter 59, 91–3.

Pushkina, N.N. & Barsky, V.E. (1989) Electron microscopy of chromosomes from barley endosperm. *Tsitologiya* (Russian) 1180–5.

Pushkina, N.N., Ananiev, E.V., Barsky, V.E. & Yakovleva, E.Yu (1989) Light and electron microscopic investigation of the structure and regularities of formation of polytene chromosomes in antipodal cells of Hordeum vulgare. *Tsitologiya* (Russian) 31, 1029–34.

Rimpau, I., Smith, D.B. & Flavell, R.B. (1980) Sequence organization in barley and oats chromosomes revealed by interspecies DNA/DNA hybridization. *Heredity* 44, 131–49.

Saghai-Maroof, M.A., Solima, K.M., Jorgensen, R.A. & Allard, R.W. (1984) Ribosomal DNA spacer-length polymorphisms in barley: Mendelian inheritance, chromosomal location and population dynamics. *Proceedings of the National Academy of Sciences (USA)* 81, 8014–18.

Saiki, R.K., Sharf, F., Faloona, K.B., Mullis, G.T., Horn, H.A. & Arnheim, N.A. (1985) Enzymatic amplification of beta-globin genomic sequences and restriction site analysis for diagnostics of sickle cell anemia. *Science* 230, 1350–4.

Schwartz, D.C. & Cantor, C.R. (1984) Separation of yeast chromosome-sized DNAs by pulsed field gel electrophoresis. *Cell* 37, 67–75.

Sivolap, Y.M., Verbitskaya, T.G. & Gornostayeva, L.N. (1983) Size and distribution of repeated nucleotide sequences in barley genome. *Cytologiya i Genetika (Kiev)* (Russian) 17, 28–34.

Sonina, N.V., Lushnikova, A.A., Tikhonov, A. P. & Ananiev, E.V. (1989) Dialect-I, species-specific repeated DNA sequence from barley (Hordeum vulgare). *Theoretical and Applied Genetics* 78, 589–93.

Stinchcomb, D.T., Struhl, K., Davis, R.W. (1979) Isolation and characterization of a yeast chromosomal replicator. *Nature* 282, 39–41.

Tautz, D. (1989) Hypervariability of single sequences as a general source for polymorphic DNA markers. *Nucleic Acids Research* 17, 6463–71.

Vakhitov, V.A. & Gimalov, F.R. (1988) Polymorphism of the length of 5S RNA genes in genomes of different wheat and Aegilops species. *Genetika* (Russian) 24, 1850–6.

Vant Hof, J., Kuniyuki, A. & Bjerknes, C.A. (1978) The size and number of replication families of chromosomal DNA of Arabidopsis thaliana. *Chromosoma* 68, 269–285.

Vanyushin, B.F., Kadyrova, D.Kh., Karimov, Kh.Kh. & Belozersky, A.N. (1971) Minor bases in DNA of higher plants. *Biokhimiya* (Russian) 36, 1251–8.

Wettstein-Knowles, P. von (1989) Facets of the barley genome. Science for Plant breeding. *Proceedings of the XII International Congress of EUCARPIA*. Elsevier Scientific Publishers, Berlin, 16, 107–24.

Young, M.W. (1979) A fluid component of the Drosophila genome. *Proceedings of the National Academy of Sciences (USA)* 76, 6274–8.

III

Analysis of Metabolism and Development

8

Molecular Analysis of Barley Chloroplast Proteins

Jens Sigurd Okkels

Introduction

The chloroplast carries out a number of processes essential for the survival of the plant; some of these are listed in Table 8.1. The main characteristic of the chloroplast is its ability to convert light energy to chemical energy by photosynthesis. Nearly all living organisms sustain life directly or indirectly from this energy transformation. Photosynthesis also contributes to the ecological balance of the Earth by recycling CO_2 respired by non-photosynthetic organisms and H_2O into carbohydrates and O_2.

The chloroplast is bounded by a double membrane (outer and inner envelope) which separates: (i) the internal lamellar system (the thylakoid membranes consisting of grana stacks and stroma lamellae); (ii) the stroma (soluble matrix of the chloroplast); and (iii) the lumen (the internal space of the thylakoid membranes) from the cytoplasm. An electron micrograph of a barley chloroplast showing some of the different compartments and membrane structures is presented in Fig. 8.1.

Barley is a major agricultural crop plant. In addition it serves as a good model plant for molecular analysis of the chloroplast. A large number of chloroplasts can conveniently be obtained shortly after sowing (e.g. from 5-day-old seedlings). The relatively large carbohydrate and protein reserves of the barley grain permit germination and growth in the dark for more than a week. Etioplasts as well as RNA can therefore be obtained from the dark-grown seedlings (Baumgartner *et al.*, 1989). The etiolated plant material is ideal for light-induction studies (Klein *et al.*, 1988; Okkels *et al.*, 1990). Mutants in chloroplast proteins have been isolated from barley (e.g. Simpson & von Wettstein, 1980; Henningsen & Stummann, 1982). Some are spontaneous mutants whereas others have been generated by mutagen treatments like X-ray irradiation or treatment with chemical compounds. The growth of homozygous recessive lethal barley mutants is also supported by the seed reserves. Leaf growth in monocots such as barley occurs primarily by cell division in the leaf basal meristem and elongation of the cells near the leaf base. Consequently, a single barley leaf contains

Table 8.1. Processes that are located in the chloroplast.

Process	Localization
Photosynthesis	
Photosynthesis light reactions	Thylakoid membrane
PS II, cyt *b6–f*, PS I, ATP synthase	
Photosynthesis dark reactions	Stroma
Rubisco and Calvin cycles	
Biosynthetic pathways (complete or partial)	
Chlorophyll synthesis	Stroma
Fatty acid synthesis	Stroma
Galactolipid synthesis	Stroma/envelope
Isoprenoid synthesis	Stroma/envelope
Amino acid synthesis	Stroma
Nitrite reduction	Stroma
Sulphate reduction	Stroma
Betaine synthesis	Stroma
Starch synthesis	Stroma
Transport	
Protein uptake	Envelope
Active transport of metabolites	Envelope
Chloroplast protein synthesis and reproduction	
Transcription and translation	Stroma/stroma membrane
Replication of the chloroplast genome	Stroma

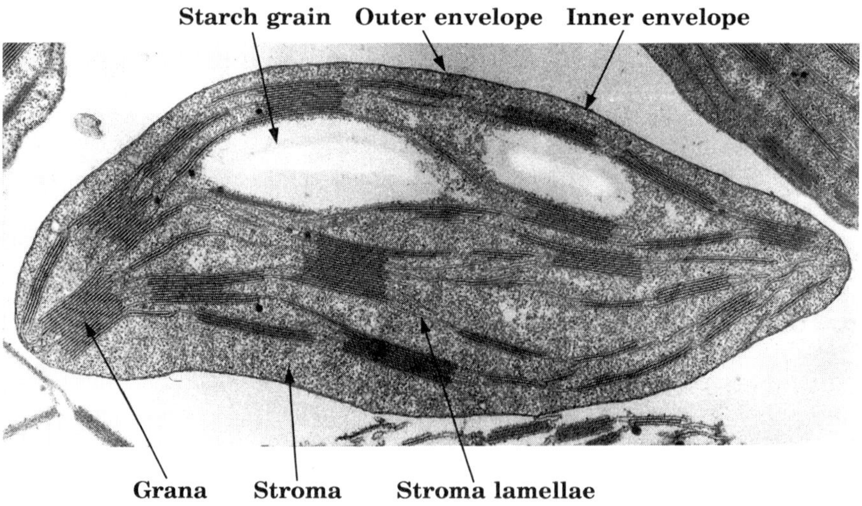

Fig. 8.1. Electron micrograph of a barley chloroplast. The different membrane systems and compartments are indicated with arrows.

proplastids in the meristematic cells at the leaf base and progressively older mature chloroplasts towards the leaf tip (Robertson & Laetsch, 1974; Baumgartner *et al.*, 1989), providing easily accessible material for studies of chloroplast development (e.g. Klein & Mullet, 1987). Molecular analyses of chloroplast proteins in barley are limited by the lack of a reliable transformation system in barley (see Chapter 26 this volume).

The complete sequencing of the chloroplast genomes of *Marchantia polymorpha* (Ohyama *et al.*, 1986), tobacco (Shinozaki *et al.*, 1986a, b) and rice (Hiratsuka *et al.*, 1989) has revealed that the chloroplast genome encodes approximately 130 different proteins and that the proteins from the various species are related. The genes located on the chloroplast genome mainly encode proteins involved in photosynthesis, besides those required for chloroplast gene expression and chloroplast genome replication. High levels of plastid gene expression may be required only during chloroplast biogenesis and not in non-photosynthetic plastids such as proplastids, amyloplasts or chromoplasts (Mullet, 1988). This possibility is consistent with the lack of plastid genes involved in most of the plastid-localized biosynthetic pathways. Enzymes for these biosynthetic pathways are directed to the plastids but probably without the need for concomitant build-up of chloroplast transcriptional and translational capacity.

The majority of the chloroplast proteins are encoded by the nuclear genome. The nuclear-encoded proteins all have a transit peptide which directs the cytoplasmatically synthesized proteins to its correct location in the chloroplast. In one or several steps the transit peptide is cleaved off providing the mature protein (e.g. Smeekens *et al.*, 1990). Experiments with protein uptake of isolated etioplasts or chloroplasts using *in vitro* translated proteins are common for studies of protein import, protein integration and assembly of multicomponent complexes (e.g. Chitnis *et al.*, 1988a; Hageman *et al.*, 1990).

In this chapter recent developments in our understanding of barley chloroplast proteins from the application of molecular biology are summarized, with emphasis on photosynthesis and especially on photosystem I.

Photosynthesis

Energy-rich NADPH and ATP molecules are required for the reduction of CO_2 to carbohydrates by the biochemical pathway, known as the Calvin cycle. The generation of NADPH and ATP within the chloroplast involves the co-operative interaction of four distinct multisubunit complexes, photosystem II (PS II), cytochrome (cyt) *b6–f*, photosystem I (PS I), and ATP synthase. The first three complexes catalyse the conversion of light energy to chemical energy, resulting in the photo-oxidation of water, the photoreduction of $NADP^+$, and the generation of a proton gradient which is utilized by the fourth complex, the ATP synthase, to produce ATP. In all four complexes, the reactions involve movement of electrons or protons across the thylakoid membrane. Therefore, all the complexes have substantial parts of their structures buried in the thylakoid membrane, and are oriented in a specific manner. A model of the electron and proton transporting complexes in the thylakoid membrane is shown in Fig. 8.2. Freeze-fracture studies and immunogold labelling analyses of barley chloroplasts have shown that the PS II complexes are located mainly in the grana stacks (Vallon *et al.*, 1987; Simpson *et al.*, 1989). In contrast, PS I and ATP synthase complexes are found only in the stroma lamellae, and in the exposed regions of the grana stacks (Simpson, 1983). The cyt *b6–f* complex is thought to be concentrated along the exposed regions of the grana stacks, as supported by fractionation studies with barley using a French press/phase-partitioning procedure (Henry & Møller, 1981) and a preparation of PS II membrane sheets with Triton X-100 (Møller *et al.*, 1983).

The nomenclature proposed by Hallick (1989) for genes of thylakoid membrane

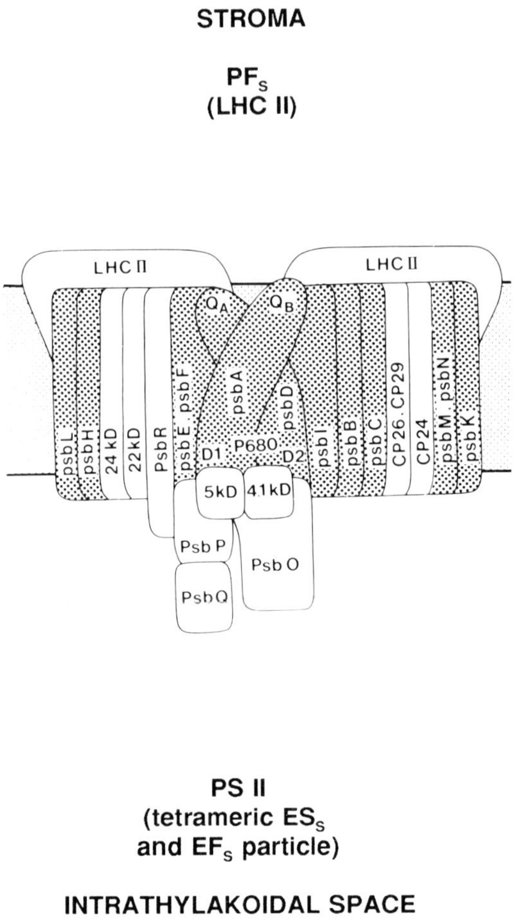

Fig. 8.2. Model of Photosystem II. The model shows the gene name and actual molecular mass, where known, of the PS II subunits. The shaded subunits are encoded by the chloroplast genome and the clear by the nuclear genome. (Courtesy of Simpson & von Wettstein, Carlsberg Research Laboratory.)

polypeptides is used in the following account.

Photosystem II

PS II is a light-driven multicomponent enzyme that oxidizes H_2O and reduces plastoquinone (PQ) generating protons, oxygen and reduced PQ (PQH_2). The PS II complex consists of an integral membrane core complex carrying the P680 reaction centre and the electron acceptors, and an extrinsic water-oxidizing complex. The reaction centre is the site where the electrons, originating from the water oxidation, are excitated by the light-energized electrons channelled to the reaction centre by the light-harvesting complexes. Figure 8.3 shows a model of PS II with the light-harvesting complex.

PS II core complex

The core complex consists of at least six polypeptides encoded by the chloroplast genome in four separate operons containing *psbA*, *psbB*, *psbD–psbC* and *psbE–psbF*, respectively. The operons *psbB* and *psbE–psbF* contain additional genes. The *psbA* (Boyer & Mullet, 1988; Efimov *et al.*, 1988a), *psbD–psbC* (Oliver & Poulsen, 1984; Efimov *et al.*, 1988b; Neumann, 1988; Reverdatto *et al.*, 1989a), *psbE–psbF* (Krupinska, 1988; Chakhmakhcheva *et al.*,

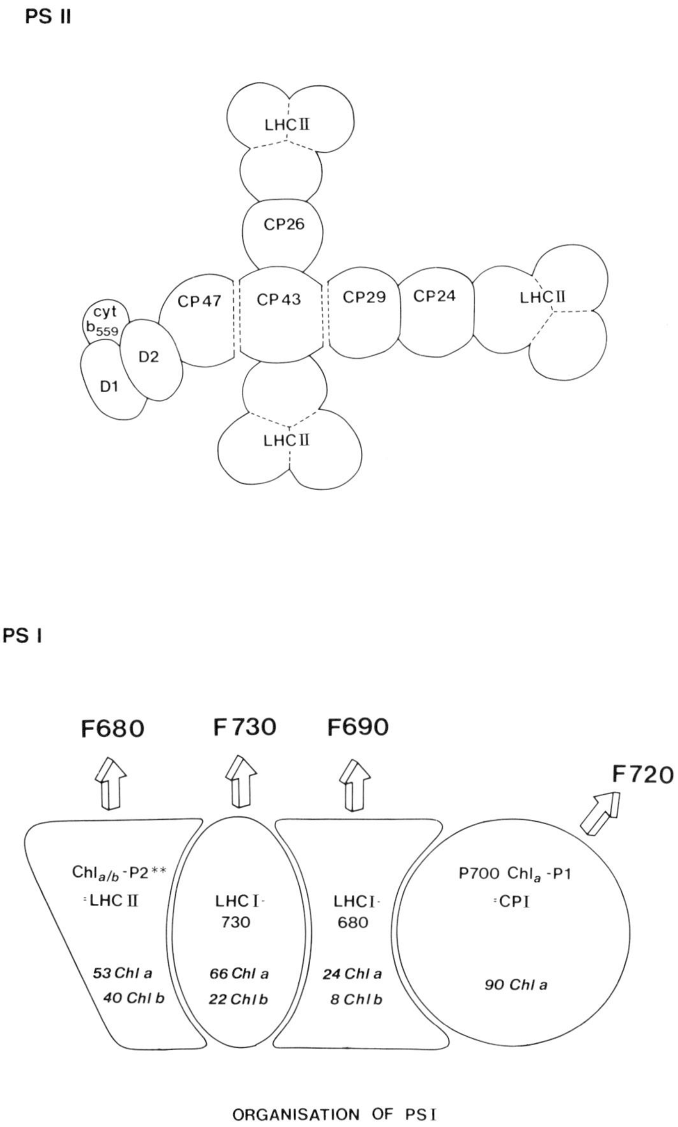

Fig. 8.3. Photosystems I and II light-harvesting complexes of barley. The model shows the relative distribution of chlorophyll *a* and *b* among the different chlorophyll–protein complexes. The amount of LHC-II shown is the maximum that can be functionally attached. *In vivo,* light conditions will determine the amount of LHC-II attached to the PS I and PS II. (Courtesy of Simpson & von Wettstein, Carlsberg Research Laboratory.)

1989) and *psbB* (Reverdatto *et al.*, 1989b) operons have been sequenced in barley. The *psbA* and *psbD* genes encode the subunits D1 and D2, respectively. D1 and D2 form a heteroduplex which binds the reaction centre P680, four chlorophyll *a* molecules, two pheophytins, one betacarotene, two quinones and one non-haem iron. Two of the chlorophylls probably form the reaction centre P680. Recent evidence indicates the participation of the two accessory chlorophylls with primary electron reactions. The

D1 and D2 subunits show sequence similarity to the purple bacterial reaction centre (RC) proteins L and M (Michel *et al.*, 1986). The structure of the purple bacterial RC has been determined to atomic resolution by X-ray diffraction studies of RC crystals from *Rhodopseudomonas viridis* (Deisenhofer *et al.*, 1985) and *Rhodobacter sphaeroides* (Allen *et al.*, 1987a, b). Regions of L and M with sequence similarity to D1 and D2 provide the principal binding sites for the primary donor bacteriochlorophylls, pheophytin, the non-haem iron and the quinones (Trebst, 1985). Site-directed mutagenesis and deletion studies in cyanobacteria (Yu & Vermaas, 1990) and in mutants of *Chlamydomonas reinhardtii* have given further information on the subunits of PS II and the amino acid residues involved in binding the different prosthetic groups (for reviews see Vermaas *et al.*, 1988; Rochaix & Erickson, 1988; Matoo *et al.*, 1989). An extensive model for the reaction centre of PS II has been developed based on these studies and the detailed structure of the bacterial RC (e.g. Marder & Barber, 1989; Mattoo *et al.*, 1989).

The *psbE* and *psbF* genes encode the two subunits of cyt b559. The cyt b559 heterodimer binds a single haem which is probably co-ordinated by a single histidine residue from each subunit. The function of cyt b559 in PS II is uncertain, but it may be involved in the electron flow around P680 by accepting electrons from the Q_B plastoquinone (Falkowski *et al.*, 1986).

The antennae chlorophyll proteins CP47 and CP43 of the PS II core complex are encoded by the *psbB* and *psbC* genes. Most of the chlorophyll in the PS II core is bound to these two chlorophyll proteins which each binds 10–20 molecules of chlorophyll *a*. The histidine residues in the transmembrane portions of CP43 and CP47 may be responsible for the ligation of chlorophyll. Deletion studies in *Synechocystis* sp. show that CP47 may also be needed for stable integration of D1 and D2 and for PS II activity (Vermaas *et al.*, 1988).

The water-oxidizing complex

In contrast to the reaction centre of purple bacteria, photosystem II catalyses the photooxidation of water. This is the major source of oxygen on the Earth. The water oxidation involves accumulation of four oxidizing equivalents in a manganese cluster (S-states), the release of which results in the generation of one molecule of oxygen from two molecules of water. Chloride and calcium ions are also involved in the catalytic site (for a review see Rutherford, 1989; Ghanotakis & Yocum, 1990). Four polypeptides of 33, 23, 16 and 10 kDa have been implicated in water oxidation, denoted OEE1 (oxygen-evolving enhancer protein 1), OEE2, OEE3, and OEE4, respectively. All four are encoded by nuclear genes *PsbO, PsbP, PsbQ*, and *PsbR*, respectively, and cDNA clones for these polypeptides have been sequenced in spinach (Jansen *et al.*, 1987; Tyagi *et al.*, 1987; Lautner *et al.*, 1988). The polypeptides are peripherally bound on the luminal side of the thylakoid membrane. Sequences for the polypeptides have so far not been obtained from barley. The 33 kDa polypeptide and the carboxyl ends of the D1/D2 subunits may be involved in forming a pocket securing a high local concentration of manganese around the water oxidation site. The 33 kDa polypeptide is an obligatory requirement for PS II activity *in vivo* (Mayfield *et al.*, 1987a). The 23 kDa polypeptide is to some extent dispensable at high calcium concentrations but its importance *in vivo* is illustrated by the fact that a *Chlamydomonas* mutant which lacks the protein maintains only very reduced levels of light-induced oxygen evolution (Mayfield *et al.*, 1987b). The 23 kDa protein is probably mainly involved in the binding of chloride and calcium ions (Rutherford, 1989). Only limited information is available about the 16 kDa protein. Antisense RNA inhibiting the expression of the 10 kDa polypeptide in transgenic potato plants reduced the level of the 10 kDa polypeptide to 1–3% compared with the wild-type control (Stockhaus *et al.*, 1990). This drastic reduction caused no phenotypic differences compared to wild-type with respect to growth rate or ultrastructure of the chloroplasts and no influence on the accumulation of the other PS II polypeptides. However, the reoxidation of Q_A^- was retarded and a general disorder may have been introduced into the PS II complex.

Additional PS II polypeptides

Smaller chloroplast-encoded subunits of the PS II complex or the oxygen-evolving complex have been isolated and given the gene designations *psbH* to *psbN*. The gene designation *psbG* has been left out of the compilation of genes for components for photosystem II (Hallick, 1989). The regions of the barley chloroplast genome containing *psbH* (Reverdatto *et al.*, 1989b), *psbK* and *psbI* (Sexton *et al.*, 1990), and *psbL* (ORF 38) (Chakhmakhcheva *et al.*, 1989) have been sequenced. The functions of these subunits remain uncertain.

Cytochrome *b6–f* complex

The cyt *b6–f* complex functions in the linear transport of electrons between PS II and PS I, as well as in cyclic electron flow around PS I. The complex contains cyt *f* which carries one c-type haem (33–38 kDa), cyt *b-563* which carries two b-type haems (19.5–23 kDa), the Rieske protein (18–20 kDa) which carries a 2Fe–2S cluster, subunit IV (15–17 kDa) and subunit V (4 kDa). The last two do not carry redox centres. The Rieske protein is encoded by the nuclear gene *PetC*, while the others are products of the chloroplast genes *petA* (cyt *f*), *petB* (cyt *b-563*), *petD* (subunit IV) and *petG* (subunit V) (for a review see Willey & Gray, 1988, and for subunit V see Haley & Bogorad, 1989). Sequence analyses indicate that cyt *b-563* (*petB*) is homologous to the N-terminal part of the mitochondrial cyt *b* which has a molecular mass of 35 kDa, and subunit IV (*petD*) is homologous to the C-terminal part (Widger *et al.*, 1984). In addition the comparison of the deduced amino acid sequences of *petB* with the mitochondrial cyt *b* sequences revealed four conserved histidine residues suggested to be involved in haem-binding. In barley, *petB* and *petD* have been sequenced (Reverdatto *et al.*, 1989b) and both contain one intron as has been found in other plants.

The nucleotide sequence of *petA* encodes a protein with a putative haem-binding site, Cys–Ala–Asn–Cys–His, located near the N-terminus (Willey & Gray, 1988). In pea, the *petA* gene is co-transcribed with an upstream open reading frame of 231 amino acid residues which may encode a *b*-type cytochrome (Willey & Gray, 1988).

The deduced amino acid sequences of the nuclear encoded *PetC* from spinach and pea show two highly conserved cysteine- and histidine-containing sequences near the C-terminus of the polypeptide which may be involved in co-ordinating the iron–sulphur cluster.

Plastocyanin and ferredoxin are encoded by the nuclear genes *PetE* and *PetF* respectively. A cDNA clone encoding plastocyanin has been isolated and sequenced from barley (Nielsen & Gausing, 1987).

Photosystem I

PS I mediates electron transport from reduced plastocyanin in the thylakoid lumen to oxidized ferredoxin in the stroma. The core complex of PS I is defined as the simplest pigment–protein complex which photoreduces $NADP^+$ in the presence of reduced plastocyanin, ferredoxin and ferredoxin–$NADP^+$:oxidoreductase (Bengis & Nelson, 1977; Reilly & Nelson, 1988). Twelve protein subunits have been identified in the PS I core complex and their primary structures have been determined from different species (for recent reviews see Scheller & Møller, 1990; Golbeck & Bryant, 1991). Five of the subunits are chloroplast encoded and seven are nuclear encoded. Bengis & Nelson (1977) originally designated the PS I subunits with roman numerals (I–VI) according to their electrophoretic mobilities on SDS-polyacrylamide gels. Owing to: (i) the increasing number of

Table 8.2. Subunits of the Photosystem I core complex. The table lists the polypeptides of PS I preparations from barley using the nomenclature proposed by Schantz & Bogorad (1988) (1) and the nomenclature of Bengis & Nelson (1977) (2). The localization of the corresponding genes in the chloroplast (C) or the nuclear (N) genomes is indicated. Apparent molecular masses are based on electrophoretic mobilities in SDS-polyacrylamide gels (Høj *et al.*, 1987; Scheller *et al.*, 1989a). Calculated molecular masses are based on amino acid or nucleotide sequencing data (Okkels *et al.*, 1988, 1989, 1990; Scheller *et al.*, 1989b, c and unpublished data). nd, not determined in barley.

Gene		Polypeptide subunit		Molecular mass (kDa)	
		(1)	(2)	Apparent	Calculated
psaA	(C)	PSI-A	I	82	nd
psaB	(C)	PSI-B	I	82	nd
psaC	(C)	PSI-C	VII, VIb	9	8.8
PsaD	(N)	PSI-D	II	18	17.6
PsaE	(N)	PSI-E	IV	16	10.8
PsaF	(N)	PSI-F	III	15	nd
PsaG	(N)	PSI-G	–	9	nd
PsaH	(N)	PSI-H	VI, VIa	9.5	10.2
psaI	(C)	PSI-I	–	1.5	4.0
psaJ	(C)	PSI-J	–	nd	nd
PsaK	(N)	PSI-K	–	nd	nd
PsaL	(N)	PSI-L	V	14	18.0
Unknown		Unknown	–	4	nd

recognized PS I subunits; (ii) large species-dependent differences between apparent and actual molecular masses; and (iii) co-migration of subunits, the use of the classical nomenclature has caused some confusion. The genes for the PS I subunits have been denoted *psaA*, *psaB*, etc. with the alphabetic order referring to the sequential order of their identification from different species (Hallick, 1989). In barley the nuclear-encoded genes are denoted with a capital P (eg. *PsaH*; Okkels *et al.*, 1989). The corresponding polypeptides may therefore conveniently be denoted PSI-A, PSI-B, etc. (Schantz & Bogorad, 1988; Møller *et al.*, 1990). The two nomenclature systems and the identified genes are shown in Table 8.2.

The chloroplast-encoded subunits

PS I contains a number of electron acceptors, identified using spectrophotophysical and biochemical techniques. The reaction centre P700 and the acceptors A_0, A_1 and X are all located in the pigment–protein complex known as P700–chl *a*-protein 1 (CP1) (Høj & Møller, 1986; Golbeck, 1987; Lagoutte & Mathis, 1989). This complex is composed of the PSI-A and PSI-B subunits which both have approximate molecular masses of 82 kDa. The PSI-A and PSI-B subunits are present in equimolar amounts (Fish & Bogorad, 1986; Bruce & Malkin, 1988), and it may be presumed that CP1 is composed of a heterodimer of the two subunits. The *psaA* and *psaB* genes have been sequenced in a number of species and the sequence identity between the species is high (for a review see Golbeck & Bryant, 1991). The deduced amino acid sequences reveal a leucine zipper motif conserved between the two subunits, very similar to the leucine zipper of DNA-binding polypeptides, suggesting that the two subunits are held together by this leucine zipper (Webber & Malkin, 1990). The *psaA* and *psaB* genes have not been sequenced in barley but clones have been isolated (Poulsen, 1983; Okkels, unpublished results). In barley, the mRNAs for the two genes are present in both dark- and light-grown seedlings whereas the polypeptides are only detected in the light-grown seedlings (Klein & Mullet, 1986; Klein *et al.*, 1988). Eihacker *et al.* (1990) have shown that *de novo* synthesis of chlorophyll *a* is necessary and sufficient to trigger accumulation of CP1 in barley. These results suggest translational or post-translational control of the expression of CP1. The accumulation of the D1 subunit of PS II in barley seems to result from stabilization of the polypeptide upon chlorophyll *a* synthesis (Mullet *et al.*, 1990), suggesting post-translational control of the expression of D1.

The PSI-A/PSI-B heterodimer of barley binds four iron atoms per P700 (Høj & Møller, 1986) and four acid-labile sulphur atoms per P700 (Høj *et al.*, 1987). Centre X has been identified as an iron–sulphur cluster by EPR and Mössbauer spectroscopy (Evans *et al.*, 1981) and requires four cysteine residues as ligands. In all species examined, the *psaA* and *psaB* genes both specify a completely conserved stretch of 12 amino acid residues: Phe–Pro–Cys–Asp–Gly–Pro–Gly–Arg–Gly–Gly–Thr–Cys. The cysteine residues in these conserved regions are almost certainly the ligands of centre X (Høj *et al.*, 1987) because the PSI-B polypeptide contains no additional conserved cysteine residues and only one additional cysteine residue is conserved in PSI-A. The stoichiometry between most of the PS I subunits and P700 has been determined as 1:1 by amino acid analyses and by the specific Coomassie Brilliant Blue R250 binding capacity of the PS I subunits (Scheller *et al.*, 1989a). From the measured contents of iron and acid-labile sulphide in PS I preparations per P700 reaction centre and from the cysteine contents of PSI-A and PSI-B, centre X was concluded to be a [4Fe–4S] cluster bridged between the PSI-A and PSI-B polypeptides of the heterodimer (Scheller *et al.*, 1989a). The conserved 12-residue stretch containing the cysteines is located in a loop connecting two hydrophobic helices. Centre X may donate electrons to the stromal side of the thylakoid membrane and therefore the loop must also be positioned on the stromal side. It still remains to be demonstrated whether centre X is directly involved in the linear electron transport through PS I. Alternatively, the main function of centre X may be to link covalently the two reaction centre polypeptides in a similar manner to the Fe atom of the quinone complex of PS II. A few cases where iron–sulphur clusters serve a structural role rather than a function as an electron carrier have been reported in the literature (Beinert, 1990).

The PSI-A and PSI-B polypeptides contain 38 and 36 conserved histidine residues, respectively. A heterodimer of the two 82 kDa polypeptides thus contains 74 conserved histidine residues of which 52 are predicted to be positioned in membrane-spanning regions. Histidine residues have been implicated in the co-ordination of chlorophyll and haem groups in polypeptides of the photosynthetic apparatus. The conserved histidine residues may provide binding sites for most of the approximately 90 chlorophyll *a* molecules associated with CP1. Very few histidine residues are found in the remaining PS I subunits (Scheller *et al.*, 1989a; Scheller & Møller, 1990).

Like the CP1 reaction centre protein of PS I, the reaction centre proteins of purple bacteria and of PS II also consist of two homologous but non-identical subunits. In the PS II-type reaction centres, the heterodimer binds the reaction centre (a chlorophyll dimer), the primary acceptors (pheophytin), and the secondary acceptors (quinones Q_A and Q_B) (Mattoo *et al.*, 1989). This set of electron transfer components is structurally similar to the reaction centre P700 (probably a chlorophyll dimer), A_0 (probably a chlorophyll molecule), and A_1 (phylloquinone) of PS I. In spite of these analogies between PS I and PS II, the amino acid sequence homology between the reaction centre proteins of PS I and PS II is low (Robert & Moenne-Loccoz, 1989).

The complete amino acid sequence of the 9 kDa PSI-C subunit of barley has been obtained (Scheller *et al.*, 1989b). This subunit carries the iron–sulphur centres A and B (Høj *et al.*, 1987). The amino acid sequence is very conserved between cyanobacteria and higher plants with a pattern of cysteine residues which is characteristic for proteins carrying two [4Fe–4S] clusters (Høj *et al.*, 1987). Alignment analysis indicates that the tertiary structure for the PSI-C polypeptide is almost identical to the structure of soluble 2[4Fe–4S] ferredoxins from bacteria with the PSI-C subunit having an extra loop of ten residues near the centre of the polypeptide chain and 14 additional residues at the C-terminal end (Dunn & Gray, 1988; Oh-Oka *et al.*, 1988; Scheller *et al.*, 1989b). The C-terminus is quite hydrophilic and the least conserved region of the polypeptide. The ten residue loop near the centre is more hydrophobic and contains charged residues which could serve to anchor the polypeptide to the PS I complex. The EPR signal of centre A and B in spinach is shifted compared with that in cyanobacteria (Mehari *et al.*, 1991). This shift may be related to the difference between Trp–His in spinach and Gly–Ala in cyanobacteria at amino acid position 69–70 (Mehari *et al.*, 1991). Since barley and other monocots contain Gly–Pro at the same position, it would be interesting to test if the EPR signal in barley is also shifted compared with both spinach and cyanobacteria.

A small 4.0 kDa polypeptide encoded by the chloroplast *psaI* gene has been identified in barley (Scheller *et al.*, 1989c). The PSI-I subunit is very hydrophobic and tightly bound to CP1 at conditions otherwise used to dissociate the low-molecular-mass polypeptides from CP1. The PSI-I polypeptide is predicted to contain a hydrophobic α-helix which may span the membrane. The polypeptide has sequence similarity with helix E of the D2 reaction centre protein of PS II. The identity is 31% when comparing PSI-I and D2 from barley. This similarity, together with the strong association of PSI-I to CP1, may suggest that PSI-I participates in binding P700, A_0 or A_1.

A 4.1 kDa PS I polypeptide has been identified in *Synechococcus vulcanus*, spinach and pea (Koike *et al.*, 1989; Ikeuchi *et al.*, 1990). A similar polypeptide has not been identified in barley but the sequenced chloroplast genomes of *Marchantia polymorpha*, tobacco and rice all contain an open reading frame encoding an almost identical polypeptide. The gene has been designated *psaJ*. The corresponding PSI-J polypeptide is predicted to contain a single membrane-spanning segment.

The presence of an additional chloroplast-encoded subunit of a molecular mass of 14–16 kDa has been reported in several papers (Nechushtai *et al.*, 1981; Mullet *et al.*, 1982; Smith & Gray 1984; Obokata, 1986) but not substantiated with sequence data and now seems unlikely (Okkels *et al.*, 1991).

The nuclear-encoded subunits

A 20 or 22 kDa subunit from spinach can be chemically cross-linked to ferredoxin whereby PS I activity is lost (Zanetti & Merati, 1987; Zilber & Malkin, 1988). Similar experiments with barley have demonstrated that PSI-D (subunit II) is the subunit interacting with ferredoxin (Andersen *et al.*, 1990). PSI-D is a peripheral membrane protein on the stromal side of the thylakoid membrane (Fig. 8.4), and has highly positively charged domains. PSI-D may enable the negatively charged ferredoxin to react with PS I in spite of the overall negative surface charge of the thylakoid membrane. Chitnis *et al.* (1989a) specifically inactivated the *psaD* gene in *Synechocystis* sp. PCC 6803. The mutant cyanobacteria contained a PS I complex consisting of the PSI-A, PSI-B, PSI-C and PSI-F polypeptides. It was inefficient in photosynthetic CO_2 fixation and required a reduced carbon source in the growth medium. The abnormal PS I complex was unable to reduce ferredoxin but could still carry out some electron transport from plastocyanin to methylviologen. These results are consistent with the conclusion from cross-linking experiments that the PSI-D polypeptide is a docking site for ferredoxin.

Early experiments by Bengis & Nelson

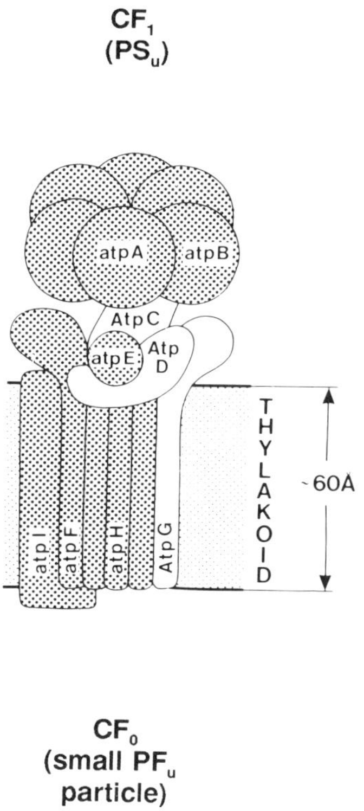

Fig. 8.4. Model of the chloroplast coupling factor (ATP synthase). The membrane-embedded CF_0 and the extrinsic CF_1 are indicated with the name of the individual polypeptide and their apparent molecular masses. (Courtesy of Simpson & von Wettstein, Carlsberg Research Laboratory.)

(1977) showed that the removal of subunit III (PSI-F) from *Beta vulgaris* var. *cicla* (Swiss chard) correlated with the loss of electron donation from plastocyanin to PS I. Recently, a 19 kDa subunit from spinach and a 20.5 kDa subunit from maize have been chemically cross-linked to plastocyanin, with loss of steady-state PS I activity (Wynn & Malkin 1988; Wynn *et al.*, 1989). A PS I preparation in which plastocyanin had been cross-linked to the PSI-F subunit showed the same fast flash-induced reduction kinetics of $P700^+$ as observed without cross-linking (Hippler *et al.*, 1989). This shows that plastocyanin was cross-linked very near to P700 and in the proper orientation for electron transfer. PSI-F has many positive charges and by analogy with the ferredoxin-binding of PSI-D, its function could be to counteract negative charge repulsion between plastocyanin and the PS I complex (Wynn & Malkin, 1988). Regions of PSI-F are highly conserved between species. In addition to providing positive charges, the polypeptide might bring plastocyanin into the proper orientation for efficient electron transfer (Hippler *et al.*, 1989). However, the presence of PSI-F was not essential for $NADP^+$-photoreducing activity of barley PS I preparations when measured using continuous illumination (Scheller *et al.*, 1990a).

PSI-E is a peripheral membrane protein on the stromal side of PS I near to PSI-D (Scheller *et al.*, 1988; Andersen *et al.*, 1990). The transit peptide of PSI-E (Münch *et al.*, 1988; Okkels *et al.*, 1988; Franzén *et al.*, 1989a) also indicates that this subunit is located on the stromal side. The apparent molecular mass of the PSI-E subunit in barley is significantly higher than the molecular mass deduced from the nucleotide sequence (Okkels *et al.*, 1988). However, the actual molecular mass of the isolated barley polypeptide as determined by plasma desorption mass spectrometry corresponded to the deduced molecular mass, indicating that the PSI-E subunit is not post-translationally modified apart from cleavage at the maturation site (Scheller *et al.*, 1990b). To study the function of the PSI-E subunit, Chitnis *et al.* (1989b) have specifically inactivated the *PsaE* gene in *Synechocystis* sp. PCC 6803. The mutant contained structurally and functionally normal PS I complexes except for the absence of PSI-E and a somewhat increased content of PSI-D. The growth rate of the mutant was almost unaffected by the lack of the PSI-E polypeptide, and it is possible that the PSI-D polypeptide can functionally substitute for the PSI-E polypeptide. In barley, the N-terminal regions of the PSI-D and PSI-E polypeptides show sequence homology (Scheller *et al.*, 1988).

Recently a cDNA clone encoding a novel PS I subunit has been isolated and characterized in barley (Okkels *et al.*, 1991). The gene for this subunit has been designated *PsaL*. The cDNA clone encodes a precursor polypeptide of 22 kDa which is processed to a hydrophobic 18 kDa mature polypeptide with two or three possible membrane-spanning regions. The polypeptide is tightly bound to CP1. An N-terminal sequence from the thermophilic cyanobacterium *S. vulcanus* shows high sequence identity to the N-terminus of barley PSI-L, indicating that this region is highly conserved between cyanobacteria and higher plants.

The *PsaG* genes of spinach (Steppuhn *et al.*, 1988) and *C. reinhardtii* (Franzén *et al.*, 1989b) have been isolated as cDNAs and sequenced. In addition, an N-terminal sequence from pea (Dunn *et al.*, 1988) and internal amino acid sequences from barley (Okkels *et al.*, 1991) have been determined. Steppuhn *et al.* (1988) assigned the *PsaG* gene product (PSI-G) as subunit V according to the nomenclature of Bengis & Nelson (1977). However, results from barley and other plants show that subunit V is equivalent to PSI-L, whereas PSI-G co-migrates with PSI-H (subunit VI) in several different SDS-polyacrylamide gel electrophoresis systems (Okkels *et al.*, 1991). N-terminal amino acid sequences from *Synechococcus* sp. PCC 7002 (Rhiel & Bryant, 1988) and *Synechococcus vulcanus* (Koike *et al.*, 1989) have been suggested to correspond to PSI-G (Scheller & Møller, 1990; Golbeck & Bryant, 1991). As mentioned above, the *S. vulcanus* sequence has been shown to be equivalent to the PSI-L subunit and not the PSI-G subunit (Okkels *et al.*, 1991). The N-terminal sequence of *Synechococcus* sp. PCC 7002 shows similarity to both the PSI-G and PSI-L subunits which may be due to co-purification and co-sequencing of the two subunits (Okkels *et al.*, 1991).

The primary structure of the PSI-H subunit has been determined in barley, *C. reinhardtii* and spinach (Franzén *et al.*, 1989b; Okkels *et al.*, 1989; Steppuhn *et al.*, 1989). No homologous polypeptide has been identified in cyanobacteria. One possible function of the PSI-H polypeptide could be to mediate binding of the LHC I polypeptides to the PS I core complex. The different organization of the antenna pigment system in the cyanobacteria could explain the absence of the PSI-H subunit in these organisms. A genomic clone, COS5, with a light-inducible promoter has been characterized in rice (de Pater *et al.*, 1990). Comparison with the sequence of PSI-H of barley (Okkels *et al.*, 1989) reveals that the clone, which contains two introns, encodes the PSI-H subunit and represents the first isolated genomic clone of a nuclear-encoded PS I polypeptide of a higher plant.

The PS I preparation from barley contains low-molecular-mass polypeptides in addition to the chloroplast-encoded PSI-I subunit. The complete nucleotide sequence of a nuclear-encoded subunit (PSI-K) with an apparent molecular mass of 5 kDa has been determined in *C. reinhardtii* (Franzén *et al.*, 1989b), and in spinach (Wynn & Malkin, 1990). The conclusion that the *C. reinhardtii* polypeptide contains one membrane-spanning α-helix was substantiated by heat-treatment experiments in spinach, demonstrating that the PSI-K subunit is closely associated to CP1 (Hoshina *et al.*, 1989). The possibility that one or more of the low-molecular-mass polypeptides of PS I are directly involved in the binding of the PS I electron transfer components needs to be further examined.

The primary structures of most of the PS I subunits have been determined. From sequence information now available, it should be easier to compare PS I preparations obtained in different laboratories and to identify homologous subunits from different species. The future challenge will be to elucidate the organization and function of the various subunits within the PS I complex. Until well-diffracting crystals of PS I are available, our knowledge of the structure and function of PS I may be improved by cross-linking studies, reconstitution experiments and genetically engineered deletions and mutations. Using antibodies for the barley PS I subunits, and the isolated cDNA clones encoding barley PS I subunits, it is now possible to analyse at the molecular level, the previously isolated and characterized barley PS I mutants (Machold *et al.*, 1979; Hiller *et al.*, 1980; Møller *et al.*, 1981). A model indicating a possible structure of the PS I complex is shown in Fig. 8.4. A general picture of a symmetrical organization of the polypeptides within PS I and with chloroplast-encoded polypeptides binding the reaction centre as well as the primary electron acceptors is becoming apparent.

The light-harvesting complex

The light-harvesting complexes (LHCs) are primarily responsible for absorbing incident radiant energy and directing the excitation energy to the core complexes of PS I and PS II. However, LHCs are not essential for photosynthetic electron transfer to take place in the core complexes.

There is evidence for at least six chlorophyll *a*/*b* binding LHCs in higher plants. Two of them, LHCI-680 and LHCI-730, are specific for PS I. Three chlorophyll *a*/*b* proteins occur as minor components and are specifically associated with PS II and denoted CP-24 (LHC-IId) (Dunahay & Staehelin, 1986; Bassi & Simpson, 1987b), CP-29 (LHC-IIa) (Machold *et al.*, 1979; Camm & Green, 1980) and CP-26 (LHC-IIc) (Bassi *et al.*, 1987). Finally, the abundant LHC-II (LHC-IIb) is associated with both photosystems. A model of the composition and organization of the chlorophyll-proteins of PS I and PS II is presented in Fig. 8.5.

In barley and maize, three polypeptides have been found associated with LHCI-730 and two with LHCI-680 with molecular masses ranging from 21 to 25 kDa (Bassi *et al.*, 1985; Bassi & Simpson, 1987a). The LHC-I polypeptides are nuclear encoded and three gene families (Types I, II and III) have been identified in tomato (Pichersky *et al.*, 1989). The overall amino acid sequence identity of the LHC-I polypeptides to LHC-II polypeptides is only 30%, but two large regions have around 50% sequence identity (Pichersky *et al.*, 1987; Stayton *et al.*, 1987; Pichersky & Green, 1990). A PS I type II chlorophyll *a*/*b* protein has been reported to be present in the LHCI-680 component of LHCI (Stayton *et al.*, 1987; Pichersky *et al.*, 1988).

CP-24 and LHCI-680 may play similar roles as connecting antennae in PS I and PS II. However, they have unique features and distinct thylakoid membrane locations indicating that they are two different complexes (Paolo *et al.*, 1990). Recently, nuclear genes for CP24 (Schwartz & Pichersky, 1990; Spangfort *et al.*, 1990) and CP29 (Pichersky & Green, 1990) have been isolated. The CP24 gene products deduced from the nucleotide sequences are more than 60% divergent from all other chlorophyll *a*/*b* binding proteins, including PS I type II of LHC-680. In addition it has been reported that PS I type II chlorophyll *a*/*b* binding protein is not present in PS II (Schwartz & Pichersky, 1990), and CP24 is found in PS II but not in PS I (Spangfort *et al.*, 1990), substantiating that LHC-680 and CP24 are two separate and non-exhangeable light-harvesting complexes.

LHC-II (LHC-IIb) carries almost half of the chlorophyll and constitutes one-third of the total protein of the thylakoid membranes. A proportion (30%) of LHC-II can be phosphorylated and is involved in the energy-transfer distribution between PS I and PS II. The remaining non-phosphorylated proportion of LHC-II is tightly bound to PS II (Islam, 1987; Bassi *et al.*, 1988). The LHC-II complex is composed of at least three polypeptides of molecular masses ranging from 25 to 29 kDa. In all plant species examined, the LHC-II polypeptides are encoded by two families of nuclear genes, types I and II. In a number of species, at least ten genes encode the type I LHC-II polypeptides (Castresana *et al.*, 1987; Chitnis & Thornber, 1988). In a few cases like *Arabidopsis thaliana* (which has a very small genome size) a smaller number have so far been identified (Leutwiler *et al.*, 1986). A genomic clone encoding a type I LHC-II polypeptide has been isolated from barley and used for assembly studies of LHC-II in barley etiochloroplasts (Chitnis *et al.*, 1988). In these studies the precursor for the barley LHC-II polypeptide is taken up by the etiochloroplasts and inserted into the thylakoids and only then processed to its mature form. Furthermore, the precursor is inserted primarily in the stromal lamellae and subsequently transferred to the grana where it is assembled into the pigmented LHC-II holoprotein (Yalovsky *et al.*, 1990). The LHC-II type I genes contain no introns whereas the type II genes contain one intron, are less numerous, and code for slightly shorter polypeptides (Chitnis & Thornber, 1988). The amino acid sequence identity between the two types is approximately 70–80%.

Three cDNA clones have been isolated encoding early light-inducible proteins of barley chloroplasts (Grimm *et al.*, 1989). They are two multigene families yielding mature polypeptides with molecular masses of 17 and 21 kDa. Two domains of the early light-inducible proteins show similarity with two transmembrane domains of all known light-harvesting chlorophyll *a*/*b* binding proteins. It has been suggested that the early light-inducible proteins are pigment-free substitutes for light-harvesting polypeptides in the assembly of PS I and PS II during early development of etioplasts to chloroplasts.

ATP synthase

The H^+-ATP synthase of the thylakoid membrane mediates the coupling of the proton gradient generated by the photosynthetic electron transfer and photophosphorylation (light-driven ATP synthesis). The ATP synthase has an F_1F_0 structure analogous to the proton-translocating ATP synthases of bacteria and mitochondria. The F_0 component (CF_0) forms a transmembrane proton channel whereas the F_1 component (CF_1) protrudes into the stroma and couples ATP synthesis to proton translocation across the thylakoid membrane. CF_1 contains five subunits designated α, β, γ, δ and ϵ. The α and β subunits are nucleotide-binding polypeptides (Merchant & Selman, 1985), clustered around the γ subunit (Akey *et al.*, 1983). The ϵ subunit is required for ATP synthesis and to prevent proton leakage through CF_0 (Richter *et al.*, 1984). The genes for the α, β, and ϵ subunits are on the chloroplast genome while those for the γ and δ subunits are encoded in the nucleus. The *atp*BE operon, encoding the β and ϵ subunits has been sequenced in barley (Zurawski & Clegg, 1984). The two coding regions overlap with four bp such that the UGA stop codon of *atp*B is preceded by an A to form the initiator codon of *atp*E as in most higher plants (Hudson & Mason, 1988). That one ϵ and three β subunits are required per CF_1 complex suggests translation of *atp*E is not tightly coupled to translation of *atp*B.

CF_0 consists of four subunits designated I, II, III and IV, where the subunit II is nuclear-encoded and the other three chloroplast-located (for a review see Hudson & Mason, 1988). The amino acid sequence of subunit III has been determined in spinach (Sebald & Wachter, 1980) showing that it is equivalent to the *E. coli* subunit *c*, which is closely associated with proton conductance. None of the genes encoding CF_0 subunits has been sequenced in barley. A model of the ATP-synthase is shown in Fig. 8.6.

Rubisco

The ATP and NADPH generated by the light reactions of photosynthesis are consumed in the dark reactions, the major portion in the reductive pentose phosphate pathway (the Calvin cycle) resulting in the fixation of CO_2 into various compounds. The Calvin cycle is catalysed by a number of soluble enzymes in the stroma, the primary one being the abundant enzyme ribulose-1,5-bisphosphate carboxylase (Rubisco). Rubisco in higher plants and cyanobacteria consists of eight large subunits of approximately 55 kDa, and eight small subunits of approximately 14 kDa. The large subunit contains the catalytic site of the enzyme and is encoded by the chloroplast gene *rbcL*, while the small subunit gene (*RbcS*) is nucleus-encoded in higher plants and algae. Most of the *rbcL* gene has been sequenced in barley and the 5′-end of its mRNA was determined (Poulsen, 1981; Poulsen, 1984; Zurawski *et al.*, 1984).

Comparative analysis of *rbcL* promoter regions from barley and other species revealed two elements analogous to the prokaryotic consensus '−35' and '−10' sequences (Zurawski & Clegg, 1987). From maximum likelihood analysis of the *rbcL* DNA sequences, Zurawski & Clegg (1987) constructed an unrooted tree that depicts the phylogenetic relationships between barley, other flowering plants, green algae and cyanobacteria.

A Rubisco large subunit-binding protein has been purified from pea (Hemmingsen & Ellis, 1986) and barley (Mushgrove *et al.*, 1987) and is required for the assembly of Rubisco from its subunits (Ellis & van der Vies, 1988; Ellis, 1990). The protein has been shown to bind other unrelated proteins as well (Lubben *et al.*, 1989). The apparent molecular mass of the purified binding protein from pea is 720 kDa as determined by gel filtration on Sephacryl columns. This complex consists of six 61 kDa and six 60 kDa subunits (for a review see Roy, 1989).

The term chaperonin has been coined for a new class of proteins with similar properties to the Rubisco large subunit-binding protein. The chaperonins belong to the general class of proteins termed molecular chaperones, which prevent the formation of 'incorrect' structures that may be produced by the transient appearance of hydrophobic or charged domains during the synthesis of a polypeptide (Ellis, 1987). Molecular chaperones are widespread in nature (Ellis, 1990) and the different types are highly conserved between the species (Hemmingsen *et al.*, 1988).

Biosynthetic pathways

Lipid biosynthesis

Chloroplast membranes have a glycerolipid composition unlike that of other plant cell membranes. The thylakoid and inner envelope membranes are rich in galactolipids (80–90%) compared with the other membranes of the plant cell which contain only phospholipids. The outer envelope of the chloroplast has an intermediate lipid composition. The lipid composition of barley plasma membrane (Rochester *et al.*, 1987; Brown & DuPont, 1989) as well as of barley PS I and PS II vesicles (Henry *et al.*, 1982) has been determined.

The first stage in lipid biosynthesis involves the synthesis of long fatty acids, reported to occur exclusively in the stroma of chloroplasts (Ohlrogge *et al.*, 1979; for a review see Hatwood, 1988). The initial step in the pathway of fatty acid biosynthesis in the chloroplast is the conversion of acetyl-coenzyme A to malonyl SCoA catalysed by acetyl-SCoA carboxylases. Subsequently, a multi-enzyme complex (fatty acid synthetase) containing an acyl-carrier protein (ACP) adds acetyl groups to the growing chain. Isolation of the ACP from barley chloroplasts revealed two ACPs, ACP I and II (Høj & Svendsen 1984). Two cDNA clones for ACP have been isolated from barley (Hansen, 1987), one encoding the chloroplast-located ACP I type polypeptide and the other cDNA clone encoding a new type of ACP, ACP III, with unknown location. Both ACP I and ACP III seem to be light-induced. Different β-ketoacyl-ACP synthase activities carry out the condensation of acetate units to an acyl primer. A β-ketoacyl-ACP synthase I has been purified and characterized from barley chloroplasts using the antibiotic cerulenin labelled with tritium (Siggaard-Andersen, 1990). Using the obtained partial amino acid sequences, cDNA clones have been isolated from barley encoding β-ketoacyl-ACP synthase I (Kauppinen, 1990).

The major products of fatty acid synthesis in chloroplasts are palmitic acid ($C_{16:0}$), stearic acid ($C_{18:0}$) and oleic acid ($C_{18:1}$) which can either be exported to the endoplasmic reticulum for eukaryotic lipid synthesis, or used for diacylglycerol synthesis. Diacylglycerol synthesis involves two sequential additions of fatty acids from activated thioesters to glycerol-3-P, forming phosphatidic acid. The phosphate group is removed from phosphatidic acid producing diacylglycerol, which is the substrate for the addition of the polar head group. For galactolipid synthesis, a galactosyl residue is transferred from UDP-galactose to diacylglycerol to form monogalactosyldiacylglycerol from which digalactosyldiacylglycerol is synthesized. These galactosylation reactions occur in the chloroplast envelope membranes. The fatty acids need to be further desaturated to produce the polyunsaturated fatty acids found in the galactolipids.

The isoprenoids, also known as terpenoids, form a diverse group of compounds with a common structural unit, the isoprene unit. Many biological functions have been attributed to members of this family, including pigments (carotenoids), hormones (derived from sterols), attractants and repellants, and phytoalexins. All isoprenoids are synthesized from isopentenyl diphosphate (the isoprene unit). The plastid appears to be an important site for isoprenoid formation within the plant cell (for a review see Klenig, 1989).

Chlorophyll biosynthesis

All steps in biosynthesis of chlorophyll in higher plants and algae occur within the plastids. Many barley mutants with altered chlorophyll synthesis have been isolated and characterized by von Wettstein and co-workers (e.g. Gough, 1972; for a review see Henningsen & Stummann, 1982). Some of them seem to contain mutated structural genes (some of the *viridis* and *xantha* mutants) whereas others (the *tigrina* mutants) have characteristics of mutated regulatory genes.

Chlorophyll is synthesized from the general tetrapyrrole precursor δ-aminolevulinic acid (ALA). In higher plants, algae and many bacteria (including cyanobacteria) ALA is synthesized from tRNA-activated glutamate in a three-step catalysis. The three enzymes, the glutamyl synthetase, glutamate reductase and glutamate-1-semialdehyde aminotransferase, are encoded by nuclear genes, while the $tRNA^{Glu}$, used to activate the glutamate, is encoded by the chloroplast genome. The barley chloroplast region containing the $tRNA^{Glu}$ gene (*trnE-(UUC)*) has been sequenced (Berry-Lowe, 1987). The *trnE-(UUC)* gene in *Synechocystis* sp. PC6803 functions both in ALA synthesis and in normal protein synthesis (Schneegurt *et al.*, 1988). Apparently, the $tRNA^{Glu}$ is processed from a precursor RNA which also contains the *trnY-(GUA)* and *trnD-(GUC)* coding regions.

A cDNA clone encoding glutamate-1-semialdehyde aminotransferase has recently been isolated and sequenced from barley (Grimm, 1990) and *Synechococcus* 6301 (Grimm *et al.*, 1991). The barley cDNA clone encodes a precursor polypeptide of 49.5 kDa with a transit peptide of 34 amino acids for chloroplast import. The cDNA clone was expressed in *E. coli* and the glutamate-1-semialdehyde aminotransferase was shown to be active in assays of bacterial cell extracts.

Through a series of biosynthetic steps eight ALA molecules are converted to protochlorophyllide. The next step is the light-dependent conversion of protochlorophyllide to chlorophyllide. This reaction is catalysed by the enzyme NADPH–protochlorophyllide oxidoreductase (Apel *et al.*, 1980). The 388 residues of the precursor for NADPH–protochlorophyllide oxidoreductase in barley has been deduced from the nucleotide sequence of a cloned full-length cDNA clone (Schulz *et al.*, 1989). The transit peptide contains 74 amino acids and is at least partially responsible for directing the polypeptide to the chloroplast.

Chlorophyll *a* is formed by esterification of chlorophyllide with a precursor of the long-chain alcohol phytol. In the absence of evidence to the contrary chlorophyll *b* is generally thought to be formed from chlorophyll *a* by oxidation of the methyl side-group at carbon 3 (ring II of the tetrapyrol system) to an aldehyde side-group. However, the existence of chlorophyllide *b* has been reported, and recently chlorophyll *b* has been shown to be formed from chlorophyllide *b* as well as from chlorophyll *a* in a mutant of *Scenedesmus obliquus* (Kotzabasis & Senger, 1990).

Other biosynthetic pathways

Starch is the major form in which photosynthetically produced carbohydrates are stored in plants. Starch in chloroplasts is usually transitory and accumulates during the light period and is used in the dark. Storage starch accumulates in reserve organs during one phase of the plant's life cycle and is used at another time. The major starch synthesis in reserve organs takes place in the proplastid-derived amyloplasts, but in some species also in chloroplast-derived plastids (Smith *et al.*, 1990). Starch is composed primarily of amylose (linear polymer of [1→4]-linked α-D-glucopyranosyl units) and amylopectin (branched polymer primarily of [1→4]-linked α-D-glucopyranosyl units with branches from [1→6] linkages). The *waxy* mutants identified in several plants including barley produce starch granules in the endosperm containing 100% amylopectin. These mutants lack the major starch granule-bound starch synthase activity. The gene encoded by the *waxy* loci has been sequenced in barley (Rohde *et al.*, 1988).

Nitrite reductase, the second step in the reduction of nitrate to ammonia, is also located in the chloroplasts. Ammonium is generated in plants by the primary nitrogen assimilation, but also as a product of certain metabolic reactions, including photorespiration, phenyl-propanoid metabolism and deamination of amino acids. The ammonium released in photorespirating leaves is reassimilated in a photorespiratory nitrogen cycle involving glutamine synthetase and ferredoxin-dependent glutamate synthase. Green leaves contain two types of glutamine synthetase, one in the cytosol (GS_1) and one in the chloroplast (GS_2). In barley, cDNA clones encoding GS_2 have been isolated (Baiama *et al.*, 1989; Freeman *et al.*, 1990) and used for characterization of deficient GS_2 barley mutants. The azide-induced barley GS_2 mutants fell into three classes: class I, in which absence of cross-reacting material was correlated with low or undetectable levels of GS_2; class II, which had normal or increased levels of GS_2 mRNA but very little GS_2 protein; and class III, which had significant amounts of GS_2 protein but little or no GS_2 activity (Freeman *et al.*, 1990).

Plants can synthesize all protein amino acids including those essential for humans (for a review see Matthews *et al.*, 1988). The chloroplast is the main, if not the sole, site for the synthesis of the aspartate-derived amino acids in plant leaves. For example, isolated intact chloroplasts carry out the light-driven synthesis of lysine, threonine, and isoleucine from labelled aspartic acid and malic acid (Mills *et al.*, 1990). This can be considered a photosynthetic process since ATP and NADPH produced in light are thought to be used directly to drive the synthesis. The shikimate pathway leads to the production of tryptophan, phenylalanine and tyrosine, as well as hormones, coumarins, quinones, flavins and lignin. Both plastidic and cytosolic forms have been reported for several of the enzymes of the shikimate pathways. 5-Enolpyruvylshikimate-3-phosphate synthase (EPSPS) catalyses the sixth step in the shikimate pathway. EPSPS is nuclear-encoded and mainly, if not solely, chloroplast localized (Smart & Amrhein, 1987). It is the primary target of the herbicid glyphosate (Roundup®). Using isolated genes encoding glyphosate-resistant EPSPS forms, transgenic plants have been obtained conferring glyphosate-resistance.

Many plants accumulate betaine for osmotic adjustment when subjected to salt stress. Betaine is synthesized by a two-step oxidation of choline in the chloroplast. The accumulation of betaine and the induction of betaine aldehyde dehydrogenase, which catalyses the last step in the synthesis of betaine, have been analysed in salt-stressed barley leaves (Arakawa *et al.*, 1990).

Several biosynthetic pathways in addition to those mentioned above are located in the chloroplast. In some cases all steps take place in the chloroplast whereas in others only some do so.

Concluding remarks

Molecular biology, particularly molecular genetics, is the newest and the most powerful approach in modern chloroplast research. Development of molecular biology techniques has provided new methods to solve old problems in many biological disciplines. The molecular analysis of barley chloroplast proteins has so far given valuable information on genes and molecules. In future studies, molecular biology has great potential for studying the complex interactions between genes and molecules and the biochemical pathways in the whole organism.

The molecular analysis of herbicide resistance has revealed the site of action of many herbicides and made it possible to construct herbicide-resistant crop plants. This might also become possible in barley. Some long-term applied goals could be to improve the efficiency of photosynthesis by improving the Rubisco enzyme, the light-harvesting with new photoreceptors and the temperature range. The molecular analysis of photosynthesis may contribute to improving solar cells by synthesis of molecules mimicking photosynthesis. Synthetic molecules containing the chemical groups of porphyrin rings, carotenoids and quinones on the same molecule have already been created by mimicking the photosynthetic reaction centres. These molecules can store about half of the photon's energy as chemical potential energy by charge separation and keeping the charges separated for up to 55 μs. If these molecules are positioned across a membrane, the charges might drive chemical reactions which store energy for later use or contribute to an electrical current. Hydrogen production by cyanobacteria and chloroplasts has previously been demonstrated and may be improved further by studies of PS I, ferredoxin and hydrogenase interactions.

Studies of the chloroplast have revealed many other useful results both for basic research and practical applications, and will in the future probably reveal many more secrets.

Acknowledgements

Professor Birger Lindberg Møller, Dr Henrik V. Scheller and Finn Okkels are thanked for a careful review of the manuscript. Professor Diter von Wettstein and Dr David Simpson are thanked for providing updated figures and suggestions, and Dr Penny von Wettstein-Knowles, Dr Bernhard Grimm, and Dr Simon Gough for advice and preprints.

References

Akey, C.W., Crepean, R.H., Dunn, S.D., McCarty, R.E. & Edelstein, S.J. (1983) Electron microscopy and single molecule averaging of subunit-deficient F_1-ATPases from *Escherichia coli* and spinach chloroplasts. *EMBO Journal* 2, 1409–13.

Allen, J.P., Feher, G., Yeates, T.O., Komiya, H. & Rees, D.C. (1987a) Structure of the reaction center from *Rhodobacter sphaeroides* R26: the cofactors. *Proceedings of the National Academy of Sciences (USA)* 84, 5730–4.

Allen, J.P., Feher, G., Yeates, T.O., Komiya, H. & Rees, D.C. (1987b) Structure of the reaction center from *Rhodobacter sphaeroides* R26: the protein subunits. *Proceedings of the National Academy of Sciences (USA)* 84, 6162–6.

Andersen, B., Koch, B., Scheller, H.V., Okkels, J.S. & Møller, B.L. (1990) Nearest neighbour analysis of the photosystem I subunits in barley and their binding of ferredoxin. In: Baltscheffsky, M. (ed.), *Current Research in Photosynthesis*, Vol. II, Kluwer, Dordrecht, pp. 671–4.

Apel, K., Santel, H.-J., Redlinger, T.E. & Falk, H. (1980) Protochlorophyllide holochrome of barley (*Hordeum vulgare* L.). Isolation and characterization of the NADP–protochlorophyllide oxidoreductase. *European Journal of Biochemistry* 111, 251–8.

Arakawa, K., Katayama, M. & Takabe, T. (1990) Levels of betaine and betaine aldehyde dehydrogenase activity in the green leaves, and etiolated leaves and roots of barley. *Plant Cell Physiology* 31, 797–803.

Baiama, S., Haegi, A., Strøman, P. & Casadro, G. (1989) Characterization of a cDNA clone for barley leaf glutamine synthetase. *Carls-*

berg Research Communications 54, 1–9.

Bassi, R. & Simpson, D.J. (1987a) Chlorophyll-protein complexes of barley photosystem I. *European Journal of Biochemistry* 163, 221–30.

Bassi, R. & Simpson, D.J. (1987b) The organization of photosystem II chlorophyll-proteins. In: Biggins, J. (ed.), *Progress in Photosynthesis Research II.* Nijhoff, Dordrecht, pp. 81–8.

Bassi, R., Machold, O. & Simpson, D.J. (1985) Chlorophyll-proteins of two photosystem I preparations from maize. *Carlsberg Research Communications* 50, 145–62.

Bassi, R., Høyer-Hansen, G., Barbato, R., Giacometti, G.M. & Simpson, D.J. (1987) Chlorophyll-proteins of the photosystem II antenna system. *Journal of Biological Chemistry* 262, 13333–41.

Bassi, R., Rigoni, F., Barbato, R. & Giacometti, G.M. (1988) Light harvesting complex (LHCII) populations in state I–state II transitions. *Biochimica et Biophysica Acta* 936, 29–32.

Baumgartner, B.J., Rapp, J.C. & Mullet, J.E. (1989) Plastid transcription activity and DNA copy number increase in barley chloroplast development. *Plant Physiology* 89, 1011–18.

Beinert, H. (1990) Recent developments in the field of iron–sulfur proteins. *FASEB Journal (American Society of Experimental Biology)* 4, 2483–91.

Bengis, C. & Nelson, N. (1977) Subunit structure of chloroplast photosystem I reaction center. *Journal of Biological Chemistry* 252, 4564–9.

Berry-Lowe, S. (1987) The chloroplast glutamate tRNA gene required for δ-aminolevulinate synthesis. *Carlsberg Research Communications* 52, 197–210.

Boyer, S.K. & Mullet, J.E. (1988) Sequence and transcript map of barley chloroplast *psbA* gene. *Nucleic Acids Research* 16, 8184.

Brown, D.J. & DuPont, F.M. (1989) Lipid composition of plasma membranes and endomembranes prepared from roots of barley (*Hordeum vulgare* L.). *Plant Physiology* 90, 955–61.

Bruce, B.D. & Malkin, R. (1988) Subunit stoichiometry of the chloroplast photosystem I complex. *Journal of Biological Chemistry* 263, 7302–8.

Camm, L.E. & Green, B.R. (1980) Fractionation of thylakoid membranes with the nonionic detergent octyl-β-D-glucopyranoside. Resolution of chlorophyll-protein complex II into two chlorophyll-protein complexes. *Plant Physiology* 66, 428–32.

Castresana, C., Staneloni, R., Malik, V.S. & Cashmore, A.R. (1987) Molecular characterization of two clusters of genes encoding the Type I CAB polypeptides of PSII in *Nicotiana plumbaginifolia. Plant Molecular Biology* 10, 117–26.

Chakhmakhcheva, O.G., Andreva, A.V., Buryakova, A.A., Reverdatto, S.V. & Efimov, V.A. (1989) Nucleotide sequence of the barley chloroplast *psbE, psbF* genes and flanking regions. *Nucleic Acids Research* 17, 2858.

Chitnis, P.R. & Thornber, J.P. (1988) The major light-harvesting complex of photosystem II: aspects of its molecular and cell biology. In: Govindjee, Bohnert, H.J., Bottomley, W., Bryant, D.A., Mullet, J.E., Ogren, W.L., Pakrasi, H. & Somerville, C.R. (eds), *Molecular Biology of Photosynthesis*. Kluwer Academic Publishers, Dordrecht, pp. 259–81.

Chitnis, P.R., Morishige, D.T., Nechushtai, R. & Thornber, J.P. (1988) Assembly of the light-harvesting chlorophyll a/b proteins in barley etiochloroplasts involves processing of the precursor on thylakoids. *Plant Molecular Biology* 11, 95–107.

Chitnis, P.R., Reilly, P.A. & Nelson, N. (1989a) Insertional inactivation of the gene encoding subunit II of photosystem I from the cyanobacterium *Synechocystis* sp. PCC 6803. *Journal of Biological Chemistry* 264, 18381–5.

Chitnis, P.R., Reilly, P.A., Miedel, M.C. & Nelson, N. (1989b) Structure and targeted mutagenesis of the gene encoding 8-kDa subunit of photosystem I from the cyanobacterium *Synechocystis* sp. PCC 6803. *Journal of Biological Chemistry* 264, 18374–80.

Deisenhofer, J., Epp, O., Miki, K., Huber, R. & Michel, H. (1985) Structure of the protein subunits in the photosynthetic reaction centre of *Rhodopseudomonas viridis* at 3 Å resolution. *Nature* 318, 618–24.

de Pater, S., Hensgens, L.A.M. & Schilperoort, R.A. (1990) Structure and expression of a light-inducible shoot-specific rice gene. *Plant Molecular Biology* 15, 399–406.

Dunahay, T.G. & Staehelin, L.A. (1986) Isolation and characterization of a new minor chlorophyll a/b complex (CP24) from spinach. *Plant Physiology* 80, 429–34.

Dunn, P.P.J. & Gray, J.C. (1988) Localization and nucleotide sequence of the gene for the 8 kDa subunit of photosystem I in pea and wheat chloroplast DNA. *Plant Molecular Biology* 11, 311–19.

Dunn, P.P.J., Packman, L.C., Pappin, D. & Gray, J.C. (1988) N-terminal amino acid sequence analysis of the subunits of pea photosystem I. *FEBS Letters* 228, 157–61.

Efimov, V.A., Andreeva, A.V., Reverdatto, S.V., Jung, R. & Chakhmakhcheva, O.G. (1988a) Nucleotide sequence of the barley chloroplast psbA gene for the Q_B protein of photosystem II. *Nucleic Acids Research* 16, 5685.

Efimov, V.A., Andreeva, A.V., Reverdatto & Chakhmakhcheva, O.G. (1988b) Nucleotide sequence of the barley chloroplast psbD gene for the D2 protein of photosystem II. *Nucleic Acids Research* 16, 5686.

Eichacker, L.A., Soll, J., Lauterbach, P., Rüdiger, W., Klein, R.R. & Mullet, J.E. (1990) *In vitro* synthesis of chlorophyll *a* in the dark triggers accumulation of chlorophyll *a* apoproteins in barley etioplasts. *Journal of Biological Chemistry* 265, 13566–71.

Ellis, R.J. (1987) Proteins as molecular chaperones. *Nature* 328, 378–9.

Ellis, R.J. (1990) Molecular chaperones. *Seminars in Cell Biology* 1 (1), 1–72.

Ellis, R.J. & van der Vies, S.M. (1988) The Rubisco subunit binding protein. *Photosynthesis Research* 16, 101–15.

Evans, E.H., Dickson, D.P.E., Johnson, C.E., Rush, J.D. & Evans, M.C.W. (1981) Mössbauer spectroscopic studies of the nature of centre X of photosystem I reaction centres from the cyanobacterium *Chlorogloea fritschii*. *European Journal of Biochemistry* 118, 82–4.

Falkowski, P.G., Fujita, Y., Ley, A. & Mauzerall, D. (1986) Evidence for cyclic electron flow around photosystem II in *Chlorella pyrenoidosa*. *Plant Physiology* 81, 310–12.

Fish, L.E. & Bogorad, L. (1986) Identification and analysis of the maize P700 chlorophyll a apoproteins PSI-A1 and PSI-A2 by high pressure liquid chromatography analysis and partial sequence determination. *Journal of Biological Chemistry* 261, 8134–9.

Franzén, L.-G., Frank, G., Zuber, H. & Rochaix, J.-D. (1989a) Isolation and characterization of cDNA clones encoding the 17.9 and 8.1 kDa subunits of photosystem I from *Chlamydomonas reinhardtii*. *Plant Molecular Biology* 12, 463–74.

Franzén, L.-G., Frank, G., Zuber, H. & Rochaix, J.-D. (1989b) Isolation and characterization of cDNA clones encoding photosystem I subunits with molecular masses 11.0, 10.0 and 8.4 kDa from *Chlamydomonas reinhardtii*. *Molecular and General Genetics* 219, 137–44.

Freeman, J., Marquez, A.J., Wallsgrove, R.M., Saarelainen, R. & Forde, B.G. (1990) Molecular analysis of barley mutants deficient in chloroplast glutamine synthetase. *Plant Molecular Biology* 14, 297–311.

Ghanotakis, D.F. & Yocum C.F. (1990) Photosystem II and the oxygen-evolving complex. *Annual Review of Plant Physiology and Plant Molecular Biology* 41, 255–76.

Golbeck, J.H. (1987) Structure, function and organization of the photosystem I reaction center complex. *Biochimica et Biophysica Acta* 895, 167–204.

Golbeck, J.H. & Bryant, D. (1991) Photosystem I. In: Lee, C.P. (ed.), *Current Topics in Bioenergetics* Vol. 16. Academic Press, New York, pp. 83–177.

Gough, S. (1972) Defective synthesis of porphyrins in barley plastids caused by mutation in nuclear genes. *Biochimica et Biophysica Acta* 286, 36–54.

Grimm, B. (1990) Primary structure of a key enzyme in plant tetrapyrrole synthesis: glutamate 1-semialdehyde aminotransferase. *Proceedings of the National Academy of Sciences (USA)* 87, 4169–73.

Grimm, B., Kruse, E. & Kloppstech, K. (1989) Transiently expressed early light-inducible thylakoid proteins share transmembrane domains with light-harvesting chlorophyll binding proteins. *Plant Molecular Biology* 13, 583–93.

Grimm, B., Bull, A. & Breu, V. (1991) Structural genes of glutamate 1-semialdehyde aminotransferase for porphyrin synthesis in cyanobacterium and *E. coli*. *Molecular and General Genetics* 225, 1–10.

Hageman, J., Baecke, C., Ebskamp, M., Pilon, R., Smeekens, S. & Weisbeek, P. (1990) Protein import and sorting inside the chloroplast are independent processes. *Plant Cell* 2, 479–94.

Haley, J. & Bogorad, L. (1989) A 4-kDa maize chloroplast polypeptide associated with the cytochrome *b6–f* complex: subunit 5, encoded by the chloroplast *petE* gene. *Proceedings of the National Academy of Sciences (USA)* 86, 1534–8.

Hallick, R.B. (1989) Proposals for the naming of chloroplast genes. II. Update to the nomenclature of genes for thylakoid membrane polypeptides. *Plant Molecular Biology Reporter* 7, 266–75.

Hansen, L. (1987) Three cDNA clones for barley

leaf acyl carrier proteins I and III. *Carlsberg Research Communications* 52, 381–92.

Harwood, J.L. (1988) Fatty acid metabolism. *Annual Review of Plant Physiology* 39, 101–38.

Hemmingsen, S.M. & Ellis, R.J. (1986) Purification and properties of ribulosebisphosphate carboxylase large subunit binding protein. *Plant Physiology* 80, 269–76.

Hemmingsen, S.M., Woolford, C., van der Vies, S.M., Tilly, K., Dennis, D.T., Georgopoulus, C.P., Hendrix, R.W. & Ellis R.J. (1988) Homologous plant and bacterial proteins chaperone oligomeric protein assembly. *Nature* 333, 330–4.

Henningsen, K.W. & Stummann, B.M. (1982) Use of mutants in the study of chloroplast biogenesis. In: Parthier, B. & Boulter, D. (eds), *Encyclopedia of Plant Physiology*, Vol. 14B. Springer-Verlag, Berlin, pp. 597–644.

Henry, L.E.A. & Møller, B.L. (1981) Polypeptide composition of an oxygen evolving photosystem II vesicle from spinach chloroplasts. *Carlsberg Research Communications* 46, 227–42.

Henry, L.E.A., Mikkelsen, J.D. & Møller, B.L. (1983) Pigment and acyl lipid composition of photosystem I and II vesicles and of photosynthetic mutants in barley. *Carlsberg Research Communications* 48, 131–48.

Hiller, R.G., Møller, B.L. & Høyer-Hansen, G. (1980) Characterization of six putative photosystem I mutants in barley. *Carlsberg Research Communications* 45, 315–28.

Hippler, M., Ratajczak, R. & Haehnel, W. (1989) Identification of the plastocyanin binding subunit of photosystem I. *FEBS Letters* 250, 280–4.

Hiratsuka, J., Shimada, H., Whittier, R., Ishibashi, T., Sakamoto, M., Mori, M., Kondo, C., Honji, Y., Sun, C.-R., Meng, B.-Y., Li, Y.-Q., Kanno, A., Nishizawa, Y., Hirai, A., Shinozaki, K. & Sugiura, M. (1989) The complete sequence of the rice (*Oryza sativa*) chloroplast genome: intermolecular recombination between distinct tRNA genes accounts for a major plastid DNA inversion during the evolution of the cereals. *Molecular and General Genetics* 217, 185–94.

Høj, P.B. & Svendsen, I. (1984) Barley chloroplasts contain two acyl carrier proteins coded for by different genes. *Carlsberg Research Communications* 49, 483–92.

Høj, P.B. & Møller, B.L. (1986) The 110-kDa reaction center protein of photosystem I, P700-chlorophyll *a*-protein 1, is an iron-sulfur protein. *Journal of Biological Chemistry* 261, 14292–300.

Høj, P.B., Svendsen, I., Scheller, H.V. & Møller, B.L. (1987) Identification of a chloroplast-encoded 9-kDa polypeptide as a 2[4Fe–4S] protein carrying centers A and B of photosystem I. *Journal of Biological Chemistry* 262, 12676–84.

Hoshina, S., Sue, S., Kunishima, N., Kamide, K., Wada, K. & Itoh, S. (1989) Characterization N-terminal sequence of a 5 kDa polypeptide in the photosystem I core complex from spinach. *FEBS Letters* 258, 305–8.

Hudson, G.S. & Mason, J.G. (1988) The chloroplast genes encoding subunits of the H^+-ATP synthase. In: Govindjee, Bohnert, H.J., Bottomley, W., Bryant, D.A., Mullet, J.E., Ogren, W.L., Pakrasi, H. & Somerville, C.R. (eds), *Molecular Biology of Photosynthesis*. Kluwer, Dordrecht, pp. 565–82.

Ikeuchi, M., Hirano, A., Hiyama, T. & Inuoe, Y. (1990) Polypeptide composition of higher plant photosystem I complex. *FEBS Letters* 263, 274–8.

Islam, K. (1987) The rate and extent of phosphorylation of the two light-harvesting chlorophyll a/b-binding protein complex (LHCII) polypeptides in isolated spinach thylakoids. *Biochimica et Biophysica Acta* 893, 333–42.

Jansen, T., Rother, C., Steppuhn, J., Reincke, H., Beyreuther, K., Jansson, C., Andersson, B. & Herrmann, R.G. (1987) Nucleotide sequence of cDNA clones encoding the complete '23 kDa' and '16 kDa' precursor proteins associated with the photosynthetic oxygen-evolving complex from spinach. *FEBS Letters* 216, 234–40.

Kauppinen, S. (1990) Molecular cloning of the gene(s) encoding barley β-ketoacyl-ACP synthase I. In: *Proceedings 9th International Symposium Plant Lipid Biochemistry, Structure and Utilization*, Wye, UK.

Klein, R.R. & Mullet, J.E. (1986) Regulation of chloroplast encoded chlorophyll-binding protein translation during higher plant chloroplast biogenesis *Journal of Biological Chemistry* 261, 11138–45.

Klein, R.R. & Mullet, J.E. (1987) Control of gene expression during higher plant chloroplast biogenesis. *Journal of Biological Chemistry* 262, 4341–8.

Klein, R.R., Gamble P.E. & Mullet, J.E. (1988) Light-dependent accumulation of radiolabeled plastid-encoded chlorophyll *a*-apoproteins requires chlorophyll *a*. *Plant Physiology* 88, 1246–56.

Klenig, H. (1989) The role of plastids in isoprenoid biosynthesis. *Annual Review of Plant Physiology and Plant Molecular Biology* 40, 39–59.

Koike, H., Ikeuchi, M., Hiyama, T. & Inoue, Y. (1989) Identification of photosystem I components from the cyanobacterium *Synechococcus vulcanus* by N-terminal sequencing. *FEBS Letters* 253, 257–63.

Kotzabasis, K. & Senger, H. (1990) Diversity of the pathways from protochlorophyllide to chlorophylls a and b. In: Baltscheffsky, M. (ed.), *Current Research in Photosynthesis*, Vol. III. Kluwer Academic Publishers, Dordrecht, pp. 881–4.

Krupinska, K. (1988) Characterization and in vitro expression of the cytochrome *b*-559 genes of barley. II. In vitro transcription and translation. *Carlsberg Research Communications* 53, 233–46.

Lagoutte, B. & Mathis, P. (1989) The photosystem I reaction center: structure and photochemistry. *Photochemistry and Photobiology* 49, 833–44.

Lautner, A., Klein, R., Ljungberg, U., Reiländer, H., Bartling, D., Andersson, B., Reinke, H., Beyreuther, K. & Herrmann, R.G. (1988) Nucleotide sequence of cDNA clones encoding the complete precursor for the '10 kDa' polypeptide of photosystem II from spinach. *Journal of Biological Chemistry* 263, 10077–81.

Leutwiler, L.S., Meyerowitz, E.M. & Tobin, E.M. (1986) Structure and expression of three light-harvesting chlorophyll a/b-binding protein genes in *Arabidopsis thaliana*. *Nucleic Acids Research* 14, 4051–76.

Lubben, T.H., Donaldsen, G.K., Viitanen, P.V. & Gatenby, A.A. (1989) Several proteins imported into chloroplasts form stable complexes with the groEL-related chloroplast molecular chaperone. *Plant Cell* 1, 1223–30.

Machold, O., Simpson, D.J. & Møller, B.L. (1979) Chlorophyll-proteins of thylakoids from wild-type and mutants of barley (*Hordeum vulgare* L.); *Carlsberg Research Communications* 44, 235–54.

Marder, J.B. & Barber, J. (1989) The molecular anatomy and function of thylakoid proteins. *Plant, Cell and Environment* 12, 595–614.

Mattoo, A.K., Marder, J.B. & Edelman, M. (1989) Dynamics of photosystem II reaction center. *Cell* 56, 241–6.

Mayfield, S.P., Bennoun, P. & Rochaix, J.-D. (1987a) Expression of the nuclear encoded OEE1 protein is required for oxygen evolution and stability of photosystem II particles in *Chlamydomonas reinhardtii*. *EMBO Journal* 6, 313–18.

Mayfield, S.P., Rahire, M., Frank, G., Zuber, H. & Rochaix, J.-D. (1987b) Expression of the nuclear gene encoding oxygen-evolving enhancer protein 2 is required for high levels of photosynthetic oxygen evolution in *Chlamydomonas reinhardtii*. *Proceedings of the National Academy of Sciences (USA)* 84, 749–53.

Matthews, B.F., Reardon, E.M., Turano, F.J. & Wilson, B.J. (1988) Amino acid biosynthesis in plants: approaching an understanding at the molecular level. *Plant Molecular Biology Reporter* 6, 137–54.

Mehari, T., Parrett, K.G., Warren, P.V. & Golbeck, J.H. (1991) Reconstitution of the iron sulfur clusters in the isolated F_A/F_B protein: ESR spectral characterization of same-species & cross-species photosystem I complexes. *Biochimica et Biophysica Acta* 1056, 139–48.

Merchant, S. & Selman, B.R. (1985) Photosynthetic ATPases: purification, properties, subunit isolation and function. *Photosynthetic Research* 6, 3–31.

Michel, H., Weyer, K.A., Gruenberg, H., Dugner, I., Oesterhelt, D. & Lottspeich, F. (1986) The 'light' and 'medium' subunits of the photosynthesis reaction centre from *Rhodopseudomonas viridis*; isolation of the genes, nucleotide and amino acid sequence. *EMBO Journal* 5, 1149–58.

Mills, W.R., Capo, S.F., Bergh, S.A. & Lassiter, C.B. (1990) Developmental variation in aspartate-family amino acid biosynthesis by isolated chloroplasts. In: Baltscheffsky, M. (ed.), *Current Research in Photosynthesis* Vol. IV. Kluwer, Dordrecht, pp. 271–4.

Møller, B.L., Nugent, J.H.A. & Evans, M.C.W. (1981) Electron paramagnetic resonance spectrometry of photosystem I mutants in barley. *Carlsberg Research Communications* 46, 373–82.

Møller, B.L., Høj, P.B. & Henry, L.E.A. (1983) Electron microscopic characteristics of photosystem II preparations and their inactivation and reactivation with respect to oxygen evolution. *Photosynthesis, Proceedings 6th International Congress III* 3, 203–6.

Møller, B.L., Scheller, H.V., Okkels, J.S., Koch, B., Andersen, B., Nielsen, H.L., Olsen, I.,

Halkier B.A. & Høj, P.B. (1990) Chloroplast encoded photosystem I polypeptides of barley. In: Baltscheffsky, M. (ed.), *Current Research in Photosynthesis*, Vol. II. Kluwer Academic Publishers, Dordrecht, pp. 523–30.

Mullet, J.E. (1988) Chloroplast development and gene expression. *Annual Review of Plant Physiology* 39, 475–502.

Mullet, J.E., Grossman, A.R. & Chua, N.-H. (1982) Synthesis and assembly of the polypeptide subunits of photosystem I. *Cold Spring Harbor Symposium on Quantitative Biology* 46, 979–84.

Mullet, J.E., Klein, P.G. & Klein, R.R. (1990) Chlorophyll regulates accumulation of the plastid-encoded chlorophyll apoproteins CP43 and D1 by increasing apoprotein stability. *Proceedings of the National Academy of Sciences (USA)* 87, 4038–42.

Münch, S., Ljungberg, U., Steppuhn, J., Schneiderbauer, A., Nechushtai, R., Beyreuter, K. & Herrmann, R.G. (1988) Nucleotide sequences of cDNAs encoding the entire precursor polypeptides of subunits II and III of the photosystem I reaction center from spinach. *Current Genetics* 14, 511–18.

Mushgrove, J.E., Johnson, R.A. & Ellis, R.J. (1987) Dissociation of the ribulosebisphosphate-carboxylase large-subunit binding protein into dissimilar subunits. *European Journal of Biochemistry* 163, 529–34.

Nechushtai, R., Nelson, N., Mattoo, A.K. & Edelman, M. (1981) Site of synthesis of subunits to photosystem I reaction center and the proton-ATPase in *Spirodela*. *FEBS Letters* 125, 115–19.

Neumann, E. (1988) Primary structure of barley genes encoding quinone and chlorophyll *a* binding proteins of photosystem II. *Carlsberg Research Communications*, 53, 259–75.

Neilsen, P.S. & Gausing, K. (1987) The precursor of barley plastocyanin. *FEBS Letters* 225, 159–62.

Obokata, J. (1986) Synthesis of a photosystem I polypeptide of 15 kilodaltons in isolated etiochloroplasts of wheat. *Plant Physiology* 81, 705–7.

Ohlrogge, J.B., Kuhn, D.N. & Stumpf, P.K. (1979) Subcellular localization of acyl carrier protein in leaf protoplasts of *Spinacia oleracea*. *Proceedings of the National Academy of Sciences (USA)* 76, 1194–8.

Oh-Oka, H., Takahashi, Y., Kuriyama, K. Saeki, K. & Matsubara, H. (1988) The protein responsible for center A/B in spinach photosystem I: isolation with iron–sulfur cluster(s) and complete sequence analysis. *Journal of Biochemistry* 103: 962–8.

Ohyama, K., Fukuzawa, H., Kohchi, T., Shirai, H., Sano, T., Sano, S., Umesono, K., Shiki, Y., Takeuchi, M., Chang, Z., Aota, S.-i., Inokuchi, H. & Ozeki, H. (1986) Chloroplast gene organization deduced from complete sequence of liverwort *Marchantia polymorpha* chloroplast DNA. *Nature* 322, 572–4.

Okkels, J.S., Jepsen, L.B., Hønberg, L.S., Lehmbeck, J., Scheller, H.V., Brandt, P., Høyer-Hansen, G., Stummann, B., Henningsen, K.W., Wettstein, D.v. & Møller, B.L. (1988) A cDNA clone encoding a 10.8 kDa photosystem I polypeptide of barley. *FEBS Letters* 237, 108–12.

Okkels, J.S., Scheller, H.V., Jepsen, L.B. & Møller, B.L. (1989) A cDNA clone encoding the precursor for a 10.2 kDa photosystem I polypeptide of barley. *FEBS Letters* 250, 575–9.

Okkels, J.S., Scheller, H.V., Jepsen, L.B., Andersen, B. & Møller B.L. (1990) Characterization of a cDNA clone for the *PsaH* gene from barley and mRNA level of PS I genes in light-induced barley seedlings. In: Baltscheffsky, M. (ed.), *Current Research in Photosynthesis*, Vol. III. Kluwer Academic Publishers, Dordrecht, pp. 613–16.

Okkels, J.S., Scheller, H.V., Svendsen, I. & Møller, B.L. (1991) Isolation and characterization of a cDNA clone encoding an 18-kDa hydrophobic photosystem I subunit (PSI-L) from barley (*Hordeum vulgare* L.). *Journal of Biological Chemistry* 266, 6767–73.

Oliver, R.P. & Poulsen C. (1984) Structure of a heavily transcribed region of barley chloroplast DNA. Transfer RNA genes for serine (UGA), glycine (GCC, UCC), formylmethionine and threonine (GGU). *Carlsberg Research Communications* 49, 647–73.

Paolo, M.L.D., Peruffo, A.D.B. & Bassi, R. (1990) Immunological studies on chlorophyll-a/b proteins and their distribution in the thylakoid membrane domains. *Planta* 181, 275–86.

Pichersky, E. & Green, B.R. (1990) The extended family of chlorophyll a/b-binding proteins of PSI and PSII. In: Baltscheffsky, M. (ed.), *Current Research in Photosynthesis*, Vol. III. Kluwer Academic Publishers, Dordrecht, pp. 553–6.

Pichersky, E., Hoffman, N.E., Bernatzky, R., Piechulla B., Tanksley S.D. & Cashmore, A.R. (1987) Molecular characterization and genetic mapping of DNA sequences encoding the Type I chlorophyll a/b-binding polypeptide of photosystem I in *Lycopersicon esculentum* (tomato). *Plant Molecular Biology* 9, 205–16.

Pichersky, E., Tanksley, S.D., Piechulla, B., Stayton, M.M. & Dunsmuir, P. (1988) Nucleotide sequence and chromosomal location of *Cab*-7, the tomato gene encoding the Type II chlorophyll a/b-binding polypeptide of photosystem I. *Plant Molecular Biology* 11, 69–71.

Pichersky, E., Brock, T.G., Nguyen, D., Hoffman, N.E., Piechulla, B., Tanksley, S.D. & Green, B.R. (1989) A new member of the CAB gene family: structure, expression and chromosomal location of *Cab*-8, the tomato gene encoding the Type III chlorophyll a/b-binding polypeptide of photosystem I. *Plant Molecular Biology* 12, 257–70.

Poulsen, C. (1981) Comments on the structure and function of the large subunit of the enzyme ribulose bisphosphate carboxylase-oxygenase. *Carlsberg Research Communications* 46, 259–78.

Poulsen, C. (1983) The barley chloroplast genome: physical structure and transcriptional activity in vivo. *Carlsberg Research Communications* 48, 57–80.

Poulsen, C. (1984) Two mRNA species differing by 258 nucleotides at the 5′end are formed from the barley chloroplast *rbcL* gene. *Carlsberg Research Communications* 49, 89–104.

Reilly, P. & Nelson, N. (1988) Photosystem I complex. *Photosynthesis Research* 19, 73–84.

Reverdatto, S.V., Andreeva, A.V., Buryakova, A.A., Chakhmakhcheva, O.G. & Efimov, V.A. (1989a) Nucleotide sequence of the barley *psbC* gene. *Nucleic Acids Research* 17, 1422.

Reverdatto, S.V., Andreeva, A.V., Buryakova, A.A., Chakhmakhcheva, O.G. & Efimov, V.A. (1989b) Nucleotide sequence of the 5.2 kbp barley chloroplast DNA fragment, containing *psbH*-psbH–petB–petD gene cluster. *Nucleic Acids Research* 17, 2859–60.

Rhiel, E. & Bryant, D.A. (1988) Preliminary results concerning the *psaC, psaD*, and *psaF* genes and their products in the cyanobacteria *Synechococcus* sp. PCC 7002 and *Nostoc* sp. PCC 8009. In: Stevens, S.E. & Bryant D.A. (eds), *Light-Energy Transduction in Photosynthesis: Higher Plants and Bacterial Models*. The American Society of Plant Physiologists, Rockville, MD, USA, pp. 320–3.

Richter, M.L., Patrie, W.J. & McCarty, R.E. (1984) Preparation of the ε subunit and ε subunit-deficient chloroplast coupling factor 1 in reconstitutively active forms. *Journal of Biological Chemistry* 259, 7371–3.

Robert, B. & Moenne-Loccoz, P. (1989) Un site possible pour l'accepteur primaire d'électrons du photosystème 1. *Comptes Rendûs des Sciences, Paris* 308, Série III, 407–9.

Robertson, D. & Laetsch W.M. (1974) Structure and function of developing barley plastids. *Plant Physiology* 54, 148–59.

Rochaix, J.-D. & Erickson, J. (1988) Function and assembly of photosystem II: genetic and molecular analysis. *Trends in Biochemical Sciences* 13, 56–9.

Rochester, C.P., Kjellbom, P., Andersson, B. & Larsson C. (1987) Composition of plasma membranes from light-grown barley (*Hordeum vulgare*) leaves: identification of cerebroside as a major component. *Archives of Biochemistry and Biophysics* 255, 385–91.

Rohde, W., Becker, D. & Salamini, F. (1988) Structural analysis of the *waxy* locus from *Hordeum vulgare*. *Nucleic Acids Research* 16, 7185–6.

Roy, H. (1989) Rubisco assembly: a model system for studying the mechanism of chaperonin action. *Plant Cell* 1, 1035–42.

Rutherford, A.W. (1989) Photosystem II, the water-splitting enzyme. *Trends in Biochemical Sciences* 14, 227–32.

Schantz, R. & Bogorad, L. (1988) Maize chloroplast genes *ndh*D, *ndh*E, and *psa*C. Sequences, transcripts and transcript pools. *Plant Molecular Biology* 11, 239–47.

Scheller, H.V. & Møller, B.L. (1990) Photosystem I polypeptides. *Physiologia Plantarum* 78, 484–94.

Scheller, H.V., Høj, P.B., Svendsen, I. & Møller, B.L. (1988) Partial amino acid sequences of two nuclear-encoded photosystem I polypeptides from barley. *Biochimica et Biophysica Acta* 933, 501–5.

Scheller, H.V., Svendsen, I. & Møller, B.L. (1989a) Subunit composition of photosystem I and identification of center X as a [4Fe–4S] cluster. *Journal of Biological Chemistry* 264, 6929–34.

Scheller, H.V., Svendsen, I. & Møller, B.L. (1989b) Amino acid sequence of the 9-kDa iron–sulfur protein of photosystem I in barley. *Carlsberg Research Communications* 54, 11–15.

Scheller, H.V., Okkels, J.S., Høj. P.B., Svendsen, I., Roepstorff, P. & Møller, B.L. (1989c) The primary structure of a 4.0-kDa photosystem I polypeptide encoded by the chloroplast *psaI* gene. *Journal of Biological Chemistry* 264, 18402–6.

Scheller, H.V., Andersen, B., Okkels, J.S., Svendsen, I. & Møller, B.L. (1990a) Photosystem I in barley: subunit PSI-F is not essential for the interaction with plastocyanin. In: Baltscheffsky, M. (ed.), *Current Research in Photosynthesis*, Vol. II. Kluwer Academic Publishers, Dordrecht, pp. 679–82.

Scheller, H.V., Okkels, J.S., Roepstorff, P., Jepsen, L.B. & Møller, B.L. (1990b) Characterization of a cDNA clone for the *PsaE* gene from barley and plasma desorption mass spectrometry of the corresponding photosystem I polypeptide PSI-E. In: Baltscheffsky, M. (ed.), *Current Research in Photosynthesis*, Vol. III. Kluwer Academic Publishers, Dordrecht, pp. 609–12.

Schneegurt, M.A., Rieble, S. & Beale, S.I. (1988) The tRNA required for *in vitro* delta-aminolevulinic acid formation from glutamate in *Synechocystis* extracts. *Plant Physiology*, 88, 1358–66.

Schultz, R., Steinmüller, K., Klaas, M., Forreiter, C., Rasmussen, S., Hiller, C. & Apel, K. (1989) Nucleotide sequence of a cDNA coding for the NADPH-protochlorophyllide oxidoreductase (PCR) of barley (*Hordeum vulgare* L.) and its expression in *Escherichia coli. Molecular and General Genetics* 217, 355–61.

Schwartz, E. & Pichersky, E. (1990) Sequence of two tomato nuclear genes encoding chlorophyll *a*/*b*-binding proteins of CP24, a PSII antenna component. *Plant Molecular Biology* 15, 157–60.

Sebald, W. & Wachter, E. (1980) Amino acid sequence of the proteolipid subunit of CF_0 (*atp*H). *Genetics* 116, 127–39.

Sexton, T.B., Jones, J.T. & Mullet, J.E. (1990) Sequence and transcriptional analysis of the barley ctDNA region upstream of *psbD–psbC* encoding *trnK*(UUU), *rps*16, *trnQ*(UUG), *psbK, psbI*, and *trnS*(GCU). *Current Genetics* 17, 445–54.

Shinozaki, K., Ohme, M., Tanaka, M., Wakasugi, T., Hayashida, N., Matsubayashi, T., Zaita, N., Chunwongse, J., Obokata, J., Yamaguchi-Shinozaki, K., Ohto, C., Torazawa, K., Meng, B.Y., Sugita, M., Deno, H., Kamogashira, T., Yamada, K., Kusuda, J., Takaiwa, F., Kato, A., Tohdoh, N., Shimada, H. & Sugiura, M. (1986a) The complete nucleotide sequence of the tobacco chloroplast genome: its gene organization and expression. *EMBO Journal* 5, 2043–9.

Shinozaki, K., Ohme, M., Tanaka, M., Wakasugi, T., Hayashida, N., Matsubayashi, T., Zaita, N., Chunwongse, J., Obokata, J., Yamaguchi-Shinozaki, K., Ohto, C., Torazawa, K., Meng, B.Y., Sugita, M., Deno, H., Kamogashira, T., Yamada, K., Kusuda, J., Takaiwa, F., Kato, A., Tohdoh, N., Shimada, H. & Sugiura, M. (1986b) The complete nucleotide sequence of the tobacco chloroplast genome. *Plant Molecular Biology Reporter* 4, 111–47.

Siggård-Andersen, M. (1990) Purification and characterization of a β-ketoacyl-ACP synthase from barley chloroplasts. In: *Proceedings 9*th *International Symposium Plant Lipid Biochemistry, Structure and Utilization*, Wye, UK.

Simpson, D. (1983) Freeze-fracture studies on barley plastid membranes. VI. Location of the P700-chlorophyll *a*-protein 1. *European Journal of Cell Biology* 31, 305–14.

Simpson, D. & von Wettstein, D. (1980) Macromolecular physiology of plastids XIV. *Viridis* mutants in barley: Genetic, fluoroscopic and ultrastructural characterization. *Carlsberg Research Communications* 45, 283–314.

Simpson, D., Vallon, O. & von Wettstein, D. (1989) Freeze-fracture studies on barley membranes. VIII. In *viridis-115*, a mutant completely lacking photosystem II, oxygen evolution enhancer 1 (OEE1) and the α-subunit of cytochrome *b*-559 accumulate in appressed thylakoids. *Biochimica et Biophysica Acta* 975, 164–74.

Smart, C.C. & Amrhein, N. (1987) Ultrastructural localization by protein A-gold immunocytochemistry of 5-enolpyruvylshikimic acid 3-phosphate synthase in a plant cell culture which overproduces the enzyme. *Planta* 170, 1–6.

Smeekens, S., Weisbeek, P. & Robinson, C. (1990) Protein transport into and within chloroplasts. *Trends in Biochemical Sciences* 15, 73–6.

Smith, A.G. & Gray, J.C. (1984) Synthesis of components of photosystem I in pea. In: Sybesma, C. (ed.), *Advances in Photosynthesis Research*, Vol. 4. Martinus Nijhoff/Dr W. Junk, The Hague, pp. 513–16.

Smith, A.M., Quinton-Tulloch, J. & Denyer, K. (1990) Characteristics of plastids responsible for starch synthesis in developing pea embryos. *Planta* 180, 517–23.

Spangfort, M., Larsson, U.K., Ljungberg, U., Ryberg, M., Andersson, B., Bartling, G.,

Wedel, N. & Herrmann, R.G. (1990) The 20 kDa apo-polypeptide of the chlorophyll a/b protein complex CP24 – characterization and complete primary amino acid sequence. In: Baltscheffsky, M. (ed.), *Current Research in Photosynthesis*, Vol. II, Kluwer Academic Publishers, Dordrecht, pp. 253–6.

Stayton, M.M., Brosio, P. & Dunsmuir, P. (1987) Characterization of a full-length petunia cDNA encoding a polypeptide of the light-harvesting complex associated with photosystem I. *Plant Molecular Biology* 10, 127–37.

Steppuhn, J., Hermans, J., Nechushtai, R., Ljungberg, U., Thümmler, F., Lottspeich, F. & Herrmann, R.G. (1988) Nucleotide sequence of cDNA clones encoding the entire precursor polypeptides for subunits IV and V of the photosystem I reaction center from spinach. *FEBS Letters* 237, 218–24.

Steppuhn, J., Hermans, J., Nechushtai, R., Ljungberg, U., Herrmann, G.S. & Herrmann, R.G. (1989) Nucleotide sequences of cDNA clones encoding the entire precursor polypeptide for subunit VI and of the plastome-encoded gene for subunit VII of the photosystem I reaction center from spinach. *Current Genetics* 16, 99–108.

Stockhaus, J., Höfer, M., Renger, G., Westhoff, P., Wydrzynski, T. & Willmitzer, L. (1990) Anti-sense RNA efficiently inhibits formation of the 10 kd polypeptide of photosystem II in transgenic potato plants: analysis of the role of the 10 kd protein. *EMBO Journal* 9, 3013–21.

Tyagi, A., Hermans, J., Steppuhn, J., Jansson, Ch., Vater, F. & Herrmann, R.G. (1987) Nucleotide sequence of cDNA clones encoding the complete '33 kDa' precursor protein associated with the photosynthetic oxygen-evolving complex from spinach. *Molecular and General Genetics* 207, 288–93.

Trebst, A. (1985) The topology of the plastoquinone and herbicide binding peptides of photosystem II in the thylakoid membrane. *Zeitschrift für Naturforschung C* 41, 240–5.

Vallon, O., Høyer-Hansen, G. & Simpson, D.J. (1987) Photosystem II and cytochrome *b*-559 in the stroma lamellae of barley chloroplasts. *Carlsberg Research Communications* 52, 405–21.

Vermaas, W.F.J., Ikeuchi, M. & Inoue, Y. (1988) Protein composition of the photosystem II core complex in genetically engineered mutants of the cyanobacterium *Synechocystis* sp. PCC 6803. In: Govindjee, Bohnert, H.J., Bottomley, W., Bryant, D.A., Mullet, J.E., Ogren, W.L., Pakrasi, H. & Somerville, C.R. (eds), *Molecular Biology of Photosynthesis*. Kluwer Academic Publishers, Dordrecht, pp. 389–405.

Webber, A.N. & Malkin, R. (1990) Photosystem I reaction-centre proteins contain leucine zipper motifs. *FEBS Letters* 264, 1–4.

Widger, W.R., Cramer, W.A., Herrmann, R.G. & Trebst, A. (1984) Sequence homology and structural similarity between cytochrome b of mitochondrial complex II and the chloroplast b_6-f complex: position of the cytochrome b hemes in the membrane. *Proceedings of the National Academy of Sciences (USA)* 81, 674–8.

Willey, D.L. & Gray, J.C. (1988) Synthesis and assembly of the cytochrome *b–f* complex in higher plants. In: Govindjee, Bohnert, H.J., Bottomley, W., Bryant, D.A., Mullet, J.E., Ogren, W.L., Pakrasi, H. & Somerville, C.R. (eds), *Molecular Biology of Photosynthesis*. Kluwer, Dordrecht, pp. 497–516.

Wynn, R.M. & Malkin, R. (1988) Interaction of plastocyanin with photosystem I: a chemical cross-linking study of the polypeptide that binds plastocyanin. *Biochemistry* 27, 5863–9.

Wynn, R.M. & Malkin, R. (1990) The photosystem I 5.5 kDa subunit (the *psaK* product). *FEBS Letters* 262, 45–8.

Wynn, R.M., Luong, C. & Malkin, R. (1989) Maize photosystem I. Identification of the subunit which binds plastocyanin. *Plant Physiology* 91, 445–9.

Yalovsky, S., Schuster, G. & Nechushtai, R. (1990) The apoprotein precursor of the major light-harvesting complex of photosystem II (LHCIIb) is inserted primarily into stromal lamellae and subsequently migrates to the grana. *Plant Molecular Biology* 14, 753–64.

Yu, J. & Vermass, W.F.J. (1990) Transcript levels and synthesis of photosystem II components in cyanobacterial mutants with inactivated photosystem II genes. *Plant Cell* 2, 315–22.

Zanetti, G. & Merati, G. (1987) Interaction between photosystem I and ferredoxin. Identification by chemical cross-linking of the polypeptide which binds ferredoxin. *European Journal of Biochemistry* 169, 143–6.

Zilber, A.L. & Malkin, R. (1988) Ferredoxin cross-links to a 22 kD subunit of photosystem I. *Plant Physiology* 88, 810–14.

Zurawski, G. & Clegg, M.T. (1984) The barley chloroplast DNA *atpBE, trnM2*, and *trnV1*

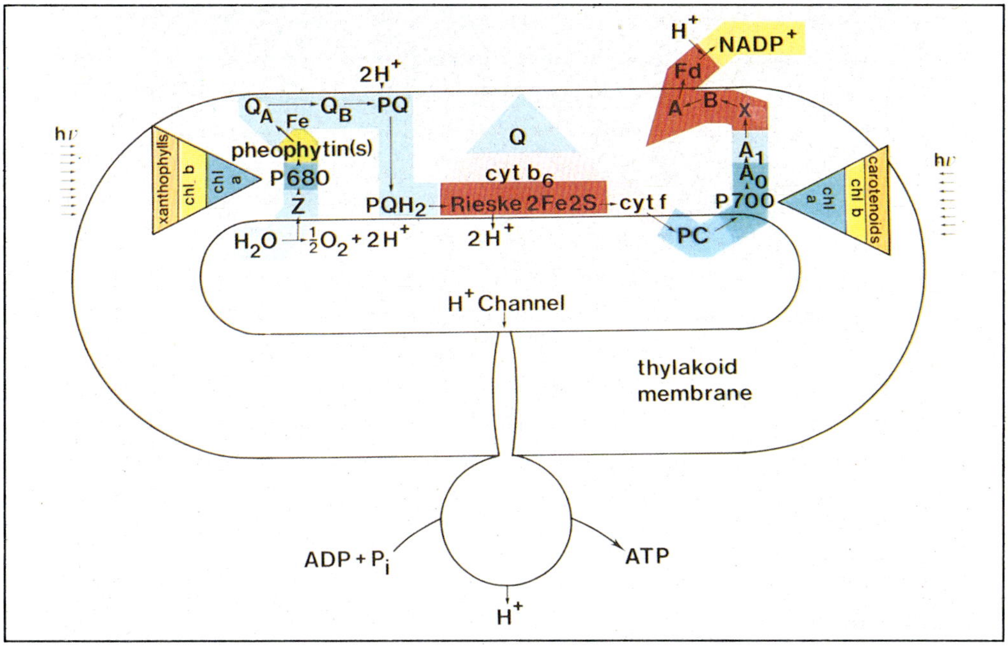

PLATE 1: A schematical model of the thylakoid membrane.

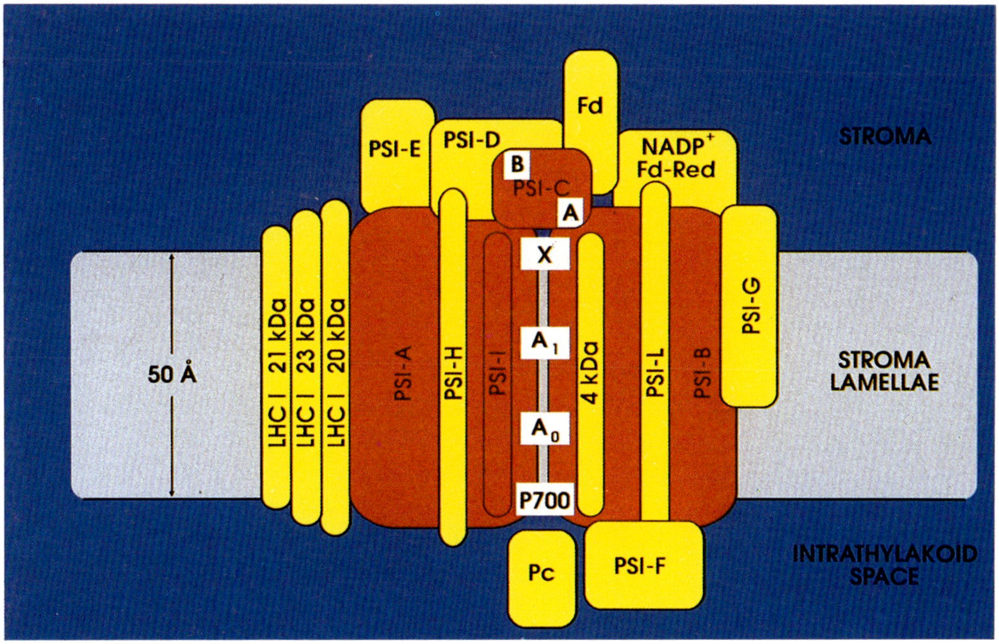

PLATE 2: Model of PS I. The model shows the subunits of the PS I core that have been identified in barley. LHC-I polypeptides and the soluble proteins plastocyanin (PC), Ferredoxin (Fd) and $NADP^+$-ferredoxin-oxidoreductase are also shown. Subunits encoded by the nuclear genome are coloured yellow whereas those encoded by the chloroplast genome are red.

loci. *Nucleic Acids Research* 12, 2549–59.

Zurawski, G. & Clegg, M.T. (1987) Evolution of higher-plant chloroplast DNA-encoded genes: implications for structure-function and phylogenetic studies. In: Briggs W.R., Russell, L.J. & Walbot V. (eds), *Annual Review of Plant Physiology* 38, 391–418.

Zurawski, G., Clegg, M.T. & Brown, A.H.D. (1984) The nature of nucleotide sequence divergence between barley and maize chloroplast DNA. *Genetics* 106, 735–49.

9

Analysis of Barley Metabolism using Mutant Genes

P.J. Lea, R.D. Blackwell & R.A. Azevedo

Introduction

In a review article written in 1979, Miflin & Lea (1980) pointed out that much of the information on biochemical pathways has arisen from the use of mutants of bacteria and yeast lacking key enzymes. They complained at that time that only a limited number of such mutants were available for study in higher plants. In the last ten years there have been rapid advances in this area and *Arabidopsis* and barley have proved to be very suitable plant material. In this chapter we describe work carried out on mutants of ammonia assimilation, photorespiration and amino acid biosynthesis selected in barley. The research projects were initiated at Rothamsted Experimental Station by S.W.J. Bright, A.J. Keys and B.J. Miflin, collaboration with whom we gratefully acknowledge.

Photorespiration

In 1979, Somerville & Ogren devised a simple screen for plants that lacked enzymes of photorespiration. This screen has been used and extended to study metabolism in barley (Blackwell *et al.*, 1988a). Second generation (M_2) plants from mutated seed can be grown in high levels of CO_2 (0.7%) where the CO_2 concentration competitively inhibits the oxygenase activity of Rubisco and thus prevents the operation of the photorespiration cycle (Fig. 9.1). Any plants with mutations in this pathway would therefore grow normally. However, when placed in air, any block in the pathway would operate and cause visible damage to the leaves. Returning these mutant plants to high CO_2 allows them to recover and grow normally.

Research into mutants of the photorespiratory pathway in barley at Lancaster and Rothamsted has produced mutants in activities of the following enzymes: Rubisco deficiency, phosphoglycolate phosphatase, catalase, glutamate:glyoxylate aminotransferase, glycine decarboxylase, serine:glyoxylate aminotransferase, NADH-dependent hydroxypyruvate reductase, ferredoxin-dependent glutamate synthase, glutamine synthetase and a mutant of chloro-

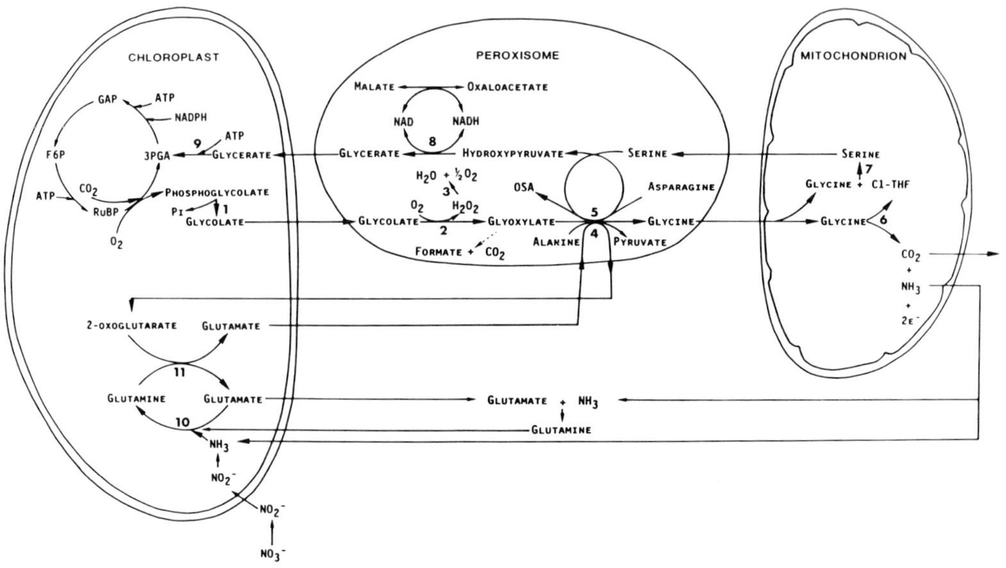

Fig. 9.1. The photorespiratory carbon and nitrogen cycle in C_3 plants. 1, phosphoglycolate phosphatase; 2, glycolate oxidase; 3, catalase; 4, glutamate:glyoxylate aminotransferase; 5, serine:glyoxylate aminotransferase; 6, glycine decarboxylase; 7, serine transhydroxymethylase; 8, hydroxypyruvate reductase; 9, glycerate kinase; 10, glutamine synthetase; 11, glutamate synthase.

plastic 2-oxoglutarate uptake. The following sections deal with these mutants more thoroughly.

Rubisco-deficient mutant

The screen devised by Somerville & Ogren (1979), as well as allowing selection of mutants with defects in the photorespiratory carbon and nitrogen cycle, also permits the selection of mutants which require additional CO_2 for growth because of other defects in their metabolism. Isolation of a mutant of *Arabidopsis thaliana* that lacked activation *in vivo* of Rubisco allowed the identification of a previously unknown enzyme characterized as Rubisco activase (Somerville *et al.*, 1982), an enzyme since found to be present in many species (Salvucci *et al.*, 1987). Evidence has been produced to show that micromolar concentrations of glyoxylate will inhibit Rubisco activation and deactivate activated Rubisco in intact, lysed and reconstituted chloroplasts (Chastain & Ogren, 1989; Campbell & Ogren, 1990). Although this mutant has not been characterized in barley, a different type of CO_2-requiring mutant has been selected and partially characterized (Hall *et al.*, 1986). In air the mutant did not die, but grew more slowly than the wild-type.

In wild-type barley, Rubisco normally accounts for over 50% of the total protein of leaves, however, in the mutant only 10% of the total protein was present as Rubisco. The specific activity of the Rubisco was also reduced to 25% of the wild-type. Comparison of kinetic constants from the wild-type and mutant revealed no differences. The difference in specific activity of Rubisco in leaves of mutant and wild-type was more marked in recently emerged leaves, indicating a lower rate of Rubisco synthesis in the mutant. ^{35}S-labelling studies supported the idea of impaired Rubisco synthesis, where the ratio of incorporation of label into the large subunit compared to the small subunit was 3.7 in the wild-type and only 1.9 for the mutant. Levels of soluble free amino acids were shown to be much higher in the mutant, especially for arginine and asparagine, but it has not been resolved whether the mutation affects all chloroplastic proteins or just the large subunit of Rubisco.

Phosphoglycolate phosphatase

The isolation of mutants of *A. thaliana* (Somerville & Ogren, 1979) and barley (Hall *et al.*, 1987; Blackwell *et al.*, 1988a; Lea & Blackwell, 1990) that lack phosphoglycolate phosphatase activity has provided the proof that phosphoglycolate was the first product of the photorespiratory pathway and hence the source of photorespiratory glycolate and CO_2.

The lack of carbon flow through the photorespiratory pathway for the phosphoglycolate phosphatase (*pgp*) barley mutant, LaPr 85/36, was shown by the decrease in levels of [^{14}C] recovered in glycine and serine when plants were fed $^{14}CO_2$ and the total absence of glycine and serine from the amino acid pool when plants were transferred to air (Lea & Blackwell, 1990). LaPr 85/36, more than any other mutant found to date, was unable to maintain wild-type rates of CO_2 fixation. When plants were photosynthetically induced in 1% O_2 and 0.035% CO_2 the fixation rate for the mutant was only 45% that of the wild-type and transfer to air caused a decline to only 7% of its own rate in 1% O_2, compared with a fall to only 67% for the wild-type. It has been suggested that the accumulation of any phosphoglycolate would inhibit triose phosphate isomerase activity and hence prevent production of dihydroxyacetone phosphate in the Calvin cycle.

Phosphoglycolate phosphatase has a role of recycling photosynthetically important phosphate within the chloroplast under photorespiratory conditions. Mutants like this therefore offer a chance to investigate this effect further (Sivak & Rowell, 1987). It is possible that LaPr 85/36 may sequester phosphate, even under conditions that are assumed to be largely non-photorespiratory. Light response curves measured for wild-type and LaPr 85/36 showed that the mutant was saturated at lower light intensities, in either low or high CO_2 concentrations (Sivak, 1987). Photosynthesis is limited by saturating CO_2, saturating light and phosphate supply (Sivak & Walker, 1986) and consequently with a lower light saturation it is possible that the mutant was subject to phosphate limitation. Feeding phosphate during light induction and after photorespiration had commenced, increased the CO_2 fixation rate in 1% O_2 from 7 to 22%, indicating that phosphate supply was limiting fixation in air.

Extracts from wild-type barley leaves subjected to native PAGE and stained for phosphoglycolate phosphatase activity showed two bands (Lea & Blackwell, 1990) in agreement with that found for *Phaseolus vulgaris* leaves (Vérin-Vergeau *et al.*, 1979). For LaPr 85/36 there were no bands of activity. Denaturing SDS gels of soluble protein stained with Coomassie blue gave similar banding patterns for both wild-type and mutant extracts, except that one band was missing for LaPr 85/36 at 18–20 kDa. This was very close to the subunit molecular weight estimated by Christeller & Tolbert (1978) of 19–22.4 kDa, although subunits of phosphoglycolate phosphatase in *Nicotiana rustica* have been shown to have molecular weights of 32 kDa (Belanger & Ogren, 1985).

Catalase

The first photorespiratory mutant characterized in barley lacked catalase (*cat2*) activity (Kendall *et al.*, 1983). Hydrogen peroxide is produced in the peroxisome from the oxidation of glycolate and it is the function of catalase to detoxify this potentially harmful metabolite. Activity staining after starch gel electrophoresis of wild-type barley extracts showed two bands of catalase activity, neither of which was present in similar extracts of the mutant, suggesting that both isoenzymes were controlled by the same gene (Kendall *et al.*, 1983). Leaves of *cat2* assimilated CO_2 at 60–80% of the wild-type irrespective of the CO_2 or O_2 levels in the surrounding atmosphere. In contrast to most other photorespiratory mutants there was no indication of a fall in this rate even when leaves were left in air for 6 h (Kendall *et al.*, 1983).

Illuminated chloroplasts can reduce hydrogen peroxide which uses glutathione and ascorbate as intermediate electron carriers (Anderson *et al.*, 1983; Foyer *et al.*, 1983). Exposure to air caused a five–tenfold increase in total glutathione content of mutant leaves (Smith *et al.*, 1984) suggesting that glutathione synthesis was protecting the leaf from peroxide toxicity, and that cellular damage only occurred where the glutathione level declined (Parker & Lea, 1983). Almost all of the increase in glutathione in a chloroplast fraction isolated from *cat2* was due to oxidized glutathione while the level of reduced glutathione remained relatively constant (Smith *et al.*, 1985). It was suggested that the eventual death of *cat2* in photorespiratory conditions was caused by the peroxide produced during photorespiration which accumulated to a level that caused peroxidation of lipids with the resulting disruption of membrane functions. Consistent with this hypothesis, transfer to air caused unusual electron opaque inclusions in the peroxisomes (Parker & Lea, 1983) and after several days cell necrosis occurred. This was associated with severely damaged plastids, and interestingly the damage was similar to that detected following application of peroxide-generating herbicides (Harvey & Frazer, 1980; Pallett & Dodge, 1980).

Metabolism of [^{14}C]glycolate by *cat2* in air was similar to that of the wild-type (Kendall *et al.*, 1983). This contrasted with the proposal of some workers that CO_2 could be produced in photorespiration by non-enzymic oxidation of glycolate by hydrogen peroxide (Grodzinski & Butt, 1976; Grodzinski & Woodrow, 1981; Walton & Butt, 1981). The *cat2* plant confirms that the major role of catalase in leaves of C_3 is to detoxify hydrogen peroxide liberated in glycolate oxidation (Tolbert, 1979).

In a study of catalase levels in *Nicotiana*, Havir & McHale (1989) showed a rapid decline (12 h) in CAT-1 activity in high CO_2 but levels of CAT-3 increased above control levels when exposed to high CO_2 for 2–3 days. This loss of CAT-1 activity was initially presumed to be related to the suppression of photorespiration by high CO_2. The use of low O_2 to suppress photorespiration in *Nicotiana* did not, however, reduce catalase activity suggesting that regulation was not directly related to photorespiratory metabolism. This effect of CO_2 on catalase activity suggests an indirect role for catalase such as control of NADP levels as shown in germinating seeds (Hendricks & Taylorson, 1975). In contrast, catalase from barley still undergoes suppression in low O_2 (Fair *et al.*, 1973). When extracts of *cat2* barley plants were chromatofocused, there were two peaks of catalatic activity, the first of which coincided with the main peak of peroxidatic activity (Havir, 1990). While this leading peak of peroxidatic activity was approximately the same in the mutant and wild-type extracts, the second peak of catalatic activity and the peroxidatic activity associated with it was all but eliminated in the mutant. When wild-type barley seeds were germinated and grown in the dark, the catalase profile was similar to that of the *cat2* mutant, i.e. the typical catalase was reduced by almost 70%. Thus in barley as well as spinach, tobacco and maize there is evidence for two types of catalase distinguishable by their rates of peroxidatic/catalatic activity.

Zelitch (1989a) has suggested that if hydrogen peroxide was not decomposed rapidly enough by peroxisomal catalase then there may be chemical decarboxylation of the vulnerable keto acids glyoxylate and hydroxypyruvate. Studies with a mutant of *Nicotiana* with enhanced catalase activity (40% more than wild-type; Zelitch, 1989b) have shown that for plants fed [1-^{14}C]glycolate the high catalase mutant discs released 19% less $^{14}CO_2$ per [1-^{14}C]glycolate metabolized than the wild-type, indicating that the mutant had produced less photorespiratory CO_2. This suggests that catalase activity could play a role in the regulation of photorespiration by removal of hydrogen peroxide, which would otherwise decarboxylate the keto acids.

Aminotransferases

The photorespiratory pathway includes two aminotransferase steps transferring amino groups to glyoxylate to produce glycine. Mutants in both of these steps have been isolated. For serine:glyoxylate aminotransferase (SGAT) there are mutants in both *A. thaliana* (Somerville & Ogren, 1980a), barley (Murray *et al.*, 1987) and *Nicotiana* (McHale *et al.*, 1988) and for glutamate:glyoxylate aminotransferase (GGAT) a barley mutant has been identified (Lea & Blackwell, 1990).

Although SGAT-deficient barley, LaPr 85/84, was recovered as a strong growing plant with a large leaf area, the rate of CO_2 fixation fell to 20–30% of the wild-type rate on transfer to photorespiratory conditions. The presence of wild-type amounts of GGAT in LaPr 85/84 should allow half the flow of carbon through to glycine and serine and hence the production of only half the wild-type levels of ammonia and CO_2. This was the case for ammonia production (Murray *et al.*, 1987) but the rate of CO_2 release when the plant was fed [1-^{14}C]glyoxylate was consistently higher than a similarly treated wild-type leaf. Without knowledge of pool sizes of organic and amino acids, it is difficult to draw conclusions about this enhanced rate. It has been postulated that in the absence of aminodonors, glyoxylate can be oxidized to produce formate and CO_2 (Walton & Butt, 1981; Zelitch, 1989a, b) and this could be the case as all of the amino groups become bound up in serine (Murray *et al.*, 1987).

It has been shown that the SGAT enzyme also catalyses asparagine:glyoxylate and serine:pyruvate aminotransferase reactions (Rehfeld & Tolbert, 1972; Ireland & Joy, 1983) and LaPr 85/84 lacked all three enzyme activities. The GGAT enzyme also has glutamate:pyruvate and alanine:glyoxylate aminotransferase activities and LaPr 85/84 possessed wild-type levels of all three activities. It is well established that the cycling of N during photorespiration should not be considered a closed system (Betsche, 1983; Ta & Joy, 1986; see later section on GS/glutamate synthase) and that asparagine and alanine in particular can act as aminodonors for glycine synthesis. In LaPr 85/84, however, asparagine did not accumulate in air indicating that in barley at least asparagine is less important than serine as a source of N.

Although only one peroxisomal form of the SGAT enzyme exists in spinach leaves (Rehfeld & Tolbert, 1972), analysis of wild-type barley extracts after ion-exchange chromatography revealed two major peaks of SGAT activity (Murray *et al.*, 1987). When wild-type extracts were subjected to SDS-PAGE and Western blotting with antisera raised against SGAT from cucumber cotyledons (Hondred *et al.*, 1985), two bands of cross-reaction were visible at 50 and 31 kDa. It is not clear whether these bands correspond to the SGAT A and B found in cucumber with molecular weights of 47 and 41 kDa, respectively, or whether the cucumber antisera react differently with barley extracts. When extracts of LaPr 85/84 were subjected to ion-exchange chromatography no peaks of activity were detected, but following SDS-PAGE and Western blot analysis the same two cross-reacting bands were present, indicating wild-type levels of enzyme protein.

A barley mutant which lacked GGAT has also been isolated and partially characterized. Extracts from the mutant showed wild-type levels of hydroxypyruvate reductase, SGAT and glycerate kinase activities but lacked GGAT activity as shown by a thin-layer chromatography assay for aminotransferases (Ireland & Joy, 1990) and the accumulation of glutamate in the soluble amino acid pool after exposure to air. This plant has since died and no further characterization was possible.

Glycine decarboxylase

Glycine produced during photorespiration from the transamination of glyoxylate is metabolized via a complex series of mitochondrial reactions with the subsequent incorporation of an additional carbon atom to produce serine. The reaction sequence, well established in animals (Kikuchi, 1973) but only recently characterized in plants (Walker & Oliver, 1986; Bourguignon *et al.*, 1988), involves two stages and uses two molecules of glycine. The first reaction, the glycine decarboxylase system, catalyses the oxidation of glycine to yield CO_2, ammonia and 5,10-methylene tetrahydrofolate. The mitochondrial isoenzyme of serine transhydroxymethylase (STHM) then transfers the C1 group from the methylene tetrahydrofolate to a second molecule of glycine to yield serine. Characterization of the complex from spinach (Neuburger *et al.*, 1986) and pea (Walker & Oliver, 1986) has demonstrated that at least four proteins are involved in this oxidation of glycine. These are designated P, H, T and L in plants and appear to be functionally equivalent to the P1, P2, P3 and P4 proteins isolated from bacteria.

Three barley plants have been isolated that accumulate glycine on exposure to air (Lea *et al.*, 1984; Blackwell *et al.*, 1990). One, RPr 83/202, has since died but full characterization has been carried out for the other two LaPr 85/55 and LaPr 87/30 (Blackwell *et al.*, 1990). The magnitude of glycine accumulation was different for the two mutants; LaPr 85/55 had a glycine content of 2.5 μmol g^{-1} FW (fresh weight) after 3 h in air, just over half that found in LaPr 87/30 (4.6 μmol g^{-1} FW) but at least six times higher than the wild-type (0.4 μmol g^{-1} FW).

When $^{14}CO_2$ was fed to leaves of the two mutants, after photosynthetic induction, at least 50% of the total ^{14}C recovered was found in glycine. One mutant LaPr 87/30 was almost totally unable to metabolize glycine, as shown by feeding experiments with [^{14}C]glycine, where over 90% of the supplied label remained in glycine even after 2 h. Although LaPr 85/55 accumulated glycine in the short term (82% of total counts recovered), after 2 h most of the radioactivity was found in the sugar fraction with only 15% in glycine, a pattern similar to that found for the wild-type. Assays for STHM activity showed that the levels in the wild-type and barley mutants were similar at 4–5 nmol min^{-1} mg^{-1} protein, indicating that the mutation was in the glycine decarboxylation step.

Measurements of CO_2 efflux, glycine bicarbonate exchange (equivalent to glycine decarboxylase activity) and ammonia release for LaPr 85/55 showed levels 80–90% of the wild-type rates, suggesting that the mutation was not in the glycine decarboxylase complex. The most probable explanation for these results taken with the glycine accumulation in the short term, but not sustained in the long term, was that the mutation affected a glycine transport system (Walker *et al.*, 1982). Levels of P, H, T and L proteins in LaPr 85/55 as shown by SDS-PAGE/Western blotting with antisera raised against the four proteins from pea, were shown to be the same as the wild-type (Blackwell *et al.*, 1990), enhancing the theory that the mutation was not in the glycine decarboxylase protein but in a transporter.

LaPr 87/30 was shown to have levels of CO_2 efflux of less than 40% wild-type values and glycine bicarbonate exchange and ammonia release of less than 15% wild-type rates. This combination of reduced rates suggested that the mutation could be in either the P, H or T proteins as these are involved in CO_2 and ammonia release. Levels of $^{14}CO_2$ efflux from [^{14}C]glycine feeding to LaPr 87/30 were higher than expected when glycine decarboxylase activity was less than 10% of the wild-type. This could be due to $^{14}CO_2$ release (Zelitch, 1989a, b) from glyoxylate, produced by the back-transamination of accumulated [^{14}C]glycine, which would occur when there was a depletion of aminodonors caused by the block at glycine decarboxylase. Immunovisualization of SDS-PAGE/Western blots for LaPr 87/30 showed wild-type levels of cross-reactivity to T and L proteins but only minor cross-reaction (<10%) for P and H antisera. This absence of major cross-reaction to P and H antisera is consistent with the other results, indicating that a combination of P and H proteins catalyse glycine-bicarbonate exchange and release CO_2 from glycine in the decarboxylation step. This could imply that the two genes are co-ordinately regulated or that one protein is unstable or undetectable in the absence of the other.

CO_2 fixation rates for both mutants fell to 30–40% of the wild-type rate when leaves were transferred to photorespiratory conditions (Blackwell *et al.*, 1990). It has been suggested that it was not glycine itself that inhibited photosynthesis but an accumulation of all amino groups as glycine (Somerville & Ogren, 1981, 1982; Somerville & Somerville, 1983). This is strengthened by the virtual disappearance of glutamate, the aminodonor for transamination of glyoxylate in the peroxisome, which occurs in both mutants in air (Blackwell *et al.*, 1990). Using the equivalent mutant in *A. thaliana*, Somerville & Ogren (1982) demonstrated a rapid release of $^{14}CO_2$ from [^{14}C]glycolate for wild-type plants, but for mutant plants a release of only 5% of wild-type rates with no evidence that a significant amount of CO_2 was evolved from the non-enzymic decarboxylation of glyoxylate (Grodzinski & Butt, 1976). It is the isolation of mutants such as those lacking glycine decarboxylase activity, which are unable to live in air yet carry out normal photosynthesis, that provides conclusive proof that this is the only reaction that generates CO_2 during photorespiration.

Hydroxypyruvate reductase

An air-sensitive barley plant, LaPr 88/29, has been identified that has less than 4% wild-type NADH-hydroxypyruvate reductase (HPR) activity (Murray *et al.*, 1989; Kleczkowski *et al.*, 1990), while rates of NADPH-HPR, NADH-glyoxylate reductase (GR) and NADPH-GR were only slightly reduced (Table 9.1). Thus under the assay conditions used (conditions which minimized non-specific NADH-HPR activity and optimized NADPH-HPR and NADH-GR) NADH-HPR activities only contributed slightly to overall NADPH-HPR and GRs in leaf extracts.

Table 9.1. Activities (μmol NADH min^{-1} mg^{-1} protein) of hydroxypyruvate reductases (HPRs) and glyoxylate reductases (GRs) in leaf extracts of wild-type and LaPr 88/29 barley.

Reaction	Activity	
	Wild-type	LaPr 88/29
NADH–HPR	0.652	0.023
NADPH–HPR	0.068	0.054
NADH–GR	0.015	0.013
NADPH–GR	0.059	0.054

Hydroxypyruvate is produced in the peroxisome (Fig. 9.1) from serine, which accumulated in the mutant. After only 3 h exposure to air over 86% of the total extractable amino acids were present as serine and glycine. When $^{14}CO_2$ was fed to LaPr 88/29, four times as much label was recovered in serine compared with the wild-type, but there was only a slight reduction in the amount of label found in sucrose (85% of the wild-type). This was the first indication that the mutation had not caused the major disruption that was expected of a plant lacking activity of a photorespiratory pathway enzyme. Measurements of the CO_2 fixation rate showed that in 21% O_2 and 0.035% CO_2 LaPr 88/29 was able to maintain rates at approximately 75% of those of the wild-type (Murray *et al.*, 1989). This was in complete contrast to other plants that had mutations in the main photorespiratory pathway, where fixation rates fall to less than 30% of the wild-type within a few minutes of transfer to photorespiratory conditions (Blackwell *et al.*, 1988a). Apparently LaPr 88/29 could metabolize hydroxypyruvate by another route. The final piece of evidence that this plant was able to bypass the lesion in the peroxisome came from feeding L-[U-^{14}C]serine, where the label recovered in sucrose from LaPr 88/29 was over 40% of that found in the wild-type.

When NADPH-HPR was initially purified and characterized (Kleczkowski & Randall, 1988) it was only possible to speculate that its function was similar to its NADH counterpart, i.e. recovery of carbon skeletons in photorespiration. The subsequent report of its cytosolic location (Kleczkowski *et al.*, 1988) and previous knowledge of the ability of hydroxypyruvate to leak out of the peroxisome (Anderson & Butt, 1986; Liang & Huang, 1983) suggested that this was how the mutant was bypassing the lesion in the peroxisome.

On the basis of estimated rates of photorespiratory CO_2 evolution in wild-type barley of 7 μmol m^{-2} s^{-1} (or 0.028 μmol min^{-1} mg^{-1} protein) NADPH-HPR appears to have sufficient activity (from NADPH-HPR rates in the mutant, Table 9.1) to deal with the whole flux of carbon through the pathway. Although NADPH-HPR has relatively high rates in leaf extracts, its contribution to the photorespiratory pathway *in vivo* is still uncertain, especially as the mutant plants show a lower rate of CO_2 fixation than the wild-type and they accumulate serine. This suggests that NADPH-HPR activity *in vivo* may be limited, perhaps by substrate availability or some unknown regulatory mechanism.

Immunoblots using spinach NADH-HPR polyclonal antisera (Kleczkowski *et al.*, 1986) against SDS/Western blots showed two cross-reacting bands at 38 and 43 kDa for wild-type barley extracts but only one band at 38 kDa for LaPr 88/29 (Kleczkowski *et al.*, 1990). These are similar to the previously established subunit molecular masses for NADPH-HPR and NADH-HPR (38 and 41 kDa respectively, Kleczkowski *et al.*, 1986). Although antisera against spinach NADH-HPR cross-reacted with protein bands to both barley NADH-HPR and NADPH-HPR the immunoreaction was much stronger for the NADH-HPR protein. Immunoprecipitation by the antibodies of crude leaf protein from wild-type barley resulted in over a 95% decrease of NADH-HPR activity but only a 30% decrease of the NADPH-HPR rate. The corresponding activities for LaPr 88/29 were decreased by merely 5–10% each.

It has been possible to fractionate crude leaf extracts with ammonium sulphate (Kleczkowski *et al.*, 1986), with most of the NADH-HPR activity being found in the 30–45% fraction, while NADPH-HPR activity is found in a 45–60% fraction. Studies of substrate kinetics by Kleczkowski *et al.* (1990) using wild-type and LaPr 88/29 barley have shown that the apparent K_m value (hydroxypyruvate) for NADH-HPR from the 30–45% fraction was 0.12 mM, a value typical for plant NADH-HPR (Kohn & Warren, 1970; Tolbert *et al.*, 1970; Anderson & Emes, 1985; Husic & Tolbert, 1987; Kleczkowski & Edwards, 1989). Barley NADH-HPR could also use NADPH as a cofactor with hydroxypyruvate as a second substrate with a K_m of 10 mM, which is similar to that for maize and spinach (Tolbert *et al.*, 1970; Kleczkowski & Edwards, 1989). The 45–60% fraction from wild-type contains only a small percentage of leaf NADH-HPR, but the enzyme greatly contributed to estimates of apparent K_m values of NADPH-HPR for hydroxypyruvate (Table 9.2).

Table 9.2. Apparent K_m (mM) for hydroxypyruvate of hydroxypyruvate reductase (HPR) activities from wild-type and LaPr 88/29 barley assayed from the 45–60% fraction.

Reaction	K_m	
	Wild-type	LaPr 88/29
NADH–HPR	0.13	0.22
NADPH–HPR	0.08	0.71

Presence of NADH-HPR lowered the apparent K_m and decreased the extent of substrate inhibition of NADPH-HPR with hydroxypyruvate. This substrate inhibition for NADPH-HPR was previously demonstrated from maize and spinach (Kleczkowski & Randall, 1988; Kleczkowski & Edwards, 1989). The K_m of 0.7 mM for NADPH-HPR activity from LaPr 88/29 is similar to that for the purified spinach enzyme (0.8 mM, Kleczkowski & Randall, 1988), while the K_m of 0.08 mM for the activity from the wild-type was comparable with the values of 0.04–0.12 mM determined for partially purified enzymes of spinach and other plants (Kleczkowski *et al.*, 1988; Kleczkowski & Edwards, 1989). Thus, the use of this mutant has allowed the unequivocal demonstration that the differences in K_m values between partially purified and purified NADPH-HPR results from interference by NADH-HPR.

Glutamine synthetase

The first clear indication that there were two distinct forms of glutamine synthetase (GS) in plant leaves came from the work of Mann *et al.* (1979, 1980) using barley. GS_1, which is localized in the cytosol, accounts for 10–15% of the total activity in green leaves, whilst GS_2 present solely in the chloroplast accounts for the remainder. The two enzymes can be distinguished by their thermal stabilities, sensitivity to sulphydryl reagents, pH optima and affinities for glutamate and ATP (Mann *et al.*, 1979; Acaster & Weitzmann, 1985).

The isoenzymes differ in their immunological characteristics. Specific antisera raised against the barley GS_1 protein cross-react with the cytosolic enzyme from the leaves, roots, nodules and seeds of a range of plants (Hirel *et al.*, 1984), but only to a lesser extent with the chloroplast GS_2 isoenzyme.

The complete amino acid sequences of cytosolic subunits from alfalfa, *N. plumbaginifolia*, *P. vulgaris* and pea have been deduced from nucleotide sequences of cloned cDNAs (see Forde & Cullimore, 1989, for full details). In a similar manner the primary structures of the chloroplast GS_2 polypeptides from *P. vulgaris*, pea and maize have been established. cDNA clones of GS_2 have been shown to encode a polypeptide that is 3–4 kDa larger than the mature chloroplast subunit. The small peptide is removed by a processing peptidase upon uptake of the GS_2 polypeptide into the chloroplast. There appears to be little similarity between the amino acid sequences of the transit peptides of the GS_2 polypeptides and those of other chloroplast-located proteins.

An interesting feature of the structure of the GS_2 isoenzyme is that all those examined so far have two conserved cysteine residues in addition to the three that are common to both the cytosolic and chloroplastic subunits. These additions may contribute to the different kinetic properties of the chloroplastic and cytosolic isoenzymes (Forde & Cullimore, 1989).

Two cDNA clones for GS_2 have been isolated from a barley leaf cDNA library by Freeman *et al.* (1990). Although neither of the clones was full length, one (pcHvGS6), contained an N-terminal extension of 55 residues. This extension was of similar length to the chloroplast transit peptides described above, although there was only limited sequence homology.

A partial GS_2 clone (pGS8) has also been isolated from barley by Baima *et al.* (1989). Over the region of overlap with pcHvGS6 (consisting of 1181 nucleotides) there was only one single nucleotide discrepancy, a result that could easily be explained by varietal differences or sequencing error. Preliminary Southern blot analysis of wheat–barley addition lines using pcHvGS6 as a probe indicated that there is only a single copy of the GS_2 gene located on the long arm of barley chromosome 2 (Cannell *et al.*, 1989). These observations are consistent with the genetic evidence that the chloroplast GS-deficient barley mutants described in the next section are all allelic.

Using the selection scheme devised by Somerville & Ogren (1979), a number of mutant lines of barley deficient in chloroplast GS_2 have been isolated and characterized (Blackwell *et al.*, 1987; Wallsgrove *et al.*, 1987). Following mass screening of air-sensitive barley plants a wide variation in the total GS activity was determined (Fig. 9.2). Plants containing less than 40% of the wild-type GS activity accumulate ammonia in the leaves following exposure to air. The maximum rate of ammonia evolution has been calculated as 120 μmol h^{-1} g^{-1} fresh weight (Lea *et al.*, 1989). This rate should theoretically be equivalent to the

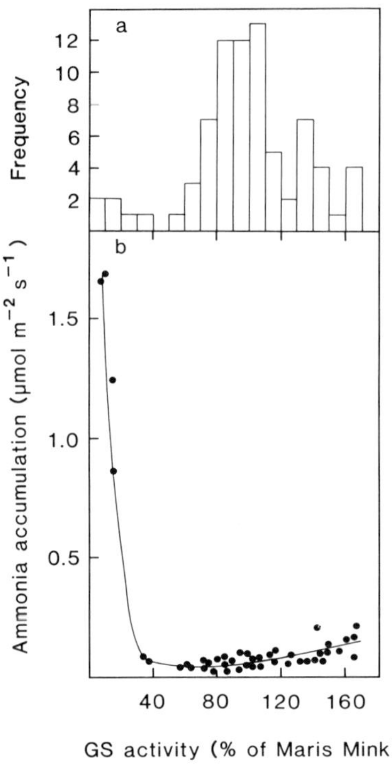

Fig. 9.2. Variation in total glutamine synthetase activity (a), and rate of ammonia accumulation during the initial 30 min in air (b) for M_2 air-sensitive plants.

rate of photorespiratory CO_2 loss from the conversion of glycine to serine and is in good agreement with the current estimates of photorespiration (40% of net rate of CO_2 fixation) in air obtained by other methods (Somerville & Somerville, 1983; Gerbaud & André, 1987).

Following the increase in ammonia levels in the air-exposed leaves, there is a dramatic decrease in the key amino acids of the photorespiratory nitrogen cycle (Blackwell *et al.*, 1988b), with only glutamate being readily detectable after one hour. The rate of CO_2 assimilation decreased slowly during this period. It was originally thought that ammonia had a direct effect on the photosynthetic rate by uncoupling electron transport. However, experiments using direct feeding of ammonium salts (Ikeda *et al.*, 1984), GS inhibitors (Lea & Ridley, 1989; Lea, 1991) and chlorophyll *a* fluorescence (Sivak *et al.*, 1988) have shown that this is most unlikely. A possible explanation for the inhibition of the photosynthetic rate in mutants lacking the enzymes of ammonia assimilation will be given in the next section.

Using antisera raised against *Phaseolus vulgaris* 'nodule specific' GS (Cullimore & Miflin, 1984), Western blot analysis has revealed that whilst barley GS mutants analysed so far have normal levels of GS_1 protein, they contain varied amounts of chloroplastic GS_2 protein (Wallsgrove *et al.*, 1987; Lea *et al.*, 1991). Freeman *et al.* (1990) have extended the analysis of Rothamsted GS mutants by employing the pcHvGS6 clone to determine mRNA levels (Fig. 9.3). Four mutant lines (designated Class I) contained low amounts of GS_2 protein and low or undetectable levels of GS_2 mRNA. One class (Class II) was unusual in that it had normal levels of GS_2 mRNA but no cross-reacting GS_2 protein. Freeman *et al.* (1990) suggested that the aberrant protein was unable to form the normal octameric enzyme or was not transported into the chloroplast, and was therefore susceptible to rapid proteolysis. Mutants in Class III contained variable amounts of GS_2 protein and mRNA. The apparent stability of the protein in these mutants suggests that the subunits are able to assemble into the holoenzyme *in vivo*, but have no catalytic activity due to alteration near the active site.

Glutamate synthase

Glutamate synthase is present in higher plants in two forms, ferredoxin-and NADH-dependent. The ferredoxin-dependent enzyme predominates in green leaves, whilst the NADH-dependent form is found in roots and is particularly active in nitrogen-fixing nodules (Lea *et al.*, 1990).

Ferredoxin-dependent glutamate synthase has been purified to homogeneity from barley leaves and has been shown to account for 1% of the total soluble protein (Marquez *et al.*, 1988). The enzyme is a monomeric iron–sulphur flavoprotein of

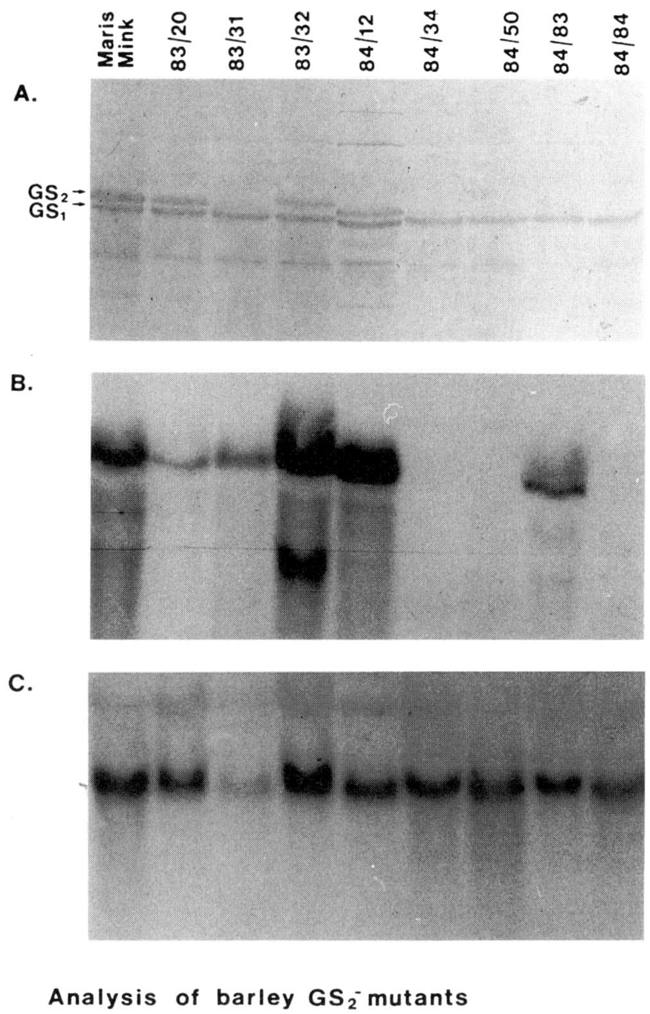

Fig. 9.3. Analysis of GS_2 mRNA and protein in leaves of Maris Mink and eight GS_2^- mutants of barley. A, Western blot of leaf extracts following SDS-PAGE. B, Northern blot of total leaf RNA after electrophoresis in a formaldehyde–agarose gel. The blot was probed with a 260 bp ^{32}P-labelled DNA fragment from the 3′-untranslated region of pcHvGS6, and the hybridizing RNA species were detected by autoradiography. C, The same Northern blot as in B was stripped of radioactivity and rehybridized using the cDNA insert from pW9 (Broglie *et al.*, 1983) as a control for the amount of RNA in each track (from Freeman *et al.*, 1990).

molecular weight 154 kDa, and appears to be localized solely in the chloroplasts of green leaves and in the plastids of other tissues. Full details of the kinetic properties of barley ferredoxin-dependent glutamate synthase in comparison with the enzyme from other sources have been given by Marquez *et al.* (1988).

In 1980, Somerville & Ogren (1980b) reported the first isolation of a mutant of *A. thaliana* deficient in ferredoxin-dependent glutamate synthase, and similar mutants have since been isolated from barley (Kendall *et al.*, 1986; Blackwell *et al.*, 1987, 1988b). The chloroplastic ferredoxin-dependent enzyme is virtually absent in the leaves and roots of the barley mutants. However, the NADH-dependent enzyme (5% of the leaf ferredoxin-dependent activity) is present at wild-type levels in all mutant lines so far tested. In contrast to the extensive work carried out on GS, little is known about the genes regulating glutamate synthase synthesis. However, work with the mutant lines of barley has already shown that the ferredoxin-dependent and NADH-dependent enzymes are under separate gene control and that the chloroplast ferredoxin-dependent enzyme is coded for by a single nuclear gene (Kendall *et al.*, 1986; Blackwell *et al.*, 1988b).

Following exposure to air the glutamate synthase-deficient plants accumulated large quantities of glutamine and small amounts of ammonia (Kendall *et al.*, 1986; Blackwell *et al.*, 1988b), the levels of all other amino acids were greatly reduced. Ten minutes after the transfer to air, the rate of photosynthetic CO_2 fixation had fallen to 10% of the rate of the wild-type plant. This loss of photosynthetic capacity is much faster than seen with the GS-deficient plant.

To gain a further insight into the role of the enzymes of ammonia assimilation in photosynthesis and photorespiration, double mutants of barley have been constructed by standard genetic crossing techniques. These double mutants lack both GS and glutamate synthase and also exhibit a dramatic fall in the photosynthetic rate following exposure to air. Low levels of glutamine and glutamate are present in the leaves which contain elevated levels of ammonia (Blackwell *et al.*, 1988b).

$^{15}NO_3^-$ and $^{15}NH_4^+$ feeding to both excised barley leaves and intact seedlings have shown that the mutant plants are capable of carrying out near-normal amino acid biosynthesis under non-photorespiratory conditions. However, following exposure to air, ammonia assimilation is blocked totally in the GS mutant and there is a massive accumulation of ^{15}N in both the amino and amide position of glutamine in the glutamate synthase deficient mutant (Lea *et al.*, 1991). A particularly important result of both the amino acid analysis and ^{15}N analysis of the mutant barley is the distinct lack of the amino acids glycine, serine, asparagine and alanine. Feeding asparagine and alanine via the xylem stream of mutant leaves has been shown to prevent the predicted fall in photosynthetic rate following exposure to air (Lea *et al.*, 1989).

It has been suggested that in the absence of suitable amino acid donors to carry out the conversion of glyoxylate to glycine, glyoxylate would leak out of the peroxisome into other organelles in the leaf cell. As already seen, glyoxylate is known to inhibit the activation of Rubisco at micromolar concentrations (Chastain & Ogren, 1989; Campbell & Ogren, 1990). Further analysis of the oxoacid content of the mutant leaves is required to establish their precise role in the photorespiratory process.

Mutants (*dct*) of *A. thaliana* and barley have also been isolated that are unable to transport 2-oxoglutarate into the chloroplast (Somerville & Ogren, 1983; Wallsgrove *et al.*, 1986). The mutant plants are very similar to glutamate synthase-deficient lines, although the symptoms are less exaggerated. The normal growth and development of these mutants at high levels of CO_2 has demonstrated that there is a particular 2-oxoglutarate carrier required for the synthesis of glutamate during photorespiration. For *A. thaliana* a chloroplast membrane polypeptide has been identified that is missing in the *dct* mutant (Somerville & Somerville, 1985). Two and three translocator models have been proposed to account for the transport of carboxylic and amino acids in and out of the chloroplast (Woo *et al.*, 1987; Yu & Woo, 1988).

Amino acid metabolism

Cereals are an excellent source of carbohydrates, although they are also characterized by their low-quality proteins which are deficient in certain essential amino acids. Feeding experiments with pigs have shown cereals to be deficient in lysine and threonine (Fuller *et al.*, 1979). Cereals are either consumed directly by humans or are fed to livestock for future consumption. For non-ruminant animals, essential amino acids must be supplied in the diet for protein synthesis and growth to occur.

Over the last 25 years, many investigators have explored different ways of obtaining cereal varieties with genetic alterations that could lead to an increase in some of the amino acids in the seeds. Breeding programmes with spontaneous mutants have been used in attempts to obtain new varieties with higher levels of amino acids with special attention to lysine. Induced mutants with high levels of specific amino acids have also been isolated, but they exhibited low productivity and high susceptibility to pathogens compared with their normal counterparts (Elmore, 1970; Koie *et al.*, 1976; Doll & Koie, 1978; Nelson, 1980).

The last decade has been marked by the use of mutagenesis, isolation and selection of mutants *in vitro* in plants, which has led to a much better understanding of amino acid biosynthesis. In this section, we discuss some of the mutant genes isolated and identified in barley that have altered regulation of the aspartic acid pathway.

Aspartic acid pathway

The synthesis of some amino acids is regulated by feedback as demonstrated for bacteria (Umbarger, 1978) and higher plants (Miflin & Lea, 1982). The essential amino acids lysine, threonine, methionine and isoleucine are derived from the aspartate family of amino acids. Since most cereals are deficient in at least one of these amino acids, this pathway has received great attention in recent years. Biochemical, molecular and genetic studies have led to a better understanding of the regulation of the aspartic acid pathway, showing that many branch-point enzymes are positively or negatively regulated by feedback (Madison &

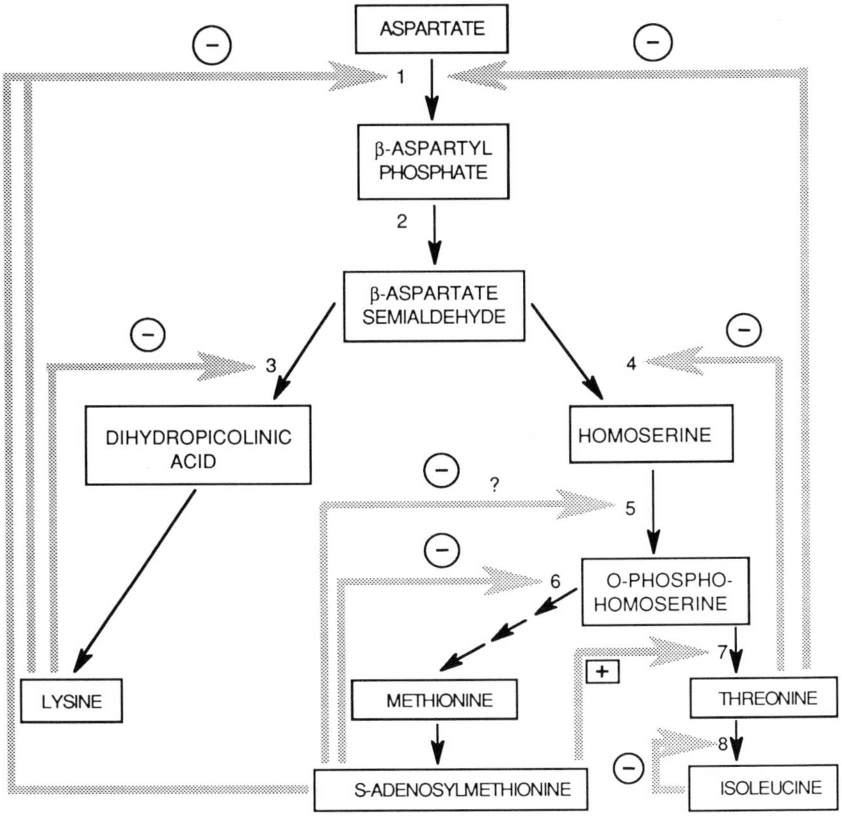

Fig. 9.4. Aspartic acid metabolic pathway. The points of regulation are indicated by arrows. (–) indicates feedback inhibition, (+) indicates enzyme activation. 1, aspartate kinase; 2, aspartate semialdehyde dehydrogenase; 3, dihydropicolinate synthase; 4, homoserine dehydrogenase; 5, homoserine kinase; 6, cystathionine γ-synthase; 7, threonine synthase; 8, threonine dehydratase.

Thompson, 1976; Bryan, 1980; Rognes *et al.*, 1980; Sainis *et al.*, 1981; Wallsgrove & Mazelis, 1981; Thompson *et al.*, 1982; Bryan, 1990a).

A scheme of the aspartic acid pathway in higher plants with its regulatory points is presented in Fig. 9.4. Aspartate kinase (AK) is the first enzyme of the pathway which converts aspartic acid to β-aspartyl-phosphate. AK isoenzymes have been identified in micro-organisms (Stadman *et al.*, 1961; Patte *et al.*, 1967) and in plants (Bryan, 1980; Rognes *et al.*, 1980; Giovanelli *et al.*, 1989a) which can be regulated by feedback inhibition, by lysine or threonine (Rognes *et al.*, 1980; Relton *et al.*, 1988). The sensitivity to lysine is increased in the presence of *S*-adenosylmethionine (Adomet) (Rognes *et al.*, 1980; Giovanelli *et al.*, 1989a). Lysine can also inhibit the enzyme dihydrodipicolinate synthase (DHPS) (Wallsgrove & Mazelis, 1981) whereas threonine inhibits some forms of homoserine dehydrogenase (Sainis *et al.*, 1981; Bryan, 1990a) which may be subject to environmentally mediated regulation *in vivo* (Bryan, 1990b). Adomet exerts an antagonistic effect on this metabolic pathway by its action of intensifying the lysine inhibition of AK activity (Bryan, 1980) and yet on the other hand, stimulating the activity of threonine synthase (Madison & Thompson, 1976). Both methionine and Adomet suppress the synthesis of cystathionine synthase in some plants (Thompson *et al.*, 1982). Isoleucine strongly inhibits its own synthesis, probably at threonine dehydratase, without causing a compensatory reduction in threonine synthesis (Bryan, 1980; Kirchner & Bryan, 1985; Giovanelli *et al.*, 1989b). Giovanelli *et al.* (1989a, b) have recently proposed a new regulatory structure for the pathway where the step catalysed by AK is not normally an important site for regulation of entry of 4-carbon units into the aspartate family of amino acids.

Many research groups have examined the regulation of the key points of the aspartic acid pathway by characterizing mutants with altered regulation of some enzymes. Several studies have showed that enzymes in these regulatory mutants have a changed pattern of feedback inhibition (Hibberd *et al.*, 1980; Bright *et al.*, 1982c; Cattoir-Reynaerts *et al.*, 1983) and also contain enhanced levels of end-product amino acids (Bright *et al.*, 1982b, c). This system has been demonstrated in micro-organisms (Demain, 1970; Eveleigh, 1981) and in some plants (Widholm, 1972a, b; Carlson, 1973; Widholm, 1976; Cattoir-Reynaerts *et al.*, 1981; Negrutiu *et al.*, 1981; Bright *et al.*, 1982b, c; Hibberd & Green, 1982).

Selection system

The selection of biochemical mutants has became a much easier task with the advance of tissue culture and plant regeneration *in vitro*. Such mutants can be selected in a Petri dish or small conical flask containing millions of cells growing in a selective medium with combinations of amino acid end-products or amino acid analogues (Widholm, 1976; Bright *et al.*, 1979a; Hibberd *et al.*, 1980; Matthews *et al.*, 1980; Schaeffer & Sharpe, 1981; Bright *et al.*, 1982c; Cattoir-Reynaerts *et al.*, 1983; Negrutiu *et al.*, 1984; Boyes & Vasil, 1987). The amino acid analogues (e.g. aminoethylcysteine, AEC) normally substitute for amino acids in their effects on the regulation of enzymes and may also substitute for them in protein synthesis (Bright *et al.*, 1979a, b). The growth of the cultures is inhibited, and the mutant cells which eventually do grow in these conditions may contain enzymes with altered feedback regulatory characteristics.

Positive results from the use of this system have been obtained in several plants. Millimolar concentrations of lysine plus threonine can inhibit the growth of a plant culture by inhibiting AK and/or homoserine dehydrogenase activities (Bright *et al.*, 1978a, b; Hibberd *et al.*, 1980). Consequently, the cells are starved of methionine and fail to grow (Bright *et al.*, 1978b; Bryan, 1980; Miflin & Lea, 1982; Giovanelli *et al.*, 1989b). This can be demonstrated when the growth of the culture is re-established by the addition of methionine, homoserine or homocysteine to the medium at low concentrations. AEC, a lysine analogue, inhibits DHPS activity, preventing the growth of a culture by the lack of lysine (Bright *et al.*, 1979a). If a mutant plant contained a higher level of soluble lysine, it could be resistant to AEC because of the dilution effect on the analogue of excess of lysine.

Although cell cultures *in vitro* or protoplasts have been used for selection of mutants with altered amino acid metabolism, another technique for the selection of biochemical mutants has been used very successfully in barley. The system involves the use of mature barley embryos which are dissected from M_2 seeds derived from M_1 seeds treated with a mutagenic agent, normally sodium azide (a complete review of mutagenesis is presented in Chapter 11). These embryos are grown on a sterile selective medium where plants which can grow normally are selected and grown to maturity for further identification at the biochemical and genetical levels. A final agronomical analysis is carried out according to the characteristics shown by the previous analyses.

The main advantage of this system is that the embryos rely entirely on the medium for nutrients and synthesize their own amino acids, and are consequently more sensitive to the inhibitor than whole seeds in which the endosperm can provide the amino acids. The main disadvantage is that the system is labour intensive and requires a large amount of space and hence fewer individuals can be screened (Miflin *et al.*, 1982).

Aspartate kinase in barley

AK has been very well characterized in different tissues of several plants including barley (Aarnes, 1974; Aarnes & Rognes, 1974; Aarnes, 1977; Davies & Miflin, 1977; Henke & Wahnbaeck, 1977; Shewry & Miflin, 1977; Bright *et al.*, 1978a; Henke & Wahnbaeck, 1979; Lea *et al.*, 1979; Bright *et al.*, 1982c; Giovanelli *et al.*, 1989a, b). Attempts to purify AK to homogeneity have been made in two laboratories (Relton *et al.*, 1988; Dotson *et al.*, 1989) with the focus on molecular characterization.

In barley plants, three peaks of AK activity can be eluted by ion-exchange chromatography (Bright *et al.*, 1982c; Fig. 9.5). The peaks of AK activity were characterized as three independent isoenzymes and labelled AKI, AKII and AKIII in order of elution (Bright *et al.*, 1982c; Rognes *et al.*, 1983). The feedback regulatory properties of these three peaks were determined.

AKI was characterized as a minor component varying from 15–18% of the total AK activity. This form bound weakly to the ion exchanger and could be eluted with 75 mM KCl. AKI was shown to be threonine-sensitive, since this peak was 60–65% inhibited by threonine at 1 mM. Lysine or lysine plus Adomet had little or no

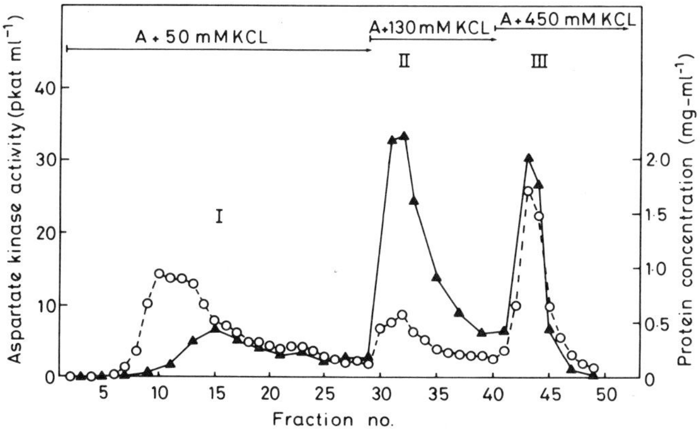

Fig. 9.5. Chromatography of aspartate kinase from sensitive plants derived from R2501. 15 ml (58 mg protein) of enzyme was applied to a 2.3 × 10 cm column of DE 52 and eluted stepwise with KCl buffer as indicated. Fractions (2.5 ml) were collected and assayed for aspartate kinase (▲) and protein content (○) (from Bright *et al.*, 1982c).

effect at low concentrations, with an inhibition between 3 and 5% by 10 mM lysine or 10 mM lysine plus 0.8 mM Adomet (Bright *et al.*, 1982c; Rognes *et al.*, 1983). AKI activity, was stimulated by 5 mM alanine or isoleucine. These amino acids could also alleviate the inhibition caused by threonine at low concentrations (Rognes *et al.*, 1983).

The second and third peaks of aspartate kinase activity were eluted from the ion-exchange chromatography column at concentrations of 90–130 mM and 220 mM KCl respectively (Bright *et al.*, 1982c). Higher concentrations of KCl did not elute any further AK activity (Rognes *et al.*, 1983). The distribution of AK activity between peaks II and III was 32 and 53% of the total, indicating a ratio of activity AKII/AKIII 0.60–0.67 (Rognes *et al.*, 1983; Arruda *et al.*, 1984). Investigations of the feedback sensitivity of the AKII and AKIII peaks showed that they could be inhibited by lysine or lysine plus Adomet (Bright *et al.*, 1982c; Rognes *et al.*, 1983). Lysine at 10 mM inhibited AKII activity by 82–87% and AKIII activity by 85–93% (Bright *et al.*, 1982c; Rognes *et al.*, 1983).

Mutants of barley resistant to lysine plus threonine inhibition were isolated and their genetics, enzymology and amino acid composition examined (Bright *et al.*, 1982c; Rognes *et al.*, 1983; Arruda *et al.*, 1984). Biochemical analysis of these mutants has shown their consistent association with alteration of the regulatory properties of the lysine-sensitive AKII and AKIII (Bright *et al.*, 1982c; Rognes *et al.*, 1983; Arruda *et al.*, 1984).

The three AK isoenzymes from Rothamsted (R) mutants R2501, 3004 and 3202 were studied in detail and their sensitivity to lysine and threonine compared with wild-type barley (Bright *et al.*, 1982c; Rognes *et al.*, 1983; Arruda *et al.*, 1984).

The profiles of elution of AKI, AKII and AKIII from the ion-exchange chromatography were similar to the wild-type (Bright *et al.*, 1982c; Rognes *et al.*, 1983; Arruda *et al.*, 1984). The AKI peak of activity was virtually unchanged in terms of feedback inhibition by threonine in all mutants, but R2501 showed a reduction of this peak from 15% in the wild-type barley to 9% of the total AK activity (Arruda *et al.*, 1984). However, AKII and AKIII were not only altered in their total AK activity distribution but also in their regulatory properties. The ratio of activities AKII/AKIII was different for all the mutants, varying from 0.35 for R3202 to 1.17 for R3004 compared with 0.60 for the wild-type (Bright *et al.*, 1982c; Rognes *et al.*, 1983; Arruda *et al.*, 1984).

The AKII and AKIII activities were both inhibited by lysine or lysine plus Adomet in wild-type barley. The AKII peak in R3202 did not show inhibition by lysine, even at very high concentrations (10 mM) (Bright *et al.*, 1982c; Rognes *et al.*, 1983). Threonine alone or in combination with lysine also showed no inhibitory effect (Rognes *et al.*, 1983). In R3004, the AKII peak was 85% inhibited by lysine at 10 mM, similar to that for wild-type barley (82%; Rognes *et al.*, 1983). In homozygous R2501 plants this peak was completely insensitive to inhibition by lysine at 10 mM but strongly inhibited (60%) in the presence of 10 mM lysine plus 0.8 mM Adomet (Arruda *et al.*, 1984). Previously, Bright *et al.* (1982c) analysed the AKII peak in heterozygous plants of the same mutant, and observed an intermediate lysine inhibition, which was enhanced by the addition of Adomet. The two different genotypes also gave different ratios of AKII/AKIII activity of 1.80 in homozygous plants (Arruda *et al.*, 1984) and 1.12 in heterozygous plants (Rognes *et al.*, 1983). Once more, the availability of different genotype combinations proved to be very useful for the genetic and biochemical characterization of a mutation.

The third peak of AK activity was shown to be altered in R3004. AKIII activity was less sensitive to lysine inhibition with a maximum inhibition at 3.5 mM lysine, compared to 0.36 mM for wild-type barley. Addition of 0.5 mM Adomet increased the inhibition in the presence of 10 mM lysine from 60 to 82%. For R3202 and R2501, the AKIII peak showed no difference in inhibition by lysine or lysine plus Adomet when

Table 9.3. Distribution of three forms of asparatate kinase (AK) extracted from a wild-type, single and double mutants of barley.

Plant type	Distribution of AK activity (% of total)			Ratio II/III	Inhibition by 10 mM lysine (%)		$I^{lys}0.5$		Gene designation	
	AKI	AKII	AKIII		AKII	AKIII	AKII	AKIII	*Ltr1a/b*	*Ltr2*
Bomi	15	32	53	0.60	82	93	0.40	0.36	+/+	+/+
R3004	17	46	37	1.17	85	60	0.30	3.50	+/+	*Ltr2/Ltr2*
R3202	17	21	62	0.37	0	93	–[a]	0.32	*Ltr1b/Ltr1*	+/+
R2501[b]	9	48	43	1.12	42	90	10.0	0.40	*Lt1a/+*	+/+
R2501[c]	9	59	32	1.80	0	93	–[a]	0.36	*Ltr1a/Ltr1*	+/+
Double	nd	nd	nd	d	0	39[e]	nd	nd	*Ltr1b/Ltr1*	*Ltr2/Ltr2*

[a] No effect of lysine
[b] Heterozygous resistant plants
[c] Homozygous resistant plants
[d] Similar to R3202
[e] Inhibition by 5 mM lysine
nd Not determined
*Data from Bright *et al.* (1982), Rognes *et al.* (1983) and Arruda *et al.* (1984).

compared with the wild-type (Rognes *et al.*, 1983; Arruda *et al.*, 1984).

Crossing of wild-type barley and the mutants was carried out to produce all possible combinations for genetic characterization of the resistance to lysine plus threonine (Table 9.3). Analysis of the F_2 generation suggested that resistance to lysine plus threonine in R2501 and R3202 was due to dominant mutations in a gene designated *ltr1*, causing altered feedback sensitivity of AKII and in R3004 was due to another dominant mutation in a gene designated *ltr2* which decreases lysine inhibition of the AKIII. These genes, *ltr1* and *ltr2*, are likely to be structural loci for AKII and AKIII (Bright *et al.*, 1982b).

For further confirmation of these results, a double mutant between R3202 (*ltr1*) and R3004 (*ltr2*) was constructed, which contained AKII insensitive to lysine inhibition and AKIII with a decreased sensitivity to lysine (Arruda *et al.*, 1984). An additive effect of the two genes was observed, as the double mutant was able to grow better in the presence of lysine (8 mM), threonine (8 mM) and arginine (1 mM) than either of its parents (Arruda *et al.*, 1984). Confirmation has been obtained from the analysis of both AK forms in roots and endosperm of the double mutant (Arruda *et al.*, 1984).

The absence of feedback inhibition of AK in these mutants allows sufficient flux of carbon for synthesis of methionine, thus allowing growth. A further effect of altered feedback regulation of some enzymes of the pathway could be overproduction of certain amino acid end-products that may produce cereal plants with improved nutritional grain quality (Green & Phillips, 1974; Miflin, 1976). Some of the mutants described here showed increased levels of threonine, lysine and methionine (Bright *et al.*, 1982b, c, 1984; Rognes *et al.*, 1983). Such increases of soluble threonine varied from tenfold in R3004 and 15-fold in heterozygous R2501 grains (Bright *et al.*, 1982b) to 18-fold in R3004, 50-fold in R2501 (homozygous grains) and 70-fold in the latest mutant identified, R2506 (Bright *et al.*, 1984). Soluble lysine increased by 15–20-fold and methionine 3–6-fold in these mutants (Bright *et al.*, 1984; Table 9.4).

Barley mutants resistant to AEC (Bright *et al.*, 1979a) were also obtained. Two mutants designated R906 and R4407 were selected by their ability to grow in the presence of 0.25 mM AEC. Genetic analysis was

Table 9.4. Soluble threonine, lysine and methionine content of mature grains from wild-type, single and double mutants of barley resistant to lysine plus threonine*.

Plant	Threonine	Lysine	Methionine	Total
Bomi	118	84	24	10 400
R2501	6129	1266	149	28 200
R2506	9032	1620	72	20 400
Bomi	127	79	19	6900
R3004	2376	107	23	12 400
R3202	125	109	19	11 700
R3004 × R3202 (double)	689	129	21	10 100

Amino acid content in nmol g^{-1} dry weight
* Data from Bright *et al.* (1984).

carried out which showed that the mutations are determined by two allelic genes *aec1a* and *aec1b*. These mutants were isolated in an attempt to select a mutation that could lead to an increase in the level of lysine. Meanwhile, biochemical analysis showed that resistance to AEC in these mutants was due to reduced uptake of AEC in the roots (Bright *et al.*, 1979a). These mutants were of little use for a study of the aspartic acid pathway, although they may help in elucidating the relationship between leaf and root uptake systems in defining the number of carrier systems, and in establishing their relevance to multiple kinetic phases of uptake (Soldal & Nissen, 1978; Borstlap, 1981; Harrington & Henke, 1981; Lien & Rognes, 1982; Van Bel *et al.*, 1982).

The selection of biochemical mutants has been claimed as an important tool to improve the nutritional quality of some plants (Green & Phillips, 1974; Miflin *et al.*, 1982). It is also an important approach to understanding the regulation of amino acid biosynthesis in plants. Many mutants were isolated and characterized in several plants, however, the effect of these mutants was only studied individually. It may be useful to combine the naturally occurring mutants with those selected *in vitro* to obtain additional information on amino acid and protein biosynthesis in cereals. It, for example, would be interesting to cross the double mutant described above with other barley mutants which lead to increased levels of lysine. What would be the effect of the double mutation with an overproduction of threonine on the overproduction of lysine by an altered composition of the protein fractions? Some interesting results have been already obtained in maize (Azevedo *et al.*, 1990).

Mutants of proline biosynthesis

Proline is synthesized from glutamate either directly or through glutamic semialdehyde and Δ1-pyrroline-5-carboxylic acid or indirectly through ornithine and arginine (Thompson, 1980). It is still not clear which pathway operates *in vivo* in higher plants (Bryan, 1990a). Only the last enzyme involved with proline synthesis, pyrroline-5-carboxylate reductase, has been studied in any detail (Krueger *et al.*, 1986). In nitrogen-fixing nodules the enzyme is thought to be involved in the production of NADP (Kohl *et al.*, 1988), and in pea leaves the enzyme is localized solely in the chloroplast (Raypati *et al.*, 1989).

The accumulation of proline following the onset of drought or the addition of increased levels of salts has been demonstrated in numerous organisms varying from bacteria to algae to higher plants (Aspinall & Paleg, 1981, Hanson & Hitz, 1982). The role of proline is unclear, but it has been suggested that the imino acid acts as an osmoticum and seems to stabilize enzymes in an active form. The accumulation has been shown to be a result of increased *de novo* synthesis as well as a decreased rate of breakdown (Rhodes *et al.*, 1986). A mutant of maize that has a proline requirement for growth was shown to have an increased rate of proline degradation (Dierks-Ventling & Tonelli, 1982).

Proline-accumulating mutants

Proline-accumulating cells of micro-organisms (Csonka, 1981) and plant cells in culture (Widholm, 1976; Cella *et al.*, 1982) have been identified by their ability to grow in toxic concentrations of proline analogues such as azetidine-2-carboxylic acid. Using a barley embryo system identical to that described in the previous section for lysine and threonine resistance, Kueh & Bright (1981) selected resistance to trans-4-hydroxyproline (Hyp). The inhibitory effects of this analogue were completely relieved by proline, particularly in the roots, a result that was not obtained with azetidine-2-carboxylic acid. Four mutant plants were isolated, of which R5201, R6102 and R6902 were shown to contain a 3–6-fold concentration of soluble proline in the leaf when compared with the parent cultivar Maris Mink. The mutations were shown to be allelic in the *Hyp1* locus (Kueh & Bright, 1981, 1982). In spite of a suggestion by Haglund (1980), that increased proline levels promoted attack by insects, the mutant R5201 was no more sensitive to locusts, slugs or aphids than the parental wild-type plants (Bright *et al.*, 1982a). In the presence of Hyp the fourth mutant (R6901) possessed very long roots, but not improved leaf growth when compared with the parental line.

The mutant R5201 could convert more of [^{14}C]glutamate to proline than the wild-type. Exogenously applied proline inhibited the incorporation of radioactivity into proline derived from glutamate, although this inhibition was less in the mutant R5201. The data confirmed the results originally obtained with maize roots (Oaks *et al.*, 1970) that proline can act as a feedback inhibitor of its own synthesis. It would appear that despite the initial higher soluble proline content in the mutant, feedback inhibition is relaxed but not absent (Kueh *et al.*, 1984).

When the wild-type and mutant (R5201) leaves were subjected to severe water stress in 40% (w/v) polyethylene glycol, proline accumulated more rapidly in R5201. However, as more water was lost from the mutant leaves during the experiment, it was not possible to positively attribute the increased rate of proline synthesis to the *Hyp1* mutation (Kueh & Bright, 1981).

Plants of R5201, R6102 and Maris Mink were grown to maturity with a liquid irrigation medium containing sodium chloride

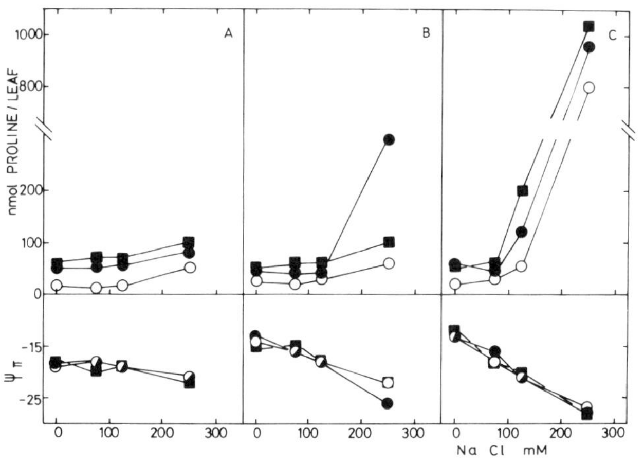

Fig. 9.6. Soluble proline content (nmol/8 cm leaf) and osmotic potential (ψ_π) of leaves of Maris Mink (○), R5201 (●) and R6902 (■) grown on NaCl at 0–250 mM concentrations after (A) 3 weeks, (B) 6 weeks and (C) 10 weeks. Each reading is an average of three samples. (From Bright *et al.*, 1984.)

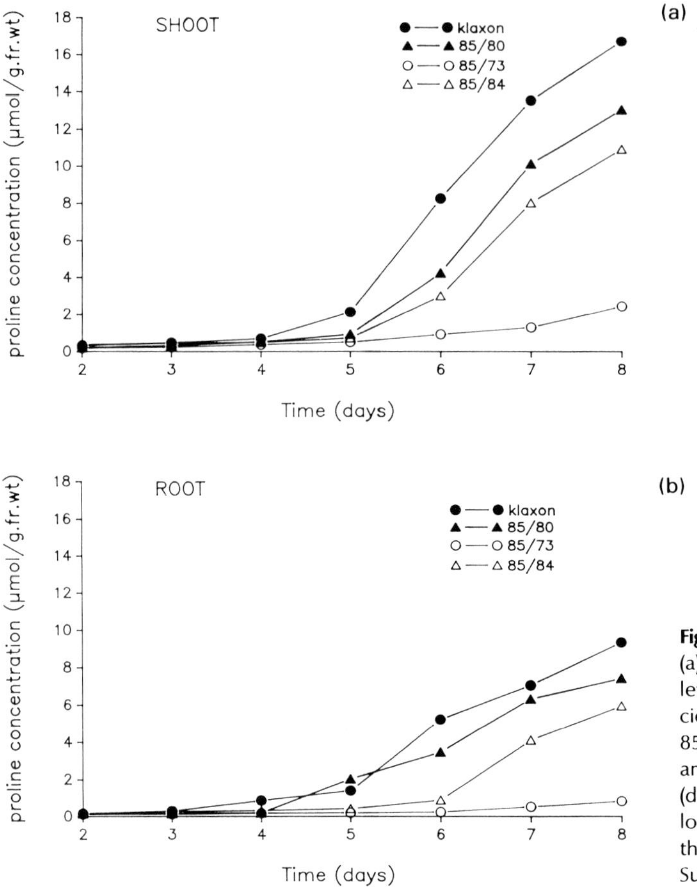

Fig. 9.7. Proline content of shoot (a) and root (b) of wild-type barley (cv. Klaxon), LaPr 85/80 (deficient in glutamine synthetase), LaPr 85/84 (deficient in serine:glyoxylate aminotransferase) and LaPr 85/73 (deficient in glutamate synthase) following the onset of soil drying by the withholding of water. (From Al-Sulaiti *et al.*, 1989.)

(0, 75, 125 or 250 mM). The proline and osmotic potential of leaves were measured after 3, 6 and 10 weeks of growth. The mutants and parent responded similarly to increasing salt by reductions in growth and leaf osmotic potential. At a clear threshold, each genotype showed the characteristic salt-induced increase in proline concentration. Under all conditions, the mutant leaves always contained slightly more proline than the parent (Fig. 9.6). However, the main conclusions were that the elevated levels of proline in the mutants under non-stress conditions were small in comparison with the stress-induced increases in proline content. It is perhaps not surprising that the response of the mutants to stress conditions is little different from the parent (Bright *et al.*, 1984).

Mutants containing reduced levels of proline

It has been shown in the previous section that the mutant lacking ferredoxin-dependent glutamate synthase (LaPr 85/73) contains a low level of glutamate. It was predicted that such a mutant would be unable to synthesize proline in large quantities.

Wild-type plants and the photorespiratory mutants lacking glutamate synthase (LaPr 85/73), glutamine synthetase (LaPr 85/80) and serine:glyoxylate aminotransferase (LaPr 85/84) were subject to a gradual increase in water deficit by soil drying. Below a water potential of –1.3 MPa in the leaf and –1.0 MPa in the root there was a rapid accumulation of proline in the wild-type and to a lesser extent in two of the mutants (Fig. 9.7). The glutamate synthase-deficient mutant lacked the ability to accumulate proline and showed premature symptoms of stress following soil drying. Unlike the other three lines, the glutamate synthase-deficient mutant exhibited a decrease in root to shoot ratio following the soil drying treatment (Al-Sulati *et al.*, 1989, 1990).

Conclusions

The work described in this chapter, clearly identifies the value of mutants in the study of metabolism, a fact well known by fungal geneticists for over 50 years. It is perhaps surprising that it has taken plant biochemistry such a long time to catch up. Mutants of *Arabidopsis* are now being used as recipients of cloned genes and for the isolation of unknown genes by complementation. It is to be hoped that the techniques for the transformation of barley will soon be as readily available as those for *Arabidopsis* and tobacco. This will then allow for full exploitation of the mutants of barley now available.

Acknowledgements

R.D.B. and P.J.L. are grateful to the AFRC, for continued support of this work. R.A.A. acknowledges the receipt of financial assistance from the CNPq, Brazil.

References

Aarnes, H. (1974) Aspartate kinase from some higher plants and algae. *Physiologia Plantarum* 32, 400–2.

Aarnes, H. (1977) A lysine-sensitive aspartate kinase and two molecular forms of homoserine dehydrogenase from barley seedlings. *Plant Science Letters* 9, 137–45.

Aarnes, H. & Rognes, S.E. (1974) Threonine-sensitive aspartate kinase and homoserine dehydrogenase from *Pisum sativum*. *Phytochemistry* 13, 2717–24.

Acaster, M.A. & Weitzman, P.D.J. (1985) Kinetic analysis of glutamine synthetase from various plants. *FEBS Letters* 189, 241–4.

Al-Sulaiti, A., Davies, W.J. & Lea, P.J. (1989) Variations in free proline accumulation in response to water deficit in barley photorespiratory mutants. *Meeting on Plant and Soil Nitrogen Metabolism*, AFRC, University of Lancaster, p. 85.

Al-Sulaiti, A., Lea, P.J. & Davies, W.J. (1990) Effects of soil drying on morphogenetic and physiological responses of barley photorespiratory mutants. *British Society for Plant Growth Regulation Monograph* 21, 347–50.

Anderson, I.W. & Butt, V.S. (1986) Permeability of lettuce leaf peroxisomes to photorespiratory metabolites. *Biochemical Society Transactions* 14, 106–7.

Anderson, I.W. & Emes, M.J. (1985) Purification and some properties of hydroxypyruvate reductase from *Lemna minor* L. *International Journal of Biochemistry* 17, 1085–9.

Anderson, J.W., Foyer, C.H. & Walker, D.A. (1983) Light-dependent reduction of hydrogen peroxide in intact spinach chloroplasts. *Biochimica et Biophysica Acta* 724, 69–74.

Arruda, P., Bright, S.W.J., Kueh, J.S.H., Lee, P.J. & Rognes, S.E. (1984) Regulation of aspartate kinase isoenzymes in barley mutants resistant to lysine plus threonine: construction and analysis of combinations of the *Lt1a*, *Lt1b* and *Lt2* mutant genes. *Plant Physiology* 76, 442–6.

Aspinall, D. & Paleg, L.G. (1981) Proline accumulation: physiological aspects. In: Paleg, L.G. and Aspinall, D. (eds), *The Physiology and Biochemistry of Drought Resistance in Plants*. Academic Press, New York, pp. 206–42.

Azevedo, R.A., Arana, J.L. & Arruda, P.A. (1990) Biochemical genetics of the interaction of the lysine plus threonine resistant mutant *Ltr*1* with opaque-2 maize mutant. *Plant Science* 70, 81–90.

Baima, S., Haegi, A., Stroman, P. & Casardoro, G. (1989) Characterisation of a cDNA clone for barley leaf glutamine synthetase. *Carlsberg Research Communications* 54, 1–9.

Belanger, F.C. & Ogren, W.L. (1985) Phosphoglycolate phosphatase: purification and preparation of antibodies. *Plant Physiology* 77, 601S.

Betsche, T. (1983) Aminotransfer from alanine and glutamate to glycine and serine during photorespiration in oat leaves. *Plant Physiology* 71, 961–5.

Blackwell, R.D., Murray, A.J.S. & Lea, P.J. (1987) Inhibition of photosynthesis in barley with decreased levels of glutamine synthetase activity. *Journal of Experimental Botany* 38, 1799–1809.

Blackwell, R.D., Murray, A.J.S., Lea, P.J., Kendall, A.C., Hall, N.P., Turner, J.C. & Wallsgrove, R.M. (1988a) The value of mutants unable to carry out photorespiration. *Photosynthesis Research* 16, 155–76.

Blackwell, R.D., Murray, A.J.S., Lea, P.J. & Joy, K.W. (1988b) Sucrose synthesis and N metabolism in a photorespiratory mutant of barley deficient in both chloroplastic glutamine synthetase and ferredoxin dependent glutamate synthase. *Journal of Experimental Botany* 39, 845–58.

Blackwell, R.D., Murray, A.J.S. & Lea, P.J. (1990) Photorespiratory mutants of the mitochondrial conversion of glycine to serine. *Plant Physiology* 94, 1316–22.

Borstlap, A.C. (1981) Invalidity of the multiphasic concept of ion absorption in plants. *Plant Cell and Environment* 4, 189–95.

Bourguignon, J., Neuburger, M. & Douce, R. (1988) Resolution and characterization of the glycine cleavage reaction in pea leaf mitochondria. Properties of the forward reaction catalysed by glycine decarboxylase and serine hydroxymethyltranferase. *Biochemical Journal* 255, 169–78.

Boyes, C.J. & Vasil, I.K. (1987) '*In vitro*' selection for tolerance to S-(2-aminoethyl)L-cysteine and overproduction of lysine by embriogenic calli and regenerated plants of *Pennisetum americanum* (L.) K. Schum. *Plant Science* 50, 195–203.

Bright, S.W.J., Shewry, P.R. & Miflin, B.J. (1978a) Aspartate kinase and the synthesis of aspartate derived amino acids in wheat. *Planta* 139, 119–25.

Bright, S.W.J., Wood, E.A. & Miflin, B.J. (1978b) The effect of aspartate derived amino acids (lysine, threonine and methionine) on the growth of excised embryos of wheat and barley. *Planta* 139, 113–18.

Bright, S.W.J., Norbury, P.B. & Miflin, B.J. (1979a) Isolation of a recessive barley mutant resistant to S-(2-aminoethyl)L-cysteine. *Theoretical and Applied Genetics* 55, 1–4.

Bright, S.W.J., Featherstone, L.C. & Miflin, B.J. (1979b) Lysine metabolism in a barley mutant resistant to S-(2-aminoethyl)cysteine. *Planta* 146, 629–33.

Bright, S.W.J., Lea, P.J., Kueh, J.S.H., Woodcock, C., Holloman, D.W. & Scott, G.C. (1982a) Proline content does not influence pest and disease susceptibility of barley. *Nature* 295, 592–3.

Bright, S.W.J., Kueh, J.S.H., Franklin, J., Rognes, S.E. & Miflin, B.J. (1982b) Two genes for threonine accumulation in barley seeds. *Nature* 299, 278–9.

Bright, S.W.J., Miflin, B.J. & Rognes, S.E. (1982c) Threonine accumulation in the seed of a barley mutant with an altered aspartate kinase. *Biochemical Genetics* 20, 229–43.

Bright, S.W.J., Lea, P.J., Arruda, P., Hall, N.P., Kendall, A.C., Keys, A.J., Kueh, J.S.H., Parker, M.L., Rognes, S.E., Turner, J.C., Wallsgrove, R.M. & Miflin, B.J. (1984) Manipulation of key pathways in photorespiration and amino acid metabolism by mutation and selection. In: Lea, P.J. & Stewart, G.R. (eds), *The Genetic Manipulation of Plants and its Application to Agriculture. Annual Proceedings of the Phytochemical Society of Europe*, Vol. 23. Clarendon Press, Oxford, pp. 141–69.

Broglie, R., Coruzzi, G., Lamppa, G., Keith, B. & Chua, N-H. (1983) Structural analysis of nuclear genes coding for the precursor to the small subunit of wheat ribulose-1,5-bisphosphate carboxylase. *Biotechnology* 1, 55–61.

Bryan, J.K. (1980) Synthesis of the aspartate family and branched chain amino acids. In: Miflin, B.J. (ed.), *The Biochemistry of Plants*, Vol. 5. *Amino Acids and their Derivatives*. Academic, London, pp. 403–52.

Bryan, J.K. (1990a) Advances in the biochemistry of amino acid biosynthesis. In: Miflin, B.J. & Lea, P.J. (eds), *The Biochemistry of Plants*, Vol. 16. Academic, London, pp. 161–95.

Bryan, J.K. (1990b) Differential regulation of maize homoserine dehydrogenase under physiological conditions. *Plant Physiology* 92, 785–91.

Campbell, W.J. & Ogren, W.L. (1990) Glyoxylate inhibition of ribulose-bisphosphate carboxylase/oxygenase activation in intact, lysed and reconstituted chloroplasts. *Photosynthesis Research* 23, 257–68.

Cannell, M., Mulcahy, M., Karp, A. & Shewry, P.R. (1989) Chromosomal assignment of characterised cDNAs and genes using RFLP analysis of ditelosomic addition lines. *Barley Genetics Newsletter*.

Carlson, P.S. (1973) Methionine sulphoximine-resistant mutants of tobacco. *Science* 180, 1366–8.

Cattoir-Reynaerts, A., Degryse, E., Negrutiu, I., Aerts, M. & Jacobs, M. (1981) Effects of aspartate-derived amino acids on growth of *Arabidopsis* plants and callus. *Zeitschrift fur Pflanzenphysiologie* 101, 67–74.

Cattoir-Reynaerts, A., Degryse, E., Verbruggen, I. & Jacobs, M. (1983) Isolation and characterization of a lysine plus threonine resistant variant in *Daucus carota. Biochemie und Physiologie der Pflanzen* 178, 81–90.

Cella, R., Parisi, B. & Nielsen, E. (1982) Characterisation of a carrot cell line resistant to azetidine-2-carboxylic acid. *Plant Science Letters* 24, 125–35.

Chastain, C.J. & Ogren, W.L. (1989) Glyoxylate inhibition of ribulose-bisphosphate carboxylase/oxygenase activation state in vivo. *Plant Cell Physiology* 30, 937–44.

Christeller, J.T. & Tolbert, N.E. (1978) Phosphoglycolate phosphatase. Purification and properties. *Journal of Biological Chemistry* 253, 1780–5.

Csonka, L.N. (1981) Proline overproduction results in enhanced osmotolerance in *Salmonella typhimurium. Molecular and General Genetics* 182, 82–6.

Cullimore, J.V. & Miflin, B.J. (1984) Immunological studies on glutamine synthetase using antisera raised to the two plant forms of the enzyme from *Phaseolus* root nodules. *Planta* 157, 245–53.

Davies, H.M. & Miflin, B.J. (1977) Aspartate kinase from carrot cell suspension culture. *Plant Science Letters* 9, 323–32.

Demain, A.L. (1970) Overproduction of microbial metabolites and enzymes due to alteration of regulation. *Advances in Biochemical Engineering* 1, 113–42.

Dierks-Ventling, C. & Tonelli, C. (1982) Metabolism of proline, glutamate and ornithine in proline mutant root tips of *Zea mays* (L.). *Plant Physiology* 69, 130–4.

Doll, H. & Koie, B. (1978) Influence of the high lysine gene from barley mutant 1508 on grain, carbodydrate and protein yield. In: *Seed Protein Improvement by Nuclear Techniques*. IAEA, Vienna, pp. 107.

Dotson, S.B., Sommers, D.A. & Gengenbach, B.G. (1989) Purification and characterization of lysine-sensitive aspartate kinase from maize cell cultures. *Plant Physiology* 91, 1602–8.

Elmore, C.D. (1970) 'Some effects of opaque-2 on kernel characteristics and the performance of counterpart maize (*Zea mays* L.) hybrids.' PhD Thesis, University of Illinois, USA.

Eveleigh, D.E. (1981) The microbiological production of industrial chemicals. *Scientific American* 245, 120–30.

Fair, P., Tew, J. & Cresswell, C.F. (1973) Enzyme activities associated with carbon dioxide exchange in illuminated leaves of *Hordeum vulgare* L. II. Effects of external concentrations of carbon dioxide and oxygen. *Annals of Botany* 37, 1035–9.

Forde, B.G. & Cullimore, J.V. (1989) The molecular biology of glutamine synthetase in higher plants. In: Miflin, B.J. (ed.), *Oxford Surveys of Plant Molecular and Cell Biology*, Vol. 6. Oxford University Press, Oxford, pp. 247–96.

Foyer, C.H., Rowell, J. & Walker, D.A. (1983) Measurement of the ascorbate content of spinach leaf protoplasts during illumination. *Planta* 157, 239–44.

Freeman, J., Marquez, A.J., Wallsgrove, R.M., Saarelainen, R. & Forde, B.G. (1990) Molecular analysis of barley mutants deficient in chloroplast glutamine synthetase. *Plant Molecular Biology* 14, 297–311.

Fuller, M.F., Livingstone, R.M., Baird, B.A. & Atkinson, T. (1979) The optimal amino acid supplementation of barley for the growing pig. I. Response of nitrogen metabolism to progressive supplementation. *British Journal of Nutrition* 41, 321–32.

Gerbaud, A. & André, M. (1987) An evaluation of recycling in measurements of photorespiration. *Plant Physiology* 83, 933–7.

Giovanelli, J., Mudd, S.H. & Datko, A.H. (1989a) Aspartokinase of *Lemna paucicostata* Hegelm. 6746. *Plant Physiology* 90, 1577–83.

Giovanelli, J., Mudd, S.H. & Datko, A.H. (1989b) Regulatory structure of the biosynthetic pathway for the aspartate family of amino acids in *Lemna paucicostata* Hegelm. 6746, with special reference to the role of aspartokinase. *Plant Physiology* 90, 1584–99.

Green, C.E. & Phillips, R.L. (1974) Potential selection system for mutants with increased lysine, threonine and methionine in cereal crops. *Crop Science* 14, 827–30.

Grodzinski, B. & Butt, V.S. (1976) Hydrogen peroxide production and the release of carbon dioxide during glycolate oxidation in leaf peroxisomes. *Planta* 128, 225–31.

Grodzinski, B. & Woodrow, L. (1981) Serine synthesis and CO_2 release from intermediates of the glycolate pathway. In: Akoyunoglou, G.

(ed.), *Photosynthesis*, Vol. 4. *Regulation of Carbon Metabolism*. Balaban International Science Services, Philadelphia, pp. 551–9.

Haglund, B.M. (1980) Proline and valine-cues which stimulate grasshopper herbivory during drought stress? *Nature* 288, 697–8.

Hall, N.P., Kendall, A.C., Turner, J.C., Wallsgrove, R.M. & Keys, A.J. (1986) A barley mutant deficient in RuBP carboxylase. *Plant Physiology* 80, 281S.

Hall, N.P., Kendall, A.C., Lea, P.J., Turner, J.C. & Wallsgrove, R.M. (1987) Characteristics of a photorespiratory mutant of barley (*Hordeum vulgare* L.) deficient in phosphoglycolate phosphatase. *Photosynthesis Research* 11, 89–96.

Hanson, A.D. & Hitz, W.D. (1982) Metabolic responses of mesophytes to plant water deficits. *Annual Review of Plant Physiology* 32, 139–69.

Harrington, H.M. & Henke, R.R. (1981) Amino acid transport into cultured tobacco cells. I. Lysine transport. *Plant Physiology* 67, 373–8.

Harvey, B.M.R. & Frazer, T.W. (1980) Paraquat tolerant and susceptible perennial ryegrasses: effects of paraquat treatment on carbon dioxide uptake and ultrastructure of photosynthetic cells. *Plant Cell and Environment* 3, 107–17.

Havir, E.A. (1990) Plant catalases: demonstration and characteristics of a catalase isozyme with enhanced peroxidatic activity. In: Zelitch, I. (ed.), *Perspectives in Biochemical and Genetic Regulation of Photosynthesis*. Wiley-Liss, New York, pp. 271–83.

Havir, E.A. & McHale, N.A. (1989) Regulation of catalase activity in leaves of *Nicotiana sylvestris* by high CO_2. *Plant Physiology* 89, 952–7.

Hendricks, S.B. & Taylorson, R.B. (1975) Breaking of seed dormancy by catalase inhibition. *Proceedings of the National Academy of Sciences (USA)* 72, 306–9.

Henke, R.R. & Wahnbaeck, R. (1977) β-Aspartokinase from developing endosperm of maize. *Biochemical and Biophysical Research Communications* 70, 38–44.

Henke, R.R. & Wahnbaeck-Spencer, R. (1979) Variations in β-aspartate kinase activity during the development of maize endosperm. *FEBS Letters* 99, 113–15.

Hibberd, K.A. & Green, C.E. (1982) Inheritance and expression of lysine plus threonine resistant selected in maize tissue culture. *Proceedings of the National Academy of Sciences (USA)* 79, 559–63.

Hibberd, K.A. Walter, T., Green, C.E. & Gengenbach, B.G. (1980) Selection and characterisation of a feedback-insensitive tissue culture of maize. *Planta* 148, 183–8.

Hirel, B., McNally, S.F., Gadal, P., Sumar, N. & Stewart, G.R. (1984) Cytosolic glutamine synthetase in higher plants; a comparative immunological study. *European Journal of Biochemistry* 138, 63–6.

Hondred, D., Hunter, J.McC., Keith, R., Titus, D.E. & Becker, W.M. (1985) Isolation of serine:glyoxylate aminotransferase from cucumber cotyledons. *Plant Physiology* 79, 95–102.

Husic, D.W. & Tolbert, N.E. (1987) NADH-hydroxypyruvate reductase and NADPH-glyoxylate reductase in algae: partial purification and characterisation from *Chlamydomonas reinhardtii*. *Archives of Biochemistry and Biophysics* 252, 396–408.

Ikeda, M., Ogren, W.L. & Hageman, R.H. (1984) Effect of methionine sulphoximine on photosynthetic carbon metabolism in wheat (*Triticum vulgare*) leaves. *Plant Cell Physiology* 25, 447–52.

Ireland, R.J. & Joy, K.W. (1983) Purification and properties of an asparagine aminotransferase from *Pisum sativum* leaves. *Archives of Biochemistry and Biophysics* 223, 291–6.

Ireland, R.J. & Joy, K.W. (1990) Aminotransferases. In: Lea, P.J. (ed.), *Methods in Plant Biochemistry*, Vol. 3. Academic, London, pp. 277–86.

Kendall, A.C., Keys, A.J., Turner, J.C., Lea, P.J. & Miflin, B.J. (1983) The isolation and characterisation of a catalase-deficient mutant of barley (*Hordeum vulgare*, L.). *Planta* 159, 505–11.

Kendall, A.C., Wallsgrove, R.M., Hall, N.P., Turner, J.C. & Lea, P.J. (1986) Carbon and nitrogen metabolism in barley mutants lacking ferredoxin dependent glutamate synthase. *Planta* 168, 316–23.

Kikuchi, G. (1973) The glycine cleavage system: composition, reaction mechanism and physiological significance. *Molecular and Cellular Biochemistry* 1, 169–87.

Kirchner, S.C. & Bryan, J.K. (1985) Isolation and characterization of two forms of maize threonine dehydratase. *Plant Physiology* 77, 109S.

Kleczkowski, L.A. & Edwards, G.E. (1989) Identification of hydroxy pyruvate and glyoxylate reductases in maize leaves. *Plant Physiology* 91, 278–86.

Kleczkowski, L.A. & Randall, D.D. (1988)

Purification and characterization of a novel NADPH (NADH)-dependent hydroxypyruvate reductase from spinach leaves. Comparison of immunological properties of leaf hydroxypyruvate reductases. *Biochemical Journal* 250, 145–52.

Kleczkowski, L.A., Randall, D.D. & Blevins, D.G. (1986) Purification and characterization of a novel NADPH(NADH)-dependent glyoxylate reductase from spinach leaves. Comparison of immunological properties of leaf glyoxylate reductase and hydroxypyruvate reductase. *Biochemical Journal* 239, 653–9.

Kleczkowski, L.A., Givan, C.V., Hodgson, J.M. & Randall, D.D. (1988) Subcellular location of NADPH-hydroxypyruvate reductase activity in leaf protoplasts of *Pisum sativum* and its role in photorespiratory metabolism. *Plant Physiology* 88, 1182–5.

Kleczkowski, L.A., Edwards, G.E., Blackwell, R.D. & Lea, P.J. (1990) Enzymology of the reduction of hydroxypyruvate and glyoxylate in a mutant of barley lacking peroxisomal hydroxypyruvate reductase activity. *Plant Physiology* 94, 819–25.

Kohl, D.H., Schubert, K.R., Carter, M.B., Hagedorn, C.H. & Shearer, G. (1988) Proline metabolism in nitrogen fixing root nodules: energy transfer and regulation of purine synthesis. *Proceedings of the National Academy of Sciences (USA)* 85, 2036–40.

Kohn, L.D. and Warren, W.A. (1970) The kinetic properties of spinach leaf glyoxylic acid reductase. *Journal of Biological Chemistry* 245, 3831–9.

Koie, B., Doll, H. & Kreis, M. (1976) Evaluation of a high-lysine barley gene using chromosome-double monopolids. *Genetika* 8, 177–82.

Kreuger, R., Jager, H.J., Hintz & Pahlich, E. (1986) Purification to homogeneity of pyrroline-5-carboxylate reductase of barley. *Plant Physiology* 80, 142–4.

Kueh, J.S. & Bright, S.W.J. (1981) Proline accumulation in a barley mutant resistant to trans-4-hydroxy-L-proline. *Planta* 153, 166–71.

Kueh, J.S. & Bright, S.W.J. (1982) Biochemical and genetical analysis of three proline accumulating mutants. *Plant Science Letters* 27, 233–41.

Kueh, J.S., Hill, J.M., Smith, S.J. & Bright, S,W.J. (1984) Proline biosynthesis in a proline accumulating barley mutant. *Phytochemistry* 23, 2207–10.

Lea, P.J. (1991) The inhibition of ammonia assimilation – a mechanism of herbicide action. In: Baker, N.R. (ed.), *Topics in Photosynthesis*, Vol. 10. Elsevier, Amsterdam, pp. 267–98.

Lea, P.J. & Blackwell, R.D. (1990) Genetic regulation of the photorespiratory pathway. In: Zelitch, I. (ed.), *Perspectives in Biochemical and Genetic Regulation of Photosynthesis*. Wiley-Liss, New York, pp. 301–18.

Lea, P.J. & Ridley, S.M. (1989) Glutamine synthetase and its inhibition. In: Dodge, A.D. (ed.), *Herbicides and Plant Metabolism*. Cambridge University Press, Cambridge, pp. 137–70.

Lea, P.J., Mills, W.R. & Miflin, B.J. (1979) The isolation of a lysine-sensitive aspartate kinase from pea leaves and its involvement in homoserine biosynthesis in isolated chloroplasts. *FEBS Letters* 98, 165–8.

Lea, P.J., Kendall, A.C., Keys, A.J. & Turner, J.C. (1984) The isolation of a photorespiratory mutant of barley unable to convert glycine to serine. *Plant Physiology* 75, 881S.

Lea, P.J., Blackwell, R.D., Murray, A.J.S. & Joy, K.W. (1989) The use of mutants lacking glutamine synthetase and glutamate synthase to study their role in plant nitrogen metabolism. In: Poulton, J.E., Romeo, J.T. & Conn, E.E. (eds), *Plant Nitrogen Metabolism*. Plenum, New York, pp. 157–89.

Lea, P.J., Robinson, S.A. & Stewart, G.R. (1990) The enzymology and metabolism of glutamine, glutamate and asparagine. In: Miflin, B.J. & Lea, P.J. (eds), *The Biochemistry of Plants*. Academic, New York, pp. 257–76.

Lea, P.J., Blackwell, R.D. & Joy, K.W. (1991) Ammonia assimilation in higher plants. In: Pilbeam, D.J. & Mengel, K. (eds), *Nitrogen Metabolism of Plants*. Oxford University Press, Oxford.

Liang, Z. & Huang, A.H.C. (1983) Metabolism of glycolate and glyoxylate in intact spinach leaf peroxisomes. *Plant Physiology* 73, 147–52.

Lien, R. & Rognes, S.E. (1982) Salt effect on amino acid transport into barley leaf cells. *Biochemie und Physiologie der Pflanzen* 177, 54–60.

Madison, J.T. & Thompson, J.F. (1976) Threonine synthetase from higher plants: stimulation by S-adenosylmethionine and inhibition by cysteine. *Biochemical and Biophysical Research Communications* 71, 684–91.

Mann, A.F., Fentem, P.A. & Stewart, G.R.

(1979) Identification of two forms of glutamine synthetase in barley. *Biochemical and Biophysical Research Communications* 88, 515–21.

Mann, A.F., Fentem, P.A. & Stewart, G.R. (1980) Tissue localisation of barley (*Hordeum vulgare*) glutamine synthetase isoenzymes. *FEBS Letters* 110, 265–7.

Marquez, A.J., Avila, C., Forde, B.G. & Wallsgrove, R.M. (1988) Ferredoxin-glutamate synthase from barley leaves: rapid purification and partial characterisation. *Plant Physiology and Biochemistry* 26, 645–51.

Matthews, B.F., Shye, S.C.H. & Widholm, J.M. (1980) Mechanism of resistance of a selected carrot cell suspension culture to S-(2-aminoethyl)L-cysteine. *Zeitschrift fur Pflanzenphysiologie* 96, 453–63.

McHale, N.A., Havir, E.A. & Zelitch, I. (1988) A mutant of *Nicotiana sylvestris* deficient in serine:glyoxylate aminotransferase activity. Callus induction and photorespiratory toxicity in regenerated plants. *Theoretical and Applied Genetics* 76, 71–5.

Miflin, B.J. (1976) In: *Genetic Improvement of Seed Proteins* (National Academic Sciences) Washington, DC, USA, pp. 135–58.

Miflin, B.J. & Lea, P.J. (1982) Ammonia assimilation and amino acid metabolism. In: Boulter, D. & Parthier, B. (eds), *Encyclopedia of Plant Physiology New Series*, Vol. 14A. Springer-Verlag, Berlin, pp. 1–64.

Miflin, B.J., Bright, S.W.J., Rognes, S.E. & Kueh, J.S.H. (1982) Amino acids, nutrition and stress: the role of biochemical mutants in solving problems of crop quality. In: Kosuge, T., Meredith, C.P. & Hollaender, A. (eds), *Genetic Engineering of Plants. An Agricultural Perspective*. Plenum Press, New York, pp. 391–414.

Murray, A.J.S., Blackwell, R.D., Joy, K.W. & Lea, P.J. (1987) Photorespiratory N donors, aminotransferase specificity and photosynthesis in a mutant of barley deficient in serine:glyoxylate aminotransferase activity. *Planta* 172, 106–13.

Murray, A.J.S., Blackwell, R.D. & Lea, P.J. (1989) Metabolism of hydroxypyruvate in a mutant of barley lacking NADH-dependent hydroxypyruvate reductase, an important photorespiratory enzyme activity. *Plant Physiology* 91, 395–400.

Negrutiu, I., Cattoir-Reynaerts, A., Verbruggen, I. & Jacobs, M. (1981) Lysine overproduction in an S-aminoethylcysteine resistant mutant, isolated in protoplast culture of *Nicotiana sylvestris*. *Archives Internationales de Physiologie et de Biochimie* 89, 188–9.

Negrutiu, I., Cattoir-Reynaerts, A., Verbruggen, I. & Jacobs, M. (1984) Lysine overproducer mutants with an altered dihydropicolinate synthase from protoplast culture of *Nicotiana sylvestris* (Spegazzini and Comes). *Theoretical and Applied Genetics* 68, 11–20.

Nelson, O.E. (1980) Genetic control of polysaccharide and storage protein synthesis in the endosperm of barley, maize and sorghum. In: Pomeranz, Y. (ed.), *Advances in Cereal Science and Technology*, Vol. 3. American Association of Cereal Chemists, St Paul, Minnesotta, USA, pp. 41.

Neuburger, M., Bourguignon, J. & Douce, R. (1986) Isolation of a large complex from the matrix of pea leaf mitochondria involved in the rapid transformation of glycine into serine. *FEBS Letters* 207, 18–22.

Oaks, A., Mitchell, D.J., Barnard, R.A. & Johnson, F.T. (1970) The regulation of proline biosynthesis in maize roots. *Canadian Journal of Botany* 48, 2249–58.

Pallett, K.E. & Dodge, A.D. (1980) Studies into the action of some photosynthetic inhibitor herbicides. *Journal of Experimental Botany* 31, 1051–66.

Parker, M.L. & Lea, P.J. (1983) Ultrastructure of the mesophyll cells of leaves of a catalase-deficient mutant of barley (*Hordeum vulgare* L.). *Planta* 159, 512–17.

Patte, J.C., Le Bras, G. & Cohen, G.N. (1967) Regulation by methionine of the synthesis of a third aspartate kinase and a second homoserine dehydrogenase in *Escherichia coli* K12. *Biochimica et Biophysica Acta* 136, 245–57.

Rayapati, P.J., Stewart, C.R. & Hack, E. (1989) Pyrroline-5-carboxylate reductase is in pea (*Pisum sativum*) leaf chloroplasts. *Plant Physiology* 91, 581–6.

Rehfeld, D.W. & Tolbert, N.E. (1972) Aminotransferases in peroxisomes from spinach leaves. *Journal of Biological Chemistry* 247, 4803–11.

Relton, J.M., Bonner, P.L.R., Wallsgrove, R.M. & Lea, P.J. (1988) Physical and kinetic properties of lysine-sensitive aspartate kinase purified from carrot cell suspension culture. *Biochimica et Biophysica Acta* 953, 48–60.

Rhodes, D., Handa, S. & Bressan, R.A. (1986) Metabolic changes associated with adaption of plant cells to water stress. *Plant Physiology* 82, 890–903.

Rognes, S.E., Lea, P.J. & Miflin, B.J. (1980) S-Adenosylmethionine, a novel regulator of aspartate kinase. *Nature* 287, 357–8.

Rognes, S.E., Bright, S.W.J. & Miflin, B.J. (1983) Feedback-insensitive aspartate kinase isoenzymes in barley mutants resistant to lysine plus threonine. *Planta* 157, 32–8.

Sainis, J.K., Mayne, R.G., Wallsgrove, R.M., Lea, P.J. & Miflin, B.J. (1981) Localization and characterization of homoserine dehydrogenase isolated from barley and pea leaves. *Planta* 152, 491–6.

Salvucci, M.E., Werneke, J.M., Ogren, W.L. & Portis, Jr. A.R. (1987) Purification and species distribution of Rubisco activase. *Plant Physiology* 84, 930–6.

Schaeffer, G.W. & Sharpe, F.T. (1981) Lysine in seed protein from S-aminoethyl-L-cysteine resistant anther derived tissue culture of rice. *In Vitro* 17, 345–52.

Shewry, P.R. and Miflin, B.J. (1977) Properties and regulation of aspartate kinase from barley seedlings. *Plant Physiology* 56, 69–73.

Sivak, M.N. (1987) Photosynthesis in barley photorespiratory mutants lacking phosphoglycolate phosphatase activity. *Research Institute for Photosynthesis Annual Report*, Sheffield, pp. 67–9.

Sivak, M.N. & Rowell, J. (1987) Some effects of oxygen on photosynthesis by photorespiratory mutants of barley. *Research Insitute for Photosynthesis Annual Report*, Sheffield, pp. 63–7.

Sivak, M.N. & Walker, D.A. (1986) Photosynthesis in vivo can be limited by phosphate supply. *New Phytologist* 102, 499–512.

Sivak, M.N., Lea, P.J., Blackwell, R.D., Murray, A.J.S., Hall, N.P., Kendall, A.C., Turner, J.C. & Wallsgrove, R.M. (1988) Some effects of oxygen on photosynthesis by photorespiratory mutants of barley (*Hordeum vulgare*). *Journal of Experimental Botany* 39, 655–66.

Smith, I.K., Kendall, A.C., Keys, A.J., Turner, J.C. & Lea, P.J. (1984) Increased levels of glutathione in catalase deficient mutants of barley. *Plant Science Letters* 37, 29–33.

Smith, I.K., Kendall, A.C., Keys, A.J., Turner, J.C. & Lea, P.J. (1985) The regulation of the biosynthesis of glutathione in leaves of barley (*Hordeum vulgare* L.). *Plant Science* 41, 11–17.

Soldal, T. & Nissen, P. (1978) Multiphasic uptake of amino acids by barley roots. *Physiologia Plantarum* 43, 181–8.

Somerville, C.R. & Ogren, W.L. (1979) A phosphoglycolate phosphatase-deficient mutant of *Arabidopsis*. *Nature* 280, 833–6.

Somerville, C.R. & Ogren, W.L. (1980a) Photorespiration mutants of *Arabidopsis thaliana* deficient in serine-glyoxylate aminotransferase activity. *Proceedings of the National Academy of Sciences (USA)* 77, 2684–7.

Somerville, C.R. & Ogren, W.L. (1980b) The inhibition of photosynthesis in *Arabidopsis* mutants lacking glutamate synthase activity. *Nature* 286, 257–9.

Somerville, C.R. & Ogren, W.L. (1981) Photorespiration-deficient mutants of *Arabidopsis thaliana* lacking mitochondrial serine transhydroxymethylase activity. *Plant Physiology* 67, 666–71.

Somerville, C.R. & Ogren, W.L. (1982) Mutants of the cruciferous plant *Arabidopsis thaliana* lacking glycine decarboxylase activity. *Biochemical Journal* 202, 373–80.

Somerville, S.C. & Ogren, W.L. (1983) An *Arabidopsis thaliana* mutant defective in chloroplast dicarboxylate transport. *Proceedings of the National Academy of Sciences (USA)* 80, 1290–4.

Somerville, S.C. & Somerville, C.R. (1983) Effects of oxygen and carbon dioxide on photorespiratory flux determined from glycine accumulation in a mutant of *Arabidopsis*. *Journal of Experimental Botany* 34, 415–21.

Somerville, S.C. & Somerville, C.R. (1985) A mutant of *Arabidopsis thaliana* deficient in chloroplast dicarboxylate transport is missing an envelope protein. *Plant Science Letters* 37, 317–20.

Somerville, C.R., Portis, A.R. & Ogren, W.L. (1982) A mutant of *Arabidopsis* which lacks activation of RuBP carboxylase *in vivo*. *Plant Physiology* 70, 381–7.

Stadman, E.R., Cohen, E.N., Le-Bras, G. & Robinchon-Szulmajster, H. (1961) Feedback inhibition and repression of aspartokinase activity in *Escherichia coli* and *Saccharomyces cerevisiae*. *Journal of Biological Chemistry* 236, 2033–8.

Ta, T.C. & Joy, K.W. (1986) Metabolism of some amino acids in relation to the photorespiratory nitrogen cycle of pea leaves. *Planta* 168, 118–22.

Thompson, G.A., Datko, A.H., Mudd, S.H. & Giovanelli, J. (1982) Methionine biosynthesis in *Lemna*. Studies on the regulation of cystathionine-synthase, *O*-phosphohomoserine sulfhydrylase, and *O*-acethylserine sulfhydrylase. *Plant Physiology* 69, 1077–83.

Thompson, J.F. (1980) Arginine synthesis,

pro-line synthesis and related processes. In: Miflin, B.J. (ed.), *The Biochemistry of Plants*, Vol. 5. Academic, New York, pp. 375–402.

Tolbert, N.E. (1979) Glycolate metabolism by higher plants and algae. In: Gibbs, M. & Latzko, E. (eds), *Encyclopaedia of Plant Physiology*, Vol. 6. Springer-Verlag, Berlin, pp. 338–53.

Tolbert, N.E., Yamazaki, R.K. & Oeser, A. (1970) Localization and properties of hydroxypyruvate and glyoxylate reductases in spinach leaf particles. *Journal of Biological Chemistry* 245, 5129–36.

Umbarger, H.E. (1978) Amino acid biosynthesis and its regulation. *Annual Review of Biochemistry* 47, 533–606.

Van Bel, A.J.E., Borstlap, A.C., Van Pinxteren-Bazuine, A. & Ammerlaan, A. (1982) Analysis of valine uptake by *Commelina* mesophyll cells in a biphasic active and a diffusional component. *Planta* 155, 335–41.

Vérin-Vergeau, C., Baldy, P. & Cavalié, G. (1979) La phosphoglycolate phosphatase des feuilles de haricot: deux formes isofonctionelles. *Phytochemistry* 18, 1279–82.

Walker, G.H., Sarojini, G. & Oliver, D.J. (1982) Identification of a glycine transporter from pea leaf mitochondria. *Biochemical and Biophysical Research Communications* 107, 856–861.

Walker, J.L. & Oliver, D.J. (1986) Glycine decarboxylase multienzyme complex. Purification and partial characterisation from pea leaf mitochondria. *Journal of Biological Chemistry* 261, 2214–21.

Wallsgrove, R.M. & Mazelis, M. (1981) Spinach leaf dihydrodipicolinate synthase: partial purification and characterization. *Phytochemistry* 20, 2651–5.

Wallsgrove, R.M., Kendall, A.C., Hall, N.P., Turner, J.C. & Lea, P.J. (1986) Carbon and nitrogen metabolism in a barley mutant with impaired chloroplast dicarboxylate transport. *Planta* 168, 324–9.

Wallsgrove, R.M., Turner, J.C., Hall, N.P., Kendall, A.C. & Bright, S.W.J. (1987) Barley mutants lacking chloroplast glutamine synthetase-biochemical and genetical analysis. *Plant Physiology* 83, 155–8.

Walton, N.J. & Butt, V.S. (1981) Metabolism and decarboxylation of glycolate and serine in leaf peroxisomes. *Planta* 153, 225–31.

Widholm, J.M. (1972a) Cultured *Nicotiana tabacum* cells with an altered anthranilate synthetase which is less sensitive to feedback inhibition. *Biochimica et Biophysica Acta* 261, 52–8.

Widholm, J.M. (1972b) Anthranilate synthetase from 5-methyltryptophan susceptible and resistant cultured *Daucus carota* cells. *Biochimica et Biophysica Acta* 279, 48–57.

Widholm, J.M. (1976) Selection and characterization of cultured carrot and tobacco cells resistant to lysine, methionine and proline analogs. *Canadian Journal of Botany* 54, 1523–7.

Woo, K.C., Flugge, U.I. & Heldt, H.W. (1987) A two translocator model for the transport of 2-oxoglutarate and glutamate in chloroplasts during ammonia assimilation in the light. *Plant Physiology* 84, 624–32.

Yu, J. & Woo, K.C. (1988) Glutamine transport and the role of the glutamine translocator in chloroplasts. *Plant Physiology* 88, 1045–54.

Zelitch, I. (1989a) Oxygen-resistant photosynthesis in tobacco plants selected for oxygen-resistant growth. In: Zelitch, I. (ed.), *Perspectives in Biochemical and Genetic Regulation of Photosynthesis.* Wiley-Liss, New York, pp. 239–51.

Zelitch, I. (1989b) Selection and characterisation of tobacco plants with novel O_2-resistant photosynthesis. *Plant Physiology* 90, 1457–64.

10

Molecular Analysis of Nitrate Metabolism

Andris Kleinhofs & Robert L. Warner

Most higher plants, including barley, can use either nitrate or ammonium as a nitrogen source. Ammonium, however, is rapidly oxidized in well-aerated soils by autotrophic nitrifying bacteria thus forcing most plants to assimilate nitrate for their nitrogen needs. Nitrate assimilation is a fundamental process of higher plants and algae, as well as many fungi and bacteria. On a global scale, the plant kingdom obtains ten- to 100-fold more reduced N through nitrate assimilation than from biological fixation of molecular nitrogen (Guerrero *et al*., 1981; Paul & Clark, 1988). The uncertainty seems to be primarily due to the inability to estimate accurately the total amount of nitrate reduced by plants annually. Regardless of which figure is used, the value of nitrate assimilation to plant productivity and therefore to all organisms who depend upon plants for their reduced nitrogen is immense.

The acquisition, reduction and assimilation of nitrate are highly regulated in higher plants. In the absence of nitrate, most plants have little or no capacity for uptake or assimilation of nitrate. Upon exposure to nitrate, the proteins and enzymes required for nitrate uptake and reduction are rapidly acquired. The proteins involved in nitrate uptake and transport are yet to be characterized. Reduction of nitrate to ammonia is catalysed by two well-characterized enzymes, nitrate reductase (NR) and nitrite reductase (NiR). Ammonia is assimilated by the glutamate synthase cycle (Wallsgrove *et al*., 1983; & Chapter 9, this volume).

Most of our early knowledge of nitrate assimilation was derived from physiological and more recently biochemical studies. During the past decade, significant progress has been made with genetic and molecular analyses. This is particularly true for nitrate reductase which has become a model system for the application of genetic, biochemical and molecular tools to the study of a complex system. Substantial progress has also been made in molecular analysis of nitrite reductase. Unfortunately, comparable progress in the molecular studies of nitrate uptake and transport has not yet materialized. The nitrate reduction pathway has been extensively reviewed (Caboche & Rouze, 1990; Campbell & Kinghorn, 1990; Crawford & Campbell, 1990; Kleinhofs & Warner, 1990; Solomonson & Barber, 1990).

Nitrate uptake and transport

Transport of nitrate from the external medium across the plasma membrane into the cells of roots is a thermodynamically uphill and highly regulated process (Clarkson, 1986). Nitrate uptake depends on metabolic energy, but the rate is affected by transporter induction, nitrate status of the cell and the external medium, and the presence of other ions in the external medium. Although the physiological aspects of nitrate uptake have been extensively studied, little is known of the transport mechanism(s) or of the molecular aspects of nitrate transporters.

Influx of nitrate into root cells is generally assumed to be against an electrochemical gradient via an active transport mechanism (Clarkson, 1986). In barley, nitrate uptake is inhibited by anaerobic conditions, uncouplers and inhibitors of oxidative phosphorylation (Rao & Rains, 1976a) and low temperatures (Bloom & Chapin, 1981). These effects, however, are much greater at low nitrate concentrations than at high concentrations indicating passive influx may occur at high external nitrate concentrations (Glass *et al.*, 1990). For inward diffusion of nitrate, the concentration of nitrate in the cytoplasm would have to be very low and/or the external concentration very high to overcome the plasma membrane electrical potential difference of about -150 mV. Under physiological conditions, concentrations of nitrate in roots are commonly one to two orders of magnitude greater than the nitrate concentration in the external medium. However, most of the cellular nitrate is located in the vacuoles (Granstedt & Huffaker, 1982; Belton *et al.*, 1985; Martinoia *et al.*, 1986). Nitrate concentrations in the cytoplasm have not been firmly established, although some estimates of cytoplasmic nitrate (Belton *et al.*, 1985; Lee & Clarkson, 1986) indicate the concentrations are too high for passive diffusion to play a significant role in nitrate influx into roots (Clarkson, 1986). However, cytoplasmic nitrate concentrations would be predicted to be low in nitrate-starved roots and perhaps under conditions where fluxes to other pools (transport to vacuole and/or xylem, reduction by NR) are high (Siddiqi *et al.*, 1990). Under these conditions passive influx of nitrate could occur even when external nitrate concentrations are relatively low (Glass *et al.*, 1990). Better estimates of cytoplasmic nitrate concentrations will be required before the potential contribution of passive influx to total nitrate uptake by plants can be estimated.

Transporter induction

In most higher plants the rate of nitrate transport increases from a low 'constitutive' level to a higher 'induced' level after roots are exposed to nitrate for several hours (Clarkson, 1986). In barley, induction by nitrate results in a two- to five-fold increase in the rate of nitrate influx requiring up to 10 h to reach full induction (Chantarotwong *et al.*, 1976; Glass *et al.*, 1985; Lee & Drew, 1986; Ward *et al.*, 1986; Behl *et al.*, 1988; Siddiqi *et al.*, 1989; Warner & Huffaker, 1989). Protein synthesis inhibitors prevent this induction (Jackson *et al.*, 1973; Rao & Rains, 1976b; Behl *et al.*, 1988) which indicates the increase in transport rate is probably due to *de novo* synthesis of nitrate transporter protein(s). External nitrate concentrations as low as 10 μM are adequate to induce the nitrate transporter (Deane-Drummond & Glass, 1983a), although higher concentrations are required for maximum induction in barley (Siddiqi *et al.*, 1989). Induction and maintenance of the nitrate transporter probably depends on the presence of nitrate in the cytoplasm. Removal of seedlings from nitrate generally results in decay of the nitrate-induced transporter in roots (Clarkson, 1986; Siddiqi *et al.*, 1989), although an elevated level of transporter activity has been observed in barley seedlings after several days of nitrogen starvation (Lee & Drew, 1986; Lee & Rudge, 1986). Whether this was due to continued induction of the transporter by nitrate mobilized from the vacuole (Siddiqi *et al.*, 1989), or to an elevation of the constitutive transporter by a starvation response (Clarkson, 1986) is not known.

Although considerable circumstantial evidence has accumulated supporting the existence of nitrate-specific transporters, none has been isolated or identified in higher plants. Correlations between nitrate transport activity and nitrate reductase (NR) activity (Rao & Rains, 1976b) have led to the hypothesis that NR functions both in the reduction and transport of nitrate (Butz & Jackson, 1977). Recently, a nitrate reductase has been recovered from the plasma membrane of maize roots (Ward *et al.*, 1989) and from barley roots (Ward *et al.*, 1988). Inhibition of nitrate uptake by anti-NR (*Chlorella*) IgG fragments has led to the suggestion that the plasma membrane NR is involved in nitrate transport by barley roots (Ward *et al.*, 1988). On the other hand, mutants defective in the NADH and/or the NAD(P)H NR structural genes had normal nitrate transport activity and kinetics (Warner & Huffaker, 1989). The NADH NR gene mutant used in these studies totally lacks NADH NR protein, ruling out the possibility that the inactive enzyme could retain transport function. The same information has not been obtained for the NAD(P)H NR mutant. The relationship of the membrane-bound NR to the soluble NRs in roots is not known. The NR structural gene mutants could provide insight to this relationship in the future.

Transport kinetics

Kinetic characterization of nitrate transport over a wide range of nitrate concentrations has revealed two or more concentration-dependent uptake phases. Multiple phases have been interpreted as evidence for multiple transporters (Rao & Rains, 1976a; Doddema & Telkamp, 1979) and as a single multiphasic transporter (Breteler & Nissen, 1982). Although enzyme kinetics have often been used to characterize ion transport, the interpretation of these results has been controversial (Borstlap, 1981, 1983; Nissen & Nissen, 1983). Kinetic analysis of nitrate uptake by higher plants could be confounded by the time-dependent induction of the transporter activity and by the simultaneous efflux of nitrate from the cytoplasm to the external medium (Clarkson, 1986; Jackson *et al.*, 1986). Simultaneous efflux may be as great as 70% of the influx rate (Clarkson, 1986).

The absence of suitable N isotopes or nitrate analogues contributes to the difficulty in characterizing nitrate influx kinetics. As an isotope, ^{15}N is not well suited for short-term kinetic studies because it is difficult to detect in small increments (Clarkson, 1986). Chlorate is an analogue of nitrate and has been reported to have the same influx kinetics into barley roots as nitrate (Deane-Drummond & Glass, 1982; Deane-Drummond & Thayer, 1986). These results, however, should be interpreted with caution because the apparent K_ms for influx of both ions were much greater than are generally observed for influx or net uptake of nitrate by barley roots. As an analogue of nitrate, chlorate is limited by the toxicity of chlorite which is produced from the reduction of chlorate to chlorite by nitrate reductase, and by the possibility that it may not be a legitimate substitute for nitrate under all circumstances (Glass *et al.*, 1985). For example, chlorate does not substitute for nitrate as an inducer of the nitrate transporter in maize roots (McClure *et al.*, 1986). Although it has a half-life of only 10 min, ^{13}N has proven to be a useful isotope for short-term nitrate influx studies (Lee & Clarkson, 1986; Lee & Drew, 1986; Siddiqi *et al.*, 1989, 1990; Glass *et al.*, 1990).

Based upon concentration-dependent uptake rates, barley has been shown to have at least two uptake systems; one or more low concentration saturable systems, and a non-saturable system (Rao & Rains, 1976a; Siddiqi *et al.*, 1990). In addition, the low concentration system may have two distinct saturable components (Lee & Drew, 1986; Warner & Huffaker, 1989). The saturable component of the 'constitutive' transporter has a lower apparent K_m for nitrate influx than the 'inducible' transporter (Lee & Drew, 1986; Siddiqi *et al.*, 1990), but it is not clear if these differences are due to two different saturable transporters (Lee & Drew, 1986) or to different expression and characteristics of a single transporter (Siddiqi *et al.*, 1990). The high-concentration non-saturable system also appears to be constitutive (Siddiqi *et al.*, 1990) and distinct from low-concentration saturable system(s). Thus, at least two mechanisms for nitrate transport operate in barley (Glass *et al.*, 1990). The saturation-type kinetics observed at low nitrate concentrations are consistent with an active transport mechanism while the linear component operating at high nitrate concentrations is consistent with a channel-mediated passive transport mechanism.

Influence of ammonium on uptake

Nitrate uptake is often inhibited by ammonium (Haynes & Goh, 1978; Clarkson, 1986; Jackson *et al.*, 1986). In some cases, products of ammonium assimilation have been proposed to inhibit or repress the induction of the nitrate transporter (Clarkson, 1986; Jackson *et al.*, 1986), while in others the inhibition by ammonium is much too rapid to be explained by product inhibition. In barley, inhibition of nitrate influx occurs within minutes of addition of ammonium, and is reversed as soon as ammonium is removed from the external medium (Lee & Drew, 1989; Warner & Huffaker, 1989). It has been proposed that ammonium inhibits net nitrate influx in barley by enhancing nitrate efflux (Deane-Drummond & Glass, 1983b), however, others have been unable to demonstrate enhanced efflux due to ammonium (Lee & Clarkson, 1986; Lee & Drew, 1989). Thus ammonium appears to have a direct effect on nitrate influx. Although ammonium generally causes an inhibition of nitrate transport, in barley nitrate influx is enhanced in some cultivars and inhibited in others (Bloom & Finazzo, 1986). In longer-term experiments ammonium may stimulate nitrate transport activity in barley (Bloom & Sukrapanna, 1990). Without better information of the rate of transporter turnover, it is difficult to distinguish between direct effects on transporter activity and indirect effects on transporter induction.

Nitrate reductase

Nitrate reductase, the first enzyme in the nitrate reduction pathway, catalyses the two electron reduction of nitrate to nitrite. In eukaryotes, the electrons are donated by either NADH or NADPH. The NADH-specific NR (EC 1.6.6.1) is found in barley and in most higher plants and numerous eukaryotic algae. The NADPH-specific NR (EC 1.6.6.3) is found in fungi, but has not been reported from higher plants and will not be considered here. The NAD(P)H-bispecific NR (EC 1.6.6.2) occurs in some higher plants, including barley and algae.

The NADH-specific NR from barley was the first higher plant NR to be purified in intact form. The holoenzyme has a molecular weight of *ca.* 220 000 and consists of two identical subunits of about 110 000 Da (Kuo *et al.*, 1980, 1982). Similar results have been reported for the barley leaf and root NR (Oji *et al.*, 1985, 1989). Studies with NRs from many different plants, algae, and fungi, show that they share common properties of holoenzyme and subunit size, prosthetic groups, and functional domains (Wray & Kinghorn, 1989; Kleinhofs & Warner, 1990).

The NAD(P)H-bispecific NR has been reported to occur in a number of higher plant species, including barley (Dailey *et al.*, 1982a; Harker *et al.*, 1986), rice (Shen *et al.*, 1976), maize (Redinbaugh & Campbell, 1981), soybean (Streit *et al.*, 1987), and *Erythrina senegalensis* (Stewart & Orebamjo, 1979). The physical properties of the NAD(P)H NR have not been well characterized, probably due to its unstable nature and consequent difficulties with purification. In wild-type barley, the NAD(P)H NR is found in the roots along with the NADH NR while in the shoots only the NADH NR can be demonstrated (Dailey *et al.*, 1982b; Warner *et al.*, 1987). When the NADH NR activity is lost due to a mutation in the structural gene, the NAD(P)H NR is induced or derepressed in the shoot (Dailey *et al.*, 1982a). This was exploited to purify partially and characterize the barley NAD(P)H NR and demonstrate a subunit molecular weight of approximately 100 000, slightly smaller than the barley NADH NR (Harker *et al.*, 1986).

All eukaryotic NRs that have been investigated contain three prosthetic groups, FAD, cytochrome b_{557}, and molybdenum in approximately one mol each per mol subunit (Giri & Ramadoss, 1979; Gewitz *et al.*, 1981; Redinbaugh & Campbell, 1985; Solomonson & McCreery, 1986). Molybdenum is associated with an organic moiety, molybdopterin, to form the molybdenum–molybdopterin complex which is probably the active molybdenum cofactor (MoCo) (Rajagopalan, 1989). Recently the structure of the molybdopterin has been identified as a 6-alkylpterin with a four carbon side-chain containing an enedithiol on C1′ and C2′, a secondary alcohol on C3′, and a phosphorylated primary alcohol on C4′ (Johnson & Rajagopalan, 1982; Johnson *et al.*, 1984; Kramer *et al.*, 1987). The sulphur atoms on the first two carbons of the side-chain probably serve as ligands for the molybdenum molecule (Johnson *et al.*, 1984; Wahl *et al.*, 1984; Kramer *et al.*, 1987). The NR MoCo is functionally interchangeable with the MoCo from all other molybdoenzymes, except nitrogenase (Pienkos *et al.*, 1977). Low levels of MoCo are present in uninduced wild-type barley seedlings. Addition of nitrate resulted in several-fold induction of measurable MoCo that was exactly correlated with NR induction in barley (Narayanan *et al.*, 1984).

Involvement of a *b*-type cytochrome in the *Neurospora crassa* assimilatory NR was first recognized by Garrett & Nason (1967). They observed that the absorption spectrum of the oxidized NR had a maximum at 412–13 nm which, upon reduction, shifted to reveal maxima at 557, 528, and 423 nm, in common with other cytochrome b_2 or b_5 enzymes. These data indicated that a *b*-type cytochrome is associated with NR and is involved with the electron transfer. The b_5-type cytochrome spectrum has been identified in all other eukaryotic NRs examined including barley (Somers *et al.*, 1982). The FAD moiety of higher plant NRs appears to be tightly bound and exogenous FAD is not commonly required for maximal activity of the NADH NR, but does show a positive effect on the NAD(P)H NR.

The intracellular localization of NR has been controversial. NRs from higher plants, algae and fungi are considered to be soluble enzymes located in the cytoplasm (Beevers & Hageman, 1980). Fractionation studies demonstrated that in leaves (Dalling *et al.*, 1972a; Wallsgrove *et al.*, 1979), roots (Dalling *et al.*, 1972b; Emes & Fowler, 1979; Suzuki *et al.*, 1981) and *Chlamydomonas reinhardtii* cells (Fischer & Klein, 1988) NR is located in the cytoplasm. Recently, however, immunolocalization has placed NR in the pyrenoid of several species of algae (Lopez-Ruiz *et al.*, 1985) and the chloroplast of spinach (Kamachi *et al.*, 1987). In contrast, immunogold localization detected NR in the cytoplasm of maize but not in the chloroplast (Vaughn & Campbell, 1988). Although no immunological localization studies have been conducted with the barley NADH NR, based on its solubility properties, it appears likely that it is localized in the cytoplasm or very loosely associated with the chloroplast. Intracellular localization studies with the NAD(P)H NR have not been successful due to the instability of the enzyme.

Particulate NR has been reported in barley roots (Miflin, 1967, 1970). Although bacterial contamination of roots may be a source of particulate NR under low oxygen conditions (Blevins *et al.*, 1976), recent reports indicate the presence of NR associated with the plasma membranes of maize and barley roots (Ward *et al.*, 1988, 1989) and *Chlorella* (Tischner *et al.*, 1989). The relationship of the plasma membrane NR to the soluble NRs has not been determined.

Many NR genes have now been cloned and sequenced. All publications describing cloned higher plant NR genes appear to be of the NADH type and all show high degree of identity. Biochemical, genetic and molecular data suggest that the overall structure and function among the higher plant NADH NRs is similar, that they evolved from a single ancestral gene and can be represented with a single model. Models based on studies with the *Chlorella* and spinach NRs have been published (Solomonson *et al.*, 1986; Kubo *et al.*, 1988). Recently these models have been further modified (Solomonson & Barber, 1990). Detailed structural studies have not been conducted with the barley NR due to difficulties in isolating large quantities of pure NR protein. However, the primary sequence homology suggests that we can expect the barley NR structure to be very similar to other NRs.

The higher plant NR is composed of two identical subunits. Each monomer is composed of three catalytic domains held together by protease-sensitive hinge regions. The catalytic domains are defined by the three prosthetic groups, partial activities of each domain and homology with animal proteins, i.e. cytochrome b_5 reductase (FAD domain), cytochrome b_5 or b_2 (heme domain), and sulphite oxidase (molybdenum co-factor domain) (Crawford *et al.*, 1988). The electron donor site is defined by the carboxy end of the protein, while the nitrate reduction site is towards the amino terminus of the protein. The electron flow proceeds from the physiological reductant through the three prosthetic group domains to nitrate (Fig. 10.1).

Besides NAD(P)H-dependent nitrate reduction, all eukaryotic NRs can also carry out several partial activities which are classified as dehydrogenase and nitrate-reducing activities. The dehydrogenase activities involve the NAD(P)H-dependent reduction of other electron acceptors, including cytochrome *c*, dichlorophenolindophenol, and ferricyanide, while the nitrate-reducing activities include nitrate reduction to nitrite by several non-physiological electron donors such as reduced viologens, flavins and bromophenol blue (Fig. 10.1). These partial activities can be differentially inactivated by heat, inhibitors, antibodies and mutations, and partially separated by limited proteolysis, indicating that they reside on different domains of the enzyme.

There are no detailed biochemical studies of the NAD(P)H NR structure probably due to the difficulties in isolating the protein in intact and pure form. Preliminary results suggest that the basic organization

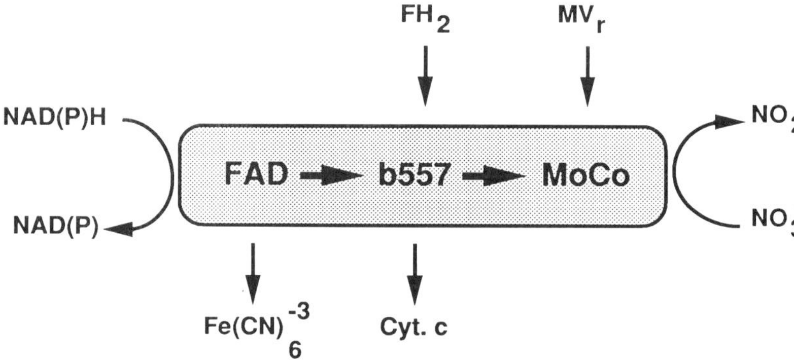

Fig. 10.1. Electron transport chain and partial activities of barley nitrate reductase. NAD(P)H dependent reduction of ferricyanide [Fe $(CN)_6^{-3}$], cytochrome *c* (Cyt. c), or dichlorophenolindophenol (DCPIP) are classified as dehydrogenase (or diaphorase) activities. These result from the reduction of ferricyanide involving the FAD domain or the reduction of cytochrome *c* or DCPIP involving the FAD and HAEM domains. Artificial electron donor nitrate reducing activities involve electrons donated from reduced flavin (FH_2) compounds at the HAEM domain or electrons donated by reduced viologens (MV_r) or bromophenol blue (BPB_r) at the MoCo domain.

of domains and structure may be similar to the NADH NR. Cloning and molecular analysis of this gene will help to resolve these questions.

Genetic analysis

Extensive genetic analysis of NR-deficient mutants has been carried out in higher plants. The first higher plant NR-deficient mutants were isolated in *Arabidopsis thaliana* (Oostindier-Braaksma & Feenstra, 1973), followed by identification of NR-deficient mutants in tobacco (Mendel & Muller, 1976, 1979; Muller & Grafe, 1978) and barley (Tokarev & Shumny, 1977; Warner *et al.*, 1977). Extensive genetic and biochemical analyses have been carried out with the barley mutants as well as those isolated in *Nicotiana*. The results show that the higher plant nitrate assimilation genetic system is similar to that found in *Aspergillus nidulans*, except for the lack of regulatory gene mutants in plants.

Two methods in use for selecting higher plant NR-deficient mutants are chlorate resistance and screening for NR deficiency by a rapid *in vivo* assay of individual segregating M_2 seedlings. The chlorate selection technique has the advantage of speed and ability to handle large populations, while the *in vivo* assay technique is more reliable when dealing with species that may have more than one NR structural gene. The barley NR-deficient mutants were selected by the *in vivo* assay method in our laboratory because we were unsuccessful with chlorate resistance selection. This is presumably because barley has two NRs which appear to be able to compensate partially for each other. Recently we have developed genetic stocks that are deficient in either or both the NADH NR and the NAD(P)H NR. These stocks have been mutagenized and successfully screened for chlorate resistant mutants.

In barley, 22 NR-deficient mutants have been identified as alleles of the *nar1* locus (Table 10.1). This locus has been rigorously identified as the NADH NR structural gene based on: (i) variation in NR-associated activities of different *nar1* alleles (Kleinhofs *et al.*, 1989); (ii) the presence in some *nar1* mutants of modified NR antigen (Kuo *et al.*, 1981; Somers *et al.*, 1983); (iii) failure of functional MoCo to reconstitute NR activity *in vitro* from extracts of *nar1* mutants (Narayanan *et al.*, 1984); (iv) a single peptide

Table 10.1. Barley nitrate reductase genes, their function, and phenotypes.

Gene	Function	Alleles	Phenotypes
Nar1	NADH NR Apoenzyme	22	CRM^- (10) $Diaphorase^+$ (10) FH_2 NR^+ (2)
Nar7	NAD(P)H NR Apoenzyme	1	NAD(P)H NR^-
Nar2	Molybdenum co-factor	3	$XADH^-$, viable (1) $XADH^-$, lethal (2)
Nar3	Molybdenum co-factor	3	$XADH^-$, viable (1) $XADH^-$, lethal (2)
Nar4	Molybdenum co-factor	1	$XADH^-$, lethal
Nar5	Molybdenum co-factor	3	$XADH^-$, viable
Nar6	Molybdenum co-factor	1	$XADH^-$, lethal
Nar8	Molybdenum co-factor	1	$XADH^-$, lethal

difference in Cleveland maps of wild-type and mutant *nar1d* NR protein (Kuo *et al.*, 1984); and (v) co-segregation of the NR-deficient *nar1* phenotype with restriction fragment length polymorphism detected by a barley NADH NR cDNA clone (Kleinhofs, unpublished).

The NR-deficient barley mutants that are alleles of the *nar1* locus all have low levels of NR activity. This activity is primarily due to a NAD(P)H-bispecific NR found in barley (Dailey *et al.*, 1982a). In wild-type barley the NAD(P)H NR activity is found in roots, but not in leaves. In the *nar1* mutants, the NAD(P)H NR activity is apparently derepressed for expression in the leaves. A single NAD(P)H NR-deficient mutant has been isolated, partially characterized and designated *nar7w* (Warner *et al.*, 1987). The *nar7* locus is presumed to be the NAD(P)H NR structural gene since it is deficient in NAD(P)H NR activity, but normal with respect to NADH NR and xanthine dehydrogenase activities (Table 10.1). Neither the *nar1* nor the *nar7* mutants have a lethal phenotype, indicating that the two NRs can substitute for one another in nitrate reduction. The double mutant has only trace NR activity, is lethal under our field conditions where nitrate is the principal form of nitrogen available, but in solution culture is able to assimilate approximately 15% of the wild-type nitrate levels (Warner & Huffaker, 1989). These data suggest that either the NADH or the NAD(P)H or both NR mutants may be leaky or that there may be other enzymes in barley capable of reducing low levels of nitrate, but not capable of sustaining reasonable growth rates.

The *nar1* locus has been mapped to the short arm of chromosome 6 (Kleinhofs *et al.*, 1988). The putative NAD(P)H NR gene clone maps to the long arm of chromosome 6 (Kleinhofs, unpublished). This is in agreement with our earlier observation that the *nar1* and *nar7* gene loci are not linked (Warner *et al.*, 1987). Identity of the *nar7* locus with the NAD(P)H NR gene clone remains to be determined.

Other monocots such as *Zea*, *Triticum* and *Oryza* also have NADH and NAD(P)H NRs and therefore must have at least two different NR genes. *Triticum aestivum* and *Aegilops umbellulata* have two genes per haploid genome with high homology to the barley NADH NR clone. It is not likely that one of these represents the NAD(P)H NR because both hybridize strongly to the barley NADH NR clone and additional, weakly hybridizing bands are present that are more likely to represent the NAD(P)H NR gene(s). Thus hexaploid wheats have at least six genes encoding the NADH NR and probably at least three genes encoding the NAD(P)H NR. The NADH NR genes map to chromosomes 6 and 7, except for one gene which appears to have been translocated to 4B from 6B (Kleinhofs *et al.*, 1988). The occurrence of two NADH NR genes per haploid genome in *Triticum* is surprising since the closely related *Hordeum* appear to have only one NADH NR gene per haploid genome. This suggests that the *Triticum* NADH NR genes may have arisen by gene duplication after the divergence of *Hordeum* and *Triticum*.

In addition to the structural gene mutants, NR activity is also lost due to mutations in genes controlling the molybdenum cofactor. These mutations are identified by the pleiotropic loss of NR and xanthine dehydrogenase activities. Detailed analysis of the genetics of MoCo mutants has been carried out in *Nicotiana* and *Hordeum* where six different genes involved with MoCo synthesis and function have been identified (Gabard *et al.*, 1988; Kleinhofs *et al.*, 1989 and unpublished). In barley, these mutants are designated *nar 2, 3, 4, 5, 6* and *8* (Table 10.1). All MoCo mutants examined have nearly wild-type or elevated cytochrome *c* reductase activities, indicating that this NR-associated partial activity does not involve the MoCo and does not require dimerization of the NR subunits (MacDonald *et al.*, 1974; Narayanan *et al.*, 1983, 1984; De Vries *et al.*, 1986; Mendel *et al.*, 1986; Kleinhofs *et al.*, 1989).

Molecular analysis

The first eukaryotic NR cloned was from barley (Cheng *et al.*, 1986). This gene most probably encodes the NADH NR and all other NR genes described to date are probably also of the NADH type (Schnorr *et al.*, 1991). Several other higher plant, algal and fungal NRs have now been cloned including from squash (Crawford *et al.*, 1986), tobacco (Calza *et al.*, 1987; Vaucheret *et al.*, 1989a, b), *Arabidopsis* (Cheng *et al.*, 1988; Crawford *et al.*, 1988), rice (Choi *et al.*, 1989; Hamat *et al.*, 1989), tomato (Daniel-Vedele *et al.*, 1989), maize (Gowri & Campbell, 1989), spinach (Prosser & Lazerus, 1990), *Aspergillus nidulans* (Kinghorn & Campbell, 1989), *Neurospora crassa* (Fu & Marzluff, 1987), and *Chlamydomonas reinhardtii* (Fernandez *et al.*, 1989). Derived amino acid sequence comparisons indicate high homology among the NRs from various sources (reviewed in Caboche & Rouze, 1990; Kleinhofs & Warner, 1990; Solomonson & Barber, 1990).

Barley NADH NR genomic sequence analysis indicates the presence of a single intron of 1.6 kb and two exons of 1170 and 1578 bp (Fig. 10.2). The exons do not correspond to the functional domains. In other higher plant NRs three introns and four exons were observed. The three introns are located in the same place in the tobacco, tomato and rice NR genes. The barley NADH NR gene intron location corresponds to the second intron in these species. The intron sizes are very different. The conservation of the NR intron locations between these very divergent higher plant species suggests that this may be a general phenomenon. The intron locations and sizes in the *A. nidulans* NADPH NR are not the same as in the plant NRs (Kinghorn & Campbell, 1989).

A putative NAD(P)H NR gene clone was recently obtained in our laboratory (Miyazaki *et al.*, in press). The coding region of this gene is slightly smaller (891 vs 915 amino acids (aa)) than the NADH NR, in agreement with the estimated molecular weight of 100 kDa for the NAD(P)H NR subunit (Harker *et al.*, 1986). The deduced amino acid sequence has good homology to the NADH NR gene (71%), but there is considerable variation in the homology of specific regions (Table 10.2). The NAD(P)H

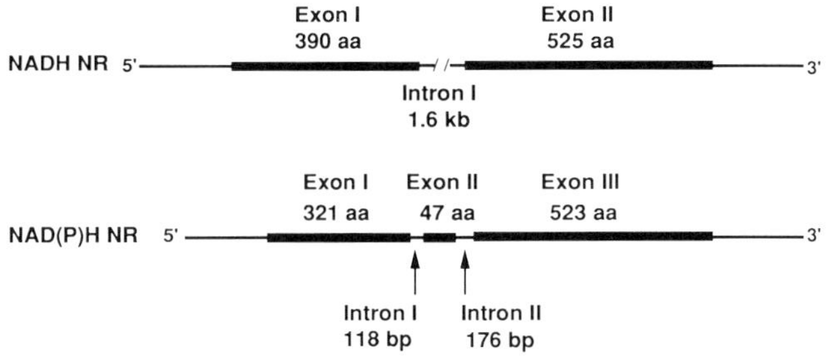

Fig. 10.2. Organization of the barley NADH and NAD(P)H nitrate reductase genes. The NAD(P)H NR coding region is slightly smaller than the NADH NR coding region (891 vs 915 aa). The NAD(P)H NR gene has two small introns and three exons while the NADH NR gene has only one large intron and two exons. The NADH NR gene intron is in the same position as Intron II in the NAD(P)H NR gene and other higher plant NR genes. A third intron observed in tobacco, tomato, and rice NR genes and located approximately 70 aa towards the 3′ end from Intron II, is not present in either barley NR gene.

Table 10.2. Identity comparison of barley NADH and NAD(P)H nitrate reductase (NR) genes and their functional domains.

Region	NADH NR (bp or aa)	NAD(P)H NR (bp or aa)	Identity (%)
5′ leader	455 bp	455 bp	36
N-terminus	110 aa	89 aa	40
MoCo domain	398 aa	397 aa	77
Hinge region	50 aa	49 aa	72
Haem domain	58 aa	58 aa	88
Hinge region	36 aa	35 aa	30
FAD domain	263 aa	263 aa	67
3′ trailer	240 bp	240 bp	40
Introns			
I vs I	1625 bp	118 bp	44
I vs II	1625 bp	176 bp	47

NR gene has two small introns, 118 and 176 bp, and three exons of 964, 141, and 1568 bp (Fig. 10.2). The introns are located in the same positions as the first two of the tobacco NADH NR gene introns.

Codon usage analysis of the barley NRs showed a remarkable preference for G or C in the third position. This is in accordance with other observations of genes from Monocotyledonae. Dicotyledonae genes show a much more balanced codon usage (Murray *et al.*, 1989).

Work to identify amino acid residues involved in binding of the prosthetic groups and the electron donors and acceptors is just beginning. The available NR-derived amino acid sequences and the homology with better-studied animal proteins allow us to make some comparisons and speculations in this area. A single cysteine residue reacting with *N*-ethyl maleimide was proposed to be involved in NADH-binding in *Chlorella* NR (Barber & Solomonson, 1986). A single cysteine residue, cys-283, was also shown to be involved in NADH-binding in microsomal NADH cytochrome b_5 reductase (Hackett *et al.*, 1986). Comparison of the *Arabidopsis* NR sequence to the microsomal NADH cytochrome b_5 reductase showed that cys-283 was not conserved between the two proteins, but that a nearby cysteine, cys-273 in cytochrome b_5 reductase and cys-889 in *Arabidopsis* NR (cys-887 in barley NADH NR), was conserved (Crawford *et al.*, 1988). Analysis of other available NR amino acid sequences showed that cys-887 (barley numbering) was conserved in all plant NRs, including the barley NAD(P)H NR (cys-863) and also the *Aspergillus* NADPH NR. Thus it seems probable that this cysteine residue plays an important role in NADH and NADPH-binding in NR. Another residue involved in NADH-binding in cytochrome b_5 reductase is lys-110 (Hackett *et al.*, 1988). This residue corresponds to lys-725 in the barley NADH NR sequence and lys-701 in the barley NAD(P)H NR. This lysine residue is also conserved in all other available NR sequences, suggesting that it serves an important role in NR function.

Arginine and histidine have also been implicated in NR functions (Pan & Nason, 1978; Baijal & Sane, 1987, 1988; Solomonson & Barber, 1989). The role of arginine was implied by the inhibition of NR activities by the arginine-specific reagents phenylglyoxal and butanedione. The role of histidine was indicated by the inhibition of diaphorase, but not nitrate-reducing activities by diethylpyrocarbonate, a histidine-specific agent (Baijal & Sane, 1987). The presence of NADH resulted in protection from inactivation with these chemical reagents. The identification of the individual residues involved will require future studies with site-specific mutagenesis.

Nitrite reductase

Several different enzymes can catalyse the six-electron reduction of nitrite to ammonia. These are: (i) assimilatory nitrite reductases (NiRs) that function in biosynthetic nitrate assimilation of higher plants, algae and fungi; (ii) ammonia-forming dissimilatory NiRs involved in anaerobic nitrate respiration of diverse bacteria; and (iii) assimilatory and dissimilatory sulphite reductases that probably reduce nitrite fortuitously due to the intrinsic capability of siroheme-containing proteins to catalyse the six-electron reduction of nitrite (Murphy *et al.*, 1974; Schmutz & Brunold, 1985). We will only consider the assimilatory NiR involved in biosynthetic assimilation of nitrate.

In photosynthetic tissues, the physiological electron donor is believed to be ferredoxin. In non-photosynthetic tissues the electron donor is a ferredoxin-like protein which is reduced by a pyridine nucleotide reductase with electrons from the pentose phosphate pathway (Suzuki *et al.*, 1985).

Nitrite reductase is localized in chloroplasts of green leaves and algae (Dalling *et al.*, 1972a; Wallsgrove *et al.*, 1979; Beevers & Hageman, 1980; Fischer & Klein, 1988) and the proplastid of roots (Dalling *et al.*, 1972b; Emes & Fowler, 1979; Suzuki *et al.*, 1981). It has been reported that some NiR may be present in the cytoplasm (Fischer & Klein, 1988).

NiRs have been purified and studied from barley (Serra *et al.*, 1982; Ip *et al.*, 1990) and a variety of other higher plants (Lancaster *et al.*, 1979; Beevers & Hageman, 1980 and references therein; Gupta *et al.*, 1984; Small & Gray, 1984; Ida & Mikami, 1986). The barley NiR, in common with other higher plant NiRs, is a monomeric polypeptide with molecular weight ca. 63 000, containing a siroheme prosthetic group and a [4Fe–4S] cluster at its active centre (Serra *et al.*, 1982; Ip *et al.*, 1990). The siroheme has been identified as an iron tetrahydroporphyrin of the isobacteriochlorin type with eight carboxylic acid-containing side-chains (Murphy *et al.*, 1974). There is no evidence to suggest that the leaf and root enzymes are different proteins (Hirasawa *et al.*, 1984; Bowsher *et al.*, 1988).

Recent evidence suggests that the *ca.* 60 000 molecular weight NiR may be a modified form of a larger, native NiR. A putative native NiR (mol. wt 85 000) isolated from spinach has 3–4-fold higher specific activity with reduced ferredoxin than with reduced methyl viologen and can be separated into two protein components of 61 000 and 24 000 molecular weight. The 61 kDa fragment retains high rates of methyl viologen, but not ferredoxin NiR activity. The 24 kDa fragment has no enzymatic activity, but contains a ferredoxin-binding site (Hirasawa & Tamura, 1980; Hirasawa-Soga & Tamura, 1981; Hirasawa-Soga *et al.*, 1982; Hirasawa *et al.*, 1984; Hirasawa & Knaff, 1985). Similar results have been reported with the *Chlamydomonas reinhardtii* nitrite reductase, where a native molecular weight of 86 000 was observed with 63 and 25 kDa subunits (Romero *et al.*, 1987). The number of sirohemes per mol of enzyme remains doubtful with most earlier reports indicating one siroheme mol per mol of enzyme, but Hirasawa *et al.* (1987) reported a value approaching 2 mol mol^{-1} of native spinach ferredoxin NiR. Perhaps the second siroheme is associated with the 24 kDa subunit.

Genetic analysis

Three NiR insoforms have been identified in tetraploid wild oat and two of them were shown to be governed by a single Mendelian locus with co-dominant alleles clearly demonstrating the nuclear nature of NiR (Heath-Pagliuso *et al.*, 1984). Southern blot analysis of the maize genome indicates the presence of two NiR genes per haploid genome (Lahners *et al.*, 1988). This is in accord with the report of two NiR isozymes in maize (Dalling *et al.*, 1973;

Kutscherra *et al.*, 1987). In spinach only a single gene per haploid genome was observed (E. Back, unpublished, quoted from Lahners *et al.*, 1988).

Nitrite reductase-deficient higher plant mutants have not been reported to date. Owing to the high toxicity of nitrite to plant cells, it is probable that such mutants would be lethal and thus not suitable for isolation in cell cultures. In whole plant systems, lethal mutants can be maintained as heterozygotes.

Molecular analysis

Nitrite reductase cDNA clones have been isolated and analysed from spinach (Back *et al.*, 1988), maize (Lahners *et al.*, 1988), and rice (Matsui *et al.*, 1990). The precursor protein for spinach nitrite reductase is 594 amino acids and contains a 32 amino acid leader sequence at the N-terminal end of the mature protein. The 32 amino acid leader most probably serves as the transit peptide for transfer of the nuclear-encoded protein from the cytoplasm to the chloroplast. The cDNA nucleotide sequences from spinach and maize NiR are 66% identical and the deduced amino acid sequences are 75% identical or 86% similar when conserved amino acids are included. Only partial rice NiR cDNA clones were reported (Matsui *et al.*, 1990). The available nucleotide sequence identity was 89% with maize and 66% with spinach NiR cDNA. The derived amino acid sequence identity was 90% with maize and 70% with spinach NiR. An extremely strong bias toward G/C use in the third codon position is observed for the maize and rice NiR. In the maize and rice gene, 98% of the codons end in C or G while in the spinach gene only 44% codons end in C or G (Lahners *et al.*, 1988; Matsui *et al.*, 1990). This codon usage bias appears to be a monocot trait which can be expected to extend to the barley NiR as already noted for the barley and rice NRs.

The four cysteines at positions 473, 479, 514 and 518 (spinach numbering) are believed to be involved in the binding of the 4Fe–4S center and siroheme co-factors (McRee *et al.*, 1986; Ostrowski *et al.*, 1987). These cysteines and the regions around them are conserved in the spinach and maize sequences (Lahners *et al.*, 1988).

Nitrate assimilation

NR was considered by Beevers & Hageman (1969) to be the rate-limiting step in the regulation of nitrate assimilation in higher plants. Nitrate, light and a number of other environmental factors are known to induce or modulate the level of NR activity in higher plants (Beevers & Hageman, 1969; Guerrero *et al.*, 1981). Barley seedlings have little or no NR activity until provided with nitrate (Somers *et al.*, 1983; Aslam *et al.*, 1987; Warner & Huffaker, 1989). Other forms of N such as nitrite have been reported to induce NR and NiR in some plant species including barley, however, the active agent is probably nitrate and not nitrite (Aslam *et al.*, 1987; Aslam & Huffaker, 1989). This is due to the oxidation of nitrite to nitrate in plant tissue and to significant nitrate contamination in many sources of nitrite. Reduced N has also been reported to induce NR; however, in barley Aslam *et al.* (1987) were unable to detect induction of NR by ammonium or urea. Since very low concentrations of nitrate are required for the induction of NR, caution should be exercised to avoid nitrate contamination including low rates of reduced N oxidation to nitrate.

The increase in NR activity corresponds to an increase in NR CRM (cross-reacting material) indicating *de novo* synthesis of the enzyme (Somers *et al.*, 1983). Upon removal of the seedlings from nitrate, both NR activity and NR CRM declined, indicating NR is rapidly induced in the presence and turned over in the absence of nitrate. Much of the regulation appears to be at the transcription level (Melzer *et al.*, 1989). Barley NR mRNA was absent in nitrate-starved seedlings, but increased very rapidly in both roots and leaves upon exposure to nitrate (Melzer *et al.*, 1989).

Light has long been known to be important in the regulation of NR and nitrate assimilation in higher plants (Beevers & Hageman, 1969). Light effects on NR are complex and may involve photoreceptors such as phytochrome and substrates produced by photosynthesis (Duke & Duke, 1984). In barley, low levels of NR are induced by nitrate, but light is required for the induction and maintenance of high NR levels (Aslam *et al.*, 1976; Somers *et al.*, 1983). Exogenous glucose can partially substitute for light (Aslam *et al.*, 1976). Fluctuations of NR activity in light and darkness correspond to changes in NR CRM indicating light modulates the level of NR protein (Somers *et al.*, 1983). Regulation of NR by light appears to be at the transcriptional level (Melzer *et al.*, 1989). In green leaves of barley seedlings, NR mRNA accumulation was enhanced by fluence-dependent white light; while in etiolated seedlings NR mRNA was also modulated by red/far red light in a manner consistent with regulation by phytochrome (Melzer *et al.*, 1989).

Although NR is a necessary component of nitrate assimilation in higher plants, other parameters may affect the level of assimilation *in vivo*. While light has direct effects on NR, light may have other indirect effects on *in vivo* nitrate assimilation. While controversy has existed in the past, it is now clear that nitrate assimilation occurs in leaves in both light and dark, provided the carbohydrate status is adequate. In barley the capacity to assimilate nitrate decreases more rapidly in leaves than in roots when seedlings are placed in darkness or CO_2 free atmosphere (Aslam *et al.*, 1979; Aslam & Huffaker, 1984) which is probably due to a much more rapid carbohydrate depletion in leaves than in roots (Aslam & Huffaker, 1982). While both nitrate uptake and assimilation are affected by carbohydrate depletion, decreases in nitrate assimilation occur sooner and to a greater extent than uptake rates. Significant quantities of photosynthesis or respiration appear to be coupled to nitrate assimilation in barley that can be detected by CO_2 and O_2 fluxes (Bloom *et al.*, 1989). Energy demands for nitrate reduction could influence plant growth processes at low light conditions but probably not at high light conditions (Bloom *et al.*, 1989).

The relative contributions of leaves and roots to nitrate assimilation have been rather difficult to determine. In part this is due to the lack of suitable N isotopes and reduced N cycling in plants. The proportion of reduced N compounds in xylem fluids has been used as an estimate of nitrate reduction in roots (Pate, 1973). In barley about 60% of the xylem fluid N was reduced N compounds. This approach probably overestimates nitrate reduction in roots due to reduced N cycling from the shoot, which is substantial in barley (Gojon *et al.*, 1986). Estimates made from ^{15}N studies indicate that nitrate reduction in roots is much greater during the first 24 h period (65%) following exposure of N-starved seedlings to nitrate than in the subsequent 24 h period (20%) (Gojon *et al.*, 1986).

The relative proportion of shoot and root nitrate reduction is also affected by light and carbohydrate status of the barley seedling (Aslam & Huffaker, 1982). Based upon nitrate depletion rates, Aslam & Huffaker (1982) concluded that roots accounted for about 40% of total seedling nitrate reduction in the dark and about 20% in the light. Similar results were obtained using ^{15}N and split root techniques (Gojon *et al.*, 1986; Oji *et al.*, 1989). After the initial exposure periods, about 10–20% of barley seedling nitrate is reduced in roots in the light; and about 30–50% during darkness.

Barley roots have two NR isozymes; NADH-specific and NAD(P)H-bispecific (Warner *et al.*, 1987). Both NRs appear to contribute toward nitrate reduction in barley seedlings during the first 24 h after exposure of N-starved seedlings to nitrate (Warner & Huffaker, 1989). Mutants deficient in one or both of the NR isozymes have reduced capacity to assimilate nitrate, although not in proportion to the reduction in NR activity. Nitrate assimilation was reduced 35% in mutants deficient in the NADH NR, 15% in mutants deficient in the NAD(P)H NR, and 85% in double mutants. These results indicate that the

NADH and NAD(P)H NRs are responsible for most of the nitrate reduction in barley seedlings, and that the NAD(P)H NR can compensate to a considerable extent for the absence of the NADH NR. In part this compensation is accomplished by expression of the NAD(P)H NR in leaves (Dailey *et al.*, 1982a; Harker *et al.*, 1986). The NAD(P)H NR isozyme is not present in leaves of the wild-type barley (Dailey *et al.*, 1982b). The NAD(P)H NR appeared to be necessary for maximum nitrate assimilation in the wild-type, indicating that it contributes to nitrate assimilation in the roots, at least during the first 24 h after exposure of N-starved seedlings to nitrate. Perhaps in succeeding periods when the contribution of the roots to total seedling nitrate reduction is smaller, this isozyme may be less important.

Summary and concluding remarks

In higher plants the assimilation of nitrate involves the acquisition and subsequent reduction of nitrate to ammonia by nitrate and nitrite reductases. Although our understanding of the biochemistry, genetics and the molecular aspects of nitrate assimilation has advanced in recent years, a great deal remains to be learned, especially about regulation and nitrate transport. Genetic and molecular studies have revealed that while only a single gene is required for the apoenzyme of nitrate reductase, multiple genes are present in some species with at least two in barley. Additional complexity is provided by the multigenic control of the nitrate reductase molybdenum co-factor. Genetic analyses of nitrite reductase in higher plants are limited, but molecular analysis has indicated that multiple nitrite reductase genes may exist in some species.

Nitrate, light and perhaps reduced nitrogen play important roles in the regulation of uptake and reduction of nitrate to ammonia. Nitrate reductase, nitrite reductase and nitrate transporters are all induced by nitrate, but little is known about the genetics or molecular aspects of their regulation. It is interesting to note that genes with regulatory functions have been identified in fungi but not in higher plants.

Acknowledgements

Research in our laboratories is supported in part by US Department of Agriculture Grant No. 89-37262-4363, National Science Foundation Grant No. DCB-8904551, and the Washington Technology Center. We wish to acknowledge our students and post-doctoral associates, past and present, for their many contributions to and discussions about nitrate assimilation.

References

Aslam, M. & Huffaker, R.C. (1982) *In vivo* nitrate reduction in roots and shoots of barley (*Hordeum vulgare* L.) seedlings in light and darkness. *Plant Physiology* 70, 1009–13.

Aslam, M. & Huffaker, R.C. (1984) Dependency of nitrate reduction on soluble carbohydrates in primary leaves of barley under aerobic conditions. *Plant Physiology* 75, 623–8.

Aslam, M. & Huffaker, R.C. (1989) Role of nitrate and nitrite in the induction of nitrite reductase in leaves of barley seedlings. *Plant Physiology* 91, 1152–6.

Aslam, M., Oaks, A. & Huffaker, R.C. (1976) Effect of light and glucose on the induction of nitrate reductase and on the distribution of nitrate in etiolated barley leaves. *Plant Physiology* 58, 588–91.

Aslam, M., Huffaker, R.C., Rains, D.W. & Rao, K.P. (1979) Influence of light and ambient carbon dioxide concentrations on nitrate assimilation by intact barley seedlings. *Plant Physiology* 63, 1205–9.

Aslam, M., Rosichan, J.L. & Huffaker, R.C. (1987) Comparative induction of nitrate reductase by nitrate and nitrite in barley leaves. *Plant Physiology* 83, 579–84.

Back, E., Burkhart, W., Moyer, M., Privalle, L. & Rothstein, S. (1988) Isolation of cDNA clones coding for spinach nitrite reductase: complete sequence and nitrate induction. *Molecular and General Genetics* 212, 20–6.

Baijal, M. & Sane, P.V. (1987) Chemical modification of nitrate reductase from *Amaranthus*. *Indian Journal of Biochemistry and Biophysics* 24, 75–9.

Baijal, M. & Sane, P.V. (1988) Arginine residue(s) at the active site(s) of the nitrate reductase complex from *Amaranthus*. *Phytochemistry* 27, 1969–72.

Barber, M.J. & Solomonson, L.P. (1986) The role of the essential sulfhydryl group in assimilatory NADH:nitrate reductase of *Chlorella*. *Journal of Biological Chemistry* 261, 4562–7.

Beevers, L. & Hageman, R.H. (1969) Nitrate reduction in higher plants. *Annual Review of Plant Physiology* 20, 495–522.

Beevers, L. & Hageman, R.H. (1980) Nitrate and nitrite reduction. In: Stumpf, P.K. & Conn, E.E. (eds), *The Biochemistry of Plants*, Miflin, B.J. (ed.), *Amino Acids and Derivatives*, Vol. 5. Academic Press, New York, pp. 115–68.

Behl, R., Tischner, R. & Raschke, K. (1988) Induction of a high-capacity nitrate-uptake mechanism in barley roots prompted by nitrate uptake through a constitutive low-capacity mechanism. *Planta* 176, 235–40.

Belton, P.S., Lee, R.B. & Ratcliffe, R.G. (1985) A ^{14}N nuclear magnetic resonance study of inorganic nitrogen metabolism in barley, maize and pea roots. *Journal of Experimental Botany* 36, 190–210.

Blevins, D.G., Lowe, R.H. & Staples, L. (1976) Nitrate reductase in barley roots under sterile, low oxygen conditions. *Plant Physiology* 57, 458–9.

Bloom, A.J. & Chapin, F.S. (1981) Differences in steady-state net ammonium and nitrate influx by cold- and warm-adapted barley varieties. *Plant Physiology* 68, 1064–7.

Bloom, A.J. & Finazzo, J. (1986) The influence of ammonium and chloride on potassium and nitrate absorption by barley roots depends on time of exposure and cultivar. *Plant Physiology* 81, 67–9.

Bloom, A.J. & Sukrapanna, S.S. (1990) Effects of exposure to ammonium and transplant shock upon the induction of nitrate absorption. *Plant Physiology* 94, 85–90.

Bloom, A.J., Caldwell, R.M., Finazzo, J., Warner, R.L. & Weissbart, J. (1989) Oxygen and carbon dioxide fluxes from barley shoots depend on nitrate assimilation. *Plant Physiology* 91, 352–6.

Borstlap, A.C. (1981) Invalidity of the multiphasic concept of ion absorption in plants. *Plant, Cell and Environment* 4, 189–95.

Borstlap, A.C. (1983) The use of model-fitting in the interpretation of 'dual' uptake isotherms. *Plant, Cell and Environment* 6, 407–16.

Bowsher, C.G., Emes, M.J., Cammack, R. & Hucklesby, D.P. (1988) Purification and properties of nitrite reductase from roots of pea (*Pisum sativum* cv. Meteor). *Planta* 175, 334–40.

Breteler, H. & Nissen, P. (1982) Effect of exogenous and endogenous nitrate concentration on nitrate utilization by dwarf bean. *Plant Physiology* 70, 754–9.

Butz, R.G. & Jackson, W.A. (1977) A mechanism for nitrate transport and reduction. *Phytochemistry* 16, 409–17.

Caboche, M. & Rouze, P. (1990) Nitrate reductase: a target for molecular and cellular studies in higher plants. *Trends in Genetics* 6, 187–92.

Calza, R., Huttner, E., Vincentz, M., Rouze, P., Galangau, F., Vaucheret, H., Cherel, I., Meyer C., Kronenberger, J. & Caboche, M. (1987) Cloning of DNA fragments complementary to tobacco nitrate reductase mRNA and encoding epitopes common to the nitrate reductases from higher plants. *Molecular and General Genetics* 209, 552–62.

Campbell, W.H. & Kinghorn, J.R. (1990) Functional domains of assimilatory nitrate reductases and nitrite reductases. *Trends in Biochemical Sciences* 15, 315–19.

Chantarotwong, W., Huffaker, R.C., Miller, B.L. & Granstedt, R.C. (1976) *In vivo* nitrate reduction in relation to nitrate uptake, nitrate content, and *in vitro* nitrate reductase activity in intact barley seedlings. *Plant Physiology* 57, 519–22.

Cheng, C.L., Dewdney, J., Kleinhofs, A. & Goodman, H.M. (1986) Cloning and nitrate induction of nitrate reductase mRNA. *Proceedings of the National Academy of Sciences (USA)* 83, 6825–8.

Cheng, C.L., Dewdney, J., Nam, H.G., den Boer, B.G.W. & Goodman, H.M. (1988) A new locus (NIA1) in *Arabidopsis thaliana* encoding nitrate reductase. *EMBO Journal* 7, 3309–14.

Choi, H.K., Kleinhofs, A. & An, G. (1989) Nucleotide sequence of rice nitrate reductase genes. *Plant Molecular Biology* 13, 731–3.

Clarkson, D.T. (1986) Regulation of the absorption and release of nitrate by plant cells: a review of current ideas and methodology. In: Lambers, H., Neeteson, J.J. & Stulen, I. (eds), *Fundamental, Ecological and Agricultural Aspects of Nitrogen Metabolism in Higher Plants*. Martinus Nijhoff, Boston, pp. 3–27.

Crawford, N.M. & Campbell, W.H. (1990) Fertile fields. *Plant Cell* 2, 829–35.

Crawford, N.M., Campbell, W.H. & Davis, R.W. (1986) Nitrate reductase from squash: cDNA cloning and nitrate regulation. *Proceedings of the National Academy of Sciences (USA)* 83, 8073–6.

Crawford, N.M., Smith, M., Bellissimo, D. & Davis, R.W. (1988) Sequence and nitrate regulation of the *Arabidopsis thaliana* mRNA encoding nitrate reductase, a metaloflavoprotein with three functional domains. *Proceedings of the National Academy of Sciences (USA)* 85, 5006–10.

Dailey, F.A., Warner, R.L., Somers, D.A. & Kleinhofs, A. (1982a) Characteristics of a nitrate reductase in a mutant deficient in NADH nitrate reductase. *Plant Physiology* 69, 1200–4.

Dailey, F.A., Kuo, T. & Warner, R.L. (1982b) Pyridine nucleotide specificity of barley nitrate reductase. *Plant Physiology* 69, 1196–9.

Dalling, M.J., Tolbert, N.E. & Hageman, R.H. (1972a) Intracellular location of nitrate reductase and nitrite reductase I. Spinach and tobacco leaves. *Biochimica et Biophysica Acta* 283, 505–12.

Dalling, M.J., Tolbert, N.E. & Hageman, R.H. (1972b) Intracellular location of nitrate reductase and nitrite reductase II. Wheat roots. *Biochimica et Biophysica Acta* 283, 513–19.

Dalling, M.J., Hucklesby, D.P. & Hageman, R.H. (1973) A comparison of nitrite reductase enzymes from green leaves, scutella, and roots of corn (*Zea mays* L.). *Plant Physiology* 51, 481–4.

Daniel-Vedele, F., Dorbe, M.F., Caboche, M. & Rouze, P. (1989) Cloning and analysis of the tomato nitrate reductase-encoding gene: protein domain structure and amino acid homologies in higher plants. *Gene* 85, 371–80.

Deane-Drummond, C.E. & Glass, A.D.M. (1982) Nitrate uptake into barley (*Hordeum vulgare*) plants. A new approach using $^{36}ClO_3^-$ as an analog for NO_3^-. *Plant Physiology* 70, 50–4.

Deane-Drummond, C.E. & Glass, A.D.M. (1983a) Short term studies of nitrate uptake into barley plants using ion-specific electrodes and $^{36}ClO_3^{-1}$. I. Control of net uptake by NO_3^- efflux. *Plant Physiology* 73, 100–4.

Deane-Drummond, C.E. & Glass, A.D.M. (1983b) Short term studies of nitrate uptake into barley plants using ion-specific electrodes and $^{36}ClO_3^{-1}$. II. Regulation of NO_3^- by NH_4^+. *Plant Physiology* 73, 105–10.

Deane-Drummond, C.E. & Thayer, J.R. (1986) Nitrate transport characteristics in *Hordeum vulgare* L. seedlings using three different tracer techniques. *Journal of Experimental Botany* 37, 429–39.

De Vries, S.E., Dirks, R., Mendel, R.R., Schaart, J.G. & Feenstra, W.J. (1986) Biochemical characterization of some nitrate reductase deficient mutants of *Nicotiana plumbaginifolia*. *Plant Science* 44, 105–10.

Doddema, H. & Telkamp, G.P. (1979) Uptake of nitrate by mutants of *Arabidopsis thaliana*, disturbed in uptake or reduction of nitrate. II. Kinetics. *Physiologia Plantarum* 45, 332–8.

Duke, S.H. & Duke, S.O. (1984) Light control of extractable nitrate reductase activity in higher plants. *Physiologia Plantarum* 62, 485–93.

Emes, M.J. & Fowler, M.W. (1979) The intracellular location of the enzymes of nitrate assimilation in the apices of seedling pea roots. *Planta* 144, 249–53.

Fernandez, E., Schnell, R., Ranum, L.P.W., Hussey, S.C., Silflow, C.D. & Lefebvre, P.A. (1989) Isolation and characterization of the nitrate reductase structural gene of *Chlamydomonas reinhardtii*. *Proceedings of the National Academy of Sciences (USA)* 86, 6449–53.

Fischer, P. & Klein, U. (1988) Localization of nitrogen-assimilation enzymes in the chloroplast of *Chlamydomonas reinhardtii*. *Plant Physiology* 88, 947–52.

Fu, Y.-H. & Marzluf, G.A. (1987) Molecular cloning and analysis of the regulation of *Nit-3*, the structural gene for nitrate reductase in *Neurospora crassa*. *Proceedings of the National Academy of Sciences (USA)* 84, 8243–7.

Gabard, J., Pelsy, F., Marion-Poll, A., Caboche, M., Saalbach, I., Grafe, R. & Muller, A.J. (1988) Isolation and characterization of *Nicotiana plumbaginifolia* nitrate reductase mutants: evidence for six complementation groups among 70 classified molybdenum cofactor deficient mutants. *Molecular and General Genetics* 213, 206–13.

Garrett, R.H. & Nason, A. (1967) Involvement of a b-type cytochrome in the assimilatory nitrate reductase of *Neurospora crassa*. *Proceedings of the National Academy of Sciences (USA)* 58, 1603–10.

Gewitz, H.-S., Piefke, J. & Vennesland, B. (1981) Purification and characterization of demolybdo nitrate reductase (NADH-Cytochrome c oxidoreductase) of *Chlorella vul-*

garis. Journal of Biological Chemistry 256, 11527–31.
Giri, L. & Ramadoss, C.S. (1979) Physical studies on assimilatory nitrate reductase from *Chlorella vulgaris. Journal of Biological Chemistry* 254, 11703–12.
Glass, A.D.M., Thompson, R.G. & Bordeleau, L. (1985) Regulation of NO_3^- influx into barley. Studies using $^{13}NO_3^-$. *Plant Physiology* 77, 379–81.
Glass, A.D.M., Siddiqi, M.Y., Ruth, T.J. & Rufty, T.W. (1990) Studies of the uptake of nitrate in barley. II. Energetics. *Plant Physiology* 93, 1585–9.
Gojon, A., Soussana, J.F., Passama, L. & Robin, P. (1986) Nitrate reduction in roots and shoots of barley (*Hordeum vulgare* L.) and corn (*Zea mays* L.) seedlings. I. ^{15}N study. *Plant Physiology* 82, 254–60.
Gowri, G. & Campbell, W.H. (1989) cDNA clones for corn leaf NADH:nitrate reductase and chloroplast $NAD(P)^+$:glyceraldehyde-3-phosphate dehydrogenase. Characterization of the clones and analysis of the expression of the genes in leaves as influenced by nitrate in the light and dark. *Plant Physiology* 90, 792–8.
Granstedt, R.C. & Huffaker, R.C. (1982) Identification of the leaf vacuole as a major nitrate storage pool. *Plant Physiology* 70, 410–13.
Guerrero, M.G., Vega, J.M. & Losada, M. (1981) The assimilatory nitrate-reducing system and its regulation. *Annual Review of Plant Physiology* 32, 169–204.
Gupta, S.C., Fletcher, J. & Beevers, L. (1984) Purification and characterization of nitrite reductase from cell suspension culture of Paul's Scarlet rose and its cross reactivity to antiserum prepared against pea leaf nitrite reductase. *Zeitschrift für Pflanzenphysiologie* 114, 321–9.
Hackett, C.S., Novoa, W.B., Ozols, J. & Strittmatter, P. (1986) Identification of essential cysteine residue of NADH-cytochrome b_5 reductase. *Journal of Biological Chemistry* 261, 9854–7.
Hackett, C.S., Novoa, W.B., Kensil, C.R. & Strittmatter, P. (1988) NADH binding to cytochrome b_5 reductase blocks the acetylation of lysine 110. *Journal of Biological Chemistry* 263, 7539–43.
Hamat, H.B., Kleinhofs, A. & Warner, R.L. (1989) Nitrate reductase induction and molecular characterization in rice (*Oryza sativa*). *Molecular and General Genetics* 218, 93–8.
Harker, A.R., Narayanan, K.R., Warner, R.L. & Kleinhofs, A. (1986) NAD(P)H bispecific nitrate reductase in barley leaves: partial purification and characterization. *Phytochemistry* 25, 1275–9.
Haynes, R.J. & Goh, K.M. (1978) Ammonium and nitrate nutrition of plants. *Biological Reviews* 53, 465–510.
Heath-Pagliuso, S., Huffaker, R.C. & Allard, R.W. (1984) Inheritance of nitrite reductase and regulation of nitrate reductase, nitrite reductase, and glutamine synthetase isozymes. *Plant Physiology* 76, 353–8.
Hirasawa, M. & Knaff, D.B. (1985) Interaction of ferredoxin-linked nitrite reductase with ferredoxin. *Biochimica et Biophysica Acta* 830, 173–80.
Hirasawa, M. & Tamura, G. (1980) Ferredoxin-dependent nitrite reductase from spinach leaves. *Agricultural and Biological Chemistry* 44, 749–58.
Hirasawa, M., Fukushima, K., Tamura, G. & Knaff, D.B. (1984) Immunochemical characterization of nitrite reductases from spinach leaves, spinach roots and other higher plants. *Biochimica et Biophysica Acta* 791, 145–54.
Hirasawa, M., Shaw, R.W., Palmer, G. & Knaff, D.B. (1987) Prosthetic group content and ligand-binding properties of spinach nitrite reductase. *Journal of Biological Chemistry* 262, 12428–33.
Hirasawa-Soga, M. & Tamura, G. (1981) Some properties of ferredoxin-nitrite reductase from *Spinacia oleracea. Agricultural and Biological Chemistry* 45, 1615–20.
Hirasawa-Soga, M., Horie, S. & Tamura, G. (1982) Further characterization of ferredoxin nitrite reductase and the relationship between the enzyme and methyl viologen-dependent nitrite reductase. *Agricultural and Biological Chemistry* 46, 1319–28.
Ida, S. & Mikami, B. (1986) Spinach ferredoxin-nitrite reductase: a purification procedure and characterization of chemical properties. *Biochimica et Biophysica Acta* 871, 167–76.
Ip, S.-M., Kerr, J., Ingledew, W.J. & Wray, J.L. (1990) Purification and characterization of barley leaf nitrite reductase. *Plant Science* 66, 155–65.
Jackson, W.A., Flesher, D. & Hageman, R.H. (1973) Nitrate uptake by dark-grown corn seedlings. Some characteristics of apparent induction. *Plant Physiology* 51, 120–7.
Jackson, W.A., Pan, W.L., Moll, R.H. & Kamprath, E.J. (1986) Uptake, translocation, and reduction of nitrate. In: Neyra, C.A.

(ed.), *Biochemical Basis of Plant Breeding*. Vol. II. *Nitrogen Metabolism*. CRC Press, Boca Raton, FL, USA, pp. 73–108.

Johnson, J.L. & Rajagopalan, K.V. (1982) Structural and metabolic relationship between the molybdenum cofactor and urothione. *Proceedings of the National Academy of Sciences (USA)* 79, 6856–60.

Johnson, J.L., Hainline, B.E., Rajagopalan, K.V. & Arison, B.H. (1984) The pterin component of the molybdenum cofactor. Structural characterization of two fluorescent derivatives. *Journal of Biological Chemistry* 259, 5414–22.

Kamachi, K., Amemiya, Y., Ogura, N. & Nakagawa, H. (1987) Immunogold localization of nitrate reductase in spinach (*Spinacia oleracea*) leaves. *Plant and Cell Physiology* 28, 333–8.

Kinghorn, J.R. & Campbell, E.I. (1989) Amino acid sequence relationships between bacterial, fungal, and plant nitrate reductase and nitrite reductase proteins. In: Wray, J.L. & Kinghorn, J.R. (eds), *Molecular and Genetic Aspects of Nitrate Assimilation*. Oxford Science Publications, Oxford, pp. 387–403.

Kleinhofs, A. & Warner, R.L. (1990) Advances in nitrate assimilation. In: Stumpf, P.K. & Conn, E.E. (eds), *The Biochemistry of Plants*, Miflin, B.J. & Lea, P.J. (eds), *Intermediary Nitrogen Metabolism*, Vol. 16. Academic, New York, pp. 89–120.

Kleinhofs, A., Chao, S. & Sharp, P.J. (1988) Mapping of nitrate reductase genes in barley and wheat. In: Miller, T.E. & Koebner, R.M.D. (eds), *Proceedings of the Seventh International Wheat Genetics Symposium*, Vol. 1. Institute of Plant Science Research, Cambridge, pp. 541–6.

Kleinhofs, A., Warner, R.L., Lawrence, J.M., Melzer, J.M., Jeter, J.M. & Kudrna, D.A. (1989) Molecular genetics of nitrate reductase in barley. In: Wray, J.L. & Kinghorn, J.R.(eds), *Molecular and Genetic Aspects of Nitrate Assimilation*. Oxford Science Publications, Oxford, pp. 197–211.

Kramer, S.P., Johnson, J.L., Ribeiro, A.A., Millington, D.S. & Rajagopalan, K.V. (1987) The structure of the molybdenum cofactor. Characterization of di-(carboxamidomethyl) molybdopterin from sulfite oxidase and xanthine oxidase. *Journal of Biological Chemistry* 262, 16357–63.

Kubo, Y., Ogura, N. & Nakagawa, H. (1988) Limited proteolysis of the nitrate reductase from spinach leaves. *Journal of Biological Chemistry* 263, 19684–9.

Kuo, T., Kleinhofs, A. & Warner, R.L. (1980) Purification and partial characterization of nitrate reductase from barley leaves. *Plant Science Letters* 17, 371–81.

Kuo, T.M., Kleinhofs, A., Somers, D.A. & Warner, R.L. (1981) Antigenicity of nitrate reductase-deficient mutants in *Hordeum vulgare* L. *Molecular and General Genetics* 181, 20–3.

Kuo, T.M., Somers, D.A., Kleinhofs, A. & Warner, R.L. (1982) NADH-nitrate reductase in barley leaves: identification and amino acid composition of subunit protein. *Biochimica et Biophysica Acta* 708, 75–81.

Kuo, T.M., Kleinhofs, A., Somers, D.A. & Warner, R.L. (1984) Nitrate reductase-deficient mutants in barley: enzyme stability and peptide mapping. *Phytochemistry* 23, 229–32.

Kutscherra, M., Jost, W. & Schlee, D. (1987) Isoenzymes of nitrite reductase in higher plants — occurrence, purification, properties, and alteration during ontogenesis. *Journal of Plant Physiology* 129, 383–93.

Lahners, K., Kramer, V., Back, E., Privalle, L. & Rothstein, S. (1988) Molecular cloning of complementary DNA encoding maize nitrite reductase. *Plant Physiology* 88, 741–6.

Lancaster, J.R., Vega, J.M., Kamin, H., Orme-Johnson, N.R., Orme-Johnson, W.H., Kreuger, R.J. & Siegel, L.M. (1979) Identification of the iron–sulfur center of spinach ferredoxin–nitrite reductase as a tetranuclear center, and preliminary EPR studies of mechanism. *Journal of Biological Chemistry* 254, 1268–72.

Lee, R.B. & Clarkson, D.T. (1986) Nitrogen-13 studies of nitrate fluxes in barley roots. I. Compartmental analysis from measurements of N^{13} efflux. *Journal of Experimental Botany* 37, 1753–67.

Lee, R.B. & Drew, M.C. (1986) Nitrogen-13 studies of nitrate fluxes in barley roots. II. Effect of plant N-status on the kinetic parameters of nitrate influx. *Journal of Experimental Botany* 37, 1768–79.

Lee, R.B. & Drew, M.C. (1989) Rapid, reversible inhibition of nitrate influx in barley by ammonium. *Journal of Experimental Botany* 40, 741–52.

Lee, R.B. & Rudge, K.A. (1986) Effects of nitrogen deficiency on the absorption of nitrate and ammonium by barley plants. *Annals of Botany* 57, 471–86.

Lopez-Ruiz, A., Roldan, J.M., Verbelen, J.P. & Diez, J. (1985) Nitrate reductase from *Monoraphidium braunii*. *Plant Physiology* 78, 614–18.

MacDonald, D.W., Cove, D.J. & Coddington, A. (1974) Cytochrome-c reductases from wild-type and mutant strains of *Aspergillus nidulans*. *Molecular and General Genetics* 128, 187–99.

Martinoia, E., Schramm, M.J., Kaiser, G., Kaiser, W.M. & Heber, U. (1986) Transport of anions in isolated barley vacuoles. I. Permeability to anions and evidence for a Cl^--uptake system. *Plant Physiology* 80, 895–901.

Matsui, J., Takeba, G. & Ida, S. (1990) Molecular cloning and partial amino acid sequence of rice ferredoxin-nitrite reductase. *Agricultural and Biological Chemistry* 54, 3069–71.

McClure, P.R., Omholt, T.E. & Pace, G.M. (1986) Anion uptake in maize roots: interactions between chlorate and nitrate. *Physiologia Plantarum* 68, 107–12.

McRee, D.E., Richardson, D.C., Richardson, J.S. & Siegel, L.M. (1986) The heme and Fe4S4 cluster in the crystallographic structure of *Eschericia coli* sulfite reductase. *Journal of Biological Chemistry* 261, 10277–81.

Melzer, J.M., Kleinhofs, A. & Warner, R.L. (1989) Nitrate reductase regulation: effects of nitrate and light on nitrate reductase mRNA accumulation. *Molecular and General Genetics* 217, 341–6.

Mendel, R.R. & Muller, A.J. (1976) A common genetic determinant of xanthine dehydrogenase and nitrate reductase in *Nicotiana tabacum*. *Biochemie und Physiologie der Pflanzen* 170, 538–41.

Mendel, R.R. & Muller, A.J. (1979) Nitrate reductase-deficient mutant cell lines of *Nicotiana tabacum*. Further biochemical characterization. *Molecular and General Genetics* 177, 145–53.

Mendel, R.R., Marton, L. & Muller, A.J. (1986) Comparative biochemical characterization of mutants at the nitrate reductase/molybdenum cofactor loci *cnx* A, *cnx* B, and *cnx* C of *Nicotiana plumbaginifolia*. *Plant Science* 43, 125–9.

Miflin, B.J. (1967) Distribution of nitrate and nitrite reductase in barley. *Nature* 214, 1133–4.

Miflin, B.J. (1970) Studies on the sub-cellular location of particulate nitrate and nitrite reductase, glutamic dehydrogenase and other enzymes in barley roots. *Planta* 93, 160–70.

Miyazaki, J., Juricek, M., Angelis, K., Schnorr, K.M., Kleinhofs, A. & Warner, R.L. Characterization and sequence of a novel nitrate reductase from barley *Molecular and General Genetics* (in press).

Muller, A.J. & Grafe, R. (1978) Isolation and characterization of cell lines of *Nicotiana tabacum* lacking nitrate reductase. *Molecular and General Genetics* 161, 67–76.

Murphy, M.J., Siegel, L.M., Tove, S.R. & Kamin, H. (1974) Siroheme: a new prosthetic group participating in six-electron reduction reactions catalyzed by both sulfite and nitrite reductases. *Proceedings of the National Academy of Sciences (USA)* 71, 612–16.

Murray, E.E., Lotzer, J. & Eberle, M. (1989) Codon usage in plant genes. *Nucleic Acids Research* 17, 447–98.

Narayanan, K.R., Somers, D.A., Kleinhofs, A. & Warner, R.L. (1983) Nature of cytochrome c reductase in nitrate reductase-deficient mutants in barley. *Molecular and General Genetics* 190, 222–6.

Narayanan, K.R., Muller, A.J., Kleinhofs, A. & Warner, R.L. (1984) In vitro reconstitution of NADH:nitrate reductase in nitrate reductase-deficient mutants of barley. *Molecular and General Genetics* 197, 358–62.

Nissen, P. & Nissen, O. (1983) Validity of the multiphasic concept of ion absorption in plants. *Phsiologia Plantarum* 57, 47–56.

Oji, Y., Hamano, T., Ryoma, Y., Miki, Y., Wakiuchi, N. & Okamoto, S. (1985) Purification and characterization of NADH-nitrate reductase from leaves of 2-row barley, and its activity as affected by some metabolites. *Journal of Plant Physiology* 119, 247–56.

Oji, Y., Otani, Y., Hosomi, Y., Wakiuchi, N. & Shiga, H. (1989) Nitrate reduction in root and shoot and exchange of reduced nitrogen between organs in two-row barley seedlings under light-dark cycles. *Planta* 179, 359–66.

Oostindier-Braaksma, F.J. & Feenstra, W.J. (1973) Isolation and characterization of chlorate-resistant mutants of *Arabidopsis thaliana*. *Mutation Research* 19, 175–85.

Ostrowski, J., Back, E.W., Madden, J., Siegel, L.M. & Kredich, N.M. (1987) Comparison of deduced amino acid sequences of α and β subunits of *S. typhimurium* NADPH-sulfite reductase (SR) with those of spinach nitrite reductase (NR) and cytochrome P-450 oxidoreductase (P450R). *Federation Proceedings* 46, 2231, Abstract 1780.

Pan, S.-S. & Nason, A. (1978) Purification and characterization of homogeneous assimilatory reduced nicotinamide adenine dinucleotide phosphate-nitrate reductase from *Neurospora crassa. Biochimica et Biophysica Acta* 523, 297–313.

Pate, J.S. (1973) Uptake, assimilation and transport of nitrogen compounds by plants. *Soil Biology and Biochemistry* 5, 109–19.

Paul, E.A. & Clark, F.E. (1988) Return of nitrogen to soil: biological nitrogen fixation. In: *Soil Microbiology and Biochemistry*. Academic Press, New York, Ch. 10.

Pienkos, P.T., Shah, V.K. & Brill, W.J. (1977) Molybdenum cofactors from molybdoenzymes and *in vitro* reconstitution of nitrogenase and nitrate reductase. *Proceedings of the National Academy of Sciences (USA)* 74, 5468–71.

Prosser, I.M. & Lazarus, C.M. (1990) Nucleotide sequence of a spinach nitrate reductase cDNA. *Plant Molecular Biology* 15, 187–90.

Rajagopalan, K.V. (1989) Chemistry and biology of the molybdenum cofactor. In: Wray, J.L. & Kinghorn, J.R. (eds), *Molecular and Genetic Aspects of Nitrate Assimilation*. Oxford Science Publications, Oxford, pp. 212–26.

Rao, K.P. & Rains, D.W. (1976a) Nitrate absorption by barley. I. Kinetics and energetics. *Plant Physiology* 57, 55–8.

Rao, K.P. & Rains, D.W. (1976b) Nitrate adsorption by barley. II. Influence of nitrate reductase activity. *Plant Physiology* 57, 59–62.

Redinbaugh, M.G. & Campbell, W.H. (1981) Purification and characterization of NAD(P)H:nitrate reductase and NADH: nitrate reductase from corn roots. *Plant Physiology* 68, 115–20.

Redinbaugh, M.G. & Campbell, W.H. (1985) Quaternary structure and composition of squash NADH:nitrate reductase. *Journal of Biological Chemistry* 260, 3380–5.

Romero, L.C., Galvan, F. & Vega, J.M. (1987) Purification and properties of the siroheme-containing ferredoxin-nitrite reductase from *Chlamydomonas reinhardtii. Biochimica et Biophysica Acta* 914, 55–63.

Schmutz, D. & Brunold, C. (1985) Localization of nitrite and sulfite reductase in bundle sheath and mesophyll cells of maize leaves. *Physiologia Plantarum* 64, 523–8.

Schnorr, K.M., Juricek, M., Huang, C., Culley, D. & Kleinhofs, A. (1991) Analysis of barley nitrate reductase cDNA and genomic clones. *Molecular and General Genetics* 227, 411–6 .

Serra, J.L., Ibarlucea, J.M., Arizmendi, J.M. & Llama, M.J. (1982) Purification and properties of the assimilatory nitrite reductase from barley *Hordeum vulgare* leaves. *Biochemical Journal* 201, 167–70.

Shen, T.-C., Funkhouser, E.A. & Guerrero, M.G. (1976) NADH- and NAD(P)H-nitrate reductases in rice seedlings. *Plant Physiology* 58, 292–4.

Siddiqi, M.Y., Glass, A.D.M., Ruth, T.J. & Fernando, M. (1989) Studies of the regulation of nitrate influx by barley seedlings using $^{13}NO_3^{-1}$. *Plant Physiology* 90, 806–13.

Siddiqi, M.Y., Glass, A.D.M., Ruth, T.J. & Rufty, T.W. (1990) Studies of the uptake of nitrate in barley. I. Kinetics of $^{13}NO_3^-$ influx. *Plant Physiology* 93, 1426–32.

Small, I.S. & Gray, J.C. (1984) Synthesis of wheat leaf nitrite reductase *de novo* following induction with nitrate and light. *European Journal of Biochemistry* 145, 291–7.

Solomonson, L.P. & Barber, M.J. (1989) Essential arginyl groups in molybdoenzymes. *Journal of Cell Biology* 107, 844a.

Solomonson, L.P. & Barber, M.J. (1990) Assimilatory nitrate reductase: Functional properties and regulation. *Annual Review of Plant Physiology and Plant Molecular Biology* 41, 225–53.

Solomonson, L.P. & McCreery, M.J. (1986) Radiation inactivation of assimilatory NADH: nitrate reductase from *Chlorella*: catalytic and physical sizes of functional units. *Journal of Biological Chemistry* 261, 806–10.

Solomonson, L.P., Barber, M.J., Robbins, A.P. & Oaks, A. (1986) Functional domains of assimilatory NADH:nitrate reductase from *Chlorella. Journal of Biological Chemistry* 261, 11290–4.

Somers, D.A., Kuo, T., Kleinhofs, A. & Warner, R.L. (1982) Barley nitrate reductase contains a functional cytochrome b-557. *Plant Science Letters* 24, 261–5.

Somers, D.A., Kuo, T.M., Kleinhofs, A., Warner, R.L. & Oaks, A. (1983) Synthesis and degradation of barley nitrate reductase. *Plant Physiology* 72, 949–52.

Stewart, G.R. & Orebamjo, T.O. (1979) Some unusual characteristics of nitrate reduction in *Erythrina senegalensis* D C. *New Phytologist* 83, 311–19.

Streit, L., Martin, B.A. & Harper, J.E. (1987) A method for the separation and partial purification of the three forms of nitrate reductase present in wild-type soybean leaves. *Plant Physiology* 84, 654–7.

Suzuki, A., Gadal, P. & Oaks, A. (1981) Intracellular distribution of enzymes associated with nitrogen assimilation in roots. *Planta* 151, 457–61.

Suzuki, A., Oaks, A., Jacquot, J.P., Vidal, J. & Gadal, P. (1985) An electron transport system in maize roots for reactions of glutamate synthase and nitrite reductase. *Plant Physiology* 78, 374–8.

Tischner, R., Ward, M.R. & Huffaker, R.C. (1989) Evidence for a plasma membrane-bound nitrate reductase involved in nitrate uptake of *Chlorella sorokiniana*. *Planta* 178, 19–24.

Tokarev, B.I. & Shumny, V.K. (1977) Detection of barley mutants with low level nitrate reductase activity after seed treatment with ethylmethane-sulphonate. *Genetika* (Moskva) 13, 2097–103.

Vaucheret, H., Kronenberger, J., Rouze, P. & Caboche, M. (1989a) Complete nucleotide sequence of the two homeologous tobacco nitrate reductase genes. *Plant Molecular Biology* 12, 597–600.

Vaucheret, H., Vincentz, M., Kronenberger, J., Caboche, M. & Rouze, P. (1989b) Molecular cloning and characterization of the two homeologous genes coding for nitrate reductase in tobacco. *Molecular and General Genetics* 216, 10–15.

Vaughn, K.C. & Campbell, W.H. (1988) Immunogold localization of nitrate reductase in maize leaves. *Plant Physiology* 88, 1354–7.

Wahl, R.C., Hageman, R.V. & Rajagopalan, K.V. (1984) The relationship of Mo, molybdopterin, and cyanolyzable sulfur in the Mo cofactor. *Archives of Biochemistry and Biophysics* 230, 264–73.

Wallsgrove, R.M., Lea, P.J. & Miflin, B.J. (1979) Distribution of the enzymes of nitrogen assimilation within the pea leaf cell. *Plant Physiology* 63, 232–6.

Wallsgrove, R.M., Keys, A.J., Lea, P.J. & Miflin, B.J. (1983) Photosynthesis, photorespiration and nitrogen metabolism. *Plant, Cell and Environment* 6, 301–9.

Ward, M.R., Aslam, M. & Huffaker, R.C. (1986) Enhancement of nitrate uptake and growth of barley seedlings by calcium under saline conditions. *Plant Physiology* 80, 520–4.

Ward, M.R., Tischner, R. & Huffaker, R.C. (1988) Inhibition of nitrate transport by anti-nitrate reductase IgG fragments and the identification of plasma membrane associated nitrate reductase in roots of barley seedlings. *Plant Physiology* 88, 1141–5.

Ward, M.R., Grimes, H.D. & Huffaker, R.C. (1989) Latent nitrate reductase activity is associated with the plasma membrane of corn roots. *Planta* 177, 470–5.

Warner, R.L. & Huffaker, R.C. (1989) Nitrate transport is independent of NADH and NAD(P)H nitrate reductases in barley seedlings. *Plant Physiology* 91, 947–53.

Warner, R.L., Lin, C.J. & Kleinhofs, A. (1977) Nitrate reductase-deficient mutants in barley. *Nature* 269, 406–7.

Warner, R.L., Narayanan, K.R. & Kleinhofs, A. (1987) Inheritance and expression of NAD(P)H nitrate reductase in barley. *Theoretical and Applied Genetics* 74, 714–17.

Wray, J.L. & Kinghorn, J.R. (eds) (1989) *Molecular and Genetic Aspects of Nitrate Assimilation*. Oxford Science Publications, Oxford.

11

Genetics of Barley Development: Mutant Phenotypes and Molecular Aspects

Gerd Bossinger, Wolfgang Rohde, Udda Lundqvist & Francesco Salamini

This chapter outlines the phytomer concept of barley development. Developmental mutants are classified by their deviation from a basic metameric scheme of the plant. Barley morphology and organogenesis of barley can be related to development within the large monocotyledonous family Poaceae. Developmental patterns of other members of the family help to establish and clarify developmental homologies required for the definition of a phytomeric model of plant organization. Recent progress in the molecular analysis of genes involved in the morphogenesis of the flower organs of dicotyledonous plants and their possible relationship to similar homoeotic genes in barley are discussed.

Morphology

Recent reviews of morphology and development in barley and other grasses are: Briggs, 1978; Dahlgren *et al.*, 1985; Deitzer, 1985; Kirby & Appleyard, 1987. Grasses undergo a life cycle of alternating generations specified by meiosis and syngamy. The sporophyte coincides with the diplophase and the gametophyte with the haplophase (Fig. 11.1). In seed formation a second fertilization event originates the endosperm, an auxiliary generation which lacks reproductive organs. Morphological properties peculiar to the grasses have been summarized by Barnard (1964) and include:

- cylindrical jointed stems with short basal internodes
- branches of successive orders arranged in alternating planes at right angles
- long narrow leaves with parallel veins and sheathing bases
- distichous phyllotaxis
- a fibrous root system consisting mainly of adventitious roots arising from nodes of the stem
- flowers of an unusual and characteristic structure
- albuminous seeds and caryopsis-type fruits
- embryos of a characteristic morphology.

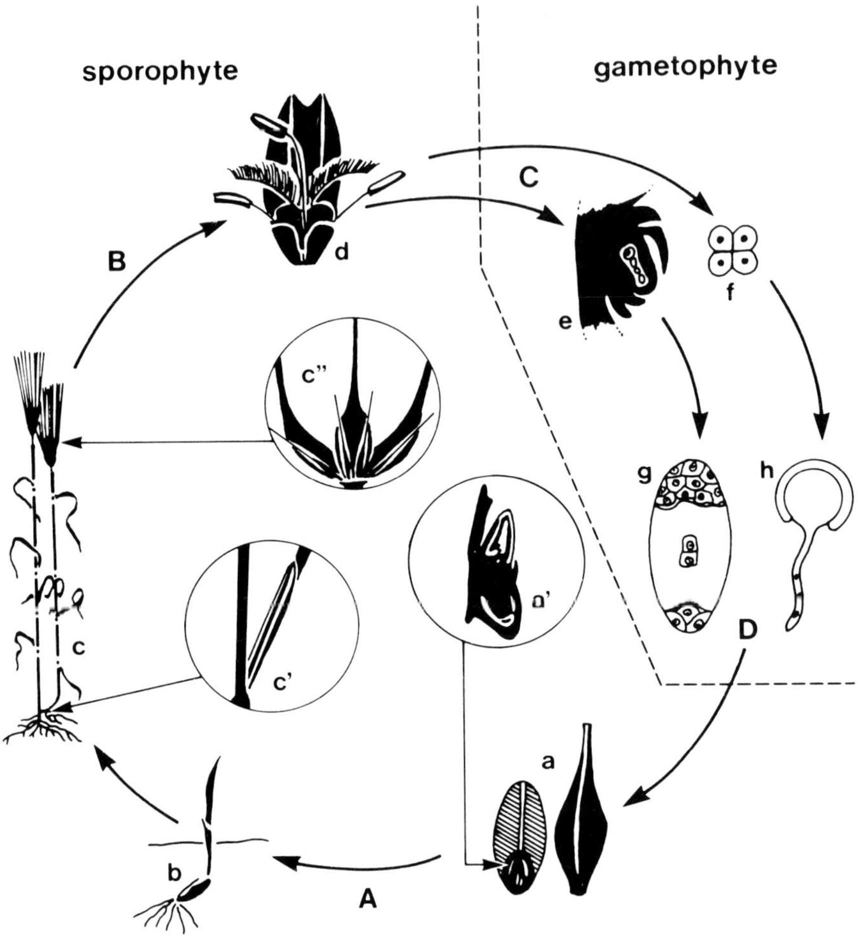

Fig. 11.1. Life cycle of barley. a, caryopsis; a′, embryo; A, germination; b, seedling; B, flowering; c, mature plant; c′, branch with prophyll; c″, spikelets; C, meiosis; d, floret; D, syngamy; e, ovule; f, pollen tetrad; g, organized embryo sac; h, pollen with tube.

Vegetative morphology

The root

The root system of grasses including barley is adventitious. Roots produced during germination from the embryo below the scutellar node are termed seminal roots (Fig. 11.1b), and may persist for months until nodal roots originating from the lower nodes of the shoot take over. The root system is fibrous and only rarely forms tubers or thorns with solid spikes. In some xerophytic grasses long and densely growing root hairs may develop a persistent 'rhizosheath' which functions for protection and uptake. Excess water inhibits root growth of most grasses, but some species may require water saturation.

The culm: nodes, internodes, branching

The primary shoot originates from the embryo, while lateral shoots develop from vegetative buds at a leaf axil or at the axil of an older lateral shoot (Gould & Shaw,

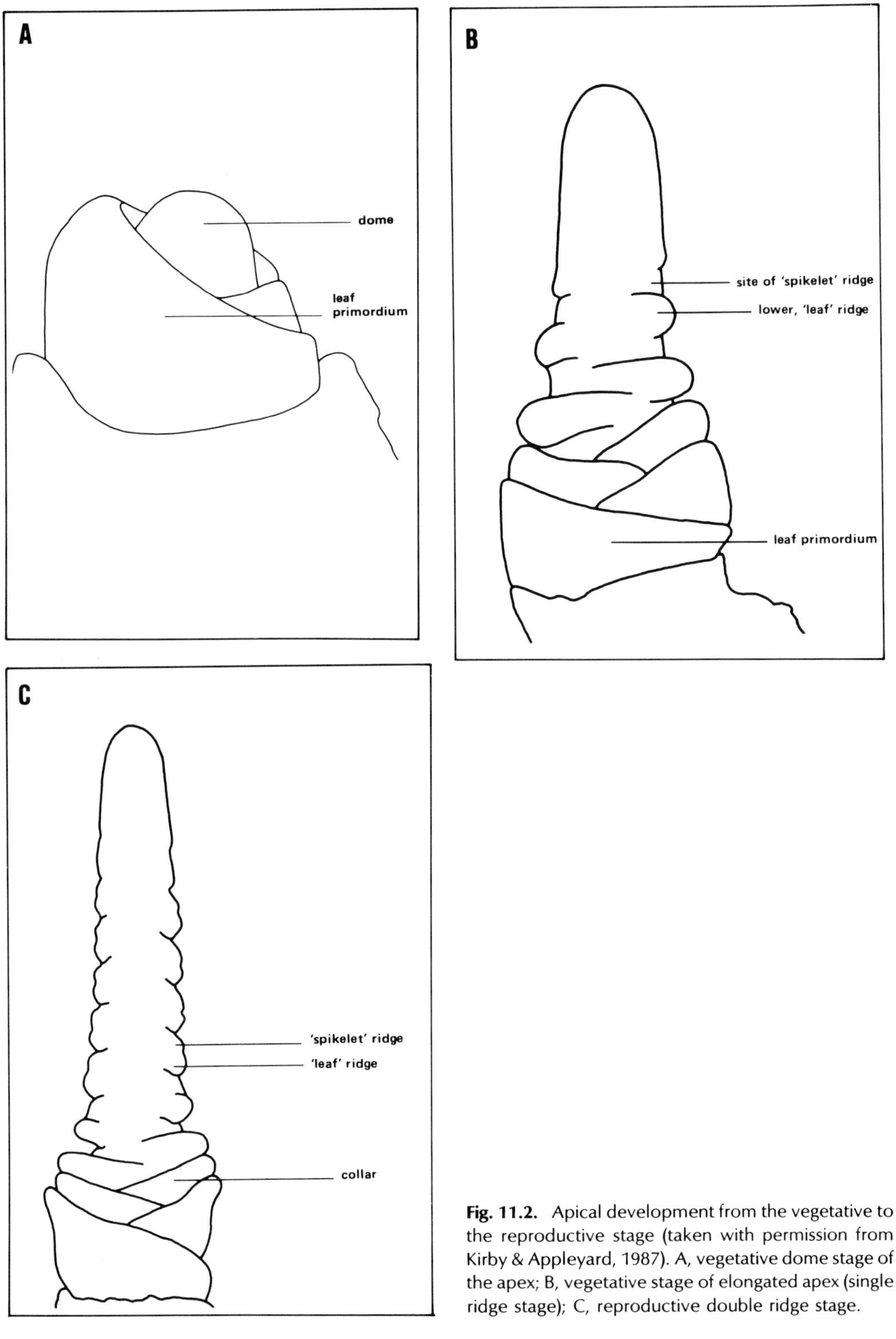

Fig. 11.2. Apical development from the vegetative to the reproductive stage (taken with permission from Kirby & Appleyard, 1987). A, vegetative dome stage of the apex; B, vegetative stage of elongated apex (single ridge stage); C, reproductive double ridge stage.

1983). In several grasses the stem segments between two successive nodes (internodes) remain short until the induction of flowering. The complex of aerial internodes is known as the culm. Meristematic tissue (intercalary meristem) is located at the basis of each internode and is involved in its elongation during development (Arber, 1934). Lateral buds at ground level (Fig. 11.1c) provide branching points for the formation of tillers. In some species of the Poaceae (but not in barley, where they invariably produce inflorescences) both basal and aerial branches can remain vegetative or terminate in an inflorescence. Lateral buds are protected by the prophyll which is the first leaf of the lateral shoot. They originate at or slightly above the nodal ridge of leaf insertion (Clark & Fisher, 1988). Bud initials frequently remain dormant; when developing they give rise to aerial culms without ramifications, which is typical of the grass family.

The leaf

Leaves of the Poaceae consist of a basal sheath which surrounds the culm, and a lamina (or blade) at varying stages of development. A ligule is almost invariably present between the sheath and blade, and, is either membranous or consists of a rim of hairs and originates from the epidermis. Close to the ligule at the blade base two projections form the auricles, and stipules may also be present.

Floral morphology

Apical development in barley

Floral initiation marks the differentiation of the apical shoot from the vegetative to the generative stage and is generally considered to have occurred when the apex is at the 'double ridge' stage (Fig. 11.2C) and the barley plant has developed three to six leaves. Before this the vegetative shoot apex forms a meristematic dome enclosed by two or more leaf primordia (Fig. 11.2A). Later on the apex becomes more elongated and accumulates many more leaf primordia which partly form prominent single ridges (Fig. 11.2B). These are the regions above which the spikelets will develop and form 'double ridges' (Fig. 11.2C). Scanning electron micrographs of these events are shown in Fig. 11.3 for barley. According to Kirby &

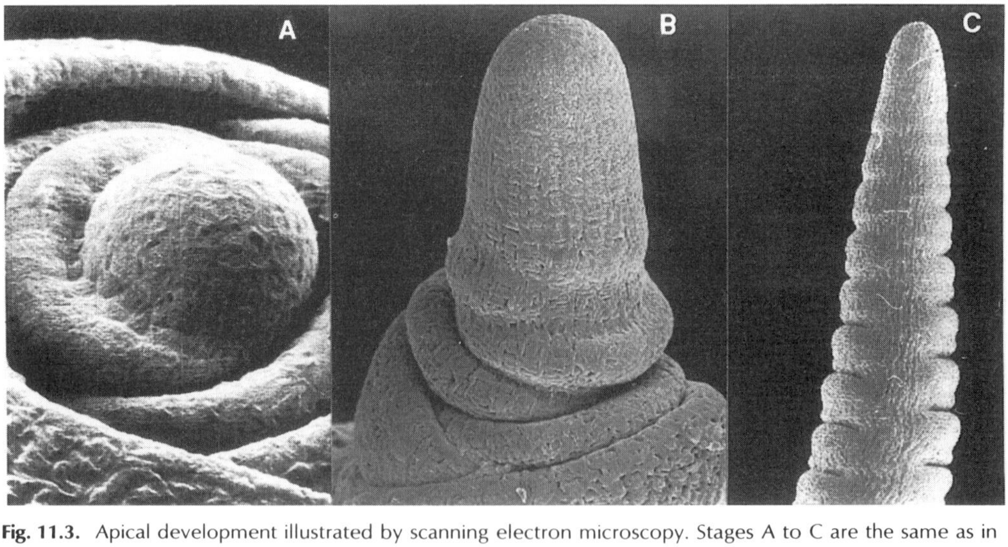

Fig. 11.3. Apical development illustrated by scanning electron microscopy. Stages A to C are the same as in Fig. 11.2. A, apex of the barley mutant *viv4*.

Appleyard (1987) further stages of barley spikelet development can be distinguished following the 'double ridge' stage. At the 'triple mound' stage the spikelet ridge will form three mounds, the precursors of the median and the two lateral spikelets which in two-row barley are sterile. In further stages the primordia of the glume, lemma, stamen and awn are separately distinguishable. At the white anther stage all of the floral organs are present, although not fully developed as shown in Fig. 11.4 for a two-row barley variety (see also Fig. 11.14F for the six-row variety *K* Atlas). At anthesis the bumps on top of the carpel have differentiated into styles and stigmas (Fig. 11.5A, B), and together with the two lodicules and three stamens these floral parts are enclosed by lemma and palea (Fig. 11.5A). Two sterile (subtending) glumes are present at the base of each floret; they are lateral appendages of the inflorescence axis and have a protective function for the spikelet (uniflowered in barley). During anthesis the lodicules swell up in open-flowering, but not in closed-flowering barley varieties, forcing the palea and lemma apart. As the filaments increase in length at the time of lodicule swelling, the pollen-containing anthers can emerge from the floret and release their pollen.

The grass inflorescence and spikelet

The inflorescence is the upper portion of the shoot separated from the culm by the collar and its subtending internode, the neck. The characteristic appearance of the inflorescence of many grasses results from 'telescoping' and from precocious branching of the axes of successive nodes. This produces a panicle of spikelets with tertiary and quarternary branches. Telescoping of lower-

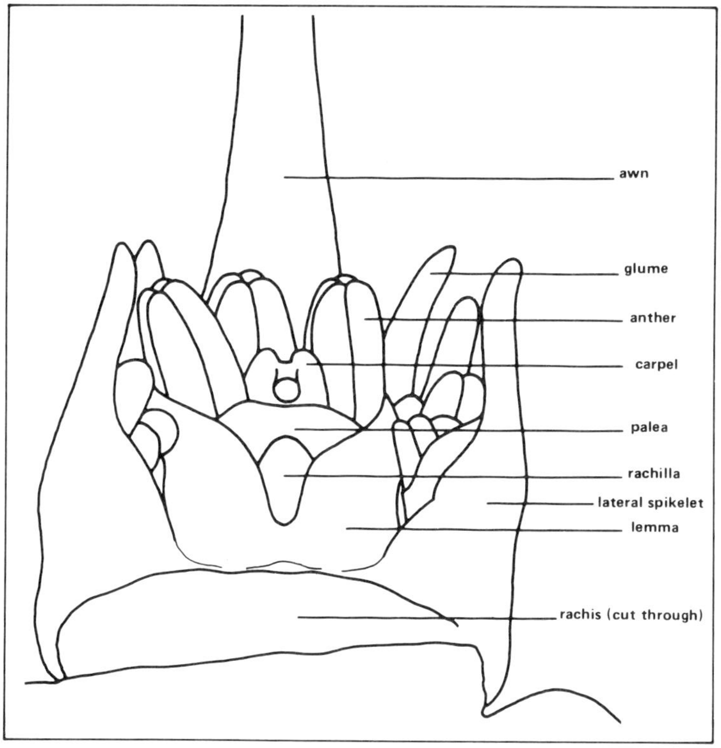

Fig. 11.4. White anther stage of the spikelet from a two-row barley (taken with permission from Kirby & Appleyard, 1987). Cutting was made through the rachis.

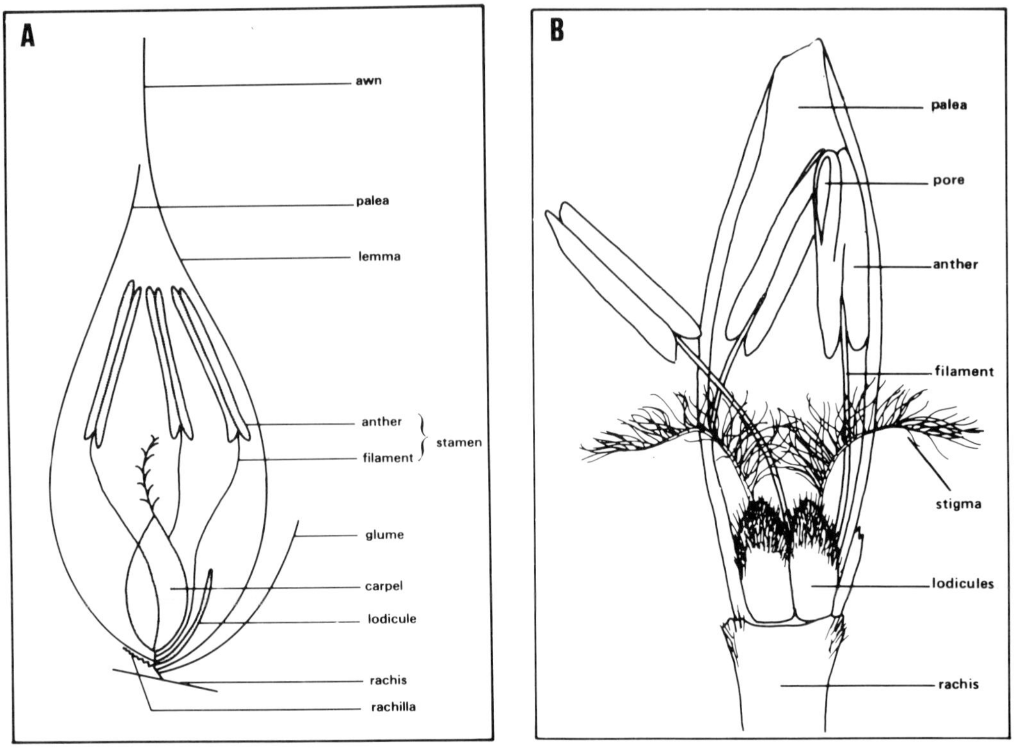

Fig. 11.5. Anthesis (diagrams taken with permission from Kirby & Appleyard, 1987). A, Diagram of the floral parts in the median spikelet (only one lodicule and one glume have been drawn). B, Anthesis stimulated by the removal of the lemma.

order branches may result in two or more branches emerging from a single node. This increase of branches at a single node may have arisen from natural selection for an increased number of spikelets per panicle, particularly in species with one floret per spikelet (Stebbins, 1982). Only in few genera or species are the inflorescence branches subtended by bracts. The plane of symmetry for each branch is at right angles to that of the parent axis; thus planes of symmetry alternate in the branches of successive orders (Arber, 1934).

The unit that forms the inflorescence is the spikelet. Its axis, the rachilla, bears overlapping bracts in a distichous order. From the axils of the bracts the reproductive shoots, the florets, develop. The two proximal bracts of the spikelet do not develop lateral shoots, but they remain as such and are known as sterile or subtending glumes. Their presence is perhaps the most characteristic feature of the grass spikelet (Harlan, 1982). The lower of the two glumes being the first leaf after a ramification is not bekeeled as expected (Clifford, 1988). Subtending glumes are not always sterile, but rarely may develop florets and spikelets from their axils. Branched spikelets provide a link to the pseudospikelets of many bamboos, which not only bear distal lemmas as in typical spikelets, but also develop branches or buds in the axils of their proximal bracts. This results in a congested branched head which bears a series of pseudospikelets of various orders, and the inflorescence becomes indeterminate. In contrast the organ is functionally determined when spikelets terminate the inflorescence branches (Dahlgren *et al.*, 1985).

The floral axis of grasses

The floral axis of the Poaceae originates in the axil of a lemma and generally bears a two-keeled palea, two to three lodicules, stamens and the pistil. These organs together with the lemma form the floret (Dahlgren *et al.*, 1985). The palea, which is inserted on the floral axis in an adaxial position, represents the first bract and has a two-keeled foliar structure closely homologous to the prophyll of the vegetative lateral shoots (Williams, 1975). This suggests that the structures are together homologous with the first bekeeled leaves present at ramification. The two palea midribs may be cilate, puberulent or rarely awn-tipped (Gould & Shaw, 1983). According to Arber (1934) the outer whorl of the perianth is missing in the Poaceae, while the inner whorl is represented by the lodicules. These organs (two of which are present in barley) are distal to the palea and – small as they are – enclosed by the two bracts, the palea and the lemma. As lodicules are inserted on the floral axis, they can be considered to represent reduced and transformed leaves, although their position is in conflict with the distichous order of other leaves. Similar considerations apply to their plane of insertion (Clifford, 1988). The stamens, three of which are present in barley as in most grasses, are composed of the pollen-containing anthers and filaments which elongate at anthesis. The stamens are located in a cylindrical arrangement above the lodicules, and according to Surkov (1961) they follow a distichous arrangement on the floral axis like other organs of the grass flower. This means that their arrangement is linear and that they are not present as a whorl. This is substantiated by the time course of differentiation for the three stamens, the degree of transformation into ovaries as a reaction to environmental conditions as well as the morphology of the mature organs which are different for each particular stamen (Clifford, 1988).

The gynoecium is the last organ present distally on the floral axis. The uniloculate pistil contains a single ovule near the posterior wall of the ovary near its base. The style bears two stigmatic lobes abundantly endowed by hairs, while the pistil may have different types of appendages. After initiation of carpel formation, the apex enclosed by the carpel develops at the tip to become the ovule (stachysporous or 'cauline' ovule; Clifford, 1988; see also below). The 'foliar' nature of the carpel is evident as deduced from its histogenesis. Most ovaries of the grass family have a pair of lateral vascular bundles and some an abaxial one (Clifford, 1988). Since one bundle is connected to the ovule, it seems plausible that in ancestral types each of the bundles was supporting a separate ovule (Gould & Shaw, 1983). It is then not unreasonable to assume that the single ovary of grasses consists of the only surviving carpel out of the original (in an evolutionary sense) three carpels, and therefore is interpreted as a reduced form of the tricapellary ovary typical for monocotyledons (Dahlgren *et al.*, 1985).

Histogenesis

Vegetative histogenesis

The vegetative shoot apex

During embryogenesis several grasses differ in the extent of meristematic activity as shown by the number of leaf primordia. In the mature embryo the shoot apex may have already initiated up to six leaf primordia enclosed by the coleoptile. The primary root is present as a slightly elongated organ and several adventitious roots are initiated (Steeves & Sussex, 1989). Embryogenesis is terminated by maturation and dessication of the seed. After germination, plant growth is resumed by the distal extremity of the embryo axis, the shoot apex. This vegetative shoot apex is surrounded by younger leaf primordia separated by intercalary meristems. Not all of these primordia differentiate immediately into tillers upon germination and shoot apex growth.

According to Klekowski (1988) plant

meristems can be classified as 'stochastic' with transitory apical initial cells and as 'structured' with permanent initial cells. Apical meristems which are not stratified into layers may be either structured (one single apical cell as in many lower plants such as ferns) or stochastic (gymnosperms). All stratified apical meristems are assumed to be structured and composed of semi-autonomous components with limited cell displacement. Most angiosperms belong to this category, and their meristems are organized in a 'tunica/corpus' system (Popham, 1951). This important concept reflects the patterns of cell division within the meristem (Steeves & Sussex, 1989). The tunica corresponds to one or more peripheral layers above the youngest leaf primordium and divides primarily according to anticlinal plane (tangential or longitudinal) forming a mantle (tunica) around the corpus. Cells of this latter tissue also divide periclinally. In the angiosperms the tunica and corpus presumably originate from few cells, if not a single permanent apical cell, the daughter cells of which generate the shoulders of the meristem (Rogers & Bonnett, 1989). More typically, the apical meristems are composed of three cell layers, the two tunica cell layers LI (dermatogen) and LII (hypodermis) and the corpus layer LIII (Satina *et al.*, 1940). In the dicotyledons LI generates the epidermis only, while in monocots it may also contribute to more internal layers of the shoot. In these species LII-derived tissues can occupy a substantial part of the plant body, while LIII forms the bulk internal part of the shoot (Poethig, 1989).

The grasses may have one (Panicoideae, Chloridoideae) or two (Bambusoideae, Oryzoideae, Arundinoideae, Pooideae) tunica layers enclosing the corpus (Clark & Fisher, 1988), although Barnard (1964) maintains that grasses invariably contain a two-layered tunica. This structural organization of the shoot apex is maintained until floral initiation. Two questions may be raised in this context. Firstly, does the vegetative apex stay in its standard tunica/corpus equilibrium state under appropriate conditions, and, secondly, does the meristem already contain the initial cells for the floral organs at this stage? The answer to the first question is straightforward. Depending on environmental conditions the shoot apices of several grasses can be maintained in the vegetative state (Halevy, 1985). It is then concluded that component meristems have the property to reproduce themselves vegetatively in a basipetal direction at any level between the dome (Fig. 11.2A and 11.3A) and the first leaf primordium. This supports the concept that cellular differentiation within the shoot apex is organized as a relatively stable process as discussed for maize by Steffensen (1968). The second question on cell commitment is not easily answered, as cell lineage and cell fate are only loosely connected in plants (Poethig, 1989). In addition, cells may originate in one layer and invade a second, where they differentiate according to the new position (Stewart, 1978). Despite this, tentative meristem fate maps have been derived as for example in maize. In these, at least two primordial cells with the potential to form the tassel are present as early as 10 days after fertilization (Poethig, 1989). It may be concluded that potentially cell-based morphogenetic information is already present during germination in the apical dome above the first leaf primordium (Steffensen, 1968), although cells in similar positions may have different fates during apical development (Poethig, 1989).

The organization of the leaf primordium, internode and axillary bud

The first sign of leaf primordium formation in the vegetative apex is the appearance of a protuberance in a region of the dome slightly above the preceding leaf primordium. This rather local protuberance transforms quickly into a crescent and then into a collar or ring-like structure, the disc of leaf insertion (Sharman, 1942), which encircles the apex in a manner typical of grasses. The circumferential growth of this new primordium terminates where it meets the central distal file of cells of the preceding primordium. As a result the leaves of most mature grasses are arranged on alternating sides of the culm generating a distichous phyllotaxis. During growth the leaf primordium assumes a hood-shaped structure which finally encloses the apex together with younger leaf primordia. It is around this time that the ligule is being formed by periclinal divisions within two or three cell layers of the adaxial epidermis (Freeling *et al.*, 1988). Also the edges of the leaf primordium base may grow on top of each other resulting in the sheaths of mature leaves that totally enclose the culm.

Initiation of a new leaf primordium is brought about by a change in the plane of cell division rather than by an increase of the division rate (Steeves & Sussex, 1989). As for all organs developing from the apical meristem, an understanding of leaf organogenesis is dependent on information about the apex layer where the first periclinal divisions take place and on the number and identity of layers that participate as a founder meristem in the development of the primordium. A general view which is valid for all grasses is not possible, even though the basic organogenetic processes are quite similar for the family members (Barnard, 1964). For six different tribes of the Poaceae that have a two-layered tunica, Barnard (1957) concluded that only dermatogen LI and hypodermis LII, but not the corpus LIII, contribute to leaf primordium formation, and that cell division first takes place in the hypodermis. This view is challenged by studies on monocotyledons with a one-layered tunica (*Zea*, *Erharta*) where the corpus is also shown to contribute to leaf histogenesis. Generally foliar histogenesis in grasses is characterized (Stewart & Dermen, 1979) by the meristematic cells that participate in histogenesis being distributed transversally in the apex and forming the founder cells (Poethig, 1984). In all cases the dermatogen layer LI divides periclinally, and can also generate mesophyll cells of the mature leaf in addition to the external layers. In one-layered tunica species, but not in those with a two-layered tunica, the corpus makes a substantial contribution to leaf formation. There is indication that in species with a two-layered tunica the first cell division during leaf primordium initiation takes place in the LII layer, while in one-layered species this may occur in the dermatogen LI. Lastly, apical growth of the leaf primordium is of minor importance in relation to the bulk of mitoses that take place during leaf formation; intercalary growth is in this sense of primary relevance (Sharman, 1942).

The origin of the internode from the leaf primordium has been extensively described for *Zea mays* (Sharman, 1942). During organization of the leaf primordium an insertion ring is formed in transversal direction. Meristematic activity at its lateral borders continues to adapt the basal part of the leaf primordium to the rapidly increasing apex diameter. This occurs as soon as new primordia are formed and the older ones start to differentiate. The cells at the lower base generate a concentric structure by active divisions. Here the cell layers whose progenies will later form all internode tissues start to divide. At the same time in the third primordium from the top, cells at the centre of the insertion ring divide in a transverse plane and generate a meristem. This basal layer of meristematic cells represents the intercalary meristem that separates two leaf primordia and is later actively involved in the elongation of the internode (Evans & Grover, 1940). During later stages of stem differentiation, the node and the upper region of the internode reach their final state, while the intercalary meristem stays active at the node base.

The axillary bud is located at the basal end of the internode next to the developing upper leaf, although the mature leaf is separated from the bud by the internode (Sharman, 1942). A precise histogenesis of the axillary bud has been given by Sharman (1945) for *Agropyron repens*, a species representative of the grasses with a two-layered tunica. In contrast to leaf primordium histogenesis, that of the axillary bud is clearly different and can be summarized as follows. Meristematic activity is first observed in the corpus, but does not spread around the apex. Instead it remains restricted to a distinct region from where it proceeds in an outward direction. As soon as a new shoot apex is organized in the bud its growth becomes apical, while in the leaf, apical growth (if present at all) ends at an early developmental stage. At the position where the lateral bud apex grows outward the tunica cell layers of dermatogen and hypodermis divide only anticlinally, so that the two tunica layers of the bud are derived directly from those of the main shoot. These important differences between leaf and axillary bud formation are summarized in the terminology used to describe the origin of organs. All organs derived from an organized apex are designated to be of 'cauline' origin when their primordia formation follows a bud-type initiation. When they are organized according to leaf primordium histogenesis they are designated to be of 'foliar' origin (Williams, 1975).

Floral histogenesis

Floral initiation in most grasses is determined by vernalization and photoperiod (Halevy, 1985). The morphogenetic event central to the development of the inflorescence is the formation of spikelet primordia. The first indication that the vegetative shoot apex enters a transition phase is the rapid elongation of the apical dome. Leaf initials are formed more rapidly than before, but progressively fail to develop. This inhibition continues up to the stage when the axillary buds appear. At this time the foliar primordia cease to grow and lateral growing points become prominent producing the double ridges discussed earlier (see also Fig. 11.2C and 11.3C). The double-ridge primordia are characteristic of all early reproductive apices of the grasses. These structures alternate along rachis, first or higher order branches and rachillae (Evans & Grover, 1940). A strong correlation exists between the appearance of the laterally growing organs and the elongation of the vegetative culm internodes. Apparently the stimulus activating the intercalary meristem is generated within the developing inflorescence (Nicholls, 1974). Other phenomena also indicate the existence of allometric correlations among differentiating organs. In barley, for example, the formation of new lateral primordia is inhibited when stamen initials appear (Nicholls, 1974).

The lateral primordia of the first order branches are still vegetative buds which may generate a secondary or tertiary branching system before differentiation of the spikelet primordia. In fact, each secondary or higher order protuberance may elongate and organize other primordia along its length. The degree of branching differs according to species, varieties and races and, together with the extent of internodal elongation within the inflorescence, is responsible for the shape of the mature inflorescence (Evans & Grover, 1940). The branch indicated as rachilla develops a first and second ridge. Their growth is not repressed and they will develop into two leaf-like primordia forming the two sterile glumes of the mature spikelet. Other leaf-like organs, the lemmas, develop on the same spikelet axis and bear the florets in their axils. These consist of an axis (the floral axis) and lateral appendages: the palea, lodicules, stamens, carpel (Figs 11.4, 11.5, 11.14).

In *Triticum* and *Phalaris* the transition of the apex from the vegetative to the reproductive stage has been studied in detail (Holt, 1954; Barnard, 1955). In wheat this transition occurs when five to seven leaves are present on the main axis. At this stage the initials of the seven youngest leaves form crescent ridges around the axis. During differentiation to the double-ridge stage small bulges appear in the axils of the 13th and 14th leaf primordia followed by new leaf primordia (which do not develop, see earlier) with associated axillary primordia. These axillary primordia develop directly into spikelets. The formation of spikelet primordia is first observed as cell divisions in the subhypodermal layer LIII between two leaf primordia. Usually three cells arranged in a vertical plane are involved, with the central cell dividing strictly periclinally, while the plane of division for the other two is at a slight angle to this direction. Further divisions of the incoming bud result in a bulge appearing on the surface of the developing rachilla. At the same time in the periphery of this region several hypodermal LII cells divide through inclined cell walls. All other divisions in the hypodermal layer are anticlinal as those in the dermatogen LI. Continued divisions of the daughter cells originating from the two central LIII cells result in the protrusion of the axillary primordium which thereby assumes a dome-like structure. The meristematic activity gradually becomes restricted to the distal region which serves as the growing point. In summary, the first and higher orders of inflorescence branches as well as spikelets have a cauline-type origin, and corpus cells and their lineages play a major role in this process. Further evidence for the cauline origin of the spikelet primordia has been obtained for other members of the grass family.

In contrast to the inflorescence axis or to the first or higher order branches, the spikelet axis (the rachilla) produces leaf primordia which are not repressed in the course of their development. In wheat these primordia originate from a well-structured apex (two tunica layers and the corpus) and follow a foliar type of histogenesis which involves only dermatogen LI and hypodermis LII (Barnard, 1955). The first two primordia will not generate a floret in their axis. Instead they form two sterile glumes located at the spikelet base and performing a protective function. In the axils of the third and following lemmas a flower primordium develops essentially in the same manner as the spikelet primordium.

Before differentiation into floral organs the flower primordium possesses the same histological traits as the apical meristems of the spike and the spikelet. In wheat the dermatogen LI and the hypodermis LII are derived from the corresponding cell layers of the spikelet, while the corpus cells originate from subhypodermal LIII initial cells (Barnard, 1955). The mature floret, the primordium of which originated in a lemma axil, uses the lemma inserted on the rachilla to protect its abaxial side, while in the adaxial position the palea develops to perform a similar function. Development of the palea primordium is similar to that of other foliar structures.

Stamen primordia are of cauline origin (Bonnett, 1953; Barnard, 1955, 1957; Williams, 1975). In wheat (Barnard, 1957) periclinal divisions take place first in the subhypodermal layer and only then in the hypodermis. The hypodermal cells divide periclinally and more frequently than during the formation of spikelet and flower primordia. Cells of the dermatogen form cell walls only in anticlinal direction. In a recent study, Dawe & Freeling (1990) pointed out the dual nature of the stamens: they initiate from a small set of cells that later on behave like apical meristems. However, this early apical growth is followed by extensive intercalary growth and expansion typical for the leaf primordia. As in leaves each cell in the founder population would form an approximately equal portion of the mature anther.

The initiation of lodicule primordia, which in wheat (Barnard, 1955) originate at the same time as stamen primordia, is of foliar mode. Periclinal divisions are first observed in the hypodermis LII from where they extend transversely along the abaxial surface of the floret primordium. The dermatogen later also divides periclinally, but only in the area between the anterior and lateral stamen initials. While Barnard (1957) and Holt (1954) localize the lodicule origin in the tunica layers, Bonnett (1953) describes lodicules (and stamens) in maize as branch-like structures. As reviewed by Guedes & Dupuy (1976), lodicules in higher plants should be considered as true appendages of the perianth. However, in the grass flower these organs have also been regarded as prophylls: the dorsal one (if present at all) as a true prophyll, the anterior pair as a split second prophyll. Although believed to have a foliar origin, lodicules are peltate organs frequently endowed with a short stalk. As peltation is unknown for bracts and prophylls of monocotyledons, grass lodicules may be suspected to be of a different origin, possibly representing staminodes, staminal filaments or fused perigonal structures. This assumption implies their assimilation to a bud-like peltate organ (Clifford, 1988).

The histogenesis of the carpel and ovule has been followed by Aziz (1972) for wheat. One to three hypodermal and dermatogen cells positioned on the anterior side of the floral axis divide periclinally followed by lateral growth similar to that of the leaf primordium. Lateral expansion rapidly leads to formation of a ring around the apex, and the newly formed primordium starts to enclose the locule of the ovary. The opening at the top reduces with time without closing completely, and this stylar channel is still present in the mature gynoecium. Stigma primordia soon start to develop in the anterior half of the gynoecium. After gynoecium initiation the floral apex develops into an ovule primordium, first in a vertical direction and later in the opposite direction. At this stage the primordia for two integuments of the ovule are initiated from the dermatogen cell layer. Both carpel and integument primordia are thus of foliar origin, and Aziz (1972) concludes that the carpel is a sin-

Table 11.1. Temporal sequence of organ initiation on the floral axis (Clifford, 1988). Organs: A, palea; B, lodicules; C, stamens; D, ovary.

Temporal sequence	Grass genera
A → B → C → D	*Avena*[10], *Bambusa*[1], *Hordeum*[2], *Lolium*[1], *Zea*[4], *Zizania*[7]
A → C → B → D	*Eleusine*[9], *Oryzopsis*[3], *Panicum*[9], *Pennisetum*[9], *Setaria*[9]
A → C → D → B	*Triticum*[8], *Oryza*[6]
C → B → A → D	*Bromus*[1]
C → B → D → A	*Bromus*[1]
C → A → B → D	*Dactylis*[5]
C → A → D → B	*Phalaris*[5]

[1]Barnard (1957); [2]Kamra & Nilan (1959); [3]Mehlenbacher (1970); [4]Bonnett (1953); [5]Holt (1954); [6]Juliano & Aldama (1937); [7]Weir & Dale (1960); [8]Barnard (1955); [9]Khosla (1964); [10]Bonnett (1961).

Table 11.2. Organ histogenesis of grasses with (a) two-layered tunica and (b) one-layered tunica.

Organ	a			b		Growth type		
	Tunica LI	LII	Corpus LIII	Tunica LI	Corpus LII	Foliar	Cauline	Mixed
Coleoptile	?	?		1	2	X		
Prophyll		1		1	2	X		
Axillary bud		2	1	2	1		X	
Leaf	2	1		1	2	X		
First-order branch		2(?)	1	2	1		X	
Rachilla		2(?)	1	2	1		X	
Glumes	2	1		1	2	X		
Lemma	2	1		1	2	X		
Floral axis		2	1	2	1		X	
Palea	1	2		1	2	X		
Stamens		2	1	2(?)	1(?)			X(?)
Lodicules	2(?)	1		2(?)	1(?)			X(?)
Carpel	1	1		1	2(?)	X		
Integuments	1			1		X		
Ovule			1		1		X	

Numbers refer to cell layers of the first (1) and further (2) periclinal divisions.

gle entity and not a syncarpous structure, and this view is shared by Esau (1965) for the grass gynoecium, while Arber (1934) describes it as tricapellary.

The temporal sequence of events which takes place during histogenesis of grass organs from germination up to spikelet formation and which we have described above, may be generalized for the grass family according to Clifford (1988) in Table 11.1. In a given species the precise knowledge of the timing of differentiation events during plant growth may help the interpretation of mutation-induced aberrations in the development of floral organs. Table 11.2 summarizes the histogenesis of primordia for all organs proper to the grass plant. Periclinal cell division is an event common to all listed primordium initiations. The table shows the cell layer in which this first cell division occurs, where the subsequent divisions take place and what type of growth is followed by the developing organ.

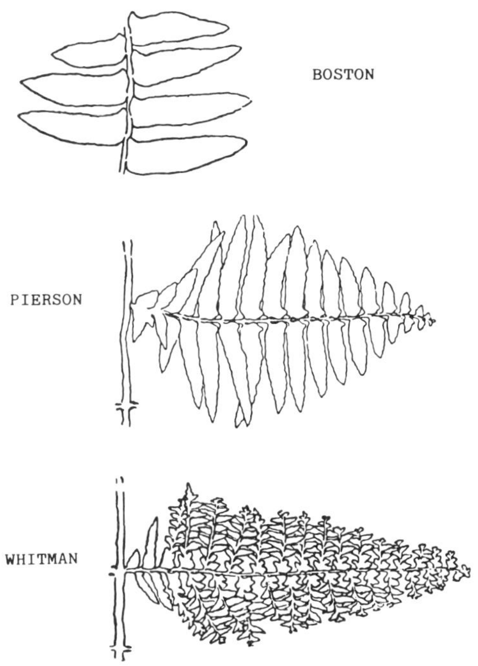

Fig. 11.6. Mutants Pierson and Whitman of the Boston fern (redrawn from Leavitt, 1909).

The phytomer concept

Bateson (1894) was the first to describe the meristic (merism: phenomenon of repetition of parts) organization of a (animal) body plan with new forms arising through variation in the destination of the repeating units. Previous observations (Bateson & Bateson, 1891) led to the term 'homoeosis' to describe a similar mode of variation in higher plants, in that one member of a meristic series could assume the form or other characters of another member. Leavitt (1905) proposed the principle of 'morphic translocation' to indicate that organs can be readily modified by translocation of parts or characters. One example illustrating the validity of this principle is the development of the compound fern leaf from a simple one (Leavitt, 1905; Fig. 11.6). This illustrates how a morphic stimulus can be translocated by mutation along the main midrib of an originally simple pinna to induce the *de novo* formation of new elementary pinnae and finally yield a fern leaf with a compound structure. Leavitt (1909) later included this principle under Bateson's homoeosis, but it is evident that Leavitt's concept is particularly useful for plants (for instance to define the class of addition mutants).

Developmental mutants of a given grass plant (as exemplified for barley) can be classified according to the disruptive effect of mutations on the meristic organization of the plant body. They will be treated in the following section, defining first the nature of the elementary unit and its variations in morphology which exist in the different parts of the plant body.

The phytomer as the morphological unit in the organization of the plant

The vegetative phytomer: phytomer type 1

The concept of the plant body as being formed as an aggregate of repetitious units has been discussed for some time, as summarized by Rutishauser & Sattler

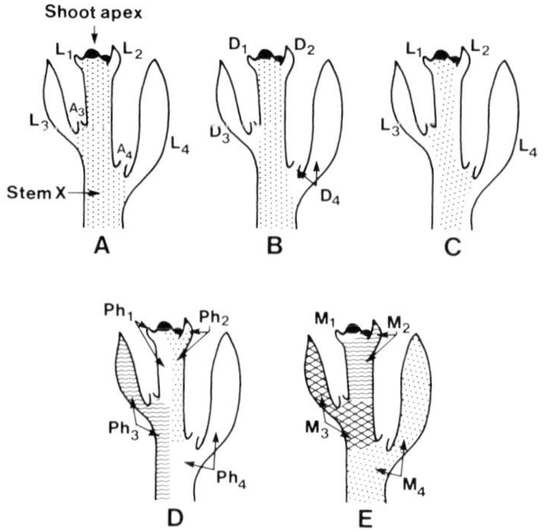

Fig. 11.7. Models of the metameric structure of vascular plants (Rutishauser & Sattler, 1985). A, stem and leaf model (the two organs belong to separate units); B, fertile leaf model (each leaf is associated with the bud in its axil); C, leaf skin model (only the external part of the stem is associated with a leaf unit); D, phytonic model (*sensu stricto*, the leaf continues into the stem); E, metameric model (a leaf and a stem segment are a simple unit). X, stem; L, leaf; A, axillary bud; D, leaf/bud primordium; Ph, phyton; M, metamer.

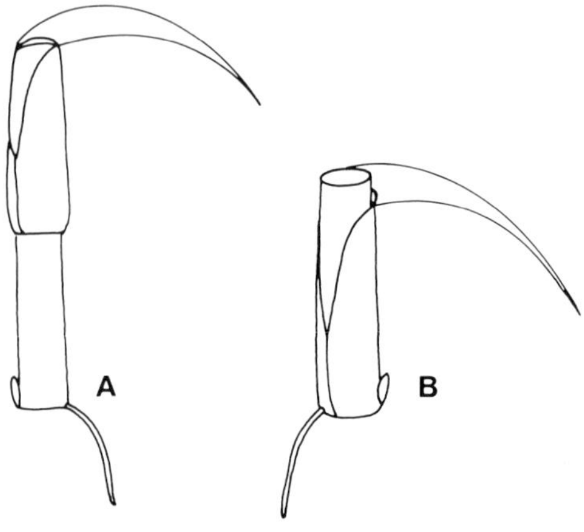

Fig. 11.8. Phytomer models drawn according to Clark & Fisher (1988). A, classical phytomer according to Weatherwax (1923) and Sharman (1942); B, structure of the phytomer based on vascular and physiological associations.

(1985). The classical 'stem and leaf' model (Fig. 11.7A) with its two variants, the 'fertile leaf' (Fig. 11.7B) and the 'leaf skin' model (Fig. 11.7C), accept the two structural units, leaf and stem, as being repeated in the plant. According to the two 'phytonic' models (Fig. 11.7D, E) each leaf or leaf whorl is the unit which extends and includes part of the stem. Lastly, in the 'metameric (phytomeric)' model the stem region associated with a leaf or leaf whorl is the internode. Monocotyledons are extremely suitable for the development of plant models based on phytonic units, and the grasses are perfect structures in terms of the phytonic theory. Their leaves are in distichous arrangement so that all lateral organs are arranged in two rows. Moreover, the alternating arrangement of leaves along the culm associates a node and an internode with each leaf. In addition the existence of intercalary meristems at the bases of the internodes further supports the view that each leaf and its associated internode form a developmentally independent unit. This 'vegetative' phytomer (or 'Sharman' phytomer) has been discussed by several authors (Weatherwax, 1923; Sharman, 1942; Clark & Fisher, 1988) and is depicted in Fig. 11.8. The phytomer concept of Sharman (1942) was mainly derived by consideration of the histogenesis of the internode and lateral bud. These are associated with the upper leaf rather than with the leaf where they are located or embraced respectively (Fig. 11.8A).

Variants of this vegetative phytomer (type 1 phytomer) are encountered in the vegetative parts of grass species as well as in the inflorescence along the rachis and the primary or higher order branches. The segments of the floral axis, for example, have been defined by Reeves (1950) as phytomers of the rachis. The basic unit of this organ is repeated, with two nodal parts divided by the internode and with leaves and roots suppressed (Fig. 11.9). Except for the first phytomeric units present at the inflorescence ramifications, all phytomers present in linear segments of primary and higher order branches display a phytomeric structure very similar to that of the rachis.

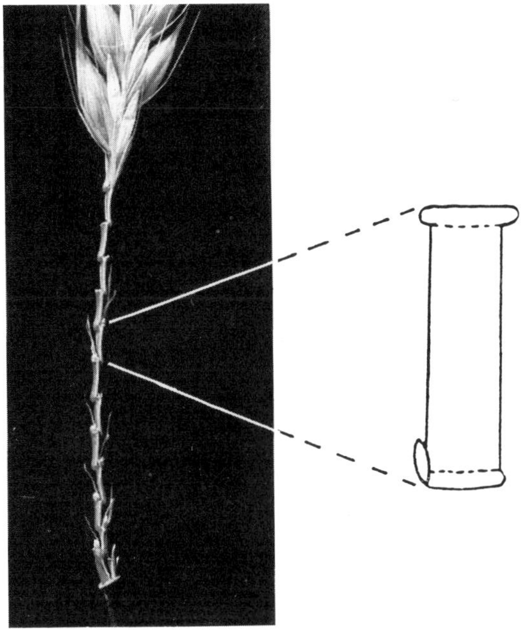

Fig. 11.9. Phytomer of the barley rachis. Left, rachis with spikelets partially removed; right, diagram of the type 1 phytomer variant present in the rachis.

The type 2 phytomer: a special structure present at branching

The prophyll is the first organized structure which develops from an axillary bud and encloses the bud except for a small opening at the tip. In doing so it behaves like a leaf sheath with an important exception: the prophyll – like the coleoptile, the preleaf of the grass seedling – contains two prominent veins, but no blade. The two keels corresponding to two main midribs are clearly visible in the mature prophyll (Fig. 11.10C). This dual-leaf nature of the prophyll is also evident at early stages, as shown for a prophyll primordium (Fig. 11.10A, B). The prophyll may, therefore, be considered as a fusion of two leaves; this situation is shown in Fig. 11.10D. Evidence for this origin of the prophyll is its capacity to stimulate the formation of two buds in its axils as observed for maize (Collins, 1924). The prophyll thus represents a fused phytomer which we will call type 2.

The coleoptile is the second organ of grass plants that can be interpreted as a type 2 phytomer. A characteristic feature of the seedling within the Poaceae (with exceptions) is the formation of the mesocotyl between the nodes of the scutellum and the coleoptile. The coleoptile, a sheath enclosing the plumule, is inserted on this node and equipped with two midribs and two vascular bundles located opposite of each other (Arber, 1934). Thus, the coleoptile is clearly of foliar origin, and it is considered to be the first leaf of the shoot with homology to the prophyll (Gould & Shaw, 1983). Studies of a large number of grasses have verified this (Jacques-Felix, 1957; Guignard, 1962) and led to the conclusion that the mesocotyl is the first internode of the plant, the scutellum a protocorm or a cotyledon and the plumule a bud of the protocorm or the terminal bud which has acquired a lateral position. The underlying morphogenetic event corresponds to the formation on the protocorm (proembryo) of a newly organized meristem from a tissue with a different organization. This mode of histogenesis is homologous to the formation of a lateral bud and favours the interpretation that the coleoptile – being present at ramifications like the prophyll and other organs – originates from the fusion of two leaves and therefore represents a type 2 phytomer.

The phytomeric structure of the grass spikelet and floret

Considering the metameric structure of plant organs, spikelet and floret are particularly difficult to analyse on the basis of repeated phytomer units. It has, however, to be emphasized that already according to Goethe (1790) 'the flowers which develop from lateral buds are to be regarded as entire plants, which are associated to the mother plant as the mother plant is associated to the soil'. His theory of the flower – floral parts are derived from basic foliar structures and located on the stem in a helical arrangement – is still accepted today (Gifford & Foster, 1989). The concept of

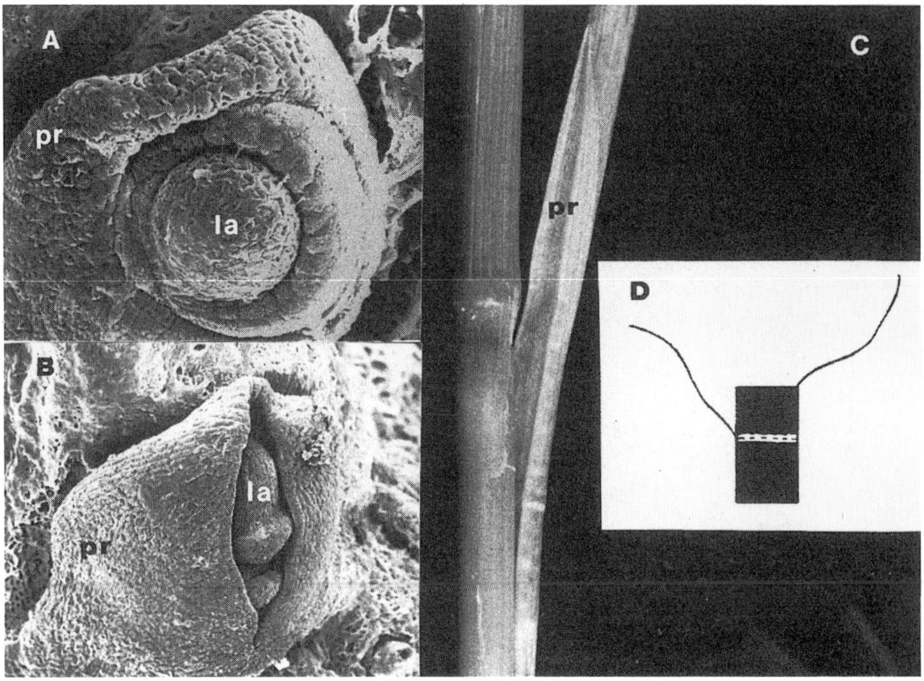

Fig. 11.10. The barley prophyll. A, Early stage of prophyll primordium; B, late stage of prophyll primordium; C, mature prophyll; D, phytomeric structure (type 2 phytomer) of the prophyll; pr, prophyll; la, lateral apex.

homology that Goethe introduced is in modern botany based on structural similarities of organs regardless of their function (Gifford & Foster, 1989). In our context homology between organs of grass species will be discussed by reference to the cell layers that undergo the first periclinal divisions, to the number of contributing cell layers, to the growth type of the primordia (apical, intercalary or mixed), to the presence of founder cells in insertion rings or of initial cells organizing a bud-like structure, and to anatomical and morphological criteria supporting the fusion of organs.

In the grass spikelet the first two organs appearing on the rachilla axis are the two sterile glumes which are leaf-like in terms of anatomy and origin. According to general consensus (summarized in Clifford, 1988) glumes are considered to be homologous with leaves of vegetative units as well as with the other bracts on the rachilla: 'As the lower of the two subtending glumes is often the first leaf on the spikelet axis it may be expected, at least in some species, to resemble other prophylls in being two keeled'. Although this is not the case, support for the prophyll-like nature of the two glumes comes from other observations. There are their foliar origin, their position after branching, the absence of nodes between the two glumes, the rare formation of vegetative shoots, spikelets or florets in their axils, the lack of an organ on the spikelet axis homologous to the prophyll, and lastly the frequent fusion of two glumes. This is depicted in Fig. 11.11 for three grasses including barley. It is clearly visible that the glume primordia of barley (Fig. 11.11C) are located on one side of the spikelet and fused at their base to form a single structure. It is therefore reasonable to conclude that in grasses the two subtending glumes can be interpreted as a unique organ

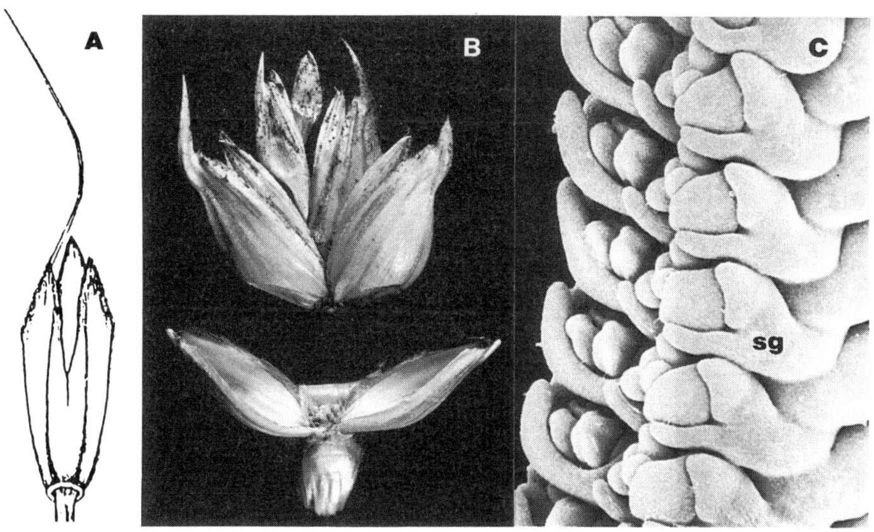

Fig. 11.11. Fusion of subtending glumes in grasses. A, *Alopecurus myosuroides* (Gould & Shaw, 1983); B, *Triticum aestivum*, in the bottom the glumes have been separated from the residual part of the rachilla; C, lateral view by scanning electron microscopy (SEM) of a young inflorescence of *Hordeum vulgare*; the two subtending glumes are fused at the base. sg, subtending glumes.

and thus correspond to a phytomer type 2.

The fertile glumes (lemmas) follow the two subtending glumes along the rachilla axis. These lemmas constitute – together with their short subtending internodes – a part of the rachilla. Each lemma bears a floret in its axil and protects it from one side (abaxial), while the adaxial side is protected by a second bract, the palea, which is inserted on the floral axis (Fig. 11.12). The palea, the first organ after a ramification, is a two-keeled organ as expected. For both lemma and palea the histogenetic origin is similar to that of the leaves. By homology, however, the lemma is associated directly with leaves (type 1 phytomer) and the palea (type 2 phytomer) to prophyll, coleoptile and subtending glumes.

The linear units of the floral axis as well as lodicules, stamens and carpel are the last constitutents of the spikelet that we wish to interpret with respect to the phytomer concept. The precise interpretation of the linear units present along the axis is difficult due to their compression. We propose to regard them as homologous to similar units present in the rachis, that is to a variant of the type 1 phytomer with a repressed leaf. In contrast, the terminal phytomer on the floral axis, the carpel, is histologically and morphologically well characterized (Aziz, 1972; Clifford, 1988). It is foliar in origin, developing as a single leaf and collinear with the preceding phytomers: it may therefore be considered as a special terminal type 1 phytomer.

Branching of the floral axis results in very specialized structures (lodicules, stamens) that are not easily interpreted. Stamen primordia are of cauline origin, but display a dual nature as discussed before (Dawe & Freeling, 1990). The apical growth is soon followed by extensive intercalary meristematic activity resulting in an expansion typical of leaf primordia. The anther can be regarded as a microsporophyll structure of foliar origin, but the filament (and possibly the central, connective tissue between pairs of sporangia) may maintain a cauline origin (Wilson, 1942). This mixed foliar/cauline origin is reflected in the transformation of stamens into ovaries, for example in wheat (Fisher, 1972). The foliar nature of the anther in the transformed organs is clearly

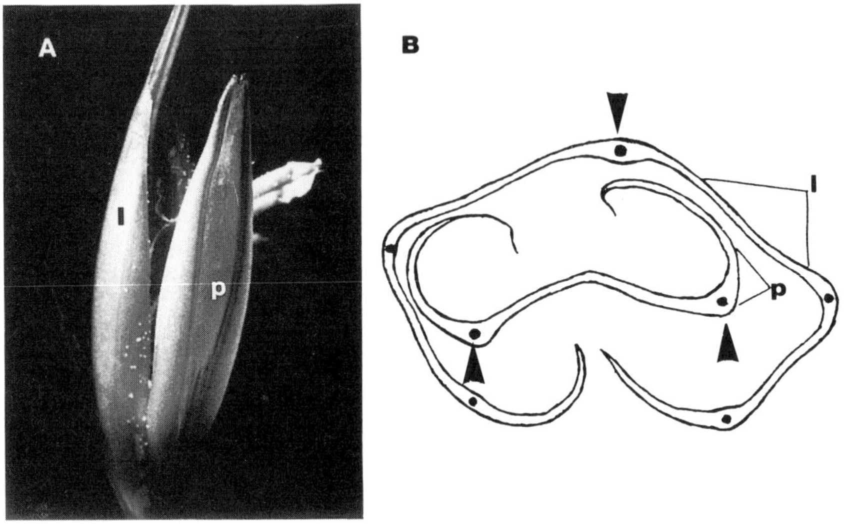

Fig. 11.12. Arrangement of lemma and palea in barley. A, single floret; B, diagram of a cross-section redrawn from Reid (1985); the arrowheads point out major midribs. l, lemma; p, palea.

evident, as they produce normal pollen while containing an ovule at their base. The known developmental flexibility of stamens and lodicules (phyllody, petalody, interconversion, carpellody) (see also the barley mutant *lax-a*, Fig. 11.17) is easily explained by referring to the part of these organs which is of foliar origin. If one extends the concept of a fused phytomer being always present at ramification to the floret axis in the Poaceae, then the homology of stamens and lodicules as being prophyll-like structures is logical. In this sense the two organs are interpreted as solitary variants of the type 2 phytomer. They are termed solitary, in contrast to any other two-keeled organ of the grass plant, because the lateral buds that develop into stamens and lodicules will not produce further growth units.

The phytomeric structure of the barley plant

Based on previous reviews of barley morphology and development (Bonnett, 1966; Briggs, 1978; Deitzer, 1985; Kirby & Appleyard, 1987) as well as on our own observations, we wish to describe in this section the phytomeric structure of the mature barley plant. For this and for further characterization of mutant classes (see later) the barley plant is divided into six distinct developmental regions, A to F (Table 11.3), and specific phytomeric units are assigned to each region (Fig. 11.13). Regions A and B are considered to represent vegetative regions, while regions C to F are associated with reproductive regions.

In the mature plant the lower internodes from which the lateral shoots arise represent the tiller region (region A) in which both type 1 and type 2 phytomers are present (Fig. 11.13). This region comprises the coleoptile and the first internode of the culm. The mesocotyl in barley does not elongate, but the first internode takes over its function pushing upward the apex of the main shoot. While both phytomer types in this region are able to form roots, buds only develop from type 1 phytomers. Leaves are formed as coleoptile and prophylls from the type 2 phytomers, whereas type 1 phytomers differentiate into normal leaves with sheath, blade, auricles and ligule. Only the first, and less frequently the second, internode elongates, if the plane of tillering has to be elevated to the soil surface as a consequence of deep sowing. In this event phytomers

Table 11.3. The phytomeric structure and the six developmental regions of the barley plant.

Region	Organ	Phytomer Type 2	Phytomer Type 1
F			
(floret)	Internode	–	–
	Leaf	+ (p, l, a)	+ (o)
	Bud	–	+
	Root	–	–
E			
(rachilla)	Internode	–	–
	Leaf	+ (sg)	+ (le)
	Bud	–	+
	Root	–	–
D			
(first-order	Internode	–	–
branch)	Leaf	–	–
	Bud	–	+
	Root	–	–
C			
(rachis)	Internode		–
	Leaf		–
	Bud		+
	Root		–
B	Internode		+
(culm)	Leaf		+
	Bud		–
	Root		–
A			
(tiller region	Internode	–	–
plus me and co)	Leaf	+ (co, pr)	+
	Bud	–	+
	Root	+ (?)	+

a, stamen; co, coleoptile; l, lodicule; le, lemma; me, mesocotyl; o, pistil; p, palea; pr, prophyll; sg, sterile glumes; –, organ repressed; +, organ present.

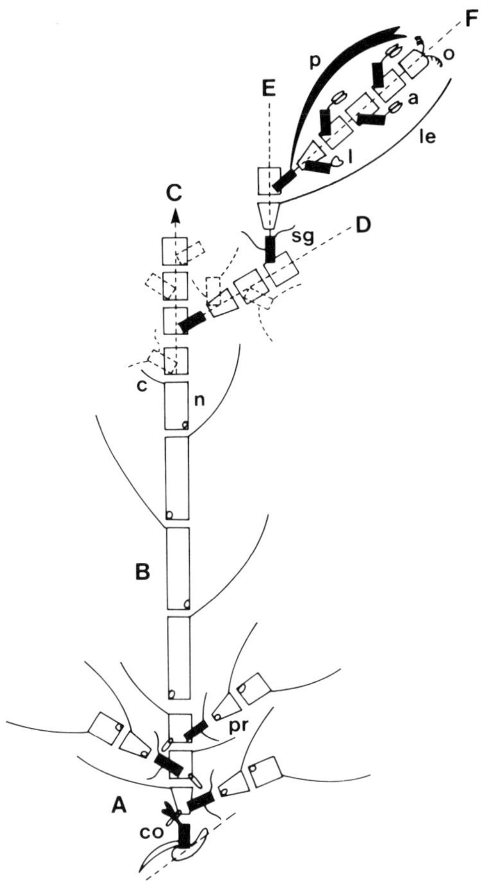

Fig. 11.13. Metameric structure of the barley plant. A to E designate the six developmental regions described in Table 11.3. a, stamen; c, collar; co, coleoptile, l, lodicule; le, lemma; n, neck; o, pistil; p, palea; pr, prophyll; sg, sterile glumes.

with long internodes will not form tillers.

The culm (region B) is formed by a number of type 1 phytomers with long internodes and fully developed leaves. In this region of the barley plant, the phytomers do not produce lateral buds nor roots.

The barley ear is separated from the culm by the collar, a rudimentary leaf that belongs to its subtending internode, the neck. The ear may be subdivided into the four regions C (rachis), D (first-order branch), E (rachilla) and F (floral axis). In the rachis, the ear continuation of the main shoot, the development of leaves is repressed in all phytomers (lower ridge of the double ridge stage; Figs 11.2C and 11.3C). The internodes are hardly elongated, and roots are not present at all. The rachis is composed of type 1 phytomers.

Lateral buds initiating from the rachis represent region D, the first-order branch region, which is similar in structure to the rachis. Leaves or roots are not present in this region, and internode formation is completely repressed. Buds of this region generate the axis of the individual spikelets, the rachilla.

The phytomers of the rachilla (region E) are compressed as are most phytomers in the different regions of the ear. The first metameric unit (type 2 phytomer) bears leaves, the subtending (or sterile) glumes, which originate from a fused organ as demonstrated in barley by scanning electron microscopy (Fig. 11.14A). The two phytomers (type 1) further along the rachilla produce leaf-like structures: the first forms the lemma, while the second stays rudimentary and is visible as the rachilla tip (Fig. 11.14B, F). The bud of this second phytomer grows to produce the last axis of the mature barley plant, the floral axis.

This axis (region F) is composed of linear phytomers (type 1) with compressed internodes. They are protected by the bekeeled palea, the leafy product of the first phytomer (type 2) of this axis. The lateral buds of the type 1 phytomers differentiate into lodicules or anthers (Fig. 11.14H). The floret axis ends in the carpel which is the leaf of the last phytomer (type 1) on the floral axis. The carpel encloses the apex tip which will form the ovule (Fig. 11.14F, G).

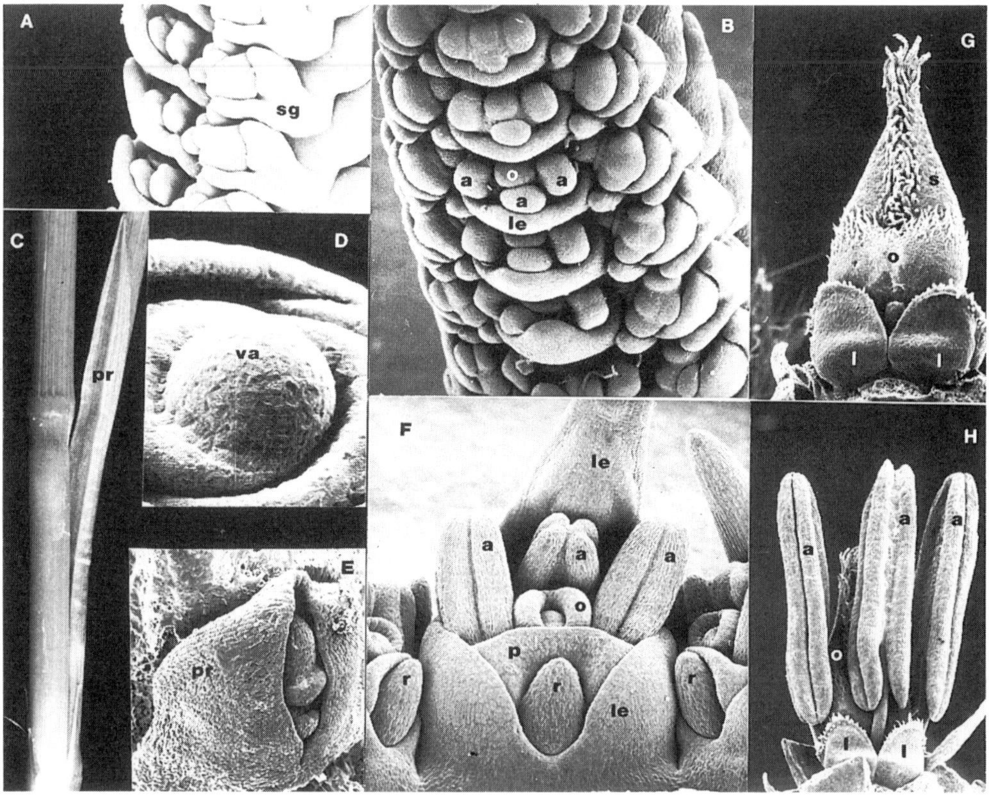

Fig. 11.14. Developmentally important regions of the barley plant. A, lateral view of a young barley ear by SEM; B, front view of a young barley ear; C, mature prophyll as seen in a mutant with lateral ramifications on the stem; D, vegetative apex (of *viv4*); E, prophyll primordium; F, white anther stage of the floret (see also Fig. 11.4); G, barley floret with stamens removed; H, barley floret. a, anthers; l, lodicule; le, lemma; o, carpel; p, palea; pr, prophyll; r, rachilla; s, style; sg, sterile glumes; va, vegetative apex.

Developmental mutants of barley

The basic metameric structure of the barley plant discussed in the preceding section(s) may be altered by mutations affecting one or several of the developmental regions defined before. The developmental mutants available (Table 11.4) may be interpreted in various terms and classified into distinct groups on the basis of the mutant phenotype conditioned. 'Transitional' mutants are those which transfer the characteristics of one developmental region to a second one. Within a particular region single organs produced from phytomers may be modified ('organ modification') and single phytomers or complex structures may be added ('additions') or may be repressed ('reductions'). We will briefly discuss the four different mutant classes and present a few examples for clarification.

Table 11.4. Morphological mutants of barley.

Symbol	Name
abr	Accordian basal rachis internodes
acr	Accordian rachis
adp	Awned palea
*ag**	Agropyroides
als	Absent lower laterals
ari	Breviaristatum
aur	Exauriculum
*bb**	Brodblädig
*be**	Branched ear
bikini	Bikini
bir	Branching inflorescent rachilla
br	Brachytic dwarf
bra	Bracteatum
bra-a,-b	Semibracteatum
brc^{+}	Branched
cal	Calcaroides
cl	Curly laterals
*clh**	Curly leaf dwarf
*cu**	Curly
cud	Curly dwarf
*D**	Sterile dwarf
den^{+}	Densinodosum
dwf	Vegetative dwarf
e	Wide outer glume
ert	Erectoides
*fer**	Few roots
gra	Granum
$int\text{-}l^{81}$	Intermedium l^{81}
K	Hooded
*L**	Lax spike
lax	Laxatum
lax-a	Laxatum five anthers
lb	Long basal rachis internode
l_c	Dense spike
L_c	Lax spike
lel^{+}	Leafy lemma
lep	Macrolepis
*Lfb**	Leafy bract
*Lfl**	Leafless

Table 11.4. *cont.*

Symbol	Name
li	Ligule and auricle less
lig	Exligulum
lin	Lesser internode number
lk	Short awn
Lk	Awnless
lnt	Low number of tillers
*lo**	Small lodicules
lr	Reduced lateral spikelet appendage on the lemma
mgl⁺	Many glumes on the laterals
*min**	Minute
*mnb**	Multinoded branched
*mo**	Multi-ovary
mul	Multiflorous
*nld**	Narrow leafed dwarf
op	Opposite spikelets
*ovl**	Ovaryless (male)
*rlb**	Reduced leaf blade
*rvl**	Revoluted leaf
*sb**	Subnodal bract
sid	Single elongated internode dwarf
sk	Subjacent hood
*so**	Stalked ovary
tar⁺	Triaristatum
*tg**	Third inner glume
tr	Triple-awned lemma
trd	Third outer glume
*tric**	Tricompositlora
u4	Unbranched style
uc	Uniculm
uz	'Uzu' (semibrachytic)
viv⁺	Viviparoides
V^t	Deficiens
*wlh**	Broad leaf

* Classification only according to literature studies. ⁺The cited symbol has not been published before. For references see Søgard & Wettstein-Knowles (1987) and Bossinger (1990).

Transitions

Transitions occur within the vegetative (A, B) and the reproductive regions (C–F) as well as between these regions (Table 11.5). The barley mutant *uniculm* (*uc2*) for example has lost the capacity for tillering and grows a single culm with elongated basal internodes. Thus the phytomers of the tillering region A take over characteristics of the culm region B and the mutation is interpreted as a transition between the two vegetative regions. The recessive mutant *densinodosum* (*den*) is a barley mutant representative for mutants with mutations in more than one developmental region. The mutant contains more and shorter internodes than the wild-type (Fig. 11.15A), and the metameric units may form branches in the lower as well as in the upper part of the culm (Fig. 11.15B). Branching from the lower part of the spike may result in the formation of additional ears (Fig. 11.15C). Transitions have thus occurred from B to A

Table 11.5. Transitional mutants.

(a) Vegetative → vegetative	
A → B	*fer, lnt, uc*
B → A	*br, cud, D, den, dwf, mnb, nld, sid, viv, uz*
(b) Vegetative → reproductive	
B → C	*Lfl*
(c) Reproductive → vegetative	
complex	*int-l*^{81}, *viv*
C → A	*den, mnb, viv*
C → B	*abr, acr, bra, den, int-l*^{81}, *L, lax, lb,* L_c, *mnb, sb, trd, viv*
D → B	*int-l*^{81}, *viv*
E → B	*int-l*^{81}, *mgl, viv*
F → B	*adp, so, viv*
(d) Reproductive → reproductive	
D → C	*int-l*^{81}
D → E	*ag, tg*
E → C	*be, bir, brc*
F → C	*adp*
F → D	*tric*

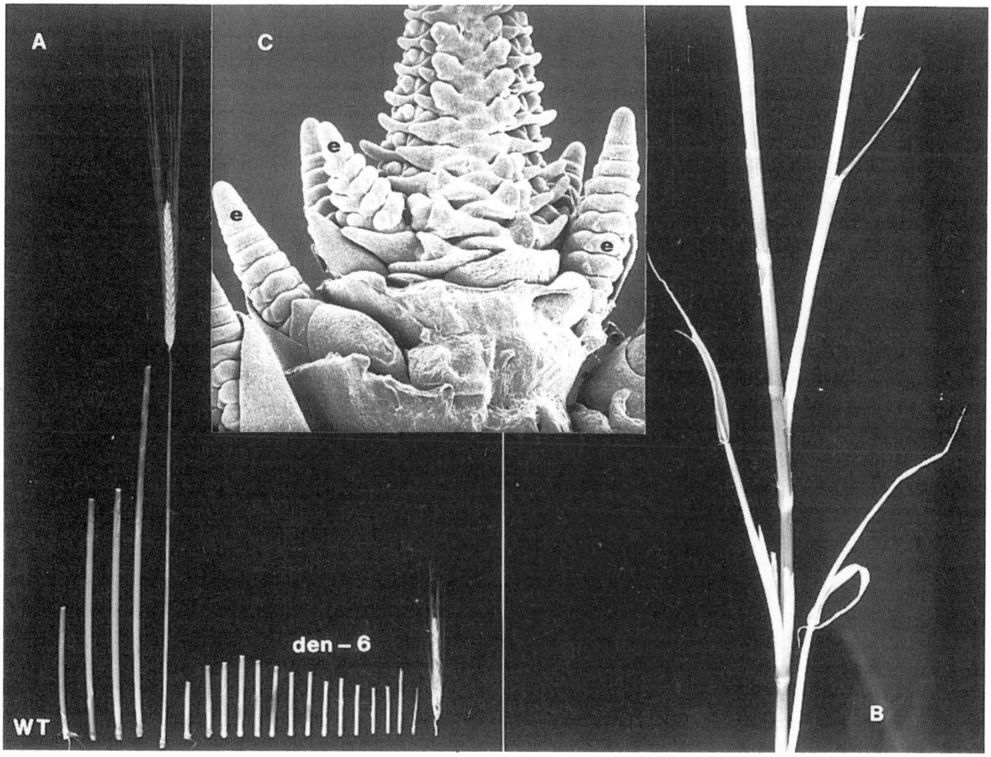

Fig. 11.15. The recessive mutant *densinodosum* (*den*) as a representative of transitions. A, internodes of wild-type barley versus the *den* 6 mutant; B, branching in the culm of *den* 6; C, branching in the lower part of the ear of *den* 6. e, extra ear.

(shortening of culm internodes, capacity for branching), from C to B (elongation of lower ear internodes), and from C to A (capacity for branching) as summarized in Table 11.5.

Organ modifications

Mutants with modified organs are less complicated than the transitional mutants. Here single organs of a phytomer are present in the mutant phenotype, though in a modified form (Table 11.6). In the mutant *liguleless* (*li*) the ligule and auricles are missing at the leaf. In the recessive mutant *leafy lemma* (*lel*) the lemma is converted to a small leaf with sheath and blade (Fig. 11.16B). The *laxatum-a* (*lax-a*) mutant *lax-a*[8] (Fig. 11.17) also exhibits modification of the lemma, but more prominent is the conversion of the two lodicules into anthers. These additional anthers (Fig. 11.17D) contain only two microsporangia as opposed to four in the three anthers proper present in both the wild-type and the mutant (Fig. 11.17C). External conditions may further result in the development of the extra anthers into carpelloid anthers in that they may carry stigma-like protrusions (Fig. 11.17E, F) or be transformed into extra pistils (Fig. 11.17G, H).

Table 11.6. Organ modification mutants.

(a) Modified leaves	
leaves:	*aur, bb, bra-a, −b, Ltl, li, lig, nld, ovl, rlb, rvl, wlh*
collar:	*bra-b, Lfb*
glumes:	*e, lep*
lemma:	*ari, cal, K, lax-a, lel, Lk, lk, lr, ovl, sk, tr, Tr2*
palea:	*adp, tar, tric*
(b) Modified internodes	
	abr, acr, den, ert, L, l_c, lb, mnb, op
(c) Modified ovules	
	ovl, u4
(d) Modified lodicules	
	int-l^{81}, lax-a, lo, mo-5
(e) Modified anthers	
	int-l^{81}, mo
(f) All organs modified	
	Bikini, cl, clh, cu, cud

Addition of phytomers

In the mutant *branched* (*brc*) branching in the inflorescence gives rise to additional ears, as particularly evident for the mutant *brc 5* (Fig. 11.18). The rachilla (region E) has been transformed into a rachis (region C) and this represents a transitional event. Addition of single florets as opposed to ears in *brc* is evident with other mutants like *intermedium* (*int-l^{81}*) or *multiflorous* (*mul*) (Table 11.7). In the dominant *Hooded* (*K*) mutants an additional flower is formed as a branch from the lemma.

Reduction of phytomers

In contrast to the class characterized by the addition of phytomers, the mutants of this group show a repression in the formation of structures, single phytomers or groups of phytomers (Table 11.8). One example is the previously mentioned mutant *uniculm* (*uc*)

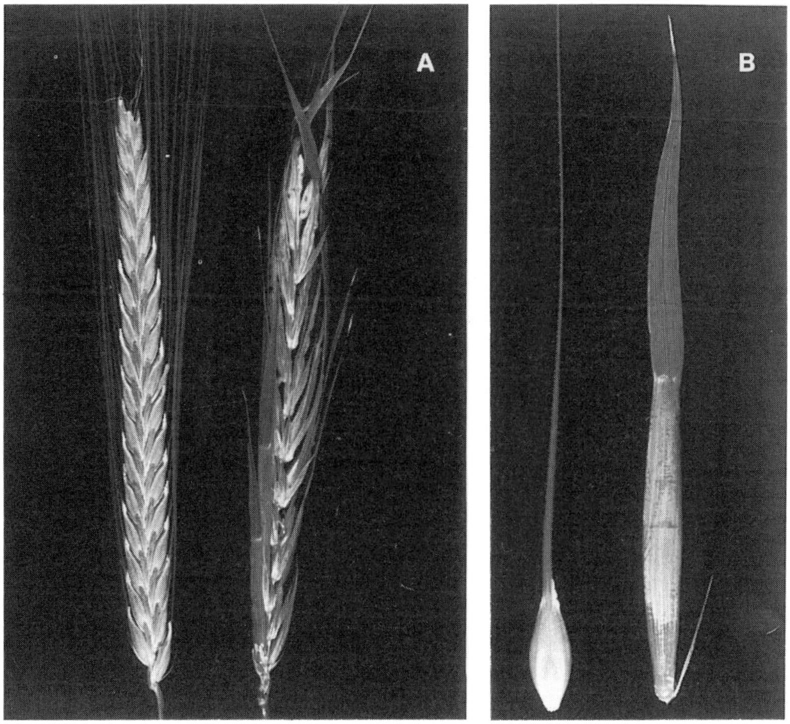

Fig. 11.16. The recessive mutant *leafy lemma* (*lel*) as a representative of a single organ modification. A, ears of wild-type (left) and *lel* mutant (right); B, lemma of wild-type (left) and *lel* mutant (right). This mutant can however be also interpreted as E → B transition (the lemma assumes characters of a vegetative leaf).

Table 11.7. Mutants by the addition of phytomers.

Addition of:	
(a) tillers	*den, gra, mnb*
(b) ears	*be, bir, brc,* $int\text{-}l^{81}$
(c) triple spikelets	*tric*
(d) spikelets	*ag,* $int\text{-}l^{81}$, *K, mul*
(e) vegetative phytomers	*mgl, viv*
(f) groups of various phytomers	*adp*

Table 11.8. Mutants by the reduction of phytomers.

Reduction of:	
(a) tillers	*lnt, uc, uz*
(b) spikelets	*ag, als,* V^t
(c) phytomers in B and C	*lin*
no reproductive phytomers	*min*

where tillering is totally suppressed. The dominant mutation *Deficiens* (V^t) suppresses the formation of lateral spikelets to various degress (Fig. 11.19). While in the wild-type lateral spikelet differentiation can be observed very early during apical development (Fig. 11.19B), the corresponding primordia in the mutant (Fig. 11.19C, la) remain in an undifferentiated state.

Molecular aspects of development

The advent of molecular genetics with its various techniques to identify and isolate genes has stimulated the study of the molecular basis of developmental processes, for example pattern formation in the *Drosophila* embryo (Ingham, 1988). As an extension of the classical analysis of mutant phenotypes, the molecular approaches to differentiation include gene tagging through transposon insertion, differential screening of subtracted cDNA libraries, chromosome walking, chromosome mapping of isolated genes by restric-

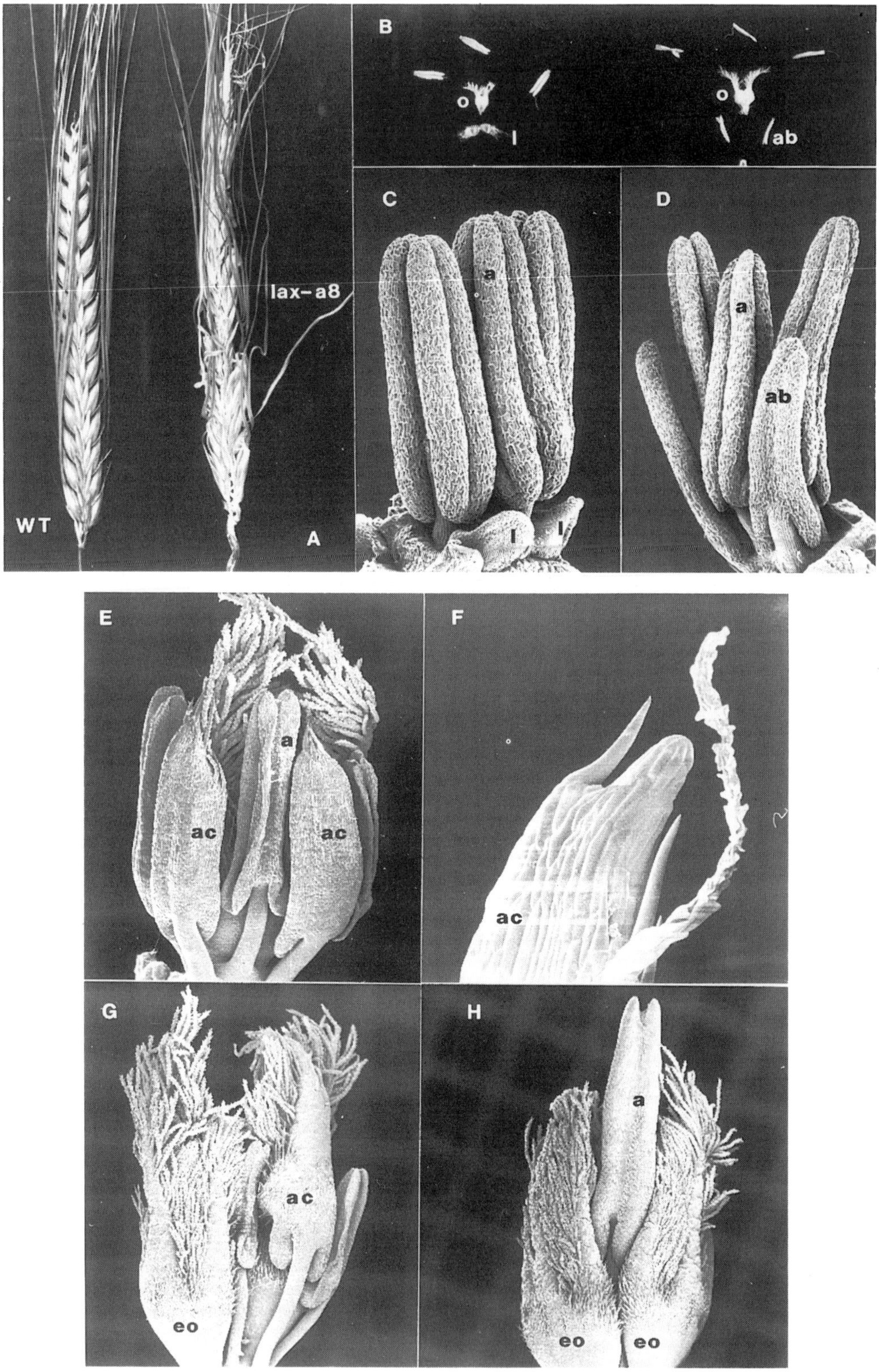
WT
lax-a8
A
B
o
l
o
ab
C
a
l
l
D
a
ab
E
a
ac
ac
F
ac
G
ac
eo
H
a
eo
eo

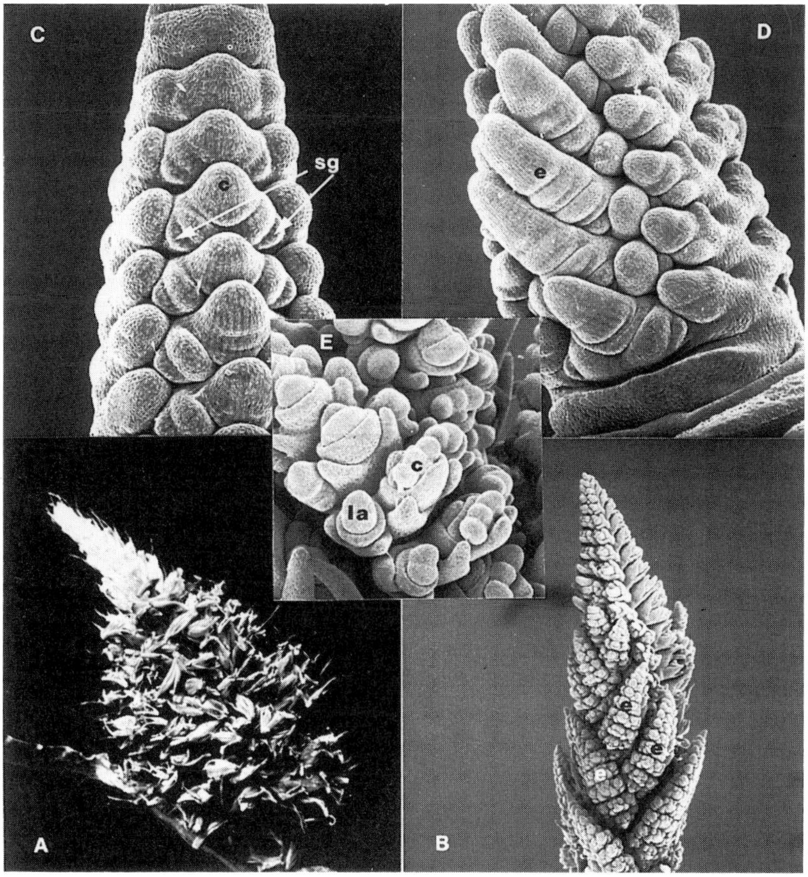

Fig. 11.18. The recessive mutant *branched* (*brc*) as a representative of phytomer addition. A to E represent various stages of ear development in the mutant *branched-5*. c, primordium of a central spikelet; e, extra ear; la, primordium of a lateral spikelet; sg, sterile glume primordium.

tion fragment length polymorphism (RFLP) analysis and complementation of mutant phenotypes by transformation with the wild-type gene(s). As far as plant development is concerned, the year 1990 will be considered a milestone due to the discovery of genes regulating floral organogenesis in *Antirrhinum majus* by Sommer, Schwarz-Sommer and their colleagues (Sommer *et al.*, 1990; Schwarz-Sommer *et al.*, 1990, in press), and in *Arabidopsis thaliana* by Yanofsky *et al.* (1990).

Model systems for flower development in dicotyledons

In *Antirrhinum majus* morphogenetic processes after floral induction give rise to a

---

Fig. 11.17. The recessive mutant *laxatum-a* (*lax-a*) as a representative of modifications in several organs. A, ears of wild-type (left) and *lax-a* ^{8} mutant (right); B, dissected florets of wild-type (left) and mutant (right); C, SEM of anthers of wild-type barley; D, SEM of anthers of the *lax-a* ^{8} mutant; E–H, different floral abnormalities observed with the *lax-a* ^{8} mutant under field conditions. a, anther; ab, anther with two microsporangia; ac, carpelloid anther; eo, extra carpel; l, lodicule; o, carpel.

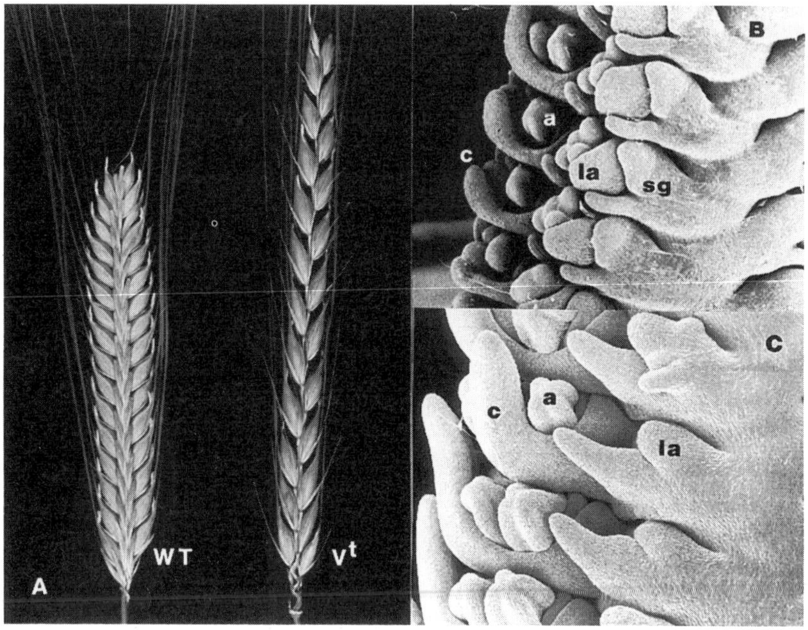

Fig. 11.19. The dominant mutant *Deficiens* (V^t) as a representative of phytomer reduction. A, ears of wild-type (WT) and mutant (V^t); B, lateral view by SEM of young wild-type inflorescence; C, lateral view by SEM of young mutant inflorescence. a, anther; c, central spikelet; la, lateral spikelet; sg, sterile glume.

mature flower of zygomorphic symmetry. Development of floral organs proceeds in four distinct whorls starting with the formation of the five sepal primordia in whorl 1, the five petals in whorl 2, the five stamens in whorl 3 and the carpel in whorl 4 (Schwarz-Sommer *et al.*, in press; Coen, in press). Available flower mutants controlling flower organ identity (homoeotic mutations) have been assigned to three classes: type 1 genes affect the perianth (whorls 1 and 2), type 2 genes control the formation of the reproductive organs in whorls 3 and 4, and type 3 genes control organ identity in whorls 2 and 3. The discovery that specific genes are expressed in overlapping whorls provides a basis for models postulating the existence of only two developmental pathways sufficient to provide each whorl with its characteristic set of functions (Schwarz-Sommer *et al.*, in press; Coen, in press).

Deficiens (*Def*), a homoeotic type 3 gene, results in a phenotype with sepaloid petals and carpelloid stamens. The gene has been cloned by differential screening of subtracted cDNA libraries and its expression correlates with the effect of a particular *def* allele (*deficiens*^{globifera}, *def A*) (Sommer *et al.*, 1990). The DEF A protein as well as AG, the product of the *Agamous* (*Ag*) gene from *Arabidopsis* (Yanofsky *et al.*, 1990), are putative transcriptional activators with domains for DNA-binding and dimerization homologous to those of two well-characterized *trans*-acting factors, SRF (serum response factor) of mammals and MCM1 (minichromosome maintenance gene product) of yeast. Both *S*RF and *M*CM1 participate in processes of cell differentiation as do *D*EF A and *A*G; their conserved region has been termed the 'MADS' domain (Schwarz-Sommer *et al.*, 1990).

Def A is one of three morphoalleles of *Def* known to condition mutated flowers in *Antirrhinum*. The two other alleles *deficiens*^{chlorantha} and *deficiens*^{nicotianoides} were also isolated and shown to exhibit mutations in the promoter region and in the DNA-binding domain, respectively, of the wild-type *Def* gene (Schwarz-Sommer *et al.*, in press). In addition *Def* specifies only one out of 15 *Def*- and *Ag*-related MADS-box containing genes that are either expressed only in the flower or in both the vegetative and floral organs. Of the flower-specific genes, two could be assigned to the genetic loci *Globosa* and *Squamosa*. It can therefore be concluded that the gene families encoding MADS-box containing proteins are fundamental to plant development.

Transcriptional factors from other eukaryotes are also very often members of families (Ziff, 1990). Control of the activity and specificity of such highly related activators of transcription could be imagined to occur *via* divergent effector domains, and heterodimerization or post-translational modifications like phosphorylation are among the mechanisms being discussed in this context. DEF A not only contains a conserved domain for dimerization, but is also a phosphorylated protein (Schwarz-Sommer *et al.*, 1990, in press). One prominent example of how phosphorylation can modulate specificity is the proto-oncogene c-*Myb* (see later) which encodes a transcriptional activator with a regulatory function during cell lineage differentiation in vertebrates. Phosphorylation of the c-MYB oncoprotein abolishes recognition of its target sequence (Lüscher, *et al.*, 1990). Similar mechanisms can be envisaged to control the activity of MADS-box containing genes during plant development.

Molecular aspects of barley development

The understanding of the molecular basis of plant development is just at its beginnings. It would appear, however, that transcriptional activators are one class of regulatory proteins that play an important role in the development of both vegetative and reproductive organs of the plant. In fact cross-kingdom homologies of transcriptional regulators are not restricted to the ones discussed above. Further regulatory plant genes are known to exist which show striking homology to transcriptional regulators belonging to various proto-oncogene families of the animal kingdom (reviewed by Birchler, 1990). These genes are involved in differentiation, as mutation gives rise to undifferentiated cell growth (Bishop, 1987). Regulatory genes of the anthocyanin biosynthesis pathway in maize show homology to the c-*Myc* (*R*, *B*) and c-*Myb* (*C*, *Pl*) proto-oncogenes, and the *Opaque-2* gene, a positive regulator of the 22 kDa class of zeins, encodes a protein with a leucine zipper homologous to the corresponding domains in the c-JUN and c-FOS oncoproteins. In this context it should be noted that the serum response factor SRF (with homology to DEF A and AG) induces the transcription of c-*Fos* and c-*Jun* genes.

In addition to *C* and *Pl*, two c-*Myb*-related genes were isolated in maize as well as six in barley (Marocco *et al.*, 1989; Rohde *et al.*, 1989; Rohde, unpublished results). Although the functions of these barley genes remain obscure, they represent a family of transcriptional activators highly conserved with respect to the DNA-binding domain. In the animal kingdom the c-*Myb* genes are predominantly expressed in immature haematopoietic tissue and apparently participate in the differentiation of this cell lineage (for a recent review see Shen-Ong, 1990). It is, however, doubtful whether the barley *Myb*-related genes, by analogy to the animal c-*Myb* genes, would be involved in differentiation, although initial *in situ* hybridization studies indicate their predominant expression in meristematic tissues (Rohde, unpublished results).

The barley mutation *calcaroides 21* (*cal 21*) induces the formation of extra ligules and auricles, the appearance of knots on the leaf surface and the formation of a hood-like appendage on the awn. The mutant shows phenotypic similarities to the *Kn-1* mutant of maize (for a review see Freeling *et al.*, 1988). However, as the awn is not present in the maize lemmas, hood-like structures are not visible in *Kn-1* mutants. *Kn-1* has been cloned (Hake *et al.*, 1989) and shown to display homology to the homoeodomain of two transcriptional activators, the yeast *MAT* locus and the *gooseberry* gene of *Drosophila*.

Other recessive barley mutants (see earlier) display homoeotic-like changes resembling those described for *Antirrhinum* suggesting that MADS-box containing genes may be involved in the normal development of the barley reproductive organs. In fact clones from flower-specific barley cDNA libraries that cross-hybridize with the conserved MADS-box sequence of *Def A* have been identified (Rohde, unpublished results).

The target genes for the transcriptional regulators encoded by developmental genes like *Def A* and *Ag* as well as the sequence of events leading from external stimuli through transmembrane signalling and the cytoplasm to the nucleus (signal transduction) resulting in the induction of *Def A* and *Ag* regulatory genes remain to be determined. An important aspect of cell division and differentiation (e.g. the stimulus active on the initial cells in the subhypodermal cell layer leading to the formation of spikelet primordia by periclinal divisions; see earlier) is the role of the cell wall in plant morphogenesis (Cooper, 1988). In addition to well-characterized structural polypeptides of the cell wall like the hydroxyproline-rich extensins (reviewed by Cooper, 1988), the gene for a glycine-rich protein (GRP) of barley has recently been characterized (Rohde *et al.*, 1990). The protein is apparently directed through the plasmalemma into the cell wall by an N-terminal transit peptide and is there immobilized. Studies on a similar *Grp* gene of bean indicated that its expression is developmentally regulated and specifically observed in protoxylem cells of the vascular system (Keller *et al.*, 1989). Apparently GRP is a component necessary for the stabilization of wall structures once they have matured by differentiation. Expression of a *Grp* gene encoding a glycine-rich protein similar to the one described in barley is suppressed in the *Antirrhinum* mutant *def-A* (Sommer, personal communication). *Grp* could therefore represent a target gene for DEF A or DEF A-induced transcriptional activators during late stages of protoxylem development.

Acknowledgements

We wish to thank Drs Hans Sommer and Zsuzsanna Schwarz-Sommer for sharing unpublished results and for providing preprints of their work.

This research was carried out in part with the financial support of the 'Piano nazionale biotechnologie', Ministry of Agriculture, Italy, project 'Mutasioni delle sviluppo'. We thank Dr M. Stanca and Mrs D. Pagani, Instituto Sperimentale per la Cerealicoltura, Fiorenzuola, Italy, for their assistance in reproducing the genetic material.

References

Arber, A. (1934) *The Gramineae. A Study of Cereal, Bamboo and Grasses*. Cambridge University Press, Cambridge.

Aziz, P. (1972) Histogenesis of the carpel in *Triticum aestivum* L. *Botanical Gazette* 133, 376–86.

Barnard, C. (1955) Histogenesis of the inflorescence and flower of *Triticum aestivum* L. *Australian Journal of Botany* 3, 1–20.

Barnard, C. (1957) Floral histogenesis in the monocotyledons. I. The *Gramineae*. *Australian Journal of Botany* 5, 1–20.

Barnard, C. (1964) Form and structure. In: Barnard, C. (ed.), *Grasses and Grasslands*. MacMillan, London, pp. 47–72.

Bateson, W. (1894) *Materials for the Study of Variation Treated with Especial Regard to Discontinuity in the Origin of Species*. MacMillan, London.

Bateson, W. & Bateson, A. (1891) On variations in the floral symmetry of certain plants having irregular corollas. *Journal of the Linnean Society, Botany* 28, 386–424.

Birchler, J. (1990) Amaizing results. *Trends in Genetics* 6, 231–3.

Bishop, J.M. (1987) The molecular genetics of cancer. *Science* 235, 305–11.

Bonnett, O.T. (1953) Developmental morphology of the vegetative and floral shoots of maize. *Illinois Experimental Station Bulletin* 568, 1–47.

Bonnett, O.T. (1961) The oat plant: its history and development. *University of Illinois Agricultural Experiment Station Bulletin* 672.

Bonnett, O.T. (1966) Inflorescences of maize, wheat, rye, barley, and oats: their initiation and development. *University of Illinois Agricultural Experiment Station Bulletin* 568, 1–47.

Bossinger, G. (1990) Klassifizierung von Entwicklungsmutanten der Gerste anhand einer Interpretation des Pflanzenaufbaus der Poaceae aus Phytomeren. Dissertation, Rheinische Friedrich-Wilhelms-Universität Bonn.

Briggs, D.E. (1978) *Barley*. Chapman & Hall, London.

Clark, L.G. & Fisher, J.B. (1988) Vegetative morphology of grasses. In: Soderstrom, T.R., Hilu, K.W., Campbell, C.S. & Barkworth, M.E. (eds), *Grass Systematics and Evolution*. Smithsonian Institution Press, Washington DC, USA, pp. 21–30.

Clifford, H.T. (1988) Spikelet and floral morphology. In: Soderstrom, T.R., Hilu, K.W., Campbell, C.S. & Barkworth, M.E. (eds), *Grass Systematics and Evolution*. Smithsonian Institution Press, Washington DC, USA, pp. 21–30.

Coen, E.S. The role of homeotic genes in flower development and evolution. *Annual Review of Plant Physiology* (in press).

Collins, G.N. (1924) The prophyllum of grasses. *Botanical Gazette* 78, 353–4.

Cooper, J.B. (1988) Cell wall extensin genes. In: Verma, D.P.S. & Goldberg, R.B. (eds), *Temporal and Spatial Regulation of Plant Genes*. Springer-Verlag, Vienna, pp. 235–52.

Dahlgren, R., Clifford, H.T. & Yeo, P.F. (1985) *The Families of the Monocotyledons. Structure, Evolution and Taxonomy*. Springer-Verlag, New York.

Dawe, R.K. & Freeling, M. (1990) Clonal analysis of the cell lineages in the male flower of maize. *Developmental Biology* 142, 233–45.

Deitzer, G.F. (1985) *Hordeum vulgare*. In: Halevy, A.H. (ed.), *Handbook of Flowering*, Vol. 2. CRC Press, Boca Raton, FL, pp. 150–66.

Esau, K. (1965) *Plant Anatomy*, 2nd edn. Wiley, New York.

Evans, M.W. & Grover, F.O. (1940) Developmental morphology of the growing point of the shoot and the inflorescence of grasses. *Journal of Agricultural Research* 61, 481–520.

Fisher, J.E. (1972) The transformation of stamens to ovaries and of ovaries to inflorescences in *Triticum aestivum* L. under short-day treatment. *Botanical Gazette* 133, 78–85.

Freeling, M., Bongard-Pierce, D.K., Harberd, N., Land, B. & Hake, S. (1988) Genes involved in the patterns of maize leaf cell division. In: Verma, D.P.S. & Goldberg, R.B. (eds), *Temporal and Spatial Regulation of Plant Genes*. Springer-Verlag, Vienna, pp. 41–62.

Gifford, E.M. & Foster, A.S. (1989) *Morphology and Evolution of Vascular Plants*, 3rd edn. Freeman, New York.

Goethe, J.W.v. (1790) *Versuch die Metamorphosen der Pflanzen zu erklären*. Gotha.

Gould, F.W. & Shaw, R.B. (1983) *Grass Systematics*, 2nd edn. Texas A & M University Press, College Station.

Guedes, M. & Dupuy, P. (1976) Comparative morphology of lodicules in grasses. *Botanical Journal of the Linnean Society* 73, 317–31.

Guignard, J.L. (1962) Recherches sur d'embryogenie des Graminees: rapports des Graminees avec les autres monocotyledones. *Annales des Sciences Naturelles, Botanique, Series 12* 2, 491–610.

Hake, S., Vollbrecht, E. & Freeling, M. (1989) Cloning *Knotted*, the dominant morphological mutant in maize using Ds2 as a transposon tag. *EMBO Journal* 8, 15–22.

Halevy, H.A. (ed.) (1985) *CRC Handbook of Flowering*. Boca Raton, FL, USA.

Harlan, J.R. (1982) Human interference with grass systematics. In: Estes, J.R., Tyrl, R.J. & Brunken, J.N. (eds), *Grasses and Grasslands*. University of Oklahoma Press, Norman, pp. 37–50.

Holt, I.V. (1954) Initiation and development of the inflorescences of *Phalaris arundinacea* L. and *Dactylis glomerata* L. *Iowa State College Journal of Science* 28, 603–21.

Ingham, P.W. (1988) The molecular genetics of embryonic pattern formation in *Drosophila*. *Nature* 335, 25–34.

Jacques-Felix, H. (1957) Sur une interpretation nouvelle de l'embryon des Graminees. La nature axillaire de la gemmule. *Comptes Rendus Hebdomadoires de l' Academie des Sciences* 245, 1260–3.

Juliano, J.B. & Aldama, M.J. (1937) Morphology of *Oryza sativa* L. *Philippine Agriculturalist* (University of Philippines Publication Series A) 26, 1–134.

Kamra, O.P. & Nilan, R.A. (1959) Multiovary in barley. Floral anatomy and embryo sac development. *Journal of Heredity* 50, 159–65.

Keller, B., Templeton, M.D. & Lamb, C.J.

(1989) Specific localization of a plant cell wall glycine-rich protein in protoxylem cells of the vascular system. *Proceedings of the National Academy of Sciences (USA)* 86, 1529–33.

Khosla, S. (1946) Developmental morphology in some Indian millets. *Proceedings of the Indian Academy of Sciences* 971, 1–20.

Kirby, E.J.M. & Appleyard, M. (1987) *Cereal Development Guide*, 2nd edn. Arable Unit, Stoneleigh, Warwickshire.

Klekowski, E.J.Jr (1988) *Mutation, Developmental Selection and Plant Evolution*. Columbia University Press, New York, USA.

Leavitt, R.O. (1905) On translocation of characters in plants. *Rhodora* 7, 14–31.

Leavitt, R.O. (1909) A vegetative mutant and the principle of Homeosis in plants. *Botanical Gazette* 47, 30–68.

Lüscher, B., Christenson, E., Litchfield, D. W., Krebs, E.G. & Eisenman, R.N. (1990) Myb DNA-binding inhibited by phosphorylation at a site deleted during oncogenic activation. *Nature* 344, 517–22.

Marocco, A., Wissenbach, M., Becker, D., Paz-Ares, J., Saedler, H., Salamini, F. & Rohde, W. (1989) Multiple genes are transcribed in *Hordeum vulgare* and *Zea mays* that carry the DNA-binding domain of the *myb* oncoproteins. *Molecular and General Genetics* 216, 183–7.

Mehlenbacher, L.E.Jr (1970) Floret development, embryology and systematic position of *Oryzopsis hendersonii* (*Gramineae*). *Canadian Journal of Botany* 48, 1741–58.

Nicholls, P.B. (1974) Interrelationship between meristematic regions of developing inflorescences of four cereal species. *Annals of Botany* 38, 827–37.

Poethig, R.S. (1984) Cellular parameters in leaf morphogenesis in maize and tobacco. In: White, R.A. & Dickison, W.C. (eds), *Contemporary Problems in Plant Anatomy*. Academic, New York, USA, pp. 235–59.

Poethig, R.S. (1989) Genetic mosaics and cell lineage analysis in plants. *Trends in Genetics* 5, 273–7.

Popham, R.A. (1951) Principle types of vegetative shoot apex organization in vascular plants. *Ohio Journal of Science* 51, 249–70.

Reeves, R.G. (1950) Morphology of the ear and tassel of maize. *American Journal of Botany* 44, 756–68.

Reid, D.A. (1985) Morphology and anatomy of the barley plant. In: Rasmusson, D.C. (ed.), *Barley*. American Society of Agronomy, Series: Agronomy No. 26, pp. 73–101.

Rogers, S.O. & Bonnett, H.T. (1989) Evidence for apical initial cells in the vegetative shoot apex of *Hedera helix* cv. goldheart. *American Journal of Botany* 76, 539–45.

Rohde, W., Marocco, A., Wissenbach, M., Barzen, E., Kristiansen, K. & Salamini, F. (1989) Anthocyanin biosynthesis in barley: characterization of structural and putative regulatory genes. In: Styles, D.E., Gavazzi, G. A. & Racchi, M.L. (eds), *The Genetics of Flavonoids*. Edizioni Unicopli, Milan, Italy, pp. 79–95.

Rohde, W., Rosch, K., Kröger, K. & Salamini, F. (1990) Nucleotide sequence of a *Hordeum vulgare* gene encoding a glycine-rich protein with homology to vertebrate cytokeratins. *Plant Molecular Biology* 14, 1057–9.

Rutishauser, R. & Sattler, R. (1985) Complementarity and heuristic value of contrasting models in structural botany. *Botanische Jahrbücher für Systematik, Pflanzengeschichte und Pflanzengeographie* 107, 415–55.

Satina, S.A., Blakeslee, A.F. & Avery, A.G. (1940) Demonstration of the three germ layers in the shoot apex of *Datura* by means of induced polyploidy in periclinal chimeras. *American Journal of Botany* 27, 895–905.

Schwarz-Sommer, Zs., Huijser, P., Nacken, W., Saedler, H. & Sommer, H. (1990) Genetic control of flower development by homeotic genes in *Antirrhinum majus*. *Science* 250, 931–6.

Schwarz-Sommer, Zs., Saedler, H. & Sommer, H. Homeotic genes in the genetic control of flower morphogenesis in *Antirrhinum majus*. In: Russo, V.E.A., Brody, S., Cove, D., Ottolenghi, S. & Ruvkun, G.B. (eds), *Molecular Genetics of Differentiation*. Springer-Verlag, Berlin (in press).

Sharman, B.C. (1942) Developmental anatomy of the shoot of *Zea mays* L. *Annals of Botany, N.S.* 6, 245–82.

Sharman, B.C. (1945) Leaf and bud initiation in the *Gramineae*. *Botanical Gazette* 106, 269–89.

Shen-Ong, G.L.C. (1990) The *myb* oncogene. *Biochimica et Biophysica Acta* 1032, 39–52.

Søgard, B. & von Wettstein-Knowles, P. (1987) Barley: genes and chromosomes. *Carlsberg Research Communications* 52, 123–96.

Sommer, H., Beltran, J., Huijser, P., Pape, H., Lönnig, W., Saedler, H & Schwarz-Sommer, Zs. (1990) *Deficiens*, a homeotic gene involved in the control of flower morphogenesis in *Antirrhinum majus*: the protein shows homology to transcription factors. *EMBO Journal* 9, 605–13.

Stebbins, G.L. (1982) Major trends of evolution in the *Poaceae* and their possible significance. In: Estes, J.R., Tyrl, R.J. & Brunken, J.N. (eds), *Grasses and Grasslands*. University of Oklahoma Press, Norman, pp. 3–36.

Steeves, T.A. & Sussex, I.M. (1989) *Patterns in Plant Development*. Cambridge University Press, Cambridge.

Steffensen, D.M. (1968) A reconstruction of cell development in the shoot apex of maize. *American Journal of Botany* 55, 354–69.

Stewart, R.N. (1978) Ontogeny of the primary body in chimeral forms of higher plants. In: Subtleny, S. & Sussex, I.M. (eds), *The Clonal Basis of Development*. Academic, New York, USA, pp. 131–60.

Stewart, R.N. & Dermen, H. (1979) Ontogeny in monocotyledons as revealed by studies of the developmental anatomy of periclinal chloroplast chimeras. *American Journal of Botany* 66, 47–58.

Surkov, V.A. (1961) Ontogeny and morphological nature of floral parts in the grasses. *Botanicheskiy Zhurnal* 46, 1134–43.

Weatherwax, P. (1923) *The Story of the Maize Plant*. University of Chicago Press, Chicago, USA.

Weir, C.E. & Dale, H.M. (1960) A developmental study of wild rice, *Zizania aquatica* L. *Canadian Journal of Botany* 38, 719–39.

Williams, R.F. (1975) *The Shoot Apex and Leaf Growth*. Cambridge University Press, Cambridge.

Wilson, C.L. (1942) The telome theory and the origin of the stamen. *American Journal of Botany* 29, 759–64.

Yanofsky, M.F., Ma, H., Bowman, J.L., Drews, G.N., Feldmann, K.A. & Meyerowitz, E.M. (1990) AGAMOUS: an *Arabidopsis* homeotic gene whose product resembles transcription factors. *Nature* 346, 35–9.

Ziff, E.B. (1990) Transcription factors: a new family gathers at the cAMP response site. *Trends in Genetics* 6, 69–72.

12

The *Slender* Mutation of Barley

Christopher J. Pollock, Helen J. Ougham & John L. Stoddart

Research on mutants in animals and micro-organisms has provided valuable information on the regulation of development and metabolism. Work with *Arabidopsis thaliana* indicates that the study of plant mutants can be equally rewarding. Unfortunately some areas of plant biology cannot be studied in such a system. These include the characteristic patterns of leaf growth, senescence and carbohydrate metabolism observed in temperate Gramineae. Such processes have, however, been studied either using genotypic (multicomponent) variation or, in some cases, specific mutations. These mutations are spread among a range of temperate cereals and grass species rather than concentrated in one genome, so the full benefit of such material cannot yet be realized, but valuable information can be gained from studying such mutants in isolation (Stoddart, 1987; Thomas, 1987).

A range of stature mutants have been described for temperate cereals (Stoddart, 1987 and references therein). Most of these mutants are genetic dwarfs but overgrowth mutants have also been isolated. Whilst many of the physiological and biochemical studies on such material have been concerned with perturbations in plant growth regulator metabolism, it has become apparent recently that, by virtue of changes in their temperature responses, such plants can also have value in fundamental studies on the factors regulating growth and environmental responses. One of the most useful mutations in this area has been the *slender* mutation of barley (Cb 3014) first described by Foster (1977). This material was isolated from the cultivar Herta following treatment with diethyl sulphate and is visually distinct by virtue of its rapid rate of extension growth, increased height and lax appearance. In this chapter, we summarize research on *slender* relevant to understanding the nature of the mutation and its consequences for the regulation of growth, plant growth regulator metabolism and gene expression.

Leaf growth and the effects of cold

The phenotype of the mutation

In the early stages of growth, the principal manifestation of the phenotype is increased

Fig. 12.1. Normal (N) and *slender* (S) barley seedlings after 7 days growth at 10°C (PPFD 300 μmol $m^{-2}s^{-1}$).

leaf size (Fig. 12.1). A detailed description of the *slender* mutant Cb 3014 (Table 12.1) shows that increased final leaf size is associated with an increase in epidermal cell length and a decline in cross-sectional area, with a consequent increase in the number of cells per mm^2. Measurements of the dynamics of cell growth show that the extension zone of both normal and *slender* genotypes lies within the first 3 mm of the insertion point but that absolute extension rates within this zone are substantially higher in the mutant material (Table 12.2). There are also differences in the overall leaf growth characteristics, since the increased final leaf length is associated with a somewhat longer duration of growth (Table 12.3). The final result in the mutant is a leaf which is larger, but with a dry weight to fresh weight ratio, specific chlorophyll content, specific protein content, total dry matter and maximum photosynthetic rate per

Table 12.1. Cell size data from mature leaves of *slender* and normal barley.

	Normal	*Slender*
Cell length[a] (mm)	0.74	1.44
Cell cross-sectional area (μm^2)	739	466
Cells mm^{-2} (transverse section)	92	137
Final leaf length (cm)	17.8	29.8

[a]Measurements were made on interstomatal rows from the upper surface of fully mature second leaves. Data cited by Stoddart, 1987.

Table 12.2. Absolute extension rates in different sections of the first leaf of *slender* and normal seedlings.

Section	Absolute extension rate (mm d^{-1})	
	Normal	*Slender*
0–3 mm	14.1	25.4
3–6 mm	2.0	3.3
6–9 mm	1.1	1.6
9–12 mm	0.2	0.8
12–15 mm	0.2	0.5
15–18 mm	0.0	0.0
18–21 mm	0.5	0.1

Plants were marked by piercing with a fine wire at 3 mm intervals midway through the period of leaf extension of the first leaf. Displacement of the marks was measured 24 h later.

Table 12.3. Growth parameters for first leaf extension in normal and *slender* barley seedlings.

	Normal	*Slender*
Duration of growth (d)	6.1 ± 0.7	7.9 ± 0.7
Maximum length (mm)	112 ± 5	220 ± 8
Absolute growth rate (mm d^{-1})	18.3 ±1.7	28.0 ± 1.7
Relative growth rate (mm d^{-1})	0.49 ± 0.06	0.38 ± 0.03

Leaf length/time data were analysed using the techniques described by Thomas (1983).

leaf which are not significantly different from leaves of normal plants (E.J. Lloyd, J.L. Stoddart & C.J. Pollock, unpublished observations).

The greater leaf size of the mutation is even more apparent if the plants are grown at low temperature. Localized cooling and high-resolution auxanometers can be used to determine directly the effects of temperature upon the growth of temperate grasses and cereals (Stoddart *et al.*, 1986). The localized basal extension zone and the linear nature of leaf extension in grasses facilitates such techniques and it can be shown that the extension zone itself is the site of temperature perception. The response to temperature change is extremely rapid and, provided a lower limit is not exceeded, fully reversible (Stoddart *et al.*, 1986; Pollock & Eagles, 1988).

When such studies were carried out comparing *slender* and normal genotypes (Stoddart & Lloyd, 1986), a remarkable difference was observed. The normal form ceased growth at around 5°C, a temperature similar to that observed in other temperate grasses (Stoddart *et al.*, 1986) and re-warming led to a full resumption of leaf extension. The *slender* genotype, however, continued to grow down to −5°C, at which point damage to the extension zone occurred and resumption of growth was subsequently impaired (Stoddart & Lloyd, 1986). Thus, it appears that a recessive mutation carried at a single locus can modify drastically the growth/temperature response of plants. Clearly this provides a most useful system for studying the short-term response of plants to temperature change. The *slender* mutation must be presumed to act at the level of cell extension rather than of cell division because of the rapidity of the response (Pollock & Eagles, 1988; Thomas *et al.*, 1989). Cold affects the duration of the cell cycle in leaves (Francis & Barlow, 1988) but there is no information on whether or not this process is differentially affected in the normal and *slender* genotypes.

The physiology of cell extension

Continuation of cell extension requires a supply of undivided cells and the development of cell turgor in excess of the yield threshold (Tomos, 1985). It is believed that the extensibility of the primary wall is high initially and falls as the cell reaches full size and that such changes are associated with an irreversible increase in rigidity, probably caused by increased covalent linkage between adjacent polymer chains (Van Volkenburgh *et al.*, 1983). Measurements of cell turgor may be made directly using a pressure probe (Hürsken *et al.*, 1978). Thomas *et al.* (1989) used this technique to measure extension zone turgor in developing leaves of *Lolium temulentum*. At the same time localized cooling was applied to the extension zone and rates of extension were measured using displacement transducers. The surgical techniques required to allow insertion of the probe were shown not to affect leaf extension rate and it was also demonstrated that the turgor pressure of both cortical and epidermal cells was effectively unchanged in the temperature range 20–0°C; over this span, growth rate fell

Table 12.4. Extension and mesophyll cell turgor in seedling leaves of normal and *slender* barley.

Meristem temperature (°C)	Growth rate (μm min^{-1})		Turgor (MPa)	
	Normal	*Slender*	Normal	*Slender*
20	11.2 ± 3.2	29.0 ± 3.8	0.65 ± 0.07	0.69 ± 0.05
15	6.5 ± 4.4	21.8 ± 1.4	0.68 ± 0.05	0.71 ± 0.05
10	4.1 ± 2.0	11.3 ± 1.9	0.62 ± 0.11	0.65 ± 0.10
5	2.5 ± 1.4	8.4 ± 0.8	0.65 ± 0.06	0.69 ± 0.09
2	1.7 ± 2.8	6.4 ± 2.2	0.63 ± 0.06	0.71 ± 0.08
0	No detectable growth	3.8 ± 1.6	0.63 ± 0.09	0.68 ± 0.11

almost 30-fold. It was concluded in this study that the direct, reversible, effects of temperature on leaf extension were manifested not by changes in turgor pressure but by changes in wall extensibility. The availability of the *slender* mutation and of the normal genotype provided an ideal way of testing this hypothesis. The turgor/growth rate/temperature interactions of the two forms can be compared directly, as can the rheological properties of the extension zone (using the extensiometer described by Van Volkenburgh *et al.*, 1983). Extension zone turgor and rates of leaf extension were measured over a range of temperatures using both genotypes (Pollock *et al.*, 1990). As in the case of *L. temulentum*, turgor was independent of temperature whereas extension rate declined during cooling (Table 12.4). Differences in extension rate between normal and *slender* were not associated with significant differences in turgor, further supporting the hypothesis that wall properties determine the gross temperature response. Evidence consistent with this suggestion was obtained by measuring wall extensibility. Measurements were made of the displacement produced per unit load applied and the plastic component was assessed by calculating the hysteresis between strain and relaxation curves. This is equivalent to the work lost as heat and would be zero for a perfect elastic solid. Comparisons between mature leaves of both normal and mutant genotypes showed no significant differences, with both behaving like rigid elastic solids (Table 12.5). In contrast, extending leaves showed marked plastic deformation, with those from the *slender* genotype being noticeably less rigid. The differences were lost when seedlings were killed before measurement (Table 12.5). Normal seedlings appeared to become more plastic during this treatment, suggesting that some

Table 12.5. Physical properties of extending and mature leaves from normal and *slender* seedlings of *Hordeum vulgare*. Data are means of four determinations ± SD.

Tissue	Condition	Genotype	Extensibility (μm extended g^{-1} applied load)	Load/relaxation hysteresis under 10 g applied load (μJ)
Mature leaf	Living	Normal	25 ± 3	1.1 ± 0.3
		slender	32 ± 4	1.2 ± 0.3
Extending leaf	Living	Normal	116 ± 12	3.3 ± 0.5
		slender	330 ± 28	8.4 ± 0.9
Extending leaf	Killed	Normal	470 ± 31	7.2 ± 0.6
		slender	480 ± 39	7.9 ± 0.7

constraining character present in living cells for normal leaves had been lost in the mutant, resulting in greater extensibility. It is suggested that, among other things, this difference could be manifested in an altered temperature response (Pollock *et al.*, 1990).

Cell wall metabolism

Recently, studies have concentrated upon establishing whether the biophysical and physiological differences between genotypes described above are detectable at the chemical or enzymological level. Orthodox chemical analysis of wall carbohydrates prepared from within the extension zone showed no significant differences between genotypes. Both normal and mutant material showed a monosaccharide distribution characteristic of monocot primary cell walls (McNeil *et al.*, 1980; Pollock *et al.*, 1990). However, the dynamic state of living extending cells, where chemical linkages are undergoing rapid turnover (Labavitch, 1981), might mean that measurements of the kind described above would be unable to detect differences which could profoundly affect cell extensibility.

Although measurement of wall turnover is complex, and there is no good coherent model for the process, the availability of the *slender* mutant simplifies the experimental approach, since any differences from the normal form must be presumed to have some connection, albeit tenuous, with the primary genetic lesion. Thus, three physiological indices which relate to the processes of wall metabolism were all shown to differ in the mutant form. Autolytic release of sugars was examined during incubation of buffered wall extracts. This technique has been used previously to study auxin-mediated changes in metabolism in the extension zones of oat (Heyn, 1986). In our study, the normal genotypes showed higher levels of autolysis, mainly attributed to the release of arabinose and arabinose-rich oligosaccharides, suggesting that the mutation altered the interaction between specific chemical structures and the enzymes which act upon them (Pollock *et al.*, 1990). Further differences were detected when radioactivity was measured in the component monosaccharides of the cell walls within the extension zone following administration of very short pulses of $^{14}CO_2$ to the autotrophic tips of the growing first leaf. Material extracted at the end of a 5-minute pulse showed more radioactivity in acid-hydrolysable glucose in preparations from normal seedlings than in preparations from the *slender* genotype. As the length of the chase period was increased, the differences became smaller and radioactivity appeared in other sugars such as arabinose, xylose and fucose (Pollock *et al.*, 1990). Finally, measurement of release of sugars following treatment with purified hydrolytic enzymes showed differences in the release of uronic acids between the two genotypes. The gross chemical analysis described shows that both genotypes contained about 12% total uronic acids, so differences in enzymatic release can be attributed to differences in accessibility of the material to the purified enzymes. Again this is consistent with the association of the mutation with altered wall architecture (Pollock *et al.*, 1990).

As yet studies such as those described have only been able to provide supporting evidence for the central hypothesis concerning the nature of the *slender* mutation. The recent development of techniques to allow incorporation of specific radiolabelled precursors into wall carbohydrates and proteins without affecting extension rates should allow more precise characterization of wall metabolism in the two genotypes and facilitate studies at a range of temperatures. In this way biochemical and physiological studies can complement the molecular biological approaches which are described below.

Plant growth regulators and the expression of mutant phenotypes

The role of endogenous gibberellins in the growth of *slender* plants

There is general agreement from genetic, physiological and biochemical studies that

Table 12.6. Endogenous gibberellin (GA) concentrations (pg g^{-1} fresh wt) in leaf sheath segments of 8-day-old wild-type and *slender* barley seedlings.

Genotype		GA_{19}	GA_{20}	GA_1	GA_3	GA_8
Wild-type	[*Sln/Sln* + *Sln/sln*]	3577	719	226	250	5607
Slender	[*sln/sln*]	6319	353	21	43	777

Sample size = 19 g fresh wt. GA_3 resolved from interfering $M^{+}536$ compound on an open tubular column coated with a SE52/54 liquid phase.

stature control in cereals is mediated *via* gibberellins (for review see Stoddart, 1987).

Reduced stature mutants can be categorized as either biosynthesis dwarfs (with GA supply blocked at one or more points in the synthetic pathway) or response dwarfs (with disabled receptor or response chain function). The former are exemplified by the '*d*' series of maize dwarfs (Phinney, 1984) and the latter by *Rht* wheat mutants (Gale *et al.*, 1975). Biosynthesis dwarfs have depleted endogenous pools for GAs downstream of the blockage site and can be restored to the normal phenotype by GA applications, whilst response mutants have active GA pool sizes which exceed those in wild-types by ten times, or more, and are insensitive to exogenous hormone treatments.

Overgrowth mutations, such as *slender* barley, can be explained either by GA overproduction or potentiated receptor function. Studies with the *la cry*^{s} slender pea mutation (Potts *et al.*, 1985) have shown greatly reduced pool sizes for GA_1 and its immediate precursors, suggesting that the elevated elongation rate results from changed response characteristics, rather than enhanced GA supply.

Recent studies on GA pool sizes in wild-type and *slender* with GC-MS quantitation by deuterated internal standards, are illustrated by the data shown in Table 12.6 (Croker *et al.*, 1990).

These data show clearly that *slender* barley does not overproduce active GAs (GA_1 and GA_3) but, in contrast, contains larger amounts of the biologically inactive C_{20} precursor, GA_{19}. Croker *et al.* contend that these data identify the GA_{19} oxidative conversion to GA_{20} as being rate-limiting and postulate a feedback regulation by active GAs. The possibility that the unexpected occurrence of GA_3 in all extracts could be artefactual was ruled out by rigorous exclusion of possible sources of contamination.

The inference from the GA pool size data that overgrowth in *slender* was not due to enhanced supplies of active GAs was confirmed using the GA biosynthesis inhibitor PP333 (paclobutrazol) which blocks at the *ent*-kaurene oxidase step.

At 20°C, *slender* seedlings showed an increase in leaf sheath length of 63% when compared with wild-type but lamina length was not significantly different. Application of a soil drench of 10 μM paclobutrazol reduced sheath length by 45% but had no comparable effect on *slender* seedlings. Growth rates of emerging third leaves were progressively reduced by inhibitor concentrations up to 100 μM but *slender* segregants showed no response to the inhibitor over the same range of applications. In 10 μM treated seedlings (showing a 37% sheath length reduction in wild-types) the concentrations of all measured GAs were reduced

Table 12.7. GA pool sizes in leaf sheath segments of wild-type and *slender* barley seedlings 7 days after treatment with 10 μM paclobutrazol. Data expressed as percentages of untreated controls. Figures in parentheses are treated growth rates as a percentage of controls.

Genotype		GA_{19}	GA_{20}	GA_1	GA_3	GA_8
Wild-type	[63]	7.5	12.8	6.6	25.6	26.2
Slender	[99]	2.6	15.6	38.1	41.9	60.2

by inhibitor treatment, as indicated in Table 12.7.

Reduced growth in wild-type seedlings was associated with large reductions in the pool sizes of the biologically active GAs, suggesting a causal relationship. In contrast, more than halving the GA_1 and GA_3 levels in *slender* had no measurable effect upon leaf sheath length.

Taken overall, the paclobutrazol data indicate that elongation growth in *slender* seedlings is independent of GA supply.

The effects of the *slender* mutation on enzyme synthesis by aleurone layers

In addition to the growth enhancement, already discussed, the *slender* locus has been shown to have major effects upon hydrolase synthesis by isolated aleurone layers (Chandler, 1988). Typically, in barley, this tissue is three to four cells thick and is located between the external layers (husk + epidermis) and the outer surface of the endosperm. At the proximal end it is contiguous with the scutellum. The individual cells are extremely thick-walled (thus resisting dehydration) and are packed with protein bodies. The cereal aleurone responds to GAs originating in the embryo/scutellum by the synthesis and secretion of α-amylase and a range of other hydrolases, with regulation being operative at the mRNA level (Zwar & Hooley, 1986).

Embryo-less half-grains of Cb 3014 wild-type were found to be highly responsive to added GA_3 and this effect could be negated by the inclusion of abscisic acid (ABA) in the incubation medium. On the other hand, after 48 h, *slender* genotype half-grains, maintained in buffer only, secreted as much α-amylase as wild-type exposed to 1.0 μM GA_3.

Further studies on the ABA responsiveness of wild-type and *slender* aleurones have suggested that there are distinct action pathways for GAs and ABA in this tissue (Chandler, 1988).

The genetics and molecular genetics of the mutation

Possible sites of action of the mutation

The inheritance of the *slender* phenotype obeys the pattern expected for an alteration in a single nuclear gene. The mutant (*sln*) allele is recessive, so that plants heterozygous at this locus (*Sln*/*sln*) are phenotypically identical to wild-type plants. The most straightforward model which accounts for a recessive mutation conferring GA insensitivity and a *slender* phenotype is to postulate the existence of a 'repressor' molecule which modulates gene expression. The normal form of this repressor acts to constrain extension growth (and other GA-responsive properties); upon addition of exogenous GA, which may bind with the repressor, the latter is inactivated and growth rate is enhanced. In plants homozygous for the mutant allele, either no repressor is synthesized, or the repressor molecule is altered so that it behaves as though it were continuously saturated with GA; in other words, it is inactive.

The alternative hypothesis, that the *slender* locus encodes an activator molecule which requires GA as a co-factor to promote extension growth, is less plausible because of the recessive nature of the mutation: a heterozygote would require only one copy of the gene for an altered, GA-independent activator in order to exhibit at least a partial alteration in growth properties, whereas plants heterozygous for a gene encoding an altered *repressor* would still produce the normal repressor which would continue to constrain growth unless GA were added. Since heterozygotes are phenotypically normal, a repressor is more likely. The term 'repressor' should be taken here to imply a general rather than a specific function.

It is, of course, quite likely that the product of the *slender* locus is a DNA-binding protein which normally inhibits transcription of GA-responsive genes. Such a mechanism would closely resemble the way in which many bacterial operons are regulated, and Scott *et al.* (1990) have suggested that the behaviour of *slender* mutants in barley, pea and tomato may be regarded as analogous to the *GAL* gene regulation system in yeast (Johnston, 1987). Normally, the *GAL80* gene produces an inhibitor of galactose-utilization gene induction; in the presence of the inducer galactose, this inhibitor is inactivated. In cells homozygous for a mutant allele of the *GAL80* gene, functional inhibitor is not produced, so galactose utilization genes are expressed whether galactose is present in the medium or not. According to this model, the *Sln* locus would correspond to the *GAL80* gene, and GA would be equivalent to galactose. Plants homozygous for the *sln* allele would produce no repressor, or an inactive repressor, for one of several reasons.

1. The altered gene is not transcribed; an analogous system is the *viviparous-1* abscisic acid-insensitive mutant of maize, in which the mRNA encoded by the *Vp1* locus and normally synthesized in the developing seed is not detectable in plants homozygous for the mutant *vp1* allele (McCarty *et al.*, 1989).
2. The altered gene produces an untranslatable mRNA.
3. The altered gene produces an RNA which is very rapidly degraded.
4. The translation product of the altered RNA is ineffective as a DNA binding protein or other repressor.
5. The translation product of the altered RNA is very rapidly degraded.

It is also possible to postulate a repressor-type function which would operate post-transcriptionally, for example by regulating initiation of translation in response to GA levels and environmental factors, but data discussed next suggest that in the case of *slender*, the lesion is manifested primarily at the transcriptional level.

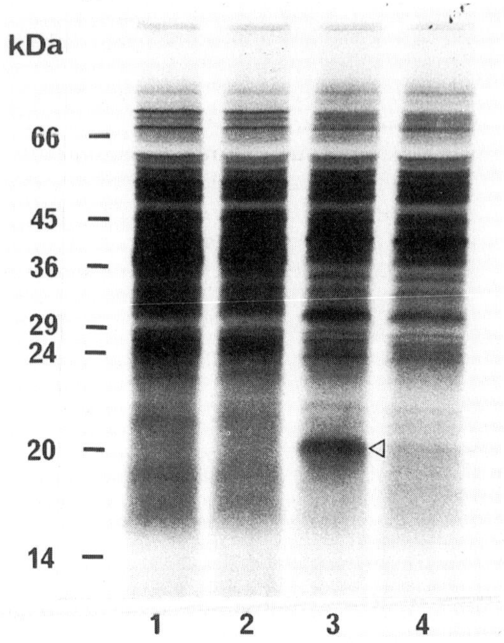

Fig. 12.2. Products of translation of total RNA in a rabbit reticulocyte lysate cell-free system, in the presence of [^{35}S]methionine. Translation products were separated by SDS-polyacrylamide gel electrophoresis on a 12.5% slab gel, and visualized by fluorography. Extension zone RNA was extracted from the basal 10% of the first and second leaf of wild-type and *slender* seedlings (grown at 20°C and harvested 5 days after sowing); root RNA was extracted from 1 cm root tip sections from the same seedlings. Track: 1, wild-type root RNA; 2, *slender* root RNA; 3, wild-type leaf extension zone RNA; 4, *slender* leaf extension zone RNA.

Altered patterns of gene expression in leaves and roots

Slender barley plants have been shown to exhibit patterns of gene expression different from those of normal plants. Figure 12.2 shows the profile of polypeptides synthesized *in vivo* from RNA extracted from the extension zone of normal wild-type and *slender* barley primary leaves, and, for comparison, from roots of the same seedlings (Schünmann, P. & Ougham, H.J., unpublished). A major band (molecular weight about 20 kDa) is consistently observed in translation products of wild-type extension zone RNA, while it is absent or present at very low abundance in *slender* translation products, and those of wild-type root tissue. This band is most obvious when RNA from young, rapidly extending tissue is used (the stage at which the behaviours of normal and *slender* plants differ most markedly), though older leaf base material also manifests this difference. It is tempting to speculate that this band results from translation of the product of transcription of the *Sln* locus (i.e. it is the putative repressor molecule), but its abundance makes this unlikely, since most regulatory proteins constitute a tiny fraction of the cellular protein complement and their mRNAs would not be expected to yield abundant translation products. Rather, the differentially expressed band is likely to represent an RNA encoding an enzymic or structural protein (e.g. a cell wall protein or enzyme associated with cross-linking) which as a consequence of GA-sensitive regulatory processes accumulates in normal but not in *slender* extension zone tissue. It is anticipated that it will be possible to characterize this and other, less abundant, differentially expressed gene products using cDNA libraries based on wild-type and *slender* extension zone mRNA; and by examining the common features of these products and the genes which encode them, to determine the mechanisms whereby the *slender* locus exerts its effects on extension growth, temperature response and other GA-responsive properties.

Table 12.8. Secretion of α-amylase by wild-type, normal and *slender* barley non-embryo half-seeds.

Genotype	α-Amylase production[1]	Percentage of half-seeds secreting α-amylase[2]
Wild-type (*Sln/Sln/Sln*)	0.23	0
Normal (*Sln/Sln/sln*, *Sln/sln/sln*)	0.17	0
slender (*sln/sln/sln*)	51	21 ± 2

[1]α-amylase activity (A_{620} units (15 min)$^{-1}$ (half grain)$^{-1}$) produced by endosperm half-grains imbibed for 3 days. Assay based on release of chromogen from starch analogue, using extract from 10–12 half-grains per assay. From Chandler (1988).

[2]Secretion of α-amylase (determined by production of cleared zone on starch/agar plates flooded with iodine) by individual half-seeds imbibed for 3 days. Half-seeds from individual parent plants were examined separately; the variation observed is within rather than between parent plants.

Altered patterns of gene expression in aleurone layers

To investigate whether, like extension growth, secretion of α-amylase by aleurone cells displays GA-independence in *slender* barley, Chandler (1988) and Lanahan & Ho (1988) examined levels of the enzyme and its mRNA in embryo-less half-seeds of the normal and *slender* phenotypes. In assays using 10–12 embryo-less half seeds, Chandler (1988) showed that endosperm homozygous for the *sln* (mutant) allele produced up to 3000 times as much α-amylase as did wild-type endosperm in the absence of GA; that when GA_3 was added, both genotypes produced similar amounts of α-amylase; and that triploid endosperm of heterozygous genotype (*Sln/Sln/sln* or *Sln/sln/sln*) behaved in the same way as wild-type endosperm (Table 12.8), confirming that the phenotype is fully recessive with respect to α-amylase secretion as well as to extension growth. Secretion of proteases and nucleases was also shown by Lanahan & Ho (1988) to be GA-independent in *slender* embryo-less half-seeds.

Recent observations have revealed unexpected complexity in the pattern of amylase secretion by *slender* aleurone layers. Lanahan & Ho (1988) examined secretion from individual half-seeds by placing them on a starch–agar solid medium for 2 days, then flooding the plate with iodine solution so that where a seed had secreted amylase a large colourless, clear zone was visible against a blue–black background. In this study, one-quarter of the progeny seeds from selfed plants heterozygous for the *sln* and *Sln* alleles showed GA-independent amylase secretion from the embryo-less half-seed, and this phenotype was shown to co-segregate with the *slender* phenotype in plants grown from the corresponding embryo half-seeds. However, we have recently observed (Ougham, H.J. & Evans, M.Ll., unpublished) that while all embryo-less halves which secrete α-amylase correspond to phenotypically *slender* embryo halves, not all *slender* embryo-less halves secrete the enzyme. Studies on pooled seed resulting from selfing a large number of heterozygous (*Sln*/*sln*) plants show that only 15–30% of *slender* embryo-less half-seeds produce a large clear zone on starch–iodine plates (Table 12.8), and that this percentage is not affected by varying the incubation period on the plates, or by pre-imbibing the half-seeds for up to 24 h. Such a result might suggest the presence of one or more additional genetic loci modifying the GA-independent secretion of amylase, but further experiments on progeny seeds from individual (selfed) parent plants do not lend strong support to this hypothesis, since progeny of every plant but one so far examined also yield a minority (less than 30%) of *slender* half-seeds which secrete amylase, and a majority, whose embryos germinate to give phenotypically identical *slender* plants, which do not. If the genetic explanation were valid, so that a *sln*/*sln* plant might additionally be homozygous or heterozygous at one or more alleles modifying the *slender* phenotype, one would expect *slender* half seeds from different parents to exhibit one of three 'secretor' phenotype patterns: all-secretor, no-secretor, and a class containing a proportion of secretors which would depend upon the dominance or otherwise of the modifier allele(s), and the degree of linkage of such allele(s) with the *slender* locus. The apparently random distribution of secretors among the population suggests that a factor such as the developmental stage of the seed, or the conditions under which it matured, rather than its genetic composition, may determine whether or not the amylase is secreted. This would account for the apparent discrepancy between the behaviour of *slender* material produced in the UK and that used by Lanahan & Ho (1988) in the USA.

Secretion of α-amylase is regulated at the transcriptional level. Aleurone cells produce a range of α-amylase isoforms (typically 7) which can be divided into two groups on the basis of their isoelectric points. mRNAs for the high and low pI families have been characterized (Chandler *et al.*, 1987). During imbibition, mRNAs for α-amylases of both the high- and the low-pI family accumulated in *slender* embryo-less half-seeds, in the presence or absence of GA_3, according to the same pattern as that seen in wild-type half-seeds supplied with exogenous GA_3 (Chandler, 1988). Unsupplemented wild-type half-seeds failed to accumulate mRNAs for either pI family. In at least this respect, then, the product of the *slender* locus regulates gene expression for both families at the transcriptional level. In contrast, use of a probe representing an ABA-inducible mRNA indicated that *slender* and wild-type seedlings exhibited identical patterns of ABA-regulated gene expression at the transcriptional level, and in both genotypes ABA prevented α-amylase synthesis. Thus, where expression of a gene is subject to regulation by both GA and ABA, the *sln*/*sln* genotype abolishes normal regulation by GA but does not affect transcriptional-level regulation by ABA.

Concluding remarks

The effects of the *sln* locus on the extension growth of vegetative tissues and enzyme secretion in germinating seeds seem, superficially, to be disparate responses. Nevertheless, because they originate from the same single-gene recessive mutation, they must be capable of rationalization down to a common biochemical action. The overgrowth effects are associated with increased cell wall plasticity, suggesting that wall linkage turnover (possibly associated with secretion into the wall of mediating enzymes) could be an end-point action of expression of *sln*. If this were the case, there would be a functional conformity with hydrolase synthesis by the aleurone layer. In each tissue response there is a clear indication of the removal of a repressive factor, presumably overcome by endogenous GAs in the 'normal', or wild-type, situation. As indicated elsewhere in this chapter there appear to be multiple consequences, at the polypeptide level, of the action of the mutant allele suggesting that the GA-sensitive locus may have regulatory function.

Associated major changes in endogenous GA levels seem not to be major causative influences and probably reflect the operation of feedback controls within the biosynthetic pathway. Essentially, the data indicate that tissues containing the homozygous *slender* mutation have escaped from GA control, either because the primary hormone receptor has been locked in the 'on' state or because processes in the response chain are no longer negatively regulated. Enhanced low-temperature growth is consistent with such suggestions.

As a final point, the behaviour of mutant aleurone layers tends to reinforce the contention that these cells are incapable of internally synthesizing GAs and points to water and ABA as the major initiating factors. In the wild-type, GAs synthesized remotely by the embryo would act to derepress expression of the genes for hydrolytic enzymes.

References

Chandler, P.M. (1988) Hormonal regulation of gene expression in the *slender* mutant of barley (*Hordeum vulgare* L.). *Planta* 175, 115–20.

Chandler, P.M., Jacobsen, J.V., Zwar, J.A., Ariffin, Z. & Huiet, L. (1987) Control of α-amylase gene expression by gibberellin and abscisic acid in barley aleurone. In: Fox, J.E. & Jacobs, M. (eds), *Molecular Biology of Plant Growth Control* (UCLA Symposium on Molecular and Cell Biology). Alan R Liss, New York, pp. 22–33.

Croker, S.J., Hedden, P., Lenton, J.R. & Stoddart, J.L. (1990) Quantification of gibberellins in normal and *slender* barley seedlings. *Plant Physiology* 94, 194–200.

Foster, C.A. (1977) *Slender*: an accelerated extension growth mutant of barley. *Barley Genetics Newsletter* 7, 24–7.

Francis, D. & Barlow, P.W. (1988) Temperature and the cell cycle. In: Long, S.P. & Woodward, F.I. (eds), *Plants and Temperature*, Symp. Soc. Exp. Biol., Vol. 42, pp. 181–202.

Gale, M.D., Law, C.N., Marshall, G.A. & Worland, A.J. (1975) The genetic control of gibberellic acid insensitivity and coleoptile length in a 'dwarf' wheat. *Heredity* 34, 393–9.

Heyn, A.N.J. (1986) A gas chromatographic analysis of the release of arabinose from coleoptile cell walls under the influence of auxin and dextranase preparation. *Plant Science* 45, 77–82.

Hürsken, D., Steudle, E. & Zimmermann, U. (1978) Pressure probe technique for measuring water relations of cells in higher plants. *Plant Physiology* 61, 158–63.

Johnston, M. (1987) A model fungal gene regulatory mechanism: the *GAL* genes of *Saccharomyces cerevisiae*. *Microbiological Reviews* 51, 458–76.

Labavitch, J.M. (1981) Cell wall turnover in plant development. *Annual Review of Plant Physiology* 32, 385–406.

Lanahan, M.B. & Ho, T.-H.D (1988) *Slender* barley: A constitutive gibberellin-response mutant. *Planta* 175, 107–14.

McCarty, D.R., Carson, C.B., Stinard, P.S. & Robertson, D.S. (1989) Molecular analysis of *viviparous-1*: an abscisic acid-insensitive mutant of maize. *Plant Cell* 1, 523–32.

McNeil, M., Darvill, A.G. & Albersheim, P. (1980) Structure of plant cell walls. X. Rhamnogalacturonan I, a structurally complex pectic polysaccharide in the walls of suspension-cultured sycamore cells. *Plant Physiology* 66, 1128–34.

Phinney, B.O. (1984) Gibberellin A_1, dwarfism and the control of shoot elongation in higher plants. In: Crozier, A. & Hillman, J.R. (eds), *The Biosynthesis and Metabolism of Plant Hormones*. Cambridge University Press, Cambridge, UK, pp. 17–41.

Pollock, C.J. & Eagles, C.F. (1988) Low temperature and the growth of plants. In: Woodward, F.I. & Long, S.P. (eds), *Plants and Temperature*. Cambridge University Press, Cambridge, UK, pp. 157–80.

Pollock, C.J., Tomos, A.D., Thomas, A., Smith, C.J., Lloyd, E.J. & Stoddart, J.L. (1990) Extension growth in a barley mutant with reduced sensitivity to low temperature. *New Phytologist* 115, 617–23.

Potts, W.C., Reid, J.B. & Murfet, I.C. (1985) Internode length in *Pisum*. Gibberellins and the *slender* phenotype. *Plant Physiology* 63, 357–64.

Scott, I.M., Horgan, R., Ougham, H.J. & Jones, M.G. (1990) The use of mutants in plant growth regulator research. In: Ormrod, J.C. & Jeffcoat, B. (eds), *Current & Future*

Methods in Plant Growth Regulator Research (British Society For Plant Growth Regulation Monograph No. 19).

Stoddart, J.L. (1987) Genetic and hormonal regulation of stature. In: Thomas, H. & Grierson, D. (eds), *Developmental Mutants in Higher Plants*. Cambridge University Press, Cambridge, UK, pp. 155–80.

Stoddart, J.L. & Lloyd, E.J. (1986) Modification by gibberellin of the growth–temperature relationship in normal and mutant genotypes of several cereals. *Planta* 167, 364–8.

Stoddart, J.L., Thomas, H., Lloyd, E.J. & Pollock, C.J. (1986) The use of a temperature-profiled position transducer for the study of low-temperature growth in gramineae. *Planta* 167, 359–63.

Thomas, H. (1983) Analysis of the response of leaf extension to chilling temperatures in *Lolium temulentum* seedlings. *Physiologia Plantarum* 57, 509–13.

Thomas, H. (1987) Foliar senescence mutants. In: Thomas, H. & Grierson, D. (eds), *Developmental Mutants in Higher Plants*. Cambridge University Press, Cambridge, UK, pp. 245–65.

Thomas, A., Tomos, A.D., Stoddart, J.L., Thomas, H. & Pollock, C.J. (1989) Cell expansion rate, temperature and turgor pressure in growing leaves of *Lolium temulentum*. *New Phytologist* 112, 1–5.

Tomos, A.D. (1985) The physical limitations of leaf cell expansion. In: Baker, N.R., Davies, W.J. & Ong, C.K. (eds), *Control of Leaf Growth*. Society for Experimental Biology Seminar Series No. 27, Cambridge University Press, Cambridge, pp. 1–33.

Van Volkenburgh, E., Hunt, S. & Davies, W.J. (1983) A simple instrument for measuring cell wall extensibility. *Annals of Botany* 51, 669–72.

Zwar, J.A. & Hooley, R. (1986) Hormonal regulation of α-amylase gene transcription in wild oat (*Avena fatua* L.) aleurone protoplasts. *Plant Physiology* 80, 459–63.

13

Biochemical and Molecular Studies of Stress Tolerance in Barley

A.M. Stanca, V. Terzi & L. Cattivelli

Plants respond to abiotic environmental stress actively by turning on metabolic pathways or by modifying gene expression, the final aim being to survive the stress. This, however, brings about a decrease in plant yield. Identification of the metabolic processes involved and the regulatory genes and active areas is indispensable in producing stress resistant plants.

Even species that are only distantly related have strong similarities in their molecular responses to abiotic stress. This makes it possible to extrapolate findings derived from only a few model plants to other species.

All evidence to date indicates that plants subjected to abiotic stress undergo degradational processes characterized by an increase in the free fatty acid content, degradation of phospholipids, peroxidation resulting in the accumulation of free radicals, acidification of the cytoplasm, accumulation of toxic metabolites and the breakdown of complex structures of the cytoplasm and cell organelles. At the same time the plant, stimulated by the critical environmental conditions, expresses new mRNAs that encode stress proteins (Sachs & Ho, 1986). Some of these proteins appear to play an important role in critical conditions and are expressed co-ordinately during different kinds of stress (Czarnecka *et al.*, 1984; Heikkila *et al.*, 1984; Uemura & Steponkus, 1989). Examples are the heat shock 70 genes which are transcribed in various stress situations (Winter *et al.*, 1988) and the thionins, which enter into action during biotic and abiotic stress (Fischer *et al.*, 1989; Grill *et al.*, 1989). It is thought that the plant hormone ABA is an important mediator in the signal transduction pathway of stress response, low temperature stress as well as desiccation or osmotic stress (Skriver & Mundy, 1990).

Heat-shock response

Barley responds to a rise in temperature by modifying gene expression to inhibit the synthesis of constitutive proteins present at normal growth temperature in favour of heat-shock proteins (HSPs). These polypeptides have been extremely well conserved in evolution: they are, in part, common to

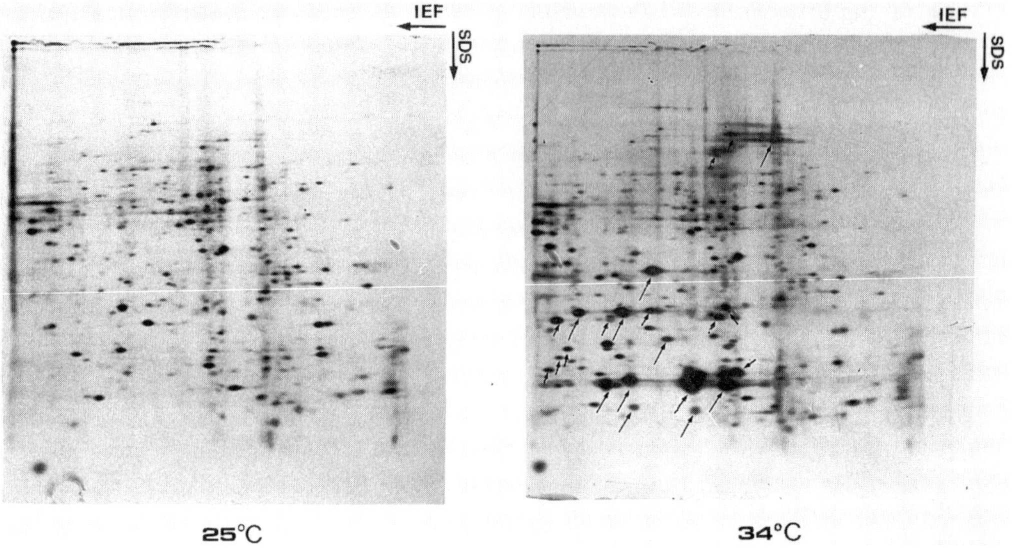

Fig. 13.1. Two-dimensional analysis of *in vitro* translation products obtained from poly(A) RNAs isolated from control (25°C) and heat-shocked (2 h at 34°C) barley shoots. The arrows indicate the proteins corresponding to the heat-shock induced mRNAs.

plants, animals and bacteria (Lindquist & Craig, 1988; Neumann *et al.*, 1989). Some of the HSPs are presumed to be 'chaperonines', proteins involved in the assembly of complex proteins (Ostermann *et al.*, 1989), in the translocation of mitochondrial and microsomal proteins, and in protection against degradational processes (Lindquist & Craig, 1988). The co-ordinated induction at high temperatures of a specific set of mRNAs that are translated into three classes of HSPs with high (70–100 kDa), medium (30–50 kDa) and low (18–20 kDa) molecular weights (Fig. 13.1) has been confirmed in barley (Marmiroli *et al.*, 1989), maize (Cooper & Ho, 1983), wheat (Necchi *et al.*, 1987; Zivy, 1987), sorghum (Ougham & Stoddart, 1986) and rice (Fourrè & Lhoest, 1989).

In barley, tissue-specific induction of HSPs and variation related to both the growth stage and the genotype occur (Marmiroli *et al.*, 1989). The variability in the HSP pattern among cultivars was the basis for a genetic analysis of the inheritance of HSPs in barley, which indicates a complex regulatory system (Terzi *et al.*, 1990).

A cDNA clone coding for a 18 kDa HS protein was isolated and sequenced. It is characterized by an open reading frame of 480 bp and encodes a sequence of 160 amino acids (Marmiroli *et al.*, 1990). The derived amino acid sequence was 73% and 41% homologous to small HSPs of soybean and maize, respectively. Homology with the small HSPs of *Drosophila* and *Caenorhabditis* was only about 20%, but hydropathic profiles indicated a preserved amino acid domain between residues 16 and 92 of HSP 18 of barley. The function of this cytoplasmic protein is unknown.

Response to low, non-freezing temperatures

The biochemical and molecular changes induced in plant tissues by low temperature form the basis of an adaptive process known as *hardening*, by which tissues improve their tolerance to cold (Levitt, 1980). Genetic analysis of the factors involved in cold hardening of cereals has demonstrated that it is controlled by many genes, with an additive effect (Brule-Babel & Fowler, 1989), although some studies in wheat have identified a locus for frost resistance on chromosome 5A closely linked to the locus that controls vernalization (*Vrn1*) (Sutka & Snape, 1989).

Studies on the physiological mechanisms responsible for acclimatization to cold have focused mainly on modification of: (i) membranes; (ii) enzyme activities and the content of soluble components; (iii) gene expression and analysis of cold-regulated genes. Although there are no specific studies of barley, in many other species the ABA is able to induce cold-hardening as well as modifications in gene expression correlated with cold-hardening (Mohapatra *et al.*, 1988; Lång *et al.*, 1989).

Membrane modifications

In cereals low-temperature stress apparently induces changes in the lipid content of membranes. Studies in barley (cv. Huron) have shown slight decreases in total lipids and in phospholipids during hardening. In rye (cv. Puma), the same experimental conditions increased the lipid content (Cloutier, 1987; Lynch & Steponkus, 1987). Parallel results for an increase in sterols in wheat during a period of cold hardening were obtained by Willemot (1980). In wheat, barley and rye, the role of linolenic acid in cold acclimatization has been demonstrated using a chemical substance (BASF 13–338) that reduces the linolenic acid content in polar lipids (St John *et al.*, 1979). See Steponkus (1984) for a more comprehensive review.

Modifications of enzyme activities and of content of soluble components

An increase in soluble sugars in response to a temperature decrease is common to all cereals (Levitt, 1980). After exposure of wheat to 4°C, the plant shows an immediate increase in the activity of the enzyme sucrose synthase, with a consequent increase in the sucrose content (Calderon & Pontis, 1985).

To evaluate the relationship between carbohydrate level and freezing tolerance, the response of three different barley varieties was compared. Carbohydrate accumulation occurs during hardening in winter cereals because the demand for assimilate is reduced, due to slower growth at lower temperatures. Although the total carbohydrate level increased upon cold stress in all three varieties tested, maximum resistance did not correspond to the higher sugar content (Livingston *et al.*, 1989). In addition, the relationship between the degree of freezing tolerance and the total carbohydrate content differed for the three cultivars examined. These findings indicate wide genetic variability for the allocation of carbohydrates with a cryoprotective function (Livingston *et al.*, 1989).

Modification of gene expression and analysis of cold-regulated genes

Following the exposure of barley seedlings to cold (+1 to +5°C), new mRNAs are induced and lead to the production of a specific set of cold-regulated proteins (Marmiroli *et al.*, 1986; Hughes & Pearce, 1988; Cattivelli & Bartels, 1989). Electrophoretic analysis of the polypeptides obtained by *in vitro* translation of the cold-induced mRNAs showed about 10–12 cold-induced polypeptides with M_r 18–30 kDa, and two cold-induced polypeptides of 45 and 75 kDa. Similar cold-induced proteins have been identified in wheat (Perras & Sarhan, 1989).

In barley these cold-related mRNAs are characterized by different induction kinetics: the mRNA coding for the 75 kDa polypeptide is expressed a few hours after exposure to cold, whereas the other cold-regulated mRNAs are expressed only after a few (1–3) days (Cattivelli & Bartels, 1989). Similar differences in the induction kinetics of cold-regulated mRNAs have been reported in *Arabidopsis* (Gilmour *et al.*, 1988).

Modified expression of some mRNAs transcribed at normal temperature has also been found after exposure to low temperatures. In rice, for example, low temperature causes a dramatic reduction in the mRNAs coding for two subunits of Rubisco (Hahn & Walbot, 1989).

The function of cold-induced proteins is at present unknown; their synthesis and accumulation in plant tissues has been correlated with the acclimatization of plants to cold (Hughes & Pearce, 1988). It has been shown that a protein fraction induced during hardening in spinach can protect the membrane against mechanical freeze–thaw damage *in vitro* (Hincha *et al.*, 1990).

Some genes whose expression in barley is regulated by low temperatures have been cloned and sequenced (Fig. 13.2) (Cattivelli & Bartels, 1990). The protein sequences from the open reading frames of two of these cDNAs are characterized by a strongly basic domain containing numerous arginine residues. No homology was found between the sequences of these cDNAs and other known sequences, and their functions remain unknown.

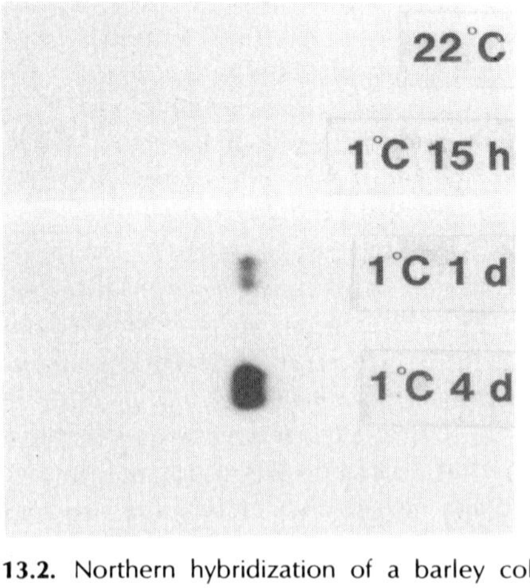

Fig. 13.2. Northern hybridization of a barley cold-regulated cDNA clone with a filter carrying poly(A) RNAs isolated from barley shoots after different temperature treatments (22°C; 1°C for 15 h; 1°C for 1 and 4 days). The mRNA corresponding to the cDNA shows a clear cold-inducibility being expressed only after exposure to low temperature for at least 24 h.

Analysis of the expression patterns of five different cold-regulated cDNAs demonstrated that some mRNAs homologous to isolated sequences are expressed tissue-specifically. One of them was expressed only in leaf tissues and not in root tissues, while other isolated cDNAs corresponded to sequences expressed in all tissues of the plant in response to low temperatures. Cold-regulated mRNAs homologous to cDNAs isolated from barley were also induced in wheat and rye seedlings during a similar period of hardening (Cattivelli & Bartels, 1990). These findings support the hypothesis that the molecular response of barley and related cereals to cold stress is a complex mechanism involving many genes, conserved during evolution among related species (Guy, 1990; Thomashow, 1990).

Drought stress response

Drought is the most important abiotic stress for barley, which is frequently grown in environments where drought is recurrent. Compared with other cereals, barley is well adapted to such environmental conditions through water-use efficiency (WUE).

There have been numerous studies on the biochemical and molecular modifications induced in barley by water shortage, by increased osmotic potential in the external medium (osmotic stress), or by both stresses. The response to these stress conditions is very similar: an increased level of ABA, the induction of a specific set of genes, and the accumulation of unusual metabolites (proline, polyamines, etc.), as well as changes in the activities of some enzymes. Most of these changes appear to be controlled by ABA, which is responsible for activating a wide set of genes (Skriver & Mundy, 1990).

In wheat, ABA is increased by up to ten-fold following dehydration. Similarly, osmotic stress induced by polyethylene glycol (PEG) causes ABA to accumulate to up to five times its basal level (Cammue *et al.*, 1989). Barley ABA-mutants, which produce only low levels of ABA in response to water stress (Walker-Simmons *et al.*, 1989), and ABA-insensitive mutants (Raskin & Ladyman, 1988) will permit studies on interactions between water stress, ABA accumulation, and the physiological and molecular response of the plant.

Proline accumulation is one of the most studied parameters of stress response (Hanson *et al.*, 1979). Proline accumulates in the leaves of barley treated with PEG (Riazi *et al.*, 1985) or with ABA alone (Stewart *et al.*, 1986) or in combination with salts, such as NaCl or KCl (Pesci, 1988, 1989); however, there is evidence of proline accumulation with PEG even in the presence of inhibitors of ABA synthesis (Stewart & Voetberg, 1987).

Other metabolites have been reported to change in response to water or osmotic stress. Accumulation of betaine and polyamines in barley leaves following water or osmotic stress has been reported (Hanson & Nelsen, 1978; Turner & Stewart, 1986, 1988), as well as decreased activity of nitrate reductase in barley and wheat plants (Smirnoff *et al.*, 1985; Larsson *et al.*, 1989), and induction of an endogenous inhibitor of α-amylase in barley (Robertson *et al.*, 1989).

Recently genes have been isolated from maize (Close *et al.*, 1989; Vilardell *et al.*, 1990), rice (Mundy & Chua, 1988; Yamaguchi-Shinozaki *et al.*, 1989) and barley (Close *et al.*, 1989) that are expressed abundantly in mature embryos, are ABA responsive and are induced in seedlings upon water or salt stress. The proteins encoded by these genes are extremely hydrophilic, rich in glycine with M_r of 15–25 kDa. They are characterized by repetitive sequences rich in lysine and by a region of uninterrupted serine residues. Owing to their ABA inducibility and to their homology at the nucleotide and amino acid levels they have all been included in a class of genes called *rab* (*R*esponsive to *AB*A).

Some other genes with similar expression patterns, but not related to *rab* genes, have recently been reported (Gomez *et al.*, 1988; Marcotte *et al.*, 1988).

One step towards understanding the regulation of these genes is that ABA-responsive elements have recently been identified in the promoter region of the early methionine (*Em*) gene (an embryo-expressed and ABA-regulated gene) of wheat and of a rice *rab* gene (Marcotte *et al.*, 1989; Mundy *et al.*, 1990).

Other genes induced by ABA are known in barley; their expression is possibly related to the acquisition of desiccation tolerance during embryogenesis. This may be of particular interest in the study of drought tolerance, because it represents a natural process during which the embryo becomes capable of resisting severe dehydration. Barley embryos excised 12 days after pollination (DAP) and grown *in vitro* germinate precociously but do not tolerate dehydration, while embryos excised 12 DAP and grown on a substrate containing ABA, or 16 DAP embryos, can withstand dehydration. The acquisition of this ability has been correlated with the appearance of specific mRNAs (Bartels *et al.*, 1988).

ABA-induced cDNAs have also been isolated from mature wheat embryos (Williamson & Quatrano, 1988) and from barley aleurone (Hong *et al.*, 1988). However, up to now it has not been demonstrated that these genes are involved in stress tolerance response. The study of drought resistance has revealed molecular complexity, but also homology among botanically related species in the responses induced.

Salt stress response

While not a halophyte, barley tolerates high salt concentrations in the soil. Exposure of the barley plants to high salt concentrations, however, alters some biochemical parameters of the cells. A plant growing in soil with high salt concentration must have a transmembrane transport system that can maintain a K^+/Na^+ ratio above the external ratio (Cheeseman, 1988). In barley there is evidence of a NaCl-induced mechanism that extrudes Na^+ based on a Na^+/H^+ antiport (Garbarino & Du Pont, 1988). Accumulation of proline has been observed in barley leaf segments treated with ABA and KCl or NaCl (Pesci, 1989), and of glycine betaine in wheat plants subjected to salt stress (McDonnel & Wyn Jones, 1988). To elucidate the effect of proline and glycine betaine accumulation on salt tolerance in barley, embryos were cultured in high salt conditions. Lone *et al.* (1987) observed that the exogenous addition of proline and glycine betaine to barley embryo cultures favours shoot lengthening. Barley mutants that accumulate a high level of proline have been isolated (Kueh & Bright, 1981). However, the comparison of the effects of exogenous supplied proline with the enhanced levels of endogenous proline in the mutant line R5201 suggested that the increased proline accumulation in the mutant is too low to have a significant physiological effect (Lone *et al.*, 1987). Much attention has been devoted to identifying specific genes induced in response to NaCl.

In barley roots, *in vivo* and *in vitro* protein-labelling has identified quantitative alterations of the expression of proteins and mRNAs (Hurkman & Tanaka, 1987; Ramagopal, 1987; Hurkman *et al.*, 1988), but did not present clear evidence of *de novo* induction of specific genes.

However, specific gene induction has been demonstrated in rice roots, where a gene expressed in an organ-specific fashion in response to drought stress has been isolated (Claes *et al.*, 1990).

Response to hypoxic conditions

In barley, as in all higher plants, a lack of oxygen inhibits aerobic respiration and stimulates fermentative metabolism with the accumulation of the end-products of glycolytic processes, such as ethanol, lactic acid, organic acids and amino acids (Davies, 1980). At the same time a switch in gene expression occurs, with the induction of anaerobic proteins, mainly glycolytic enzymes (Lazlo & St Lawrence, 1983; Kelly & Freeling, 1984a, b; Hoffman *et al.*, 1986). In particular, both alcohol dehydrogenase (ADH) and lactate dehydrogenase (LDH) are strongly induced in barley tissues upon anaerobiosis; these are involved in the reduction of pyruvate to ethanol and lactic acid, respectively (Hoffman *et al.*, 1986; Good & Crosby, 1989a). The isoenzymes ADH_2 and ADH_3, which are specifically induced under stress, have K_m values for acetaldehyde which are lower than that of the constitutively expressed form, ADH_1 (Mayne & Lea, 1984; Fagerstedt & Crawford, 1986). They show tissue-specificity and are induced to different levels in roots and aleurone (Hanson & Brown, 1984; Good & Crosby, 1989a). Alternatively, pyruvic acid can be reduced to lactic acid by LDH, whose isoenzymatic forms have induction kinetics essentially similar to ADH (Good & Crosby, 1989a) and appear to be encoded by two genes (*Ldh1* and *Ldh2*) (Hoffman *et al.*, 1986). The finding that LDH concentrations increase up to 20 times under stress supports the hypothesis of a specific, but still unclear, function for this enzyme. This is in contrast with the hypothesis of Davies (Davies *et al.*, 1974) and Roberts (Roberts *et al.*, 1984) for a short, self-limiting induction of LDH during the initial phase of stress. This would result in a decrease of intracellular pH, making it optimal for the catalysis of pyruvate to acetaldehyde by pyruvate dehydrogenase (PDH).

Glutamate - pyruvate - aminotransferase (GPT), which is characterized by anaerobic induction kinetics similar to those ofADH and LDH, increased its activity more than four times in barley root tissues. The single enzymatic form induced is tissue-specific, being present in the roots but not in leaf tissue, and is independent of the nitrogen level of the substrate (Good & Crosby, 1989b).

The ADH gene system in higher plants, and in maize in particular, is one of the best-characterized from the molecular and genetic point of view. In barley, two ADH loci (*Adh1* and *Adh2*) are tightly linked on chromosome 4, whereas the third locus, *Adh3*, is independent (Harberd & Edwards, 1982, 1983; Hanson & Brown, 1984). Two complete and one partial genomic clones corresponding to the three loci for the ADH isoenzymes have been isolated (Trick *et al.*, 1988), using a cDNA for maize *Adh1* as probe. Two of the genes (*Adh2* and *Adh3*) are strictly homologous to the *Adh2* gene of maize, while the third gene cloned (*Adh1-Mg*), was found, by comparing the sequences and expression pattern, to be a null-allele mutant related to the *Adh1* gene of maize. This suggests that the present three-gene ADH system evolved from a single ancestral gene by duplication and subsequent transposition. *Adh2* and *Adh3* have diverged considerably in their 5′ regions upstream of the ATG translation initiation codon, with the exception of a consensus region, TC(A/G) TCA(G/C)CAA, located between −70 and −80 bp, and a second region −33−34 bp upstream to the start of translation. These regions may have a regulatory role (Trick *et al.*, 1988).

Heavy metal response

Toxic concentrations of heavy metals have wide-ranging effects on cell biochemistry, maintenance of membrane potential, cell growth, and chloroplast function. Siegel & Haug (1983) hypothesize that aluminium ions interfere with calmodulin-dependent formation of membrane potential in barley root plasma membrane vesicles. The ability of some barley genotypes to tolerate toxic concentrations of metals may be due to exclusion or detoxification. Tolerant cultivars appear to have a greater capacity to maintain normal ion flow and membrane potential in the presence of aluminum (Miyasaka *et al.*, 1989), triggering a more active exclusion mechanism (Zhang & Taylor, 1989). Some peptides and proteins probably act in this way. In fact, higher plants accumulate metallothionins (MTs) designated as class III MTs in response to heavy metal stress. These are peptides not encoded by genes but derived from the metabolism of glutathione and are characterized by activity in binding metal ions (Grill *et al.*, 1989). Three MT classes are known: MTsI, MTsII, MTsIII. Class I and class II MTs are polypeptides encoded by specific genes. They are widespread in animals and fungi but have to date been found in only one higher plant: *Mimulus guttatus* (De Miranda *et al.*, 1990). However, barley does have a class of genes encoding proteins called thionins (García-Olmedo *et al.*, 1989) which can be induced by heavy metal stress. Fisher *et al.* (1989) have shown that mRNAs coding for thionins are induced in barley leaf tissues following treatment with heavy metal salts. These are cysteine-rich membrane proteins highly homologous to the hordothionins of the endosperm (Bohlman & Apel, 1987). The thionins of leaf tissues are encoded by a multigene family consisting of 50–100 genes on chromosome 6 (Bohlman *et al.*, 1988). These are toxic proteins that are negatively regulated by light (Gausing, 1987) and are also induced in response to pathogen attack. A co-ordinate expression of thionins and pathogenesis-related proteins (PR) has also been demonstrated. In particular, high chromium concentrations induce the accumulation of chitinase and putrescine, which are also synthesized in pathological conditions (Jacobsen *et al.*, 1990).

It has also been reported that specific cadmium-binding proteins (Cd-BP) may be induced and play a role by slowing the absorption and translocation of cadmium, as may other still unclear physiological mechanisms (Rauser, 1986).

Conclusion

Barley is the fourth most important food cereal. Since most agriculture is conducted under limiting soil and climatic conditions, studies of barley resistance to abiotic stress are important. In particular, a better understanding is needed of regulatory mechanisms involved in co-ordinated expression of stress-regulated genes and also the identification of genes with key functions. This will improve understanding of the molecular basis of stress response and the mechanisms of plant resistance and susceptibility.

Knowledge of gene function will help in developing new genotypes aimed at reducing the wide gap between the actual and potential yield of barley.

Acknowledgements

We thank Dr D. Bartels for critically reading the manuscript. The work was supported by Italian Ministry of Agriculture 'Progetto: Tecnologie Avanzate Applicate alle Piante'.

References

Bartels, D., Singh, M. & Salamini, F. (1988) Onset of desiccation tolerance during development of the barley embryo. *Planta* 175, 485–92.

Bohlmann, H. & Apel, K. (1987) Isolation and characterization of cDNAs coding for leaf-specific thionins closely related to the endosperm-specific hordothionin of barley (*Hordeum vulgare L.*). *Molecular and General Genetics* 207, 446–54.

Bohlmann, H., Clausen, S., Behnke, S., Giese, H., Hiller, C., Reimann-Philipp, U., Schrader, G., Barkholt, V. & Apel, K. (1988) Leaf-specific thionins of barley – a novel class of cell wall proteins toxic to plant-pathogenic fungi and possibly involved in the defence mechanism of plants. *EMBO Journal* 7, 1559–65.

Brule-Babel, A.L. & Fowler, D.B. (1989) Genetic control of cold hardiness and vernalization requirement in rye. *Genome* 32, 19–23.

Calderon, P. & Pontis, H.G. (1985) Increase of sucrose synthase activity in wheat plants after a chilling shock. *Plant Science* 42, 173–6.

Cammue, B.P.A., Broekaert, W.F., Kellens, J.T.C., Raikhel, N.V. & Peumans W.J. (1989) Stress-induced accumulation of wheat germ agglutinin and abscisic acid in roots of wheat seedlings. *Plant Physiology* 91, 1432–5.

Cattivelli, L. & Bartels, D. (1989) Cold-induced mRNAs accumulate with different kinetics in barley coleoptiles. *Planta* 178, 184–8.

Cattivelli, L. & Bartels, D. (1990) Molecular cloning and characterization of cold-regulated genes in barley. *Plant Physiology* 93, 1504–10.

Cheeseman, J.M. (1988) Mechanisms of salinity tolerance in plants. *Plant Physiology* 87, 547–50.

Claes, B., Dekeyser, R., Villarroel, R., Van den Bulcke, M., Bauw, G., Van Montagu, M. & Caplan, A. (1990) Characterization of a rice gene showing organ-specific expression in response to salt stress and drought. *Plant Cell* 2, 19–27.

Close, T.J., Kortt, A.A. & Chandler, P.M. (1989) A cDNA-based comparison of dehydration-induced proteins (dehydrins) in barley and corn. *Plant Molecular Biology* 13, 95–108.

Cloutier, Y. (1987) Lipid and protein changes in cold- and drought-hardened cereals. *Phytoprotection* 68, 87–96.

Cooper, P. & Ho, T.-H.D. (1983) Heat shock proteins in maize. *Plant Physiology* 71, 215–22.

Czarnecka, E., Edleman, L., Schöffl, F. & Key, J.L. (1984) Comparative analysis of physical stress responses in soybean seedlings using cloned heat shock cDNAs. *Plant Molecular Biology* 3, 45–58.

Davies, D.D. (1980) Anaerobic metabolism and the production of organic acids. In: Davies, D.D. (ed.), *The Biochemistry of Plants*, Vol. 2. Academic Press, New York, pp. 581–611.

Davies, D.D., Grego, S. & Kenworthy, P. (1974) The control of the production of lactate and ethanol by higher plants. *Planta* 118, 297–310.

De Miranda, J.R., Thomas, M.A., Thurman, D.A. & Tomsett, A.B. (1990) Metallothionein genes from the flowering plant *Mimulus guttatus*. *FEBS Letters* 260, 277–80.

Fagerstedt, K.V. & Crawford, R.M.M. (1986) Changing kinetic properties of barley alcohol dehydrogenase during hypoxic conditions. *Journal of Experimental Botany* 37, 857–64.

Fischer, R., Behnke, S. & Apel, K. (1989) The effect of chemical stress on the polypeptide composition of the intercellular fluid of barley leaves. *Planta* 178, 61–8.

Fourrè, J.L. & Lhoest, J. (1989) Protein synthesis and modification by heat in rice cell culture. *Plant Science* 61, 69–74.

Garbarino, J. & DuPont, F.M. (1988) NaCl induces a Na^+/H^+ antiport in tonoplast vesicles from barley roots. *Plant Physiology* 86, 231–6.

García-Olmedo, F., Rodriguez-Palenzuela, P., Hernández-Lucas, C., Ponz, F., Maraña, C., Carmona, M.J., Lopez-Fando, J., Fernandez,

J.A. & Carbonero, P. (1989) The thionins: a protein family that includes purothionins, viscotoxins and crambins. In: Miflin, B.J. (ed.), *Oxford Surveys of Plant Molecular & Cell Biology*, Vol. 6. Oxford University Press, pp. 31–60.

Gausing, K. (1987) Thionin genes specifically expressed in barley leaves. *Planta* 171, 241–6.

Gilmour, S.J., Hajela, R.K. & Thomashow, M.F. (1988) Cold acclimation in *Arabidopsis thaliana*. *Plant Physiology* 87, 745–50.

Gómez, J., Sánchez-Martínez, D., Stiefel, V., Rigau, J., Puigdomènech, P. & Pagès, M. (1988) A gene induced by the plant hormone abscisic acid in response to water stress encodes a glycine-rich protein. *Nature* 334, 262–4.

Good, A.G. & Crosby, W.L. (1989a) Induction of alcohol dehydrogenase and lactate dehydrogenase in hypoxically induced barley. *Plant Physiology* 90, 860–6.

Good, A.G. & Crosby, W.L. (1989b) Anaerobic induction of alanine aminotransferase in barley root tissue. *Plant Physiology* 90, 1305–9.

Grill, E., Löffler, S., Winnacker, E.-L. & Zenk, M.H. (1989) Phytochelatins, the heavy-metal-binding peptides of plants, are synthesized from glutathione by a specific glutamylcysteine dipeptiolyl transpeptidase (phytochelatin synthase). *Proceedings of the National Academy of Sciences (USA)* 86, 6838–42.

Guy, C.L. (1990) Cold acclimation and freezing stress tolerance: role of protein metabolism. *Annual Review of Plant Physiology and Plant Molecular Biology* 41, 187–223.

Hahn, M. & Walbot, V. (1989) Effects of cold-treatment on protein synthesis and mRNA levels in rice leaves. *Plant Physiology* 91, 930–8.

Hanson, A.D. & Nelsen, C.E. (1978) Betaine accumulation and [^{14}C] formate metabolism in water-stressed barley leaves. *Plant Physiology* 62, 305–12.

Hanson, A.D. & Brown, A.D.H. (1984) Three alcohol dehydrogenase genes in wild and cultivated barley *Hordeum vulgare*. Characterization of the products of variant alleles. *Biochemical Genetics* 22, 495–515.

Hanson, A.D., Nelsen, C.E., Pedersen, A.R. & Everson, E.H. (1979) Capacity for proline accumulation during water stress in barley and its implications for breeding for drought resistance. *Crop Science* 19, 489–93.

Harberd, N.P. & Edwards, K.J.R. (1982) A mutational analysis of the alcohol dehydrogenase system in barley. *Heredity* 48, 187–95.

Harberd, N.P. & Edwards, K.J.R. (1983) Further studies on the alcohol dehydrogenases in barley: evidence for a third alcohol dehydrogenase locus and data on the effect of an alcohol dehydrogenase-1 null mutation in homozygous and in heterozygous condition. *Genetical Research* 41, 109–16.

Heikkila, J.J., Papp, J.E.T., Schultz, G.A. & Bewley, J.D. (1984) Induction of heat shock protein messenger RNA in maize mesocotyls by water stress, abscisic acid and wounding. *Plant Physiology* 74, 270–4.

Hincha, D.K., Heber, U. & Schmitt, J.M. (1990) Proteins from frost-hardy leaves protect thylakoids against mechanical freeze-thaw damage in vitro. *Planta* 180, 416–19.

Hoffman, N.E., Bent, A.F. & Hanson, A.D. (1986) Induction of lactate dehydrogenase isozymes by oxygen deficit in barley root tissue. *Plant Physiology* 82, 658–63.

Hong, B., Uknes, S.J. & Ho, T.-H.D. (1988) Cloning and characterization of a cDNA encoding a mRNA rapidly-induced by ABA in barley aleurone layers. *Plant Molecular Biology* 11, 495–506.

Hughes, M.A. & Pearce, R.S. (1988) Low temperature treatment of barley plants causes altered gene expression in shoot meristems. *Journal of Experimental Botany* 39, 1461–7.

Hurkman, W.J. & Tanaka, C.K. (1987) The effects of salt on the pattern of protein synthesis in barley roots. *Plant Physiology* 83, 517–24.

Hurkman, W.J., Tanaka, C.K. & DuPont, F.M. (1988) The effects of salt stress on polypeptides in membrane fractions from barley roots. *Plant Physiology* 88, 1263–73.

Jacobsen, S., Hauschild, M. & Kragh, K.M. (1990) Comparison of chitinase and putrescine induction in barley leaves by Cr (VI). *Physiologia Plantarum* 79, A 120.

Kelley, P.N. & Freeling, M. (1984a) Anaerobic expression of maize glucose phosphate isomerase I. *Journal of Biological Chemistry* 259, 673–7.

Kelley, P.M. & Freeling, M. (1984b) Anaerobic expression of maize fructose 1,6-diphosphate aldolase. *Journal of Biological Chemistry* 259, 14180–3.

Kueh, J.S.H. & Bright, S.W.I. (1981) Proline accumulation in a barley mutant resistant to *trans*-4-hydroxy-L-proline. *Planta* 153, 166–71.

Lång, V., Heino, P. & Palva, E.T. (1989) Low temperature acclimation and treatment with exogenous abscisic acid induce common polypeptides in *Arabidopsis thaliana* (L.) Heynh. *Theoretical and Applied Genetics* 77, 729–34.

Larsson, M., Larsson, C.-M., Whitford, P.N. & Clarkson, D.T. (1989) Influence of osmotic stress on nitrate reductase activity in wheat (*Triticum aestivum* L.) and the role of abscisic acid. *Journal of Experimental Botany* 40, 1265–71.

Lazlo, A. & St Lawrence, P. (1983) Parallel induction and synthesis of PDC and ADH in anoxic maize root. *Molecular and General Genetics* 192, 110–17.

Levitt, J. (1980) *Responses of Plants to Environmental Stresses*, Vol. 1, *Chilling, Freezing, and High Temperature Stresses*, 2nd edn. Academic Press, New York.

Lindquist, S. & Craig, E.A. (1988) The heat-shock proteins. *Annual Review of Genetics* 22, 631–77.

Livingston, III, D.P., Olien, C.R. & Freed, R.D. (1989) Sugar composition and freezing tolerance in barley crowns at varying carbohydrate levels. *Crop Science* 29, 1266–70.

Lone, M.I., Kueh, J.S.H., Wyn Jones, R.G. & Bright, S.W.J. (1987) Influence of proline and glycinebetaine on salt tolerance of cultured barley embryos. *Journal of Experimental Botany* 38, 479–90.

Lynch, D.V. & Steponkus, P.L. (1987) Plasma membrane lipid alterations associated with cold acclimation of winter rye seedlings (*Secale cereale* L. cv Puma). *Plant Physiology* 83, 761–7.

Marcotte, W.R. Jr, Bayley, C.C. & Quatrano, R.S. (1988) Regulation of a wheat promoter by abscisic acid in rice protoplasts. *Nature* 335, 454–7.

Marcotte, W.R. Jr, Russell, S.H. & Quatrano R.S. (1989) Abscisic acid-responsive sequences from the Em gene of wheat. *Plant Cell* 1, 969–76.

Marmiroli, N., Terzi, V., Odoardi Stanca, M., Lorenzoni, C. & Stanca, A.M. (1986) Protein synthesis during cold shock in barley tissues. *Theoretical and Applied Genetics* 73, 190–6.

Marmiroli, N., Lorenzoni, C., Cattivelli, L., Stanca, A.M. & Terzi, V. (1989) Induction of heat shock proteins and acquisition of thermotolerance in barley (*Hordeum vulgare* L.). Variations associated with growth habit and plant development. *Journal of Plant Physiology* 135, 267–73.

Marmiroli, N., Cattivelli, L., Raho, G., Terzi, V. & Perrotta, C. (1990) Genetic and molecular analysis of genes involved in heat shock response in higher plants. In: Amaldi, F., De Felice, M., Di Fonzo, N., Merlin, P., Polsinelli, M. & Viotti, A. (eds), *Recombinant DNA Technologies in Industry and Agriculture: State-of-the-Art in Italy*. Amity, Milan, Italy, pp. 193–218.

Mayne, R.G. & Lea, P.J. (1984) Alcohol dehydrogenase in *Hordeum vulgare*: changes in isoenzyme levels under hypoxia. *Plant Science Letters* 37, 73–8.

McDonnell, E. & Wyn Jones, R.G. (1988) Glycinebetaine biosynthesis and accumulation in unstressed and salt-stressed wheat. *Journal of Experimental Botany* 39, 421–30.

Miyasaka, S.C., Kochian L.V., Shaff, J.E. & Foy, C.D. (1989) Mechanisms of aluminum tolerance in wheat. *Plant Physiology* 91, 1188–96.

Mohapatra, S.S., Poole, R.J. & Dhindsa, R.S. (1988) Abscisic acid-regulated gene expression in relation to freezing tolerance in alfalfa. *Plant Physiology* 87, 468–73.

Mundy, J. & Chua, N.-H. (1988) Abscisic acid and water-stress induce the expression of a novel rice gene. *EMBO Journal* 7, 2279–86.

Mundy, J., Yamaguchi-Shinozaki, K. & Chua, N.-H. (1990) Nuclear proteins bind conserved elements in the abscisic acid-responsive promoter of a rice *rab* gene. *Proceedings of the National Academy of Sciences (USA)* 87, 1406–10.

Necchi, A., Pogna, N.E. & Mapelli, S. (1987) Early and late heat shock proteins in wheat and other cereal species. *Plant Physiology* 84, 1378–84.

Neumann, D., Nover, L., Parthier, B., Rieger, R., Scharf, K.-D., Wallghiehn, R. & Nieden, U.Z. (1989) Heat shock and other stress response systems of plants. *Biologisches Zentralblatt* 108, 1–146.

Ostermann, J., Horwich, A.L., Neupert, W. & Hartl, F.-U. (1989) Protein folding in mitochondria requires complex formation with hsp60 and ATP hydrolysis. *Nature* 341, 125–30.

Ougham, H.J. & Stoddart, J.L. (1986) Synthesis of heat-shock protein and acquisition of thermotolerance in high-temperature tolerant and high-temperature susceptible lines of sorghum. *Plant Science* 44, 163–7.

Perras, M. & Sarhan, F. (1989) Synthesis of freezing tolerance proteins in leaves, crown, and roots during cold acclimation of wheat. *Plant Physiology* 89, 577–85.

Pesci, P. (1988) Ion fluxes and abscisic acid-induced proline accumulation in barley leaf segments. *Plant Physiology* 86, 927–30.

Pesci, P. (1989) Involvement of Cl^- in the increase in proline induced by ABA and stimulated by potassium chloride in barley leaf segments. *Plant Physiology* 89, 1226–30.

Ramagopal, S. (1987) Salinity stress induced tissue-specific proteins in barley seedlings. *Plant Physiology* 84, 324–31.

Raskin, I. & Ladyman, J.A.R. (1988) Isolation and characterization of a barley mutant with abscisic-acid-insensitive stomata. *Planta* 173, 73–8.

Rauser, W.E. (1986) The amount of cadmium associated with Cd-binding protein in roots of *Agrostis gigantea*, maize and tomato. *Plant Science* 43, 85–91.

Riazi, A., Matsuda, K. & Arslan, A. (1985) Water-stress induced changes in concentrations of proline and other solutes in growing regions of young barley leaves. *Journal of Experimental Botany* 36, 1716–25.

Roberts, J.K.M., Callis, J., Wemmer, D., Walbot, V. & Jardetzky, O. (1984) Mechanism of cytoplasmic pH regulation in hypoxic maize root tips and its role in survival under hypoxia. *Proceedings of the National Academy of Sciences (USA)* 81, 3379–83.

Robertson, M., Walker-Simmons, M., Munro, D. & Hill, R.D. (1989) Induction of α-amylase inhibitor synthesis in barley embryos and young seedlings by abscisic acid and dehydration stress. *Plant Physiology* 91, 415–20.

Sachs, M.M. & Ho, T.-H.D. (1986) Alteration of gene expression during environmental stress in plants. *Annual Review of Plant Physiology* 37, 363–76.

Siegel, N. & Haug, A. (1983) Calmodulin-dependent formation of membrane potential in barley root plasma membrane vesicles: a biochemical model of aluminum toxicity in plants. *Physiologia Plantarum* 59, 285–91.

Skriver, K. & Mundy, J. (1990) Gene expression in response to abscisic acid and osmotic stress. *Plant Cell* 2, 503–12.

Smirnoff, N., Winslow, M.D. & Stewart, G.R. (1985) Nitrate reductase activity in leaves of barley (*Hordeum vulgare*) and durum wheat (*Triticum durum*) during field and rapidly applied water deficits. *Journal of Experimental Botany* 36, 1200–8.

St John, J.B., Christiansen, M.N., Ashworth, E.N. & Gentner, W.A. (1979) Effect of BASF 13-338, a substituted pyridazinone, on linolenic acid levels and winterhardiness of cereals. *Crop Science* 19, 65–9.

Steponkus, P.L. (1984) Role of the plasma membrane in freezing injury and cold acclimation. *Annual Review of Plant Physiology* 35, 543–84.

Stewart, C.R. & Voetberg, G. (1987) Abscisic acid accumulation is not required for proline accumulation in wilted leaves. *Plant Physiology* 83, 747–9.

Stewart, C.R., Voetberg, G. & Rayapati, P.J. (1986) The effects of benzyladenine, cycloheximide, and cordycepin on wilting-induced abscisic acid and proline accumulations and abscisic acid- and salt-induced proline accumulation in barley leaves. *Plant Physiology* 82, 703–7.

Sutka, J. & Snape, J.W. (1989) Location of a gene for frost resistance on chromosome 5A of wheat. *Euphytica* 42, 41–4.

Terzi, V., Cattivelli, L., Gatti, A., Stanca, A.M., Lorenzoni, C. & Marmiroli, N. (1990) Physiological and molecular studies of heat and drought tolerance in barley. *Bulletin Sociètè Botanique de France*, 137, Actualites Botaniques (1), 73–80.

Thomashow, M.F. (1990) Molecular genetics of cold acclimation in higher plants. In: Scandalios, J.G. (ed.), *Advances in Genetics*, Vol. 28. Academic Press, Inc., pp. 99–131.

Trick, M., Dennis, E.S., Edwards, K.J.R. & Peacock, W.J. (1988) Molecular analysis of the alcohol dehydrogenase gene family of barley. *Plant Molecular Biology* 11, 147–60.

Turner, L.B. & Stewart, G.R. (1986) The effect of water stress upon polyamine levels in barley (*Hordeum vulgare* L.) leaves. *Journal of Experimental Botany* 37, 170–7.

Turner, L.B. & Stewart, G.R. (1988) Factors affecting polyamine accumulation in barley (*Hordeum vulgare* L.) leaf sections during osmotic stress. *Journal of Experimental Botany* 39, 311–16.

Uemura, M. & Steponkus, P.L. (1989) Parallel effects of freezing and osmotic stress on the ATPase activity and protein composition of the plasma membrane of winter rye seedlings. *Plant Physiology* 91, 961–9.

Vilardell, J., Goday, A., Freire, M.A., Torrent, M., Martínez, M.C., Torné, J.M. & Pagès, M. (1990) Gene sequence, developmental expression, and protein phosphorylation of RAB-17 in maize. *Plant Molecular Biology* 14, 423–32.

Walker-Simmons, M., Kudrna, D.A. & Warner, R.L. (1989) Reduced accumulation of ABA during water stress in a molybdenum cofactor mutant of barley. *Plant Physiology* 90, 728–33.

Willemot, C. (1980) Sterols in hardening winter wheat. *Biochemistry* 19, 1071–3.

Williamson, J.D. & Quatrano, R.S. (1988) ABA-regulation of two classes of embryo-specific sequences in mature wheat embryos. *Plant Physiology* 86, 208–15.

Winter, J., Wright, R., Duck, N., Gasser, C., Fraley, R. & Shah, D. (1988) The inibition of petunia hsp 70 mRNA processing during $CdCl_2$ stress. *Molecular and General Genetics* 211, 315–19.

Yamaguchi-Shinozaki, K., Mundy, J. & Chua, N.-H. (1989) Four tightly linked *rab* genes are differentially expressed in rice. *Plant Molecular Biology* 14, 29–39.

Zhang, G. & Taylor, G.J. (1989) Kinetics of aluminum uptake by excised roots of aluminum-tolerant and aluminum-sensitive cultivars of *Triticum aestivum* L. *Plant Physiology* 91, 1094–9.

Zivy, M. (1987) Genetic variability for heat shock proteins in common wheat. *Theoretical and Applied Genetics* 74, 209–13.

IV

Seed Development, Composition, Germination and Utilization

14
Grain Structure and Composition

C.M. Duffus & M.P. Cochrane

Grain development

In the UK the spring barley crop is harvested 5–6 months after the grain is sown, but it is not until about the end of the third month that anthesis (shedding of pollen) takes place, and the development of the new generation begins. By this stage many of the factors affecting the final yield of the crop have already been determined. These include the number of tillers, the total leaf area, the amount of assimilate stored in the stem, and the number of spikelets which have been produced in each ear (Cottrell *et al.*, 1982). Environmental conditions before, and at anthesis, determine the number of florets which are successfully fertilized and thus undergo further development.

In wheat, there is only one spikelet at each node, and each spikelet consists of about four florets, but in barley, there is only one floret in each spikelet and at each node there are three spikelets, of which one is fertile in two-row barley, two are fertile in four-row barley, and all are fertile in six-row barley. In each floret the ovary and stamens are enclosed in two flowering glumes. The lower dorsal glume (lemma) slightly overlaps the upper ventral glume (palea). In many cultivars the lemma extends into a long, thin, somewhat barbed awn. The lemma and palea are more or less fully developed at anthesis and their green tissue contributes photosynthate to the carbon economy of the ear. In non-hulled or naked cultivars such as Himalaya, the flowering glumes separate completely from the rest of the grain at harvest, but in hulled or covered cultivars such as those grown in Europe and N. America, the lemma and palea adhere to the underlying pericarp, forming the husk. Gaines *et al.* (1985) have described a layer of material which appears on the surface of the pericarp soon after anthesis, and cements the pericarp to the glumes when the two organs eventually come into contact. The glumes do not adhere to the pericarp at the apical (brush) end or over the embryo.

The inflorescence of barley is a terminal spike which emerges from the leaf sheath or 'boot' of the flag leaf as the terminal internode of the stem elongates. Both genotypic and environmental factors influence flowering. The inflorescence may or may not be fully emerged from the boot at anthesis, and similarly, the florets in the emerged ears may or may not be open when the pollen is shed. Barley is thus usually, but not always, self-fertilized. The opening of the floret is brought about by the swelling of a pair of lodicules which lie between the glumes and the ovary. In the mature grain, these organs almost enclose the embryo end of the caryopsis and staining with phloroglucinol indicates that they contain lignified tissue. During imbibition they may have a function in gas exchange, water availability or the physical protection of the embryo.

The ovary is mature at anthesis. It is approximately 1 mm from the narrow base to the broad, hairy apex which is divided into two lobes each bearing a large feathery stigma. The inner cells of the ovary wall are chlorenchymatous but the numerous outer layers of parenchymatous cells mask the green colour, making the ovary appear whitish. Attached to the inside of the ovary wall along the ventral side are two integuments which enclose the nucellus except at the small gap known as the micropyle. The embryo sac, formed as the result of a meiotic division in a nucellar cell and thus containing haploid nuclei, is embedded in the nucellus at the micropylar end. Fertilization takes place within an hour of pollination (Pope, 1937) and results in the formation of a diploid zygote and a triploid primary endosperm nucleus. Also in the embryo sac are two synergid nuclei close to the zygote at the end of the embryo sac nearest the micropyle, and an indefinite number (up to 100) of antipodal cells at the chalazal side of the embryo sac (Cass & Jensen, 1970). Conditions experienced by the plant before anthesis have been shown to influence the subsequent development of the grain (Scott *et al.*, 1983) but there is as yet no evidence to show whether environmental factors cause differences in, for example, the amount of reserve material present in the nucellus, or the numbers or degree of polyploidy of the antipodal cells (Bennett *et al.*, 1975).

All barley grains go through the same stages of development. A period of slow growth follows fertilization and then grain weight increases rapidly. When dry matter accumulation ceases, the grain is said to have reached physiological maturity but it must undergo the dehydration phase of development before, at a moisture content of 15% or less, it is said to have reached full maturity or harvest-ripeness. The rate at which these changes take place depends both on the genotype and the environmental conditions during grain growth. Temperature appears to be a particularly important factor (Hoshikawa, 1962). Thus, when comparing observations from different laboratories it is necessary to take into account variations due to cultivar and to climate, and also differences in the interpretation of the term 'grain maturity'. Much confusion can be avoided if grain age is expressed as a fraction of the period between anthesis and harvest-ripeness, or, if reference is made to a definite stage of development such as the attainment of maximum dry weight.

Pericarp

The ovary contains a single ovule enclosed in integuments from which the testa or seed-coat is formed. The wall of the ovary develops into the pericarp of the one-seeded fruit which is classed as a caryopsis because at maturity the testa is fused to the pericarp. At fertilization, the cells of the ovary wall are more or less isodiametric but as the growth of the grain proceeds, the cells of the outer layers elongate in a plane parallel to the long axis of the grain while the cells of the chlorenchymatous inner layers, after a short period of cell division, elongate in a plane at right angles to the long axis of the grain and are thus called cross-cells. The cells of the inner epidermis do not undergo cell division after fertilization but they elongate considerably in the same plane as the cells of the outer pericarp. As the diameter of the grain increases, these tube cells become widely separated and eventually crushed, except over the embryo where they and some distended cells of the cross-cell layer form a tissue similar to the spongy mesophyll of a leaf (Cochrane & Duffus, 1983a). A cuticular layer covers the outer wall of the outer epidermis of the pericarp, and up to about ten stomata are present among the hairs at the apical end of the grain in the ventral furrow or crease (Cochrane & Duffus, 1979).

At anthesis, starch granules are present in the non-chlorenchymatous cells of the ovary wall (Krauss, 1933). MacGregor & Dushnicky (1989a) reported that in cv. Bonanza grown in Manitoba, starch granules could be detected in the outer layers of the developing pericarp up to about 17 days after anthesis. Most of the starch granules in this tissue are round and 2–3 μm in diameter. Alpha-amylase 1 is found in the outer pericarp during the period when the granules are present and it would seem likely that starch breakdown products are used in cell elongation. While the outer cell layers are elongating, the inner non-chlorenchymatous layers are degenerating, leaving an air space outside the cross cells. The outer pericarp thus becomes a thin layer of cells, devoid of starch and containing very little cytoplasm. This tissue is known as the transparent layer.

The cross-cell layer is two to three cells thick, each cell having a large nucleus and numerous chloroplasts. The cells are interconnected by numerous plasmodesmata (Cochrane & Duffus, 1979). The starch granules in this layer are lenticular and are found in the chloroplasts (Fig. 14.1) which suggests that the photosynthetic activity of these cells is more than sufficient for their metabolism. Using ^{14}C-labelled carbon dioxide, Watson & Duffus (1988) showed that immature caryopses of barley were able to fix externally supplied carbon dioxide and that the assimilate thus produced was transported to the endosperm. Previous work (Nutbeam & Duffus, 1976) had demonstrated that C4 photosynthesis takes place in isolated green layers and that intact immature caryopses are capable of net oxygen evolution in the light (Nutbeam & Duffus, 1978). The outer pericarp with its cuticular layer and fewer stomata appears to constitute some barrier to gas exchange with the atmosphere. It would seem likely there-

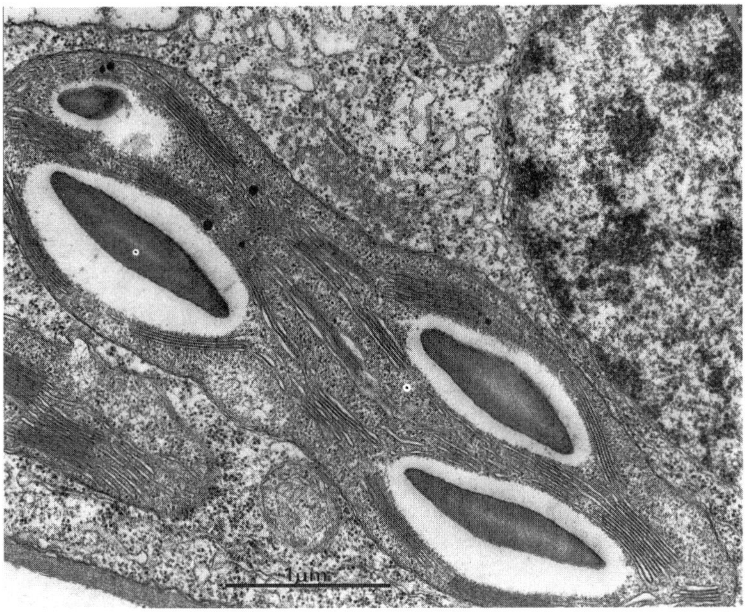

Fig. 14.1. Chloroplast in the cross-cell layer in a transverse section of the pericarp of a caryopsis of cv. Julia. The grain was from a plant grown at 10°C from anthesis, and it was fixed in glutaraldehyde at a stage of development equivalent to 25 days after anthesis in a grain development period of 60 days.

fore that much of the carbon dioxide fixed by the green layer would come from respiration in the inner tissues of the grain. It has not yet been established under what conditions and to what extent the fixation of atmospheric carbon dioxide by the pericarp contributes to the accumulation of carbohydrate in the developing grain. There is some evidence that the transparent layer reduces the rate of oxygen loss from the pericarp (Nutbeam & Duffus, 1978) and some of the oxygen thus trapped in the space between the cross-cells and the transparent layer may subsequently be used in endosperm respiration.

Towards the end of the grain-filling period, starch is no longer found in the inner pericarp, the chlorophyll in the cross cells breaks down and the caryopsis becomes yellow, first on the dorsal side and the flanks and then finally at the crease. During the dehydration phase, the pericarp becomes compressed between the glumes which adhere to it on the outside (in hulled cultivars) and the testa which adheres to it on the inside. The description of the pericarp just given, refers to the dorsal and lateral regions of the grain. On the ventral side is the crease, groove or furrow in which lies the main vascular bundle of the pericarp. The vascular tissue is just one part of a highly integrated set of tissues which apparently control the movement of assimilates, water and other substances into, and perhaps also out of, the developing endosperm. These will be considered together later.

Testa

The outermost of the two integuments which form the outer layers of the ovule at anthesis degenerates, and it is the inner integument which develops into the testa or seed coat. The testa is composed of two layers of cells which surround the nucellus except at the micropyle where there is a small gap, and on the ventral side where a larger gap is occupied by chalazal cells. The two layers of testa cells have very different characteristics (Fig. 14.2). The outer layer is more thin-walled and has large nuclei

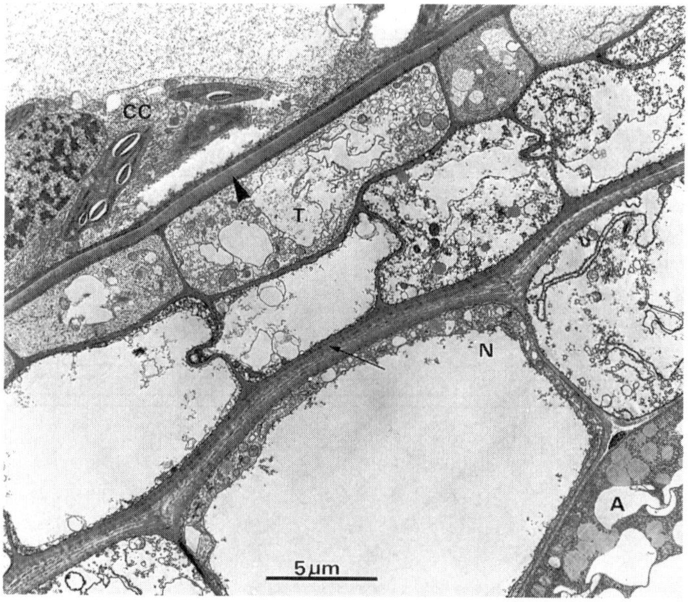

Fig. 14.2. Transverse section cut at mid-grain showing cross-cells (CC), testa (T), nucellus (N), aleurone (A), cuticular layer of nucellus (arrow), and cuticular layer of testa (arrow head). Caryopsis as in Fig. 14.1.

and dense cytoplasm, perhaps associated with the formation of the thick cuticular layer which lies between the testa and the pericarp. The cell walls of the inner layer are thicker and many of the anticlinal walls appear to be buckled. Polyphenolic material accumulates in these cells and is particularly abundant in the cells on either side of the chalaza (Cochrane, 1983). As the grain matures, the cytoplasm of both layers degenerates and the anticlinal walls of the outer layer become buckled, leaving three layers of periclinal walls close together. The cuticular layer of the outermost periclinal walls is fused to the pericarp, and the innermost periclinal wall is fused to the cuticular layer of the nucellus (Freeman & Palmer, 1984).

Nucellus

Fertilization initiates endosperm and embryo formation within the embryo sac and, as these tissues develop, the starch-filled cells surrounding the embryo sac, and then the remaining parenchymatous cells in the dorsal and lateral parts of the nucellus, degenerate (Krauss, 1933) leaving material which Norstog (1974) described as nucellar lysate. The epidermis of the nucellus persists as a layer of highly vacuolated cells between the testa and the endosperm (Fig. 14.2). The outermost wall of the nucellar epidermis is relatively thick compared with that of the testa, and is covered with a cuticular layer which is much thinner than that of the testa but has a similar structure (Cochrane & Duffus, 1979; Freeman & Palmer, 1984). In the mature grain, the nucellar epidermis is visible only as a layer of cell wall material which, being unpigmented, is often referred to as the hyaline layer. Using scanning electron microscopy, Evers & Reed (1988) have detected in mature wheat caryopses an amorphous layer between the nucellar epidermis and the aleurone. This layer, which appears to be derived from the nucellar lysate, varies in thickness over the surface of the grain and may well be responsible for the adherence of the aleurone to the hyaline layer.

The development of the nucellar tissue on the ventral side of the caryopsis differs from that described above and will be considered in greater detail below.

Endosperm

Early development

The triploid primary endosperm nucleus is suspended in a strand of cytoplasm in the very large and highly vacuolate central cell within the embryo sac. In plants growing at 20°C the first division of the primary endosperm nucleus is completed within about 7 h of pollination (Bennett *et al.*, 1975) and further nuclear divisions follow quite rapidly, the mean nuclear doubling time being initially about 4 h. The nuclei lie outside a large central vacuole and as many as 2000 are formed before wall formation takes place and the coenocytic endosperm becomes cellular. The cellularization of the endosperm coincides with the disappearance of the antipodal cells. Cell division continues throughout the endosperm but the nuclear doubling time is much greater than in the coenocytic phase. After a few days, cell division is confined to the outermost layers of cells. These cells have many of the characteristics of the meristematic cells found in other parts of the plant. They are brick-shaped and thin-walled with large nuclei and dense cytoplasm. However, by 15 days after anthesis in a developmental period of 60 days from anthesis to harvest ripeness, they contain lipid droplets and numerous small vacuoles, and thus appear to be partially differentiated into aleurone cells. They continue to undergo both anticlinal and periclinal divisions for a further 10–15 days (Cochrane & Duffus, 1981). The inner of the two daughter cells resulting from a periclinal division may simultaneously dedifferentiate and differentiate to become a starchy endosperm cell (Cochrane & Duffus, 1982).

Because starchy endosperm cells next to the embryo are continuously being broken down (MacGregor & Dushnicky, 1989b), total endosperm cell number may reach a maximum before cell division ceases. It has been reported that in wheat, the number of endosperm nuclei falls after the maximum is reached (Briarty *et al.*, 1979; Gleadow *et al.*, 1982). Chojecki *et al.* (1986a) concluded that there is no satisfactory method for obtaining absolute cell counts on large numbers of endosperms, but, by estimating the mean DNA per cell and per endosperm, they were able to compare endosperm cell numbers in different genotypes of wheat. It was shown that in two cultivars the maximum endosperm cell number was reached in 24 days from anthesis, which was about two-fifths of the total grain development time. This is apparently somewhat later in development than was indicated by previous reports of the cessation of endosperm cell division 11–13 days after anthesis (Brocklehurst *et al.*, 1978) and 16 days after anthesis (Evers, 1970). However, the results of Briarty *et al.* (1979) showing that maximum endosperm cell number in wheat was reached 16–20 days after anthesis in a developmental period of 52 days are, in fact, comparable with those of Chojecki *et al.* (1986a) and, notwithstanding errors inherent in cell counting methods, and genotypic variation within species (Gleadow *et al.*, 1982), they are remarkably similar to the values obtained for barley (Cochrane & Duffus, 1981; Kvaale & Olsen, 1986). It should be emphasized that estimates of the maximum endosperm cell number cannot be used to determine the duration of cell division in the endosperm. Light microscopy would seem to be the best method available, and for this, sections from the apex and base of the grain should be examined as well as those from mid-grain.

Grain weight is one of the major components of yield in cereals and there is some evidence to suggest that final grain weight may be a function of endosperm cell number (Brocklehurst, 1977; Gleadow *et al.*, 1982; Cochrane & Duffus, 1983b; Chojecki *et al.*, 1986a). It is therefore of interest to know what factors control the rate and duration of endosperm cell division and, in barley, what determines whether a cell differentiates into a starchy endosperm cell or into an aleurone cell.

Starchy endosperm

The innermost cells of the starchy endosperm are irregular in shape, having been formed during the cellularization of the coenocytic endosperm. Except on the ventral side, these cells are surrounded by cells which are prismatic in shape and lie roughly in columns, indicating that they were formed as a result of cell divisions in the peripheral layers. Between the aleurone and the prismatic cells are the last-formed starchy endosperm cells. These are much less elongated in the radial plane and are known as the subaleurone. In plants growing at

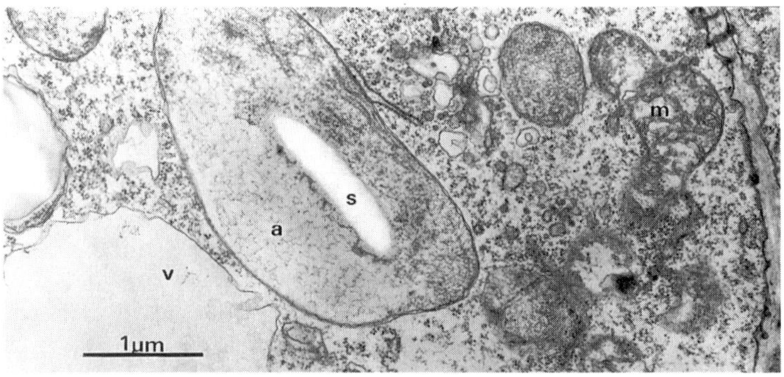

Fig. 14.3. Part of a subaleurone cell of a field-grown caryopsis of cv. Midas fixed in glutaraldehyde 19 days after anthesis in a grain development period of 60 days, showing amyloplast (a), starch granule (s), vacuole (v) and mitochondrion (m).

20°C, numerous starch granules have been observed in the endosperm within about 24 h of the start of the cell wall formation (Bennett *et al.*, 1975). Starch granules form in plastids which have tubuli but little or no internal membrane structure (Fig. 14.3). Usually lenticular with an equatorial groove, a single granule grows until it eventually almost fills an amyloplast. These A-type granules may reach a diameter of 30 μm. About 2 weeks after fertilization, small, round starch granules are observed in the older endosperm cells (May & Buttrose, 1959). As these B-type granules grow they retain their spherical shape and they do not exceed 10 μm in diameter. Experimental work described by Williams & Duffus (1977) has provided additional evidence that the large and small granules are initiated at different stages of endosperm development. UDP-($U^{14}C$)-glucose was supplied to barley plants at 0, 10, 14 or 24 days after anthesis and endosperms from each set of plants were analysed 25 days after anthesis. Radioactivity was found in large granules recovered from each set of plants whereas little if any radioactivity was found in small granules in endosperms from plants supplied with the labelled starch precursor before 14 days after anthesis.

May & Buttrose (1959) found that in mature grains of cv. Prior grown in South Australia there were ten times as many B-type granules as A-type granules but that 90% of the starch volume was made up of A-type granules. Similar results were obtained from analyses of mature grains of barley cv. Glacier grown in field conditions in Scotland. The large A-type granules accounted for 91% of the total volume of starch, but only 8% of the total number of starch granules (McDonald *et al.*, 1991). At a certain stage of development, the A-type amyloplasts develop protuberances or buds (Fig. 14.4) in which B-type granules develop (Buttrose, 1960). Until serial sections are studied it will not be possible to determine whether the outgrowths separate from the parent A-type amyloplasts and form B-type amyloplasts or whether the B-type granules remain enclosed in the A-type amyloplast (Parker, 1985). Using stereological analysis, Hughes & Briarty (1976) found that during the development

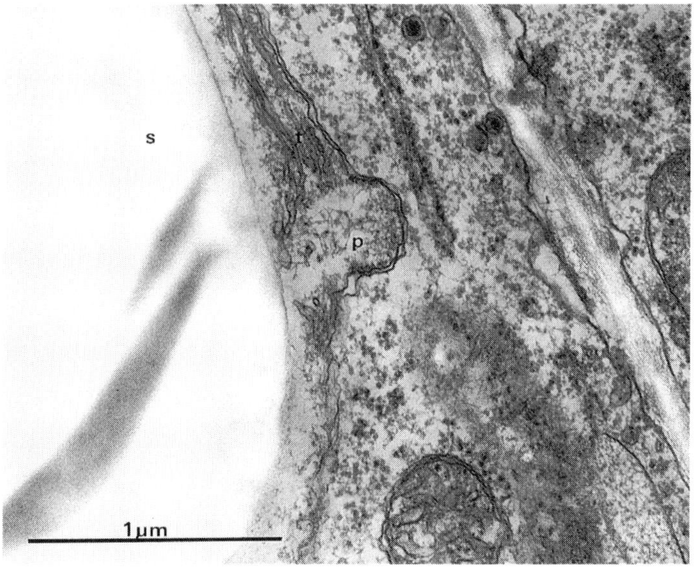

Fig. 14.4. Part of an A-type amyloplast in an inner starchy endosperm cell showing starch granule (s), protuberance (p), and tubuli (t). Caryopsis as in Fig. 14.3.

of the starchy endosperm in wheat, plastid replication ceased before cell division. As a result, the last-formed cells, i.e. the outermost starchy endosperm cells, had fewer amyloplasts per cell than the inner endosperm cells. No comparable analysis of the development of the starchy endosperm in barley has been reported. Experiments designed to investigate the effect of temperature on starch accumulation indicate that the control of plastid replication is different from the control of cell division. Hoshikawa (1962) showed that endosperm cell number in wheat was independent of temperature in the range 20–30°C but the number of small granules was considerably less at the higher temperature of grain growth. Similarly, MacLeod & Duffus (1988a) were unable to detect any correlation between temperature and endosperm cell number in barley over the range 15–30°C but they found that grains from plants grown 2–3 days pre-anthesis at 30°C (day) and 25°C (night) accumulated less starch than grains grown at 20°C (day) and 15°C (night). The difference was attributable to reductions in the numbers rather than the sizes of both A-type and B-type granules. Using their own data and those from Gleadow *et al.* (1982), Chojecki *et al.* (1986b) have suggested that the A-type starch granule volume and the number of A-type granules per endosperm cell are major determinants of endosperm cell weight in wheat, and are under separate genetic control.

As the starchy endosperm develops, there is a steady increase in the volume of the A-type granules and the number and volume of the B-type granules, except in those cells lying next to the embryo. The contents of these cells disappear, and the walls are crushed to form what is known as the intermediate layer, the depleted layer or the crushed cell layer, of the mature grain. Early in embryo development, the cells which are crushed are apparently devoid of starch, but later the cells from which the contents are digested have well-formed starch granules. These starch granules are apparently hydrolysed by α-amylase 2 which is present in this part of the endosperm during development, rather than by α-amylase 1 which is characteristic of the pericarp and the embryo (MacGregor & Dushnicky, 1989b). The presence in mature grains of some granules with holes characteristic of amylolytic attack does not necessarily indicate that preharvest sprouting has taken place. It is possible that in the final stages of grain maturation, a similar process of dissolution of cell contents and crushing of cell wall material takes place in the starchy endosperm cells at the crease, during the formation of the sheaf which lies between the two flanks of the starchy endosperm in the dry grain.

The other main type of storage material accumulated in the starchy endosperm is protein. Unlike starch, it is not formed or stored in a self-replicating organelle. It is detected in the differentiating starchy endosperm cells of young grains as small uniform deposits in very large vacuoles (Cameron-Mills & von Wettstein, 1980). In older cells nearer the centre of the endosperm, protein aggregates are heterogeneous in size and content, and are surrounded by one or several layers of membranes (Fig. 14.5). During grain maturation and dehydration the integrity of these membranes is apparently lost, and the storage protein forms a

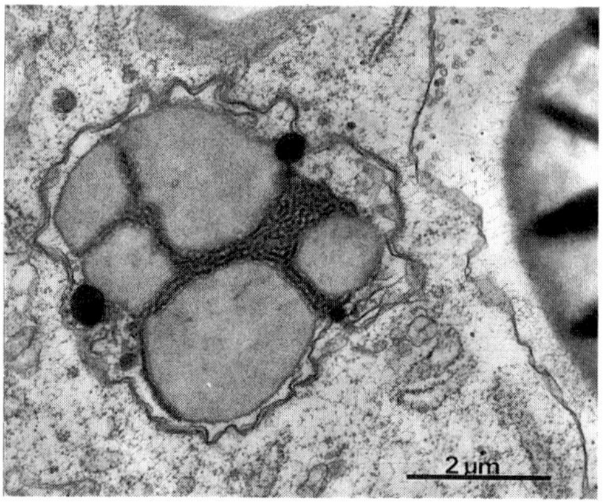

Fig. 14.5. Membrane-bound heterogeneous deposit of protein in an inner starchy endosperm cell of a caryopsis of cv. Midas fixed in glutaraldehyde 25 days after anthesis in a grain development period of 60 days.

matrix in which amyloplasts are embedded. In the absence of any evidence of morphogenetic continuity between these different forms of protein deposits and their enclosing membranes, it is perhaps misleading to use the term 'protein body' in relation to storage protein deposition in the starchy endosperm.

Aleurone

In the mature barley grain the starchy endosperm is surrounded, except at the crease, by a layer of aleurone tissue, typically two to three cells thick, but up to six cells thick in some cultivars. The differentiation of these cells takes place in two stages. Until cell division in the peripheral layers ceases, the cells which are brick-shaped, remain relatively thin-walled, and their numerous vacuoles contain only small electron-dense globoids. Lipid droplets are present and also plastids which may contain small starch granules. When cell division ceases, the walls of the outer cell layers thicken, thus clearly distinguishing the aleurone from the subaleurone (Hoshikawa, 1984). A second type of inclusion, identified by Jacobsen *et al.* (1971) as a protein–carbohydrate complex, appears in the vacuoles, in addition to more of the phytin-containing globoids. These inclusions become embedded in the protein matrix which eventually fills each vacuole or aleurone grain. Lipid droplets increase in number, surrounding the nucleus and each aleurone grain, and also lining the inside of the plasma membrane. The dense cytoplasm contains numerous mitochondria and some plastids. Storage protein deposition in the aleurone differs from that in the starchy endosperm in two respects. It takes place at a later stage in grain development, and it occurs in aleurone grains which are vacuolar in origin and also retain their integrity from the early stages of aleurone differentiation through seed maturation and dehydration, and into the first stages of germination.

The DNA content of aleurone nuclei exhibits considerable heterogeneity throughout grain development. Keown *et al.* (1977) found that the mean value doubled from the beginning of aleurone differentiation to grain maturity, and that at all stages a considerable proportion of the cells had more than the triploid ($3n$) amount of DNA. The cell walls of mature aleurone vary in thickness from 2 to 5 μm and consist of a fibrillar secondary wall and a less-structured primary wall. The walls are composed mainly of heteroxylans and 1,3:1,4-β-glucans, and they also contain phenolic acids (Bacic & Stone, 1981a, b). Deposits of $\beta(1\rightarrow3)$-D-glucan (callose) are found in the walls at the aleurone–subaleurone interface (Fulcher *et al.*, 1977; Bacic & Stone, 1981b).

In addition to the 'normal aleurone' described above, 'groove aleurone' is found on the ventral side at the crease, and 'germ aleurone' lies between the embryo and the testa. The germ aleurone in the mature grain is a single layer of thick-walled, somewhat flattened cells. It adheres tightly to the adaxial (concave) surface of the scutellum (Negbi, 1984) but is not present over the micropylar end of the coleorhiza. Groove aleurone cells do not pass through a meristematic phase and, early in development, they are irregular in shape, thick-walled and large in comparison with normal aleurone. Their failure to divide may be responsible for the formation of the groove or crease in the mature grain (Evers, 1970) but their other unusual characteristics are probably related to the role of the crease in the transport of water and assimilates. These are described below.

Embryo

Following fertilization, the first division in the zygote is later than the first division of the primary endosperm nucleus and, as the mean nuclear doubling time in the embryo is, for the first few days, greater than that in the endosperm, and cell division is not accompanied by cell growth (Bennett *et al.*, 1975), the development of the embryo lags behind that of the endosperm. Early in development, part of the embryo is recognized as the suspensor which attaches the embryo to the nucellus, but, at the 20-celled stage the embryo lies free in the endosperm and the suspensor is no longer functional (Norstog, 1972). It may be, however, that the suspensor is the morphological precursor of the embryonic appendage referred to by Collins (1918). Symons *et al.* (1984) have described a similar group of thin-walled cells at the tip of the coleorhiza in wheat, and they have attributed to this coleorhizal papilla the function of the splitting of the seed coat during germination.

The morphology of the developing embryo has been described in detail by Merry (1941). He observed the differentiation of the scutellum as early as 8 days after fertilization, and root and leaf primordia were identified about 2 days later. As root primordia develop they become enclosed in the coleorhiza, and, similarly, the leaf primordia and shoot apex are surrounded by the coleoptile. All are joined to the scutellum at the central nodal region of the embryo. The abaxial surface of the scutellum is made up of the columnar cells of the scutellar epithelium which become closely associated with the crushed cell walls of the intermediate layer of the starchy endosperm. Vascular strands in the scutellum extend from the apex to the node but remain in the provascular state until about 18 h after the start of imbibition (MacLeod & Palmer, 1966). The cell walls of the scutellum strongly resemble those of the aleurone, being composed mainly of hemicelluloses and exhibiting a strong autofluorescence attributable to the phenolic compound, ferulic acid (Smart & O'Brien, 1979). Analyses of embryo growth in cv. Julia have shown that the maximum dry weight of the embryo is not reached until three-quarters of the way through grain development. After this, the total carbohydrate content falls and then the protein content reaches a maximum (Duffus & Rosie, 1975). During embryo growth, numerous plastids contain small starch granules, but at harvest-ripeness very little starch is present in the embryo. Storage protein accumulates in single membrane-bound organelles similar to aleurone grains but having only the globoid, phytin-containing type of inclusion. The other main storage material accumulated in the embryo is lipid, numerous droplets being associated with protein deposits and the plasma membrane as in aleurone cells. By 24 h after the start of imbibition the amount of lipid is noticeably reduced and the numerous plastids in the cells of the scutellum each contain several small starch granules (Palmer, 1989).

The crease region

The vascular tissue of the developing grain is embedded in a wedge of chlorenchymatous cells along the ventral side of the pericarp. It appears in transverse section as an arc of phloem outside a group of xylem vessels and their accompanying xylem parenchyma cells. As grain-filling proceeds, the flanks of the grain bulge out on the ventral side, leaving the intervening tissues in a hollow known as the crease, furrow or groove. Along the length of the grain in this crease region the testa is discontinuous, and the gap between the ends of the testa is occupied by the chalaza. Radially elongated cells with few vacuoles separate the xylem from the chalaza. Inside the testa of the immature grain, the nucellus is present as a single layer, the nucellar epidermis, but inside the chalazal tissue, early in development, the nucellar tissue proliferates to form the nucellar projection which extends into the endosperm cavity.

The endosperm cells, i.e. the groove aleurone cells which lie opposite the nucellar projection, are unlike any of the other cells of the endosperm. Early in development they have irregularly shaped walls, massively thickened with what appears to be callose, and their dense cytoplasm is fragmented by a complex and extensive endomembrane system. Within a few days, the cytoplasm becomes less fragmented and the nature of the wall-thickening changes. The widely spaced cells, and the homogeneous material which separates them, form the lining of the dorsal side of the endosperm cavity during the grain-filling period. The endosperm cells inside the groove aleurone have wall ingrowths characteristic of transfer cells (Cochrane & Duffus, 1980; Hoshikawa, 1983). Transfer cells are also found in the nucellar projection (Cochrane & Duffus, 1980; Felker *et al.*, 1984a). As in the groove aleurone, one to two layers of thick-walled cells separate these transfer cells from the endosperm cavity and, in very young caryopses, the walls of these cells are thickened with callose.

The morphology of the tissues of the crease region thus appears to make it possible for assimilate to move from the phloem to the transfer cells of the nucellar projection in the symplast, and then enter the apoplast. The apoplastic pathway would be through firstly, the thick walls of the innermost cells of the nucellar projection, then the endosperm cavity containing mucilaginous material (Krauss, 1933; Cochrane & Duffus, 1980), the thick walls of the groove aleurone and finally, the wall projections of the endosperm transfer cells from which assimilate could move into the symplast of the endosperm. There are numerous plasmodesmata between the cells of the endosperm but the proportion of assimilate which moves through the endosperm in the symplast has not been established. The temporary deposition of comparatively large amounts of callose in the tissues next to the endosperm cavity may be a means of stabilizing morphology during a period of rapid growth, and it may also be a means of storing carbohydrate until the starch-synthesizing system of the endosperm is fully functional.

In the final stages of grain maturation, dry matter accumulation ceases and the water content of the grain decreases rapidly. Dehydration does in fact begin before maximum dry matter is attained, and this is apparently made possible by changes which take place in the cytochemistry and morphology of the chalaza and nucellar projection towards the end of the grain-filling period (Cochrane, 1983). About 35 days after anthesis in a developmental time scale of 60 days, the primary cell wall of each chalazal cell becomes lignified. This lignified layer is separated from the plasma membrane by the deposition of a layer of suberin, inside which a thick and irregular layer of wall material is laid down. At this stage of development, plasmodesmata still provide symplastic connections within the chalaza and between the chalaza and the nucellar projection.

The cell wall modifications in the chalaza have been interpreted as providing a route for the movement of water, but not sucrose, out of the endosperm in the lignified zone, while the movement of water from the lignified zone into the cytoplasm of the chalazal cells is prevented by the layer of suberin. Thus, symplastic movement of assimilate into the nucellar projection can take place while water is being lost from the endosperm. From about 15 days after anthesis, increasing amounts of phenolic material can be detected in the vacuoles of the chalazal cells (Fig. 14.6). The amount of phenolic material increases, and eventually by 50 days after anthesis, phenolic material appears to occupy most of the lumen of the somewhat distorted cells, and only a few organelles remain intact (Cochrane, 1983; Felker *et al.*, 1984a; Lingle & Chevalier, 1985). Experiments in which detached ears were supplied with [U-^{14}C] sucrose indicated that in grains at this stage of development, some assimilate was reaching the endosperm (Cochrane, 1983, 1985). By 55 days, however, disorganization of chalazal tissue was complete and uptake of sucrose had ceased. In the *seg* 1 shrunken endosperm mutant of cv. Betzes, Felker *et al.* (1984a) found that the phenolic material was not confined to the central vacuole of the chalazal cells and so caused disorganization of the cytoplasm at an early stage of grain-filling. The symplastic pathway for assimilate movement was blocked and starch synthesis ceased. The duration of dry matter accumulation was thus determined by a developmental defect in maternal tissues. It has yet to be established whether, during normal grain development, the embryo and endosperm exert any influence on the timing of morphological

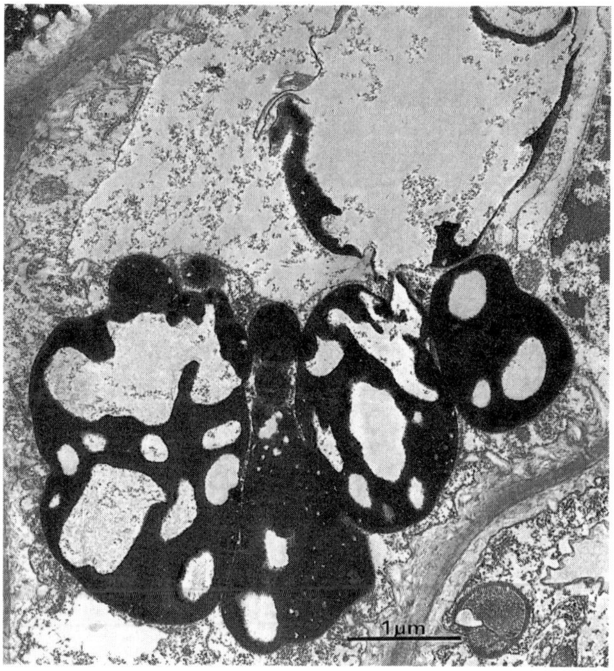

Fig. 14.6. Tannin-containing vacuoles in a chalazal cell in a transverse section cut at mid-grain from a caryopsis of cv. Midas fixed in glutaraldehyde 42 days after anthesis in a grain development period of 60 days. The tissue was not post-fixed in osmium tetroxide.

changes in maternal tissues, or whether the duration of grain-filling is entirely under the control of the maternal plant.

Throughout most of grain development, the amount of water used to transport the assimilate needed for starch and protein synthesis is much greater than that accumulating in the grain. It is not yet known how the water content of the developing grain is controlled but Jenner *et al.* (1988) have recently reported evidence, obtained from nuclear magnetic resonance micro-imaging, of water flow in certain localized regions of the wheat grain. Prior to the final phase of water loss, the cells of the crease aleurone and some neighbouring cells become crushed to form a sheaf of cell wall material stretching at mid-grain from the chalaza to the centre of the starchy endosperm where it terminates in a delta-like structure. A broad apoplastic route for water loss thus exists in the endosperm, and some pericarp cells at the crease which might be capable of bringing about active water loss have been identified (Cochrane, 1983). Grain growth is more or less synchronous along the ear, but dehydration proceeds from the apex to the base of the ear and appears to be accompanied by cytochemical changes in the vascular tissue of the rachis (Cochrane, 1985). Further investigations are needed to establish whether this is another way in which maternal tissues control grain maturation.

Grain composition

Nutrient supply

The identity and concentration of the nutrients entering the developing endosperm tissue are so far unknown. The main carbon source for grain biosynthetic processes is generally believed to be sucrose (Jenner, 1982; Duffus, in press), the evidence for this observation being partly based initially on results which demonstrated that sucrose is the major transport material in higher plants (Porter, 1962). A firmer indication of which nutrients are available to the developing grain has come from analysis of peduncle sieve tube contents (Fisher, 1987) from wheat plants during grain filling. The total contributions of sucrose, amino acids and potassium accounted for 75–90% of the total exudate osmolarity, with sucrose always accounting for most of the solutes present. Glutamine was by far the most abundant amino acid with alanine, asparagine, serine, threonine and valine next in importance. It may not be reasonable to assume that the nutrients entering the wheat grain are the same as those entering the barley grain. Furthermore, the composition of the nutrient supply entering the developing endosperm cells is not necessarily the same as that found in the phloem since the route through the crease and nucellar projection may involve some metabolic rearrangement. Radiolabelling experiments with developing barley cv. Betzes have shown, however, that sucrose is not metabolized during transfer from the phloem to the developing endosperm

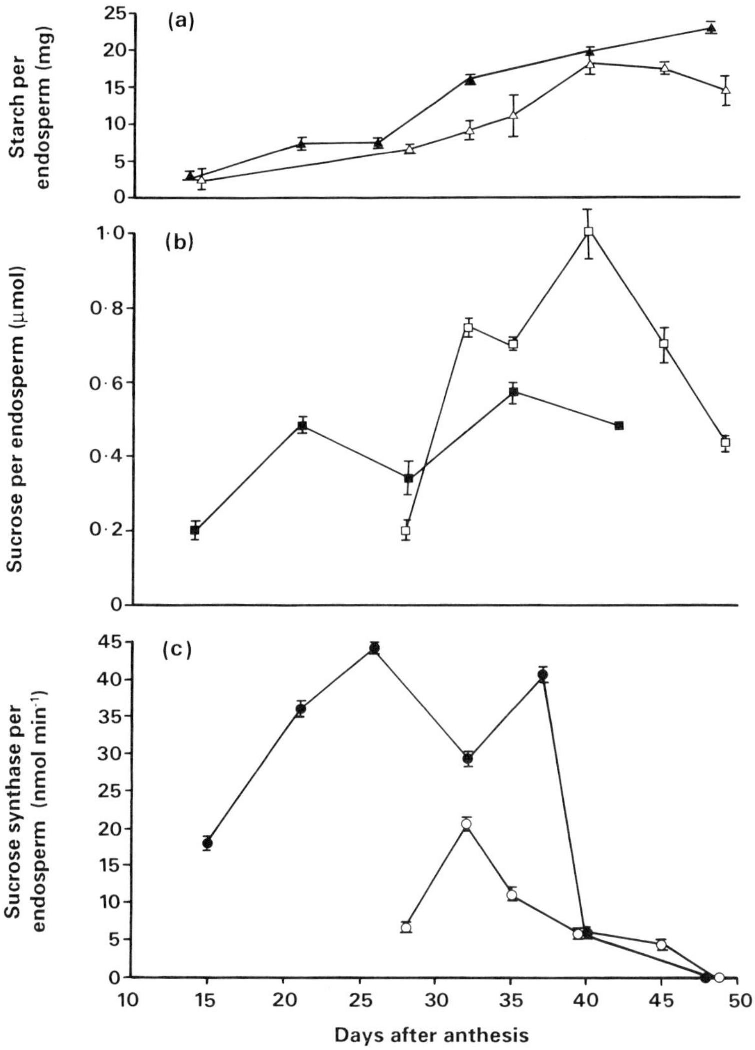

Fig. 14.7. Changes in (a) starch content, (b) sucrose content and (c) sucrose synthase activity of endosperms during grain development at 20/15°C and 30/25°C in cv. Triumph. ▲, ■, ●, 20/15°C; △,□,○, 30/25°C. ▲, △, starch; ■, □, sucrose; ●, ○, sucrose synthase activity. Bar markers indicate standard error of the mean (MacLeod & Duffus, 1988b).

(Felker *et al.*, 1984b). In these experiments, [fructose-U-^{14}C]sucrose was supplied to barley spikes, 14 days after anthesis. Most of the radioactivity (over 96%) recovered from the endosperms at all sampling times up to 3 h was in the sucrose fraction with no apparent increase in the percentage of labelled hexoses. There was no initial or progressive randomization of label among hexose moieties of sucrose as compared with the free space sampled after one hour of incubation. These results suggest that hydrolysis of sucrose is not required for uptake into the developing barley endosperm. It was also shown that sucrose was rapidly absorbed by the endosperm without first accumulating in the endosperm cavity free space. Similar results to those described by Felker *et al.* (1984b) for barley had earlier been reported for wheat (Jenner, 1973).

Starch is the principal component of yield in barley and contributes between 60 and 75% of final grain dry weight. The major source of carbon for starch biosynthetic is considered to be sucrose, although some contribution may be made from carbon skeletons derived from amino acids not used in protein synthesis. Sucrose levels appear to be maintained throughout grain development and there is much evidence to suggest that sucrose supply does not limit dry matter accumulation or starch deposition. Thus, experiments in which detached ears of barley aged between 22 and 56 days after anthesis were supplied with [U-^{14}C]sucrose for 4 h, showed that radioactivity moved into the grains at all stages of development up to the final dehydration stage (Cochrane, 1985). Since sucrose continued to reach the endosperm after the end of grain filling it was thought unlikely that lack of assimilate was responsible for the cessation of starch synthesis in the barley endosperm. Sucrose has been found to be the predominant free sugar in all tissues of the developing barley grain with very little free glucose or fructose present (Chevalier & Lingle, 1983).

Further evidence that the rate and duration of starch synthesis in barley endosperm are limited by biosynthetic processes within the endosperm rather than by the availability of sucrose comes from work described by Lingle & Chevalier (1984). In these experiments, [U-^{14}C]sucrose was introduced to specific kernels through the awn of the subtending lemma during the linear phase of dry matter accumulation which extended from 8 to 24 days after heading. Since significant amounts of label were present in the kernel, including endosperm and pericarp tissues, on day 33 and on subsequent dates, it was concluded that starch synthesis was not limited by sucrose supply.

That sucrose supply may not be limiting even under environmental conditions in which starch deposition is reduced, has been shown by MacLeod & Duffus (1988b). In endosperms from whole plants grown at 30/25°C, starch accumulation was less than in endosperms from plants grown at 20/15°C (Fig. 14.7). Since sucrose concentrations in the endosperms grown at the higher temperature were not reduced, but were, in fact, increased, it was concluded that the reduction in starch deposition was not a result of a decrease in the supply of sucrose.

Starch biosynthesis

Sucrose metabolism

Most of the sucrose entering the developing barley endosperm is thought to be used in starch biosynthesis. Some will enter the pathways of respiratory metabolism to supply energy for grain biosynthesis reactions. A further portion will be used in the biosynthesis of non-starch polysaccharides, lipids and even amino acids. Incoming sucrose is thought to be first converted to UDP-glucose and fructose in the developing barley endosperm by UDP-dependent sucrose synthase. Invertase activity was found to be low and its presence may well be due to contaminating maternal tissues in the endosperm extracts (Baxter & Duffus, 1973a; Chevalier & Lingle, 1983). The involvement of sucrose synthase in regulating starch deposition through its effect on sucrose metabolism is borne out by the effect on its activity of elevated temperatures during grain growth. When endosperms developed at 30/25°C, sucrose concentration was greater and sucrose synthase activity and starch deposition were less than in endosperms which developed at 20/15°C (Fig. 14.7) (MacLeod & Duffus, 1988b). This suggests that the increased sucrose levels were a result of decreased sucrose synthase activity and that starch deposition was reduced as a result of the inhibition of sucrose utilization.

Since it appears that the greater part of α-(1 → 4)-D-glucan synthesis is catalysed by an ADP-glucose-dependent starch syn-

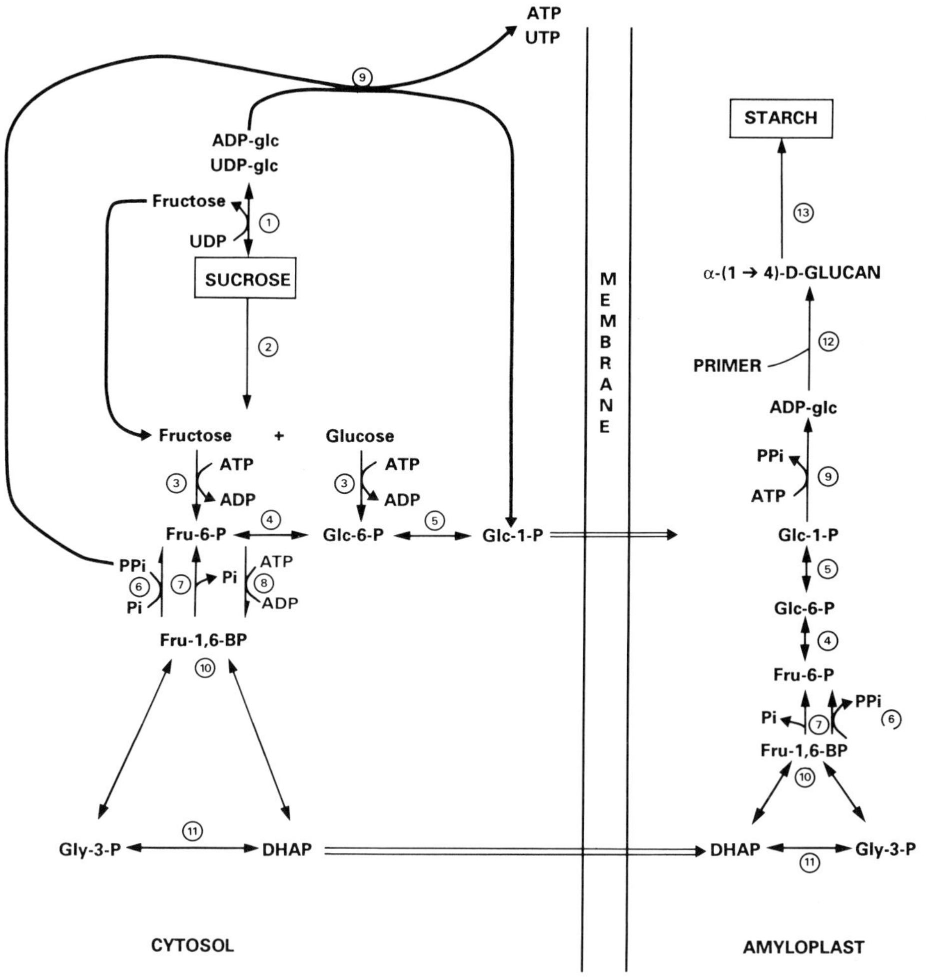

Fig. 14.8. Hypothetical scheme for conversion of sucrose to starch in developing amyloplasts. 1, sucrose synthase; 2, invertase; 3, hexokinase; 4, phosphohexoisomerase; 5, phosphoglucomutase; 6, PP_i-dependent phosphofructokinase; 7, fructose-1,6-bisphosphatase; 8, ATP-dependent phosphofructokinase; 9, ADP-glucose and UDP-glucose pyrophosphorylases; 10, aldolase; 11, triosephosphateisomerase; 12, starch synthase; 13, branching enzyme. UDP-glc, UDP-glucose; ADP-glc, ADP-glucose; Glc-6-P, glucose-6-phosphate; Glc-1-P, glucose-1-phosphate; Fru-1-P, fructose-1-phosphate; Fru-6-P, fructose-6-phosphate; Fru-1,6-BP, fructose-1,6-bisphosphate; Gly-3-P, glyceraldehyde-3-phosphate; DHAP, dihydroxyacetone phosphate; PP_i, pyrophosphate; P_i, inorganic phosphate.

thase it has been suggested that the UDP-glucose formed from the sucrose synthase reaction, together with any hexose phosphates derived from fructose, are first converted to glucose-1-phosphate and then to ADP-glucose via the enzyme ADP-glucose pyrophosphorylase (Baxter & Duffus, 1973a). Starch could then be formed in a co-ordinated reaction involving starch synthase and branching enzyme. A debranching enzyme might also be required to produce primers for the starch synthase reaction.

The form in which carbon enters the developing amyloplast has not so far been identified with certainty and thus the biochemical pathway whereby sucrose is converted to starch is not fully understood. One reason that progress has been slow is the difficulty of purifying intact, active amyloplasts in good yield for transport studies. However, Keeling *et al.* (1988) have shown that it is not essential to prepare amyloplasts in order to solve this problem. Thus, by supplying ^{13}C-labelled hexoses *in vivo* to developing wheat grains via a nick in the stem of intact wheat plants, they showed that a hexose phosphate is the most likely candidate for entry to the amyloplast. Some of the enzyme-catalysed reactions that could be involved in the conversion of sucrose to starch in immature starchy endosperm cells are shown in Fig. 14.8.

Starch deposition

The amount of endosperm starch deposited is a function of both rate and duration of grain filling. In studies with two barley cultivars differing in final grain weight, it has been shown that while rates of deposition were similar, cv. Kym attained a higher final starch content than cv. Golden Promise because it had a longer grain-filling period (Renwick & Duffus, 1987).

The starch of mature normal barley endosperms contains about 25% amylose and 75% amylopectin. How such a constant composition is achieved remains a mystery. There is a steady increase in the amylose content of total starch over the developmental period (Banks *et al.*, 1973) and so it seems likely that amylopectin synthesis predominates at early stages of endosperm development. Results from wheat (Morrison & Gadan, 1987) and barley (Kang *et al.*, 1985; McDonald *et al.*, 1991) suggest that the A-type granules have a slightly higher amylose content than the B-type granules and so the maintenance of a constant starch composition in the mature grain indicates that there is likely to be tight control of the relative sizes and/or numbers of each type present.

Lipid content of barley starches increases markedly with increase in ambient temperature during the grain-filling period (Tester *et al.*, 1991). The significance of this observation is unclear, but higher lipid content reduces the swelling potential of the starches, thus affecting food processing characteristics.

Some of the technological properties of barley starches may be correlated with granule size and composition. Thus, during mashing, the small barley starch granules, unlike the large granules, are resistant to swelling and as a consequence are less completely converted to maltose (Slack *et al.*, 1979). It was suggested that association of the small granules with hordein is responsible for their resistance to amylolysis. One objective of any barley improvement programme should be to determine the factors concerned in the development of starch with known and predictable technological characteristics. Unfortunately, the relationship between structure and function in the starches is poorly understood.

Termination of starch deposition

The processes that may be involved in the termination of starch synthesis and deposition have been investigated in developing wheat endosperms by Caley *et al.* (1990). One possibility is that water loss may trigger cessation of dry matter accumulation. An *in vitro* system was set up with detached ears in liquid culture in which water loss in the developing endosperms was induced by including polyethylene glycol in the culture medium. Starch deposition was reduced in these endosperms compared with controls with no polyethylene glycol. The results showed that starch synthase activity was not affected by the water deficit and that activity was maintained until well after the termination of starch deposition. Hence, some factor other than starch synthase activity must be involved in the termination of starch deposition. Since ADP-glucose pyrophosphorylase activity was completely inhibited under conditions of reduced water content, it was suggested that this enzyme may have a key role in the termination of starch synthesis. The involvement of other enzyme systems in this process cannot of course be ruled out.

Protein composition and synthesis

The protein content of mature barley grains is typically between 9 and 13% of total grain dry weight. The precise composition depends on the growth conditions and on the rate and timing of nitrogen fertilizer application. For example, high temperatures or water shortage increase the percentage of protein in the grain (Russell, 1986). In general there is a linear relationship between the rate of nitrogen fertilizer applied to the soil and the protein content of the grain.

The proteins of barley may be classified into four solubility groups, each of which is highly heterogeneous (Miflin *et al.*, 1983) and whose characteristics can vary with cultivar and with agronomic practice. The albumins are soluble in water in neutral or slightly acidic conditions; globulins are soluble in salt solutions but insoluble in water; glutelins are soluble in strong acid or alkali but insoluble in water, salt solutions or ethanol; prolamins are extracted by propan-1-ol (50% by volume) or propan-2-ol (55% by volume). The relative proportions of these fractions depend very much on the extraction procedures employed.

The prolamins are the major storage proteins of barley and it is this fraction which appears to exert a key influence on the technological characteristics of the harvested grain. In barley this fraction is called hordein and accounts for around 50% of total grain protein. The name prolamin relates to the high contents present of glutamine and proline which in hordein constitute about 50% of the total amino acids (Doll, 1984). In general, barley protein has a low content of the indispensable (essential) amino acids which explains why non-ruminant farm animals cannot grow and reproduce satisfactorily on a complete barley diet without amino acid supplementation. The low content of indispensable amino acids in barley protein is a direct consequence of the high content of prolamin. Not only is the lysine content low (around 1%) but the contents of threonine, valine and isoleucine are also low.

While diets can be supplemented with lysine-rich oilseed meals, the adverse effects of the hordeins during the brewing process are more difficult to overcome. Indeed, the amount of extract that can ultimately be derived from malt is inversely related to the protein content of the original grain (Bathgate, 1987). Displacement of starch by protein is not thought to be the principal cause of this problem. Rather it is thought that the constitution of the protein matrix surrounding the starch granules determines the rate of conversion of starch to soluble extract. More specifically as we saw above, this may in part be due to the aggregation of the small starch granules with hordein, thus conferring resistance to amylolysis (Slack *et al.*, 1979). Endosperm modification (degradation) during malting may also depend on the packing of small starch granules and storage proteins (Palmer, 1989). Thus loosely packed 'mealy' or 'low nitrogen' endosperms may modify more easily than 'steely' endosperms which have compact matrix areas of protein and small starch granules. A good malting barley will thus have a low protein content. At the same time, however, it should also have a high capacity for the synthesis of starch-degrading enzymes during malting.

The biochemical pathways whereby

nitrogenous substances entering the developing endosperm may be converted to amino acid precursors for protein synthesis have received little attention. From a comparison of the amino acid composition of sieve tube contents (Fisher, 1987) with the amino acid composition of mature cereal protein, it seems likely that much rearrangement of amino acids takes place before their incorporation into grain storage protein. The presence of a number of amino acid metabolizing enzymes in the developing barley endosperm, testa/pericarp and embryo has been reported (Duffus & Rosie, 1978). These include glutamate–oxaloacetate transaminase and glutamate–pyruvate transaminase.

Since hordein exerts such a major influence on the technological characteristics of the mature barley grain, there has been much recent interest in the mechanisms controlling its synthesis and deposition. Hordein consists of three major groups of polypeptides called B, C and D hordeins which account for 75–90, 10–20 and 2–4% respectively of the total fraction (Kreis *et al.*, 1987). Hordein-related mRNA species are detectable in developing endosperm some days before hordeins can be detected *in vivo* (Rahman *et al.*, 1984).

Differential rates of accumulation of the B and C hordeins were reported (Fig. 14.9) and their rates of synthesis were shown to be related to the abundance of their respective

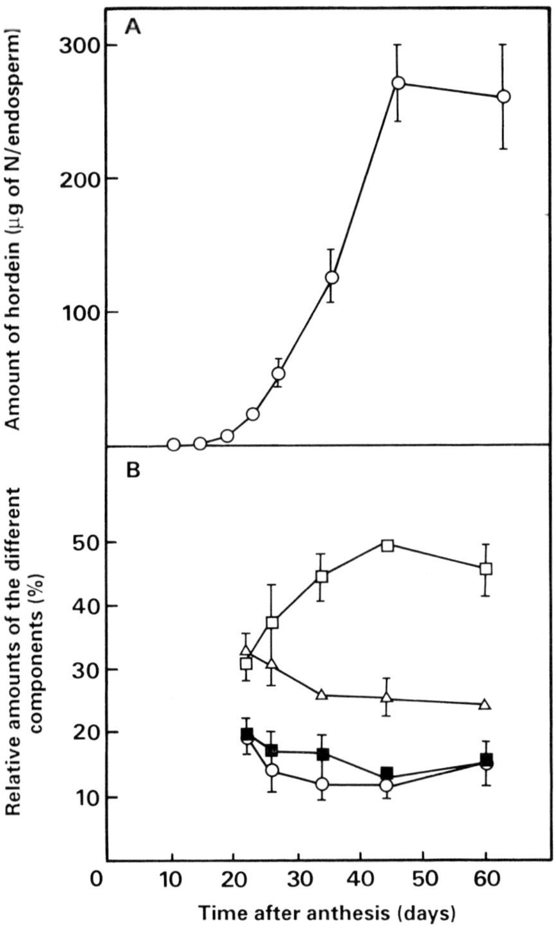

Fig. 14.9. Amount of hordein (A) and the proportions of the different components (B) at different stages of development. Endosperms were harvested from field-grown barley at various days after anthesis, and hordeins were extracted in propan-1-ol/2-mercaptoethanol/water (25:1:24, by vol). The amount of hordein (○) was determined by Kjeldahl analysis. The results are the means of four determinations. The hordeins were separated by SDS-polyacrylamide-gel electrophoresis and the relative amounts of the different B and C hordeins (□, B1; ■, B2; △, B3; ○, C1) were determined by gel scanning. Results are the mean of at least two scans of each sample from at least three gels. The error bars represent 2 S.D. The results are taken from the experiments described in detail by Rahman *et al.* (1982, 1984).

mRNA species. It was suggested that the amount of hordein mRNA present was likely to be controlled by the rate of hordein-gene transcription rather than the rate of hordein mRNA turnover. The transcriptional control of hordein gene expression has been further investigated by Sørensen *et al.* (1989). They showed that the hordein mRNAs increased three- to four-fold from 8 to 25 days after anthesis and thereafter decreased. The maximum levels of hordein mRNAs which are reached at 25 days after anthesis correlate well with the observed pattern of accumulation of the corresponding hordein polypeptides in the developing endosperm (Rahman *et al.*, 1982, 1984) (Fig. 14.9). However, the relative abundance of the B hordein mRNA later in endosperm development suggested that B hordein polypeptides are preferentially synthesized at later stages of endosperm development. From a comparison of the ratios of steady-state levels of mRNAs coding for the hordeins and several other proteins, with the ratios for the *in vitro* transcriptional activities of the corresponding genes, it was suggested that the high levels of B- and C-hordein mRNAs are the result of post-transcriptional regulation.

The proportions of the different storage proteins present and their component polypeptides are markedly influenced by growth conditions. Thus, although as we saw above, nitrogen application increases yield, it also decreases quality since the increased nitrogen content of the grain is due mainly to the deposition of additional hordein (Giese *et al.*, 1983; Giese & Hopp, 1984; Corke & Atsmon, 1988). A deficiency of sulphur, relative to nitrogen was found to decrease the amount of S-rich B-hordeins and their mRNAs, relative to that of the S-poor C hordeins (Rahman *et al.*, 1983). The results suggested that there may be two components to the effect of sulphur deficiency on hordein accumulation, one acting at the level of transcription, or mRNA degradation, and one at the level of translation.

One of the objectives of the seed scientist is to consider how far seed protein may be modified in order to improve nutritive value and technological characteristics without loss of yield, resistance to disease or germinability. It has been suggested by Doll (1984) that improvements in protein quality that do not affect grain production may only be obtained by changing the amino acid composition of prolamin or by increasing the content of storage proteins with a better nutritional value. One possibility is that the large genetic variation in the composition of barley prolamin may represent some variation in lysine content of the constituent polypeptides. However, the structural storage protein genes consist of multigene families coding for several different polypeptides. To improve protein quality by altering the gene combinations coding for better polypeptides would not be easy. Attempts to improve protein nutritional quality by replacing prolamin by more lysine-rich proteins already present in the barley seed may be more promising. The barley endosperm contains a number of lysine-rich salt-soluble proteins: chymotrypsin inhibitors CI-1 and CI-2, β-amylase, protein Z (Hejgaard & Boisen, 1980) and a protein synthesis inhibitor (Hejgaard & Bjorn, 1985). These proteins may have some of the properties that can be used to replace prolamin as a storage protein. For example, the chymotrypsin inhibitor CI-2 has been shown to be deposited in the protein bodies of developing barley endosperm cells (Rasmussen *et al.*, 1990).

Non-starch polysaccharides

The non-starch polysaccharides found in mature barley grains include fructans, β-(1 → 4)-D-glucan (cellulose), β-(1 → 3; 1 → 4)-D-glucan (β-D-glucan), arabinoxylan and glucomannan. Over 96% of total grain cellulose is present in the husk (MacLeod & Napier, 1959) with very little being present in the endosperm cell walls (Fincher & Stone, 1986). The cell walls of the starchy endosperm contain around 70% β-D-glucan and 20% arabinoxylan, whereas the aleurone cell walls contain only about 26% β-D-glucan and 67% arabinoxylan. The β-D-glucans of barley have long been a problem facing both the malting and brewing industries and the poultry feed manufacturers. In the former case, poor filtration caused by β-glucans is a major problem for brewers. In the latter case, high β-glucan content of broiler feed results in incomplete feed use, sticky droppings and even death. The β-D-glucans are linear molecules with around 30% β-(1 → 3) and 70% β-(1 → 4) linkages randomly dispersed, and are associated with firmly linked peptide sequences (Fleming & Kawakami, 1977; Forrest & Wainwright, 1977).

Changes in the content of β-D-glucans over the developmental period in intact barley kernels have been described by Aman *et al.* (1989). It was shown that β-D-glucan content at maturity depended on cultivar and growing conditions. The results suggested that β-D-glucan deposition is a relatively late event in grain development, since dry matter accumulation was nearly complete before the majority of β-D-glucan was synthesized.

Lipid synthesis and deposition

The lipid content of the mature barley grain is around 2–4% of which most is present in the embryo and aleurone. The major class of lipid is triglyceride with free fatty acids forming less than 2% of the total. Other constituents of the lipid fraction include diacylgalactosylglycerides, di-*O*-acylphospholipids and a unique class of lipids called the starch lipids. These are defined as those lipids which are present within the native starch granule (Morrison, 1988) and in barley are almost exclusively lysophospholipids. They are thought to exist as inclusion complexes of amylose. From experiments in which large, or A-type, granules were isolated at different stages of grain development, it has been shown that a linear relationship exists between granule amylose and lipid phosphorus content for two pairs of isogenic lines differing in amylose content (McDonald *et al.*, 1991). It was suggested from these results that A-type granules have a core of low amylose–low lipid starch and an outer shell of high amylose–high lipid starch. How these observations relate to the technological characteristics of starches is not fully understood. The possible effects of lipid on malting and brewing have been described by Palmer (1989). For example, it may be that lipid can reduce the susceptibility of starch granules to amylolytic digestion. There is also a possibility that products of lipid oxidation may contribute to the stale-flavour characters of beers.

Degradative processes during grain development

Intact immature grains respire, as confirmed by measurements of carbon dioxide evolution in wheat (Evans & Rawson, 1970) and oxygen uptake in barley (Nutbeam & Duffus, 1978). This information, together with evidence for the presence of glycolytic pentose phosphate pathway and mitochondrial enzyme activities in immature barley endosperm (Duffus & Rosie, 1977), suggests that the energy required to drive grain biosynthetic reactions is derived from normal respiratory processes.

As expected from observations on grain development, a number of enzymes implicated in the degradation of starch have been found in developing cereal endosperms. Of these, α-amylase has received particular attention since it is thought to be responsible for reduced grain quality arising from preharvest sprouting phenomena. This can result in economic losses in all field-grown cereal crops. The problem is probably greatest in wheat, where the high α-amylase activity of flour from sprouted grain is taken to be a cause of poor bread-making quality (Drapron & Godon, 1987). With barley, post-harvest problems arising as a result of pregermination are not strictly due to amylolytic activity but rather are related to the more complex physiological states of viability and dormancy. In barley, α-amylase activity is detectable in immature grain from a few days after anthesis. Activity reaches a maximum early in development and then declines to a low value that remains until maturity (Duffus, 1969). Much of the early activity has been shown to be present in the pericarp (MacGregor *et al.*, 1972) and it was suggested that the function of the enzyme is to hydrolyse the starch granules of the pericarp transparent layer to provide energy and raw material for the growing kernel.

More recently, Lindblom (1985) has examined the activities of starch-degrading enzymes in intact barley grains during development (Fig. 14.10). Starch-degrading activity was present in grains of four genotypes over the developmental period, considerable variation being observed between genotypes. Alpha-amylase activity was relatively low during the greater part of grain development but in all four cultivars it showed a marked increase at the time when

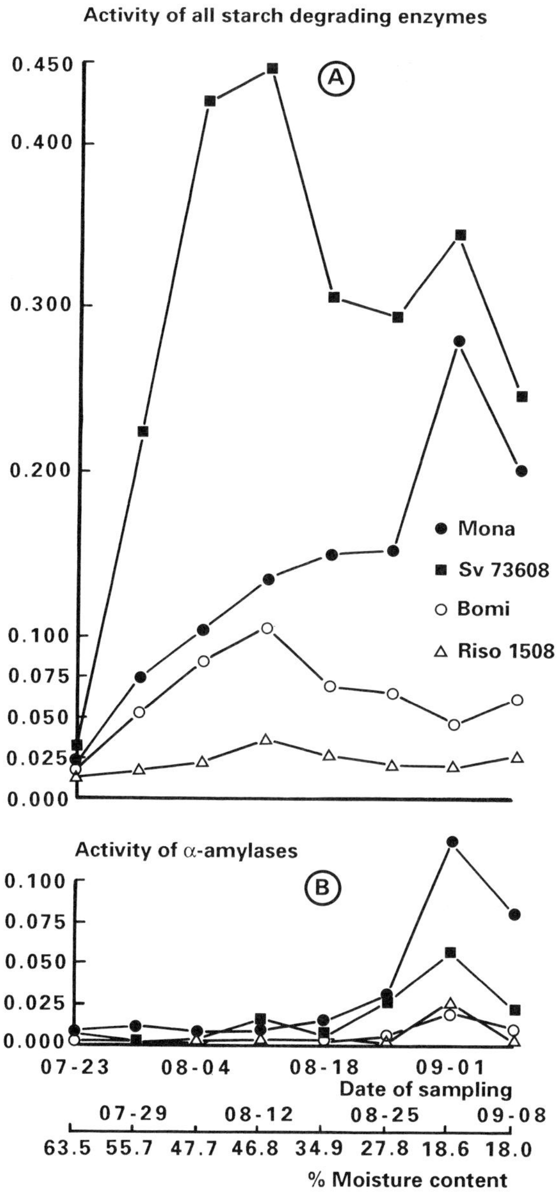

Fig. 14.10. Activities of starch-degrading enzymes in four barley genotypes at different stages of kernel development in 1980. All activities calculated on the same number of kernels. (A) All starch-degrading enzymes. (B) All α-amylases. Moisture content mean of all genotypes at date of sampling (Lindblom, 1985).

grain moisture content was falling to low levels. While increase in α-amylase activity did show some correlation with the appearance of sprouted grains this was not always the case. This activity may be equivalent to the prematurity production of α-amylase described by Gale *et al.* (1983) for wheat.

It has been shown (Robertson & Hill, 1989) that an α-amylase inhibitor accumulates during barley grain development. While its location was not reported, it may be that it functions in the control of starch degradation during endosperm development. Since total starch-degrading activity over the developmental period was shown to be much greater than that of α-amylase (Lindblom, 1985) it seems likely that some of the difference is made up by β-amylase activity (Duffus & Rosie, 1973). This is present in barley endosperm in both latent and available forms. Changes in free, bound and total β-amylase activities of intact barley kernels are shown in Fig. 14.11. From experiments where the latent form was released by incubating with protease, it was suggested that in endosperm, β-amylase is first synthesized in the soluble form, subsequently being converted to the latent form during development (Duffus & Rosie, 1973). Similar conclusions have been reached by Hara-Nishimura *et al.* (1986).

Further investigations of the β-amylases of intact immature barley grains by isoelectric focusing and chromatofocusing demonstrated the presence of eight major β-amylases (LaBerge & Marchylo, 1986). Using both enzymatic and immunochemical analysis it has been shown (Nishimura *et al.*, 1987) that the β-amylase of developing barley endosperm is not located in the protein body fraction but is found only in the cytosolic fraction. Other starch-degrading enzymes present in developing barley endosperm include starch phosphorylase (Baxter & Duffus, 1973b), α-glucosidase and debranching enzyme (Williams, 1976). The role of such hydrolytic enzymes during endosperm development is open to speculation. One possibility is that they provide a continuous source of primers for starch biosynthesis. They also appear to be involved in the degradation of endosperm cells next to the scutellum which in wheat (Smart & O'Brien, 1983) and more recently in barley (MacGregor & Dushnicky, 1989a) have been shown to be broken down during embryo growth and development. It may be then that the amylolytic activity detected in the developing endosperm is responsible for the mobilization of food reserves to be used as nutrients for the growing embryo.

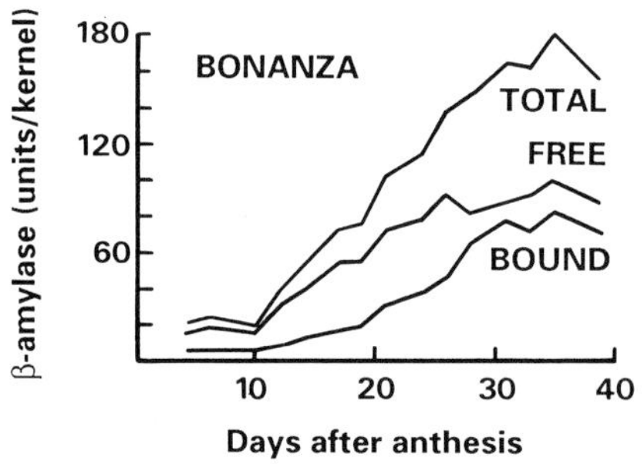

Fig. 14.11. Changes in free, bound and total β-amylase activities during kernal development and maturation of Bonanza barley (LaBerge & Marchylo, 1986).

References

Aman, P., Graham, H. & Tilly, A.-C. (1989) Content and solubility of mixed linked (1 → 3, 1 → 4)-β-D-glucan in barley and oats during kernel development and storage. *Journal of Cereal Science* 10, 45–50.

Bacic, A. & Stone, B.A. (1981a) Isolation and ultrastructure of aleurone cell walls from wheat and barley. *Australian Journal of Plant Physiology* 8, 453–74.

Bacic, A. & Stone, B.A. (1981b) Chemistry and organisation of aleurone cell wall components from wheat and barley. *Australian Journal of Plant Physiology* 8, 475–95.

Banks, W., Greenwood, C.T. & Muir, D.D. (1973) Studies on the biosynthesis of starch granules. 5. Properties of starch components of normal barley, and barley with starch of high amylose content during growth. *Die Stärke* 25, 153–7.

Bathgate, G.N. (1987) Quality requirement for malting. *Aspects of Applied Biology* 15, 18–31.

Baxter, E.D. & Duffus, C.M. (1973a) Enzymes of carbohydrate metabolism in developing *Hordeum distichum* grain. *Phytochemistry* 12, 1923–8.

Baxter, E.D. & Duffus, C.M. (1973b) Phosphorylase activity in relation to starch synthesis in developing *Hordeum distichum* grain. *Phytochemistry* 12, 2321–30.

Bennett, M.D., Smith, J.B. & Barclay, I. (1975) Early seed development in the *Triticeae*. *Philosophical Transactions of the Royal Society of London B* 272, 199–277.

Briarty, L.G., Hughes, C.E. & Evers, A.D. (1979) The developing endosperm of wheat – a stereological analysis. *Annals of Botany* 44, 641–58.

Brocklehurst, P.A. (1977) Factors controlling grain weight in wheat. *Nature* 266, 348–9.

Brocklehurst, P.A., Moss, J.P. & Williams, W. (1978) Effects of irradiance and water supply on grain development in wheat. *Annals of Applied Biology* 90, 265–76.

Buttrose, M.S. (1960) Sub-microscopic development and structure of starch granules in cereal endosperms. *Journal of Ultrastructure Research* 4, 231–57.

Caley, C.Y., Duffus, C.M. & Jeffcoat, B. (1990) The effects of elevated temperature and reduced water uptake on enzymes of starch synthesis in developing wheat grains. *Australian Journal of Plant Physiology* 17, 431–9.

Cameron-Mills, V. & von Wettstein, D. (1980) Protein body formation in the developing barley endosperm. *Carlsberg Research Communications* 45, 577–94.

Cass, D.D & Jensen, W.A. (1970) Fertilization in barley. *American Journal of Botany* 57, 62–79.

Chevalier, P. & Lingle, S.E. (1983) Sugar metabolism in developing kernels of wheat and barley. *Crop Science* 23, 272–7.

Chojecki, A.J.S., Gale, M.D. & Bayliss, M.W. (1986a) Cell production and DNA accumulation in the wheat endosperm, and their association with grain weight. *Annals of Botany* 59, 809–17.

Chojecki, A.J.S., Gale, M.D. & Bayliss, M.N. (1986b) The number and sizes of starch granules in the wheat endosperm and their association with grain weight. *Annals of Botany* 58, 819–31.

Cochrane, M.P. (1983) Morphology of the crease region in relation to assimilate uptake and water loss during caryopsis development in barley and wheat. *Australian Journal of Plant Physiology* 10, 473–91.

Cochrane, M.P. (1985) Assimilate supply and water loss in maturing barley grains. *Journal of Experimental Botany* 36, 770–82.

Cochrane, M.P. & Duffus, C.M. (1979) Morphology and ultrastructure of immature cereal grains in relation to transport. *Annals of Botany* 44, 67–72.

Cochrane, M.P. & Duffus, C.M. (1980) The nucellar projection and modified aleurone in the crease region of developing caryopses of barley (*Hordeum vulgare* L. var. *distichum*) *Protoplasma* 103, 361–75.

Cochrane, M.P. & Duffus, C.M. (1981) Endosperm cell number in barley. *Nature* 289, 399–401.

Cochrane, M.P. & Duffus, C.M. (1982) Opportunities for the regulation of grain development. In: Hawkins, A.F. & Jeffcoat, B. (eds), *Opportunities for Manipulation of Cereal Productivity*, Monograph 7. British Plant Growth Regulator Group, Wantage, pp. 167–78.

Cochrane, M.P. & Duffus, C.M. (1983a) Observations on the development of the testa and pericarp in barley. In: Kruger, J.E. & LaBerge, D.E. (eds), *Third International Symposium on Pre-Harvest Sprouting Damage in Cereals*. Westview Press, Boulder, CO, USA, pp. 154–61.

Cochrane, M.P. & Duffus, C.M. (1983b) Endosperm cell number in cultivars of barley differing in grain weight. *Annals of Applied Biology* 102, 177–81.

Collins, E.J. (1918) The structure of the integumentary system of the barley grain in relation to localized water absorption and semipermeability. *Annals of Botany* 32, 381–414.

Corke, H. & Atsmon, D. (1988) Effect of nitrogen nutrition on endosperm protein synthesis in wild and cultivated barley grown in spike culture. *Plant Physiology* 87, 523–8.

Cottrell, J.E., Dale, J.E. & Jeffcoat, B. (1982) Endogenous control of spikelet initiation and development in barley. In: Hawkins, A.F. & Jeffcoat, B. (eds), *Opportunities for Manipulation of Cereal Productivity*, Monograph 7. British Plant Growth Regulator Group, Wantage, pp. 130–9.

Doll, H. (1984) Nutritional aspects of cereal proteins and approaches to overcome their deficiencies. *Philosophical Transactions of the Royal Society of London B* 304, 373–80.

Drapron, R. & Godon, B. (1987) Role of enzymes in baking. In: Kruger, J.E., Lineback, D.

& Stauffer, C.E. (eds), *Enzymes and their Role in Cereal Technology*. American Association of Cereal Chemists, St Paul, Minnesota, USA, pp. 281–324.

Duffus, C.M. (1969) Alpha-amylase activity in the developing barley grain and its dependence on gibberellic acid. *Phytochemistry* 8, 1205–9.

Duffus, C.M. Control of grain growth and development. In: Marshall, C. & Grace, J. (eds), *Environmental Physiology and Ecology of Fruit and Seed Production*. Society for Experimental Biology, Seminar Series Symposium. Cambridge University Press, Cambridge (in press).

Duffus, C.M. & Rosie, R. (1973) Starch hydrolysing enzymes in the developing barley grain. *Planta* 109, 153–60.

Duffus, C.M. & Rosie, R. (1975) Biochemical changes during embryogeny in *Hordeum distichum*. *Phytochemistry* 14, 319–23.

Duffus, C. & Rosie, R. (1977) Carbohydrate oxidation in developing barley endosperm. *New Phytologist* 78, 391–5.

Duffus, C.M. & Rosie, R. (1978) Metabolism of ammonium ion and glutamate in relation to nitrogen supply and utilization during grain development in barley. *Plant Physiology* 61, 570–4.

Evans, L.T. & Rawson, L.T. (1970) Photosynthesis and respiration by the flag leaf and components of the ear during grain development in wheat. *Australian Journal of Biological Sciences* 23, 245–54.

Evers, A.D. (1970) Development of the endosperm of wheat. *Annals of Botany* 34, 547–55.

Evers, A.D. & Reed, M. (1988) Some novel observations by scanning electron microscopy on the seed coat and nucellus of the mature wheat grain. *Cereal Chemistry* 65, 81–5.

Felker, F.C., Peterson, D.M. & Nelson, O.E. (1984a) Development of tannin vacuoles in chalaza and seed coat of barley in relation to early chalazal necrosis in the *seg*1 mutant. *Planta* 161, 540–9.

Felker, F.C., Peterson, D.M. & Nelson, O.E. (1984b) [^{14}C]-sucrose uptake and labeling of starch in developing grains of normal and *seg*1 barley. *Plant Physiology* 74, 43–6.

Fincher, G.B. & Stone, B.A. (1986) Cell walls and their components in cereal grain technology. In: Pomeranz, Y. (ed.), *Recent Advances in Cereal Science and Technology*. American Association of Cereal Chemists, St Paul, Minnesota, USA, 8, 207–95.

Fisher, D.B. (1987) Changes in the concentration and composition of peduncle sieve tube sap during grain filling in normal and phosphate-deficient wheat plants. *Australian Journal of Plant Physiology* 14, 147–56.

Fleming, M. & Kawakami, K. (1977) Studies of the fine structure of β-D-glucans of barleys extracted at different temperatures. *Carbohydrate Research* 57, 15–23.

Forrest, I.S. & Wainwright, T. (1977) The mode of binding of β-glucans and pentosans in barley endosperm cell walls. *Journal of the Institute of Brewing* 83, 279–86.

Freeman, P.L. & Palmer, G.H. (1984) The structure of the pericarp and testa of barley. *Journal of the Institute of Brewing* 90, 88–94.

Fulcher, R.G., Setterfield, G., McCully, M.E. & Wood, P.J. (1977) Observations on the aleurone layer. II. Fluorescence microscopy of the aleurone–sub-aleurone junction with emphasis on possible β-1, 3-glucan deposits in barley. *Australian Journal of Plant Physiology* 4, 917–28.

Gaines, R.L., Bechtel, D.B. & Pomeranz, Y. (1985) A microscopic study on the development of a layer in barley that causes hull–caryopsis adherence. *Cereal Chemistry* 62, 35–40.

Gale, M.D., Flintham, J.E. & Arthur, E.D. (1983) Alpha-amylase production in the late stages of grain development – an early sprouting damage risk period? In: Kruger, J.E. & LaBerge, D.E. (eds), *Third International Symposium on Pre-Harvest Sprouting in Cereals*. Westview Press, Boulder, CO, USA, pp. 29–35.

Giese, H. & Hopp, E.H. (1984) Influence of nitrogen nutrition on the amount of hordein, protein Z and β-amylase messenger RNA in developing endosperm of barley. *Carlsberg Research Communications* 49, 365–83.

Giese, H., Anderson, D. & Doll, H. (1983) Synthesis of the major storage protein, hordein, in barley. Pulse labelling study of grain filling in liquid cultured detached spikes. *Planta* 159, 60–5.

Gleadow, R.M., Dalling, M.J. & Halloran, G.M. (1982) Variation in endosperm characteristics and nitrogen content in six wheat lines. *Australian Journal of Plant Physiology* 9, 539–51.

Hara-Nishimura, I., Nishimura, M. & Daussant, J. (1986) Conversion of free beta-amylase to bound beta-amylase on starch granules in the barley endosperm during dessication phase of seed development. *Protoplasma* 134, 149–53.

Hejgaard, J. & Bjorn, S.E. (1985) Four major basic proteins of barley grain. Purification and practical characterization. *Physiologia Plantarum* 64, 301–7.

Hejgaard, J. & Boisen, S. (1980) High lysine proteins in Hiproly barley breeding: identification, nutritional significance and new screening methods. *Hereditas* 93, 311–20.

Hoshikawa, K. (1962) Studies on the ripening of wheat grain. 4. Influence of temperature upon the development of the endosperm. *Proceedings of the Crop Science Society of Japan* 30, 228–31.

Hoshikawa, K. (1983) Development of endosperm tissue with special reference to the translocation of reserve substances in cereals. 1. Transfer cells in the endosperm of two-rowed barley (*Hordeum distichum* L. emend. LAM) *Japanese Journal of Crop Science* 52, 529–33.

Hoshikawa, K. (1984) Development of endosperm tissue with special reference to translocation of reserves in cereals. II. Modification of cell shape in the developing endosperm parenchyma, aleurone and sub-aleurone of two-rowed barley. *Japanese Journal of Crop Science* 53, 64–70.

Hughes, C.E. & Briarty, L.G. (1976) A quantitative analysis of some ultrastructural aspects of seed development. In: *Proceedings of the Fourth International Congress for Stereology*, National Bureau of Standards Special Publication 431, pp. 385–8.

Jacobsen, J.V., Knox, R.B. & Pyliotis, N.A. (1971) The structure and composition of aleurone grains in the barley aleurone layer. *Planta* 101, 189–209.

Jenner, C.F. (1973) The uptake of sucrose and its conversion to starch in detached ears of wheat. *Journal of Experimental Botany* 24, 295–306.

Jenner, C.F. (1982) Storage of Starch. In: Loewus, F.A. & Tanner, W. (eds), *Encyclopaedia of Plant Physiology, New Series (Plant Carbohydrates I)*, Vol. 13A, Springer-Verlag, Berlin, pp. 700–47.

Jenner, C.F., Xia, Y., Eccles, C.D. & Callaghan, P.T. (1988) Circulation of water within wheat grain revealed by nuclear magnetic resonance micro-imaging. *Nature* 336, 399–402.

Kang, M.Y. Sugimoto, Y., Sakamoto, S. & Fuwa, H. (1985) Developmental changes in the amylose content of endosperm starch of barley (*Hordeum vulgare* L.) during the grain filling period after anthesis. *Agricultural and Biological Chemistry* 49, 3463–6.

Keeling, P., Wood, J.R., Tyson, R.H. & Bridges, I.C. (1988) Starch biosynthesis in developing wheat grain. Evidence against the direct involvement of triose phosphates in the metabolic pathway. *Plant Physiology* 87, 311–19.

Keown, A.C., Taiz, L. & Russel, L.J. (1977) The nuclear content of developing barley aleurone cells. *American Journal of Botany* 64, 1248–53.

Krauss, L. (1933) Entwicklungsgeschichte der Fruchte von *Hordeum, Triticum, Bromus,* und *Poa* mit besonderer Berucksichtigung ihrer Samenschalen. *Jahrbücher für Wissenschaftliche Botanick* 77, 733–808.

Kreis, M., Williamson, M.S., Forde, J., Clarke, J., Buxton, B., Pywell, J., Marris, C., Gallois, P. & Shewry, P.R. (1987) Structure and regulation of expression of seed protein genes in barley. *Plant Physiology and Biochemistry* 25, 291–302.

Kvaale, E. & Olsen, O.A. (1986) Rates of cell division in developing barley endosperms. *Annals of Botany* 57, 829–33.

LaBerge, D.E. & Marchylo, B.A. (1986) Changes in beta-amylase enzymes during kernel development of barley and the effect of papain as an extractant. *Journal of American Society of Brewing Chemists* 44, 16–19.

Lindblom, H. (1985) Amylase studies in normal and high-lysine barleys. *Swedish Journal of Agricultural Research* 15, 145–51.

Lingle, S.E. & Chevalier, P. (1984) Movement and metabolism of sucrose in developing barley kernels. *Crop Science* 24, 315–19.

Lingle, S.E. & Chevalier, P. (1985) Development of the vascular tissue of the wheat and barley caryopsis as related to the rate and duration of grain filling. *Crop Science* 25, 123–8.

MacGregor, A.W. & Dushnicky, L. (1989a) Alpha-amylase in developing barley kernels – a reappraisal. *Journal of the Institute of Brewing* 95, 29–33.

MacGregor, A.W. & Dushnicky, L. (1989b) Starch degradation in endosperms of developing barley kernels. *Journal of the Institute of Brewing* 95, 321–5.

MacGregor, A.W., Gordon, A.G., Meredith, W.O.S & Lacroix, L. (1972) Site of alpha-amylase in developing barley kernels. *Journal of the Institute of Brewing* 78, 174–9.

MacLeod, A.M. & Napier, J.P. (1959) Cellulose distribution in barley. *Journal of the Institute of Brewing* 65, 188–96.

MacLeod, A.M. & Palmer, G.H. (1966) The embryo of barley in relation to modification of the endosperm. *Journal of the Institute of Brewing* 72, 580–9.

MacLeod, L.C. & Duffus, C.M. (1988a) Temperature effects on starch granules in developing barley grains. *Journal of Cereal Science* 8, 29–37.

MacLeod, L.C. & Duffus, C.M. (1988b) Reduced starch content and sucrose synthase activity in developing endosperms of barley plants grown at elevated temperatures. *Australian Journal of Plant Physiology* 15, 367–75.

McDonald, A.M.L., Stark, J.R., Morrison, W.R. & Ellis, R.P. (1991) The composition of starch granules from developing barley genotypes. *Journal of Cereal Science* 13, 93–112.

May, L.H. & Buttrose, M.S. (1959) Physiology of cereal grain. II. Starch granule formation in the developing barely kernel. *Australian Journal of Biological Sciences* 12, 146–59.

Merry, J. (1941) Studies on the embryo of *Hordeum sativum*. 1. The development of the embryo. *Bulletin of the Torrey Botanical Club* 68, 585–98.

Miflin, B.J., Field, J.M. & Shewry, P.R. (1983) Cereal storage proteins and their effect on technological properties. In: Daussant, J., Mosse, J. & Vaughan, J. (eds), *Seed Proteins*. Academic, London, pp. 255–319.

Morrison, W.R. (1988) Lipids in cereal starches: a review. *Journal of Cereal Science* 8, 1–15.

Morrison, W.R. & Gadan, H. (1987) The amylose and lipid contents of starch granules in developing wheat endosperms. *Journal of Cereal Science* 5, 263–75.

Negbi, M. (1984) The structure and function of the scutellum of the Gramineae. *Botanical Journal of the Linnean Society* 88, 205–22.

Nishimura, M., Hara-Nishimura, I., Bureau, D. & Daussant, J. (1987) Subcellular distribution of beta-amylase in developing barley endosperm. *Plant Science* 49, 117–22.

Norstog, K. (1972) Early development of the barley embryo: fine structure. *American Journal of Botany* 59, 123–32.

Norstog, K. (1974) Nucellus during early embryogeny in barley: fine structure. *Botanical Gazette* 135, 97–103.

Nutbeam, A.R. & Duffus, C.M. (1976) Evidence for C_4 photosynthesis in barley pericarp tissue. *Biochemical and Biophysical Research Communications* 70, 1198–203.

Nutbeam, A.R. & Duffus, C.M. (1978) Oxygen exchange in the pericarp green layer of immature cereal grains. *Plant Physiology* 62, 360–2.

Palmer, G.H. (1989) Cereals in malting and brewing. In: Palmer, G.H. (ed.), *Cereal Science and Technology*. Aberdeen University Press, Aberdeen, pp. 61–242.

Parker, M.L. (1985) The relationship between A-type and B-type starch granules in the developing endosperm of wheat. *Journal of Cereal Science* 3, 271–8.

Pope, M.N. (1937) The time factor in pollen tube growth and fertilization in barley. *Journal of Agricultural Research* 54, 525–9.

Porter, H.K. (1962) Synthesis of polysaccharides of higher plants. *Annual Review of Plant Physiology* 13, 303–28.

Rahman, S., Shewry, P.R. & Miflin, B.J. (1982) Differential protein accumulation during barley grain development. *Journal of Experimental Botany* 33, 717–28.

Rahman, S., Shewry, P.R., Forde, B.G., Kreis, M. & Miflin, B.J. (1983) Nutritional control of storage protein synthesis in developing grain of barley (*Hordeum vulgare* L.). *Planta* 159, 366–72.

Rahman, S., Kreis, M., Forde, B.G., Shewry, P.R. & Miflin, B.J. (1984) Hordein gene expression during development of the barley (*Hordeum vulgare*) endosperm. *Biochemical Journal* 223, 315–22.

Rasmussen, U., Munck, L. & Ullrich, S.E. (1990) Immunogold localization of chymotrypsin inhibitor-2, a lysine rich protein in developing barley endosperm. *Planta* 180, 272–7.

Renwick, F. & Duffus, C.M. (1987) Factors affecting dry weight accumulation in developing barley endosperm. *Physiologia Plantarum* 69, 141–6.

Robertson, M. & Hill, R.D. (1989) Accumulation of an endogenous alpha amylase inhibitor in barley during grain development. *Journal of Cereal Science* 9, 237–46.

Russell, G. (1986) Fertilisers and quality of wheat and barley. Proceedings No. 253. *The Fertiliser Society*, Cowcross Street, London EC1M 6BH. pp. 1–23.

Scott, W.R., Appleyard, M., Fellowes, G. & Kirby, E.J.M. (1983) Effect of genotype and position in the ear on carpel and grain growth and mature grain weight of spring barley. *Journal of Agricultural Science* 100, 383–92.

Slack, P.T., Baxter, E.D. & Wainwright, T. (1979) Inhibition by hordein of starch degradation. *Journal of the Institute of Brewing* 85, 112–14.

Smart, M.G. & O'Brien, T.P. (1979) Observations on the scutellum. II. Histochemistry and autofluorescence of the cell wall in mature grain and during germination of wheat, barley, oats and ryegrass. *Australian Journal of Botany* 27, 403–11.

Smart, M.G. & O'Brien, T.P. (1983) The development of the wheat embryo in relation to the neighbouring tissues. *Protoplasma* 114, 1–13.

Sørensen, M.B., Cameron-Mills, V. & Brandt, A. (1989) Transcriptional and post-transcriptional regulation of gene expression in developing barley endosperm. *Molecular and General Genetics* 217, 195–201.

Symons, S.J., Angold, R., Black, M. & Chapman, J.M. (1984) A possible mechanism for seed coat rupture by the embryo in germinating wheat. *Annals of Botany* 53, 765–72.

Tester, R.F., South, J.B., Morrison, W.R. & Ellis, R.P. (1991) The effects of ambient temperature during the grain-filling period on the composition and properties of starch from four barley genotypes. *Journal of Cereal Science* 13, 113–27.

Watson, P.A. & Duffus, C.M. (1988) Carbon dioxide fixation by detached cereal caryopses. *Plant Physiology* 87, 504–9.

Williams, J.M. (1976) 'The ultrastructure and biochemistry of the developing barley grain'. PhD Thesis, University of Edinburgh, Edinburgh, Scotland.

Williams, J.M. & Duffus, C.M. (1977) The development of endosperm amyloplasts during grain maturation in barley. *Journal of the Institute of Brewing* 84, 47–50.

15

The Control of Protein Synthesis in Developing Barley Seeds

Martin Kreis & Peter R. Shewry

The fertilization of the barley ovule gives rise to two tissues, the diploid embryo and triploid endosperm. The embryo differentiates further into the embryonic axis (which gives rise to the seedling) and a single cotyledon, the scutellum, which contains some storage reserves but also functions as a source of enzymes for digestion of the endosperm reserves during germination. These embryonic tissues together account for only about 3–5% of the final dry weight (Tallberg, 1982), the bulk of the grain being endosperm.

The endosperm also differentiates further, into the starchy endosperm and the aleurone. The former is the major storage tissue, and consists at maturity of dead cells distended by starch granules in a protein matrix. The aleurone is several layers of small cells surrounding the starchy endosperm, which contain some storage reserves but also remain alive during maturation and produce hydrolytic enzymes during germination.

The differentiation of these tissues during grain development is associated with different patterns of protein synthesis, making barley an excellent system to study the control of gene expression. In addition, the amount and composition of the grain proteins have important effects on nutritional and technological quality.

We have studied several groups of seed storage proteins and their genes which exhibit contrasting patterns of expression, and have potential or actual impact on the nutritional quality of the grain. These are the hordein storage proteins, chymotrypsin inhibitors CI-1 and CI-2, β-amylase and a 7S storage globulin.

Barley seed proteins and grain quality

Proteins account for about 10% of the dry weight of mature barley grain, and are usually classified into solubility groups according to the work of Osborne (1924). These are albumins (water-soluble), globulins (soluble in dilute saline), prolamins (soluble in alcohol/water mixtures) and glutelins (soluble only in dilute acid or alkali). The prolamins, called hordein in barley, are the major storage proteins. They include not only proteins which are readily soluble in aqueous alcohols (usually 50% v/v propan-1-ol), but also structurally related proteins which are only extracted in the presence of a reducing agent, due to the formation of inter-chain disulphide bonds. The albumins, globulins and glutelins consist predominantly of structural and metabolic proteins, although some storage proteins are also present (see later), and may have overlapping compositions. It is, therefore, more valid to classify barley seed proteins into only two broad groups according to function: storage and non-storage (i.e. structural and metabolic). Prolamins, the major storage proteins, account for between about 35 and 50% of the total grain nitrogen, according to the level of nitrogen fertilization (see Kirkman *et al.*, 1982).

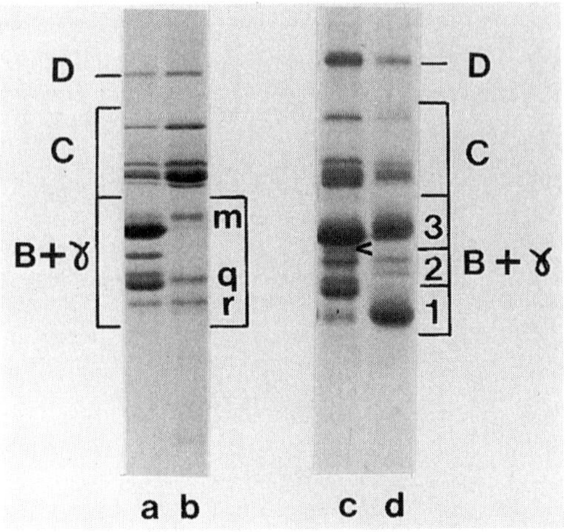

Fig. 15.1. SDS-PAGE of total hordein from: (a) cv. Carlsberg II; (b) Risø mutant 56; (c) line P12/3; (d) cv. Bomi. All the proteins present in the B + γ-hordein region of Risø 56 are γ-hordeins which are labelled according to Kreis *et al.* (1983b). 1,2 and 3 indicate the major B hordein bands present in cv. Bomi and related genotypes (e.g. cv. Sundance). The arrow in track (c) indicates a minor B hordein polypeptide encoded by the *Hor4* (*HrdG*) locus which is present in P12/3 and other lines derived from the Soviet cultivar Elgina. (From Shewry, in press.)

Hordein families and subfamilies

The hordeins are classified into four groups, or families, of polypeptides, called B, C, D and γ-hordeins (Fig. 15.1). Although these are all structurally related and presumably derived, at least in part, from the same ancestral protein, they do differ considerably in their structure and properties. The amino acid sequences of typical B, C and γ-hordeins are summarized in Fig. 15.2, which also shows the structure of a high-molecular-weight (HMW) subunit of wheat glutenin. The HMW subunits are homologous with D hordein, and have been studied in considerable detail (see Shewry *et al.*, 1989). All the prolamins contain repeated sequences, which in C hordein, γ-hordein and B hordein are based on a common motif. In C hordein these repeats probably account for about 95% of the protein, which appears to contain only a single sulphur-containing amino acid residue (methionine) and is present in the grain as monomer. B and γ-hordeins both have repetitive N-terminal domains accounting for between about a third and half of the protein, and non-repetitive C-terminal domains which contain most or all of the cysteine residues. Both probably occur in monomeric and polymeric (disulphide-stabilized) forms. B hordeins can also be classified into two subfamilies, which differ in their cyanogen bromide peptide maps and in the cross-hybridization behaviour of their cDNAs and genes (Kreis *et al.*, 1983a). These are called class I/II and class III polypeptides and correspond to the B1 and B3 bands respectively in the cultivars Sundance and Bomi (Fig. 15.1).

Less is known about the structure of D hordein, but it probably resembles the HMW subunits in consisting of a central repetitive domain based on several motifs (all of which are different to those in the other hordein groups), which is flanked by shorter non-repetitive domains.

These four groups of hordeins are synthesized at different levels. B hordein and C hordein account for about 70–80% and 10–20% respectively of the total, while D and γ-hordeins are only minor com-

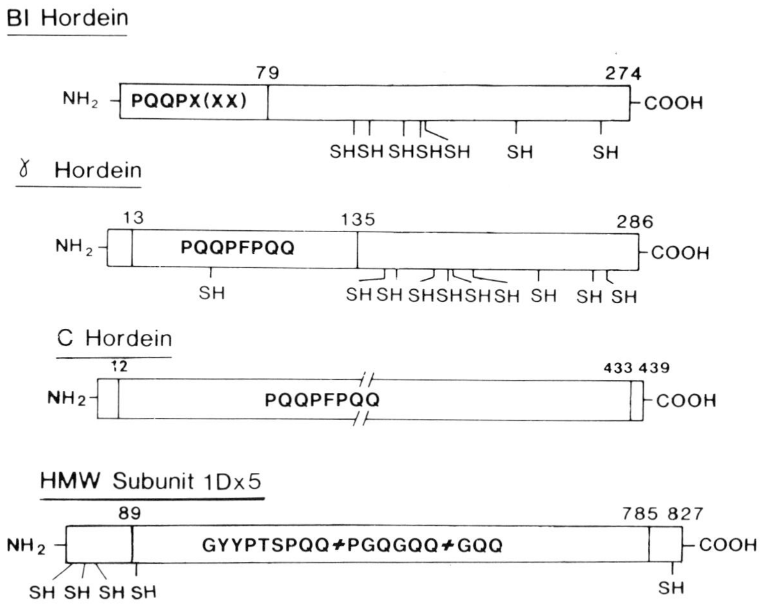

Fig. 15.2. Comparison of the amino acid sequences of the B1 hordein from cv. Sundance (Forde *et al.*, 1985), a γ-type hordein from cv. Carlsberg II (Cameron-Mills & Brandt, 1988), C hordein (see Shewry *et al.*, 1987a) and wheat HMW subunit 1Dx5 (Anderson *et al.*, 1989). All are based on the nucleotide sequences of genomic clones except for C hordein, which is a proposed structure based on data from a number of sources (see Shewry *et al.*, 1987a). SH indicates position of cysteine residue. Consensus repeat motifs are shown using standard single-letter abbreviations: G, glycine; F, phenylalanine; P, proline; Q, glutamine; S, serine; T, threonine; Y, tyrosine. X = variable.

ponents. All hordeins except D hordein are low in the essential amino acids lysine and threonine (Table 15.1), and are responsible for the poor nutritional quality of the whole grain.

Other seed proteins

CI-1 and CI-2 have M_rs of about 8800 and 9400 respectively, and contain about 9.5 and 11.5 g% lysine (Hejgaard & Boisen, 1980). Although they are present in small amounts in normal barley lines (with total activities of about 0.7 units/g grain), increased amounts (4.3–5.6 units/g) in Hiproly and derived lines contribute to the high lysine phenotype. They contain no cysteine residues and are present as monomers.

β-Amylase (which contains about 4.1 mol% lysine) (Kreis *et al.*, 1987) accounts for about 1% of the total grain proteins in most barley lines, but is also increased in Hiproly. It occurs in several isoforms with M_rs of between 54 000 and 59 700, and in free (salt-soluble) and bound (thiol-extractable) forms (see Shewry *et al.*, 1988a). The 7S globulin appears to be related to the vicilins of legumes and to α^1-globulins of cotton seeds, and is a holoprotein consisting of M_r 50 000, 40 000, 25 000 and 20 000 subunits (Yupsanis *et al.*, 1990).

Tissue distribution and subcellular location

The proteins just discussed show clear differences in their tissue specificity, as summarized in Table 15.2, and provide an excellent system to study the control of gene expression. They also differ in their subcellular

Table 15.1. Amino acid compositions (expressed as mol%) of the groups of hordein polypeptides.

	B	C	D	γ 'r'	γ 'm + q'
Asp[a]	1.4	1.0	1.3	2.9	2.8
Thr	2.1	1.0	8.1	3.1	2.1
Ser	4.7	4.6	9.7	5.5	5.7
Glu[b]	35.4	41.2	29.6	32.4	29.9
Pro	20.6	30.6	11.6	16.5	22.1
Gly	1.5	0.3	15.7	5.9	2.5
Ala	2.2	0.7	2.5	2.6	2.2
Cys	2.5	0	1.5	2.7	3.3
Val	5.6	1.0	4.5	3.7	5.5
Met	0.6	0.2	0.2	1.2	1.2
Ile	4.1	2.6	0.7	2.9	2.5
Leu	7.0	3.6	3.3	8.6	6.9
Tyr	2.5	2.3	3.9	1.7	2.1
Phe	4.8	8.8	1.4	4.7	6.1
His	2.1	1.1	3.4	2.0	1.6
Lys	0.5	0.2	1.1	1.6	1.2
Arg	2.4	0.8	1.5	2.0	2.5
% amidation of asp + glu	90	92	nd	nd	nd
Ref.	1	2	3	4	4

[a]includes Asn; [b]includes Gln.
1, total B hordein from cv. Julia (Shewry *et al.*, 1980);
2, total C hordein from cv. Julia (Shewry *et al.*, 1980);
3, D hordein from Risø mutant 1508 (Kreis *et al.*, 1984);
4, γ-hordein bands from Risø 56 (Kreis *et al.*, 1983b).
nd, not determined;
tryptophan was not determined.

Table 15.2. Tissue specificity of barley seed proteins.

	Embryo	Endosperm	
		Starchy	Aleurone
7S globulin	✓	x	✓
CI-1	x	✓	✓
CI-2	x	✓	✓
β-amylase	✓	✓	✓
B hordein	x	✓	x
C hordein	x	✓	x
D hordein	x	✓	x
γ-hordein	x	✓	x

location. The hordeins are synthesized with a signal peptide which is cleaved as the nascent polypeptide is transported into the endoplasmic reticulum (ER). They are then packaged into protein bodies, which may be of vacuolar or ER origin (probably both) (Cameron-Mills & von Wettstein, 1980; Miflin *et al.*, 1981). The chymotryptic inhibitors appear to be synthesized without a signal peptide and were therefore thought to be cytosolic (Williamson *et al.*, 1987, 1988). However, recently immunogold labelling has provided clear evidence for the presence of CI-2 in hordein-containing protein bodies (Rasmussen *et al.*, 1990), although the mechanism for this deposition is not known. In contrast, β-amylase is also synthesized without a signal peptide (Kreis *et al.*, 1987) and is principally cytosolic, although an increasing amount becomes associated with the periphery of starch granules during the later stages of development (Hara-Nishimura *et al.*, 1986), which may result in a higher proportion of the bound form. Nothing is known of the subcellular location of the 7S globulin in the aleurone and embryo cells, but it may be in the protein bodies present in these tissues. In fact, little further is known about the structure, genetics and control of expression of this protein, but its unusual distribution in embryo and aleurone makes it an attractive target for future study.

The genetic control of barley seed proteins

All four groups of hordeins appear to be encoded by structural genes on barley chromosome 5. The B and C hordein groups are encoded by single loci (*Hor2* and *Hor1* respectively) located on the short arm of the chromosome (Shewry & Miflin, 1982), restriction fragment length polymorphism (RFLP) analysis demonstrating that each consists of about 20–30 gene copies (Bunce *et al.*, 1986). Recent studies indicate that the class I/II and class III B hordeins may be encoded by spatially distinct clusters of genes within the *Hor2* locus (Shewry *et al.*, 1990). In contrast, D hordein is encoded by a single copy locus (*Hor3*) on the long arm of the same chromosome. The γ-hordeins are encoded by a small multigene family probably located at the *Hor5* (*HrdF*) locus which is closely linked to *Hor2* (Shewry *et al.*, 1990). The *Hor4* locus is present only in some lines from the USSR. It is closely linked to *Hor1* and encodes B hordein-like polypeptides (Shewry *et al.*, 1988b). RFLP analysis indicates only a low copy number (Shewry & Burgess, in press).

CI-1 and CI-2 are also encoded by small multigene families (in each case three to four genes) (Williamson *et al.*, 1987, 1988) located at loci (*Ica1* and *Ica2* respectively) on the short arm of chromosome 5 (Cannell *et al.*, in press).

β-Amylase is encoded by genes at two loci (Kreis *et al.*, 1988). The *Bmy1* locus on the long arm of chromosome 4 probably consists of two gene copies and encodes the major isoforms present in the mature grain. The product of a second and probably single copy locus (*Bmy2*) on the short arm of chromosome 2 has not been identified, but may be a minor seed form or a form expressed in vegetative tissues.

Factors affecting gene expression

The accumulation of hordein in the developing grain can be altered by various factors in three ways:

1. the total amount
2. the proportions of the different groups
3. the proportions of individual polypeptides within groups

Protein synthesis during grain development

Hordein can only be detected by SDS-PAGE at about 14 days after anthesis, when the endosperm has reached about 10% of its final dry weight (Rahman *et al.*, 1982). The subsequent rapid accumulation is accompanied by decreases in the proportions of C and class III B hordeins and an increase in class I/II B hordeins (Rahman *et al.*, 1982). Hordein related mRNAs are first detected slightly before the corresponding proteins with a three-fold greater increase in those encoding class I/II hordeins than in those encoding class III B hordeins and C hordeins (Rahman *et al.*, 1984). However, the total amounts of all mRNAs decreased after 26 days although the amounts of proteins continue to increase. This may indicate preferential translation of hordein mRNAs during late endosperm development.

Changes in isoforms of β-amylase are observed during grain development (Shewry *et al.*, 1988a). The enzyme present in young developing caryopses consists of two low M_r isoforms, which are replaced later in development by those typical of the mature grain. The young caryopsis isoforms appear to be associated with a separate population of mRNAs and may be 'green' forms present in the pericarp and/or testa. Their genetic control has not been determined. CI-1 and CI-2 show remarkably similar patterns and their mRNAs are just detected at about 14 days after anthesis, and they continue to be synthesized up to about 30 days (Rasmussen *et al.*, 1988) (Fig. 15.3).

Mineral nutrition

Increased availability of nitrogen results in more total hordein, with an increase in the proportion of C hordein (Kirkman *et al.*, 1982). The latter may result from limited availability of sulphur, as a dramatic increase in the proportion of C hordein also occurs under conditions of sulphur stress (Shewry *et al.*, 1983). Analyses of the populations of mRNAs at early and later stages of development show that they are broadly related to the levels of hordein accumulation, but sulphur deficiency also appeared to increase the efficiency of translation of C hordein mRNAs and decrease that of B hordein mRNAs (Rahman *et al.*, 1983). This could result from limited availability of cysteine and methionine.

β-Amylase is also increased in response to nitrogen fertilization, although only a small increase in the population of translatable mRNAs is observed (Giese & Hejgaard, 1984; Giese & Hopp, 1984).

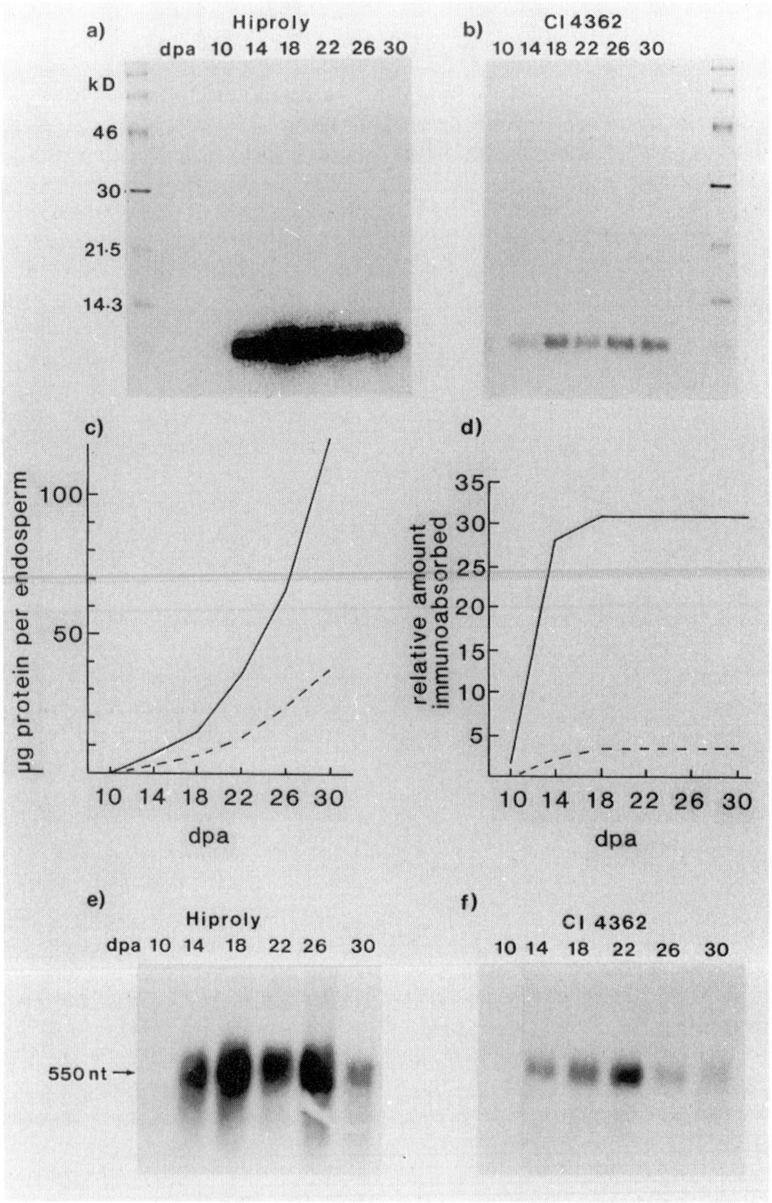

Fig. 15.3. *In vitro* synthesis and *in vivo* deposition of CI-2 (a,b). Fluorographs of SDS-PAGE analyses of immunoprecipitated CI-2 from total *in vitro* translation products of RNA prepared from endosperms of Hiproly and CI 4362 at various stages of development (dpa). CI-2 polypeptides were immunoprecipitated, separated by SDS-PAGE and visualized by fluorography. Numbers on the left indicate molecular mass in kDa. (c) Amount of CI-2 protein per endosperm of Hiproly (—) and CI 4362 (---) at various stages of development, measured by single radial immunodiffusion. (d) Relative amounts of immunoabsorbed *in vitro*-synthesized CI-2 polypeptides of Hiproly (—) and CI 4362 (----), estimated by scanning densitometry of fluorographs (a) and (b). (e,f) Northern blot analysis of poly $(A)^+$RNA fractions from developing endosperms of Hiproly and CI 4362. Denatured poly $(A)^+$ RNA (2 µg/lane) were separated on agarose gels, blotted on to membranes and hybridized to a ^{32}P-labelled cDNA for CI-2. (From Rasmussen *et al.*, 1988.)

Mutant genes

The total amount of hordein is also affected by the presence of mutant high lysine genes. A number of such genes have been identified, which are discussed in detail by Miflin & Shewry (1979) and Shewry *et al.* (1987b). All result in decreases in total hordein (by up to 60%) which are accompanied by increases in other proteins. Only three mutant lines have been studied in detail.

Hiproly was the first high lysine line to be discovered (Munck *et al.*, 1970). It originates from Ethiopia, and the high lysine phenotype is controlled by a spontaneous mutant gene, *lys1*, on chromosome 7. The original line contained about 30% more lysine but had only about 30% of the yield of normal barleys. Although the yield of Hiproly-derived lines is greater, they are not sufficiently high for commercial release. The amounts of total hordein in Hiproly and derived lines are decreased by about 10–20%, with no apparent effects on the polypeptide composition.

The high-lysine Risø mutants 56 and 1508 are derived from the mutagenesis programme at Risø National Laboratory, Denmark, and resulted from treatment with X-rays and ethylinimine respectively (see Doll, 1983).

Risø mutant 56 has a deletion of the *Hor2* locus, resulting in the complete elimination of B hordeins (Fig. 15.1) (Kreis *et al.*, 1983b). Compensatory increases in C hordein and possibly also γ-hordeins (but not D hordein) are not sufficient to prevent a reduction in total hordein of about 20–30%. Messenger RNAs for B hordeins are absent while the population encoding C hordeins is increased.

Risø mutant 1508 results from mutation of a 'regulatory' gene on chromosome 7. Total hordein is reduced to about 15% of the total grain proteins with complete absence of C hordeins. Class I/II B hordeins appear to be decreased more than class III, while D hordein is increased four-fold. Similar changes are observed in the populations of mRNAs (Kreis *et al.*, 1984).

Mutant high-lysine genes also affect the synthesis of CI-1, CI-2 and β-amylase (Hejgaard & Boisen, 1980). Three- to five-fold increases in β-amylase contribute to the increased lysine content of Hiproly and derived lines, while decreases occur in several other high-lysine mutants including Risø 1508. The differences between Hiproly, Risø 1508 and Sundance (a normal line) are associated with differences in the populations of β-amylase mRNA (Kreis *et al.*, 1987).

In contrast increases in the amounts of CI-1 and CI-2 and in the populations of their mRNAs occur in Hiproly, Risø 1508 and Risø 56. Rasmussen *et al.* (1988) have compared the patterns of synthesis of CI-1 and CI-2 in Hiproly and its normal lysine sister line CI 4362. They found that the *lys1* gene affected the level of synthesis of both proteins, but not the timing.

These studies demonstrate that, in most cases, the rates of accumulation of seed proteins are related to their mRNA populations. Although mRNA populations may be regulated by changes in mRNA stability and turnover, control is most commonly exerted at the level of gene transcription (see for example Goldberg, 1988). It is therefore logical to isolate genes encoding barley seed proteins and study the control of their transcription.

Gene isolation and characterization

Five genes for hordeins have been isolated and characterized. Three encode class I/II B hordeins (Brandt *et al.*, 1985; Forde *et al.*, 1985; Chernyshev *et al.*, 1989) and one a γ-type hordein (Cameron-Mills & Brandt, 1988). The fifth is an unexpressed gene encoding an atypical C hordein protein (Entwistle, 1988). The structures of genes encoding a B1 (class I/II) hordein from cv. Sundance, γ-hordein and C hordein are summarized in Fig. 15.4. All have 'TATA' boxes about 80–100 bp upstream of the ATG initiation codons and one or more putative polyadenylation signals between about 50 and 150 bp downstream of the

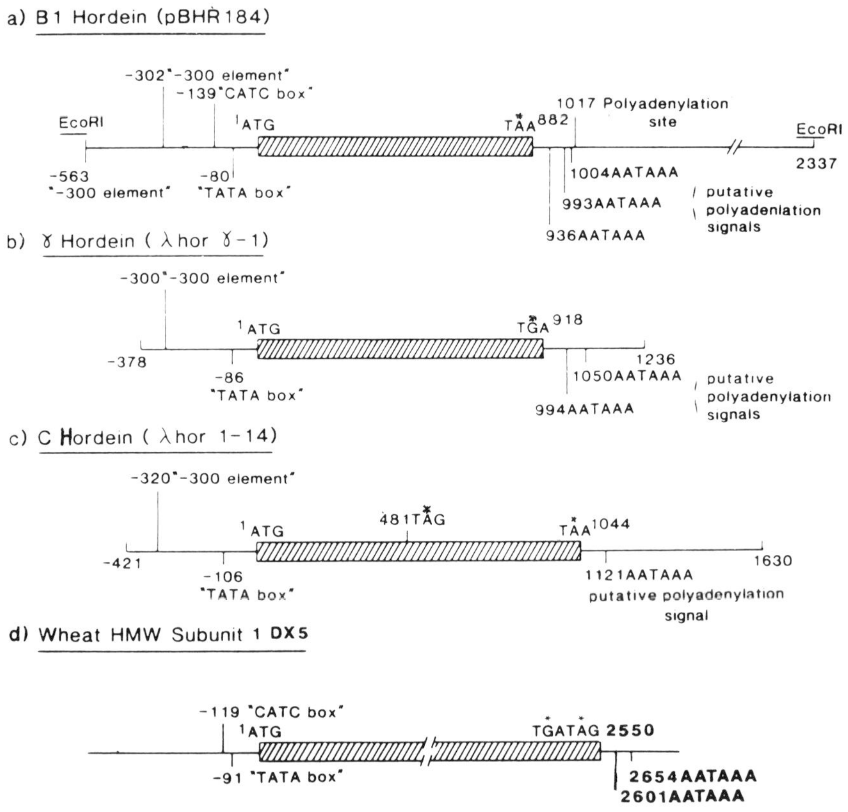

Fig. 15.4. Comparison of the structures of genes encoding B1 hordein (Forde *et al.*, 1985), γ-hordein (Cameron-Mills & Brandt, 1988), C hordein (Entwistle, 1988) and wheat HMW subunit 1Dx5 (Anderson *et al.*, 1989).

stop codon. They also all have a conserved region of about 30 bp located about 300 bp upstream of the ATG. Forde *et al.* (1985) reported that sequences related to these 30 bp regions are present in other prolamin genes, and suggested that it might have a role as an 'enhancer' in controlling tissue-specific expression. We have focused on the role of this '−300 element' in our studies of the B1 (class I/II) hordein gene from cv. Sundance (see later).

No genes for D hordein have been isolated, but genes encoding related proteins (HMW subunits of wheat) have been isolated and studied in detail (Shewry *et al.*, 1989). A typical example, encoding subunit 1Dx5, is shown in Fig. 15.4. These genes have TATA boxes and CCAT sequences at about 90 and 120 bp respectively upstream of the ATG initiation codon, but little other 5′ homology with other prolamin genes from wheat and barley. In particular, a conserved '−300 element' is not present, although several short regions with partial homology can be recognized (Halford *et al.*, 1989).

A gene for chymotrypsin inhibitor CI-2 has also been characterized (Peterson *et al.*, 1991). It contains a single 90 bp intron 18 bp upstream of the coding region, and CAT and TATA boxes. In addition the 5′ upstream region contains sequences homologous with the '−300 element' of cereal prolamins and with the 7–11–7 element, the latter being also present in the prolamin (zein and glutelin) genes of maize (Boronat *et al.*, 1986), but not in the hordein genes of barley. A gene encoding β-amylase has also been isolated and characterized (Thomas *et al.*, in preparation).

Control of gene expression

Reproducible and efficient production of transgenic cereals is not yet possible. Consequently, promoter analysis of cereal seed protein genes must be carried out using alternative systems. Two such systems have been used: transformation into heterologous species (usually tobacco or petunia) using *Agrobacterium*-based vectors, and transient expression in protoplasts. However, both these systems have limitations. With transformation the individual regenerated plants may show wide variation in the levels of gene expression, and strict tissue-specific expression is not retained for all storage protein genes (see for example Ueng *et al.*, 1988). The use of transient expression eliminates the need for time-consuming regeneration of whole plants. In addition it is possible to use cells from species that cannot be transformed, for example cereal aleurone and endosperm cells. However, the DNA is present in a 'naked' form rather than integrated into the chromosome, which may explain why the specificity of gene expression may differ from that *in planta* (see for example Huttly & Baulcombe 1989). Thus the main value of transient expression systems may be to assay promotor strength rather than tissue-specificity.

We have, therefore, used both approaches to study the expression of genes for B1 hordein, for wheat HMW subunits and for CI-2.

Expression in transgenic tobacco

A 2.9 kb *Eco*RI fragment containing the complete B1 hordein coding sequence (879 bp) flanked by 5′ (563 bp) and 3′ (1458 bp) sequences was initially transferred into tobacco using the Bin6 binary vector system (see Shewry *et al.*, 1988c). B1 hordein transcripts were detected in the seeds but not the leaves of transgenic plants, but B hordein protein was not detected. A 549 bp 5′ flanking sequence containing the transcription start site, the 'TATA' box, the 'CATC' box and the '−300 element' was therefore linked to the chloramphenicol acetyl transferase (CAT) reporter gene and transferred to tobacco using Bin19. In this case enzyme activity was detected in the endosperms (not the embryos or testas) of the developing seeds, and only from 15 days after pollination (Fig. 15.5) (Marris *et al.*, 1988). Further

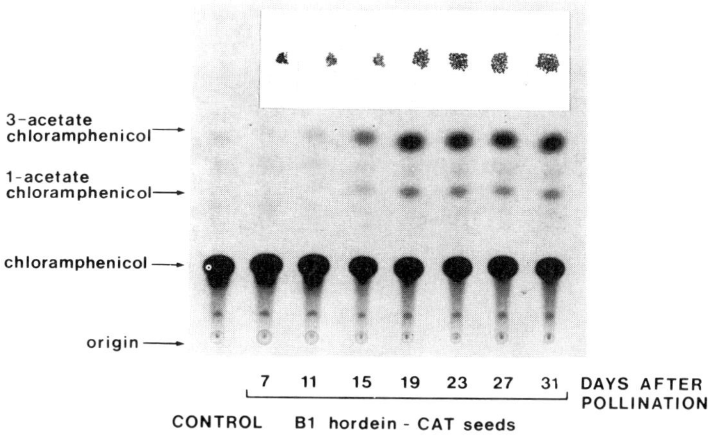

Fig. 15.5. Expression of a B1 hordein–CAT chimeric gene in seeds of transgenic tobacco. Protein was extracted from seeds between 7 and 31 days after pollination and assayed for ability to acetylate chloramphenicol (to the 1- and 3-acetyl derivates). The inset shows seeds of the various stages. (Taken in part from Marris *et al.*, 1988.)

attempts to define precise sequences responsible for tissue and developmental specificity using shorter regions of 5′ sequence fused to the β-glucuronidase (GUS) reporter gene were unsuccessful, due to low levels of expression combined with high variability between individual plants (Marris, 1990). Similar studies with CI-2 and β-amylase genes have shown that expression in seeds is conferred by sequences between 263 and 1100 bp and by about 400 bp of sequence immediately upstream of the transcription start site respectively.

Expression of HMW subunit genes in transgenic tobacco has demonstrated that only 338 bp of sequence directly upstream of the ATG initiation codon is sufficient to confer tissue-specific and developmentally regulated expression (Halford *et al.*, 1989). This region does not contain sequences with extensive homology with the '−300 element', indicating that it does not have a role in the control of HMW subunit (and presumably also D hordein) gene expression. Differential regulation of D hordein gene expression is consistent with the effects of the mutant *lys3a* gene (see earlier).

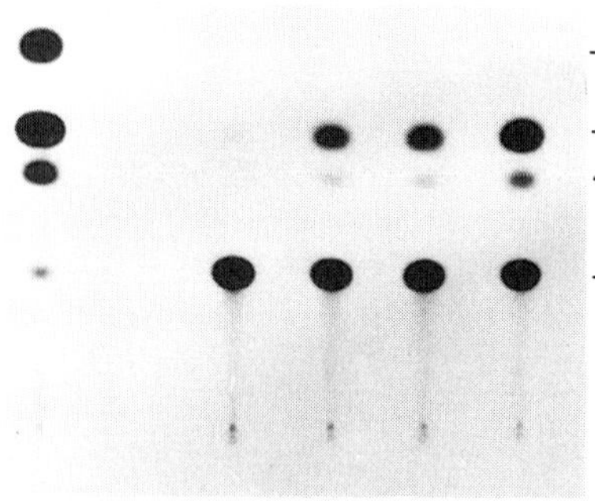

Fig. 15.6. CAT activity following introduction of pCaMV1_1CN (a construct with the 'Constitutive' CaMV 35S promoter) into starchy endosperm protoplasts isolated from 12–15 days post-anthesis barley grains of Risø mutant 1508. Lane 1, bacterial control; lane 2, PEG-treated protoplasts with no DNA; lanes 3, 4 and 5, CAT activity following introduction of pCaMV1_1CN using different salts for the PEG-induced uptake of the plasmid. In lanes 3 and 4 the protoplasts were washed in protoplast incubation medium before PEG treatment and the transformation solution contained 15 mM $MgCl_2$ (lane 3), or 50 mM $CaCl_2$ (lane 4). Before transformation the protoplasts in lane 5 were washed first in Krens' salts and then in protoplast incubation medium. The transformation solution contained 15 mM $MgCl_2$. (From Lee *et al.*, in press.)

Transient expression in protoplasts

We have studied the expression of seed protein genes in mesophyll protoplasts from tobacco and in starchy endosperm protoplasts from barley. The mesophyll protoplasts were isolated as described by Foulger & Jones (1986) and Gallois *et al.* (submitted). The methods developed for the isolation of synthetically active starchy endosperm protoplasts are described by Lee *et al.* (in press), and a typical result using the CAT reporter gene and the 'constitutive' 35S CaMV promoter is shown in Fig. 15.6.

Transient expression of β-amylase/GUS constructs in mesophyll protoplasts of tobacco showed that only 120 bp of sequence immediately upstream of the ATG initiation codon was sufficient to drive expression, while additional sequences lying between about 1100 and 6000 bp upstream increased the level of expression by about 50% (Fig. 15.7). Similar studies demonstrated that about 1100 bp of the CI-2 gene gave expression of the GUS gene in potato mesophyll protoplasts (IACR Annual Report, 1989).

Recent studies by Lee *et al.* (in press) have shown interesting differences between the transient expression of CI-2/GUS and HMW subunit/GUS constructs in protoplasts derived from tobacco mesophyll and barley starchy endosperm. Four constructs were assayed in the two types of protoplast using polyethylene glycol-mediated direct gene transfer. The pGUS construct contained the promoter-less GUS coding region, and showed the expected low level of background expression in both systems (Fig. 15.8). The CaMV–GUS construct with the CaMV 35S promoter was expressed at a fairly high level in both systems, which is consistent with the 'constitutive' promoter. The Glu–1Dx–GUS1 construct contained 688 bp of upstream region from the HMW

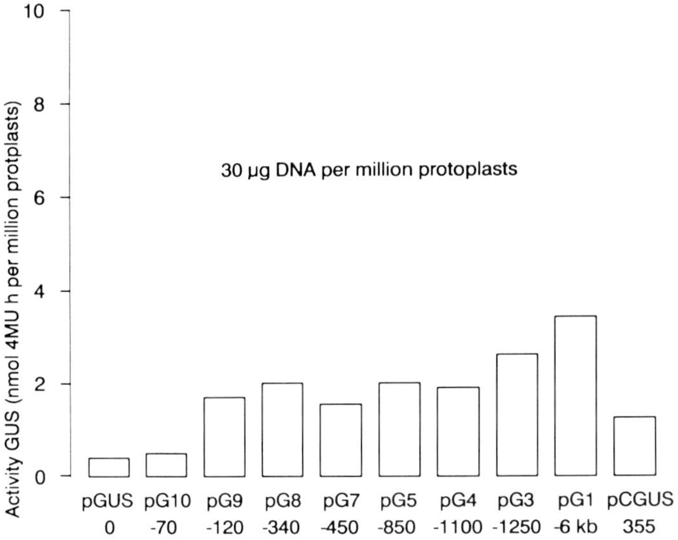

Fig. 15.7. β-Glucuronidase (GUS) activity in tobacco mesophyll protoplasts electroporated with constructs in which the GUS reporter gene is placed downstream of increasing lengths of the β-amylase promoter. pGUS and pCGUS are negative and positive controls, respectively. (From IACR Annual Report, 1989.)

subunit gene encoding subunit 1Dx5, starting 32 bp upstream of the ATG initiation codon (see Halford *et al.*, 1989). This showed a high level of expression in the endosperm protoplasts, but very low expression (similar to background) in the mesophyll protoplasts. Previous studies with an HMW subunit promoter have demonstrated expression in protoplasts derived from oat aleurone cells (Huttly & Baulcombe, 1989). The HMW subunit genes are not normally expressed in aleurone cells of developing wheat seeds which suggests that the HMW subunit genes retain their endosperm-specificity but lose the ability to discriminate between aleurone and starchy endosperm cells when expressed transiently in protoplasts.

The fourth construct studied by Lee *et al.* (in press), pCI-2–SP–GUS, was a translational fusion containing 1100 bp of the 5′ region of the CI-2 gene (relative to the ATG initiation codon). This encodes a GUS fusion protein containing 47 extra amino acids derived from CI-2 at the N-terminus, including the initiation methionine codon of CI-2. This showed significant but relatively low expression in the endosperm protoplasts (Fig. 15.8), which is consistent with the known *in vivo* strengths of the CI-2 and HMW subunit promoters (based on the relative amounts of protein synthesized). However, it gave very high expression in the mesophyll protoplasts higher than the CaMV–GUS construct. Although initially surprising, a possible reason for this result is given by comparison with the related protease inhibitors of tomato, whose synthesis is induced in vegetative tissues by wounding (Graham *et al.*, 1986). The high level of expression of the pCI-2–SP–GUS construct in mesophyll protoplasts may therefore be related to damage of the tissue during preparation of the protoplasts (unpublished results of M. Kreis and co-workers).

Conclusion from gene expression studies

The combination of transient expression in protoplasts and expression of integrated genes in transgenic tobacco have allowed us to define relatively small lengths of the 5′ upstream sequences of the B1 hordein, CI-2 and β-amylase genes which are sufficient to confer tissue-specificity and developmental regulation on CAT and GUS reporter genes.

They have also demonstrated clear differences between apparent promoter strength, most notably the HMW subunit genes of wheat appear to have very strong promoters

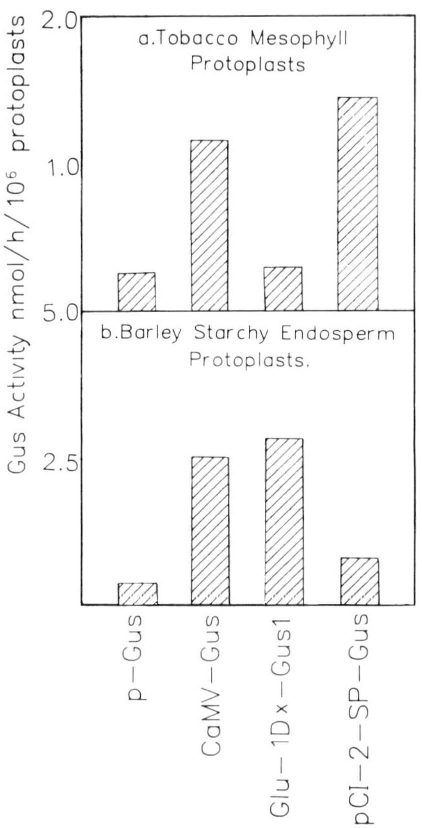

Fig. 15.8. Transient expression of genes in protoplasts derived from (a) mesophyll cells of tobacco and (b) starchy endosperm cells of Risø mutant 1508 (12–15 days post-anthesis). Constructs are: GUS control; CaMV–GUS; Glu–1Dx–GUS 1; pCI-2–SP–GUS. See text for details of constructs. (Based on Lee *et al.*, in press.)

which are ideal for obtaining high expression levels in transgenic plants.

Further studies may enable us to define even shorter sequences and, perhaps to discriminate between the sequences that confer tissue-specificity and developmental regulation. However, such studies are very labour-intensive and time-consuming, and their ultimate value for the improvement of barley by genetic engineering is open to question. The information and genes currently available should enable us to make improvements to barley feed quality as soon as methods for transformation become available. Recent success in transforming maize (Fromm *et al.*, 1990; Gordon-Kamm *et al.*, 1990) and regenerating wheat (Vasil *et al.*, 1990) indicate that this should soon be possible.

References

Anderson, O.D., Greene, F.C., Yip, R.E., Halford, N.G., Shewry, P.R. & Malpica-Romero, J.-M. (1989) Nucleotide sequences of the two high-molecular-weight glutenin genes from the D-genome of a hexaploid bread wheat, *Triticum aestivum* L. cv Cheyenne. *Nucleic Acids Research* 17, 461–2.

Boronat, A., Martinez, M.C., Reina, M., Puigdomenech, P. & Palau, J. (1986) Isolation and sequencing of a 28 kD glutelin-2 gene from maize. Common elements in the 5′ flanking regions among zein and glutelin genes. *Plant Science* 47, 95–102.

Brandt, A., Montembault, A., Cameron-Mills, V. & Rasmussen, S.K. (1985) Primary structure of a B1 hordein gene from barley. *Carlsberg Research Communications* 50, 333–45.

Bunce, N.A.C., Forde, B.G., Kreis, M. & Shewry, P.R. (1986) DNA restriction fragment length polymorphism at hordein loci: application to identifying and fingerprinting barley cultivars. *Seed Science and Technology* 14, 419–29.

Cameron-Mills, V. & Brandt, A. (1988) A γ-hordein gene. *Plant Molecular Biology* 11, 449–61.

Cameron-Mills, V. & von Wettstein, D. (1980) Endosperm morphology and protein body formation in developing wheat grain. *Carlsberg Research Communications* 45, 577–94.

Cannell, M., Karp, A., Shewry, P.R. & Issacs, P. Chromosomal assignment of genes in barley using telosomic wheat–barley addition lines *Grenome* (in press).

Chernyshev, A.K., Davletova, Sh.K., Bashkirov, V.I., Shakhmanov, N.B., Mekhedov, S.L. & Ananiev, E.V. (1989) Nucleotide sequence of B1 hordein gene of barley *Hordeum vulgare* L. *Genetica* 25, 1349–55.

Doll, H. (1983) Barley seed proteins and possibilities for their improvement. In: Gottschalk, W. & Muller, H.P. (eds), *Seed Proteins: Biochemistry, Genetics, Nutritive Value*. Martin Nijhoff, The Hague, pp. 205–23.

Entwistle, J. (1988) Primary structure of a C-hordein gene from barley. *Carlsberg Research Communications* 53, 247–58.

Forde, B.G., Heyworth, A., Pywell, J. & Kreis, M. (1985) Nucleotide sequence of a B1 hordein gene and the identification of possible upstream regulatory elements in endosperm storage protein genes from barley, wheat and maize. *Nucleic Acids Research* 13, 7327–39.

Foulger, D. & Jones, M.G.K. (1986) Improved efficiency of genotype dependent regeneration from protoplasts of important tobacco cultivars. *Plant Cell Reports* 5, 72–6.

Fromm, M.E., Morrish, F., Armstrong, C., Williams, R., Thomas, J. & Klein, T.M. (1990) Inheritance and expression of chimaeric genes in the progeny of transgenic maize plants. *Bio/Technology* 8, 833–9.

Gallois, P., Malone, R., Lindsey, K., Kreis, M. & Jones, M.G.K. Direct gene transfer into tobacco protoplasts by electroporation or PEG: analysis of transient expression and stable transformation (submitted).

Giese, H. & Hejgaard, J. (1984) Synthesis of salt-soluble proteins in barley. Pulse-labeling study of grain filling in liquid-cultured detached spikes. *Planta* 161, 172–7.

Giese, H. & Hopp, H.E. (1984) Influence of nitrogen nutrition on the amount of hordein, protein Z and β-amylase messenger RNA in developing endosperms of barley. *Carlsberg Research Communications* 49, 365–83.

Goldberg, R. (1988) Novel developmental processes. *Science* 240, 1460–7.

Gordon-Kamm, W.J., Spencer, T.M., Mangano, M.L., Adam, T.R., Daines, R.J., Start, W.G., O'Brien, J.V., Chambers, S.A., Adams, W.R., Willetts, N.G., Rice, T.B., Mackey, C.J., Krueger, R.W., Kausch, A.P. & Lamaux, P.G. (1990) Transformation of maize cells and regeneration of fertile transgenic plants. *Plant Cell* 2, 603–18.

Graham, J.S., Hall, G., Pearce, G. & Ryan, C.A. (1986) Regulation of synthesis of proteinase inhibitor I and II mRNA's in leaves of wounded tomato plants. *Planta* 169, 399–405.

Halford, N.G., Forde, J., Shewry, P.R. & Kreis, M. (1989) Functional analysis of the upstream regions of a silent and an expressed member of a family of wheat seed protein genes in transgenic tobacco. *Plant Science* 62, 207–16.

Hara-Nishimura, I., Nishimura, M. & Daussant, J. (1986) Conversion of free β-amylase to bound β-amylase on starch granules in the barley endosperm during desiccation phase of seed development. *Protoplasma* 134, 149–53.

Hejgaard, J. & Boisen, S. (1980) High lysine proteins in Hiproly barley breeding: identification, nutritional significance and new screening methods. *Hereditas* 93, 311–20.

Huttly, A. & Baulcombe, D. (1989) A wheat α-*Amy2* promoter is regulated by gibberellin in transformed oat aleurone protoplasts. *EMBO Journal* 8, 1907–13.

IACR Annual Report (1989) Institute of Arable Crops Research, Harpenden, Herts, UK.

Kirkman, M.A., Shewry, P.R. & Miflin, B.J. (1982) The effect of nitrogen nutrition on the lysine content and protein composition of barley seeds. *Journal of the Science of Food and Agriculture* 33, 115–27.

Kreis, M., Rahman, S., Forde, B.G., Pywell, J., Shewry, P.R. & Miflin, B.J. (1983a) Subfamilies of hordein mRNA encoded at the *Hor2* locus of barley. *Molecular and General Genetics* 191, 194–200.

Kreis, M., Shewry, P.R., Forde, B.G., Rahman, S. & Miflin, B.J. (1983b) Molecular analysis of a mutation conferring the high lysine phenotype on the grain of barley (*Hordeum vulgare*). *Cell* 34, 161–7.

Kreis, M., Shewry, P.R., Forde, B.G., Rahman, S., Bahramian, M.B. & Miflin, B.J. (1984) Molecular analysis of the effects of the mutant *lys3a* gene on the expression of *Hor* loci in developing endosperms of barley (*Hordeum vulgare*). *Biochemical Genetics* 22, 231–55.

Kreis, M., Williamson, M., Buxton, B., Pywell, J., Hejgaard, J. & Svendsen, I. (1987) Primary structure and differential expression of β-amylase in normal and mutant barleys. *European Journal of Biochemistry* 169, 517–25.

Kreis, M., Williamson, M.S., Shewry, P.R., Sharp, P. & Gale, M. (1988) Identification of a second locus encoding β-amylase on chromosome 2 of barley. *Genetical Research* 51, 13–16.

Lee, B.T., Murdoch, K., Topping, J., Jones, M.G.K. & Kreis, M. Transient expression of foreign genes introduced into barley endosperm protoplasts by PEG-mediated transfer or into intact endosperm tissue by microprojectile bombardment *Plant Science* (in press).

Marris, C. (1990) Regulation of the expression of a seed-protein gene from barley in transgenic tobacco plants. PhD Thesis, CNAA, UK.

Marris, C., Gallois, P., Copley, J. & Kreis, M. (1988) The 5′-flanking region of a barley B hordein gene controls tissue and development specific CAT expression in tobacco plants. *Plant Molecular Biology* 10, 359–66.

Miflin, B.J. & Shewry, P.R. (1979) The synthesis of proteins in normal and high lysine barley seeds. In: Laidman, D. & Wyn Jones, R.G. (eds), *Recent Advances in the Biochemistry of Cereals*. Academic Press, London, pp. 239–73.

Miflin, B.J., Burgess, S.R. & Shewry, P.R. (1981) The development of protein bodies in the storage tissues of seeds. *Journal of Experimental Botany* 32, 199–219.

Munck, L., Karlsson, K.-E., Hagburg, A. & Eggum, B.O. (1970) Gene for improved nutritive value in barley seed protein. *Science* 168, 985–7.

Osborne, T.B. (1924) *The Vegetable Proteins*. Longmans, Green & Co., London, 154 pp.

Peterson, D.M., Forde, J., Williamson, M.S., Rhode, W. & Kreis, M. (1991) The nucleotide sequence of a chymotrypsin inhibitor-2 gene of barley (*Hordeum vulgare*). *Plant Physiology* 96, 1389–90.

Rahman, S., Shewry, P.R. & Miflin, B.J. (1982) Differential protein accumulation during barley grain development. *Journal of Experimental Botany* 33, 717–28.

Rahman, S., Shewry, P.R., Forde, B.G., Kreis, M. & Miflin, B.J. (1983) Nutritional control of storage protein synthesis in developing grain of barley (*Hordeum vulgare* L.). *Planta* 159, 366–72.

Rahman, S., Kreis, M., Forde, B.G., Shewry, P.R. & Miflin, B.J. (1984) Hordein-gene expression during development of the barley (*Hordeum vulgare*) endosperm. *Biochemical Journal* 223, 315–22.

Rasmussen, U., Williamson, M.S., Mundy, J. & Kreis, M. (1988) Differential effects of the Hiproly *lys1* gene on the developmental synthesis of (lysine-rich) proteins from barley endosperm. *Plant Science* 55, 255–66.

Rasmussen, U., Munck, L. & Ullrich, S.E. (1990) Immunogold localisation of chymotrypsin inhibitor-2, a lysine-rich protein, in developing barley endosperm. *Planta* 180, 272–7.

Shewry, P.R. Barley seed storage proteins – structure, synthesis and deposition. In: Pilbeam, D.J. & Mengel, K. (eds), *Nitrogen Metabolism of Plants*. Oxford University Press, Oxford, UK (in press).

Shewry, P.R. & Burgess, S.R. RFLP analysis of the *Hor4* (*Hrd G*) locus encoding a B hordein-like polypeptide. *Barley Genetics Newsletter* (in press).

Shewry, P.R. & Miflin, B.J. (1982) Genes for the storage proteins of barley. *Qualitas Plantarum. Plant Foods for Human Nutrition* 31, 251–67.

Shewry, P.R., Field, J.M., Kirkman, M.A., Faulks, A.J. & Miflin, B.J. (1980) The extraction, solubility and characterization of two groups of barley storage polypeptides. *Journal of Experimental Botany* 31, 393–407.

Shewry, P.R., Franklin, J., Parmar, S., Smith, S.J. & Miflin, B.J. (1983) The effects of sulphur starvation on the amino acid and protein compositions of barley grain. *Journal of Cereal Science* 1, 21–31.

Shewry, P.R., Field, J.M. & Tatham, A.S. (1987a) The structures of cereal seed storage proteins. In: Morton, I.D. (ed.), *Cereals in a European Context*. Ellis Horwood, Chichester, UK, pp. 421–37.

Shewry, P.R., Williamson, M.S. & Kreis, M. (1987b) Effects of mutant genes on the synthesis of storage components in developing barley endosperms. In: Thomas, H. (ed.), *Mutant Genes that affect Plant Development*. Cambridge University Press, Cambridge, UK, pp. 95–118.

Shewry, P.R., Parmar, S., Buxton, B., Gale, M.D., Liu, C.J., Hejgaard, J. & Kreis, M. (1988a) Multiple molecular forms of β-amylase in seeds and vegetative tissues of barley. *Planta* 176, 127–34.

Shewry, P.R., Parmar, S., Franklin, J. & White R. (1988b) Mapping and biochemical analysis of *Hor4* (*Hrd G*), a second locus encoding B hordein seed proteins in barley *Hordeum vulgare* L. *Genetical Research* 51, 5–12.

Shewry, P.R., Tatham, A.S., Field, J.M., Forde, B.G., Clark, J., Gallois, P., Marris, C., Halford, N., Forde, J. & Kreis, M. (1988c) The structure of barley and wheat prolamins and their genes. *Biochemie und Physiologie der Pflanzen* 183, 117–27.

Shewry, P.R., Halford, N.G. & Tatham, A.S. (1989) The high molecular weight subunits of wheat, barley and rye: genetics, molecular biology, chemistry and role in wheat gluten structure and functionality. In: Miflin, B.J. (ed.), *Oxford Surveys of Plant Molecular and Cell Biology*, Vol. 6. Oxford University Press, Oxford, pp. 163–219.

Shewry, P.R., Parmar, S., Franklin, J. & Burgess, S.R. (1990) Analysis of a rare recombination event within the multigenic *Hor2* locus of barley (*Hordeum vulgare* L.) *Genetical Research* (Camb.) 55, 171–6.

Tallberg, A. (1982) Characterisation of high-lysine barley genotypes. *Hereditas* 96, 229–45.

Ueng, P., Galili, G., Sapanara, V., Goldsborough, P.B., Dube, P., Beachy, R.N. & Larkins, B.A. (1988) Expression of a maize storage protein gene in petunia plants is not restricted to seeds. *Plant Physiology* 86, 1281–5.

Vasil, V., Redway, F. & Vasil, I.K. (1990) Regeneration of plants from embryogenic suspension cultivar protoplasts of wheat (*Triticum aestivum* L.) *Bio/Technology* 8, 429–34.

Williamson, M.S., Forde, J., Buxton, B. & Kreis, M. (1987) Nucleotide sequence of barley chymotrypsin inhibitor-2 (CI-2) and its expression in normal and high-lysine barley. *European Journal of Biochemistry* 165, 99–106.

Williamson, M.S., Forde, J. & Kreis, M. (1988) Molecular cloning of two isoinhibitor forms of chymotrypsin inhibitor 1 (CI-1) from barley endosperm and their expression in normal and mutant barleys. *Plant Molecular Biology* 10, 521–35.

Yupsanis, T., Burgess, S.R., Jackson, P.J. & Shewry, P.R. (1990) Characterisation of the major protein component from aleurone cells of barley (*Hordeum vulgare* L). *Journal of Experimental Botany* 41, 385–92.

16

Trypsin/α-Amylase Inhibitors and Thionins: Possible Defence Proteins from Barley

F. García-Olmedo, G. Salcedo, R. Sanchez-Monge, C. Hernandez-Lucas, M.J. Carmona, J.J. Lopez-Fando, J.A. Fernandez, L. Gomez, J. Royo, F. García-Maroto, A. Castagnaro & P. Carbonero

Plant storage tissues in general should be good substrates for a wide range of organisms on the basis of their gross biochemical compositions. However, only a few specific taxa are able to feed on each type. One possible reason for this is that they have proteins and other molecules that are toxic or inhibitory towards heterologous systems. In the case of barley endosperm, over 10% of the total protein is represented by protein families that fall within this description. A substantial fraction of the albumins and globulins of barley belong to two of these families, the trypsin/α-amylase inhibitors and the thionins, which are the subject of the present chapter.

The first family includes inhibitors of serine proteases and of heterologous α-amylases. Some members of this family can be selectively extracted with chloroform/methanol mixtures and have been designated CM-proteins. Their apparent molecular weights are in the 12–16 kDa range. The α-amylase inhibitors can be classified according to their degree of aggregation into monomeric, dimeric and tetrameric forms. No members of this family have been found in tissues other than endosperm, although this has not been investigated in detail. These inhibitors have been found in other species of the Poaceae, including wheat, where they have been extensively studied.

The thionins are polypeptides of about 5 kDa, which have three or four disulphide bridges and are toxic to plant pathogenic micro-organisms. Although they are very abundant in endosperm, different types are found in other tissues. Members of this family have been identified in a few but wide-ranging taxa.

Comprehensive reviews of work on the trypsin/α-amylase inhibitors (García-Olmedo *et al.*, 1987) and on the thionins (García-Olmedo *et al.*, 1989, 1991) have been published recently. This chapter focuses on recent developments concerning the two protein families in barley and related species, with special emphasis on their possible role as defence proteins.

Trypsin/α-amylase inhibitors

Inhibitor types

The first member of this protein family in barley was reported by Mikola & Suolinna (1969), who isolated a 14.1 kDa protein which was active against trypsin and inactive against the endogenous proteinases from green malt. Antibodies raised against this protein did not react with inhibitors from barley embryo or from wheat endosperm (Mikola & Kirsi, 1972). A similar or identical inhibitor was purified by Boisen (1976) from a different genetic stock. The inhibitor was sequenced by Odani *et al.* (1983a), who found that it was homologous to a monomeric α-amylase inhibitor from wheat which had been previously sequenced by Kashlan & Richardson (1981). Two chloroform/methanol soluble proteins (CM-proteins) from barley endosperm, designated CMe and CMc, were found to have antitrypsin activity (Lazaro *et al.*, 1985; Barber *et al.*, 1986a). The first of these proteins (BTI-CMe) was found to be identical to the previously described trypsin inhibitor of Mikola & Suolinna (1969).

A tetrameric α-amylase inhibitor was characterized in barley endosperm by Sanchez-Monge *et al.* (1986b), who also described the presence of dimeric and monomeric inhibitors in this tissue. The subunits of the tetrameric inhibitor were identified as the previously described proteins CMa, CMb and CMd, which were known to be homologous to the previously characterized α-amylase inhibitors (Salcedo *et al.*, 1982; Shewry *et al.*, 1984; Lazaro *et al.*, 1985; Barber *et al.*, 1986a, b). The complete sequence of subunit CMd has been deduced from the corresponding cloned cDNA (Paz-Ares *et al.*, 1986; Halford *et al.*, 1988).

A dimeric α-amylase inhibitor from barley and its corresponding cDNA have been reported by Lazaro *et al.* (1988b) and a monomeric inhibitor was identified as the major barley allergen reacting with sera from patients suffering baker's asthma (Barber *et al*;, 1989; Sanchez-Monge *et al.*, unpublished). A number of cDNA clones encoding members of this family have been identified, but the corresponding proteins have not been characterized yet.

Proteins corresponding to the trypsin inhibitors and to the subunits of the tetrameric α-amylase inhibitor of *Hordeum vulgare* have been identified in *Hordeum spontaneum* and their allelic relationships have been established by genetic analysis (Salcedo *et al.*, 1984; Molina-Cano *et al.*, 1987). A similar study has been carried out for the subunits of the tetrameric inhibitor in *Hordeum chilense* (Sanchez-Monge *et al.*, 1987).

This protein family has been investigated for half a century in wheat, where all the inhibitor types seem to be present (for reviews, see Buonocore *et al.*, 1977; García-Olmedo *et al.*, 1987; Silano, 1987; Sanchez-Monge *et al.*, 1988). In addition, members of this family have been identified in rye (Lyons *et al.*, 1987), in maize (Mahoney *et al.*, 1984) and in finger-millet (Campos & Richardson, 1983).

Sequences of members of this family in barley are aligned with total or partial sequences from the corresponding proteins in other species in Fig. 16.1. Three domains, which correspond approximately to those proposed by Kreis *et al.* (1985) for the trypsin inhibitor CMe, have been used in the alignment. The reactive site of this inhibitor, Pro–Arg–Leu (P–R–L), is located at the C-terminal border of the first domain, a region which is quite variable throughout the family.

In addition to the homology relationships shown in Fig. 16.1, weaker and more elusive relationships have been proposed between this family and the 2S storage proteins from dicots (Odani *et al.*, 1983c; Shewry *et al.*, 1984), with dispersed sulphur-rich domains from cereal prolamins (Kreis *et al.*, 1985), and even with the Kazal secretory trypsin inhibitor from bovine pancreas (Odani *et al.*, 1983b).

Genomic distribution of inhibitor genes in barley and related species

The α-amylase/trypsin inhibitors are encoded by a multigene family which is dispersed over several chromosomes, both in barley and in wheat. Present knowledge of the genomic organization of this family is summarized in Table 16.1. A total of 24 members of the family have been identified in barley, wheat and rye. These can be grouped into nine types according to sequence similarity and/or *in vitro* activity. All but four of these have been characterized at the protein level and their *in vitro* activities have been determined except for protein WTI from *Triticum monococcum*, whose sequence is quite close to BTI-CMe.

Table 16.1. Genomic distribution of genes encoding α-amylase/trypsin inhibitors

Inhibitory activity (aggregation)	Protein (syn.)	Clones	Gene locus (syn.)	Chromosome no. genome (arm)
α-amylase (monomeric)	BMAI-1 (Horv1)		*lam1*	?
	WMAI-1 (0.28)	pUP-28	Imha-D1	6D(S) ab
	WMAI-2		Imha-B1	6B(S) c
α-amylase (dimeric)	BDAI-1	pUP-44	*lad1*	?
	WDAI-1 (0.53)		Idha-B1-1	3B(S) abd
	WDAI-3		Idha-B1-2	3B(S) e
	WDAI-2 (0.19)		Idha-B1	3D(S) abd
α-amylase (tetrameric)				
1st subunit	BTAI-CMa		*lat1*	1(7H) f
	WTAI-CM2		Itha-B1	7B(S) adgh
	WTAI-CM1	pCT-2	Itha-D1	7D(S) adgh
	RAI		Itha-R1	?R
2nd subunit	BTAI-CMb		*lat2*	4(4H) f
	WTAI-CM16	pCT-3	Itha-B2	4B(S) adh
	WTAI-CM17		Itha-D2	4D(S) adh
3rd subunit (2 copies)	BTAI-CMd	pUP-38	*lat3*	4(4H) f
	WTAI-CM3B	pCT-1	Itha-B3	4B(S) adh
	WTAI-CM3D		Itha-D3	?D
Trypsin	BTI-CMe	λcCMe λgCMe	*ltr1*	3(3H) fij
	WTI?			?A
	?	pCT-4		?A, B, D?
	RTI		Itr-R1	3R k
Trypsin	BTI-CMc		*ltr2*	1(7H) f
Unknown	?	pUP-23	*lah2*	?
	?	pCT-88	?	?A, B, D?
Unknown	?	pUP-13	*lah1*	?

(a) Fra-Mon *et al.*, 1984; (b) Sanchez-Monge *et al.*, 1986a; (c) Gomez *et al.*, unpublished; (d) Aragoncillo *et al.*, 1975; (e) Sanchez-Monge *et al.*, 1989; (f) Salcedo *et al.*, 1984; (g) García-Olmedo & Carbonero, 1970; (h) García-Maroto *et al.*, 1990; (i) Hejgaard *et al.*, 1984a; (j) Rodriguez-Palenzuela *et al.*, 1989; (k) Hejgaard *et al.*, 1984b.

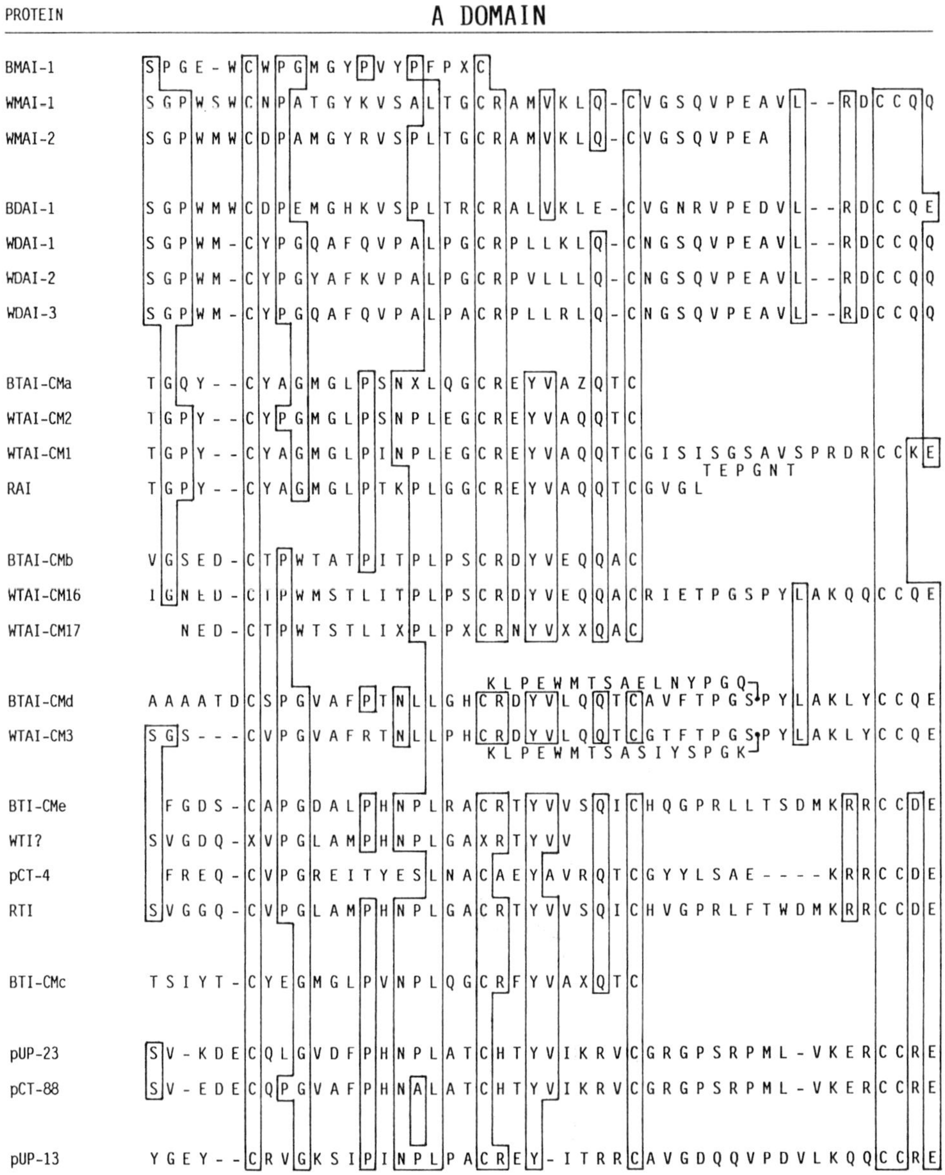

Fig. 16.1. Alignment of total or partial amino acid sequences of barley inhibitors of α-amylase and trypsin with those from wheat and rye: monomeric α-amylase inhibitors from barley (BMAI-1, ref. 1) and wheat (WMAI-1, ref. 2; WMAI-2, ref. 3); dimeric α-amylase inhibitors from barley (BDAI-1, ref. 4) and wheat (WDAI-1 and WDAI-2, ref. 5; WDAI-3, ref. 6); tetrameric α-amylase inhibitor subunits from barley (BTAI-CMa and BTAI-CMb, ref. 7; BTAI-CMd, refs 11–13), wheat (WTAI-CM2 and WTAI-CM17, ref. 8; WTAI-CM1 and WTAI-CM16, refs 8,9; WTAI-CM3, refs 9,11) and rye (RAI, ref. 10); trypsin inhibitors from barley (BTI-CMe, refs 14–16; BTI–CMc, refs 7,11), rye (RTI, ref. 10) and two related wheat proteins (WTI, ref. 11; pCT-4, ref. 17); amino acid sequences deduced from barley cDNA clones (pUP-23, ref. 18; pCT-88, ref. 17; pUP-13, ref. 12). References: (1) Barber *et al.*, 1989; (2) Kashlan & Richardson, 1981; (3) Gomez *et al.*, unpublished; (4) Lazaro *et al.*, 1988b; (5) Maeda *et al.*, 1985; (6) Sanchez-Monge *et al.*, 1989; (7) Barber *et al.*, 1986b; (8) Barber *et al.*, 1986a; (9) García-Maroto *et al.*, 1990; (10) Lyons *et al.*, 1987; (11) Shewry *et al.*, 1984; (12) Paz-Ares *et al.*, 1986;

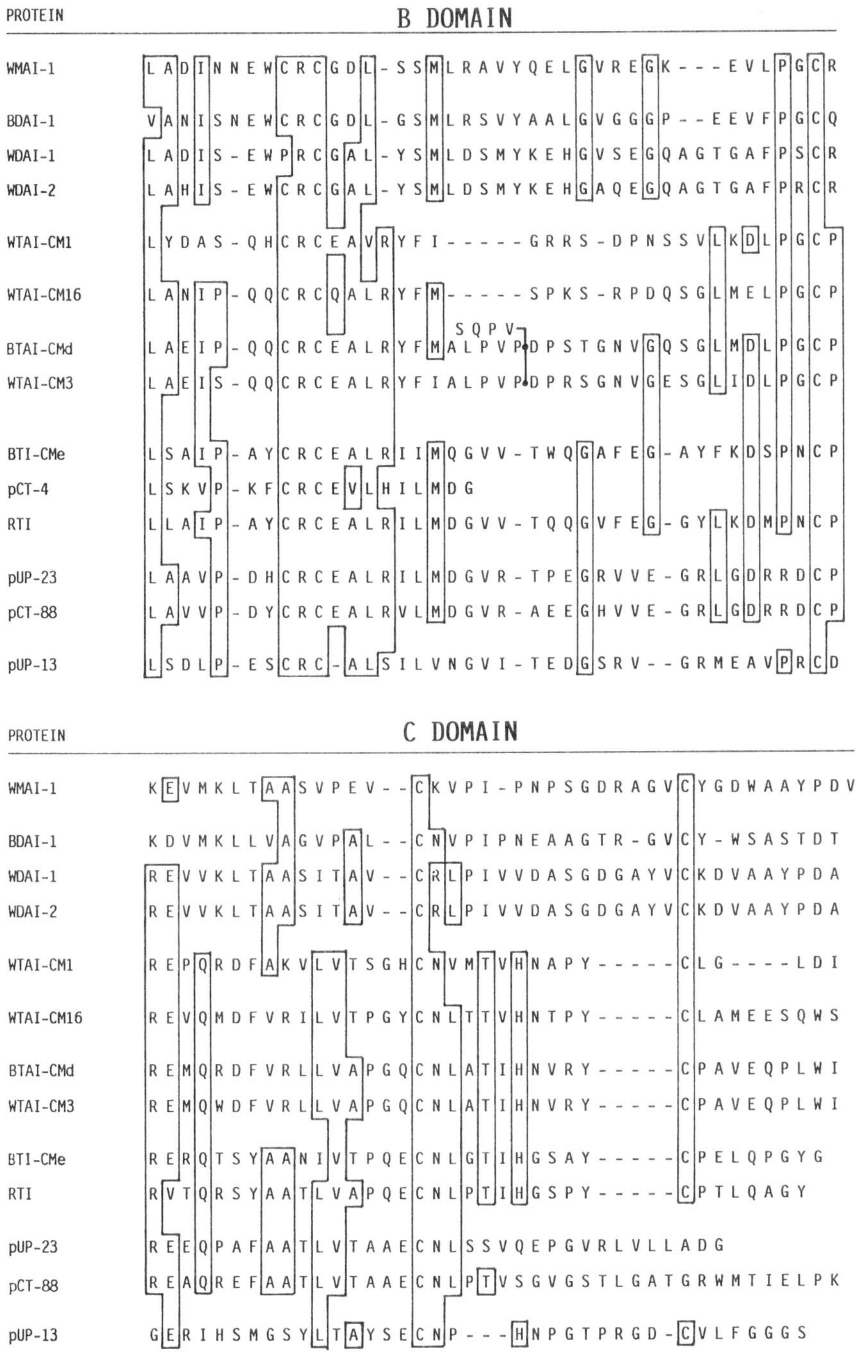

(13) Halford *et al.*, 1988; (14) Odani *et al.*, 1983b; (15) Lazaro *et al.*, 1985; (16) Rodriguez-Palenzuela *et al.*, 1989; (17) Carbonero *et al.*, unpublished; (18) Lazaro *et al.*, 1988a; (19) Gautier *et al.*, 1990. Domains A, B and C are indicated. Sequences have been grouped according to similarity and to their *in vitro* activities and chromosomal locations of the genes as indicated in Table 16.1.

The proteins corresponding to four of the cDNA clones have not been characterized yet, although the sequence deduced for one of them (pCT-4) can be aligned with that of the trypsin inhibitor BTI-CMe from barley.

The sequence homology data and the chromosomal location of the corresponding genes indicate that this dispersed gene family has originated both by translocation and by intrachromosomal duplication, and that most if not all of the dispersion must have occurred before the divergence of the barley genome from the diploid genomes included in allohexaploid wheat. An interesting example of intrachromosomal duplication is that of the *Iat1* locus encoding CMa and the *Itr2* locus encoding CMc on chromosome 1 (7H). The two proteins are more closely related to each other than to any other proteins encoded by the same genome. Nevertheless, BTAI-CMa, which is a subunit of the tetrameric α-amylase inhibitor, is even closer to the equivalent subunits in wheat (CM1 and CM2), which are encoded by chromosomes of group 7. In contrast, BTI-CMc is a trypsin inhibitor which shows much greater divergence from the other trypsin inhibitor, BTI-CMe [chromosome 3 (3H)], than from CMa (45% identical positions vs 78%).

The considerable sequence divergence found for this protein family within a given genome is in sharp contrast with the low allelic variability found for the individual members of the group both in barley (Salcedo *et al.*, 1984; Molina-Cano *et al.*, 1987; Sanchez-Monge *et al.*, 1988) and in wheat (García-Olmedo & García-Faure, 1969; García-Olmedo and Carbonero, 1970; Rodriguez-Loperena *et al.*, 1975; Salcedo *et al.*, 1978). In a wide-ranging survey of *Hordeum vulgare* cultivars and *Hordeum spontaneum* accessions analysed by two-dimensional electrophoresis (Salcedo *et al.*, 1984; Molina-Cano *et al.*, 1987), the following observations were made: BTAI-CMd, BTAI-CMa and BTI-CMc were essentially invariant throughout the two species (the last two proteins occurred in variant forms in one accession of *H. spontaneum* each); BTAI-CMb existed in variant forms in both species; and in the case of BTI-CMe, two variants, designated CMe2/CMe2′, were jointly inherited and co-dominantly expressed with respect to CMe (Salcedo *et al.*, 1984). Variant subunits of BTAI-CMa, BTAI-CMb and BTAI-CMd have been found in *Hordeum chilense* (Sanchez-Monge *et al.*, 1987).

The distribution of rare genetic variants of members of this protein family among Moroccan and non-Moroccan accessions of *Hordeum spontaneum* and among selections from several Moroccan landraces of *H. vulgare* and cultivars of the same species of wide-ranging origins, suggested that domestication of barley might have taken place in Morocco (Molina-Cano *et al.*, 1987).

Gene expression

The synthesis of α-amylase/trypsin inhibitors appears to precede that of the bulk of reserve proteins and starch. Thus Kirsi (1973) detected the barley trypsin inhibitor at about 5 days after pollination (DAP) and found that most was accumulated between 10 and 23 DAP and that the net amount was not affected by nitrogen fertilization. In a study of the *in vivo* and *in vitro* synthesis of these proteins in barley, Paz-Ares *et al.* (1983) concluded that synthesis peaked between 15 and 20 DAP. The inhibitors were synthesized by membrane-bound polysomes as precursors of higher apparent molecular weight (13–21 kDa) than the mature proteins (12–16 kDa). Accordingly, all cloned cDNAs corresponding to this protein family encode putative signal peptides that precede the N-terminal sequences of the mature proteins (Paz-Ares *et al.*, 1986; Halford *et al.*, 1988; Lazaro *et al.*, 1988b; García-Maroto *et al.*, 1990 and unpublished).

A significant homology has been found between at least some of these signal peptides and that of the precursor of the sweet protein thaumatin II (Lazaro *et al.*, 1988a).

The effects of high-lysine mutations on the expression of different genes from this family in barley have been investigated (Salcedo *et al.*, 1984; Lazaro *et al.*, 1985; Rodriguez-Palenzuela *et al.*, 1989). The most notable effect concerns the gene for BTI-CMe, which is regulated in *trans* by high-lysine loci. The inhibitor is not detected in the mutant Risø 1508 (*lys 3a*) and is present at a much lower level in Hiproly (*lys*) (Salcedo *et al.*, 1984; Rodriguez-Palenzuela *et al.*, 1989). In both cases the mutations affect loci on chromosome 7 (5H), while the *Itr1* locus encoding CMe is on chromosome 3 (3H) (see Lazaro *et al.*, 1985). Steady-state levels of the corresponding mRNA in the mutant Risø 1508 are about 1% of those in the wild-type (Rodriguez-Palenzuela *et al.*, 1989). One or two copies of the gene have been estimated to be present in both the wild-type and in the mutant, and DNA restriction patterns are identical in both stocks, so neither a change in copy number nor a major rearrangement of the structural gene account for the markedly decreased expression (Rodriguez-Palenzuela *et al.*, 1989). A single dose of the wild-type allele at the regulatory *lys3a* locus restored the expression of the structural gene (Rodriguez-Palenzuela *et al.*, 1989).

The first inhibitor gene to be cloned was that coding for BTI-CMe (Royo *et al.*, unpublished). The gene has no introns and is preceded at about 2 kb upstream of the initiation codon by the 5′ end of a retroposon-like sequence. This suggests a possible dispersal mechanism responsible for the spread of inhibitor genes over several chromosomes.

The subcellular location of the different members of this family has not been fully established. In the case of α-amylase inhibitors, indirect evidence of association with starch granules has been reported (see Buonocore *et al.*, 1977). More recent immunodetection data seem to confirm this evidence, suggesting an external location of the inhibitors in these granules (Carmona *et al.*, unpublished).

Rapid disappearance of the inhibitors with the onset of germination suggests that these proteins do not play a specific role during germination (Kirsi & Mikola, 1971; Pace *et al.*, 1978).

In vitro activities

Monomeric and dimeric inhibitors show different specificities when tested against α-amylases from human saliva and from the digestive tract of the insect *Tenebrio molitor*, which have been routinely used as standard amylases in purification procedures (Fig. 16.2). It should be noted that the barley dimeric inhibitor (BDAI-1) shows different specificity from the wheat dimeric inhibitors. Not all insect α-amylases show the same specificity towards the different types of inhibitor: i.e. the enzyme from *T. molitor* is significantly more sensitive to the wheat monomeric inhibitor than to the dimeric type and the opposite is true for the enzyme from *Leptinotarsa decemlineata* (Colorado potato beetle), while in certain other cases the enzymes are about equally sensitive to both types (Gutierrez *et al.*, 1990).

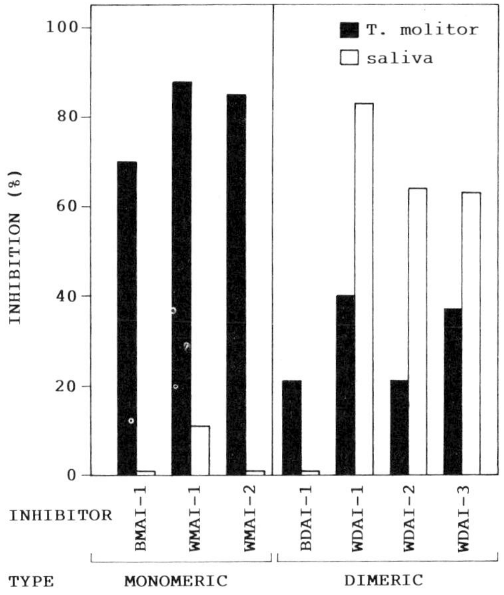

Fig. 16.2. Different specificity of monomeric and dimeric inhibitors towards α-amylases from the insect *Tenebrio molitor* and from human saliva. Inhibition was achieved by 1 μg of pure inhibitor against 1 unit of enzyme.

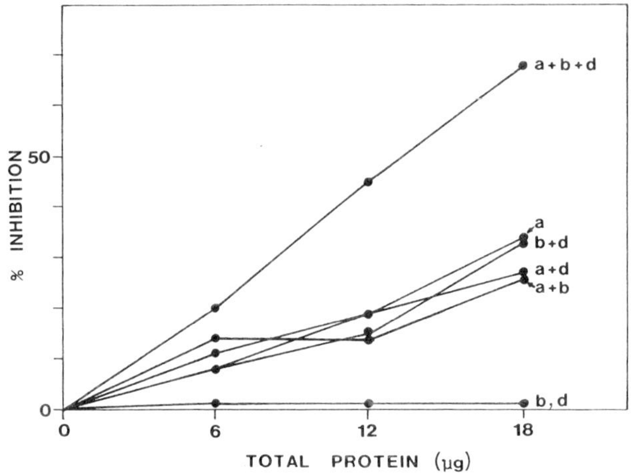

Fig. 16.3. Inhibition of α-amylase from larvae of *Tenebrio molitor* by purified proteins CMa, CMb, CMd and their mixtures. Equal amounts of each of the indicated proteins were used in the mixtures. All tests were carried out with approx. 2 units of α-amylase (1 unit defined as the amount of enzyme required to produce the reducing equivalents of 1 mg of maltose in 20 min).

The tetrameric inhibitor from barley seems to be active against the α-amylase from *T. molitor* and not against the salivary one (Sanchez-Monge *et al.*, 1986b). The subunits of this inhibitor were identified as proteins BTAI-CMa, BTAI-CMb and BTAI-CMd (two copies). Only the first protein showed some activity by itself but, as shown in Fig. 16.3, the association of the three different subunits was required for full activity (Sanchez-Monge *et al.*, 1986b). The identity of the subunits of the wheat tetrameric inhibitors was predicted on the basis of the studies of barley, and subsequently confimed by analysis of the inhibitors and by their reconstitution from the purified subunits (Gomez *et al.*, 1989).

The trypsin inhibitor BTI-CMe is inactive against chymotrypsin, papain, subtilopeptidase A, pepsin, bacterial or fungal proteinases, as well as against the endogenous proteinases from green malt (Mikola & Suolinna, 1969).

An important biological property of all these proteins is that they are the major allergens in baker's asthma, which is an important occupational disease. BMAI-1 was the first to be shown to react with sera from patients (Barber *et al.*, 1989) and subsequently subunits of both dimeric and tetrameric barley inhibitors have been identified as significant allergens (Sanchez-Monge *et al.*, unpublished). As would be expected, most of the wheat components of this family also show strong reactions (Gomez *et al.*, 1990).

Thionins

Structural types

More than 20 different thionin sequences have been determined so far, either directly by amino acid sequencing or indirectly through cDNA cloning and nucleotide sequencing (see García-Olmedo *et al.*, 1989, 1991). These sequences have been classified into five structural types (García-Olmedo *et al.*, 1991), using the algorithm of Feng & Doolittle (1987). At least three of these types, whose gross structural features are represented in Fig. 16.4, are present in barley and wheat.

Type I corresponds to that of the original purothionins isolated from wheat endosperm by Balls *et al.* (1942a, b). Two non-allelic genetic variants (α, β) of this type have been characterized in barley endosperm (Ozaki *et al.*, 1980; Hernandez-Lucas *et al.*, 1986; Ponz *et al.*, 1986). Thionins of this type have four disulphide bridges and are highly basic, with no negatively charged amino acid residues. Total chain length is 45 residues, 8 of which are in the central disulphide loop.

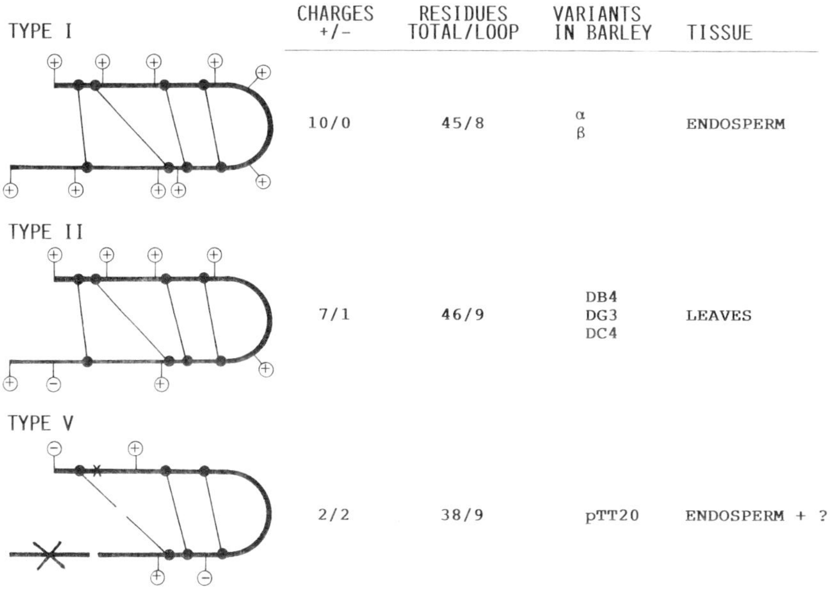

Fig. 16.4. Structural types of thionins present in barley.

Type II corresponds to thionins isolated from the leaves of the parasitic plant *Pyrularia pubera* (Vernon *et al*., 1985) and from those of barley (Bohlmann & Apel, 1987; Gausing, 1987). These thionins are also quite basic, but have some negatively charged residues. Disulphide bridges are conserved and total chain length is 46–7 residues, with 9–10 in the central loop. Three different but closely related complete sequences have been deduced from the nucleotide sequences of cDNA clones in barley.

The third thionin type in barley and wheat, type V, is quite divergent from the other types: the 2nd and 8th cysteines of types I and II are missing, through point mutation and deletion, respectively, thus disrupting the 1st and 2nd disulphide bridges and potentially allowing the formation of a new bridge between the unpaired cysteines (Castagnaro *et al*., unpublished). This new type, which is neutral, has been first identified in a cDNA library derived from developing wheat kernels and is also present in barley.

Thionin genes in barley and related species

There are 1–2 genes for type I thionins per haploid genome in barley, wheat and rye. In the last two species, these genes have been located on the long arms of the group 1 chromosomes by aneuploid analysis (Fernandez de Caleya *et al*., 1976; Sanchez-Monge *et al*., 1979; Rodriguez-Palenzuela *et al*., 1988). A similar location on chromosome 5 (1H) of barley is to be expected, but direct confirmation has not been achieved by aneuploid analysis because the critical addition line has not been obtained. Type II genes have been located on chromosome 6 (6H) of barley by Southern blot analysis of DNAs from wheat–barley addition lines (Bohlman *et al*., 1988), but there are discrepancies about their number: about 100 genes (Bohlman *et al*., 1988) or 9–11 genes (Gausing, 1987). Type V genes have been located within a few kb of the type I genes in wheat, with only one copy per haploid genome (Castagnaro *et al*., unpublished).

Gene expression

Barley endosperm thionins (Type I) are synthesized by membrane-bound polysomes as much larger precursors that undergo at least two processing steps: one co-translational and one post-translational (Ponz *et al.*, 1983). As predicted from these studies, the nucleotide sequences of the cDNAs corresponding to α and β thionins from barley endosperm were found to encode precursors with an N-terminal signal peptide followed by the sequence of the mature protein and that of a C-terminal acidic peptide, as shown in Fig. 16.5 (Hernandez-Lucas *et al.*, 1986; Ponz *et al.*, 1986). The same precursor structure has been subsequently found for type II (Bohlmann & Apel, 1987; Gausing, 1987) and for type V thionins (Castagnaro *et al.*, unpublished). It is therefore likely that all three types are processed in the same manner.

Thionin accumulation in developing endosperm appears to start and to level off before synthesis of the major storage proteins, with maximum steady-state concentrations of mRNAs 13–16 days after pollination (Ponz *et al.*, 1983; Rodriguez-Palenzuela *et al.*, 1988).

The final deposition site of type I thionins seems to be in the periphery of the protein bodies, based on recent immunogold detection by electron microscopy (Carmona *et al.*, unpublished) and on previous subcellular fractionation experiments (Carbonero *et al.*, 1980; Ponz *et al.*, 1983).

It was first claimed that leaf thionins were exclusively located in the cell wall, based on electron micrographs of immunogold-labelled thin sections, using an antibody raised against a fusion protein expressed in *E. coli* (Bohlmann *et al.*, 1988). Subsequently, these authors found that about 98% of the leaf thionins were intracellular (Reiman-Philipp *et al.*, 1989b). To reconcile the contradictory observations, they proposed that there were two distinct structural groups of leaf thionins: the cell wall and the intracellular ones (Reiman-Philipp *et al.*, 1989b). In our opinion, this claim was not sufficiently supported by their evidence.

The gene coding for α-thionin from barley endosperm (type I) has been cloned and its complete sequence has been published (Rodriguez-Palenzuela *et al.*, 1988). As represented in Fig. 16.5, the gene has two introns of 420 and 91 bp, respectively, both of which interrupt the sequence that corresponds to the C-terminal acidic peptide of the precursor. Although genomic clones of type II thionins have not been described in detail, it seems that they also have two introns (Bohlman *et al.*, 1988). More recently, it has been found that type V genes have the same intron organization as type I genes (Castagnaro *et al.*, unpublished).

Type I genes appear to be specifically expressed in barley endosperm during the cell proliferation phase (Rodriguez-Palenzuela *et al.*, 1988) and fusions of the corresponding promoter with the glucuronidase (GUS) reporter gene direct endosperm-specific expression in transgenic tobacco plants with the same temporal control (Fernandez *et al.*, unpublished). When fusions of the 35S promoter with the α-thionin gene (coding regions and introns) are expressed in transgenic tobacco, the introns are correctly spliced, in contrast with what happens when some genes from

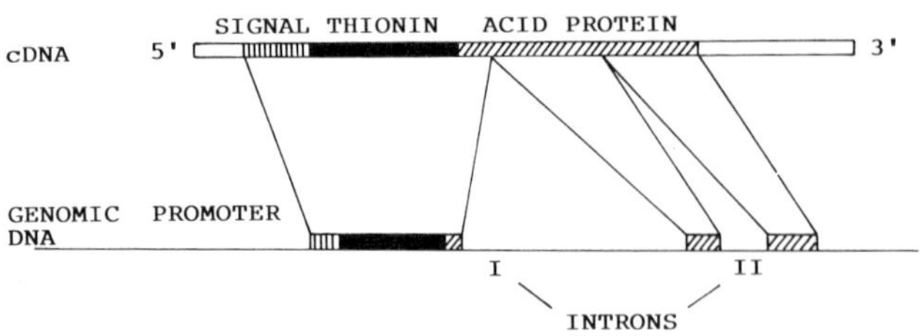

Fig. 16.5. cDNAs and genomic DNAs encoding thionin precursors.

monocotyledonous plants are expressed in a dicotyledonous species (Carmona *et al.*, unpublished).

Type II genes seem to be leaf-specific. They are expressed in the dark and are switched off upon illumination (Bohlman & Apel, 1987; Gausing, 1987; Reiman-Philipp *et al.*, 1989a). The inhibitory effect of light can be overcome by stress and pathogen-induced signals, as it has been shown that fungal infection induces a transient expression of the thionin genes in the leaves (Bohlman *et al.*, 1988; Ebrahim-Nesbat *et al.*, 1989) and that the chlorides of divalent cations (Mg^{2+}, Mn^{2+}, Cd^{2+}, Zn^{2+}) elicit a more permanent response (Fisher *et al.*, 1989).

No information is yet available concerning the developmental expression of type V genes.

In vitro properties

The toxicity of thionins towards different kinds of organisms and to cells in culture has been investigated for several decades (for a review, see García-Olmedo *et al.*, 1991). Toxicity of thionins to plant pathogens was first reported by Fernandez de Caleya *et al.* (1972). These studies were stimulated by earlier reports concerning the antimicrobial properties of these polypeptides (Stuart & Harris, 1942; Balls & Harris, 1944). Plant pathogenic bacteria of the genera *Pseudomonas*, *Xanthomonas*, *Agrobacterium*, *Erwinia* and *Corynebacterium* were found to be sensitive to wheat-endosperm thionins (type I) at concentrations ranging from 10^{-7} to 10^{-5}M, and purified genetic variants of the thionins had different activities and showed some degree of specificity (Fernandez de Caleya *et al.*, 1972). In a recent survey, the sensitivity of fungal plant pathogens to pure genetic variants of thionins from barley (types I and II) and from wheat (type I) was found to be in the 0^{-6}M–10^{-5}M range (García-Olmedo *et al.*, 1991; Molina & Fraile, unpublished). Barley thionins of types I and II were about equally effective, although for a given pathogen certain variants were somewhat more toxic than others. Bohlmann *et al.* (1988) have reported the sensitivities of two fungi, *Thielaviopsis paradoxa* and *Drechslera teres*, to both types of thionins at 5×10^{-4}M, a concentration that is orders of magnitude higher than those discussed above.

Two types of mechanism could be responsible for the toxicity of thionins: the permeabilization of the cell membrane, followed by inhibition of macromolecular synthesis (Fernandez de Caleya, 1973; Carrasco *et al.*, 1981; García-Olmedo *et al.*, 1983) or the interference with thioredoxin-mediated redox reactions (Wada & Buchanan, 1981; Johnson *et al.*, 1987).

Evidence of a possible defence role for inhibitors and thionins

Although both families are toxic or inhibitory to heterologous systems *in vitro*, their true *in vivo* functions remain unknown. In addition both show little or no intraspecific variability at given loci and their distribution is apparently limited to a few taxa, although the thionins might yet prove to be ubiquitous.

The known properties of the trypsin/α-amylase inhibitors do not lead to any obvious hypothesis about their role in plant metabolism, whereas in the case of thionins, a participation in the regulatory pathways linking light to enzyme activation/inactivation or in disulphide rearrangement mechanisms can be suggested (see García-Olmedo *et al.*, 1991). The evidence of a defence role for these proteins is increasing and merits a brief discussion here.

In the case of the inhibitors, the following aspects are to be noted: (i) different inhibitors show different specificities towards enzymes from different insects (Gutierrez *et al.*, 1990 and unpublished); (ii) considerable intra- and interspecific variation in inhibitor levels has been observed (Kirsi & Ahokas, 1983; Gomez *et al.*, 1989 and unpublished); and (iii) gene silencing has occurred in some cases (García-Maroto *et al.*, 1990). These characteristics, together with the low genetic variability at given loci, suggest that if these proteins are involved in plant defence, they are probably related to non-host resistance rather than to more specific interactions. More direct evidence for a defence role stems from observations and experiments with insect pests. Thus, insects that are able to feed on wheat endosperm have unusually high levels of α-amylase (Silano *et al.*, 1975; Gutierrez *et al.*, 1990). High inhibitor concentrations in an artificial diet were required to affect the development of larvae of *Tribolium confusum*, a storage pest of wheat products, while quite low concentrations were effective against *Callosobruchus maculatus*, a pest of legume seeds (Gatehouse *et al.*, 1986). More recently, transgenic tobacco plants carrying the structural parts of the barley gene coding for BTI-CMe or the wheat gene coding for WMAI-1 (0.28) under the control of the 35S promoter have been found to be lethal to *Agrotis ipsilon* in a leaf-disc assay (Carbonero *et al.*, unpublished).

No true specificity has yet been found among the genetic variants of thionins with respect to pathogens, which suggests that their possible defence role would be related to the general resistance barriers. Besides their *in vitro* biocidal activity, preliminary experiments with transgenic tobacco plants carrying thionin genes under the 35S promoter were less sensitive to bacterial pathogens than the non-transformed controls (Carmona *et al.*, unpublished).

From our current knowledge of the two protein families, α-amylase/trypsin inhibitors and thionins, it can be concluded that they are potentially useful for the genetic manipulation of plant resistance to pests and pathogens, independently of whether or not they are true defence proteins.

Acknowledgements

We would like to thank D. Lamoneda and J. García for technical assistance, the Fundación Ramón Areces for support of our current work on thionins. Grants ALI89-0121 and BIO88-0216 from the Comisión Interministerial de Ciencia y Tecnología supports work on the α-amylase/trypsin inhibitors.

References

Aragoncillo, C., Rodriguez-Loperena, M.A., Carbonero, P. & García-Olmedo, F. (1975) Chromosomal control of non-gliadin proteins from the 70% ethanol extract of wheat endosperm. *Theoretical and Applied Genetics* 45, 322–6.

Balls, A.K., Hale, W.S. & Harris, T.H. (1942a) A crystalline protein obtained from a lipoprotein of wheat flour. *Cereal Chemistry* 19, 279–88.

Balls, A.K., Hale, W.S. & Harris, T.H. (1942b) Further observations on a crystalline wheat protein. *Cereal Chemistry* 19, 840–4.

Balls, A.K. & Harris, T.H. (1944) The inhibitory effect of a protamine from wheat flour on the fermentation of wheat mashes. *Cereal Chemistry* 21, 74–9.

Barber, D., Sanchez-Monge, R., García-Olmedo, F., Salcedo, G. & Mendez, E. (1986a) Evolutionary implications of sequential homologies among members of the trypsin/α-amylase inhibitor family (CM-proteins) in wheat and barley. *Biochimica et Biophysica Acta* 873, 147–51.

Barber, D., Sanchez-Monge, R., Mendez, E., Lazaro, A., García-Olmedo, F. & Salcedo, G. (1986b) New α-amylase and trypsin inhibitors among the CM-proteins of barley (*Hordeum vulgare*). *Biochimica et Biophysica Acta* 869, 115–18.

Barber, D., Sanchez-Monge, R., Gomez, L., Carpizo, J., Armentia, A., Lopez-Otin, C., Juan, F. & Salcedo, G. (1989) A barley flour inhibitor of insect α-amylase is a major allergen associated with baker's asthma disease. *FEBS Letters* 248, 119–22.

Bohlmann, H. & Apel, K. (1987) Isolation and characterization of cDNAs coding for leaf-specific thionins closely related to the endosperm-specific hordothionin of barley (*Hordeum vulgare* L.). *Molecular and General Genetics* 207, 446–54.

Bohlmann, H., Clausen, S., Behnke, S., Giese, H., Hiller, C., Reiman-Philipp, U., Schrader, G., Barkholt, V. & Apel, K. (1988) Leaf-specific thionins of barley – a novel class of cell wall proteins toxic to plant-pathogenic fungi and possibly involved in the defense mechanism of plants. *EMBO Journal* 7, 1559–65.

Boisen, S. (1976) Characterization of the endosperm trypsin inhibitor in barley. *Phytochemistry* 15, 641.

Buonocore, V., Petrucci, T. & Silano, V. (1977) Wheat protein inhibitors of α-amylase. *Phytochemistry* 16, 811–20.

Campos, F.A.P. & Richardson, M. (1983) The complete amino acid sequence of the bifunctional α-amylase/trypsin inhibitor from seeds of ragi (Indian finger millet; *Eleusine coracana* Goertn.). *FEBS Letters* 152, 300–4.

Carbonero, P., García-Olmedo, F. & Hernandez-Lucas, C. (1980) External association of hordothionin with protein bodies in mature barley. *Journal of Agriculture and Food Chemistry* 28, 399–402.

Carrasco, L., Vazquez, D., Hernández-Lucas, C., Carbonero, P. & García-Olmedo, F. (1981) Thionins: plant peptides that modify membrane permeability in cultured mammalian cells. *European Journal of Biochemistry* 116, 185–9.

Ebrahim-Nesbat, F., Behnke, S., Kleinhofs, A. & Apel, K. (1989) Cultivar-related differences in the distribution of cell-wall-bound thionins in compatible and incompatible interactions between barley and powdery mildew. *Planta* 179, 203–10.

Feng, D.-F. & Doolittle, R.F. (1987) Progressive sequence alignment as a prerequisite to correct phylogenetic trees. *Journal of Molecular Evolution* 25, 351–60.

Fernandez de Caleya, R. (1973) Caracterización química y propiedades antimicrobianas de purotioninas. PhD Thesis, Universidad Politecnica de Madrid, Spain.

Fernandez de Caleya, R., Gonzalez-Pascual, B., García-Olmedo, F. & Carbonero, P. (1972) Susceptibility of phytopathogenic bacteria to wheat purothionins in vitro. *Applied Microbiology* 23, 998–1000.

Fernández de Caleya, R., Hernandez-Lucas, C., Carbonero, P. & García-Olmedo, F. (1976) Gene expression in alloploids: genetic control of lipopurothionins in wheat. *Genetics* 83, 687–99.

Fisher, R., Behnke, S. & Apel, K. (1989) the effect of chemical stress on the polypeptide composition of the intercellular fluid of barley leaves. *Planta* 178, 61–8.

Fra-Mon, P., Salcedo, G., Aragoncillo, C. & García-Olmedo, F. (1984) Chromosomal assignment of genes controlling salt-soluble proteins (albumins and globulins) in wheat and related species. *Theoretical and Applied Genetics* 69, 167–72.

García-Maroto, F., Maraña, C., Mena, M., García-Olmedo, F., & Carbonero, P. (1990) Cloning of cDNA and chromosomal location of genes encoding the three types of subunits of the wheat tetrameric inhibitors of insect α-amylase. *Plant Molecular Biology* 14, 845–53.

García-Olmedo, F. & García-Faure, R. (1969) A new method for the estimation of common wheat (*T. aestivum* L.) in pasta products. *Lebensmittel-Wissenschaft Technologie* 2, 94–6.

García-Olmedo, F. & Carbonero, P. (1970) Homoeologous protein synthesis controlled by homoeologous chromosomes in wheat. *Phytochemistry* 9, 1495–7.

García-Olmedo, F., Carbonero, P., Hernandez-Lucas, C., Paz-Ares, J., Ponz, F., Vicente, O. & Sierra, J.M. (1983) Inhibition of eukaryotic cell-free protein synthesis by thionins from wheat endosperm. *Biochimica et Biophysica Acta* 740, 52–6.

García-Olmedo, F., Salcedo, G., Sanchez-Monge, R., Gomez, L., Royo, J. & Carbonero, P. (1987) Plant proteinaceous inhibitors of proteinases and α-amylases. *Oxford Surveys of Plant Molecular and Cell Biology* 4, 275–334.

García-Olmedo, F., Rodriguez-Palenzuela, P., Hernandez-Lucas, C., Ponz, F., Maraña, C., Carmona, M.J., Lopez-Fando, J., Fernandez J.A. & Carbonero, P. (1989) The thionins: a protein family that includes purothionins, viscotoxins and crambins. *Oxford Surveys of Plant Molecular and Cell biology* 6, 31–60.

García-Olmedo, F., Carmona, M.J., Lopez-Fando, J.J., Fernandez, J.A., Castagnaro, A., Molina, A., Hernández-Lucas, C. & Carbonero, P. (1991) Characterization and analysis of thionin genes. In: Boller, Th. & Meins, F. (eds), *Genes Involved in Plant Defense* (Plant Gene Research Series). Springer-Verlag, Berlin.

Gatehouse, A.M.R., Fenton, K.A., Jepson, I. & Pavey, D.J. (1986) The effects of α-amylase inhibitors on insect storage pests: inhibition of α-amylase in vitro and effects on development in vivo. *Journal of the Science of Food and Agriculture* 37, 727–34.

Gausing, K. (1987) Thionin genes specifically expressed in barley leaves. *Planta* 171, 241–6.

Gautier, M.F., Alary, P. & Joudrier, P. (1990) Cloning and characterization of a cDNA encoding the wheat (*Triticum aestivum* Desf.) CM16 protein. *Plant Molecular Biology* 14, 313–22.

Gomez, L., Sanchez-Monge, R., García-Olmedo, F. & Salcedo, G. (1989) Wheat tetrameric inhibitors of insect α-amylases: alloploid heterosis at the molecular level. *Proceedings of the National Academy of Sciences (USA)* 86, 3242–46.

Gomez, L., Martin, E., Hernandez, D., Sanchez-Monge, R., Barber, D., del Pozo, V., Andres, B., Armentia, A., Lahoz, C., Salcedo, G. & Palomino, P. (1990) Members of the α-amylase inhibitors family from wheat endosperm are major allergens associated with baker's asthma. *FEBS Letters* 261, 85–8.

Gutierrez, C., Sanchez-Monge, R., Gomez, L., Ruiz-Tapiador, M., Castañera, P. & Salcedo, G. (1990) α-Amylase activities of agricultural insect pests are specifically affected by different inhibitor preparations from wheat and barley endosperm. *Plant Science* 72, 37–44.

Halford, N.G., Morris, N.A., Urwin, P., Williamson, M.S., Kasarda, D.D., Lew, E.J.L., Kreis, M. & Shewry, P. (1988) Molecular cloning of the barley seed protein CMd: a variant member of the α-amylase/trypsin inhibitor family of cereals. *Biochimica et Biophysica Acta* 950, 435–40.

Hejggard, J., Bjorn, S.E. & Nielsen, G. (1984a) Localization to chromosomes of structural genes for the major protease inhibitors in barley grains. *Theoretical and Applied Genetics* 68, 127–30.

Hejggard, J., Bjorn, S.E. & Nielsen, G. (1984b) Rye chromosomes carrying structural genes for the major grain protease inhibitors. *Hereditas* 101, 257–9.

Hernandez-Lucas, C., Royo, J., Paz-Ares, J., Ponz, F., García-Olmedo, F. & Carbonero, P. (1986) Polyadenylation site heterogeneity in mRNA encoding the precursor of the barley toxin α-hordothionin. *FEBS Letters* 200, 103–5.

Johnson, T.C., Wada, K., Buchanan, B.B.. & Holmgren, A. (1987) Reduction of purothionin by the wheat seed thioredoxin system. *Plant Physiology* 85, 446–51.

Kashlan, N. & Richardson, M. (1981) The complete amino acid sequence of a major wheat protein inhibitor of α-amylase. *Phytochemistry* 20, 1781–4.

Kirsi, M. (1973) Formation of proteinase inhibitors in developing barley grains. *Physiologia Plantarum* 29, 141–4.

Kirsi, M. & Ahokas, H. (1983) Trypsin inhibitor activities in the wild progenitor of barley. *Phytochemistry* 22, 2739–40.

Kirsi, M. & Mikola, J. (1971) Occurrence of proteolytic inhibitors in various tissues of barley. *Planta* 96, 281–91.

Kreis, M., Forde, B.G., Rahman, S., Miflin, B.J. & Shewry, P.R. (1985) Molecular evolution of the seed storage proteins of barley, rye and wheat. *Journal of Molecular Biology* 183, 499–502.

Lazaro, A., Barber, D., Salcedo, G., Mendez, E. & García-Olmedo, F. (1985) Differential effects of high-lysine mutations on the accumulation of individual members of a group of proteins encoded by a disperse multigene family in the endosperm of barley (*Hordeum vulgare* L.). *European Journal of Biochemistry* 149, 617–23.

Lazaro, A., Rodriguez-Palenzuela, P., Maraña, C., Carbonero, P. & García-Olmedo, F. (1988a) Signal peptide homology between the sweet protein thaumatin II and unrelated cereal α-amylase/trypsin inhibitors. *FEBS Letters* 239, 147–50.

Lazaro, A., Sanchez-Monge, R., Salcedo, G., Paz-Ares, J., Carbonero, P. & García-Olmedo, F. (1988b) A dimeric inhibitor of insect α-amylase from barley. Cloning of the cDNA and identification of the protein. *European Journal of Biochemistry* 172, 129–34.

Lyons, A., Richardson, M., Tatham, A.S. & Shewry, P.R. (1987) Characterization of homologous inhibitors of trypsin and α-amylase from seeds of rye (*Secale cereale* L.). *Biochimica et Biophysica Acta* 915, 305–13.

Maeda, K., Kakabayashi, S. & Matsubara, H. (1985) Complete amino acid sequence of an α-amylase inhibitor in wheat kernel (0.19 inhibitor). *Biochimica et Biophysica Acta* 828, 213–21.

Mahoney, W.C., Hermodson, M.A., Jones, B., Powers, D.D., Corfman, R.S. & Reeck, G.R. (1984) Amino acid sequence and secondary structural analysis of the corn inhibitor of trypsin and activated Hageman factor. *Journal of Biological Chemistry* 259, 8412–16.

Mikola, J. & Kirsi, M. (1972) Differences between endospermal and embryonal trypsin inhibitors in barley, wheat and rye. *Acta Chemica Scandinavica* 26, 787–95.

Mikola, J. & Soulinna, E.M. (1969) Purification and properties of a trypsin inhibitor from barley. *European Journal of Biochemistry* 9, 555–60.

Molina-Cano, J.L., Fra-Mon, P., Salcedo, G., Aragoncillo, C., Roca de Tagores, F. & García-Olmedo, F. (1987) Morocco as a possible domestication center for barley: biochemical and agromorphological evidence. *Theoretical and Applied Genetics* 73, 531–6.

Odani, S., Koide, T. & Ono, T. (1983a) The complete amino acid sequence of barley trypsin inhibitor. *Journal of Biological Chemistry* 258, 7998–8003.

Odani, S., Koide, T. & Ono T. (1983b) A possible evolutionary relationship between plant trypsin inhibitor, α-amylase inhibitor, and mammalian pancreatic secretory trypsin inhibitor (Kazal). *Journal of Biochemistry* 93, 1701–4.

Odani, S., Koide, T., Ono, T. & Ohnishi, K. (1983c) Structural relationships between barley (*Hordeum vulgare*) trypsin inhibitor and castor-bean (*Ricinus communis*) storage protein. *Biochemical Journal* 213, 543–5.

Ozaki, Y., Wada, K., Hase, T., Matsubara, H., Nakanishi, T. & Yoshizumi, H. (1980) Amino acid sequence of a purothionin homologue from barley flour. *Journal of Biochemistry* 87, 549–55.

Pace, W., Parlamenti, R., Rab, A., Silano, V. & Vittozzi, L. (1978) Protein α-amylase inhibitors from wheat flour. *Cereal Chemistry* 55, 244–54.

Paz-Ares, J., Ponz, F., Aragoncillo, C., Hernandez-Lucas, C., Salcedo, G., Carbonero, P. & García-Olmedo, F. (1983) In vivo and in vitro synthesis of CM-proteins (A-hordeins) from barley (*Hordeum vulgare* L.). *Planta* 157, 74–80.

Paz-Ares, J., Ponz, F., Rodriguez-Palenzuela, P., Lazaro, A., Hernandez-Lucas, C., García-Olmedo, F. & Carbonero, P. (1986) Characterization of cDNA clones of the family of trypsin/α-amylase inhibitors (CM-proteins) in barley (*Hordeum vulgare* L.). *Theoretical and Applied Genetics* 71, 842–6.

Ponz, F., Paz-Ares, J., Hernandez-Lucas, C., Carbonero, P. & García-Olmedo, F. (1983) Synthesis and processing of thionin precursors in developing endosperm from barley (*Hordeum vulgare* L.). *EMBO Journal* 2, 1035–40.

Ponz, F., Paz-Ares, J., Hernandez-Lucas, C., García-Olmedo, F. & Carbonero, P. (1986) Cloning and nucleotide sequence of a cDNA encoding the precursor of the barley toxin α-hordothionin. *European Journal of Biochemistry* 156, 131–5.

Reimann-Philipp, U., Behnke, S., Batschauer, A., Schafer, E. & Apel, K. (1989a) The effect of light on the biosynthesis of leaf-specific thionins in barley *Hordeum vulgare*. *European Journal of Biochemistry* 182, 283–9.

Reimann-Philipp, U., Schrader, G., Martinoia, E., Barkholt, V. & Apel, K. (1989b) Intracellular thionins of barley. A second group of leaf thionins closely related to but distinct from cell wall-bound. *Journal of Biological Chemistry* 264, 8978–84.

Rodriguez-Loperena, M.A., Aragoncillo, C., Carbonero, P. & García-Olmedo, F. (1975) Heterogeneity of wheat endosperm proteolipids (CM-proteins). *Phytochemistry* 14, 1219–23.

Rodriguez-Palenzuela, P., Pintor-Toro, J.A., Carbonero, P. & García-Olmedo, F. (1988) Nucleotide sequence and endosperm-specific expression of the structural gene for the toxin α-hordothionin in barley (*Hordeum vulgare* L.). *Gene* 70, 271–81.

Rodriguez-Palenzuela, P., Royo, J., Gomez, L., Sanchez-Monge, R., Salcedo, G., Molina-Cano, J.L., García-Olmedo, F. & Carbonero, P. (1989) The gene for trypsin inhibitor CMe is regulated in trans by the *lys 3a* locus in the endosperm of barley (*Hordeum vulgare* L.). *Molecular and General Genetics* 219, 474–9.

Salcedo, G., Aragoncillo, C., Rodriguez-Loperena, M.A., Carbonero, P. & García-Olmedo, F. (1978) Differential allelic expression at a locus encoding an endosperm protein in tetraploid wheat (*T. turgidum*). *Genetics* 89, 147–56.

Salcedo, G., Sanchez-Monge, R. & Aragoncillo, C. (1982) The isolation and characterization of low molecular weight hydrophobic salt-soluble proteins from barley. *Journal of Experimental Botany* 33, 1325–31.

Salcedo, G., Fra-Mon, P., Molina-Cano, J.L., Aragoncillo, C. & García-Olmedo, F. (1984) Genetics of CM-proteins (A-hordeins) in barley. *Theoretical and Applied Genetics* 68, 53–9.

Sanchez-Monge, R., Delibes, A., Hernandez-Lucas, C., Carbonero, P. & García-Olmedo, F. (1979) Homoeologous chromosomal location of the genes encoding thionins in wheat and rye. *Theoretical and Applied Genetics* 54, 61–3.

Sanchez-Monge, R., Barber, D., Mendez, E., García-Olmedo, F. & Salcedo, G. (1986a) Genes encoding α-amylase inhibitors are located in the short arms of chromosomes 3B, 3D and 6D of wheat (*Triticum aestivum*

L.). *Theoretical and Applied Genetics* 72, 108–13.

Sanchez-Monge, R., Gomez, L., García-Olmedo, F. & Salcedo, G. (1986b) A tetrameric inhibitor of insect α-amylase from barley. *FEBS Letters* 207, 105–9.

Sanchez-Monge, R., Fernandez, J.A. & Salcedo, G. (1987) Subunits of tetrameric α-amylase inhibitors of *Hordeum chilense* are encoded by genes located in chromosomes $4H^{ch}$ and $7H^{ch}$. *Theoretical and Applied Genetics* 74, 811–16.

Sanchez-Monge, R., García-Olmedo, F., Gomez, L., Royo, J., Carbonero, P. & Salcedo, G. (1988) Genetics of the α-amylase/trypsin inhibitor family in wheat and related species. *Anales Aula Dei* 19, 195–205.

Sanchez-Monge, R., Gomez, L., García-Olmedo, F. & Salcedo, G. (1989) New dimeric inhibitor of heterologous α-amylases encoded by a duplicated gene in the short arm of chromosome 3B of wheat (*Triticum aestivum* L.). *European Journal of Biochemistry* 183, 37–40.

Shewry, P.R., Lafiandra, D., Salcedo, G., Aragoncillo, C., García-Olmedo, F., Lew, E.J.-L., Dietler, M.D. & Kasarda, D.D. (1984) N-terminal amino acid sequences of chloroform/methanol-soluble proteins and albumins from endosperm of wheat, barley and related species. *FEBS Letters* 175, 359–63.

Silano, V. (1987) α-Amylase inhibitors. In: Kruger, J.E., Lineback, D. & Stanffer, C.E. (eds), *Enzymes and their Role in Cereal Technology*. AACC, St Paul, MN, USA, pp. 141–99.

Silano, V., Furia, M., Gianfreda, L., Macri, A., Palescandolo, R., Rab, A., Scardi, V., Stella, E. & Valfre, F. (1975) Inhibition of amylases from different origins by albumins from the wheat kernel. *Biochimica et Biophysica Acta* 391, 170–8.

Stuart, L.S. & Harris, T.H. (1942) Bactericidal and fungicidal properties of a crystalline protein isolated from unbleached wheat flour. *Cereal Chemistry* 19, 288–300.

Vernon, L.P., Evett, G.E., Zeikus, R.D. & Gray, W.R. (1985) A toxic thionin from *Pyrularia pubera*: purification, properties and amino acid sequence. *Archives of Biochemistry and Biophysics* 238, 18–29.

Wada, K. & Buchanan, B.B. (1981) Purothionin: a seed protein with thioredoxin activity. *FEBS Letters* 124, 237–40.

17

Nutritional Aspects of Barley Seed Structure and Composition

C.W. Newman & R.K. Newman

The structure and composition of the barley kernel was not designed as a nutrient source for animals or humans but rather to propagate the plant species. The chemical components, which are put together in a unique structural arrangement, provide the biochemistry to meet the physiological requirements for germination, emergence, survival, and growth of a seedling. Perhaps it is a coincidence, but these same chemical components can provide most of the nutrients required by other forms of life including domestic farm animals and humans.

The origin and taxonomy of *Hordeum* species have been summarized in an excellent review by von Bothmer & Jacobson (1985) and are also discussed elsewhere in this volume. Similarly, the morphology and anatomy of the barley plant and kernel were addressed by Reid (1985). Seeds of wild barley species and landrace cultivars have widely varying shapes and sizes. Natural selection dictated by survival of the species probably produced seeds containing only the components necessary for reproduction and propagation. When man began to cultivate barley as a crop some 9000 years ago (Harlan, 1971), it can be reasoned that shifts in the ratio of the components were initiated. Originally, early agriculturists probably selected plants that grew well in the local environment, but in time, selected those that were more desirable for food and beverages, and could be stored for future use. This practice continued for centuries and led to the genotypes of landrace barleys from which modern barley breeders developed high-yielding cultivars that produce plump, high-energy kernels for malting, feed and food.

Barley seed structure and the basic components are relatively constant over wide ranges of the individual components (Briggs, 1978). The effects of genotype and environment on the levels of kernel nutrients have been well-documented (Newman & McGuire, 1985; Åman & Newman, 1986; Bach Knudsen *et al.*, 1987a; Henry, 1988). The nutritional implications of the shifts in nutrient concentrations are dramatic, as will be documented in this review. Physical parameters of the seed, such as kernel weight and hectolitre weight, are closely associated with certain nutrient concentrations. These parameters are controlled to some extent by genotype but, because these measurements are directly related to nutrient content, they are heavily influenced by environmental growing conditions.

This chapter considers the structural composition and chemical components that affect the nutritional value of the barley kernel for feed and food. These include starch, dietary fibre, protein and its amino acid components, lipid constituents, and the important mineral components. Additionally, compounds which alter the energy value or have direct effects on enzyme systems of mammalian and avian species will be addressed.

Carbohydrates

Carbohydrates comprise about 80% of the barley kernel (Englyst *et al.*, 1983) and provide the major source of available energy for food and feed. The carbohydrates are generally subdivided into starches, sugars, and non-starch polysaccharides plus lignin. Although not a carbohydrate, lignin is usually included with carbohydrates because of the chemical association with these compounds.

Starches and sugars

Starch is the major carbohydrate component of the barley kernel and its concentration is inversely related to the total dietary fibre and protein contents. In most barleys, the starch is predominantly amylopectin (74–78%) in which chains of α-(1 → 4) D-glucopyranose units are branched through α-(1 → 6) linkages. The remainder of the starch is amylose (22–26%) containing straight chains of D-glucopyranose units linked α-(1 → 4) (Briggs, 1978). Waxy starch, which approaches 100% amylopectin, is produced by the recessive waxy endosperm character, *wx* (Nakao, 1950). An amylose level of 44% was discovered in a mutant of Glacier (CI 9676) and designated Glacier Ac38 by Merritt (1967). This character was shown to be under genetic control (Walker & Merritt, 1969) and thus may be expressed in other cultivars. A wide range in the percentage of starch exists across genotype, and this is also influenced by environment. Barleys planted late or grown under arid conditions will contain less starch than when produced under ideal agronomic conditions. Hull-less barley contains a greater percentage of starch than comparable covered types due to the absence of the fibrous hull. Analyses of waxy isotype cultivars indicated a slightly but significantly lower percentage of starch (4.4–8.5%) compared with isotypes of covered and hull-less barleys derived from Compana (CI 5438) and Betzes (CI 6398) (Newman *et al.*, 1990a). The reduced level of starch was accompanied by significant increases in sucrose (33–44%) and maltose (111–176%). According to MacLeod *et al.* (1953), maltose is only produced in barley during germination and Henry (1988) stated that maltose in barley is produced by the action of amylases. Åman & Newman (1986) first suggested that waxy-type barleys contain less starch and a greater quantity of sucrose, but did not measure maltose. Additionally, these authors presented data on three chemically induced mutants of a waxy short-awned hull-less barley that ranged from 21.2 to 45.9% in starch percentage and from 5.3 to 6.4% sucrose. The significant increase in sucrose in waxy types may be a reflection of altered starch synthesis; however, the presence of relatively high levels of maltose is possibly due to the high level of α-amylase activity in the waxy cultivars that was reported earlier by Goering & Eslick (1976). An observation made by Harlan (1957) that waxy barleys deposit amylodextrin instead of starch during the final stage of carbohydrate deposition may also be a partial explanation of this phenomenon. Amylodextrin may be more susceptible to amylase activity than normal starch granules in the intact kernel. This may be surmised in this case since the total percentages of starch plus mono- and disaccharides and β-glucans were essentially equal in both waxy and non-waxy types (Newman *et al.*, 1990a).

Nutritional implications

As starch is the major source of readily available energy, barleys containing higher levels of this nutrient have higher digestible energy values (Bach Knudsen *et al.*, 1987b). Starches from maize, wheat, grain sorghum and oats are reported to be almost completely hydrolysed to glucose (>95%) in the small intestine of man (Englyst & Cummings, 1986; Sandberg *et al.*, 1986) and swine (Keys & DeBarthe, 1974; Graham *et al.*, 1986a). The digestibility of barley starch anterior to the large intestine in pigs has been found in some studies to be considerably less (69 to 84%) than the 95% reported for other grain starches (Cunningham *et al.*, 1963; Keys & Debarthe, 1974; Fadel *et al.*, 1988). On the other hand, Fadel *et al.* (1989) and Graham *et al.* (1989) reported that barley starch from hull-less cultivars had digestibilities of 94.8 and 90.3% in the small intestine of swine. Graham *et al.* (1986b) also found a 96% digestibility of starch by swine in a covered barley (cv. Tellus). Covered barleys were evaluated in all of the earlier studies in which lower digestibilities were reported.

Waxy starch from maize, barley and grain sorghum in the uncooked state has been purported to be more nutritionally desirable than normal starch or high amylose starch because it is more readily digested by α-amylase *in vitro* (Sandstedt & Hites, 1968; Goering *et al.*, 1973; Sullins & Rooney, 1974; and Björck *et al.*, 1990). This has not proven true when waxy barley starch was fed to rats (Calvert *et al.*, 1976) or when whole grain waxy barleys were compared with normal barleys for ruminants (Moss *et al.*, 1980), rats and swine (Calvert *et al.*, 1977) and broiler chickens (Moss *et al.*, 1983).

In contrast to waxy starches, high-amylose barley starch was found to be less susceptible to α-amylase than normal barley starch (Pomeranz *et al.*, 1972) or to normal and waxy barley starch (Björck *et al.*, 1990). Calvert *et al.* (1976) found that rats gained at a slower rate when fed high-amylose barley starch compared with normal barley starch in purified diets. Rats fed the high-amylose starch consumed somewhat less feed compared with those fed normal starch; however, the reduced growth rate may have been due to a significant reduction in digested and retained protein observed on the high-amylose starch diets.

Protein of the high-amylose barley, Glacier Ac38, has a higher concentration of lysine and other essential amino acids than Glacier (Pomeranz *et al.*, 1972; Newman *et al.*, 1978) and consequently a higher biological value as determined with rats (Newman *et al.*, 1978) and swine (Calvert *et al.*, 1981). When the protein requirements of swine were met with supplemental protein, feed efficiency was significantly reduced in pigs fed the high-amylose barley, suggesting reduced available energy (Calvert *et al.*, 1981).

Björck *et al.* (1990) concluded that the amylose/amylopectin ratio in different barley genotypes (waxy, normal and high-amylose starch) had only a marginal influence on starch gelatinization. These authors, however, noted significant differences in *in vitro* digestibility of these starches, but all the barley starches were significantly more susceptible to α-amylolysis than a wheat starch reference. Waxy barley starch showed the highest susceptibility to α-amylolysis when tested raw, with the high-amylose starch being least susceptible. No differences in enzymatic hydrolysis were found when the barley starches were boiled. Autoclaving the starches, however, produced a concomitant decrease in *in vitro* digestibility with increased amylose content. This was explained by an increased amount of resistant starch and the formation of amylose–lipid complexes, which did not occur in autoclaved waxy starch or in any of the starches when boiled.

High-amylose rice (Goddard *et al.*, 1984) and high-amylose maize (Behall *et al.*, 1988, 1989) have been shown to have a positive lowering effect on blood glucose and insulin levels in humans. Behall *et al.* (1989) also reported a lowering of blood triglycerides and cholesterol in subjects consuming high-amylose maize. Data from the Montana Agricultural Experiment Station

(R.K. Newman, unpublished data) showed significantly higher blood glucose levels in broiler chicks fed two uncooked milling fractions (flour and red dog) from high-amylose covered and hull-less barleys compared with normal barley milling fractions and maize. These data indicate relatively high digestibility of the raw high-amylose barley starch. Serum cholesterol, however, was significantly reduced by 138–194% in chicks fed the high-amylose flour and red dog fractions, respectively. Amylose will form a stable complex *in vitro* and *in vivo* with monoacyl lipids and possibly other lipids, which is relatively resistant to enzymatic degradation (Larsson & Miezis, 1979). The high-amylose maize products evaluated by Goddard *et al.* (1984) and Behall *et al.* (1988, 1989), were cooked, while the Montana experiment evaluated raw milling fractions from high-amylose barleys. The influence on lipid metabolism may have been the result of the formation of amylose–lipid complexes or the effect of the soluble dietary fibre components present in the milling fractions. The latter is suggested in the effects observed in the Montana experiment since the starches were not cooked. Early work with maize indicated that the digestibility of raw maize starch is not directly associated with amylose content. Three genes, sugary-2 (su_2), dull (*du*), and amylose extender (*ae*) were reported to increase the amylose content of maize starch, but digestibility was depressed only when the *ae* gene was present (Sandstedt, *et al.*, 1962; Fujita *et al.*, 1989). The high-amylose trait of barley is due to a single recessive gene (Walker & Merritt, 1969). Eslick & Ullrich (1977) suggested that the gene which is located on chromosome 2 should be designated *amol* and demonstrated that low kernel weight and high lysine are associated traits. Further study is required to determine the effects of cooking on the digestibility of high-amylose barley starch, as has been observed in maize starch altered by the *ae* gene.

In his review of the carbohydrates in barley grain, Henry (1988) listed three trisaccharides, raffinose, ketose and isoketose (fructans), that occur in the kernel. The presence of stachyose and verbacose in barley has not been established. The fructans of barley range from trisaccharides to long polymers that must be regarded as polysaccharides. The level of fructans in barley is quite low, less than 1% in most instances (Åman & Newman, 1986). The presence of low-molecular-weight saccharides and fructans in barley kernels is well established, but their significance in grain quality has not been established. Maltose and sucrose are readily hydrolysed to monosaccharides by intestinal sucrase and maltase, thus adding to the energy value of starch.

Non-starch polysaccharides and lignin (NSP)

The NSP are collectively designated total dietary fibre (TDF) and are generally considered to be plant cell wall material. Although controversial, starch retrograded during certain processing methods is included in TDF in some instances due to analytical technique. In this discussion, TDF will refer to cell wall material.

The cell wall material or the structural polysaccharides of barley kernels which comprise TDF, consist primarily of cellulose microfibrils and the matrix polysaccharides, mainly β-glucan and arabinoxylans (Henry, 1988). Cellulose is composed of linear $(1\rightarrow4)$-β-glucans. The β-glucans are also linear polymers and consist of β-cellotriosyl and β-cellotetraosyl residues separated by $(1\rightarrow3)$ linkages arranged in an independent or random way along the chain (Staute *et al.*, 1983). Arabinoxylans of the starchy endosperm are polysaccharides containing $(1\rightarrow4)$-β-xylan chains with α-L-arabinofuranose residues attached through $(1\rightarrow2)$ or $(1\rightarrow3)$ linkages. Pentosans from the husk vary from those of the endosperm and aleurone cell walls. Some material consists of $(1\rightarrow4)$ β-D-xylopyranose residues variously substituted with α-L-arabinofuranose and D-glucuronic acid (Briggs *et al.*, 1981). Most of the cellulose occurs in the husk of covered barleys, comprising 3.5–4.5% of the kernel, whereas hull-less types contain ≈2% (Newman *et al.*, 1990a). The β-glucans and arabinoxylans are concentrated in the aleurone and endosperm cell wall material but in quite different ratios (Henry, 1988). Aleurone cell walls contain ≈60% arabinoxylans and ≈20% β-glucans (Bacic & Stone, 1981) whereas endosperm cell wall material contains ≈25% arabinoxylans and ≈70% β-glucans (Ballance & Manners, 1978). According to Bacic & Stone (1981), aleurone cell walls contain 23% arabinose, 44% xylose, 2% mannose, 2% galactose and 29% glucose. Ballance & Manners (1978) reported endosperm cell wall components as 10% arabinose, 13% xylose, 2.5% mannose and 74% glucose. Small percentages of glucomannans and cellulose are also present in barley cell wall material along with phenolic acids, ferulic acid (Henry, 1988) and uronic acid (Fincher, 1975; Åman & Hesselman, 1984; Henry, 1986a). In that the amount of endosperm cell wall material far exceeds that of aleurone cell wall material simply due to the size of the respective tissues, analyses of whole barley generally reflect the composition of the endosperm constituents.

Varietal or genotypic differences in the β-glucan content of barley have been well established in numerous studies, for example Bourne & Pierce (1970), Aastrup (1979), Ullrich *et al.* (1986) and Henry (1987). Åman *et al.* (1989) presented data to demonstrate that stage of grain development as well as cultivar and growing conditions has great influence on the percentage of β-glucans in barley. The dry matter in the barley kernel is largely deposited before the majority of the β-glucan in the endosperm cell wall is synthesized. Numerous studies have reported the effects of environment on β-glucan concentration in barley as cited by Newman & McGuire (1985). The report of Åman *et al.* (1989) also demonstrates this phenomenon and clearly shows a genotype × environment effect.

Powell *et al.* (1985) reported that the β-glucan content of barley is controlled by a simple additive genetic system. Significant genetic correlations were observed between β-glucan content, kernel weight, and plant height in doubled haploid lines. Early reports, as reviewed by Henry (1986a), showed much smaller concentrations of β-glucans than more recent data which reflect improved analytical techniques. Aastrup (1979) reported a range of 1.93 to 8.29% in 18 European cultivars, whereas Åman & Graham (1987) showed the average for Scandinavian barleys was 4.4% with a range of 3.0–5.6%. In an evaluation of 12 North American cultivars over a 3-year period, Hockett *et al.* (1987) found a range in β-glucan of 4.9–6.2% with a mean of 5.5%. In a study reported by Ullrich *et al.* (1986), the mean β-glucan content of Titan (CI 7055), Betzes (CI 6398) and Compana (CI 5438) barleys grown in three locations were 4.9, 5.7 and 3.8% respectively. Corresponding waxy isotypes of these cultivars contained 7.1, 7.3 and 4.7% β-glucans. Similar differences due to the waxy gene in Compana and Betzes barleys were observed by Newman *et al.* (1990a) in both covered and hull-less genotypes. Azhul, a waxy hull-less six-rowed barley developed at the Arizona Agricultural Experiment Station (R.T. Ramage, personal communication) was found to contain 10–11% β-glucans while the non-waxy parent barley contains 6–7% β-glucans (Mori, 1990). Thus the current evidence indicates that the *wx* gene mediates an increase in barley β-glucans. The mechanism of this effect is unknown.

Less attention has been given to the pentose polymer, arabinoxylan, compared with β-glucans. Henry (1986b) evaluated 17 Australian cultivars grown at three locations. The pentosan content varied from 4.4 to 7.8% and the β-glucan content from 3.4 to 5.7%. Genetic differences were found in arabinose, xylose, total pentosan and β-glucan contents and in the pentosan:β-glucan ratio. Smaller kernels had higher pentosan contents and pentosan:β-glucan ratios. Hoshimoto *et al.* (1986) reported 7.10% pentosans in a covered six-rowed barley. After pearling, the pearled barley contained 5.2% pentosans and the pearlings, 16.4%, indicating a concentration of these polymers in the outer portion of the kernel.

Nutritional implications

Barley meal suspended in water at either alkaline, neutral, or acid pH will produce a highly viscous solution. This property was used in early studies to estimate the quantity of soluble β-glucans in barley (Greenberg & Whitmore, 1974; Bendelow, 1975). Other grain constituents, however, contribute to the viscosity of extracts (Scott, 1972; Smith *et al.*, 1980). Recently it has been shown that the pentosan constituents, xylose and arabinose, are more highly correlated with barley extract viscosity than are β-glucans (R.K. Newman, unpublished data). Fengler & Marquardt (1988) determined that the principal antinutritional factor in rye grain is a highly viscous, water soluble, pentosan-rich carbohydrate. Even so, the *in vitro* viscosity of barley extracts is greatly reduced by the addition of endo-β-glucanase (Hesselman & Åman, 1986; Bengtsson *et al.*, 1990). Numerous reports, such as Classen *et al.* (1985) and Hesselman & Åman (1986), have demonstrated that enzymatic destruction of barley β-glucans *in vivo* results in improved performance of poultry. The improvement is thought to be the result of reduced intestinal viscosity created by β-glucans and this was recently confirmed by Wang *et al.* (1991). This report showed that barley β-glucans contributed to the small intestinal viscosity of chicks and the effect was related to altered lipid metabolism as well as reduced lipid and protein digestibility. The deleterious effect of β-glucans on poultry is now well recognized, but similar undesirable effects have not been consistent in swine.

The human health benefits of TDF have been well documented and reviewed (Vahouny & Kritchevsky, 1986). Soluble fibres such as β-glucans have been demonstrated to produce hypocholesterolemic effects in rats (Klopfenstein & Hoseney, 1987; Mori, 1990), chicks (Fadel *et al.*, 1987; Newman *et al.*, in press), and humans (Newman *et al.*, 1989a, b; McIntosh & Oakenfull, 1990). Bengtsson *et al.* (1990) concluded that the hypocholesterolemic response observed in a waxy barley, but not in a non-waxy barley, by Fadel *et al.* (1987) was due to the higher molecular weights and concentrations of the β-glucans in the waxy type. β-glucanase eliminates partially or completely the hypocholesterolemic effect of barley (Newman *et al.*, in press), demonstrating that the β-glucans play a major role in the observed effects. In some cases, however, the β-glucanase does not completely eliminate the effect, suggesting that other factors such as the viscosity causing pentosans, other soluble fibre components, or the tocotrienols also influence lipid metabolism. The latter will be discussed in the lipid section. The influence of other soluble fibres, (pentosans, pectins, etc.) has not been documented, but it is reasonable to assume that these compounds in barley will react in a similar fashion to that reported from other cereals or fibre sources (Vahouny & Kritchevsky, 1986). Less is known about the insoluble portion of TDF of barley, i.e. cellulose, but as with other insoluble fibre from other sources, this probably contributes to improved bowel regulation through added bulk. Cellulose is indigestible by monogastrics, contributing only a small amount of available energy through the production of volatile fatty acids in the large intestine. This amounts to less than 1% of the total available energy of cereals. In ruminants, the cellulose contributes considerably more energy and carbon skeletons through fermentation in the rumen. This is influenced by the degree of lignification and the presence of readily fermentable carbohydrates (starch and sugars) and soluble nitrogen.

Protein

Protein, the second major component of the barley grain, varies inversely with the concentration of the starch component (Briggs, 1978) and exhibits the greatest fluctuation of the major nutrients (Newman & McGuire,

1985) due to genotype and environment. The overall nutritional quality of most cereal proteins, including barley, is in general below that of legume and animal/marine tissue protein. This is due to the smaller amounts of essential amino acids, especially lysine, methionine, tryptophan, and threonine in the total protein component. Osborne (1924) classified cereal proteins by sequential extraction with water, dilute salt solutions, aqueous alcohol and alkaline extractants. These protein groups or fractions are designated respectively, albumins, globulins, prolamins and glutelins.

Albumin–globulins

The albumin and globulin fractions are often extracted together and referred to as the salt-soluble proteins. These proteins are the main constituents of the kernel embryo and aleurone layer, accounting for 15–30% of the total grain nitrogen (Shewry *et al.*, 1984). Although mainly metabolic proteins, Shewry *et al.* (1984) suggested the possibility of some storage proteins in the globulin fraction. Non-protein components, peptides, free amino acids and nitrates are associated with the salt-soluble proteins and account for 10–12% of the kernel nitrogen.

Prolamins

Osborne coined the name prolamin for this distinct group of proteins to reflect the high contents of proline and amide nitrogen (Shewry & Tatham, 1990). The prolamins of barley, called hordein after the latin generic name, *Hordeum vulgare*, are the major storage proteins, accounting for 35–50% of the total nitrogen (Kirkman *et al.*, 1982). Hordein is a complex mixture of polypeptides which have been classified on the basis of their amino acid composition and sequences into three groups, the sulphur-rich (S-rich), sulphur-poor (S-poor) and high-molecular-weight (HMW) prolamins (Shewry & Tatham, 1990). The S-rich, S-poor, and HMW prolamins of barley correspond to B and γ, C and D hordeins, respectively. A further fraction designated A hordein (Miflin *et al.*, 1981) is not now considered a true prolamin as it does not show the characteristics of these proteins (Salcedo *et al.*, 1980 as cited by Shotwell & Larkins, 1989). The S-rich B hordeins of barley account for 80% of the total fraction, followed in order of concentration by the C hordeins and the D and γ-hordeins, respectively. There are at least four loci on chromosome 5 controlling prolamin synthesis in barley. D hordein is controlled by the *Hor3* locus on the long arm and B, C and γ-hordeins are controlled by the *Hor2*, *Hor1* and *Hor5* loci, respectively, on the short arm (Shewry *et al.*, 1983, 1990).

Glutelins

Glutelins were initially thought to be a distinct group of proteins. However, it is now known that many of the component proteins are structurally related to prolamins. Although not soluble in aqueous alcohol, according to Shewry & Tatham (1990) these components may be considered as prolamins since the individual reduced subunits are alcohol-soluble and rich in prolamine and glutamine. Other components of the glutelin fraction are insoluble structural proteins associated with membranes and cell walls and correspond to matrix proteins of the cell (Miflin & Shewry, 1979; Shewry *et al.*, 1984). It has been difficult to define accurately these true barley glutelins because of their heterogeneous nature and difficulties in their complete extraction and separation from the hordeins.

Nutritional implications

Lysine comprises 5–7% of the albumin–globulin fraction and less than 1% of the prolamins (in most barleys), with levels intermediate between these two levels in the glutelin (Munck, 1972; Brandt, 1976; El-Negoumy *et al.*, 1977; Kirkman *et al.*, 1982; Shewry *et al.*, 1984). These differences mean that an increase in hordein in barley reduces the lysine percentages in the total protein, or an increase in the salt-soluble proteins and (or) glutelins will produce an increase in the concentration of lysine and other essential amino acids. At least five separate genes are known to enhance lysine content in seven barley genotypes (Tallberg, 1981). These are often referred to as 'high lysine' mutants (Shotwell & Larkins, 1989). The nutritional quality of the seed protein in the mutants is improved as a consequence of the reduced hordein content with a concomitant increase in other more lysine-rich proteins. The improved value of these proteins has been demonstrated in numerous studies as cited in Newman & McGuire (1985). A major impediment of the widespread introduction of high-lysine genotypes into conventional barley is an associated shrunken endosperm resulting in reduced kernel weight and overall yield. Some progress has been made in overcoming this problem by a conventional breeding programme to incorporate the high-lysine character of Risø 1508 into high-yielding European cultivars (Bang-Olsen *et al.*, 1987). This resulting barley, designated CA 700202, has 6.0–6.5 g lysine/100 g protein and has proven to have extremely high (>95%) biological value (BV) (Bang-Olsen *et al.*, 1987; Newman *et al.*, 1990b). As with other high-lysine barleys, the protein of CA 700202 is less digestible than that of normal cultivars. However, the increase in BV is of such magnitude that the net protein utilization is significantly increased over that of standard cultivars.

The increased BV of high-lysine barley protein has significant implications for providing high-quality protein to populations where animal protein is in short supply. Quality protein can either be provided through direct consumption of the grain or through improved animal and poultry production. The young of any species are the individuals most likely to suffer from protein quality deficiency. Recently the CA 700202 barley was effectively used to prepare weaning foods with improved energy and nutrient (lysine) density (Hansen *et al.*, 1989; Pedersen *et al.*, 1989).

Lipids

Total lipids

Barley generally contains 2–3% lipid (Welch, 1978), although there is genotypic variation. Bhatty & Rossnagel (1980) reported relatively high levels, (3.5–5.3%) in the mutant Bomi Risø barley. Parsons & Price (1974) had earlier reported a cultivar with 4.6% lipid content. Prowashonupana, a high-protein, waxy hull-less mutant of Compana has been analysed to contain up to 7% total lipid on a dry basis (Åman & Newman, 1986; Newman *et al.*, in press).

Price & Parsons (1975) compared lipid levels in several cereal grains, using Prilar as a representative barley. Total lipid of the Prilar was 3.2% and the predominant fatty acids were linoleic (55%), palmitic (21%) and oleic (18%). These investigators later reported that the endosperm of Prilar hull-less barley contained 77.1% of the lipid, the embryo 17.9% and the hull 5.0% (Price & Parsons, 1979). Other workers have reported that one-third of the lipid is in the germ, which represents less than 3% of the kernel weight (Morrison, 1978). Skarsaune *et al.* (1973) reported that free lipid was extracted from whole barley at 8–12 mg/100 g, while the husk, endosperm and aleurone contained 9–12, 4–6 and 22–36 mg respectively. The lipids of the whole barley kernel consist of 65–78% non-polar lipid, 7–13% glycolipid and 15–26% phospholipid (Parsons & Price, 1974). Demchenko *et al.* (1969) reported an average of 8% non-saponifiable matter in oil from five barley varieties.

Welch (1978) and DeMan (1985) both reported a higher total fatty acid content in two-rowed barleys compared with their six-rowed counterparts, and also that winter barleys have a higher proportion of linolenic acid than spring types. DeMan and co-workers have intensively investigated various factors related to the lipid composition of barley. Lipid composition varies during seed ripening due to metabolic processes, and the date of harvest is a source of variation for both lipid content and composition (DeMan & Cauberghe, 1988). The proportion of unsaturated fatty acids tended to be higher and saturated fatty acids lower in smaller grains compared with large grains, possibly related to different proportions of neutral lipids, phospholipids and glycolipids (DeMan & Vervenne, 1988).

Starch lipids complex

The presence of lipids associated with cereal starches is often overlooked. Morrison (1988) recently reviewed cereal starch lipids. In barley, these lipids are predominately monoacyl lysophospholipids with smaller amounts of free fatty acids, and are present inside the starch granules of the endosperm. In addition, the starch granules may acquire lipids which are mostly free fatty acids, from the endosperm non-starch lipids. The concentration of starch lipids is in proportion to the amylose content of the starch. Waxy barleys contain 300–493 mg starch lipids/100 g starch and normal varieties 675–924 mg starch lipids/100 g, while high-amylose varieties contain 1051–1219 mg starch lipids/100 g (Morrison, 1988). Barley is unique among cereal grains in that the proportion of lipid in the waxy type is higher than that found in waxy types of maize and rice.

Nutritional implications

The non-saponifiable components of barley have received recent attention for health promoting activities. These include carotenoids, tocopherols and isoprenoid products. Demchenko *et al.* (1969) reported β-carotene levels of 5.2–8.5 mg/100 g in five barley varieties. Barley contains all the known tocopherols, including tocotrienols. A summary of ranges reported by Morrison (1978) is shown in Table 17.1.

Wang *et al.* (1991b) reported α-T-3 levels in nine barleys, ranging from 0.96 to 1.73 mg/100 g. The highest value was found in a Korean hull-less barley, Bangsa, which was reported by Newman *et al.* (in press) to be very effective in lowering cholesterol in spite of relatively low β-glucan and soluble fibre content.

The tocotrienols differ from tocopherols only in three double-bonds in the isoprenoid chain. D-α-tocotrienol was identified by Qureshi *et al.* (1986) as an inhibitor of cholesterol biosynthesis in livers of experimental animals. The mechanism is due to suppression of HMG-CoA reductase, a rate-limiting enzyme in the mevalonate pathway. A second cholesterol inhibitor of barley isolated by Qureshi *et al.* (1984), γ-linolenic acid, is present only in a very low concentration, corresponding to about 5% of total fatty acids.

Wang *et al.* (1990) fed barley oil, maize oil, and margarine to hypercholesterolemic chicks. The determined α-tocotrienol contents were 558 ppm in the barley oil, 23 ppm in maize oil, and not detectable in the margarine. After 10 days, the serum cholesterol levels of chicks consuming barley oil were 172 $\pm$ 10 compared with 227 $\pm$ 15 and 262 $\pm$ 50 mg dl^{-1} for those fed maize oil and margarine respectively.

The effect of the fat-soluble components appears to be retained through the malting process as a number of studies with brewer's grains have shown a hypocholesterolemic effect (Qureshi *et al.*, 1987, 1991; Robinson & Lupton, 1990).

The use of extracted barley oil for nutritional purposes has many possibilities. The

Table 17.1. Tocopherols and tocotrienols of barley.

Tocopherols	mg/100 g barley	Tocotrienols	mg/100 g barley
α-T	0.20–0.40	α-T-3	1.10–1.30
β-T	0.04–0.40	β-T-3	0.30–0.70
γ-T	0.03–0.50	γ-T-3	0.20
δ-T	0.01–0.04	δ-T-3	0.16

oil can be extracted by hexane, similarly to wheat germ oil, and has potential as a nutritional supplement in capsule form or incorporated into food products. One particular application in medical nutritional supplements is being studied (R.K. Newman, pers. comm.). Stability of the oil is a subject which will require investigation. The presence of lipolytic enzymes in cereal grains can cause changes in the oil components resulting in oxidative rancidity (Morrison, 1988). Furthermore, the tocotrienols may be rendered ineffective due to oxidation during the oil extraction process.

The properties of the oil components in barley serve to enhance the desirability of barley as a cereal food, possibly acting in synergism with the soluble fibre components to lower serum cholesterol. In addition, there are potential physiological benefits in minor constituents such as ferulic acid (Collins, 1986) and isoprenoid products (Elson, 1989).

Minerals

Mineral elements exist in small concentrations in cereal grains, and are essential to the health and growth of plants, as well as being important nutrient sources for animals and man. The ash content of barley ranges from 2 to 3%, the major components being phosphorus, potassium and calcium, with lesser amounts of chlorine, magnesium, sulphur, and sodium (Owen *et al.*, 1977). There are wide varietal differences and smaller environmental variations (Kleese *et al.*, 1968). Covered barleys contain more calcium and silica than hull-less barleys due to the hull composition (Newman & McGuire, 1985) and the hull has also been reported to have higher concentrations of iron and nickel than the groat (Weaver *et al.*, 1981). Distribution of mineral content throughout the seed was investigated by Weaver *et al.* (1981), through milling and pearling processes. Hulls, separated with a decorticator, contained significantly greater concentrations of all elements except chromium, while the flour contained the smallest amounts. Liu *et al.* (1974) determined minerals in Himalaya hull-less barley and Atlas covered barley after separation of the grain by abrasion and milling. The germ contained the highest, and central endosperm the lowest concentrations of mineral components. The shorts fraction of roller-milled barley, which contains the aleurone layer, had higher levels of all minerals than did the flour. Liu & Pomeranz (1975) examined the localization of minerals in barley kernels by X-ray analysis. Silicon was localized mostly in the outer lemma, while potassium had the highest concentration in the aleurone but was distributed in all portions. Calcium was highest in the aleurone, with smaller amounts in the lemma and endosperm cell walls. Magnesium and phosphorus were most concentrated in the aleurone, with lesser amounts in the endosperm, and sulphur and chlorine were confined to cell walls of the aleurone and endosperm.

Nutritional implications

Phosphorus is one of the most important mineral nutrients for animals and humans, and 50–80% of the phosphorus in barley is unavailable because it is bound to phytic acid. In addition, phytic acid, because of its unique molecular structure, binds several other important cations such as iron, zinc, magnesium and calcium. In human diets, the unavailability of zinc is of particular significance (McCance & Widdowson, 1935).

Raboy *et al.* (1990) surveyed a large collection of maize kernel mutations which block phytic acid synthesis. Some mutations which affected the germ and aleurone were found to be associated with substantial reductions in kernel phytic acid. Whenever a reduction in kernel phytic acid was observed, there was a concomitant increase in kernel inorganic phosphorus. There was also a positive correlation between kernel protein and the level of phytic acid. Similar investigations should possibly be directed toward other cereal grains, including barley, as a step toward development of low phytate crops (Raboy, 1990).

Inhibitors

Phenolic compounds

Proanthocyanidins and catechins are found in the seed coat of mature barley kernels (Aastrup *et al.*, 1984), and differences in concentrations have been recorded due to genotype (Frost & Holm, 1973; Eckman, 1981). The composition of proanthocyanidins is complex and considerable work has been accomplished on their structure (Jende-Strid, 1981; Jende-Strid & Møller, 1981; Outtrup & Schaumburg, 1981). Proanthocyanidin-free barley has been produced through mutation breeding (Jende-Strid & Møller, 1981).

Nutritional implications

The phenolic compounds in barley are known to associate with proteins to form insoluble complexes which cause haze formation in beer (Jende-Strid, 1981). The suggestion by Munck (1964) that phenolic compounds in barley may inhibit nutrient utilization was confirmed *in vitro* (Eckman, 1981) and with rats (Newman *et al.*, 1984; Heintzman, 1986; Øverland, 1988). The last two reports demonstrated significantly higher protein digestibility in proanthocyanidin-free mutant barleys compared with the parent barley stock. The improvement in protein digestibility, however, was not of such a magnitude as to have a major effect on the overall feed value of the barley. It was proposed by Heintzman (1986) that the proanthocyanidins may be the causative agents in the observed darker coloured flour from milled barley compared with wheat flour. Baking tests, however, showed that the proanthocyanidin-free flour and baked products were in some cases darker than flour and products from normal barley. This was possibly due to the presence and increased concentration of other pigments produced as a result of the genetic mutations (Øverland, 1988).

Protease inhibitors

The presence and nutritional impact of these compounds in legume seed protein digestibility is well established. Barley kernels contain two main chymotrypsin inhibitors (Kirsi & Mikola, 1971; Mikola & Suolinna, 1971; Mikola & Kirsi, 1972; Boisen *et al.*, 1981). The concentration of chymotrypsin inhibitors varies with cultivar, whereas the amount of trypsin inhibitors appears to be relatively constant in different genotypes (Kirsi, 1973; Hejgaard & Boisen, 1980; Boisen *et al.*, 1981).

Nutritional implications

The two main chymotrypsin inhibitors and one of the trypsin inhibitors are pepsin-labile (Mikola & Suolinna, 1969; Boisen *et al.*, 1981) and are not considered to have any negative effect in animals. The activity of the pepsin-stable trypsin-inhibitor represents about one-third of the total inhibitor activity in most normal barleys (Pedersen & Boisen, 1982). In the mutant Risø 1508, however, its concentration is three to four times greater than in normal cultivars. Although a reduced protein digestibility has been observed in this barley (Newman & El-Negoumy, 1980), Pedersen & Boisen (1982) concluded this was not due to any protease inhibitory effect. It is generally felt that due to the low level of protease inhibitors in barley kernels, it is unlikely that they have a significant negative influence on the digestibility of protein.

Summary

The barley kernel is a complex integrated mixture of carbohydrates, proteins, lipids, minerals and other compounds. Many of these components are under direct genetic control that is often modified by environment.

Starch is the major component of the barley kernel and has the greatest overall influence on nutritional quality through its energy value. Energy value and utilization are, however, modified by the presence of non-starch polysaccharides and lignin as well as by the total percentage of starch. Variations in the ratio of the starch components amylose and amylopectin can have a profound effect on digestibility. The viscosity of water extracts of barley due to the non-starch polysaccharides, particularly β-glucans and arabinoxylans, is a negative factor for some species, particularly poultry. On the other hand, this viscosity interferes with fat absorption and is highly desirable in human diets as an anticholesterolemic factor.

Protein quality is dependent upon the concentration of essential amino acids, which is determined to a large extent by changes in the proportions of Osborne protein fractions due to genotype, environment and the interaction of the two. Lysine, the most limiting amino acid for nutritional quality in cereal grains, continues to be of interest. High-lysine barley is important to livestock feeding by reducing the need for protein supplements and to human diets in developing countries where animal protein sources are lacking.

Lipids, although comprising a small percentage of the kernel, contain minor components with potent nutritional benefits. Tocotrienols have been shown to be effective in inhibiting cholesterol synthesis, and give potential dietary applications to barley oil.

Antinutritional compounds in barley must be considered when evaluating nutritional value. Phytate, which sequesters important minerals, phenolic compounds which complex with proteins, and protease inhibitors all have negative influences on nutritional quality.

The complicating or confounding effects of processing and genetic alteration of other nutrients or non-nutrients, i.e. starch type, dietary fibre or protein, make it difficult if not impossible to separate the impact of a single component on multiple metabolic variables. Further research is necessary to identify and clarify the mechanisms that mediate these effects.

References

Aastrup, S. (1979) The effect of rain on β-glucan content in barley grains. *Carlsberg Research Communications* 44, 381–93.

Aastrup, S., Outtrup, H. & Erdal, K. (1984) Location of the proanthocyanidin in barley grain. *Carlsberg Research Communications* 49, 105–9.

Åman, P. & Graham, H. (1987) Analysis of total and insoluble mixed-linked (1 → 3), (1 → 4)-β-D-glucans in barley and oats. *Journal of Agricultural Food Chemistry* 35, 704–9.

Åman, P. & Hesselman, K. (1984) Analysis of starch and other main constituents of cereal grains. *Swedish Journal of Agricultural Research* 14, 135–9.

Åman, P. & Newman, C.W. (1986) Chemical composition of some different types of barley grown in Montana, USA. *Journal of Cereal Science* 4, 133–41.

Åman, P., Graham, H. & Tilly, A.-C. (1989) Content and solubility of mixed-linked (1 → 3), (1 → 4)-β-D-glucan in barley and oats during kernel development and storage. *Journal of Cereal Science* 10, 45–50.

Bach Knudsen, K.E., Åman, P. & Eggum, B.O. (1987a) Nutritive value of Danish-grown barley varieties. I. Carbohydrates and other major constituents. *Journal of Cereal Science* 6, 173–86.

Bach Knudsen, K.E., Eggum, B.O. & Jacobsen, I. (1987b) Nutritive value of Danish-grown barley varieties. II. Effect of carbohydrate composition on digestibility of energy and protein. *Journal of Cereal Science* 6, 187–95.

Bacic, A. & Stone, B.A. (1981) Isolation and ultrastructure of aleurone cell walls from wheat and barley. *Australian Journal of Plant Physiology* 8, 453–74.

Ballance, G.M. & Manners, D.J. (1978) Structural analysis and enzymic solubilization of barley endosperm cell walls. *Carbohydrate Research* 61, 107–18.

Bang-Olsen, K., Stilling, B. & Munck, L. (1987) Breeding for yield in high-lysine barley. In: Yasuda, S. & Konishi, T. (eds), *Barley Genetics V* (Proceedings of the 5th International Barley Genetics Symposium) Sanyo Press, Okayama, Japan pp. 865–70.

Bendelow, V.M. (1975) Determination of non-starch polysaccharides in barley breeding programmes. *Journal of the Institute of Brewing* 81, 127–30.

Behall, K.M., Scholfield, D.J. & Canary, J.J. (1988) Effect of type of starch on metabolic parameters in human subjects. *American Journal of Clinical Nutrition* 47, 428–32.

Behall, K.M., Scholfield, D.J., Yuhauiak, I. & Canary, J. (1989) Diets containing high amylose vs amylopectin starch: effects on metabolic variables in human subjects. *American Journal of Clinical Nutrition* 49, 337–44.

Bengtsson, S., Åman, P., Graham, H., Newman, C.W. & Newman, R.K. (1990) Chemical studies on mixed-linked β-glucans in hull-less barley cultivars giving different hypocholesterolemic responses in chickens. *Journal of Science of Food and Agriculture* 52, 435–45.

Bhatty, R.S. & Rossnagel, B.G. (1980) Lipids and fatty acid composition of Risø 1508 and normal barley. *Cereal Chemistry* 59, 89–91.

Björck, I., Eliasson, A.C., Drews, A., Gudmundsson, M. & Karlsson, R. (1990) Some nutritional properties of starch and dietary fiber in barley genotypes containing different levels of amylose. *Cereal Chemistry* 67, 327–33.

Boisen, S., Anderson, C.Y. & Hejgaard, J. (1981) Inhibitors of chymotrypsin and microbial serine proteases in barley grains. Isolation, partial characterization and immunological relationships of multiple molecular forms. *Physiology of Plants* 52, 167–76.

Bourne, D.T. & Pierce, J.S. (1970) β-glucan and β-glucanase in brewing. *Journal of the Institute of Brewing* 76, 328–35.

Brandt, A.B. (1976) Endosperm protein formulation during kernel development of wild type and a high-lysine barley mutant. *Cereal Chemistry* 53, 890–901.

Briggs, D.E. (1978) *Barley*. Chapman & Hall, New York, USA.

Briggs, D.E., Hough, J.S., Stevens, R. & Young, T.W. (1981) *Malting and Brewing Science*, Vol. 1. Chapman & Hall, London, Ch. 4.

Calvert, C.C., Newman, C.W., Eslick, R.F. & Goering, K.G. (1976) Comparison of waxy, high amylose and normal barley starches in rat diets. *Nutrition Reports International* 14, 56–61.

Calvert, C.C., Newman, C.W., Eslick, R.F. & Goering, K.G. (1977) Waxy vs normal barley in rat and pig diets. *Nutrition Reports International* 15, 157–64.

Calvert, C.C., Newman, C.W., El-Negoumy, A.M. & Eslick, R.F. (1981) High amylose Glacier barley in swine diets. *Nutrition Reports International* 23, 29–36.

Classen, H.L., Campbell, G.L., Rossnagel, B.G., Bhatty, R.S. & Reichert, R.P. (1985) Studies on the use of hull-less barley in chick diets: deleterious effects and methods of alleviation. *Canadian Journal of Animal Science* 65, 725–33.

Collins, F.W. (1986) Oat phenolics: structure, occurrence and function. In: Webster, F.H. (ed.), *Oats: Chemistry and Technology*. American Association of Cereal Chemists, St Paul, MN, USA.

Cunningham, H.M., Friend, D.W. & Nicholson, J.W.G. (1963) Observations on digestion in the pig using re-entrant intestinal fistula. *Canadian Journal of Animal Science* 43, 215–25.

DeMan, W. (1985) The effect of genotype and environment on the fatty acid content of barley (*Hordeum vulgare* L.) grains. *Plant, Cell and Environment* 8, 571–7.

DeMan, W. & Cauberghe, N. (1988) Changes of lipid composition in maturing barley kernels. *Phytochemistry* 27, 1639–42.

DeMan, W. & Vervenne, B. (1988) Lipid composition of barley in relation to grain size. *Phytochemistry* 27, 2037–9.

Demchenko, A.I., Olifson, L.E. & Nechaev, A.P. (1969) Chemical composition of barley oil. *Maslo-Zhir Prom* 35, 11–13.

Eckman, R. (1981) Determination of protein digestibility in barley and evaluation of the breedability for the character. In: Whitehouse, R.N.H. (ed.), *Barley Genetics IV*. (Proceedings of the 4th International Barley Genetics Symposium, Edinburgh), Edinburgh University Press, pp. 263–70.

El-Negoumy, A.M., Newman, C.W. & Moss, B.R. (1977) Chromatographic fractionation and composition of the salt soluble proteins from Hiproly (CI 3947) and Hiproly Normal (CI 4362) barleys. *Cereal Chemistry* 54, 333–44.

Elson, C.E. (1989) Potential value of isoprenoid co-products in health maintenance. *New Technologies for Value-Added Products from Protein and Co-products* (AOCS Symposium Proceedings, May 3–6, 1989), Cincinnati, OH, USA.

Englyst, H.N. & Cummings, J.H. (1986) Digestion of the polysaccharides of some cereal foods in the human small intestine. *American Journal of Clinical Nutrition* 42, 778–87.

Englyst, H.N., Anderson, V. & Cummings, J.H.

(1983) Starch and non-starch polysaccharides in some cereal foods. *Journal of the Science of Food and Agriculture* 34, 1434–40.

Eslick, R.F. & Ullrich, S.E. (1977) A characterization of three high lysine mutants in relation to their normal isotypes and their environmental response. *Barley Newsletter* 20, 39–45.

Fadel, J.G., Newman, R.K., Newman, C.W. & Barnes, A.E. (1987) Hypocholesterolemic effects of beta-glucans in different barley diets fed to broiler chicks. *Nutrition Reports International* 35, 1049–58.

Fadel, J.G., Newman, C.W., Newman, R.K. & Graham, H. (1988) Effects of extrusion cooking of barley on ileal and fecal digestibilities of dietary components in pigs. *Canadian Journal of Animal Science* 68, 891–7.

Fadel, J.G., Newman, R.K., Newman, C.W. & Graham, H. (1989) Effects of baking hull-less barley on the digestibility of dietary components as measured at the ileum and in the feces in pigs. *Journal of Nutrition* 119, 722–6.

Fengler, A.I. & Marquardt, R.R. (1988) Water-soluble pentosans from rye. II. Effects on rate of dialysis and on the retention of nutrients by the chick. *Cereal Chemistry* 65, 298–302.

Fincher, G.B. (1975) Morphology and chemical composition of barley endosperm cell walls. *Journal of the Institute of Brewing* 81, 116–22.

Frost, S. & Holm, G. (1973) A new phylogenetic approach in barley. *Hereditas* 74, 304–8.

Fujita, S., Blover, D.V., Okuno, K. & Fuwa, H. (1989) In vitro and in vivo digestion of high-amylose type starch granules. *Stärke* 4, 221–4.

Goddard, M.S., Young G. & Narcus, R. (1984) The effect of amylose content on insulin and glucose response to ingested rice. *American Journal of Clinical Nutrition* 39, 388–92.

Goering, K.J. & Eslick, R.F. (1976) Barley starch. VI. A self-liquefying waxy barley starch. *Cereal Chemistry* 53, 174–80.

Goering, K., Eslick, R.F. & DeHaas, B.W. (1973) Barley starch. V. A comparison of the properties of waxy Compana barley starch with starches of its parents. *Cereal Chemistry* 50, 322–8.

Graham, H., Hesselman, K. & Åman, P. (1986a) The influence of wheat bran and sugar beet pulp on the digestibility of dietary components in a cereal based diet. *Journal of Nutrition* 116, 242–51.

Graham, H., Hesselman, K., Jonsson, E. & Åman, P. (1986b) Influence of β-glucanase supplementation on the ileal and fecal digestibilities of a barley based diet in the pig gastrointestinal tract. *Nutrition Reports International* 34, 1089–96.

Graham, H., Fadel, J.G., Newman, C.W. & Newman, R.K. (1989) Effect of pelleting and β-glucanase supplementation on the ileal and fecal digestibilities of a barley based diet in the pig. *Journal of Animal Science* 67, 1293–8.

Greenberg, D.C. & Whitmore, E.T. (1974) A rapid method for estimating the viscosity of barley extracts. *Journal of the Institute of Brewing* 80, 31–3.

Hansen, J., Pedersen, B., Munck, L. & Eggum, B.O. (1989) Weaning foods prepared with improved energy and nutrient density prepared from germinated cereals. 1. Preparation and dietary bulk of gruels based on barley. *United Nations University, Foods and Nutrition Bulletin* 11, 40–5.

Harlan, H.V. (1957) In: Harlan, J.R. (ed.), *One Man's Life with Barley – The Memories and Observations of Harry V. Harlan*. Exposition Press, NY, pp. 168–9.

Harlan, J.R. (1971) On the origin of barley: a second look. In: Nilan, R.A. (ed.), *Barley Genetics II* (Proceedings 2nd International Barley Genetics Symposium), pp. 45–50.

Heintzman, K.B. (1986) 'The nutritional value of six barleys and their proanthocyanidin-free mutants'. Unpublished Master's Thesis, Montana State University, Bozeman, MT 59717, USA.

Hejgaard, J. & Boisen, S. (1980) High-lysine proteins in Hiproly barley breeding: identification, nutritional significance and new screening methods. *Hereditas* 93, 311–20.

Henry, R.J. (1986a) A comparison of the non-starch carbohydrates in cereal grains. *Journal of the Science of Food and Agriculture* 36, 1243–53.

Henry, R.J. (1986b) Genetic and environmental variation in the pentosan and β-glucan contents of barley, and their relation to malting quality. *Journal of Cereal Science* 4, 269–77.

Henry, R.J. (1987) Variation in the carbohydrate composition of barley. In: Yasuda, V.W. & Konishi, T. (eds), *Barley Genetics V.* (Proceedings of the 5th International Barley Genetics Symposium), Sanyo Press, Okayama, Japan, pp. 763–6.

Henry, R.J. (1988) The carbohydrates of barley grains – a review. *Journal of the Institute of Brewing* 94, 71–8.

Hesselman, K. & Åman, P. (1986) The effect of β-glucanase on the utilization of starch and nitrogen by broiler chickens fed on barley of low- or high-viscosity. *Animal Feed Science and Technology* 15, 83–93.

Hockett, E.A., McGuire, C.F., Newman C.W. & Prentice, N. (1987) The relationship of barley beta-glucan content to agronomic and quality characteristics. In: Yasuda, S. & Konishi, T. (eds), *Barley Genetics V* (Proceedings of the 5th International Barley Genetics Symposium), Sanyo Press, Okayama, Japan, pp. 851–60.

Hoshimoto, S., Shogrew, M.D., Bolte, L.C. & Pomeranz, Y. (1986) Cereal pentosans: their estimation and significance. III. Pentosans in abraded grains and milling by-products. *Cereal Chemistry* 64, 39–41.

Jende-Strid, B. (1981) Characterization of mutants in barley affecting flavonoid synthesis. In: Whitehouse, R.N.H. (ed.), *Barley Genetics IV* (Proceedings of the 4th International Barley Genetics Symposium, Edinburgh), Edinburgh University Press, pp. 631–4.

Jende-Strid, B. & Møller, L. (1981) Analysis of proanthocyanidins in wild type and mutant barley (*Hordeum vulgare* L.). *Carlsberg Research Communications* 46, 53–64.

Keys, J.E., Jr & Debarthe, J.V. (1974) Site and extent of carbohydrate, dry matter, energy and protein digestion and rate of passage of grain diets in swine. *Journal of Animal Science* 39, 57–62.

Kirkman, M.A., Shewry, R.R. & Miflin, B.J. (1982) The effect of nitrogen nutrition on the lysine content and protein composition of barley seeds. *Journal of the Science of Food and Agriculture* 3, 115–27.

Kirsi, M. (1973) Formation of proteinase inhibitors in developing barley grain. *Physiology of Plants* 29, 141.

Kirsi, M. & Mikola, J.M. (1971) Occurrence of proteolytic inhibitors in various tissues of barley. *Plants* 96, 281–91.

Kleese, R.A., Rasmussen, D.C. & Smith, L.H. (1968) Genetic and environmental variation in mineral element accumulation in barley, wheat and soybeans. *Crop Science* 8, 591–3.

Klopfenstein, C.F. & Hoseney, R.C. (1987) Cholesterol-lowering effect of β-glucan enriched bread. *Nutrition Reports International* 36, 1091–8.

Larsson, K. & Miezis, Y. (1979) On the possibility of dietary fiber formation by interaction in the intestine between starch and lipids. *Stärke* 31, 301–2.

Liu, D.J. & Pomeranz, Y. (1975) Distribution of minerals in barley at the cellular level by X-ray analysis. *Cereal Chemistry* 52, 620–9.

Liu, D.J., Robbins, G.S. & Pomeranz, Y. (1974) Composition and utilization of milled barley products. IV. Mineral components. *Cereal Chemistry* 51, 209–316.

MacLeod, A.M., Travis, D.C. & Wreay, D.G. (1953) Studies on the free sugars of the barley grain. III. Changes in sugar content during malting. *Journal of the Institute of Brewing* 54, 154–65.

McCance, R.A. & Widdowson, E.M. (1935) Phytic acid in human nutrition. *Biochemistry Journal* 29, 42694–9.

McIntosh, G.H. & Oakenfull, D. (1990) Possible health benefits from barley grain. *Chemistry in Australia*, September, 294–6.

Merritt, N.R. (1967) A new strain of barley with starch of high amylose content. *Journal of the Institute of Brewing* 73, 583–5.

Miflin, J.B. & Shewry, P.R. (1979) The biology and biochemistry of cereal prolamins. In: *Seed Protein Improvement in Cereals and Grain Legumes*. IAEA FAO GSF. STI PUB 496, Vienna, pp. 137–57.

Miflin, B.J., Burgess, S.R. & Shewry, P.R. (1981) The development of protein bodies in the storage tissues of seeds: subcellular separations of homogenates of barley, maize and wheat endosperms and of pea cotyledons. *Journal of Experimental Botany* 32, 199–219.

Mikola, J. & Kirsi, M. (1972) Differences between endospermal and embryonal trypsin inhibitors in barley, wheat and rye. *Acta Chemica Scandinavica* 26, 787–95.

Mikola, J. & Suolinna, E.M. (1969) Purification and properties of a trypsin inhibitor from barley. *European Journal of Biochemistry* 9, 555–60.

Mikola, J. & Suolinna, E.M. (1971) Purification and properties of an inhibitor of microbial alkaline proteinases from barley. *Archives of Biochemistry and Biophysics* 144, 566–75.

Mori, T. (1990) Chemical characterization and metabolic function of soluble dietary fiber from select milling fractions of a hull-less barley and its waxy starch mutant. Unpublished MS Thesis. Montana State University, Bozeman, MT 59717–0290, USA.

Morrison, W.R. (1978) Cereal lipids. In: Pomeranz, Y. (ed.) *Advances in Cereal Science and Technology*, Vol. II. American Association of Cereal Chemists, St Paul, MN, USA, pp. 221–343.

Morrison, W.R. (1988) Lipids in cereal starches: a review. *Journal of Cereal Science* 8, 1–15.

Moss, B.R., Laudert, S.B., Stockton, G., Miller, C. & Baldridge, D. (1980) Nutritional value of waxy vs normal barley for ruminants. *Nutrition Reports International* 21, 487–96.

Moss, B.R., El-Negoumy, A.M. & Newman, C.W. (1983) Nutritional value of waxy barley for broiler chickens. *Animal Feed Science and Technology* 8, 25–34.

Munck, L. (1964) The variation of nutritional value of barley. I. Variety and nitrogen fertilizer effects on chemical composition and laboratory feeding experiments. *Hereditas* 52, 1–35.

Munck, L. (1972) Improvement of nutritional value in cereals. *Hereditas* 72, 1–128.

Nakao, S. (1950) On the waxy barleys in Japan. *Seikin Ziko* 4, 111–13.

Newman, C.W. & El-Negoumy, A.M. (1980) Increased lysine in barley protein improves the biological value but depresses protein digestibility. *Proceedings of the Federation of American Societies of Experimental Biology* 39, 1115.

Newman, C.W. & McGuire, C.F. (1985) Nutritional quality of barley. In: Rasmusson, D.C. (ed.) *Barley*. Agronomy Monograph No. 26, ASA, CSA, SSSA, Madison, WI, USA, pp. 403–56.

Newman, C.W., Eslick, R.F., Calvert, C.C. & El-Negoumy, A.M. (1978) Comparative nutritive value of Glacier and high amylose Glacier barleys. *Journal of Animal Science* 47, 448–55.

Newman, C.W., Newman, R.K. & Graham, H. (1990a) Influence of the waxy starch gene on the carbohydrate composition of covered and hull-less barleys. *Cereal Foods World* 35, 835.

Newman, C.W., Øverland, M., Newman, R.K., Bang-Olsen, K. & Pedersen, B. (1990b) Protein quality of a new high-lysine barley derived from Risø 1508. *Canadian Journal of Animal Science* 70, 279–85.

Newman, R.K., Newman, C.W., El-Negoumy, A.M. & Aastrup, S. (1984) Nutritional quality of proanthocyanidin-free barley. *Nutrition Reports International* 30, 809–16.

Newman, R.K., Lewis, S.E., Newman, C.W., Boik, R.J. & Ramage, R.T. (1989a) Hypocholesterolemic effect of barley foods on healthy men. *Nutrition Reports International* 39, 749–60.

Newman, R.K., Newman, C.W. & Graham, H. (1989b) The hypocholesterolemic function of barley β-glucans. *Cereal Foods World* 34, 883–6.

Newman, R.K., Newman, C.W., Hofer, P.J. & Barnes, A.E. The influence of hull-less barley β-glucans on growth and lipid metabolism in the chick model. *Plant Foods for Human Nutrition* (in press).

Osborne, T.B. (1924) *The Vegetable Proteins*, 2nd edn. Longmans, Green & Co., London.

Outtrup, H. & Schaumburg, K. (1981) Structure elucidation of some proanthocyanidins in barley by H270MHz spectroscopy. *Carlsberg Research Communications* 46, 43–52.

Øverland, M. (1988) Protein quality of normal barley cultivars and their proanthocyanidin-free mutants. Unpublished Master's Thesis, Montana State University, Bozeman, MT 59717, USA.

Owen, B.D., Sosulski, F., Wu, K.K. & Farmer, J.J. (1977) Variation in mineral content in Saskatchewan feed grains. *Canadian Journal of Animal Science* 57, 679–87.

Parsons, J.G. & Price, P.B. (1974) Search for barley (*Hordeum vulgare* L.) with higher lipid content. *Lipids* 9, 804–8.

Pedersen, B. & Boisen, S. (1982) The nutritional effect of a pepsin stable trypsin inhibitor from barley grains. 2. *Zeitschrift für Tierphysiologie, Tierphysiologie und Futtermittelkunde* 48, 65–75.

Pedersen, B., Hansen, M., Munck, L. & Eggum, B.O. (1989) Weaning foods with improved energy and nutrient density prepared from germinated cereals. 2. Nutritional evaluation of gruels based on barley. *United Nations University, Foods and Nutrition Bulletin* 11, 46–52.

Pomeranz, Y., Eslick, R.F. & Robbins, G.S. (1972) Amino acid composition and malting and brewing performance of high amylose and hiproly barleys. *Cereal Chemistry* 49, 629–35.

Powell, W., Caligari, P.D.S., Swanston, J.S. & Jinks, J.L. (1985) Genetical investigations into β-glucan content in barley. *Theoretical and Applied Genetics* 71, 461–6.

Price, P.B. & Parsons, J.G. (1975) Lipids of seven cereal grains. *Journal of the American Oil Chemists Society* 52, 490–3.

Price, P.B. & Parsons, J.G. (1979) Distribution of lipids in embryonic axis, bran-endosperm and hull fractions of hull-less barley and hull-less oat grain. *Journal of Agricultural Food Chemistry* 27, 813–16.

Qureshi, A.A., Schnoes, H.K., Din, Z.Z. & Peterson, D.M. (1984) Determination of the structure of cholesterol inhibitor II isolated from high-protein barley flour (HPBF). *Federation Proceedings* 43, 1866 (#2626).

Qureshi, A.A., Burger, W.C., Peterson, M.D. & Elson, C.E. (1986) The structure of an inhibitor of cholesterol biosynthesis isolated from barley. *Journal of Biological Chemistry* 261, 10544–50.

Qureshi, A.A., Crenshaw, T.D., Abuirmeileh, H., Peterson, D.M. & Elson, C.E. (1987) Influence of minor plant constituents on porcine hepatic lipid metabolism. *Atherosclerosis* 64, 109–15.

Qureshi, A.A., Chaudhary, V., Weber, F.E., Choicoye, E. & Qureshi, N. (1991) Effects of brewer's grain and other cereals on lipid metabolism in chickens. *Nutrition Research* 11, 159–68.

Raboy, F., Dickinson, D.B. & Neuffer, M.G. (1990) A survey of maize kernel mutants for variation in phytic acid. *Maydica* 35, 383–90.

Raboy, V. (1990) The biochemistry and genetics of phytic acid synthesis. In: Morke, D.J., Boss, W. & Loewus, F. (eds), *Inositol Metabolism in Plants*. Wiley-Liss, New York, USA, pp. 52–73.

Reid, D.A. (1985) Morphology and anatomy of the barley plant. In: Rasmusson, D.C. (ed.), *Barley*. Agronomy Monograph No. 26, ASA, CSA, SSSA, Madison, WI, USA, pp. 73–101.

Robinson, M.C. & Lupton, J.R. (1990) The effects of barley flour and barley oil on hypercholesterolemic men and women. *Journal of the American Dietetic Association* 90 (supplement), A–104.

Sandberg, A.-S., Anderson, H., Kivistö, B. & Sandström, B. (1986) Extrusion cooking of a high-fiber cereal product. 1. Effect on digestibility and absorption of protein, fat, starch, dietary fiber and phytate in the small intestine. *British Journal of Nutrition* 55, 245–54.

Sandstedt, R.M. & Hites, B.O. (1968) Genetic variations in maize. Effects on the properties of starches. *Cereal Science Today* 13, 82–94.

Sandstedt, R.M., Strahan, D., Ueda, S. & Abbot, R.C. (1962) The digestibility of high amylose corn starches compared to that of other corn starches. The apparent effect of the *ae* gene on susceptibility to amylase action. *Cereal Chemistry* 39, 123–31.

Scott, R.W. (1972) The viscosity of worts in relation to their content of β-glucan. *Journal of the Institute of Brewing* 78, 179–86.

Shewry, P.R. & Tatham, A.S. (1990) The prolamin storage proteins of cereal seeds: structure and evaluation. *Biochemical Journal* 267, 1–12.

Shewry, P.R., Finch, R., Parmer, S., Franklin, J. & Miflin, B.J. (1983) Chromosomal location of *Hor3*, a new locus governing storage protein in barley. *Heredity* 50, 179–89.

Shewry, P.R., Bright, S.W.J., Burgess, S.R. & Miflin, B.J. (1984) Approaches to improving the nutritional value of barley seed proteins. In: *Use of Nuclear Techniques for Cereal Grain Protein Improvement*. IAEA, STI PUB 664, Vienna, pp. 227–40.

Shewry, P.R., Parmar, S., Franklin, J. & Burgess, S.R. (1990) Analysis of a rare recombination event within the multigenic *Hor2* locus of barley (*Hordeum vulgare* L.). *Genetical Research* 55, 171–6.

Shotwell, M.A. & Larkins, B.A. (1989) The biochemistry and molecular biology of seed storage proteins. In: Marcus, A. (ed.), *The Biochemistry of Plants: A Comprehensive Treatise*, Vol. 15, *Molecular Biology*. Academic, London, pp. 297–345.

Skarsaune, S.K., Banasik, I.J. & Watson, C.A. (1973) Lipids of post-harvest barley. *Proceedings of the American Society of Brewing Chemistry* 30, 94–7.

Smith, D.B., Morgan, A.G. & Gill, A.A. (1980) The major biochemical constituents of an acid extract of barley flour as used in viscosity determinations by the falling ball method. *Journal of the Institute of Brewing* 86, 113–19.

Staute, R.G., Woodward, J.R., Fincher, G.B. & Stone, B.A. (1983) Water soluble (1 → 3), (1 → 4)-β-D-glucans from barley (*Hordeum vulgare*) endosperm. III. Distribution of cellotriosyl and cellotetrasyl residues. *Carbohydrate Polymers* 3, 299–312.

Sullins, R.D. & Rooney, L.W. (1974) Microscopic evaluation of the digestibility of sorghum lines that differ in endosperm 25 characteristics. *Cereal Chemistry* 51, 134–42.

Tallberg, A. (1981) Characterization of high-lysine barley genotypes. PhD Thesis, Lund University, Lund, Sweden.

Ullrich, S.E., Clancy, J.A., Eslick, R.F. & Lance, R.C.M. (1986) β-glucan and viscosity of extracts from waxy barley. *Journal of Cereal Science* 4, 279–85.

Vahouny, G.V. & Kritchevsky, D. (eds) (1986) *Dietary Fiber Basics and Clinical Aspects*. Plenum Press, New York, USA.

von Bothmer, R. & Jacobsen, N. (1985) Origin,

taxonomy, and related species. In: Rasmusson, D.C. (ed.), *Barley*. Agronomy Monograph No. 26, ASA, CSA, SSSA, Madison, WI, USA, pp. 73–101.

Walker, J.L. & Merritt, N.R. (1969) Genetic control of abnormal starch granules and high amylose content in a mutant of Glacier barley. *Nature (London)* 221, 482–3.

Wang, L., Newman, R.K., Newman, C.W., & Hofer, P.J. (1990) Effect of barley oil on serum cholesterol in chicks. *Cereal Foods World* 35, 819.

Wang, L., Newman, R.K., Newman, C.W., Hofer, P.J. & Fengler, A.I. (1991a) Effect of barley β-glucans on intestinal viscosity and lipid metabolism in the chick model. *The Federation of American Societies for Experimental Biology Journal* 5 (no. 5) A1286.

Wang, L., Newman, R.K., Newman, C.W. & Ramage, R.T. (1991b) Effect of cultivar and environment on oil and its α-tocotrienol content in barley grain. *Sixth International Barley Genetics Symposium*, Helsingborg, Sweden.

Weaver, C.M., Chen, P.H. & Rynearson, S.L. (1981) Effects of milling on trace element and protein content of oats and barley. *Cereal Chemistry* 58, 120–4.

Welch, R.W. (1978) Genotypic variation in oil and protein in barley grain. *Journal of Science of Food and Agriculture* 29, 953–8.

18

Barley Germination: Biochemical Changes and Hormonal Control

Dennis E. Briggs

Thousands of barley varieties are available (Briggs, 1978), only a few of which are extensively cultivated under a wide range of climatic conditions. Grains of different barley samples can differ widely in size, shape, chemical composition and physiological state. These differences may account for some of the discrepancies between experimental results obtained by different research groups. For example the huskless variety 'Himalaya' is widely used in experiments but the behaviour of isolated aleurone layers from it differs from that of several British husked, malting-grade barleys. There are appreciable differences between grains from one plant, and very great differences between samples of one variety grown in different seasons (Briggs *et al.*, 1981). Thus inhomogeneity of grain samples is a problem for the technologist and experimentalist.

Experiments on barley germination are usually from the viewpoint of a seedsman who needs grain capable of generating new seedlings in the field, or the maltster who needs to produce routinely malts of specified qualities, in the shortest possible times, and in high yield (Briggs, 1978; Briggs *et al.*, 1981). Grains grown under test conditions in ways appropriate to these two outlooks behave differently and this should be kept in mind when considering experimental reports.

Grain composition

The structure of the barley grain is described elsewhere (Briggs, 1978, 1987a; Cook, 1962; Chapter 14 this volume). Commercial European barleys are husked and usually have average thousand corn dry weights of 30–45 g. In any batch, grains vary in size from those retained on a 2.8 mm slotted screen through to those which will pass through a 2.2 mm screen. Grain nitrogen contents frequently range from 1.3 to 2.5% dry weight (8.1–15.6% crude protein; except for moisture contents all proportions of grains will be given on a percentage dry weight basis).

The chemistries of the husk and pericarp (9–14%) have not been adequately investigated. The husk is lignified and is abrasive because of silica in the outer epidermis. The husk and pericarp protect the inner, living tissues of the grain and resist swelling during imbibition. They are slowly permeated by aqueous solutions (Briggs & MacDonald, 1983a). They carry variable populations of micro-organisms which may be controlled experimentally but seem to be impossible to eliminate without damaging or otherwise significantly altering the grain.

The husk and pericarp may be removed using 50% sulphuric acid. Such decorticated grain can be surface-sterilized and grown under aseptic conditions making it a favourable experimental preparation. However, in a number of respects it differs in its behaviour from whole grains (Briggs & MacDonald, 1983b; Briggs, 1987a).

The testa *plus* the nucellar cuticle, the pigment strand and the micropylar structures (1–3%) make up the selectively permeable system which divides the interior from the exterior of the grain. Salts, sugars and microbes do not pass through this layer in healthy undamaged grains, while nutrients from within do not pass out. The two cuticularized layers contain an estolide of fatty acids and hydroxy-fatty acids and have a wax containing alkanes, sterols, various esters and *n*-alkyl-resorcinols (Briggs, 1974). Proanthocyanidin polyphenols occur between the two cuticularized layers (Aastrup *et al.*, 1984). These and the *n*-alkyl-resorcinols have antimicrobial properties which may help to limit the inward invasion of micro-organisms from the surface layers.

The quiescent embryo (axis *plus* scutellum) weighs 1.3–1.7 mg, about 2.5% of the grain. It contains little or no starch but it is rich in 'crude' protein (34%), lipids (14–17%), ash (5–10%) and extremely rich in sugars [sucrose, 15%; raffinose, 5–10% and some fructosans (Briggs, 1978)]. Thus the embryo is well stocked with substances to support its initial metabolic activities and growth. The cell walls make up about 15% of the total.

The rest of the grain consists of nucellar materials and endosperm. These parts of the grain cannot be cleanly separated and in most experimental reports are grouped as the aleurone layer (including the hyaline nucellar layer between it and the testa) and the starchy endosperm (including the crushed and depleted cell layer and the sheaf-cells over the furrow which are nucellar in origin). The aleurone layer (8–15% of the grain) is usually analysed with the testa and pigment strand still attached.

The thick (3–5 μm) cell walls of the aleurone layer are 40–45% of the whole cells, and constitute a substantial reserve of polysaccharide, chiefly feruloylated and acetylated arabinoxylan (Bacic & Stone, 1981a, b). The tissue contains no starch but is rich in protein (17–20%), triacyl glycerol (20%), minerals (ca. 75% of those of the decorticated grain) and sugars (including sucrose, raffinose and probably stachyose, verbascose and fructosans). Thus, as well as being a source of hydrolytic enzymes the aleurone layer is also an important reserve tissue.

The cells of the aleurone layer, like those of the embryo, are living and, in hydrated grain, they respire and metabolize. However, they do not multiply and divide.

The starchy endosperm makes up a variable proportion of the grain (about 75%) and is the major reserve tissue. Starch is the chief component of the grain (58–65%). It is virtually confined to this tissue, and makes up around 80% of it. The starch granules fall into two populations, usually of 1.7–10 μm and 20–48 μm diameter but the sizes vary with variety. About 90% (w/w) of the starch is in large granules (Briggs, 1978). In most varieties the ratio of amylose to amylopectin is 20–30 to 80–70. The starch granules have associated protein. Polar lipids occur within the granules and appear to contribute to their structural characteristics (Morrison, 1978, 1988). In addition the starchy endosperm contains major amounts of protein (ca. 9% of the tissue). The cell walls are thin (2 μm), and are about 70% mixed-link β-glucan, 23% arabinoxylan and 6% holocellulose, together with some associated protein (Briggs, 1978; Fincher, 1975, 1989; Fincher & Stone, 1986). The arabinoxylan is feruloylated (and quite possibly acetylated), but the β-glucan is not (Ahluwalia & Fry,

1986). There are small deposits of β-(1→3) glucans in the endosperm cell walls, especially in the junction with the aleurone layer (Fulcher *et al.*, 1977; MacGregor *et al.*, 1989). It is probably the (1→3)-β-glucan rich layer that is degraded in the freshly hydrated grain or degermed grain to allow the separation of aleurone layer from the underlying starchy endosperm (Clutterbuck & Briggs, 1973; Briggs, 1987a).

There are also smaller amounts of lipids, minerals and nucleic acids in the endosperm. This tissue is dead; its cells do not respire or carry out other metabolic activities.

Morphological changes during germination

The reserves used to support the growth and development of the embryo/seedling plant during germination are initially drawn from the embryo itself, and later from the starchy endosperm and, less obviously, from the aleurone layer.

Much is inferred, but little has been experimentally demonstrated, about the ways in which the embryo's own reserves are used. A little more is known about the use of the reserves of the aleurone layer. Most is known about how components of the dead starchy endosperm are degraded to yield soluble products, such as sugars, peptides and amino acids, which diffuse to the embryo, and are taken up by the scutellum and used. The patterns of breakdown of the components of the endosperm, particularly the cell walls and the starch granules, reveal a great deal about where the enzymes catalysing the process are produced and released and the sequences in which this happens.

Endosperm breakdown has been studied in (i) decorticated grains grown aseptically with a continuous supply of water at elevated temperatures (25°C) in the dark ('wet grown'), and in (ii) entire and decorticated grains grown under malting conditions, that is steeped (soaked) under water and then germinated in cool (14–16°C), water-saturated air in the dark, *not* in contact with liquid water. The changes undergone by the grains in the two cases are strikingly different (Briggs, 1987a).

'Wet-grown' decorticated grains swell during the initial period of hydration. Subsequently, the scutellum pulls away from the starchy endosperm and this partly dissolves and is replaced by a sugary liquid. Root growth is vigorous, the coleoptile grows to its full length in a few days and the first leaf appears (Briggs, 1968a). Endosperm dissolution begins under the scutellum, then the liquid-filled space extends away from the embryo, advancing fastest under the aleurone layer and on a front roughly parallel to the face of the scutellum (Briggs & Morrall, 1984). In time, the fluid-filled space comes to extend completely around a residual mass of endosperm tissue, except above the furrow where the sheaf cells (of nucellar origin) remain intact and anchor the plug of endosperm to the testa and pigment strand. The fluid-filled space seems to enlarge both by the stretching of the surface layers, the aleurone layer and testa–nucellar cuticle, which distorts the shape of the grain, and by the continuing dissolution of the tissue of the starchy endosperm (Briggs & Morrall, 1984; Briggs, 1987a). In grains grown in this way it has been estimated that after 6 days the degradation of the starch provided 81.5% of the carbohydrate used, that had been derived from polysaccharide reserves, while the cells walls of the aleurone layer and starchy endosperm together contributed the other 18.5% (Morrall & Briggs, 1978). In the initial stages of germination, the cell walls of the starchy endosperm are degraded before the starch grains are attacked, while at the end of the germination period cell walls remain towards the apex of the grain but the cell spaces enclosed by them are devoid of starch.

In malting grain, endosperm degradation ('modification' in maltster's terms) is incomplete and is less obvious because no total dissolution of the tissue occurs. During germination the endosperm progressively softens. When the grain is dried, it is found to have lost its toughness, and become 'crunchy' and friable (Briggs *et al.*, 1981). The patterns of modification have been the subject of debate, some acrimonious, for about a century (Briggs, 1987a). Mapping endosperm modification is important both commercially to monitor the progress of malting and scientifically to follow the progress of enzyme release from the living tissues around the starchy endosperm. The comparative solidity of even the modified regions restricts the spread of enzymes (in contrast to 'wet-grown' grains in which liquid accumulates and which, therefore, permits free diffusion of enzymes).

Modification can be mapped using thin sections of grains, cut in various planes, and stained with Congo Red, Trypan Blue or Calcofluor (a fluorescent stain); these stains preferentially bind to the β-glucans in the cell walls. Alternatively, grains are dried and stuck to a wooden or glass backing. They can then be ground down with a sander to expose the interior parts, including the starchy endosperm. Modified regions can be detected by: (i) penetration of the dye methylene blue into the modified zones; (ii) fluorescence of unmodified regions treated with Calcofluor; (iii) loss of physical strength of the modified regions; (iv) presence in the modified regions of enzyme(s) which hydrolyse dibutyryl fluorescein to liberate fluorescent fluorescein (Gram, 1982a, b; Briggs, 1987a, b). The results given by the different techniques are in reasonable agreement when technical problems are allowed for (Oh & Briggs, 1989).

The first sign of endosperm degradation in malting grain is the fragmentation of the cell walls followed by their apparent disappearance in the most modified regions. Whether or not cell wall destruction is complete is discussed elsewhere but β-glucan degradation is essentially complete (Briggs, 1987a, b; Selvig *et al.*, 1986; Oh & Briggs, 1989). Subsequently the protein 'matrix' is partly degraded and the starch granules, which were more or less anchored by the protein, become 'loose'. The small starch granules are attacked first, decreasing in size as a result of a general attack all over the surfaces of the granules. Later the large starch granules appear etched and pitted and layers are selectively degraded within, giving broken granules a laminated appearance. Often attack on large, lenticular granules is first apparent at the peripheral groove.

Modification proceeds faster in decorticated grains than in entire husked grains when they are malted. Variations between grains in the rates and patterns of modification are greater in husked than in decorticated grains (Briggs & MacDonald, 1983b; Briggs, 1987a). Modification begins beneath the scutellum and, in some grains, advances slightly faster on the ventral side, around the sheaf cells. The cell walls of this tissue pull apart in some grains during steeping. This line of weakness, which permits the spread of methylene blue, may also permit hydrolytic enzymes to spread faster causing the more rapid spread of modification next to the sheaf cells. After germination for 2 or 3 days modification also occurs ahead of the region caused by enzymes from the scutellum beneath the aleurone layer, and is clearly caused by enzymes from this latter tissue (Gibbons, 1980, 1981; Briggs & MacDonald, 1983b). Endosperm modification in malting grain is normally finished in 5–6 days, in that β-glucan destruction is almost complete. In contrast to decorticated grains, the patterns of endosperm breakdown in partially malted, husked grains are rather variable, suggesting that the relative contributions of enzymes from the scutellum and aleurone layer may be variable in husked grains, but less so in decorticated grains.

Externally applied gibberellic acid, GA_3, 'reinforces' the normal pattern of endosperm breakdown, accentuating the breakdown under the aleurone layer. GA_3 must gain entry in the region of the embryo since modification still spreads from this organ. Despite claims to the contrary, the same patterns of endosperm breakdown occur in normal, abraded and tumbled grains when they are malted with applications of GA_3 (Smith & Briggs, 1979, 1980; Briggs & MacDonald, 1983b; Briggs, 1987a). In well-made malt nearly all the endosperm cell walls are degraded and consequently such malt has a very low content of β-glucan, e.g. 0.25% (Bourne *et al.*, 1982).

It has been widely assumed that the mixtures of enzymes released from the scutellum and from the aleurone layer contain the same enzymes in similar proportions. Increasingly there are indications that this is not so (see later). Oh & Briggs (1989) noticed that in colour photographs of cut grains, fluorescent because they had been stained with Calcofluor, the 'colours' of the non-fluorescent, modified zones adjacent to the scutellum and the aleurone layer were different. The simplest explanation for this is that the enzyme mixtures from the tissues differ and so degrade the endosperm tissue differently. Other data support this idea (see below).

The enzyme systems of grains

At the onset of germination the embryo mobilizes and uses its internal reserves of sugars, lipids, proteins and minerals. Little is known of the details of these processes, or how they are regulated, but the limited evidence indicates that the usual biochemical pathways are involved (Briggs, 1978). A mixture of hydrolytic enzymes is released from the scutellar epithelium into the endosperm. Hydrolysis products generated by the breakdown of the starch, the cell walls, the protein deposits and (presumably) residual nucleic acids diffuse to the embryo and are used by it to support its metabolism and growth and restore endogenous sucrose depleted in the initial stages.

A little later the aleurone layer begins to generate some hydrolytic enzymes. In response to gibberellins from the embryo (see below), the production of these enzymes may be enhanced and the production of others initiated. These enzymes are released into the starchy endosperm to assist in its degradation. However, before these endosperm-targeted enzymes are generated, the hydrated aleurone needs to mobilize its own reserves to support enzyme production. Once again, little is known of how this is achieved. The process requires oxygen and is prevented by respiratory inhibitors. The aleurone grains are progressively degraded and phytate is hydrolysed. Some of the inorganic phosphate so formed and, by analogy with wheat, other inorganic ions, are released from the aleurone in response to the gibberellins and used by the growing embryo (Briggs, 1973; Clutterbuck & Briggs, 1974; Gram, 1982a). Aleurone tissue is rich in potassium, phosphorus, magnesium and calcium (Liu & Pomeranz, 1975).

Gibberellins also trigger gluconeogenesis from the triglycerides and at least some of the sucrose is released into the pool of sugars in the endosperm. This explains why changes in the level of this sugar do not parallel changes in the levels of maltose, maltotriose and other products of starch hydrolysis (Harris, 1962). The aleurone is also capable of taking up some sugars and amino acids and, after converting them into sucrose and glutamine, releasing them again (Chrispeels *et al.*, 1973; Newman & Briggs, 1976).

The details of the enzymes that mobilize the protein reserves of the aleurone grains are uncertain, but they can be inhibited by bromate (Melcher & Varner, 1971). The effect of bromate on enzyme formation by gibberellin-triggered aleurone layers may be overcome by supplying a mixture of amino acids (Melcher & Varner, 1971; Briggs, 1973).

Substantial data are available on enzymes responsible for the hydrolysis of the major reserves of the starchy endosperm, i.e. the starch, the protein, and the major components of the cell walls, the β-glucans and the arabinoxylans. Less is known of the DNA- and RNA-degrading enzymes, phosphatases and lipid-degrading enzymes. Chitinase, which has no known substrate within the grain, has been detected; it is proposed that the enzyme attacks the cell-walls of invading fungi.

Here, most attention will be given to the enzymes which degrade starch and protein. The enzymes which degrade the cell walls are discussed in detail by Fincher (1989), Fincher & Stone (1986), and in

Chapter 20. As germination proceeds the levels of starch, cell walls and protein decline in the starchy endosperm, there is a rise in the levels of soluble, low-molecular-weight substances (sugars, oligosaccharides, amino acids and peptides), products of the hydrolysis of the macromolecular reserves, and these are used by the growing embryo. This occurs to very different degrees in wet-grown and in malted grains (Briggs 1968a; Briggs *et al.*, 1981).

Starch degradation

The enzymes involved in the degradation of starch are two groups of α-amylases, β-amylase, debranching enzyme (R-enzyme; limit dextrinase) and α-glucosidase/maltase. Phosphorylase also occurs in germinating barley. It is probably confined to the growing embryo where glucose 1-phosphate, the product of the phosphorolysis of the transient starch that forms at the onset of germination, can immediately enter the metabolic pathways. In principle, if phosphorylase were released into the starchy endosperm, inorganic phosphate released from the aleurone would be available to permit it to act. The glucose 1-phosphate produced by the phosphorolysis of the starch would almost certainly be rapidly hydrolysed by the phosphatases which are present to regenerate inorganic phosphate and release glucose (Briggs, 1973; Ashford & Gubler, 1984).

α-Amylase, which occurs as several isozymes, is essentially absent from sound, mature, ungerminated grain. It is generated entirely during germination (see below), and is the enzyme which initiates the degradation of the starch granules in the endosperm. The enzyme attacks α-(1→4) linkages within chains of amylose and amylopectin. It does not split links immediately adjacent to α-(1→6) joined glucose residues or the α(1→6)-linkages themselves.

The action patterns of these α-amylases have been extensively studied (MacGregor, 1983, 1987; Hill & MacGregor, 1988). Originally agar electrophoresis was used to detect three groups of isoenzymes of α-amylase which vary in their properties and stabilities (Frydenberg & Nielsen, 1965; Hill & MacGregor, 1988). Continuing studies have confirmed and extended these results. The occasional presence has been noted of a fourth α-amylase, 'α-amylase X'. Like the others discussed here, α-amylase X does not originate from the microflora of the grains (Niku-Paavola, 1977; MacGregor *et al.*, 1979). The low pI of this fourth isoenzyme (3.0), distinguishes it from any of the other intracellular precursor forms or secreted forms of α-amylase I (Jacobsen *et al.*, 1988). There are wide variations between the isoenzyme patterns observed in germinating grains with different genotypes (e.g. Briggs & Raynes, 1985). The isozymes can be divided into three groups: α-amylase I, α-amylase II and α-amylase III. These groups of isozymes are known by various names. They differ significantly in their properties (Table 18.1).

α-Amylase III is α-amylase II reversibly combined with the heat-labile protein barley α-amylase/subtilisin inhibitor (BASI) which reduces its activity (Weselake *et al.*, 1983, 1985a, b; MacGregor, 1983, 1987; Hill *et al.*, 1987; Hill & MacGregor, 1988). BASI, which also binds to the bacterial protease subtilisin, can be partly disassociated from its α-amylase III complex by increasing the ionic strength (Hill & MacGregor, 1988).

Quantitatively α-amylase I is a minor component of the total α-amylase found in germinating grain, but it is the most efficient at degrading starch granules (MacGregor, 1987; Hill & MacGregor, 1988). α-Amylase I is immunologically distinct from α-amylase II and the two groups of isozymes rise and fall independently during germination (Bøg-Hansen & Daussant, 1974). α-Amylase I is said to be the form preferentially formed in the scutellum but Gibbons (1979, 1980), using an immunofluorescent technique, located α-amylase II spreading from this organ.

α-Amylase II is the 'classical' malt enzyme. In the presence of calcium ions α-amylase I is more liable to inhibition by

Table 18.1. The α-amylases of germinating barley*.

Enzyme group	pI	Ca^{2+} binding	pH optima	Mobility in agarose electrophoresis pH 8.5	Ca^{2+} Hg^{2+}	EDTA	Proportion in germinated grain	Location of gene
α-Amylase I (Group A)	Low 4.1–5.3	Very strong	3.0–5.5	Rapid	Inhibit	–	Minor	Chromosome I
α-Amylase II (Group B)	High 5.3–6.6	Strong	5.0–5.4	Slower	–	Inhibit	Major	Chromosome 6
α-Amylase III (α-Amylase II + BASI)	~6.7	–	–	Slow	–	–	Major	–
[BASI] – α-amylase inhibitor	7.3	–	–	–	–	–	–	–
α-Amylase 'X'	3.0	–	–	–	–	–	Small, not always detected	–

*Data from Bertoft *et al.*, 1984; Briggs & Raynes, 1985; Hill & MacGregor, 1988; Hil *et al.*, 1987; Muthukrishnan & Chandra, 1988; McGregor, 1983, 1987.
BASI, barley α-amylase/subtilisin inhibitor.

mercuric chloride and copper ions than α-amylase II and less likely to be inhibited by calcium-chelating agents (Frydenberg & Nielsen, 1965; Bertoft *et al.*, 1984). At 70°C α-amylase I is most stable between pH 5 and 6 in the presence of calcium ions, while α-amylase II is most stable at pH 7.0 (Bertoft *et al.*, 1984). Calcium ions bind to both groups of enzymes, but more strongly to α-amylase I than II (Bush *et al.*, 1989b). Both α-amylases I and II have been crystallized (Svensson *et al.*, 1987).

For several years evidence has pointed to α-amylase being synthesized on the rough endoplasmic reticulum (RER) (Briggs, 1973). More recently it has been shown that the enzyme is transferred from the RER to the Golgi and directionally transported to the plasmalemma of the secreting cell (Fernandez & Staehelin, 1985; Gubler *et al.*, 1986). It seems uncertain whether or not the enzyme is glycosylated, but in the early stages its transport into the ER is directed by a peptide leader-sequence which is removed from the mature enzyme.

In the aleurone layer enzyme formation and release are dependent on a gibberellin (e.g. GA_3) and calcium ions.

α-Amylase release through the predominantly arabinoxylan cell walls of the aleurone layer seems to be via channels dissolved by pentosanases. Calcium ions serve some function in addition to enzyme stabilization during enzyme generation, and the production of the two families of isoenzymes have different calcium ion requirements (Deikman & Jones, 1986). Sometimes the reported optimal requirements are surprisingly high (e.g. 30 mM; Chrispeels & Varner, 1967). However, using a more specific assay for α-amylase and aleurone layers from a different variety of barley prepared in a different way, Clutterbuck & Briggs (1973) confirmed a requirement for Ca^{2+}, but no significant differences in enzyme yields were detected when the concentration of calcium sulphate was varied in the range 0.75–15 mM. The assay used in previous studies was shown to be unspecific and to detect β-amylase and α-glucosidase. It would be expected to be influenced by debranching enzyme and is not reliable (Briggs, 1967; Clutterbuck & Briggs, 1973; Smith *et al.*, 1979; Briggs *et al.*, 1980; Briggs & Cornish-Bowden, 1981).

The existence of different components in each group of α-amylase isoenzymes may be due to the presence of families of genes differing to minor extents but closely linked at each locus (Khursheed & Rogers, 1988) or to secondary modifications. Minor forms which are often detected by electrophoresis may be due to limited proteolysis of the enzymes (Hayes *et al.*, 1988), to which enzyme activity appears to be resistant. This proteolysis is blocked by PMSF (phenyl methane sulphonyl fluoride). The parent enzyme has an M_r of 44 000, the proteolysed enzyme as little as 27 000. Unlike the 'parental' form, the proteolysed enzyme failed to bind to cyclodextrin (Hayes *et al.*, 1988). In aleurone layers incubated with GA_3 distinct precursor forms of α-amylase I have been detected (Jacobsen *et al.*, 1988).

β-Amylase is without action on intact starch granules, which must first be attacked by α-amylase, or possibly α-glucosidase. β-Amylase attacks the penultimate links of starch chains, with the release of maltose (Harris, 1962; Briggs, 1973). In the quiescent grain the enzyme which is experimentally dissolved by salt solutions is highly heterogeneous; more 'latent' or initially insoluble enzyme is brought into solution by extractants containing thiols and/or papain, even boiled papain (Ford & Guthrie, 1908; Harris, 1962). Reducing agents simplify the mixture of soluble forms of β-amylase, converting the higher-molecular-weight forms to components having lower molecular weights. All the components react with a specific antibody.

Thus, extracts contain monomers and oligomers of β-amylase and at least one other protein. One dimer is a heterodimer in which a monomer unit of β-amylase is linked via a disulphide bridge to a monomer of protein Z. Protein Z has an M_r of approx. 40 000. It probably also occurs in solution as monomers, dimers and tetramers, and may constitute as much as

4% of the grain protein (Hejgaard, 1976, 1977, 1978, 1982; Hejgaard & Carlsen, 1977). It is unclear if the larger soluble forms of β-amylase are homo- or heteropolymers. In one investigation β-amylase components ranging in M_r from 56 000 to 392 000 were detected (Niku-Paavola *et al.*, 1973). As barley germinates the bound form of β-amylase becomes soluble, and the mixture of soluble forms alters so that monomeric forms come to predominate (Briggs, 1973). The old conclusion that 'barley' β-amylase monomer is converted into a smaller 'malt' β-amylase monomer by limited proteolysis has been confirmed and extended. Proteolysis and hence this transformation may be blocked by a mixture of leupeptine and pepstatin (Lundgard & Svensson, 1986). Four monomeric forms have been isolated and characterized (Lundgaard & Svensson, 1987) with M_r values of 59 700, 58 000, 56 000 and 54 000 and pI values of 5.2, 5.3, 5.5. and 5.7 respectively. The kinetic properties of all four are similar when starch is the substrate. The largest monomer is 'complete' since it has an amino acid sequence which coincides with that deduced from a cDNA clone (Kries *et al.*, 1987). The smaller forms are derived by limited proteolysis of the 'parental' form at the carboxyl-terminal end.

Apparently the heterogeneity of β-amylase may be increased by the presence of isoenzymes as the enzyme is encoded by two gene copies located on chromosome 4 and one on chromosome 2 (Kreis *et al.*, 1988; Shewry *et al.*, 1988). It is possible that the β-amylase encoded by the gene on chromosome 2 is not expressed in the grain (Kreis *et al.*, 1988).

In contrast to α-amylase, β-amylase exists preformed in quiescent barley and little or none is formed during germination. The enzyme occurs almost entirely in the starchy endosperm, where it is concentrated in the subaleurone region (Briggs, 1973). It constitutes 1–2% of the total grain protein (Kreis *et al.*, 1987). Traces of β-amylase are also present in the embryo (Laurière *et al.*, 1986).

α-Amylase and β-amylase, acting in concert, have been thought to be the main agents of starch degradation in germinating grain. Certainly this is the case when malt is mashed during the brewing, distilling or malt vinegar manufacturing processes (Briggs *et al.*, 1981). More recent evidence suggests that α-glucosidase has an important role to play in degrading starch granules (Sun & Henson, 1990). Both α- and β-amylases are specific for the α-(1→4) linkages between the glucose residues of starch. The α-(1→6) linkages are broken by debranching enzyme or possibly α-glucosidase acting on the oligosaccharide products of starch hydrolysis.

Debranching enzyme was previously regarded as two different enzymes, limit dextrinase and R-enzyme. However, it appears that only one enzyme is involved both in hydrolysing the α-(1→6) links in limit dextrins and in debranching amylopectin. The enzyme increases in amount during germination, apparently due to *de novo* synthesis in the aleurone layer (Hardie, 1975) and the release of a form bound to insoluble protein in the starchy endosperm (Serre & Laurière, 1989). A number of forms of this enzyme have been detected (MacGregor, 1987).

α-Glucosidase (maltase) occurs in the quiescent grain in substantial amounts. Not all the enzyme is soluble even in strong salt solutions (Briggs, 1973). During germination the amount of enzyme increases. Histochemistry suggested that it is practically confined to the embryo and aleurone layer in quiescent grains, but that some is released into the endosperm during germination (Briggs, 1973; Clutterbuck & Briggs, 1973). This agrees with the observed release of α-glucosidase activity from isolated barley embryos (Briggs, 1962) and aleurone layers incubated with gibberellic acid (Pollard, 1968) and has been confirmed by assays of enzyme activities in the separated parts of grains (MacGregor & Lenoire, 1987).

Four distinct enzymes, α-glucosidases I and II and maltases I and II, having different characteristics and time courses of development in germinating grains were identified by Stark & Yin (1987). In previous studies

α-glucosidase and maltase activities had been assumed to be due to one protein, so earlier work needs to be re-evaluated. Nevertheless it is interesting that α-glucosidase (by histochemical tests) and maltase (by measurements of enzyme activity) are concentrated in the living tissues. Thus they are available to degrade maltose, isomaltose and other oligosaccharides derived from the degradation of starch caused by α-amylase, β-amylase and debranching enzyme working together in the starchy endosperm. Another recent report indicates that α-glucosidase attacks starch granules at a high rate, and that it may be the 'raw starch factor', long sought, that facilites granule degradation by α-amylases (Sun & Henson, 1990).

The mobilization of the reserve proteins

The hydrolysis of the reserve proteins of germinating barley is less well understood than amylolysis. In part this is because the problem is more complex, but also because less effort has been devoted to elucidating the proteolytic systems. About a third of the endospermic reserve proteins are located in the aleurone layer and two-thirds in the starchy endosperm (Enari & Sopanen; 1986). Four stages of proteolysis can be distinguished (Enari & Sopanen, 1986; Mikola, 1983).

1. The hydrolysis of the reserve proteins of the aleurone grains or protein bodies in the embryo and aleurone layer to support initial growth and the production of hydrolytic enzymes.
2. The hydrolysis of the 'bulk' reserve proteins of the starchy endosperm, generating amino acids and peptides which are taken up by the embryo. The scutellar epithelium actively accumulates amino acids, glutamine and simple peptides (Mikola, 1983).
3. The hydrolysis of peptides by enzymes in the scutellum and in the aleurone layer.
4. The virtually complete breakdown of all the proteins remaining in the starchy endosperm and aleurone layer, which apparently autolyses, may be regarded as a final stage.

The amino acids, either taken up directly or derived from the hydrolysis of simple peptides, are used by the embryo to support its growth. Although enzyme production by the aleurone layer is virtually independent of an external supply of amino acids, some amino acids are taken up and used to synthesize glutamine, which is then released (Newman & Briggs, 1976).

Quiescent barley contains proteins which are inhibitors of its own proteases. These inhibitors decline in amount and are finally destroyed in the later stages of germination (Enari & Sopanen, 1986; Mikola & Enari, 1970). Germinating barley contains two major groups of proteinases (endopeptidases) – thiol-dependent and metal-activated enzymes. There are also small amounts of a pepstatin-inhibitable proteinase. At least three metal-activated enzymes (pH optima 5.5, 6.9 and 8.5) occur in extracts of barley; together they make up about 10% of the proteolytic activity at pH 5.4 (Mikola, 1983; Enari & Sopanen, 1986). The thiol-dependent protease activity makes up about 90% of the activity in extracts and is also due to a mixture of enzymes. Activity optima occur around pH 3.9 and 5.5 (Enari & Sopanen, 1986), so at least two enzymes are present. One purified thiol-dependent proteinase, M_r of about 30 000, has a pH optimum of 3.8 and is inhibited by leupeptin (Poulle & Jones, 1988). A different leupeptin-inhibitable protease, M_r approximately 30 000, has been described having a pI of 4.6–4.7 (two closely positioned bands) and a pH optimum of about 4.5 (Koehler & Ho, 1990). Another protease, which has three isoenzymes, M_r of 37 000, has a pH optimum of 5.0 (Koehler & Ho, 1988). Other evidence points to the presence of multiple forms of these enzymes (Burger, 1973).

In addition to the proteinases, germinating barley contains acid carboxypeptidases, neutral aminopeptidases and alkaline peptidases (dipeptidases) (Enari & Sopanen, 1986). Five acid carboxypeptidases have been detected, with complementary substrate specificites and pH optima in the range 4.8–5.6. Several of these enzymes have been purified and all are inhibited by DFP (di-isopropylfluorophosphate). Their levels change independently during germination (Enari & Sopanen, 1986; Fincher, 1989). Four neutral aminopeptidases (naphthylamidases) have been detected with pH optima in the range 5.5–7.2 (Enari & Sopanen, 1986). These enzymes are confined to the embryo and the aleurone layer, and appear to increase very little during germination. Two alkaline peptidases, a leucine aminopeptidase (pH optimum 8–10) and a dipeptidase (pH optimum 8.8) also occur in grain, possibly together with a third enzyme (Mikola, 1983).

In germinating grain the starchy endosperm contains high activities of proteinases and carboxypeptidases. The level of the proteinases is thought to limit the rate of protein degradation. The initial products of the proteolysis of hordein are comparatively large peptides (Rastogi & Oaks, 1986). The scutellum is rich in all the peptidases and the aleurone contains proteases and all the groups of peptidases (Mikola & Kolehmainen, 1972; Mikola, 1983). Thus the final degradation of the peptides produced by the hydrolysis of the reserve proteins of the starchy endosperm occurs in the living tissues of the grains. An interesting parallel is the way in which the oligosaccharides produced by α-amylase apparently are also finally degraded by α-glucosidase in the living tissues.

Cell wall degradation

Cell wall degradation is discussed more fully elsewhere (Fincher & Stone, 1986; Fincher, 1989; Chapter 20).

The cell walls of the scutellum are not degraded during germination, although the cells of the scutellar epithelium separate along the middle lamellae and the cells elongate as germination proceeds (Briggs, 1973, 1978; Gram, 1982a). The cell walls of the starchy endosperm appear to be completely degraded in well-made malt and in wet-grown grains, but the walls of the aleurone layer are only slowly and incompletely degraded.

The difference in resistance to enzymolysis between the cell walls of the aleurone layer and starchy endosperm may be due to the large proportion (ca. 70%) of β-glucan in the walls of the starchy endosperm or the high degree of acetylation and feruloylation of the arabinoxylan which makes up the major part of the cell walls of the aleurone layer. The cell walls of the embryo, the crushed and depleted cell layer and the sheaf cells are all also highly fluorescent and so all are heavily substituted, probably with ferulic acid residues. These cell walls also are either slowly degraded or are not degraded during germination (Briggs, 1987a; Fincher, 1989; Oh & Briggs, 1989). As barley aleurone cell walls are degraded ferulic acid esters, in which ferulic acid is joined to tri- and tetrasaccharides, are released, indicating that esterolytic removal of ferulic acid residues plays little part in the initial steps of wall degradation (Gubler & Ashford, 1985; Gubler *et al.*, 1985). Nothing is known of the processes of cell wall deacetylation. Pentosans make up about 7–11% of the entire grains; β-glucans about 3–6% (Lehtonen & Aokasalo, 1987a, b).

The cell walls of the aleurone layer and the starchy endosperm together constitute a significant carbohydrate reserve in the grain. The enzymes which attack the β-glucans are usually recorded as endo-β-$(1 \rightarrow 3)$-glucanase, or laminarinase, endo-β-$(1 \rightarrow 3;\ 1 \rightarrow 4)$-glucanase, or licheninase or mixed-link barley β-glucanase and endo-β-$(1 \rightarrow 4)$-glucanase or cellulase. Cellulase is largely or entirely derived from microorganisms on the husk (Hoy *et al.*, 1981; Yin *et al.*, 1989). In addition there are β-glucosidases, with cellobiase and laminaribiase activities (Briggs *et al.*, 1981). There are references to an exo-β-glucanase, but the evidence for a distinct enzyme is unconvincing. This picture is confused by the presence of isozymes. The two isozymes of mixed-link β-glucanase, for example, differ significantly in their properties and are encoded by genes located on different chromosomes (Loi *et al.*, 1988). When barley grain is malted an enzyme appears which catalyses an increase in the amount of β-glucan which comes into solution when the grain is mashed. The role of this 'β-glucan solubilase' enzyme (or enzymes) *in vivo* is unknown. There are uncertainties as to its (their) nature(s), but proposals include a carboxypeptidase or an endo-β-(1→4)-glucanase derived from the grains microflora (Bamforth & Martin, 1981; Yin & MacGregor, 1988, 1989; Bamforth, 1989; Yin *et al.*, 1989).

The enzymes known to be involved in the degradation of the arabinoxylans are arabinosidase, endo-xylanase and β-xylopyranosidase. Three (1→4)-β-D-xylan endohydrolases have been purified and characterized (Slade *et al.*, 1989). The existence of an exo-β-xylanase is doubtful (Briggs *et al.*, 1981). The esterases which presumbly remove the acetyl and feruloyl substituents from the arabinoxylans or arabinoxylan oligosaccharides appear not to have been studied.

Other grain reserves

During germination there is a massive flux of inorganic materials to the growing embryo. These are derived largely from the aleurone layer. Gibberellins trigger their release from this tissue (Jones, 1973; Clutterbuck & Briggs, 1974). The mobilization of phosphate has been studied most thoroughly. Much of the phosphate of grains is stored as phytate, predominantly the potassium and magnesium salt of phytic acid (*meso*-inositol hexaphosphate; Pomeranz, 1973; Stewart *et al.*, 1988). Curiously, despite the high affinity of phytic acid for calcium ions, little calcium is associated with the phytin deposits. This salt represents about 1% of the weight of the grain. Approximately 85% of the phytate occurs in the aleurone layer (Dalling & Bhalla, 1984). In addition, phosphate occurs bound in lipids, DNA and RNA. Since appropriate hydrolytic enzymes are released into the starchy endosperm from the adjacent living tissues as germination progresses it seems probable that the DNA and RNA are hydrolysed in this tissue as are the phospholipids present in the starch granules (Briggs, 1973; Baisted, 1981; Baisted & Stroud, 1982; Fujikura & Baisted, 1983). As the aleurone hydrates, the level of phosphatase increases and that of phytate declines (Ashford & Jacobsen, 1974; Clutterbuck & Briggs, 1974; Brookes & Martin, 1975; Katayama & Suzuki, 1980). In response to gibberellic acid, phosphatases increase and they and inorganic phosphate are released from the aleurone layer. The phosphate is used by the growing embryo (Ashford & Jacobsen, 1974; Clutterbuck & Briggs, 1974; Brookes & Martin, 1975; Bailey *et al.*, 1976; Obata, 1987).

The aleurone layer is rich in glycerides (chiefly triacyl glycerols) the levels of which decline in the hydrated tissue, apparently as they are used as respiratory substrates. It is proposed that lipase, which initiates their degradation, migrates from the protein bodies to the spherosomes in response to gibberellin, and this is blocked by ABA (Fernandez & Staehelin, 1987). However, gibberellic acid triggers a change in the metabolic pathways and the tissue releases sucrose derived from the lipids. To permit this to occur the glyoxylate pathway becomes operative and malate synthase and isocitrate lyase, the key enzymes of this pathway, appear and increase in amount (Newman & Briggs, 1976). The levels of these enzymes reached in the aleurone layers of germinating grains were higher than those reached in degermed grains dosed with gibberellic acid. The respiratory quotient (RQ) of aleurones from germinating grains can be as low as 0.3–0.4, suggesting a substantial conversion of triacylglycerols to carbohydrates (Merry & Goddard, 1941).

Regulation of the mobilization of the grain reserves

Barley needs to be hydrated to a moisture content of about 35–37% (fresh weight basis) or more to germinate completely. It will not germinate under anaerobic conditions. Almost nothing is known of the processes which regulate the metabolism of the reserves of the embryo, or of the aleurone layer until this is 'triggered' by the arrival of gibberellins and possibly other hormonal 'signal substances', from the embryo. However, much is known about the processes which regulate 'modification', the mobilization of the reserves which make up the starchy endosperm.

It has long been appreciated that isolated barley embryos grow well on nutrient media and hydrated, degermed grains undergo very minor changes when held under aseptic conditions that permit germination, and endosperm breakdown in intact grains. If embryos are replaced on degermed grains then, as shown by Brown & Morris in 1890, modification occurs (Briggs, 1973). Damage to the embryo reduces or prevents endosperm modification in reconstructed grains. Transplanting embryos to 'killed' degermed grains permitted less embryo growth than if the degermed grains were still living (Briggs, 1973), and the endosperms of 'dead' degermed grains united with living embryos did not break down in the usual way. It has been known for hundreds of years that modification spreads through the malting grain from the embryo end and progressively advances to the apex of the grain. Thus, the embryo regulates mobilization of the reserves of the endosperm, which is logical since it is the growing seedling that uses the breakdown products derived from the reserve substances.

There have been many disagreements regarding the roles of the embryo and the aleurone layer in supplying the enzymes which catalyse the dissolution of the endosperm. Early experiments showed that isolated embryos could generate enzymes which degraded starch and endosperm cell walls, but the quantities produced did not approach those in whole grains (Brown & Morris, 1890; Brown & Escombe, 1898). Degermed grains also failed to generate significant quantities of endosperm-degrading enzymes. The dilemma was resolved when Yomo (1958) showed that when embryos and endosperms were incubated together in a liquid medium vastly greater amounts of α-amylase were generated than when the grain parts were cultured separately. He demonstrated that a weak organic acid, 'AFX', moved from the embryo and triggered enzyme formation in the degermed grain. It was subsequently shown that both the embryo and the 'degermed grains' have to be living for this interaction to occur (Briggs, 1963a). Independently, Yomo and Paleg demonstrated that AFX could be replaced by gibberellic acid, GA_3, a substance derived from the fungus *Gibberella fujikuroi* (Briggs, 1973). The Japanese had added mixed fungal gibberellins to malting barley two decades earlier to hasten the malting process (Briggs, 1963b, 1973).

Before this striking discovery was known or appreciated various attempts had been made to discover if the inability of isolated embryos to generate the quantities of hydrolytic enzymes found in whole grains was due to inadequate supplies of nutrients. Briggs (1962, 1964) confirmed and extended earlier work and showed that isolated embryos cultured *in vitro* contained, and released into the culture medium, complex mixtures of enzymes. Enzymes released into the culture medium were able to hydrolyse starch, maltose, lactose, sucrose, cellobiose, melibiose, laminarin, lichenin, arabinoxylans, various phosphates and synthetic glycosides, polygalacturonic acid, DNA, RNA and various proteins and dipeptides. Peroxidase was also present.

Quantitative studies confirmed that the amounts of α-amylase produced by isolated embryos were strongly dependent on the nutrients available and enzyme levels were enhanced by the presence of calcium ions in the culture medium. This was attributed to the stabilizing effects of the calcium ions on the enzyme. The effects of the nutrients are worth noting, since they are quite different to those which occur with aleurone layers (Briggs, 1964, 1973, 1987a). In agreement with others it was noted that gibberellic acid did not increase α-amylase production in the absence of nutrients, in absolute contrast to what occurs with aleurone layers. Various nutrients enhanced α-amylase production. Of the substances tested, L-glutamate and DL-proline were the most effective, but a commercial casein hydrolysate containing a complex mixture of substances was the best nutrient. Several other commercial protein hydrolysates were found to be poor nutrients in terms of their abilities to support the production of α-amylase. This may explain why MacGregor & Marchylo (1986) failed to stimulate enzyme production by adding a particular protein hydrolysate to the culture medium. In the presence of the casein hydrolysate, but not in its absence, gibberellic acid approximately doubled the amount of α-amylase released into the medium. Thus, under these conditions, with isolated barley embryos, both gibberellin levels and supplies of nutrients limited the formation of this enzyme. The inclusion of glucose (5 mM) in the culture medium had no clear effect on the production of α-amylase by embryos cultured without casein hydrolysate, with or without gibberellic acid. In contrast glucose (5 mM) did reduce α-amylase yields from isolated embryos cultured with casein hydrolysate whether or not gibberellic acid was present. Thus, nutrient supplies had dramatic effects on the production of α-amylase by isolated embryos. However, even under the most favourable conditions the enzyme levels were much less than those encountered in entire germinating grains.

Investigations with the aleurone layer

Before gibberellins were discovered there had been many attempts to induce endosperm breakdown in degermed grains. In general these attempts were failures, but one observation is of particular interest: if degermed, hydrated grains were held, in air, for several days it became possible to peel the aleurone layer away from the starchy endosperm. This was not possible if the aleurone had been poisoned by the inclusion of toxic substances in the hydrating medium. This provided evidence for the limited release from the living aleurone layer of an enzyme able to degrade the cell wall material at the aleurone layer–starchy endosperm interface. It provided a means of preparing isolated aleurone layers (Briggs, 1973).

The dramatic discovery that gibberellins could trigger the near-total dissolution of the starchy endosperm of degermed grains incubated under aerobic conditions in liquid has diverted research efforts away from studies with whole grains and embryos. Various cytological studies convinced some early investigators that the aleurone layer was an enzyme-secreting tissue (Briggs, 1973). The pattern of endosperm breakdown in degermed grains dosed with gibberellic acid which begins next to the aleurone layer, the failure of breakdown to occur after the removal of aleurone tissue and the demonstration of the production of numerous enzymes by isolated aleurone layers all showed that the gibberellin was acting by triggering enzyme production and/or release from the aleurone tissue. Respiration is confined to the aleurone layer in degermed, surface-sterilized grains and so this tissue was expected to be metabolically responsive (Briggs, 1973). Enzymes released by 'triggered' aleurone layers include α-amylases (various isoenzymes) (1 → 3, 1 → 4,)-β-glucanase (licheninase) (1 → 3)-β-glucanase (laminarinase), proteases, peroxidase and enzymes able to cleave DNA, RNA, various disaccharides and glycosides, phosphates and peptides. Not all of these have yet been studied in detail. For a response to gibberellic acid to occur the aleurone layer had to be living, and the response was reduced or prevented by anoxia and a range of inhibitors including heavy metal ions, cyanide, fluoroacetate, amino acid analogues, puromycin, chloramphenicol and cycloheximide. In general, enzyme production was not influenced by external supplies of amino acids unless endogenous proteolysis was checked with potassium bromate, so that the supply of amino acids by the proteolysis of the reserve proteins of the tissue was prevented (Melcher & Varner, 1971).

The means by which gibberellins may trigger the responses of aleurone layers are beyond the scope of the review. However, the recent report that additions of GA_3 to isolated aleurone layers induce changes in the metabolism of phosphatidyl inositol in 30 s suggests that, as in animal cells, this may be a link between a surface hormone receptor and a chain of events that control the cell's responses (Murthy *et al.*, 1989). There is also evidence for gibberellin receptors on the surfaces of protoplasts prepared from the aleurone layers of wild oat caryopses (Hooley *et al.*, 1990).

Using inhibitors and the incorporation of $[^{14}C]$amino acids and $[^{18}O]H_2O$, α-amylases, protease and ribonuclease were found to be synthesized *de novo* in aleurone layers treated with gibberellic acid (Briggs, 1973). However, enzymes were shown to differ in their responses to different doses of GA_3 and in the time courses of their appearance in and release from the aleurone layer (Briggs, 1973, 1987a). Furthermore, some hydrolytic enzymes such as α-glucosidase occurred preformed in the aleurone layer and increased in amount following the hydration of the tissue and/or the arrival of gibberellins. Other enzymes appeared in the hydrated tissue but their release was gibberellin-dependent. Yet other enzymes were dependent on gibberellins to trigger both their formation and release. Xylanase is present in quiescent aleurone layers but subsequent *de novo* synthesis depends on the presence of gibberellins and calcium ions. It seems likely that by producing channels in the aleurone layer cell walls xylanase assists the release of α-amylase and other enzymes from this tissue (Gubler *et al.*, 1987; Benjavongkulchai & Spencer, 1989).

We have never observed significant quantities of α-amylases formed by degermed grains or isolated aleurone layers incubated under aseptic conditions. However, in some preparations of some varieties this seems to occur. Possibly, aleurone behaviour differs between samples because of varietal or manipulative differences. These complications will be discussed below.

Gibberellins in germinating grains

Some gibberellins and, with lower efficiencies, helminthosporol and helminthosporic acid and some related substances, trigger aleurone responses, such as the synthesis and release of α-amylase. Numerous claims that other compounds act as trigger-substances have not been substantiated (Briggs, 1973). Do then gibberellins or other substances, acting separately or together, trigger substances which induced aleurone responses in entire grains?

In an effort to resolve this uncertainty Groat & Briggs (1969) experimented with decorticated grains grown under malting conditions at 14.4°C and on wet paper at 25°C. The levels of gibberellin were determined by a lettuce hypocotyl extension bioassay and α-amylase was measured by a specific method. The results obtained in the two cases were substantially different. In the 'malted' grain, gibberellin levels rose sharply between germination days 1 and 2 to 0.8–0.9 ng GA_3 equivalents/grain (by bioassay), then after day 3 declined and levelled to an approximately constant value in days 4–7. In entire husked grain growing under malting conditions a similar rise and fall of endogenous gibberellin has been observed, with a peak level of ca. 1.4 ng GA_3 equivalents/grain determined by a different bioassay (Brookes & Martin, 1975). In our experiments, α-amylase rose progressively from day 2 and continued to rise at a decreasing rate to day 7. The peak rate of α-amylase formation occurred between 2.3 and 3.5 days of germination, a lag of about 20 h after the peak gibberellin level. Assuming that most α-amylase formation occurred in the aleurone layer, such a lag could be caused by the time taken for the gibberellin to move from its site of formation (in the embryo) to the aleurone and then the time for the aleurone layer (the target tissue) to respond. Experiments involving the removal of embryos at different times, followed by the incubation of degermed grains to a total of 7 days to allow enzyme development, showed that most gibberellin, or other factor inducing subsequent α-amylase formation in the aleurone, moved from the embryo between germination days 1 and 2, the time of the peak in gibberellin content. However, in days 5–7 the grains contained gibberellin that did not appear to be involved with α-amylase formation. It was tentatively concluded that this was retained in the embryo, and was connected with its growth.

Decorticated grains degermed before the movement of 'gibberellin' from the embryo to the endosperm and aleurone produced no α-amylase when subsequently held under malting conditions. However, if microdrops (1 μl) of water containing known amounts of gibberellic acid were applied to the depression previously occupied by the scutellum, before the embryo was removed, then α-amylase was formed during subsequent incubation. The amount of enzyme formed in response to a given dose of GA_3 was greater in grains degermed after steeping than in those degermed before. This supports the idea of an 'embryo factor' that moves from the embryo to the aleurone during steeping and potentiates the responses of the aleurone to gibberellins. In two experiments the doses of gibberellic acid that induced (in post-steeped degermed grains) the same levels of α-amylase as found in entire, germinating grains were 0.3 and 0.6 ng. These values are of the order of the peak levels of 'gibberellins' (0.6 ng GA_3 equivalents) found by bioassay in the endosperm of grains after 2 days of germination, the balance being located in the embryo. Within the limits set by the methods, the endogenous gibberellins qualitatively, quantitatively and in terms of their distribution in the grain at various times, 'fitted' the role of the major regulator of α-amylase levels in malted, decorticated grain. However, it is quite possible, even probable, that other substances act as modulators in this process.

For decorticated grains grown under 'wet and warm' conditions the results were entirely different (Groat & Briggs, 1969). During germination under these conditions, α-amylase levels rose, peaked and declined, as expected (Briggs, 1968a, b, c). In these grains, gibberellin levels rose *after* the initial increase in α-amylase levels. This is wholly *inconsistent* with the view that the gibberellins that were detected by the bioassay regulate α-amylase formation in grains grown under these conditions unless a redistribution of preformed gibberellins is involved. This anomaly remains unexplained and is important as in many laboratories experiments are made exclusively with 'warm and wet'-grown grains. Usually when gibberellic acid is applied at the onset of germination the grains respond by producing more α-amylase. There are differences between different varieties and samples of one variety (e.g. Briggs *et al.*, 1981; Raynes & Briggs, 1985) and in some instances the responses are nil. Some of our attempts to characterize gibberellin derivatives in grains previously dosed with GA_3 were unsatisfactory, and suggested that the techniques used were in some way inadequate (Smith & Briggs, 1980). In this connection it had been observed that bioassayable gibberellin declined in amount daily, during germination, from the surfaces of malting barley grains dosed with gibberellic acid, and disappeared in two days (Brookes & Martin, 1975). Also it has been shown that tannins can act as gibberellin antagonists (Corcoran *et al.*, 1972; Jacobson & Corcoran, 1977). The discovery that tight but reversible binding could occur between gibberellic acid and a wide range of phenolic materials may explain the 'inhibiting' effects of tannins, the apparent disappearance of GA_3 from the surface of the grains and the difficulties of separating gibberellins in crude grain extracts (Nutbeam & Briggs, 1982).

The responsiveness of aleurone layers to gibberellins

The 'responsiveness' of the aleurone layer to added gibberellin and its behaviour in the absence of gibberellin varies with the history of the grain and the ways in which degermed grains or aleurone layers are prepared or handled (Briggs, 1973, 1987a). The responsiveness of degermed grains, or aleurone layers to a given dose of gibberellin is dependent on the availability of oxygen (Briggs, 1987a). Holding prehydrated, degermed grains in air for up to 24 h before they are dosed with gibberellic acid increases their responsiveness (Briggs, 1973, 1987a). Steeping before embryo removal gives a more sensitive 'degermed grain' preparation than if degerming precedes hydration. Usually this effect has been attributed to an unidentified embryo factor that is hypothesized to diffuse from the embryo during steeping and act to increase the responsiveness of the aleurone layer (Groat & Briggs, 1969; Briggs, 1987a). However, the difference between the preparations might equally well be due to the partial removal of some necessary soluble factor, by leaching, when the degermed grains are steeped. In contrast to all our experience with British malting barleys, there are several reports of degermed Canadian barley making α-amylase in the absence of gibberellic acid and not generating more when gibberellic acid is supplied (Marchylo *et al.*, 1985, 1987).

In some instances, but not others, it has been found that isolated aleurone layers yield less α-amylase in response to gibberellic acid than the degermed grains from which they are prepared (Chrispeels & Varner, 1967; Briggs, 1973; Clutterbuck & Briggs, 1973). This might be due to damage during the separation of the aleurone layer, or to the removal of some desirable factor(s) with the starchy endosperm, or to the use of assays not specific for α-amylase and influenced by α-glucosidase and β-amylase which are associated with aleurone layer preparations (Briggs, 1973). Also some aleurones seems to generate a little α-amylase when incubated in the absence of gibberellins; others, prepared from different varieties of barley and using different techniques, do not. We have never found that aleurone layers from British malting barleys generate significant amounts of α-amylase in the absence of gibberellins, while this has been observed with some Canadian samples, and samples of Himalaya barley are reported to vary widely in their responsiveness from season to season (MacGregor, 1976; MacGregor & MacGregor, 1987).

Isolated barley aleurone protoplasts seem to be even more prone to produce α-amylase in the absence of gibberellin, suggesting that damage during their preparation may be causing a loss of metabolic control (Jacobsen *et al.*, 1985; MacGregor & MacGregor, 1987). On the other hand protoplasts from wild oat grains seem to be less damaged during their preparation (Hooley, 1982).

In some barley samples produced using unusual cultural conditions, for example those grown in greenhouses under special lighting and temperature conditions, the aleurone layers can generate substantial amounts of α-amylase and endosperm-degrading enzymes without gibberellins being supplied (Nicholls, 1980, 1983; MacGregor *et al.*, 1983; Nicholls *et al.*, 1986). Thus the details of the control of enzyme formation can differ between grain samples, even of one variety, when the growing conditions of the parent plants are different. Studies with mutant barleys have revealed lines in which aleurone layers are either insensitive or hypersensitive to gibberellic acid, and yet other lines that are insensitive to abscisic acid (ABA) (Ho *et al.*, 1980; Chandler, 1988; Lanahan & Ho, 1988). In wheat, low-temperature treatments can induce responsiveness to gibberellic acid in otherwise hormonally insensitive aleurone tissue (Singh & Paleg, 1984). Suitable heating and storage treatments improve the ability of endosperm slices of freshly harvested barley grains to respond to gibberellic acid by making more α-amylase (Crabb, 1971).

Enzyme formation by aleurone layers, in response to gibberellins, is little influenced by supplies of amino acids or low concentrations of sugars (e.g. 5 mM glucose). However, high concentrations of sugars (200–400 mM) do reduce the response, as do other agents which create high osmotic pressures (Jones & Armstrong, 1971; Armstrong & Jones, 1973). It is unclear whether this effect is significant in malting grains, but it might be in grains grown 'warm and wet' where the endosperm partly dissolves and is replaced with a sugary liquid (Briggs, 1987a). Frequently calcium ions have been supplied in culture media to stabilize α-amylases but it now seems that these ions are also needed for the secretion of this enzyme (Carr & Goodwin, 1972; Varner & Mense, 1972; Jones & Jacobsen, 1983) and some other enzymes. Varying Ca^{2+} concentrations differentially affect the formation and release of the different isoenzymes of α-amylase. The effect on synthesis seems to be post-transcriptional (Muthukrishnan & Chandra, 1988). Calcium ions promote the secretion of α-amylase from aleurone layer cell protoplasts (Bush *et al.*, 1986), and gibberellic acid influences the subcellular distribution of calcium ions (Bush & Jones, 1988; Bush *et al.*, 1989a b). In the presence of GA_3 and Ca^{2+} the endoplasmic reticulum and Golgi apparati undergo changes that correlate well with α-amylase secretion from barley aleurone protoplasts (Cornejo *et al.*, 1988). Calcium ions also influence the formation of β-glucanase in the scutellum, and the release of this enzyme and other proteins and enzymes, including α-amylase (Stuart *et al.*, 1986; Muthukrishnan & Chandra, 1988; Fincher, 1989). A difficulty with assessing the importance of Ca^{2+} *in vivo* is that stimulatory effects *in vitro* are achieved with concentrations of 10–30 mM, which seems too high to be achieved *in vivo* (Fincher, 1989). The addition of calcium ions increased the amount of β-glucanase I and, even more, β-glucanase II in the medium (Stuart *et al.*, 1986). In addition, under some conditions, aleurone layers require ferric ions to permit α-amylase synthesis (Goodwin & Carr, 1970).

When grains were degermed, after various periods of germination under malting conditions, it was found that they were progressively less able to respond (by making more α-amylase) to a dose of gibberellic acid (Smith & Briggs, 1980). It seemed that the growing embryo was the cause of this progressive loss of responsiveness in the aleurone layer. Similarly when grains were grown warm, on wet paper, they became unresponsive to gibberellic acid. Indeed while the addition of this hormone at the start of germination accelerated the formation of α-amylase, later additions of GA_3 accelerated enzyme destruction (Briggs 1968a, b, c). This occurred together with proteolysis in the starchy endosperm, a fall in the level of starch, the removal of calcium ions and a fall in the pH of the starchy endosperm from a peak of about pH 5.4–5.6 to about pH 4.0. All these factors would favour the destruction of α-amylase (Briggs, 1968a). When degermed grains are incubated in water with gibberellic acid the pH may fall to between 3 and 4 and α-amylase is largely inactivated (Briggs, 1968a). We assumed that phosphate release from the tissue caused the drop in pH (Clutterbuck & Briggs, 1974), which others have noted (Eastwell & Spencer, 1982c) and which is dependent on a living embryo in whole grains or on an exogenous supply of gibberellin if the embryo has been killed or removed (Briggs, 1968b). However, others, working with aleurone layers suspended in phosphate buffer, have attributed the fall in pH caused by GA_3-treated aleurone layers to the secretion of L-malic acid (Mikola & Virtanen, 1980). According to Hamabata *et al.* (1988) the acidification of the medium by wheat aleurone layers is independent of gibberellic acid.

Gibberellins in the grain

Gibberellins are formed in the germinating barley embryo and some are released from it. At least initially the hormone seems to be synthesized from stored precursors but at later stages of germination it may be formed '*de novo*' from mevalonic acid. The stored precursor may be *ent*-kaurene, which is converted via *ent*-kaurenol, *ent*-kaurenal, *ent*-kaurenoic acid and hydroxykaurenoic acid to active gibberellins. Various inhibitors prevent gibberellin formation in germinating grains, but they do not alter the responsiveness of aleurone tissue to gibberellins (Briggs, 1973; Gilmour & MacMillan, 1984).

There have been substantial, unexplained disagreements in the literature regarding which gibberellins are present in germinating barley grains (Briggs, 1987a). In malted grain the major gibberellins seem to be GA_1, with smaller amounts of GA_3, GA_4 and GA_7 were not detected. GA_1 and GA_3 rose, peaked and fell in parallel in malting grains (Yamada, 1982) as might be predicted from 'model' experiments with decorticated grains in which 'gibberellins' were determined by bioassays (Groat & Briggs, 1969). Other gibberellins were present in barley seedlings, where they may be formed in the growing axis. The mixture of gibberellins can be complex (Gaskin *et al.*, 1983, 1984; Briggs, 1987a). Radley (1969) demonstrated that sugars supplied to the scutellum prevented the production/release of gibberellins. In both wet-grown and malted grains it seems that the initial metabolism of the embryo, following grain hydration, causes a fall in the levels of the sugars in the scutellum and this correlates with the onset of gibberellin production and release into the endosperm. As the breakdown of the endosperm begins, a supply of hydrolysis products reaches the embryo, and the sugar reserves in the scutellum are restored. This coincides with, and is thought to be the cause of, the cessation of gibberellin production and release (Briggs, 1973, 1987a, b).

From these observations the concept of the feedback control loop, controlling gibberellin levels, arose. Sugar depletion in scutellum → gibberellin formation and release → gibberellins trigger enzyme formation and release from aleurone layer into the starchy endosperm → endosperm breakdown begins, soluble sugars released → sugars taken up by scutellum; sugar levels rise in embryo → sugars 'switch off' gibberellin formation and/or release. By cultivating embryos and degermed grain together in liquid culture it has been shown that with increasing concentrations of sucrose in the medium α-amylase formation is reduced until, at a concentration of 100 mM sucrose, enzyme formation is wholly prevented. As in entire grains, the addition of gibberellic acid to this system causes a major increase in the amount of α-amylase synthesized and the presence of sucrose (100 mM) decreases this to a relatively small extent (Briggs & Clutterbuck, 1973). Thus small concentrations of sucrose have a dramatic effect on gibberellin production while 100 mM sucrose totally prevents α-amylase formation and, presumably, endogenous gibberellin formation. However, this concentration of sucrose does not totally prevent the response of the enzyme-forming tissues to added gibberellic acid as 200–400 mM sucrose is reported to do (Jones & Armstrong, 1971). Thus, sugars cut off gibberellin production at concentrations lower than those which block the aleurone response to gibberellins, by a general inhibition of protein synthesis (Chrispeels, 1973). Gibberellic acid added to entire germinating grains can more than double the amount of α-amylase produced (Smith & Briggs, 1980). It follows that, at least under some germination conditions, if endogenous sugar levels *are* regulating the levels of α-amylase it must be at the level of gibberellin amounts rather than at the level of aleurone responsiveness, because the grain does respond to added gibberellic acid by making more enzyme. Sugars vary in their abilities to suppress α-amylase formation in 'divided grain' cultures (Smith & Briggs, 1980).

A tenaciously held view is that gibberellins are released from the apex of the scutellum on the dorsal, rounded side of the grain and that endosperm breakdown spreads from this location (Palmer & Duffus, 1986). That α-amylase formation and endosperm degradation, both of which are gibberellin dependent, occur roughly equally on the dorsal and ventral sides of the grain demonstrates absolutely that endosperm breakdown is not initiated selectively on the dorsal side of the grain, and so gibberellin release must also be general from the face of the scutellum (Briggs, 1972, 1973, 1987a, b).

One report, based on immunological assays of the gibberellins, seemed to show that aleurones from barley generated GA_4 in the presence of GA_1 and that gibberellin synthesis in the aleurone was essential for the response of the tissue (Atzorn & Weiler, 1983). This conclusion runs contrary to most others, and specifically to the experiments of Gilmour & MacMillan (1984).

Other hormones

Other non-gibberellin hormones may influence the formation of hydrolytic enzymes in germinating grain. Evidence is lacking for this, although some hormones added to isolated aleurone layers have dramatic effects on response of tissue to gibberellic acid.

Indoleacetic acid (IAA) does occur in barley (Briggs, 1978) but its role, if any, in regulating the mobilization of endosperm reserves is not established. Additions of IAA to entire germinating grains are usually damaging, and a claim that IAA enhances the response of aleurone layers to gibberellic acid has been disputed (Briggs, 1973, 1978). Cytokinins have been detected in barley (Briggs, 1978), and when added to grain growing on a wet surface root growth is reduced, coleoptile senescence is enhanced and the degradation rate of α-amylase is accelerated, presumably an effect mediated by the embryo (Briggs, 1973; Verbeek *et al.*, 1973). However, kinetin and benzylaminopurine alter the ratio of α-amylase secreted to that retained in the tissue in aleurone layers incubated with gibberellic acid (Clutterbuck & Briggs, 1973).

Abscisic acid (ABA) has been detected in barley grains and is concentrated in the embryo (Yamada, 1985). Additions to whole grains arrest germination and α-amylase formation; these effects are opposed by the addition of cytokinins (Briggs, 1973). Additions of ABA reduce sugar release from degermed grains and α-amylase formation in aleurone layers incubated with GA_3. Ethylene can reverse this ABA effect in the presence of high concentrations of GA_3 (Briggs, 1987a; Muthukrishnan & Chandra, 1988; Fincher, 1989). ABA is not a competitive inhibitor or an antagonist of general protein synthesis. Rather it directs the aleurone cells to synthesize proteins other than those whose formation is provoked by gibberellins (Ashford & Gubler, 1984; Fincher, 1989). Aleurone layers take up ABA and convert it to phaseic acid, 4′-dihydrophaseic acid, and three other polar metabolites (Dashek *et al.*, 1979).

Ethylene is regarded as a ubiquitous plant hormone but it has not been found in germinating barley. Addition of ethylene to whole grains can cause reduced root growth (Briggs, 1978). Its ability, together with an excess of GA_3, to overcome the effects of ABA on aleurone layers has been noted (Jacobsen, 1973). In further studies Tittle & Spencer (1986) noted that, while ethylene overcame the ABA-induced inhibition of α-amylase formation in aleurone layers incubated with gibberellic acid, carbon dioxide in turn overcame the effect of ethylene and restored the inhibitory effect of the ABA. Reports on the effects of ethylene are not entirely consistent. Scott & Leopold (1967) found that ethylene (10 ppm) reduced the release of reducing sugars from degermed grains dosed with GA_3. Jones (1968) noted that at some levels (e.g. 1 ppm) ethylene favoured the release of α-amylase from aleurone layers incubated with GA_3 but the total amount of enzyme formed was unchanged. However, Jacobsen (1973) found that slightly more enzyme was formed in the presence of ethylene. Ho *et al.* (1982) showed that the 'extra' enzyme which appeared in the medium in the presence of ethylene was released from the aleurone cell walls. Eastwell & Spencer (1982a, b, c) noted that the kinetics of appearance of total amounts of α-amylase, and its release, in the presence of GA_3 and various amounts of ethylene are complex. They proposed that:

(i) xylanase formation, and hence the degradation creating channels through the aleurone cell walls, was enhanced by ethylene; and (ii) this favoured α-amylase release; and (iii) enhanced protease activity contributed to losses of α-amylase in extended incubation periods. It seems that hydrolases pass readily into the walls in the regions of the plasmodesmata (Gubler, *et al.*, 1985; 1987).

Enzymes from the scutellar epithelium

The role of the scutellar epithelium in producing hydrolytic enzymes has been disputed (Macleod & Palmer, 1966; Palmer & Bathgate, 1976; Palmer, 1980, 1982; Palmer & Duffus, 1986; Hill & MacGregor, 1988). It has been argued that the enzymes released from the embryo are derived exclusively from the residual thin layer of flattened aleurone tissue that overlaps the edges of the embryo (Briggs, 1978, 1987a; Palmer & Duffus, 1986). The reasons for accepting that the scutellar epithelium is a source of hydrolases are discussed next.

The subcellular changes undergone by the cells of the scutellar epithelium during germination are similar to those undergone by the aleurone layer that has been 'triggered' with gibberellic acid (Briggs, 1973; Gram, 1982a). In particular, the reserve substances (in the oleosomes and protein bodies) are mobilized, mitochondria proliferate and develop cristae and membranes, RER appears and many well-defined Golgi bodies, typical of secretory cells, make their appearance. In gibberellin-triggered barley aleurone layers α-amylases I and II are associated with the RER and the Golgi bodies, indicating that the enzyme is transferred to the plasmalemma by the 'classical' secretory pathway (Gubler *et al.*, 1986; Zingen-Sell *et al.*, 1990). However, there are differences between the aleurone cells and the cells of the scutellar epithelium. The epithelial cells separate and elongate and starch appears in them briefly during germination. The major changes occur in the scutellar epithelium before comparable alterations in the aleurone layers in malting barley (Gram, 1982a). Brown & Morris (1890) showed that when isolated embryos were cultured *in vitro* they grew and released starch- and cell wall-degrading enzymes. Scraping away the scutellar epithelium prevented enzyme release, but not growth, strong evidence for the epithelium as a source of enzymes.

The endosperm breakdown, which begins under the full width of the scutellum and in some grains proceeds a little more rapidly in the mid-region around the sheaf cells, is initiated by enzymes from the scutellum (Briggs, 1987a). In malting grain it is 2–3 days before breakdown beneath the aleurone layer is observed (Gibbons, 1979, 1980, 1981; Briggs & MacDonald, 1983b). The breakdown of the cell walls of the starchy endosperm, chiefly caused by $(1\rightarrow3;1\rightarrow4)$-endo-$\beta$-glucanases, is followed by degradation of the protein matrix and attack on the starch granules. Oh & Briggs (1989) found that when germinating grains were cut, stained with Calcofluor, exposed to ultraviolet radiation, and photographed in colour, the degraded areas of starchy endosperm adjacent to the embryo seemed unlike those which appeared later next to the aleurone layer. They tentatively suggested that the mixture of enzymes from the scutellar epithelium differed from that from the aleurone layer, and so degraded the starchy endosperm tissue in a different way. There is independent evidence for this (see later).

The production of α-amylase by isolated embryos differs markedly from that by degermed grains and isolated aleurone layers in a number of ways. Thus enzyme production by embryos has no absolute requirement for an external supply of gibberellins – which is mandatory for aleurone preparations. When embryos are supplied with amino acids, enzyme yields are greatly increased – but amino acids are almost without influence on enzymes production by gibberellin-triggered aleurone layers. On the other hand gibberellins have little or no influence on enzyme production by isolated embryos cultured without nutrients, although this results in the production of large amounts of enzymes by isolated aleurone layers (Briggs, 1987a). However, when embryos are incubated with amino acids the addition of gibberellic acid does increase the production of α-amylase. These results strongly support the view that the peripheral aleurone tissue makes at most only a tiny contribution to α-amylase production by isolated embryos. Similarly the ability of 5 mM glucose to reduce sharply α-amylase production by isolated embryos, even in the presence of gibberellic acid, when this concentration of glucose is almost without effect on enzyme production by gibberellin-triggered aleurone tissue is further evidence for the independence of the scutellar system from any residual aleurone tissue. Further proof was provided by experiments in which 'plugs' of central scutellar tissue, from which all aleurone tissue had been removed (as checked by scanning electron microscopy), were shown to produce α-amylase (Briggs & Clutterbuck, 1973; Ranki, 1990).

Using substrate film histochemical techniques Okamoto *et al.* (1980) showed that α-amylase, proteinase and ribonuclease first originated in the scutellum of barley and some other germinating cereal grains and only later did they appear in the aleurone layer (Briggs 1987a). Histochemical studies with fluorescein dibutyrate esterase gave similar results (Jensen & Heltved, 1982). Using anti-α-amylase II antibodies in a fluorescent double-antibody technique Gibbons (1979, 1980) showed that α-amylase II was first detectable in the crushed cell layer, which is pressed against the scutellar epithelium, and then the enzyme spreads away from the scutellum into the starchy endosperm. Only after 2–3 days did α-amylase II appear to have been formed in and released from the aleurone layer. Furthermore, it seemed that initially enzyme was concentrated in the tips of the cells of the scutellar epithelium. Using antibodies active against α-amylases I and II, Ranki (1990) and Pogson *et al.* (1989) were able to show that α-amylase occurred in the cells of the scutellar epithelium after 1 day of germination, and it was detectable in the peripheral aleurone tissue as well as the 'normal' aleurone tissue under the conditions in which their grains were grown. All three tissues seemed to secrete enzyme in grains grown 'wet' at 20°C (Pogson *et al.*, 1989). Only rarely were traces of enzyme located in the scutellar parenchyma.

mRNAs coding for α-amylase, carboxypeptidase and β-glucanase were isolated from scutellar and aleurone tissues (Mundy *et al.*, 1985). Some of the mRNA in the scutellar preparations may have come from the peripheral aleurone tissue. However, using hybridization histochemistry, with [^{32}P]cDNA it was shown that the mRNAs coding for (1 → 3; 1 → 4)-β-glucanases appeared first in the scutellum and only later in the aleurone layer (McFadden *et al.*, 1988; Fincher, 1989). Barley scutellum can provide perhaps 40% of the (1 → 3; 1 → 4)-β-glucanase generated in the grain and much of this seems to be liberated in the early stages of germination (Fincher, 1989).

There is increasing evidence that the mixtures of isoenzymes released from the scutellum differ significantly from those liberated from the aleurone layer. Thus carboxypeptidase I is generated almost exclusively in the scutellum (Ranki *et al.*, 1983) and the mRNA coding for this enzyme is virtually confined to this tissue (Ranki, 1990). Similarly (1 → 3; 1 → 4)-β-glucanase I is secreted from isolated scutella together with an uncharacterized protein, termed isoenzyme III, which is not seen in intact grain, while aleurone layers secrete chiefly (1 → 3; 1 → 4)-β-glucanase II (Fincher, 1989). MacGregor & Marchylo (1986) noted that isolated barley embryos generated mainly α-amylase I, while α-amylase II is by far the major isoenzyme in germinated grains and is the main component formed by the aleurone layer. On the other hand, the scutellum and aleurone layer seemed to form the same isoenzymes in the experiments of MacGregor *et al.* (1984). Thus there is evidence that the scutellum generates essentially all the carboxypeptidase I and 40% of (1 → 3; 1 → 4)-β-glucanase in germinating barley as well as some α-amylase and other enzymes. It has often been implied or assumed that the scutellum and aleurone release 'equal mixes' of enzymes into the starchy endosperm. Clearly this is not so, and may explain the observations of Oh & Briggs (1989; see earlier).

The quantities of α-amylase generated by the two organs has also been a matter of dispute. By combining the amounts from embryos (tissue and medium) cultured *in vitro* and in the separated parts of decorticated grains grown under malting conditions it was estimated that of the total α-amylase in the grains 7% was retained in the embryo, 6.5% was released by the embryo into the endosperm, and the remaining 86.5% was derived from the aleurone layer (Briggs, 1964). Using dissected grain parts monitored by scanning electron microscopy and cultured in liquid, Briggs & Clutterbuck (1973) estimated that about 10% of the α-amylase generated arose from the scutellar tissue. Part of this would have been retained in the scutellum. Ranki & Sopanen (1984) germinated barley at 20°C. At intervals they separated the aleurone and scutellum and estimated the rates of enzyme secretion from these tissues. They concluded that 5–10% of the α-amylase appearing in the starchy endosperm originated from the embryo, and much of this was released in the early stages of germination. On the other hand it has been proposed from histochemical experiments that the scutellum generates a much higher proportion of α-amylase (Okamoto *et al.*, 1980). Gibbons (1981) suggested that as much as 50% of the enzyme in the starchy endosperm may originate from the scutellum. The problems of interpreting some of these conflicting results have been noted (Briggs, 1987a). Probably α-amylase is derived initially from the embryo, particularly the scutellum, but as germination progresses, the amount of enzyme from the aleurone layer comes to exceed substantially that from the embryo.

Final considerations

Some of the factors thought to regulate endosperm breakdown are summarized in Fig. 18.1. This scheme seems to describe and adequately account for the changes in malting grains. However, as the situation in 'warm and wet'-grown grains is not so satisfactory, the changes that occur have been less adequately explained. Less work has been carried out on whole grains compared with isolated aleurone layers, and there is little evidence for many of the interactions proposed to occur in whole grains.

Some experimental problems remain. For example, the mixed microbial populations on grains can cause irregular germination, as they compete for oxygen. Some microbes produce phytotoxins and their activities could lead to the misinterpretation of metabolic experiments. In addition, these microbes produce a range of enzymes. Many reports fail to take into consideration grain microflora. Antimicrobial agents are indispensable but can also cause difficulties. Calcium hypochlorite is less damaging than the strongly alkaline sodium hypochlorite, but other agents such as formaldehyde, bromine water, silver nitrate and mixtures of antibiotics have all been used and are more or less objectionable (Yomo, 1958; Groat & Briggs, 1969; Abdul-Baki, 1974; Briggs *et al.*, 1981; Raynes, 1983; Raynes & Briggs, 1985; Goudey *et al.*, 1987; Tittle *et al.*, 1988). We know of no 'safe' treatment that will eliminate microbes from grains unless they have first been decorticated with sulphuric acid, although with lightly contaminated samples we seem able to limit effectively microbial activities.

Figure 18.1 emphasizes the role of supplies of oxygen to the living tissues within the grain. Dormant barleys are particularly sensitive to impaired supplies of oxygen. Embryo growth, aleurone responses to gibberellins and the formation of gibberellins are all oxygen dependent (Briggs, 1978, 1987a). Entry of oxygen into the interior of grains is limited by the diffusion barriers presented by the husk and pericarp and by water-film on the surface of grains where

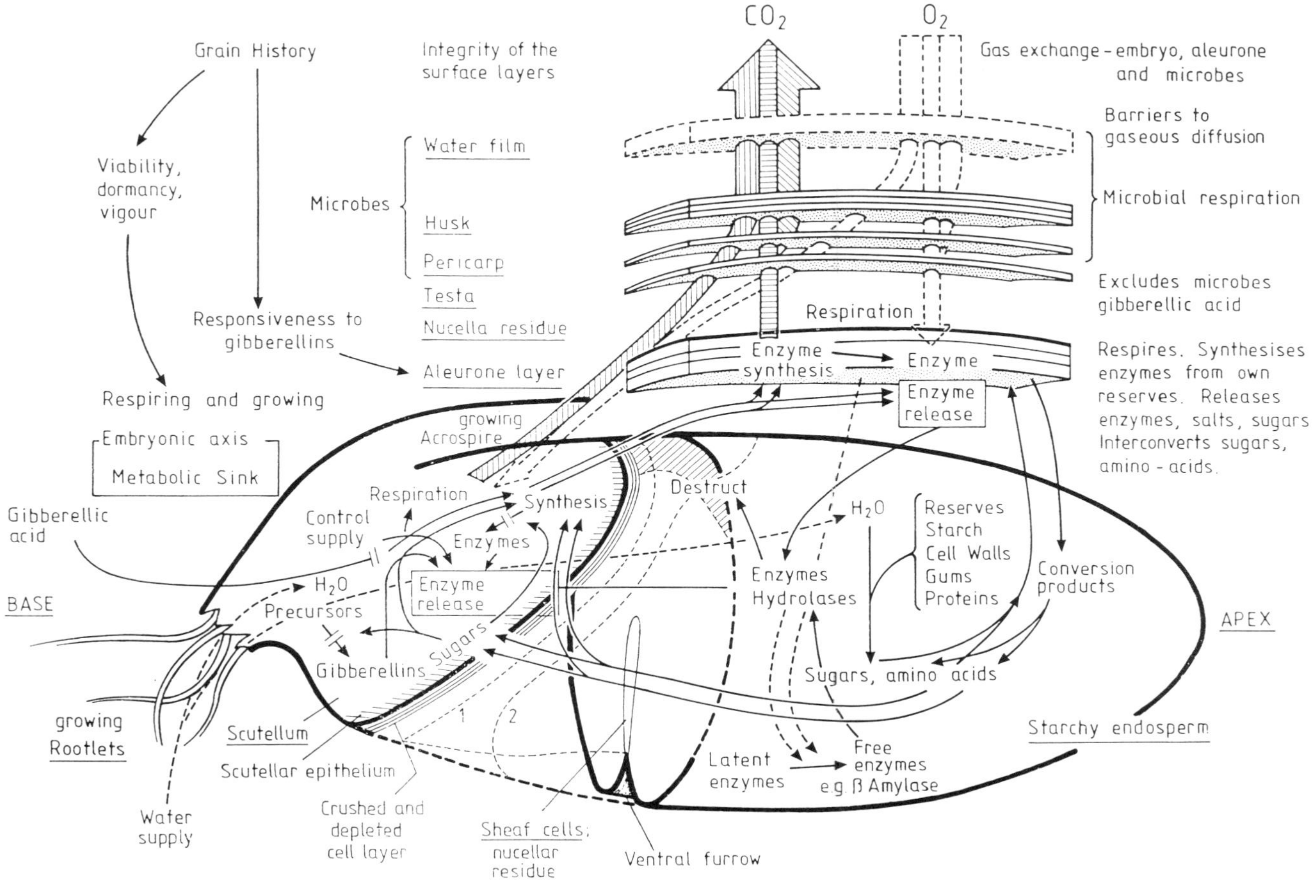

Fig. 18.1. A scheme summarizing some of the interrelationships between the parts of the barley grain, growing under malting conditions (after Briggs, 1987a). The grain must be sufficiently hydrated (e.g. 43% moisture fr. wt) and held in a humid atmosphere. Several essential processes depend upon an adequate supply of oxygen, including the germination of the embryo, gibberellin production and the metabolism of the aleurone layer. Microbes seem to cause their adverse effects by consuming oxygen and, less certainly, by producing phytotoxins, and the husk, pericarp and any film of surface water are important barriers to the diffusion of oxygen from the air to the living tissues in the grain. It is also possible to trace the feed-back loop by which it is proposed sugar levels regulate the production and/or release of gibberellins from the scutellum.

present. Furthermore microbes on or in the surface layers consume oxygen. Enhanced availability of oxygen to the living tissues may partly explain the effectiveness of limited physical treatments, such as commercial 'abrasion', in enhancing responses to gibberellic acid during malting (Smith & Briggs, 1979; Briggs, 1987a). However, Fig. 18.1 does not indicate the role of the 'embryo factor' in enhancing the responsiveness of the aleurone layer to gibberellins, nor does it indicate roles for other plant hormones which, in any case, are uncertain.

Numerous apparent discrepancies in the literature remain to be resolved. These may be partly due to the use of grains from widely differing genotypes produced under widely differing growth conditions. Widespread use of grain of the naked (huskless) variety 'Himalaya' from one source has reduced this problem. However, this variety differs considerably from commercial husked barleys. As studies are extended, important significantly different characteristics having genetics bases are likely to be found and used in plant breeding for their commercial value; for example in improving the malting characteristics of grain.

Another set of problems arise from wide variations in germination conditions used in different trials. As is well known to maltsters, apparently minor changes in germination condition can substantially alter the metabolic behaviour of germinating grain (Briggs *et al.*, 1981). Comparisons are very difficult to make between 'malted' samples of grain and those grown 'wet and warm' (Briggs, 1987a). Even the production and release of α-amylase by isolated aleurone layers triggered with gibberellic acid differs dramatically at temperatures below 19°C and above 21°C (Fadeel *et al.*, 1980). A maltster wishing to obtain maximum yields of this enzyme will germinate grain for an extended period at a low temperature (13–14°C) (Briggs *et al.*, 1981).

References

Aastrup, S., Outtrup, H. & Erdal, K. (1984) Location of the proanthocyanidins in the barley grain. *Carlsberg Research Communications* 49, 105–9.

Abdul-Baki, A.A. (1974) Hypochlorite and tissue sterilization. *Planta* 115, 373–6.

Ahluwalia, B. & Fry, S.C. (1986) Barley endosperm cell walls contain a feruloylated arabinoxylan and a non-feruloylated β-glucan. *Journal of Cereal Science* 4, 287–95.

Armstrong, J.E. & Jones, R.L. (1973) Osmotic regulation of α-amylase synthesis and polyribosome formation in aleurone cells of barley. *Journal of Cell Biology* 59, 444–55.

Ashford, A.E. & Gubler, F. (1984) Mobilization of polysaccharide reserves from the endosperm. In: Murray, D.R. (ed.), *Seed Physiology. 2. Germination and Reserve Mobilization*. Academic Press, Sydney, pp. 117–62.

Ashford, A.E. & Jacobsen, J.V. (1974) Cytochemical localization of phosphatase in barley aleurone cells: the pathway of gibberellic-acid-induced enzyme release. *Planta* 120, 81–105.

Atzorn, R. & Weiler, E.W. (1983) The role of endogenous gibberellins in the formation of α-amylase by aleurone layers of germinating barley caryopses. *Planta* 159, 289–99.

Bacic, A. & Stone, B.A. (1981a) Isolation and ultrastructure of aleurone cell walls from wheat and barley. *Australian Journal of Plant Physiology* 8, 453–74.

Bacic, A. & Stone, B.A. (1981b) Chemistry and organisation of aleurone cell wall components from wheat and barley. *Australian Journal of Plant Physiology* 8, 475–95.

Bailey, K.M., Phillips, I.D.J. & Pitt, D. (1976) Effects of gibberellic acid on the activation, synthesis and release of acid phosphatase in barley seed. *Journal of Experimental Botany* 27, 324–36.

Baisted, D.J. (1981) Turnover of starch-bound lysophosphatidylcholine in germinating barley. *Phytochemistry* 20, 985–88.

Baisted, D.J. & Stroud, F. (1982) Enhancement by gibberellic acid and asymmetric distribution of lysophospholipase in germinating barley. *Phytochemistry* 21, 2619–23.

Bamforth, C.W. (1989) β-Glucan solubilase. *Journal of the Institute of Brewing* 95, 314–15.

Bamforth, C.W. & Martin, H.L. (1981) The development of β-glucan solubilase during barley germination. *Journal of the Institute of Brewing* 87, 81–4.

Benjavongkulchai, E. & Spencer, M.S. (1989) Barley aleurone xylanase: its biosynthesis and

possible role. *Canadian Journal of Botany* 67, 297–302.
Bertoft, E., Andtfolk, C. & Kulp, S.-E. (1984) Effect of pH, temperature and calcium ions on barley malt α-amylase isoenzymes. *Journal of the Institute of Brewing* 90, 298–302.
Bøg-Hansen, T.C. & Daussant, J. (1974) Immunochemical quantitation of isoenzymes. α-Amylase isoenzymes in barley malt. *Analytical Biochemistry* 61, 522–7.
Bourne, D.T., Powlesland, T. & Wheeler, R.E. (1982) The relationship between total β-glucan of malt and malt quality. *Journal of the Institute of Brewing* 88, 371–5.
Briggs, D.E. (1962) Development of enzymes by barley embryos *in vitro*. *Journal of the Institute of Brewing* 68, 470–5.
Briggs, D.E. (1963a) Biochemistry of barley germination. Action of gibberellic acid on barley endosperm. *Journal of the Institute of Brewing* 69, 13–19.
Briggs, D.E. (1963b) Effects of gibberellic acid on barley germination and its use in malting: a review. *Journal of the Institute of Brewing* 69, 244–8.
Briggs, D.E. (1964) Origin and distribution of α-amylase in malt. *Journal of the Institute of Brewing* 70, 14–24.
Briggs, D.E. (1967) Modified assay for α-amylase in germinating barley. *Journal of the Institute of Brewing* 73, 361–70.
Briggs, D.E. (1968a) α-Amylase in germinating, decorticated barley. I. α-Amylase, conditions of growth and grain constituents. *Phytochemistry* 7, 513–29.
Briggs, D.E. (1968b) α-Amylase in germinating, decorticated barley. II. Effects of physically damaging the grain. *Phytochemistry* 7, 531–8.
Briggs, D.E. (1968c) α-Amylase in germinating, decorticated barley III. Effects of adding CCC and other chemical substances. *Phytochemistry* 7, 539–54.
Briggs, D.E. (1972) Enzyme formation, cellular breakdown and the distribution of gibberellins in the endosperm of barley. *Planta* 108, 351–8.
Briggs, D.E. (1973) Hormones and carbohydrate metabolism in germinating cereal grains. In: Milborrow, B.V. (ed.), *Biosynthesis and its Control in Plants*. Academic Press, London, pp. 219–77.
Briggs, D.E. (1974) Hydrocarbons, phenols, and sterols of the testa and pigment strand in the grain of *Hordeum distichon*. *Phytochemistry* 13, 987–96.
Briggs, D.E. (1978) *Barley*. Chapman & Hall, London, pp. 612.
Briggs, D.E. (1987a) Endosperm breakdown and its regulation in germinating barley. In: Pollock, J.R.A. (ed.), *Brewing Science 3*. Academic Press, London, pp. 441–532.
Briggs, D.E. (1987b) Modification processes in germinating barley. In: Morton, I.D. (ed.), *Cereals in a European Context: First European Conference on Food Science and Technology*. VCH/Ellis Horwood, Chichester, UK, pp. 119–36.
Briggs, D.E. & Clutterbuck, V.J. (1973) Generation of α-amylase in germinating *Hordeum distichon*. *Phytochemistry* 12, 1047–50.
Briggs, D.E. & Cornish-Bowden, A. (1981) The calculation of results of α-amylase assays. *Journal of the Institute of Brewing* 87, 222–3.
Briggs, D.E. & MacDonald, J. (1983a) The permeability of the surface layers of cereal grains and implications for tests of abrasion in barley. *Journal of the Institute of Brewing* 89, 324–32.
Briggs, D.E. & MacDonald, J. (1983b) Patterns of modification in malting barley. *Journal of the Institute of Brewing* 89, 260–73.
Briggs, D.E. & Morrall, P. (1984) Endosperm breakdown in decorticated grains grown at 25°C on a wet substratum. *Journal of the Institute of Brewing* 90, 363–70.
Briggs, D.E. & Raynes, J.G. (1985) α-Amylase isozymes of germinated barley. *Journal of Cereal Science* 3, 67–72.
Briggs, D.E., Cornish-Bowden, A. & Smith, M.T. (1980) α-Amylase assays – calculation of results. *Journal of the Institute of Brewing* 86, 162–3.
Briggs, D.E., Hough, J.S., Stevens, R. & Young, T.W. (1981) *Malting and Brewing Science*, 2nd edn, Vol. 1, *Malt and Sweet Wort*. Chapman & Hall, London, 387 pp.
Brookes, P.A. & Martin, P.A. (1975) The determination and utilization of gibberellic acid in germinating barley. *Journal of the Institute of Brewing* 81, 357–63.
Brown, H.T. & Escombe, F. (1898) On the depletion of the endosperm of *Hordeum vulgare* during germination. *Proceedings of the Royal Society of London* 63, 3–25.
Brown, H.T. & Morris. G.H. (1890) Researches on the germination of some of the Gramineae. *Journal of the Chemical Society* 57, 458–528.
Burger, W.C. (1973) Multiple forms of acidic endopeptidase from germinated barley. *Plant Physiology* 51, 1015–21.

Bush, D.S. & Jones, R.L. (1988) Cytoplasmic calcium and α-amylase secretion from barley aleurone protoplasts. *European Journal of Cell Biology* 46, 466–9.

Bush, D.S., Cornejo, M.-J., Huang, C.-N. & Jones, R.L. (1986) Ca^{2+}-stimulated secretion of α-amylase during development in barley aleurone protoplasts. *Plant Physiology* 82, 566–74.

Bush, D.S., Biswas, A.K. & Jones, R.L. (1989a) Gibberellic-acid-stimulated Ca^{2+} accumulation in endoplasmic reticulum of barley aleurone: Ca^{2+} transport and steady state levels. *Planta* 178, 411–20.

Bush, D.S., Sticher, L., van Huystee, R., Wagner, D. & Jones, R.L. (1989b) The calcium requirement for stability and enzymatic activity of two isoforms of barley aleurone α-amylase. *Journal of Biological Chemistry* 264, 19392–8.

Carr, D.J. & Goodwin, P.B. (1972) Stages during the induction of α-amylase by gibberellic acid in barley aleurone layers. II. Some properties of the pre- and post-transcription stages. In: Carr, D.J. (ed.), *Plant Growth Substances-1970*. Springer, Berlin, pp. 378–87.

Chandler, P.M. (1988) Hormonal regulation of gene expression in the 'slender' mutant of barley (*Hordeum vulgare, L.). Planta* 175, 115–20.

Chrispeels, M.J. (1973) Mechanism of osmotic regulation of hydrolase synthesis in aleurone cells of barley: inhibition of protein synthesis. *Biochemical and Biophysical Research Communications* 53, 99–104.

Chrispeels, M.J. & Varner, J.E. (1967) Gibberellic acid-enhanced synthesis and release of α-amylase and ribonuclease by isolated barley aleurone layers. *Plant Physiology* 42, 398–406.

Chrispeels, M.J., Tenner, A.J. & Johnson, K.D. (1973) Synthesis and release of sucrose by the aleurone layer of barley: regulation by gibberellic acid. *Planta* 113, 35–46.

Clutterbuck, V.J. & Briggs, D.E. (1973) Enzyme formation and release by isolated barley aleurone layers. *Phytochemistry* 12, 537–46.

Clutterbuck, V.J. & Briggs, D.E. (1974) Phosphate mobilization in grains of *Hordeum distichon*. *Phytochemistry* 13, 45–54.

Cook, A.H. (ed.) (1962) *Barley and Malt: Biology, Biochemistry, Technology*. Academic Press, London, 740 pp.

Corcoran, M.R., Geissman, T.A. & Phinney, B.O. (1972) Tannins as gibberellin antagonists. *Plant Physiology* 49, 323–30.

Cornejo, M.J., Platt-Aloia, K.A., Thomson, W.W. & Jones, R.L. (1988) Effects of GA_3 and Ca^{2+} on barley aleurone protoplasts: a freeze-fracture study. *Protoplasma* 146, 157–65.

Crabb, D. (1971) Changes in the response to gibberellic acid of barley endosperm slices during storage. *Journal of the Institute of Brewing* 77, 522–8.

Dalling, M.J. & Bhalla, P.L. (1984) Mobilization of nitrogen and phosphorus from the endosperm. In: Murray, D.R. (ed.) *Seed Physiology. 2. Germination and Reserve Mobilization*. Academic Press, Sydney, pp. 163–99.

Dashek, W.V., Singh, B.N. & Walton, B.C. (1979) Abscisic acid localization and metabolism in barley aleurone layers. *Plant Physiology* 64, 43–8.

Deikman, J. & Jones, R.L. (1986) Regulation of the accumulation of m-RNA for α-amylase isoenzymes in barley aleurone. *Plant Physiology* 80, 672–5.

Eastwell, K.C. & Spencer, M.S. (1982a) Effect of ethylene on the gibberellic acid-enhanced synthesis and release of amylase by isolated barley aleurone layers. *Plant Physiology* 69, 557–62.

Eastwell, K.C. & Spencer, M.S. (1982b) Modes of ethylene action in the release of α-amylase from barley aleurone layers. *Plant Physiology*, 69, 563–7.

Eastwell, K.C. & Spencer, M.S. (1982c) Ethylene effects on amylase activity from isolated barley aleurone layers: possible modification by proteolytic enzymes. *Plant Physiology* 70, 849–52.

Enari, T.-M. & Sopanen, J. (1986) Centenary review: mobilization of endospermal reserves during the germination of barley. *Journal of the Institute of Brewing* 92, 25–31.

Fadeel, A., Moll, B.A. & Jones, R.L. (1980) Effect of temperature on the synthesis and secretion of α-amylase in barley aleurone layers. *Plant Physiology* 66, 466–70.

Fernandez, D.E. & Staehelin, L.A. (1985) Structural organisation of ultra-rapidly frozen barley aleurone cells actively involved in protein secretion. *Planta* 165, 455–68.

Fernandez, D.E. & Staehelin, L.A. (1987) Effect of gibberellic acid on lipid degradation in barley aleurone layers. In: Fox, J.E. & Jacobs, M. (eds), *Molecular Biology of Plant Growth Control*. Alan R. Liss, New York, pp. 323–34.

Fincher, G.B. (1975) Morphology and chemical composition of barley endosperm cell walls. *Journal of the Institute of Brewing* 81, 116–22.

Fincher, G.B. (1989) Molecular and cellular biology associated with endosperm mobilization in germinating cereal grains. *Annual Review of Plant Physiology and Plant Molecular Biology* 40, 305–46.

Fincher, G.B. & Stone, B.A. (1986) Cell walls and their components in cereal grain technology. *Advances in Cereal Science and Technology* 8, 207–95.

Ford, J.S. & Guthrie, J.M. (1908) Contributions to the biochemistry of barley. *Journal of the Institute of Brewing* 14, 61–87.

Frydenberg, O. & Nielsen, G. (1965) Amylase isozymes in germinating barley seed. *Hereditas* 54, 123–39.

Fujikura, Y. & Baisted, D. (1983) Changes in starch-bound lysophospholipids and lysophospholipase in germinating Glacier and Hi-Amylose barley varieties. *Phytochemistry* 22, 865–8.

Fulcher, R.G., Setterfield, G., McCully, M.E. & Wood, P.J. (1977) Observations on the aleurone layer. II. Fluorescence microscopy of the aleurone–sub-aleurone junction with emphasis on possible β-1, 3-glucan deposits in barley. *Australian Journal of Plant Physiology* 4, 917–28.

Gaskin, P., Gilmour, S.J., Lenton, J.R., MacMillan, J. & Sponsel, V.M. (1983/1984) Endogenous gibberellins and kaurenoids identified from developing and germinating barley grain. *Journal of Plant Growth Regulation* 2, 229–42.

Gibbons, G.C. (1979) On the localization and transport of α-amylase during germination and early seedling growth of *Hordeum vulgare*. *Carlsberg Research Communications* 44, 353–66.

Gibbons, G.C. (1980) On the sequential determination of α-amylase transport and cell wall breakdown in germinating seeds of *Hordeum vulgare*. *Carlsberg Research Communications* 45, 177–84.

Gibbons, G.C. (1981) On the relative roles of the scutellum and aleurone in the production of hydrolases during germination of barley. *Carlsberg Research Communications* 46, 215–25.

Gilmour, S.J. & MacMillan, J. (1984) Effects of inhibitors of gibberellin synthesis on the induction of α-amylase in embryoless caryopses of *Hordeum vulgare* cv. Himalaya. *Planta* 162, 89–90.

Goodwin, P.B. & Carr, D.J. (1970) Ferric ion requirement for amylase synthesis by barley aleurone layers in response to gibberellic acid. *Cytobios* 2, 165–74.

Goudey, J.S., Tittle, F.L. & Spencer, M.S. (1987) Effects of hypochlorite disinfection on the response of barley aleurone layers to gibberellic acid. *Physiologia Plantarum* 69, 295–8.

Gram, N.H. (1982a) The ultrastructure of germinating barley seeds. I. Changes in the scutellum and aleurone layer in Nordal barley. *Carlsberg Research Communications* 47, 143–62.

Gram, N.H. (1982b) The ultrastructure of germinating barley seeds. II. Breakdown of starch granules and cell walls of the endosperm in three barley varieties. *Carlsberg Research Communications* 47, 173–85.

Groat, J.I. & Briggs, D.E. (1969) Gibberellins and α-amylase formation in germinating barley. *Phytochemistry* 8, 1615–27.

Gubler, F. & Ashford, A.E. (1985) Release of ferulic acid esters from barley aleurone. I. Time-course of gibberellic acid-induced release from isolated layers. *Australian Journal of Plant Physiology* 12, 297–305.

Gubler, F., Ashford, A.E., Bacic, A., Blakeney, A.B. & Stone, B.A. (1985) Release of ferulic acid esters from barley aleurone. II. Characterisation of the feruloyl compounds released in response to GA_3. *Australian Journal of Plant Physiology* 12, 307–17.

Gubler, F., Jacobsen, J.V. & Ashford, A.E. (1986) Involvement of the Golgi apparatus in the secretion of α-amylase from gibberellin-treated barley aleurone cells. *Planta* 168, 447–52.

Gubler, F., Ashford, A.E. & Jacobsen, J.V. (1987) The release of α-amylase through gibberellin-treated barley aleurone cell walls. *Planta* 172, 155–61.

Hamabata, A., Garcia-Maya, M., Romero, T. & Bernal-Lugo, I. (1988) Kinetics of the acidification capacity of aleurone layers and its effects on solubilisation of reserve substances from starchy endosperm of wheat. *Plant Physiology* 86, 643–44.

Hardie, D.G. (1975) Control of carbohydrase formation by gibberellic acid in barley endosperm. *Phytochemistry* 14, 1719–22.

Harris, G. (1962) The enzyme content and enzymic transformations in malt. In: Cook, A.H. (ed.), *Barley and Malt. Biology, Biochemistry, Technology*. Academic Press, London, pp. 583–694.

Hayes, B., Harmey, M.A. & Byrne, H. (1988) Multiple forms of α-amylase in malting barley, *var.* Emma. *Journal of the Institute of Brewing* 94, 253–9.

Hejgaard, J. (1976) Free and protein-bound β-amylases of barley grain. Characterisation by two-dimensional immunoelectrophoresis. *Physiologia Plantarum* 38, 293–9.

Hejgaard, J. (1977) Origin of a dominant beer protein: immunochemical identity with a β-amylase-associated protein from barley. *Journal of the Institute of Brewing* 83, 94–6.

Hejgaard, J. (1978) 'Free' and 'bound' β-amylases during malting of barley. Characterisation by two-dimensional immunoelectrophoresis. *Journal of Institute of Brewing* 84, 43–6.

Hejgaard, J. (1982) Purification and properties of protein-Z, a major albumin of barley endosperm. *Physiologia Plantarum* 54, 174–82.

Hejgaard, J. & Carlsen, S. (1977) Immunoelectrophoretic identification of a heterodimer β-amylase in extracts of barley grain. *Journal of the Science of Food and Agriculture* 28, 900–4.

Hill, R.D. & MacGregor, A.W. (1988) Cereal α-amylases in grain research and technology. *Advances in Cereal Science and Technology* IX, 217–61.

Hill, R.D., MacGregor, A.W., Weselake, R.J. & Daussant, J. (1987) A review of some properties of an endogenous inhibitor of cereal α-amylase. In: Mares, D.J. (ed.), *Proceedings of the 4th International Symposium on Preharvest Sprouting in Cereals*. Westview Press, Boulder, CO, USA, pp. 474–82.

Ho, T.-H.D., Shih, S.-C. & Kleinhofs, A. (1980) Screening for barley mutants with altered hormone sensitivity in their aleurone layers. *Plant Physiology* 66, 153–7.

Ho, T.-H.D., Abroms, J. & Varner, J.E. (1982) Effect of ethylene on the release of α-amylase through cell walls of barley aleurone layers. *Plant Physiology* 69, 1128–31.

Hooley, R. (1982) Protoplasts isolated from aleurone layers of wild oat (*Avena fatua*, L.) exhibit the classic response to gibberellic acid. *Planta* 154, 29–40.

Hooley, R., Beale, M.H., Smith, S.J. & MacMillan, J. (1990) Novel affinity probes for gibberellin receptors in aleurone protoplasts of *Avena fatua*. In: Pharis, R.P. & Rood, S.B. (eds), *Plant Growth Substances 1988*. Springer Verlag, Berlin, pp. 145–53.

Hoy, J.L., Macauley, B.J. & Fincher, G.B. (1981) Cellulases of plant and microbial origin in germinating barley. *Journal of the Institute of Brewing* 87, 77–80.

Jacobsen, J.V. (1973) Interaction between gibberellic acid, ethylene and abscisic acid in control of amylase synthesis in barley aleurone layers. *Plant Physiology* 51, 198–202.

Jacobson, A. & Corcoran, M.R. (1977) Tannins as gibberellin antagonists in the synthesis of α-amylase and acid phosphatase by barley seeds. *Plant Physiology* 59, 129–33.

Jacobsen, J.V., Zwar, J.A. & Chandler, P.M. (1985) Gibberellic-acid-responsive protoplasts from mature aleurone of Himalaya barley. *Planta* 163, 430–8.

Jacobsen, J.V., Bush, D.S., Sticher, L. & Jones, R.L. (1988) Evidence for precursor forms of the low isoelectric point α-amylase isozymes secreted by barley aleurone cells. *Plant Physiology* 88, 1168–74.

Jensen, S.A. & Heltved, F. (1982) Visualisation of enzyme activity in germinating cereal seeds using a lipase-sensitive fluorochrome. *Carlsberg Research Communications* 47, 297–303.

Jones, R.L. (1968) Ethylene enhanced release of α-amylase from barley aleurone cells. *Plant Physiology* 43, 442–4.

Jones, R.L. (1973) Gibberellic acid and ion release from barley aleurone tissue. *Plant Physiology* 52, 303–8.

Jones, R.L. & Armstrong, J.E. (1971) Evidence for osmotic regulation of hydrolytic enzyme production in germinating barley seeds. *Plant Physiology* 48, 137–42.

Jones, R.L. & Jacobsen, J.V. (1983) Calcium regulation of the secretion of α-amylase isoenzymes and other proteins from barley aleurone layers. *Planta* 158, 1–9.

Katayama, N. & Suzuki, H. (1980) Possible effect of gibberellin on phytate degradation in germinating barley seeds. *Plant and Cell Physiology* 21, 115–23.

Khursheed, B. & Rogers, J.C. (1988) Barley α-amylase genes. Quantitative comparisons of steady-state in m-RNA levels from individual members of the two different families expressed in aleurone cells. *Journal of Biological Chemistry* 263, 18953–60.

Koehler, S. & Ho, T.-H.D. (1988) Purification and characterisation of gibberellic acid-induced cysteine endoproteases in barley aleurone layers. *Plant Physiology* 87, 95–103.

Koehler, S. & Ho, T.-H.D. (1990) A major gibberellic acid-induced barley aleurone cysteine proteinase which digests hordein. *Plant Physiology* 94, 251–8.

Kreis, M., Williamson, M., Buxton, B., Pywell, J., Hejgaard, J. & Svendsen, I. (1987) Primary structure and differential expression of β-amylase in normal and mutant barleys. *European Journal of Biochemistry* 169, 517–25.

Kreis, M., Williamson, M.S., Shewry, P.R., Sharp, P. & Gale, M. (1988) Identification of a second locus encoding β-amylase on chromosome 2 of barley. *Genetic Research (Camb.)* 51, 13–16.

Lanahan, M.B. & Ho, T.-H.D. (1988) Slender barley: a constitutive gibberellin-responsive mutant. *Planta* 175, 107–14.

Laurière, C., Laurière, M. & Daussant, J. (1986) Immunohistochemical localization of β-amylase in resting barley seeds. *Physiologia Plantarum* 67, 383–8.

Lehtonen, M. & Aikasalo, R. (1987a) Pentosans in barley varieties. *Cereal Chemistry* 64, 133–4.

Lehtonen, M. & Aikasalo, R. (1987b) β-Glucan in two- and six-rowed barley. *Cereal Chemistry* 64, 191–3.

Liu, D.J. & Pomeranz, Y. (1975) Distribution of minerals in barley at the cellular level by x-ray analysis. *Cereal Chemistry* 52, 620–9.

Loi, L., Ahluwalia, B. & Fincher, G.B. (1988) Chromosomal location of genes encoding barley (1 → 3, 1 → 4)-β-glucan-4-glucanohydrolases *Plant Physiology* 87, 300–2.

Lundgard, R. & Svensson, B. (1986) Limited proteolysis in the carboxy terminal region of barley β-amylase. *Carlsberg Research Communications* 51, 487–91.

Lundgard, R. & Svensson, B. (1987) The four major forms of barley β-amylase. Purification, characterisation and structural relationship. *Carlsberg Research Communications* 52, 313–26.

MacGregor, A.W. (1976) A note on the formation of α-amylase in de-embryonated barley kernels. *Cereal Chemistry* 53, 792–6.

MacGregor, A.W. (1983) Cereal α-amylases: synthesis and action patterns. In: Daussant, J., Mossé, J. & Vaughn, J. (eds), *Seed Proteins*. Academic Press, London, pp. 1–34.

MacGregor, A.W. (1987) α-Amylase, limit dextrinase and α-glucosidase enzymes in barley and malt. *CRC Critical Reviews in Biotechnology* 5, 117–28.

MacGregor, A.W. & Lenoir, C. (1987) Studies on α-glucosidase in barley and malt. *Journal of the Institute of Brewing* 93, 334–7.

MacGregor, A.W. & Marchylo, B.A. (1986) α-Amylase components in excised, incubated barley embryos. *Journal of the Institute of Brewing* 92, 159–61.

MacGregor, A.W., Daussant, J. & Niku-Paavola, M.-L. (1979) Further purification and characterisation of a new amylase found in barley. *Journal of the Science of Food and Agriculture* 30, 1071–6.

MacGregor, A.W., Nicholls, P.B. & Dushnicky, L. (1983) Endosperm degradation in barley kernels that synthesize α-amylase in the absence of embryos and exogenous gibberellic acid. *Food Microstructure* 2, 13–22.

MacGregor, A.W., MacDougall, F.H., Mayer, F. & Daussant, J. (1984) Changes in the levels of α-amylase components in barley tissues during germination and early seedling growth. *Plant Physiology* 75, 203–6.

MacGregor, A.W., Ballance, G.M. & Dushnicky, L. (1989) Fluorescence miscroscopy studies on (1,3)-β-D-glucan in barley endosperm. *Food Microstructure* 8, 235–44.

MacGregor, E.A. & MacGregor, A.W. (1987) Studies of cereal α-amylase using cloned DNA. *CRC Critical Reviews in Biotechnology* 5, 129–42.

MacLeod, A.M. & Palmer, G.H. (1966) The embryo of barley in relation to modification of the endosperm. *Journal of the Institute of Brewing* 72, 580–9.

Marchylo, B.A., MacGregor, A.W. & Kruger, J.E. (1987) Quantitative analysis of multiple forms of alpha-amylase produced in germinating cereals. In: Mares, D.J. (ed.), *Proceedings of the 4th International Symposium on Pre-Harvest Sprouting in Cereals*. Westview Press, Boulder, CO, USA, pp. 522–33.

Marchylo, B.A., MacGregor, A.W. & Kruger, J.E. (1985) Production of α-amylases in germinating whole and incubating de-embryonated barley kernels. *Journal of the Institute of Brewing* 91, 161–5.

McFadden, G.I., Ahluwalia, B., Clarke, A.E. & Fincher, G.B. (1988) Expression sites and developmental regulation of genes encoding (1 → 3, 1 → 4)-β-glucanases in germinating barley. *Planta* 173, 500–8.

Melcher, U. & Varner, J.E. (1971) Protein release by barley aleurone layers. *Journal of the Institute of Brewing* 77, 456–61.

Merry, J. & Goddard, D.R. (1941) A respiratory study of barley grain and seedlings. *Proceedings of the Rochester Academy of Science* 8, 28–44.

Mikola, J. (1983) Proteinases, peptidases and inhibitors of endogenous proteinases in germinating seed. In: Daussant, J., Mossé, J. & Vaughn, J. (eds), *Seed Proteins*. Academic Press, London, pp. 35–52.

Mikola, J. & Enari, T.-M. (1970) Changes in the contents of barley proteolytic inhibitors during malting and mashing. *Journal of the Institute of Brewing* 76, 182–8.

Mikola, J. & Kolehmainen, L. (1972) Localization and activity of various peptidases in germinating barley. *Planta* 104, 167–77.

Mikola, J. & Virtanen, M. (1980) Secretion of L-malic acid by plant aleurone layers. *Plant Physiology* (Supplement), Abstract 783.

Morrall, P. & Briggs, D.E. (1978) Changes in the cell wall polysaccharides of germinating barley grains. *Phytochemistry* 17, 1495–502.

Morrison, W.R. (1978) Cereal lipids. *Advances in Cereal Chemistry* 2, 221–348.

Morrison, W.R. (1988) Lipids in cereal starches; a review. *Journal of Cereal Science* 8, 1–15.

Mundy, J., Brandt, A. & Fincher, G.B. (1985) Messenger RNAs from the scutellum and aleurone of germinating barley encode (1 → 3, 1 → 4)-β-D-glucanase, α-amylase and carboxypeptidase. *Plant Physiology* 79, 867–71.

Murthy, P.P.N., Renders, K.M. & Keranen, L.M. (1989) Phosphoinositides in barley aleurone layers and gibberellic acid-induced changes in metabolism. *Plant Physiology* 91, 1266–9.

Muthukrishnan, S. & Chandra, G.R. (1988) Regulation of the expression of hydrolase genes in cereal seeds. *Advances in Cereal Science and Technology* IX, 129–59.

Newman, J.C. & Briggs, D.E. (1976) Glyceride metabolism and gluconeogenesis in barley endosperm. *Phytochemistry* 15, 1453–8.

Nicholls, P.B. (1980) Development of responsiveness to gibberellic acid in the aleurone layer of immature wheat and barley caryopses: effect of temperature. *Australian Journal of Plant Physiology* 7, 645–53.

Nicholls, P.B. (1983) Environment, developing barley grain, and subsequent production of alpha-amylase by the aleurone layer. In: Kruger, J.E. & LaBerge, D.E. (eds), *Proceedings of the 3rd International Symposium on Pre-Harvest Sprouting in Cereals*. Westview Press, Boulder, CO, USA, pp. 147–53.

Nicholls, P.B., MacGregor, A.W. & Marchylo, B.A. (1986) Production of α-amylase isoenzymes in barley caryopses in the absence of embryos and exogenous gibberellic acid. *Australian Journal of Plant Physiology* 13, 239–47.

Niku-Paavola, M.-L. (1977) Partial characterisation of a new barley amylase. *Journal of the Science of Food and Agriculture* 28, 728–38.

Niku-Paavola, M.-L., Skakoun, A., Nummi, M. & Daussant, J. (1973). The polymorphism of barley β-amylase. *Biochimica et Biophysica Acta* 322, 181–4.

Nutbeam, A.R. & Briggs, D.E. (1982) Gibberellin-phenol interactions in plant extracts. *Phytochemistry* 21, 2217–23.

Obata, T. (1987) Transfer of information within the germinating grain. In: Mares, D.J. (ed.), *Proceedings of the 4th International Symposium on Pre-Harvest Sprouting in Cereals*. Westview Press, Boulder, CO, USA, pp. 345–57.

Oh, S.Y. & Briggs, D.E. (1989) Modification in malting barley. *Journal of the Institute of Brewing* 95, 83–8.

Okamoto, K., Kitano, H. & Akazawa, T. (1980) Biosynthesis and secretion of hydrolases in germinating cereal grains. *Plant and Cell Physiology* 21, 201–4.

Palmer, G.H. (1980) Enzyme location and endosperm modification. *Journal of the Institute of Brewing* 86, 163–4.

Palmer, G.H. (1982) A reassessment of the pattern of endosperm hydrolysis (modification) in germinating barley. *Journal of the Institute of Brewing* 88, 145–53.

Palmer, G.H. & Bathgate, G.N. (1976) Malting and brewing. *Advances in Cereal Science and Technology* 1, 237–324.

Palmer, G.H. & Duffus, J.H. (1986) Aleurone or scutellar enzymes in malting? *Journal of the Institute of Brewing* 92, 512–13.

Pogson, B.J., Ashford, A.E. & Gubler, F. (1989) Immunofluorescence localization of α-amylase in the scutellum, germ aleurone and 'normal' aleurone of germinated barley grains. *Protoplasma* 151, 128–36.

Pollard, C.J. (1968) Effects of gibberellic acid on metabolism of barley seeds. *Federation Proceedings* 27, 812 (Abstr. 3365).

Pomeranz, Y. (1973) Structure and mineral composition of cereal aleurone cells as shown by scanning electron microscopy. *Cereal Chemistry* 50, 504–11.

Poulle, M. & Jones, B.L. (1988) A proteinase from germinating barley. I. Purification and some physical properties of a 30 kD cysteine endoproteinase from green malt. *Plant Physiology* 88, 1454–60.

Radley, M. (1969) The effect of the endosperm on the formation of gibberellin by barley embryos. *Planta* 86, 218–23.

Ranki, H. (1990) Secretion of α-amylase by the epithelium of barley scutellum. *Journal of the Institute of Brewing* 96, 307–9.

Ranki, H. & Sopanen, T. (1984) Secretion of α-amylase by the aleurone layer and the scutellum of germinating barley grain. *Plant Physiology* 75, 710–15.

Ranki, H., Sopanen, T. & Mikola, I. (1983) Secretion of carboxypeptidase-I by the scutellum of germinating barley grain. *Plant Physiology* 72, S-101.

Rastogi, V. & Oaks, A. (1986) Hydrolysis of storage proteins in barley endosperms. Analysis of soluble products. *Plant Physiology* 81, 901–6.

Raynes, J.G. (1983) 'The ability of barley grains of different varieties to produce α-amylase and gibberellins during germination'. Unpublished PhD thesis, Department of Biochemistry, University of Birmingham.

Raynes, J.G. & Briggs, D.E. (1985) Genotype and the production of α-amylase in barley grains germinated in the presence and absence of gibberellic acid. *Journal of Cereal Science* 3, 55–65.

Scott, P.C. & Leopold, A.C. (1967) Opposing effects of gibberellin and ethylene. *Plant Physiology* 42, 1021–2.

Selvig, A., Aarnes, H. & Lie, S. (1986) Cell wall degradation in endosperm of barley during germination. *Journal of the Institute of Brewing* 92, 185–7.

Serre, L. & Laurière, C. (1989) Limit dextrinase in cereal seeds (a review). *Sciences des Aliments* 9, 645–63.

Shewry, P.R., Parmar, S., Buxton, B., Gale, M.D., Liu, C.J., Hejgaard, J. & Kreis, M. (1988) Multiple molecular forms of β-amylase in seeds and vegetative tissues of barley. *Planta* 176, 127–34.

Singh, S.P. & Paleg, L.G. (1984) Low temperature induction of hormonal sensitivity in genotypically gibberellic acid-insensitive aleurone tissue. *Plant Physiology* 74, 437–8.

Slade, A.M., Høj, P.B., Morrice, N.A. & Fincher, G.B. (1989) Purification and characterisation of three (1 → 4)-β-D-xylan endohydrolases from germinated barley. *European Journal of Biochemistry* 185, 533–9.

Smith, M.T. & Briggs, D.E. (1979) The mechanisms by which tumbling and abrasion alter enzyme formation in malting grain. *Journal of the Institute of Brewing* 85, 160–7.

Smith, M.T. & Briggs, D.E. (1980) Externally applied gibberellic acid and α-amylase formation in grains of barley (*Hordeum distichon*). *Phytochemistry* 19, 1025–33.

Smith, M.T., Cornish-Bowden, A. & Briggs, D.E. (1979) The determination of barley α-amylase activity. *Journal of the Institute of Brewing* 85, 157–9.

Stark, J.R. & Yin, X.S. (1987) Evidence for the presence of maltase and α-glucosidase isoenzymes in barley. *Journal of the Institute of Brewing* 93, 108–12.

Stewart, A., Nield, H. & Lott, J.N.A. (1988) An investigation of the mineral content of barley grains and seedlings. *Plant Physiology* 86, 93–7.

Stuart, I.M., Loi, L. & Fincher, G.B. (1986) Development of (1 → 3, 1 → 4)-β-D-glucan endohydrolase isoenzymes in isolated scutella and aleurone layers of barley (*Hordeum vulgare*). *Plant Physiology* 80, 310–14.

Sun, Z. & Henson, C.A. (1990) Degradation of native starch granules by barley α-glucosidases. *Plant Physiology* 94, 320–7.

Svensson, B., Gibson, R.M., Haser, R. & Astier, J.P. (1987) Crystallization of barley malt α-amylases and preliminary x-ray diffraction studies of the high pI enzyme, α-amylase-2. *Journal of Biological Chemistry* 262, 13682–4.

Tittle, F.T. & Spencer, M.S. (1986) Interaction between ethylene, CO_2 and ABA on GA-induced amylase synthesis in barley aleurone tissue. *Plant Physiology* 80, 1034–7.

Tittle, F.L., Goudey, J.S. & Spencer, M.S. (1988) Hypochlorite disinfection influences the GA_3-induced synthesis of α-amylase isozymes by isolated barley aleurone layers. *Plant Physiology* 86, 510–11.

Varner, J.E. & Mense, R.M. (1972) Characteristics of the process of enzyme release from secretory plant cells. *Plant Physiology* 49, 187–9.

Verbeek, R., Van Onckelen, H.A. & Gaspar, Th. (1973) α-Amylase in the barley grain influenced by the auxin–cytokinin interaction in the coleoptile. *Physiologia Plantarum* 29, 208–11.

Weselake, R.J., MacGregor, A.W. & Hill, R.D. (1983) An endogenous α-amylase inhibitor in barley kernels. *Plant Physiology* 72, 809–12.

Weselake, R.J., MacGregor, A.W. & Hill, R.D. (1985a) Endogenous α-amylase inhibitor in various cereals. *Cereal Chemistry* 62, 120–3.

Weselake, R.J., MacGregor, A.W. & Hill, R.D. (1985b) Effect of endogenous barley α-amylase inhibitor on the hydrolysis of starch under various conditions. *Journal of Cereal Science* 3, 249–59.

Yamada, K. (1982) Determination of endogenous gibberellins in germinating barley by combined gas chromatography – mass spectrometry. *Journal of the American Society of Brewing Chemists* 40, 18–25.

Yamada, K. (1985) Endogenous abscisic acid in barley and use of abscisic acid in malting. *Agricultural and Biological Chemistry* 49, 429–34.

Yin, X.S. & MacGregor, A.W. (1988) An approach to the identification of a β-glucan solubilase from barley. *Journal of the Institute of Brewing* 94, 327–30.

Yin, X.S. & MacGregor, A.W. (1989) Substrate specificity and nature of action of barley β-glucan solubilase. *Journal of the Institute of Brewing* 95, 105–9.

Yin, X.S., MacGregor, A.W. & Clear, R.M. (1989) Field fungi and β-glucan solubilase in barley kernels. *Journal of the Institute of Brewing* 95, 195–8.

Yomo, H. (1958) Studies on barley malt. The sterilisation of barley seed and amylase formation of separated embryos and endosperms. *Hakkô Kyôkaishi* 16, 444–8.

Zingen-Sell, I., Hillmer, S., Robinson, D.G. & Jones, R.L. (1990) Localization of α-amylase isozymes within the endomembrane system of barley aleurone. *Protoplasma* 154, 16–24.

19

Gibberellin Responses in Barley

Peter M. Chandler

The effects of gibberellins (GAs) on plant processes have largely been deduced from studies in which one particular GA (gibberellic acid, GA_3) is applied to intact plants or to portions of plants. A typical response would be an increase in stem length when treated plants are compared with controls. From observing the effects of such treatments, and from the knowledge that GAs occur naturally in plants, physiological roles for GA have been proposed. For the particular response just mentioned a role in internode expansion has been proposed.

Recently, it has become possible to examine some of these proposed roles more rigorously than previously. This results from the emergence of a number of new experimental approaches which should improve our understanding of how GAs influence plant growth and development. This will help unravel the contribution of GA responses to varietal differences of agronomic importance (e.g. culm length, photoperiod control of flowering, tendency for preharvest sprouting, and grain malting characteristics). Among these new approaches, are improved assays for GAs, a growing number of defined mutants, and molecular methods for studying GA responses.

A range of different GAs are present in plants, but their low levels and the problem of interconversion have made identification and measurement difficult. However, substantial progress has been made with the availability of doubly labelled standards, improved analytical procedures, as well as the possibility that antibody techniques may be used for the purification and identification of individual GAs (Takahashi *et al.*, 1986; Smith & MacMillan, 1989).

There has been renewed interest in studying mutants with altered GA levels or altered expression of GA responses. GA-responsive dwarf mutants have been described in several species, including barley (Suge, 1983; Hentrich *et al.*, 1985; Reid, 1987). Mutations affecting the expression of GA responses (Foster, 1977) will be useful for defining components of the response such as receptors and signal transduction pathways, although there has been little progress to date.

Nucleic acid and antibody probes for products of GA-regulated gene expression are valuable for analysing responses to GA, both spatially and temporally (McFadden *et al.*, 1988; Chandler & Jacobsen, 1991). Research is in progress to define the DNA sequences required for gene expression in response to GA (Huttly & Baulcombe, 1989; Jacobsen & Close, 1991). A natural progression of such studies will be the characterization of proteins binding to 'GA response elements' with the exciting possibility that one (or more) of these proteins may form part of a GA–receptor complex.

The challenge for our future understanding of the role(s) played by GAs in plant biology is to bring these new approaches to bear on one or a few systems. No single system has yet received the multidisciplinary input necessary for these tools to be most effectively used and for optimal progress to be made.

This chapter summarizes the major GA responses in barley, and then focuses on one system in particular and assesses the evidence for GA control of α-amylase production by aleurone *in vivo*. By highlighting the deficiencies in our knowledge in this relatively well-understood system, it is hoped that the need for a multidisciplinary approach will be demonstrated.

Effects of applied GA

The effects of applied GA_3 in barley have been well described by Briggs (1978). In general, these effects are similar to those seen in other plant species. Perhaps most obvious are: (i) stimulation of grain germination and seedling growth; and (ii) enhanced internode elongation in older plants.

Grains imbibed in GA_3 solutions will often show earlier germination, more rapid shoot elongation and higher α-amylase activity per seedling than control grains imbibed in water. This system has been extensively studied in barley from several different perspectives (e.g. hormone biology, steeping and malting technology, and seed dormancy characteristics). The role played by endogenous GA following germination will be discussed here in greater detail.

Another major effect of GA in many plant species, including barley, is stimulation of internode elongation in the culm or stem (Kaufman & Dayanandan, 1983). There are corresponding effects on leaf shape and inflorescence characteristics. These effects result, in part at least, from GA_3-stimulated cell expansion in basal regions of the internode or leaf sheath. The mechanism(s) underlying this have not been fully established, but the current view is that an increased extensibility of the cell wall may be involved (Keyes *et al.*, 1990).

Establishing the role of GA in a response

The 'spray and pray' approach used in many studies has been useful in suggesting physiological roles for GA, but there are obvious limitations to the amount of confidence which can be placed in such conclusions. One complication arises because plants already exhibit a GA response due to endogenous GA. Consequently the effect of applied GA on the response measured will depend upon the degree to which the response is already saturated by endogenous GA. For instance, many 'tall' barley cultivars show a modest (10–40%) growth stimulation when GA_3 is applied. Dwarf mutants, on the other hand, can show several-fold increases (i.e. 300–400%) in growth rate in response to GA_3. This difference presumably reflects the contribution of endogenous gibberellin to the growth of the tall varieties. Other complications in ascribing a response to GA could include genetic or environmental effects on GA sensitivity, 'non-physiological' artefacts resulting from abnormally high GA levels, or 'secondary' effects mediated by some other regulator whose level might be influenced by the applied GA.

Fortunately several approaches can be taken to improve confidence that a particular response is mediated through GA action. One approach is to minimize the complications from endogenous GA by using tissue from the plant which appears incapable of making its own GA (e.g. aleurone), or by reducing the endogenous GA level with GA synthesis inhibitors. These methods may not always be applicable – for instance, there may be no suitable tissue system for the response in question, or the specificity of an inhibitor may be a problem. Nevertheless, an enhanced response to applied GA when endogenous GA levels have been lowered, supports GA involvement in the response. Problems associated with interpretation of endogenous GA levels are described later.

A second approach towards establishing the role of GA in a particular response is to isolate mutants which are altered either in the response in question, or in an established GA response (such as internode length). Correlative evidence for or against the involvement of GA can then be obtained by comparing the expression of the two responses, and by further studies such as GA analysis. The best example of GA responses is plant height mutants (see later). In many species, including barley, dwarf mutants have been isolated which revert to the normal tall phenotype when GA is applied (Reid, 1987). This suggests that the dwarf characteristic results from a block in the GA biosynthetic pathway; in some cases, examination of the growth response to intermediates in the pathway has suggested which enzymatic step might be involved.

Mutants which are taller than normal have also been found in several plant species. In one recent study the increased internode length of a mutant of *Brassica* was attributed to higher than normal GA levels (Rood *et al.*, 1990). However, in other cases, including the *slender* mutant of barley (Foster, 1977), the explanation does not seem to be due to elevated GA levels; rather it seems that there is constitutive expression of GA responses in this mutant in the presence of near-normal GA levels (Chandler, 1988; Lanahan & Ho, 1988). Perhaps the mutation results from an alteration in a GA receptor or in some component of the GA response pathway.

So by using several different approaches the case for GA involvement in a response can be strengthened. Unfortunately details of this involvement remain obscure at present because of our ignorance concerning the mechanism(s) of GA action both in terms of the GAs involved and the details of their biological activity.

Although major advances have been made in describing the GAs which might be involved in stem elongation, some uncertainty still remains. The conclusion that GA_1 determines the extent of elongation (reviewed in Phinney, 1985) is now complicated by the fact that other active GAs are derived from GA_{20}, the precursor to GA_1 (Fujioka *et al.*, 1990). For other GA responses there is much less information as to which endogenous GA might be involved. An important limitation is the amount of tissue required for assay, since ideally GAs should be assayed in populations of responding cells rather than in whole organ systems where the responding cells might form only a small proportion of the total. Even if we were able to assay hormone levels in a selected cell population, the interpretation of such data is potentially complicated by problems of intercellular, intracellular and perhaps extracellular distribution.

In addition to the problems of knowing which GA species to assay, where to assay it, and how to interpret the result, little is known about putative GA receptors and how a GA–receptor complex might result in the responses measured. Progress in this area has been slow. Major limitations have been the lack of suitable mutants (e.g. GA-insensitive, *slender*) and neglect of the few mutants available to help establish a vital connection between (for instance) a putative receptor protein and an altered response to GA.

Action of GA in endosperm

The demonstrations in the 1960s that GA_3 could replace the embryo signal required by the endosperm for α-amylase production led to detailed investigations of the effects of GA_3 on α-amylase production by aleurone tissue and more recently by aleurone protoplasts. This active area of research has been reviewed recently (Fincher, 1989; Jones & Jacobsen, in press). After a brief summary this research is discussed here, in particular evidence for GA involvement *in vivo* (i.e. α-amylase production by aleurone in germinated grain or seedlings).

There is broad agreement that applied GA_3 results in stimulation of α-amylase synthesis in isolated aleurone layers of cv. Himalaya. This response is similar to that of the aleurone in intact grains several days after germination. It is tempting to speculate that the isolated tissue provides a useful model system for studying aleurone behaviour *in vivo*.

Increased α-amylase synthesis in layers treated with GA_3 results from higher levels of the different α-amylase mRNAs. These, in turn, result from enhanced transcription of α-amylase genes (Jacobsen & Beach, 1985). Particular DNA sequences are required for transcriptional stimulation in isolated aleurone protoplasts (Huttly & Baulcombe, 1989; Jacobsen & Close, in press). In addition a role can be postulated for enhanced stability of α-amylase mRNAs once formed, although this has not been demonstrated, nor is it known if the translational efficiency of α-amylase transcripts remains constant. There is a similar increase of α-amylase mRNA levels in the aleurone of seedlings, but whether increased transcription is responsible for this increase has not been determined.

There are two major classes of α-amylase isozymes, the high-pI and the low-pI forms. Each class is a small gene family with two or three major expressed members, but also minor species (reviewed in Jones & Jacobsen, in press). The genes for the two isozyme classes are on different chromosomes, while the genes for members of each class are linked (Brown & Jacobsen, 1982). There is a differential effect of GA_3 on the levels of high-pI mRNAs versus low-pI mRNAs in isolated aleurone layers. Both sets of α-amylase mRNAs accumulate from the time of GA_3 application to reach a maximum level at about 16 h. But whereas the low-pI mRNAs persist at this maximum level for a further 48 h at least, levels of the high-pI mRNAs decline and by 36 h are very low (Nolan & Ho, 1988; Chandler & Jacobsen, 1991). There is no satisfactory explanation for this difference in behaviour but it presumably results from differences between the two pI classes in the duration of transcriptional stimulation and/or the stability of transcripts once formed. Surprisingly, this difference between the two pI classes is not apparent in aleurone protoplasts treated with GA_3 (Chandler & Jacobsen, 1991).

The 'model' and arguments for and against

It is now clear that the same set of α-amylase isozymes and mRNAs is expressed by the aleurone of seedlings, and by isolated aleurone layers or protoplasts treated with GA_3 (Chandler & Jacobsen, 1991). This might suggest that GA_3 is the natural regulator of α-amylase production by aleurone in germinated grain, supporting the 'simple' model in which embryo-produced GA (GA_3?) triggers aleurone cells to produce α-amylase. In fact GA_3 and GA_1 have both been identified in germinated barley (Yamada, 1982). However, a number of observations indicate that the situation is likely to be more complex than suggested by this simple model.

Firstly, there are problems arising from apparent differences in the regulation of α-amylase mRNA in the aleurone of seedlings versus isolated aleurone layers. High-pI mRNAs initially predominate in the aleurone of seedlings, but then fall as the low-pI mRNAs rise and persist at high levels. From studies of the levels of α-amylase mRNAs in isolated aleurone layers we might have predicted the earlier appearance of low-pI mRNAs, based on their greater response to low GA_3 concentrations (for more detail see Chandler & Jacobsen, 1991). Although the data do not fit our expectations based on the simple model, this observation need not necessarily represent a fundamental problem. It is clear, for example, from the differences observed between α-amylase gene regulation in isolated aleurone layers and in aleurone protoplasts that the environment of the aleurone cell can influence its response to GA_3. Therefore, the differences described above in the responses of isolated aleurone layers and the aleurone of seedlings may reflect differences in the environment of the aleurone cell.

A second problem concerns endosperm half-grains which produce as much α-amylase in the absence of added GA_3 as in its presence (Marchylo *et al.*, 1985; Nicholls *et al.*, 1986). This has been observed in cv. Himalaya as well as in other cultivars. Perhaps a different signal is responsible for α-amylase production in such grain. Alternatively, through a combination of genotype and environment, α-amylase production may be independent of the usual 'signal' (constitutive production). It is also possible that the aleurone of such grains is or has been exposed to GA. Although there have been attempts to show that this explanation is not correct, it is difficult to assess the data because of problems associated with the measurement and interpretation of GA levels. For example, during development barley grains accumulate relatively high levels of a spectrum of GAs, many of which are regarded as inactive or of low activity (Gaskin *et al.*, 1984). These may be derived from active GAs of maternal origin during grain fill. Clearly there is potential for some GA activity in hydrated endosperm, and the extent of this could be influenced by either genetic or environmental factors.

More fundamental objections to the model have been raised by Trewavas (1988) who has criticized its simplicity, its apparent inability to explain data from other grass species (as well as barley), and the relevance of an experimental system (cultivated barley) in which 'endogenous controls have been crippled in the search for malting excellence'. Few would deny that the model is an oversimplification of the real situation. However, for this reviewer its value lies more in establishing a framework for making predictions which can be experimentally tested in barley than in its ability to account for a diverse set of phenomena associated with the germination of grasses.

Predictions of the model and some results

The model predicts that barley mutants with altered responses to GA, or with altered capacity to produce GA, should show corresponding changes in α-amylase production if the latter is dependent on both GA levels and a GA-response mechanism. Fortunately both types of mutants exist in barley.

The *slender* mutant of barley exhibits a phenotype normally associated with saturating levels of applied GA_3: the culm and rachis internodes are highly elongated, the leaves are long and narrow, and the plants have a pale green and limber appearance. α-Amylase production by endosperm half-grains and by the aleurone of the *slender* mutant has been investigated, and the data indicate that α-amylase is produced to about the same extent in the presence or absence of exogenous GA_3 (Chandler, 1988; Lanahan & Ho, 1988). This contrasts with the parent cultivar which showed α-amylase production only when GA_3 was added.

Recent investigation of gibberellin metabolism in the *slender* genotype found no evidence to support the possibility that *slender* might possess high endogenous GA levels (Croker *et al.*, 1990). The current view that the *slender* mutant exhibits a constitutive GA response, is supported by the observation that a similar phenotype in pea is observed even when crossed into a severely GA-deficient dwarfing background (Potts *et al.*, 1985). If we accept that the enhanced internode elongation seen in the *slender* mutant represents an altered GA response, the constitutive production of α-amylase presumably does also.

Dwarf mutants of barley have been described which have shortened internodes, and leaves which are shorter, broader and a darker green than those of the parent. These mutants are presumed to be defective in some aspect of GA synthesis since application of GA_3 restores a near-normal phenotype. We have isolated and characterized GA-responsive dwarf mutants of cv. Himalaya. Preliminary studies of some of these mutants have begun after several generations of inbreeding before the establishment of suitably backcrossed mutant stocks.

One of these mutants, M117, produced much less α-amylase following germination than did the control. Furthermore, the low level of α-amylase mRNAs found in the mutant grain was further reduced by treatment with paclobutrazol (an inhibitor of GA synthesis) but restored to high levels in the presence of both paclobutrazol and GA_3. This latter observation suggests that reduced levels of α-amylase mRNA in the mutant can be overcome through applied GA_3, suggesting in turn that the reductions result from low endogenous GA (Chandler & Mosleth, 1990).

M359 is another GA-responsive dwarf mutant of cv. Himalaya, and feeding of biosynthetic intermediates suggests a block in the GA_{20} to GA_1 conversion. Grain of this mutant was imbibed in water or in a GA_3 solution, and the mean shoot height and α-amylase content determined at different stages of seedling growth. The results (Fig. 19.1) indicate that M359 is low in α-amylase production following germination on water relative to the parent, but there is little difference in α-amylase production when germinated on a GA_3 solution. The maximum α-amylase level produced by M359 was approximately 19% of the control value when germinated on water, and 81% of the control when germinated on GA_3 solution. The difference between these values presumably reflects a deficiency in endogenous GA level in the mutant which is largely corrected by exogenous GA_3. A similar conclusion is reached when shoot heights of the two lines are compared for seedlings grown on water or on GA_3 solution.

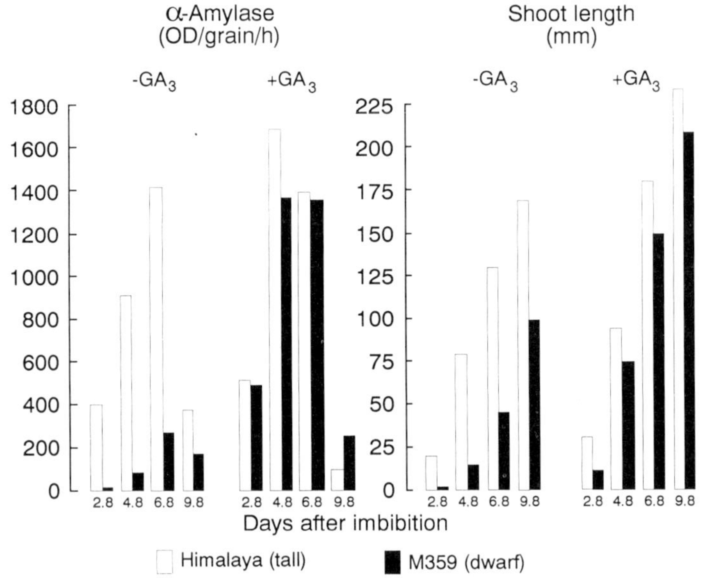

Fig. 19.1. Grain of barley cv. Himalaya or M359 (a sodium azide-induced GA_3-responsive dwarf in cv. Himalaya) was surface-sterilized and germinated under conditions described previously (Chandler & Jacobsen, 1991) either on water or on a solution of GA_3 (10^{-5} M). At the indicated times after imbibition, the average shoot length of 20 seedlings and the α-amylase activity in the grain part of 10 seedlings were determined.

The results obtained using these two dwarf mutants are most easily interpreted to indicate that the supply of GA limits α-amylase production in germinated grain of cv. Himalaya. Care must be exercised in making this interpretation, as it remains an assumption that the mutants are deficient in their ability to synthesize GA. When appropriately backcrossed lines are available their GA metabolism and status will be assessed. A further complication is that the level of one or a few distinct GAs may be critical for determining α-amylase production rather than the level of GA_1 or GA_3 which have been suggested to be important for regulating vegetative growth.

The capacity to respond to GA is an important component in α-amylase production. Across a range of barley varieties, Raynes & Briggs (1985) observed that in some cases there was almost as much α-amylase produced by grain germinating on water as by grain on a GA_3 solution. In these cases, α-amylase production may be limited by the capacity to respond to GA. In other cases, however, GA_3 treatment stimulated α-amylase production two-fold, suggesting that the level of GA was limiting rather than the response capacity.

Mutants in cv. Himalaya with altered expression of α-amylase were reported by Ho *et al.* (1980). These included grain with altered hormone (GA_3 and ABA) sensitivity as well as constitutive phenotypes. However, despite the potential value of these mutants, no follow-up studies have been reported.

One class of mutant which has not yet been characterized in any detail in barley is the GA-insensitive dwarf analogous to Rht3 in wheat, or D8 in corn. Such mutants would be valuable for a number of reasons, including their obvious use in strengthening the case for GA involvement in a response. In addition, these mutants would be extremely useful for defining the role of putative receptor proteins and components of the GA response pathway.

Applications and prospects for the future

Evidence continues to accumulate that production of α-amylase in germinated grain of barley cv. Himalaya is a GA response. It is likely that this is also true in other cultivars of barley but there are much less data. This conclusion cannot be made with any certainty until we have a better understanding of the mechanism of GA action. It may also be the case that in some circumstances GA is not the main regulator of α-amylase production. Given the variety of barley habitats and genotypes it would probably be surprising if one mechanism was predominant in all situations.

This GA-response system compares favourably with any other system in terms of understanding plant hormone action. This is despite the fact that until recently little attention had been given to whole plant (i.e. seedling) aspects. With recent analyses of GAs in germinated grain and the availability of mutants, there should now be progress in biochemical and physiological studies to match that expected from molecular studies.

Another reason for studying GA response is its relevance to both primary and secondary industries. If GA is a major regulator of α-amylase expression this information is important to breeders, farmers, bakers, maltsters and other grain users. Practical issues include a tendency for grain to show preharvest sprouting or remain dormant, early seedling vigour and extract potential during malting. It is important to establish if the genetic variation in these traits reflects altered GA activity or responses or if there are more complex explanations. This information will be relevant for attempts to genetically engineer barley when transformation becomes available.

References

Briggs, D.E. (1978) *Barley*. Chapman & Hall, London.

Brown, A.H.D. & Jacobsen, J.V. (1982) Genetic basis and natural variation of α-amylase isozymes in barley. *Genetical Research (Camb.)* 40, 315–24.

Chandler, P.M. (1988) Hormonal regulation of gene expression in the 'slender' mutant of barley (*Hordeum vulgare* L.). *Planta* 175, 115–20.

Chandler, P.M. & Jacobsen J.V. (1991) Primer extension studies on α-amylase mRNAs in barley aleurone. II. Hormonal regulation of expression. *Plant Molecular Biology* 16, 637–45.

Chandler, P.M. & Mosleth, E. (1990) Do gibberellins play an *in vivo* role in controlling α-amylase gene expression? In: Ringlund, K., Mosleth, E. & Mares, D.J. (eds), *Proceedings of the 5th International Symposium on Pre-Harvest Sprouting in Cereals*. Westview Press, Boulder, CO, USA, pp. 100–9.

Croker, S.J., Hedden, P., Lenton, J.R. & Stoddart, J.L. (1990) Comparison of gibberellins in normal and slender barley seedlings. *Plant Physiology* 94, 194–200.

Fincher, G.B. (1989) Molecular and cellular biology associated with endosperm mobilization in germinating cereal grains. *Annual Review of Plant Physiology and Plant Molecular Biology* 40, 305–46.

Foster, C.A. (1977) Slender: an accelerated extension growth mutant of barley. *Barley Genetics Newsletter* 7, 24–7.

Fujioka, S., Yamane, H., Spray, C.R., Phinney, B.O., Gaskin, P., MacMillan, J. & Takahashi, N. (1990) Gibberellin A_3 is biosynthesized from gibberellin A_{20} via gibberellin A_5 in shoots of *Zea mays* L. *Plant Physiology* 94, 127–31.

Gaskin, P., Gilmour, S.J., Lenton, J.R., MacMillan, J. & Sponsel, V.M. (1984) Endogenous gibberellins and kaurenoids identified from developing and germinating barley grain. *Journal of Plant Growth Regulation* 2, 229–42.

Hentrich, W., Bergner, Ch. & Sembdner, G. (1985) Characterization of a gibberellin sensitive dwarf mutant of barley (*Hordeum vulgare* L., *Mut Dornburg* 576). *Plant Growth Regulation* 3, 103–10.

Ho, T.-H.D., Shih S.-C. & Kleinhofs, A. (1980) Screening for barley mutants with altered hormone sensitivity in their aleurone layers. *Plant Physiology* 66, 153–7.

Huttley, A.K. & Baulcombe, D.C. (1989) A wheat α-*amy2* promoter is regulated by gibberellin in transformed oat aleurone protoplasts. *EMBO Journal* 8, 1907–13.

Jacobsen, J.V. & Beach, L.R. (1985) Control of transcription of α-amylase and rRNA genes in barley aleurone protoplasts by gibberellin and abscisic acid. *Nature* 316, 275–7.

Jacobsen, J.V. & Close, T.J. (1991) Control of transient expression of chimaeric genes by gibberellic acid and abscisic acid in protoplasts prepared from mature barley aleurone layers. *Plant Molecular Biology* 16, 713–24.

Jones, R.L. & Jacobsen, J.V. Regulation of the synthesis and transport of secreted proteins in cereal aleurone. *International Review of Cytology* (in press).

Kaufman, P.B. & Dayanandan, P. (1983) Gibberellin-induced growth in *Avena* internodes. In: Crozier, A. (ed.), *The Biochemistry and Physiology of Gibberellins*, Vol. II, Praeger, New York, USA.

Keyes, G., Sorrells, M.E. & Setter, T.L. (1990) Gibberellic acid regulates cell wall extensibility in wheat (*Triticum aestivum* L.). *Plant Physiology* 92, 242–5.

Lanahan, M.B. & Ho, D.T. (1988) Slender barley. A constitutive gibberellin-response mutant. *Planta* 175, 107–14.

Marchylo, B.A., MacGregor, A.W. & Kruger, J.E. (1985) Production of α-amylases in germinating whole and incubating de-embryonated barley kernels. *Journal of the Institute of Brewing* 91, 161–5.

McFadden, G.I., Ahluwalia, B., Clarke, A.E. & Fincher, G.B. (1988) Expression sites and developmental regulation of genes encoding (1→3, 1→4)-β-glucanases in germinated barley. *Planta* 173, 500–8.

Nicholls, P.B., MacGregor, A.W. & Marchylo, B.A. (1986) Production of α-amylase isozymes in barley caryopses in the absence of embryos and exogenous gibberellic acid. *Australian Journal of Plant Physiology* 13, 239–47.

Nolan, R.C. & Ho, T.-H.D. (1988) Hormonal regulation of gene expression in barley aleurone layers. *Planta* 174, 551–60.

Phinney, B.O. (1985) Gibberellin A_1 dwarfism and shoot elongation in higher plants. *Biologia Plantarum (Praha)* 27, 172–9.

Potts, W.C., Reid, J.B. & Murfet, I.C. (1985) Internode length in *Pisum*. Gibberellins and the slender phenotype. *Physiologia Plantarum* 63, 357–64.

Raynes, J.G. & Briggs, D.E. (1985) Genotype and the production of α-amylase in barley grains germinated in the presence and absence of gibberellic acid. *Journal of Cereal Science* 3, 55–65.

Reid, J.B. (1987) The genetic control of growth via hormones. In: Davies, P.J. (ed.) *Plant Hormones and their Role in Plant Growth and Development*. Martinus Nijhoff, Dordrecht, pp. 318–40.

Rood, S.B., Williams, P.H., Pearce, D., Murofushi, N., Mander, L.N. & Pharis, R.P. (1990) A mutant gene that increases gibberellin production in *Brassica*. *Plant Physiology* 93, 1168–74.

Smith, V.A. & MacMillan, J. (1989) An immunological approach to gibberellin purification and quantification. *Plant Physiology* 90, 1148–55.

Suge, H. (1983) Gibberellin relationships in a dwarf mutant of barley: brachytic (*br2*). *Japanese Journal of Genetics* 58, 555–66.

Takahashi, N., Yamaguchi, I. & Yamane, H. (1986) Gibberellins. In: Takahashi, N. (ed.), *Chemistry of Plant Hormones*. CRC Press, Boca Raton, FL, USA, pp. 57–151.

Trewavas, A.J. (1988) Gibberellins, amylase and germination. *BioEssays* 9, 213–14.

Yamada, K. (1982) Determination of endogenous gibberellins in germinating barley by combined gas chromatography-mass spectrometry. *Journal of American Society of Brewing Chemists* 40, 18–25.

20
Cell Wall Metabolism in Barley

Geoffrey B. Fincher

In barley, as in all plants, cell walls provide the skeletal framework and intercellular cohesion necessary for structural integrity. They also provide a balance between rigidity and flexibility that is appropriate for the functional requirements of specific tissues at different stages of development. Walls of meristematic cells are relatively thin, consisting of cellulosic microfibrils embedded in a gel-like matrix, but as tissues mature or adopt specialized roles, the walls usually thicken and additional layers may be deposited. Because cell walls surround the protoplast, they represent a potential barrier to metabolite exchange and to intercellular communication. However, the gel-like matrix of walls is hydrated and porous, and allows apoplastic diffusion of water, sucrose, amino acids, ions, phytohormones and other molecules with diameters of up to approximately 5 nm (Carpita *et al.*, 1979). In most instances, poplypeptides and other large molecules do not move freely across the wall. Indeed, the wall acts as a primary physical barrier against microbial invasion of plant cells. Since walls can contain up to 60% water, much of which is held tightly in the interstices of the gel, they may help prevent desiccation of developing tissues under conditions of mild water stress.

Cell walls are dynamic, and changes that occur during growth and differentiation involve the metabolism of both matrix phase and fibrillar components. Metabolism of the matrix phase components appears to be subject to precise positional regulation (Knox *et al.*, 1990). As might be expected, hydrolytic and oxidative enzymes responsible for wall modifications are located in the matrix (Cassab & Varner, 1988) and wall metabolism remains under the direct control of the protoplast itself, presumably through interactions between wall components and proteins of the plasma membrane (Roberts, 1990).

During primary cell wall deposition, the wall may be modified to accommodate cell elongation or expansion in a process that involves simultaneous hydrolysis or 'loosening' of wall components, and deposition of newly synthesized material. When primary wall deposition is completed, the walls may remain unmodified, as in parenchymatous tissues, or a secondary wall layer may be laid down and lignin deposited in both layers, as in sclerenchyma or other walls that need to be further thickened and strengthened. In addition, the wall may be modified in response to water, to salt or temperature stress, to wounding, or to attempted penetration by microbial pathogens. In senescing tissues or in germinated grain, cell walls may be partially or completely digested.

In this chapter, the molecular and cellular biology of cell wall metabolism in barley are considered, with emphasis on the role of (1 → 3, 1 → 4)-β-glucanases. Little is known of the synthesis and deposition of barley wall constituents, primarily because they are mostly polysaccharides for which the corresponding synthase enzymes have not been purified or characterized. However, the structural features of major components of mature walls have been defined and the enzymes that depolymerize wall polysaccharides in germinated barley grain, in particular the (1 → 3, 1 → 4)-β-glucan endohydrolases, have been studied in detail. Thus, the developmental regulation, enzymology and genetics of the (1 → 3, 1 → 4)-β-glucanases are under investigation and corresponding cDNAs and genes have been cloned. The central importance of cell wall degradation in starchy endosperm mobilization during malting, coupled with the generally undesirable consequences of residual wall polymers in the brewing process, have heightened interest in the regulation of the genes and in their potential manipulation through gene technology. In addition, wall polysaccharide fragments of both plant and microbial origin are important participants in molecular communication between plants and invading pathogenic micro-organisms (Dixon & Lamb, 1990). Because disease resistance remains one of the most important selection criteria in barley breeding programmes, the regulation of genes for enzymes that generate these wall fragments is of considerable importance. The physiology, enzymology and molecular biology of barley (1 → 3)-β-glucanases are therefore considered in relation to disease resistance, and to the evolution of gene families for enzymes involved in cell wall metabolism.

Composition and organization of barley cell walls

The plant cell wall may be considered as a reinforced, multicomponent gel in which microfibrils of non-covalently aggregated cellulose molecules are surrounded by a continuous non-covalently or covalently cross-linked gel (Fincher & Stone, 1986; Fry, 1986). Structural proteins, such as glycine-rich proteins, threonine-rich glycoproteins (Kieliszewski *et al.*, 1990) and hydroxyproline-rich glycoproteins may form a second fibrillar network within the gel matrix (Cassab & Varner, 1988). The matrix phase of the primary cell wall consists predominantly of polysaccharides with extended conformations but, in contrast to cellulose, they do not aggregate into microfibrils because of steric hindrance afforded by short side-chains on the polysaccharide backbone or by molecular kinking in the chain itself. It has been suggested that the strength, resilience, porosity and plasticity of the wall is determined by the stability and frequency of interchain polysaccharide–polysaccharide interactions (Fincher & Stone, 1986). During secondary cell wall deposition, the matrix phase of both the primary and secondary wall may become encrusted with lignin, a copolymer of coniferyl, sinapyl and *p*-hydroxycinnamyl alcohols, which is also covalently linked to the matrix phase polysaccharides themselves (Iiyama *et al.*, 1990).

Composition

Compositions of cell walls from several barley tissues are compared in Table 20.1. The walls of the aleurone and starchy endosperm of mature grain are characterized by the predominance of arabinoxylans and (1 → 3, 1 → 4)-β-glucans, which together account for up to 95% of wall polysaccharides. Pectin, xyloglucan, lignin and hydroxyproline-rich glycoproteins are not detectable. A substantial proportion of the protein found in wall preparations is of cytoplasmic origin (Fincher, 1975; Bacic &

Table 20.1. Composition of cell walls from barley.

Tissue	Neutral monosaccharide composition (%)					Protein (%)	Phenolic acids (%)	Major polysaccharide components*	References
	Ara	Xyl	Glc	Man	Gal				
Aleurone (mature grain)	24	47	26	2	2	16	1.2	71% arabinoxylan 26% (1 → 3,1 → 4)-β-glucan 2% cellulose 2% glucomannan 1% (1→ 3)-β-glucan	Bacic & Stone (1981a,b)
	(total polysaccharides)								
Starchy endosperm (mature grain)	11	11	75	3	0	5	0.05	75% (1 → 3,1 → 4)-β-glucan 20% arabinoxylan 2% cellulose 2% glucomannnan	Fincher (1975, 1976) Ballance & Manners (1978)
	(total polysaccharides)								
Coleoptiles (4 days)	23	29	43	0	6	?	?	55% cellulose 19% arabinoxylan 19% (1 → 3,1 → 4)-β-glucan ? pectin ? xyloglucan	Sakurai & Masuda (1978)
	(non-cellulosic polysaccharides)								
Stems (4th internode)	5	78	16	0.7	0.3	?	?	65% cellulose 28% heteroxylan 5% (1 → 3,1 → 4)-β-glucan	Kokubo *et al.* (1989)
	(non-cellulosic polysaccharides)								
Young leaves (top 2 cm, dark grown)	9	27	63	0	1	?	?	63% cellulose 16% (1 → 3,1 → 4)-β-glucan 11% arabinoxylan 5% pectin ? xyloglucan	N. Sakurai (pers. comm.)

*Values for polysaccharide components are estimated from compositional analyses presented in the references cited.

Stone, 1981b). However, wheat aleurone walls from which adventitious protein has been removed contain about 1% protein by weight; this integral protein is rich in proline and glycine (D. Rhodes & B.A. Stone, personal communication) and may belong to the proline- and glycine-rich protein families reported in the walls of other plants (Cassab & Varner, 1988). One would expect that a small amount of protein constitutes an integral component of barley aleurone walls. Indeed, low levels of protein are covalently linked to the $(1 \rightarrow 3, 1 \rightarrow 4)$-$\beta$-glucan in walls of starchy endosperm (Forrest & Wainwright, 1977). Walls have not been isolated from the scutellum of barley, but histochemical studies indicate that ferulic acid is present and that lignin and pectin are absent (Smart & O'Brien, 1979a, b).

Overall, the walls of the starchy endosperm and aleurone are relatively uncomplicated. The main difference between the two lies in the proportions of $(1 \rightarrow 3, 1 \rightarrow 4)$-$\beta$-glucan and arabinoxylan (Table 20.1). Very low levels of cellulose are present and the walls are not lignified. This could facilitate rapid degradation of walls during mobilization of the starchy endosperm in germinated grains (Fincher, 1989). In contrast, the walls of vegetative tissues of young barley plants have 55–65% cellulose and may contain some lignin, in keeping with the requirement of walls to provide greater structural rigidity in these tissues than is needed in the starchy endosperm. In addition to $(1 \rightarrow 3, 1 \rightarrow 4)$-$\beta$-glucans and arabinoxylans, which constitute 30–40% of the walls, pectic polysaccharides and xyloglucan may also be present (Table 20.1).

Fine structure and organization

$(1 \rightarrow 3, 1 \rightarrow 4)$-$\beta$-glucans

The $(1 \rightarrow 3, 1 \rightarrow 4)$-$\beta$-glucans of starchy endosperm and aleurone walls in barley grain are comprised of a family of polysaccharides that is heterogeneous with respect to molecular size, solubility in aqueous solvents, and molecular structure (Bacic & Stone, 1981b; Woodward & Fincher, 1983; Woodward *et al.*, 1983b). One subpopulation of the family, namely the $(1 \rightarrow 3, 1 \rightarrow 4)$-$\beta$-glucans extracted from endosperm walls with water at 40°C and precipitated with 30% saturated ammonium sulphate, has been examined in detail using chemical and enzymic procedures. These $(1 \rightarrow 3, 1 \rightarrow 4)$-$\beta$-glucans are linear polysaccharides of up to 1200 glucosyl residues, polymerized through $(1 \rightarrow 4)$ and $(1 \rightarrow 3)$ β-linkages (Woodward *et al.*, 1983a, b). The ratio of $(1 \rightarrow 4)$ to $(1 \rightarrow 3)$ linkages falls in the range 2.0–2.5 and although there is no strictly repeating structure in the polysaccharide, neither is the distribution of $(1 \rightarrow 3)$- and $(1 \rightarrow 4)$-linked β-glucosyl residues random. Thus, the $(1 \rightarrow 4)$-linked β-glucosyl units are found predominantly in blocks of two or three contiguous residues linked by single $(1 \rightarrow 3)$-β linkages; these structures account for up to 90% of the polysaccharide (Woodward *et al.*, 1983b). At a higher organizational level, however, the blocks of two and three $(1 \rightarrow 4)$-linked β-glucosyl residues are arranged in an essentially independent or random fashion (Staudte *et al.*, 1983). Approximately 10% by weight of the $(1 \rightarrow 3, 1 \rightarrow 4)$-$\beta$-glucan contains longer blocks of up to 14 adjacent $(1 \rightarrow 4)$-β-glucosyl residues (Woodward *et al.*, 1983b) and computer-assisted, theoretical conformational analysis showed not only that the longer blocks of adjacent $(1 \rightarrow 4)$ linkages are important determinants of the overall conformation of the polysaccharide, but also confirmed that contiguous $(1 \rightarrow 3)$ linkages are not present in this $(1 \rightarrow 3, 1 \rightarrow 4)$-$\beta$-glucan preparation (Buliga *et al.*, 1986).

Physicochemical methods used to define the molecular weight distribution, solution behaviour and shape of the same $(1 \rightarrow 3, 1 \rightarrow 4)$-$\beta$-glucan preparation indicate that the polysaccharide is highly asymmetric in aqueous medium, where it exists as extended, worm-like chains and forms solutions of high viscosity (Woodward *et al.*, 1983a). The high viscosity of barley $(1 \rightarrow 3, 1 \rightarrow 4)$-$\beta$-glucan solutions can cause significant technical problems in the malting and brewing industries and accounts for much of the interest in (1 → 3, 1 → 4)-β-glucan levels in barley varieties and in the potential of grain to produce rapidly (1 → 3, 1 → 4)-β-glucan hydrolases during germination.

Another barley (1 → 3, 1 → 4)-β-glucan population that has been examined in some detail is the fraction extracted from barley flour with water between 40°C and 65°C (Woodward *et al.*, 1988). This population exhibits chemical and solution properties that are similar to the fraction soluble at 40°C, but small differences in fine structure that alter the ability of chains to align into molecular aggregates are believed to be responsible for its lower solubility in aqueous medium (Woodward *et al.*, 1988).

Arabinoxylans

Arabinoxylans in walls of barley aleurone or starchy endosperm consist of a (1 → 4)-β-xylopyranosyl backbone to which single α-L-arabinofuranosyl residues are linked mostly through 03 but also through 02 of the xylosyl residues. Terminal galactosyl, glucosyl, xylosyl and glucuronopyranosyl (and its four methyl ether) residues may also be present but are quantitatively minor (Fincher, 1975; McNeil *et al.*, 1975; Ballance & Manners, 1978; Bacic & Stone, 1981b). The arabinoxylans form a heterogeneous population, with differences in degree of arabinose substitution, molecular weight, solubility and probably in the distribution of substituents along the xylan backbone. Phenolic acids, chiefly ferulic and *p*-coumaric acids, are covalently associated with arabinoxylans, the ferulic acid via ester linkages to 05 atoms of arabinosyl substituents (Gubler *et al.*, 1985). Arabinoxylans from barley endosperm have a degree of polymerization of 400–500 (Podrazky, 1964; Fincher & Stone, 1986) and although they are smaller than the (1 → 3, 1 → 4)-β-glucans, cereal arabinoxylans have an asymmetrical conformation and form solutions of high viscosity (Andrewartha *et al.*, 1979). Arabinoxylans will therefore contribute to the problems associated with high viscosity of grain extracts during the brewing process.

Molecular organization of cell wall constituents

Despite the detailed chemical and physical information available on barley (1 → 3, 1 → 4)-β-glucan fractions, cereal arabinoxylans and cellulose, the molecular organization and interactions of these components in the wall matrix and at the matrix–microfibril interface have not been clearly defined. Complex models involving covalent and non-covalent cross-linking of polysaccharide and protein constituents of plant cell walls have been proposed (Fry, 1986), together with suggestions that (1 → 3, 1 → 4)-β-glucans and arabinoxylans have the potential to form a non-covalent, three-dimensional gel network with all the properties required for cell wall function (Fincher & Stone, 1986).

Degradation of walls in germinated grain

During germination of the barley grain, hydrolytic enzymes are secreted from the aleurone and the scutellar epithelium into the starchy endosperm, where they depolymerize walls, storage proteins, starch and residual nucleic acids (Fincher, 1989). The degradation products are absorbed through the epithelial layer of the scutellum and translocated to support growth of the developing seedling. While the scutellum is involved in enzyme secretion, particularly during the early stages of germination (McFadden *et al.*, 1988), and some hydrolases pre-exist in the starchy endosperm itself in active or zymogen form, the aleurone appears to be the principal source of hydrolytic enzymes in germinating barley. Endosperm dissolution begins in the region immediately adjacent to the scutellum and progresses towards the distal end of the grain in a front that is approximately parallel to the face of the scutellum. This pattern results from the spatially and temporally co-ordinated secretion of hydrolases from the scutellum and aleurone, in a process that involves gibberellic acid and calcium ions (Deikman & Jones, 1986; Jacobsen & Chandler, 1987) but for which detailed information of precise cellular and molecular mechanisms remains elusive (Fincher, 1989).

After the synthesis of hydrolytic enzymes and their secretion across the plasma membrane of scutellar or aleurone cells, the enzymes face two physical barriers before they can reach their substrates in the starchy endosperm, namely the walls of the secretory cells themselves and the walls of starchy endosperm cells. Removal of these cell wall barriers is therefore a critically important early event that is necessary for successful germination of the grain.

Ultrastructural changes

Aleurone

In the mature barley grain, two distinct layers can be detected in aleurone walls; an inner, relatively thin layer and a thicker outer layer (Taiz & Jones, 1973; Bacic & Stone, 1981a). During germination, enzymes are released initially through channels that form in the outer wall layer (Gubler *et al.*, 1987; Fig. 20.1). Subsequently, this outer layer is degraded completely. The thin, inner layer appears to remain intact during enzyme secretion, presumably to provide continuing structural support for the aleurone protoplasts. Although the distribution of specific components between the two aleurone wall layers has not been unequivocally defined, the inner layer is probably enriched in (1 → 3, 1 → 4)-β-glucan and protein, while arabinoxylan may be the major constituent of the outer layer (B.A. Stone, pers. comm.). It remains to be demonstrated which enzymes are responsible for the degradation of the outer wall layer, although xylanases are obvious candidates (Benjavongkulchai & Spencer, 1989;

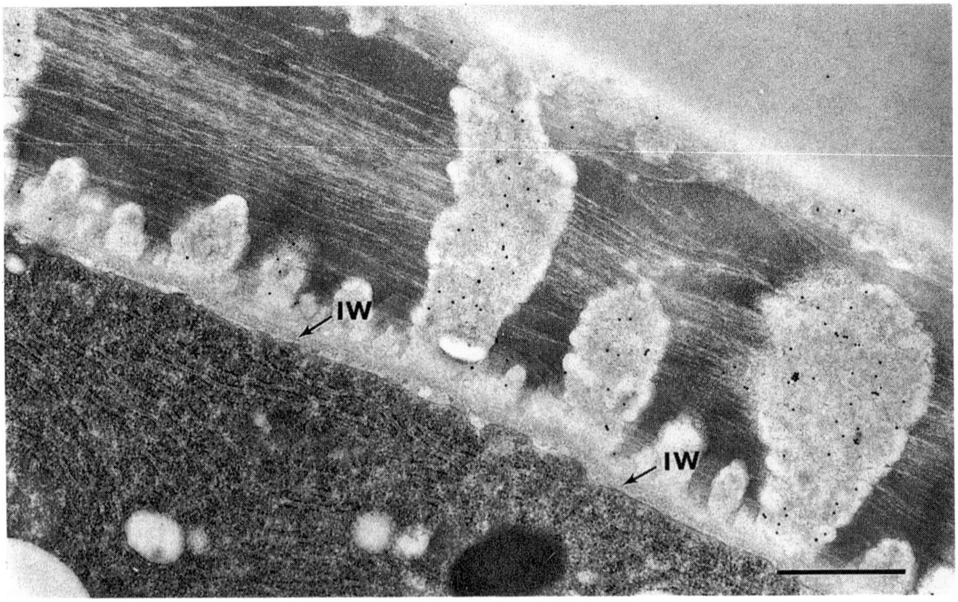

Fig. 20.1. Electron microscopy of sections from gibberellic acid-treated barley aleurone layers, probed with antibodies against α-amylase and gold-labelled protein A (from Gubler *et al.*, 1987). The gold particles are localized over large digested areas of the thick outer wall layer, indicating that the enzyme is secreted through these channels. The relatively thin, inner wall layer (IW) is retained in the digested zones, suggesting that this layer is resistant to the action of hydrolases responsible for the formation of the channels. The magnification bar is 1 μm. (The photograph was generously provided by Dr Anne Ashford.)

Fincher, 1989), how the secreted hydrolytic enzymes cross the inner wall layer, and how this inner layer remains intact in the presence of high concentrations of xylanases, (1 → 3, 1 → 4)-β-glucanases and a battery of exo- and endo-peptidases.

Scutellum

Ultrastructural studies have revealed two wall layers in cells of the scutellar epithelium (Smart & O'Brien, 1979a). During germination, individual epithelial cells become disconnected along their lateral interface to form cylindrical papillae (Nieuwdorp & Buys, 1964). This process leads to an increase in surface area of epithelial cells, which presumably enhances their ability to absorb the degradation products released by mobilization of endosperm reserves. The lateral separation of cells clearly involves partial degradation of outer wall layers, and may be analogous to aleurone wall alteration.

Starchy endosperm

During endosperm mobilization, walls of the starchy endosperm are completely degraded. Some remnants of walls are left undegraded after the initial front of wall-degrading enzymes passes, but these subsequently disappear (Selvig *et al.*, 1986). This biphasic degradation pattern would suggest temporal or spatial differences in the appearance of hydrolytic enzymes responsible for depolymerization of individual wall components.

(1 → 3, 1 → 4)-β-glucan degradation

Purification and properties of (1 → 3, 1 → 4)-β-glucanases

Two (1 → 3, 1 → 4)-β-glucan 4-glucanohydrolases (EC 3.2.1.73) can be purified from aqueous extracts of germinated barley grain (Woodward & Fincher, 1982a). The enzymes show classical endohydrolase action patterns and catalyse the hydrolysis of (1 → 4)-β-linkages in (1 → 3, 1 → 4)-β-glucans when the D-glucosyl residue is substituted at 03 (Woodward & Fincher, 1982b), as follows:

```
  ↓        ↓         ↓
---G3G4G4G3G4G4G4G3G4G4G3G---(red)
```

where G represents β-glucosyl residues, the numbers indicate the linkage positions and (red) denotes the reducing terminus of the polysaccharide chain. The enzymes therefore require adjacent (1 → 3)- and (1 → 4)-β-glucosyl residues. They do not hydrolyse (1 → 3)-β-glucans or (1 → 4)-β-glucans. The enzymes are usually assayed by their ability to decrease the viscosity of barley (1 → 3, 1 → 4)-β-glucan solutions (Woodward & Fincher, 1982a). The characteristic tri- and tetrasaccharides 3-*O*-β-cellobiosyl-D-glucose ($G4G3G_{red}$) and 3-*O*-β-cellotriosyl-D-glucose ($G4G4G3G_{red}$) are the major oligosaccharides released from (1 → 3, 1 → 4)-β-glucans of the starchy endosperm (Table 20.1).

Properties of the two barley (1 → 3, 1 → 4)-β-glucanase isoenzymes are shown in Table 20.2. The enzymes can be separated

Table 20.2. Properties of barley (1 → 3,1 → 4)-β-glucanases*

	Isoenzyme EI	Isoenzyme EII
Apparent molecular weight (SDS-PAGE)	30 000	32 000
Isoelectric point	8.5	10.6
pH optimum	4.7	4.7
Amino acids	306	306
Carbohydrate	trace (?)	4%
N-Glycosylation sites	0	1
Polyclonal antibodies	cross-react with EII	cross-react with EI
Monoclonal antibodies	specific	specific

*Values taken from Woodward *et al.*, 1982a,b; Høj *et al.*, 1990; Slakeski *et al.*, 1990.

by SDS polyacrylamide gel electrophoresis, although analysis of corresponding cDNAs shows that they both consist of 306 amino acids and that the positional identity of their amino acid sequences is 92%; most amino acid substitutions are conservative (Fincher *et al.*, 1986; Slakeski *et al.*, 1990). Differences in electrophoretic mobility of the two isoenzymes may result from differences in their isoelectric points or in their degree of glycosylation. Isoenzyme EI carries only traces of carbohydrate, while isoenzyme EII has 3.6% by weight carbohydrate, of which three residues are *N*-acetylglucosamine (Woodward & Fincher, 1982a). The cDNA sequences show that isoenzyme EII has a single, strong *N*-glycosylation site but that the site is not present in isoenzyme EI because of a nucleotide substitution in the gene at that position (Slakeski *et al.*, 1990).

Another difference between the isoenzymes that may prove to be of practical interest relates to their thermal stabilities. Isoenzyme EII is far more stable at elevated temperatures than isoenzyme EI, both in unpurified extracts of germinated barley (Loi *et al.*, 1987) or as highly purified enzyme preparations (Woodward & Fincher, 1982b). Whether the glycosylation of isoenzyme EII is responsible for its greater thermostability is not yet known.

Development of (1 → 3, 1 → 4)-β-glucanases

In ungerminated barley grain, (1 → 3, 1 → 4)-β-glucanase activity is very low or absent. After an initial lag period of 1–2 days, activity rises to a maximum 4–6 days after the initiation of germination and then declines (Ballance *et al.*, 1976; Stuart & Fincher, 1983). The separation of the two isoenzymes by SDS polyacrylamide gel electrophoresis enables the use of Western transfer procedures, coupled with monoclonal or polyclonal antibody probes, to identify and quantitate levels of the individual isoenzymes in unpurified extracts of germinated grain (Stuart & Fincher, 1983; Stuart *et al.*, 1987; Høj *et al.*, 1990). In the barley variety Clipper, isoenzymes EI and EII develop in approximately equimolar proportions, athough isoenzyme EI may appear slightly earlier than isoenzyme EII (Stuart & Fincher, 1983). Variations in absolute levels and in the ratio of individual isoenzymes are observed in other varieties (Henry, 1990).

Immunological studies with isolated scutella and aleurone layers reveal a tissue-specific difference in the development of the two (1 → 3, 1 → 4)-β-glucanase isoenzymes. Isoenzyme EI is the predominant isoenzyme secreted by isolated scutella, although this isoenzyme can also be detected in aleurone secretions. Isoenzyme EII is found exclusively in aleurone secretions (Stuart *et al.*, 1986). A third, putative (1 → 3, 1 → 4)-β-glucanase, found in secretions from scutella and tentatively designated isoenzyme III (Stuart *et al.*, 1986), is now thought to be a (1 → 3)-β-glucanase that cross-reacts with some (1 → 3, 1 → 4)-β-glucanase polyclonal antibody preparations. The level of (1 → 3, 1 → 4)-β-glucanase activity secreted per isolated scutellum is up to 40% of that secreted by a single aleurone layer (Stuart *et al.*, 1986), indicating that the scutellum is likely to be a major contributor of (1 → 3, 1 → 4)-β-glucanase during endosperm dissolution. However, it is not clear whether experiments with isolated tissue fragments accurately reflect the behaviour of the tissues *in vivo*. Using a cDNA encoding (1 → 3, 1 → 4)-β-glucanase isoenzyme EII (Fincher *et al.*, 1986), McFadden *et al.* (1988) located the corresponding mRNA transcripts in intact, germinated barley by hydridization histochemistry of grain cryosections. The cDNA probe did not differentiate between mRNAs for isoenzymes EI and EII, but it nevertheless confirmed the participation of both the scutellum and the aleurone in (1 → 3, 1 → 4)-β-glucanase synthesis *in vivo* (McFadden *et al.*, 1988). The hybridizations also revealed a temporal and spatial co-ordination of (1 → 3, 1 → 4)-β-glucanase gene regulation in intact grain that was not observed with isolated tissue fragments. Expression of (1 → 3, 1 → 4)-

β-glucanase genes is first detected in the scutellum after 1 day, when little or no expression is apparent in the aleurone. Expression is confined to the epithelial layer of the scutellum, consistent with the secretion of the enzyme into the adjacent starchy endosperm. Two days after the initiation of germination, levels of (1 → 3, 1 → 4)-β-glucanase mRNA decrease in the scutellar epithelium, but increase in the aleurone. Thereafter, (1 → 3, 1 → 4)-β-glucanase gene expression is confined to the aleurone, where it progresses from the proximal to the distal end of the grain (McFadden *et al.*, 1988).

Other enzymes in (1 → 3, 1 → 4)-β-glucan degradation

Although the (1 → 3, 1 → 4)-β-glucan endohydrolases are clearly the most important in the degradation of wall (1 → 3, 1 → 4)-β-glucan in the starchy endosperm of germinated barley, other enzymes may contribute to the depolymerization, either through hydrolysis of the polysaccharide itself or through the conversion of oligosaccharides released by (1 → 3, 1 → 4)-β-glucanase action to glucose. Relatively high levels of (1 → 3)-β-glucan endohydrolase are synthesized in germinated barley (Ballance & Svendsen, 1988; Høj *et al.*, 1988, 1989) and have been implicated in the initial degradation of cell wall (1 → 3, 1 → 4)-β-glucans (Ballance & Manners, 1978). However, pure barley (1 → 3)-β-glucanases do not hydrolyse the 40°C water-soluble (1 → 3, 1 → 4)-β-glucan (Høj *et al.*, 1988, 1989) because this substrate does not have the contiguous (1 → 3) linkages necessary for enzyme action (Woodward *et al.*, 1983b). While it is possible that other (1 → 3, 1 → 4)-β-glucan fractions from barley starchy endosperm walls do have blocks of contiguous (1 → 3) linkages that would render the polysaccharide susceptible to (1 → 3)-β-glucanase action, there is as yet no good evidence that these exist in barley (1 → 3, 1 → 4)-β-glucans or that (1 → 3)-β-glucanases function to release cell wall (1 → 3, 1 → 4)-β-glucans.

Cellulases are capable of hydrolysing barley (1 → 3, 1 → 4)-β-glucans by virtue of the adjacent (1 → 4) linkages in the substrate. However, levels of cellulase in germinating grain are low and variable, with much of the activity attributable to microorganisms that reside on the surface of the grain (Hoy *et al.*, 1981) and a function for cellulases in endosperm mobilization is not yet established.

Exo-β-glucanases capable of hydrolysing (1 → 3, 1 → 4)-β-glucan are found in extracts of germinated barley (Preece & Hoggan, 1957). An exo-β-glucanase of M_r 67 000 can rapidly degrade both (1 → 3, 1 → 4)-β-glucans and (1 → 3)-β-glucans to glucose (P.B. Høj, J. Wang & G.B. Fincher, unpublished), but its location and role in the germinated grain have not been defined. The enzyme can also hydrolyse the oligo-(1 → 3, 1 → 4)-β-glucosides released by the action of (1 → 3, 1 → 4)-β-glucan endohydrolases EI and EII, and its detection in the grain well after the initiation of germination suggests a possible role in the conversion of these oligosaccharides to glucose. The glucose would then be available for transport to the seedling. Up to 18% of glucose residues in the starchy endosperm reside in cell wall (1 → 3, 1 → 4)-β-glucan (Morall & Briggs, 1978) and one would expect an efficient enzyme system to have evolved to reclaim this glucose as an energy source for the seedling. The β-glucosidases reported in germinated barley (Manners & Marshall, 1969) might also degrade oligo-β-glucosides, but their action on tri- and tetrasaccharides would be slow and their participation in (1 → 3, 1 → 4)-β-glucan depolymerization remains to be demonstrated.

Arabinoxylan degradation

Depolymerization of wall arabinoxylans to their constituent monosaccharides is mediated by the combined action of xylan endo- and exohydrolases, xylobiase (or β-xylopyranosidase) and α-arabinofuranosidase (Preece & MacDougall, 1958), but there is very little information available on the properties of the individual enzymes involved. The ferulic acid that is esterified to aleurone wall arabinoxylan is not removed during wall hydrolysis, but is recovered as feruloylated oligosaccharides in the degradation products (Gubler *et al.*, 1985).

Three (1 → 4)-β-xylan endohydrolases (EC 3.2.1.8) have been purified 1200- to 1800-fold from extracts of grain 5 days after the initiation of germination, and additional xylanases may also be present (Slade *et al.*, 1989). Each of the three enzymes has an apparent M_r of 41 000 on SDS polyacrylamide gels and a pI value of 5.2, and all are highly unstable in dilute solution (Slade *et al.*, 1989). Sequences of the 30 NH_2-terminal amino acids are identical for each isoenzyme and although their genetic origins are not clear, it seems likely that they are the products of different genes (Slade *et al.*, 1989).

The three xylanases purified by Slade *et al.* (1989) clearly differ from a xylanase of M_r 34 000 and pI 4.6 that has been purified 29-fold from proteins secreted by gibberellic acid-treated barley aleurone layers (Benjavongkulchai & Spencer, 1986). The latter enzyme may be an additional xylan endohydrolase or a xylan exohydrolase; both classes of xylanase are secreted from isolated barley aleurone layers (Taiz & Honigman, 1976; Dashek & Chrispeels, 1977). The aleurone layer xylanase studied by Benjavongkulchai & Spencer (1989) is synthesized *de novo* in response to gibberellic acid and Ca^{2+}, and although it appears in the layers immediately, there is a long lag phase before it is released into the medium. Similarly, endoxylanases are detected in aqueous extracts of intact, germinated barley grain several days later than the (1 → 3, 1 → 4)-β-glucanases (Slade *et al.*, 1989) and this may indicate that these enzymes mediate degradation of aleurone walls, to which they remain tightly bound during channel formation and are only released when degradation of the outer layer is complete (Benjavongkulchai & Spencer, 1989; Slade *et al.*, 1989).

Structure and regulation of genes encoding barley (1 → 3, 1 → 4)-β-glucanases

Genetics

When the NH_2-terminal amino acid sequences of barley (1 → 3, 1 → 4)-β-glucanase isoenzymes EI and EII are compared, it is clear that the enzymes are the products of separate genes which probably arose by duplication of a common ancestral gene (Woodward *et al.*, 1982). Southern analyses of barley genomic DNA extracted from several cultivars and hydrolysed with several restriction enzymes have confirmed the presence of two (1 → 3, 1 → 4)-β-glucanase genes (Fincher *et al.*, 1986; Loi *et al.*, 1988; Slakeski *et al.*, 1990). The cDNA probes used in the Southern analyses detect both genes and the total copy number is estimated at less than two per haploid genome (Loi *et al.*, 1988). It seems likely, therefore, that the isoenzymes are derived from single copy, or very low copy number, genes.

Using euplasmic wheat–barley addition lines that contain a full complement of wheat chromosomes with a single additional barley chromosome (Islam & Shepherd, 1981), together with aneuploid and ditelosomic wheat lines that lack specific chromosomes or chromosome arms, it has been shown that the gene encoding barley (1 → 3, 1 → 4)-β-glucanase isoenzyme EI is probably located on the long arm of barley chromosome 5, while the isoenzyme EII gene is located on the long arm of barley chromosome 1 (Loi *et al.*, 1988; Slakeski *et al.*, 1990; Fig. 20.2).

Examination of 13 barley cultivars has revealed restriction fragment length polymorphisms (RFLPs), but these are only seen with the isoenzyme EII gene (Fig. 20.2). Within the constraints imposed by the small number of varieties analysed, it would seem that the gene for isoenzyme EI is conserved, while the isoenzyme EII gene exhibits polymorphism (Slakeski *et al.*, 1990). The RFLPs may be applied in the construction of detailed genetic maps for barley and wheat,

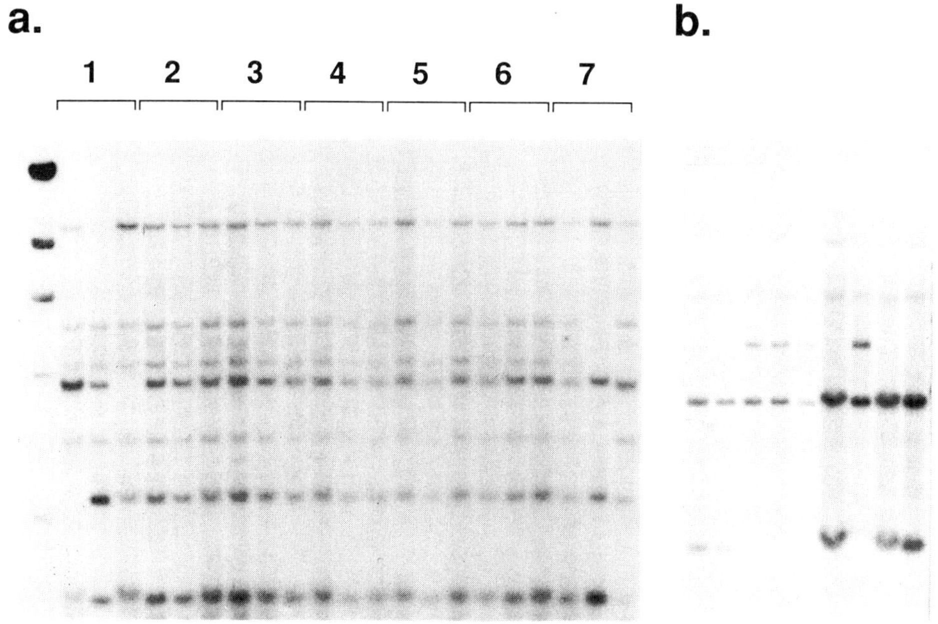

Fig. 20.2. (a) Southern analyses of wheat aneuploid line DNA digested with a restriction endonuclease and probed with a (1→3,1→4)-β-glucanase cDNA (from Slakeski *et al.*, 1990). In each aneuploid group (1–7) the three lines are nullisomic for that chromosome for the A, B, and D genomes respectively. The absence of bands from groups 1 and 7 indicates that the (1→3,1→4)-β-glucanase genes are located on these chromosomes in wheat. These are homeologous with barley chromosomes 5 and 1, respectively. It has been concluded on the basis of this and other data that the gene for barley (1→3,1→4)-β-glucanase isoenzyme EI is on the long arm of barley chromosome 5 and that the gene for (1→3,1→4)-β-glucanase isoenzyme EII is on the long arm of barley chromosome 1 (Slakeski *et al.*, 1990). (b) Southern analyses of DNA from nine barley varieties cut with a restriction enzyme and probed with a (1→3,1→4)-β-glucanase cDNA (from Slakeski *et al.*, 1990). The fragment that corresponds to the isoenzyme EI gene (3.8 kb) is conserved in these varieties, while the isoenzyme EII gene exhibits polymorphisms.

and lead to the identification of linkages between marker genes and genes of agronomic or commercial importance that might be useful in breeding programmes (Beckmann & Soller, 1986). Indeed, the ability of barley grain to synthesize high levels of (1 → 3, 1 → 4)-β-glucanases is correlated with malting performance and has been used as a selection criterion in barley breeding programmes (Stuart *et al.*, 1988).

Gene structure

The gene encoding barley (1 → 3, 1 → 4)-β-glucanase isoenzyme EI has recently been isolated and sequenced, together with near full-length cDNAs for both isoenzyme EI and EII (Slakeski *et al.*, 1990). The isoenzyme EI gene contains a single, long intron of 2514 bp inserted in the region encoding the signal peptide of the precursor polypeptide, close to the cleavage position at the NH_2-terminus of the mature enzyme (Fig. 20.3). The exon at the 5′ end of the intron is identifiable by comparison of its sequence with the 5′ sequence of a corresponding cDNA encoding isoenzyme EI (Slakeski *et al.*, 1990). However, the presence of sequence motifs for transcription, RNA processing, translation and protein processing, together with primer extension data, suggest that there may be an additional exon, with an associated promoter,

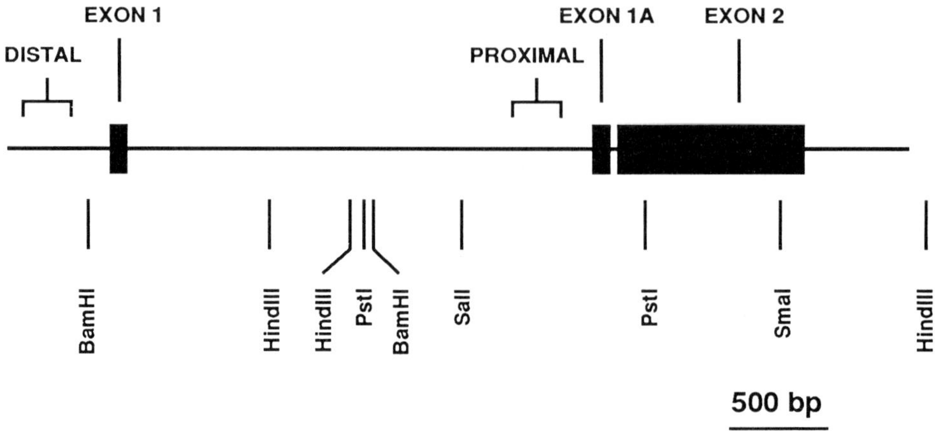

Fig. 20.3. Struture of the barley (1→3,1→4)-β-glucanase isoenzyme EI gene, showing restriction enzyme sites, the exons, and the approximate positions of the distal promoter and the putative proximal promoter (from Slakeski *et al.*, 1990).

within the large intron (Fig. 20.3). No cDNAs corresponding to transcription from this putative, proximal promoter have so far been isolated (Slakeski *et al.*, 1990). The possibility that tissue-specific and developmentally regulated transcription from the two promoters, coupled with alternative splicing of the associated intron, generates different mRNA species (Slakeski *et al.*, 1990) requires further investigation.

Within the polypeptide-coding region of both (1 → 3, 1 → 4)-β-glucanase genes a very high G/C content (67%) is attributable to the use of G or C in the third base position of more than 90% of codons (Fincher *et al.*, 1986; Slakeski *et al.*, 1990). Similar strong preferences in codon usage are found in other barley genes, in particular those that are expressed at high levels in germinating grain and are enhanced by gibberellic acid (Knox *et al.*, 1987; Whittier *et al.*, 1987; Koehler & Ho, 1990). It has been suggested that the high G/C content might contribute to mRNA stability or translational efficiency, since the rapid translation of abundant, stable mRNAs in the aleurone or scutellum of germinating cereals would clearly be advantageous in achieving rapid endosperm mobilization (Fincher, 1989; cf. Brinkmann *et al.*, 1987; Campbell & Gowri, 1990).

It is noteworthy that in the 5′ exon, codon usage is balanced with respect to the use of A/T or G/C in the wobble base position, but this changes abruptly at the position of intron insertion to yield the extreme bias mentioned above for the protein coding region of the 3′ exon (Slakeski *et al.*, 1990). This might reflect a shuffling of exons encoding specific cellular targeting signals on to the 5′ end of the coding region for the enzyme (Gilbert, 1985). Comparison of the isoenzyme EI and EII cDNAs indicates that the sequence of the signal peptide and the untranslated region immediately 5′ to it are very similar for the two genes. In contrast, the 3′ untranslated regions diverge completely, immediately after the stop codons (Slakeski *et al.*, 1990).

Regulation of expression in germinating grain

In the presence of gibberellic acid and Ca^{2+}, secretion of both (1 → 3, 1 → 4)-β-glucanase isoenzymes from isolated aleurone layers is enhanced, and isoenzyme EII may be preferentially stimulated under these conditions (Stuart *et al.*, 1986). The amounts of translatable mRNA for the enzymes in aleurone layers are also enhanced by gibberellic acid, while abscisic acid dramatically suppresses the relative abundance of (1 → 3, 1 → 4)-β-glucanase mRNA (Mundy & Fincher, 1986); the antibodies used in this work to immunoprecipate the (1 → 3, 1 → 4)-β-glucanase precursors from *in vitro* translation products of mRNA preparations did not discriminate between the individual isoenzymes. In isolated scutella, secretion of isoenzyme EI is enhanced by Ca^{2+} and possibly by gibberellic acid, but the effects of the hormone are masked by endogenous gibberellic acid in the excised scutella (Stuart *et al.*, 1986). Messenger RNAs for (1 → 3, 1 → 4)-β-glucanases have been identified in scutellar poly(A)$^+$ RNA preparations but the effects of phytohormones on their levels were not investigated (Mundy *et al.*, 1985).

The marked differences in the 3′ untranslated regions of isoenzyme EI and EII genes permit the synthesis of complementary oligonucleotides to examine specifically expression sites of the individual genes (Slakeski *et al.*, 1990). The transcription of the isoenzyme EI gene in both the scutellum and aleurone from intact, germinated grain has been confirmed, as has the transcription of the isoenzyme EII gene in the aleurone (Slakeski *et al.*, 1990).

Regulation in vegetative tissues

Leaves

During Northern analyses of (1 → 3, 1 → 4)-β-glucanase gene expression in tissues from germinated barley grain, transcription of the genes was also examined in young and mature leaves, in roots, coleoptiles and stems (Slakeski *et al.*, 1990). High levels of isoenzyme EI gene expression are detected in young leaves and are significantly higher than in the aleurone and scutellum of germinated grain (Slakeski *et al.*, 1990). No expression of isoenzyme EII occurs in young leaves at this time, neither is there expression of either (1 → 3, 1 → 4)-β-glucanase gene in mature leaves, coleoptiles, roots or stems under the growth conditions used. However, transcription of the (1 → 3, 1 → 4)-β-glucanase isoenzyme EI gene has subsequently been detected in the roots of etiolated seedlings (N. Slakeski & G.B. Fincher, unpublished data).

In the germinating grain, (1 → 3, 1 → 4)-β-glucanases clearly function to depolymerize (1 → 3, 1 → 4)-β-glucans of cell walls during endosperm mobilization. The functional significance of high levels of isoenzyme EI in young leaves is not so obvious. Time-course experiments show that (1 → 3, 1 → 4)-β-glucanase isoenzyme EI mRNA levels increase for up to 20 days in young leaves (N. Slakeski & G.B. Fincher, unpublished data). Enzyme activity in aqueous extracts of the leaves parallels the mRNA levels, indicating that the mRNA is translated into active enzyme. The relatively high levels of enzyme activity and isoenzyme EI mRNA suggest that large-scale cell wall degradation or turnover is occurring in the leaves at this time. The (1 → 3, 1 → 4)-β-glucanase in developing leaves may function in the formation of intercellular airspaces that are necessary for diffusion of carbon dioxide, oxygen and water vapour in parts of the leaf that initially have no direct access to the atmosphere. Airspaces, or aerenchyma, develop by selective cell dissolution (lysogeny) or more frequently by partial cell separation (schizogeny) (Fincher & Stone, 1981). Selective wall dissolution between palisade cells and mesophyll cells is observed during airspace formation in developing leaves of *Phaseolus vulgaris* (Jeffree *et al.*, 1986). The formation of this apoplastic compartment by partial cell separation is clearly under precise control, both with respect to the position and extent of separation, and to the ability of the plant to stop the separation before tissue disintegration occurs (Roberts, 1990). Monoclonal antibodies against barley (1 → 3, 1 → 4)-β-glucanase isoenzyme EI (Høj *et al.*, 1990) can now be used to investigate whether the enzyme functions in airspace formation.

Coleoptiles and roots

Although $(1 \rightarrow 3, 1 \rightarrow 4)$-$\beta$-glucanase mRNA was not detected in barley coleoptiles or mature roots (Slakeski *et al.*, 1990), 'loosening' of the cell wall matrix is known to occur during cell expansion (Hoson & Nevins, 1989) and there is a progressive, auxin-inducible decrease in $(1 \rightarrow 3, 1 \rightarrow 4)$-$\beta$-glucan content in barley coleoptiles with time (Sakurai & Masuda, 1978). Both of these observations would imply a potential role for $(1 \rightarrow 3, 1 \rightarrow 4)$-$\beta$-glucanases in loosening the wall matrix, but the levels of enzyme required for such loosening would probably be low.

Regulation of the two $(1 \rightarrow 3, 1 \rightarrow 4)$-$\beta$-glucanase genes in barley is under tissue-specific and hormonal control. The gene for isoenzyme EII is transcribed in the aleurone layer, probably under the influence of gibberellic acid, and at this stage can be classified as 'germination-specific'. The gene for isoenzyme EI is expressed in a number of tissues and appears to be subject to a more complex control mechanism. The isoenzyme EI gene is expressed in the aleurone layer of germinated grain, but stimulation by gibberellic acid has not been unequivocally demonstrated. Other stimuli are likely to control the expression of isoenzyme EI (but not isoenzyme EII) in the scutellum, in developing leaves and in elongating roots. Auxin and ethylene are strong candidates for the stimulus in these tissues. The availability of the isoenzyme EI gene will permit fusion of its promoter region on to reporter genes such as β-glucuronidase for the investigation of *cis*- and *trans*-acting factors that participate in the tissue-specific and hormonal regulation of the gene.

Wall degradation and malting quality

Barley quality encompasses a range of physical and chemical attributes, depending on whether the grain is to be used in the preparation of malt for brewing purposes, in the formulation of stockfeed or as a component of human foods. Currently, specifications of barley quality are tailored primarily for the malting and brewing industries, and include such parameters as grain size, dormancy, malt extract, grain protein content, development of enzymes for starch degradation in malt and $(1 \rightarrow 3, 1 \rightarrow 4)$-$\beta$-glucan content. Malt extract is a widely used quality indicator. It is an estimate of the percentage of malted grain that can be extracted with hot water. Barley breeders and growers strive to produce grain with high malt extract values, because greater extract percentages provide higher levels of materials for subsequent fermentative growth by yeast during brewing. Malt extract values are influenced both by the composition of the ungerminated barley and by the speed and extent of endosperm modification during malting. Given the central role of cell walls as a potential barrier against the free diffusion of starch- and protein-degrading enzymes from the scutellum or from the aleurone to their substrates in cells of the starchy endosperm, it is not surprising that wall composition and the ability of the grain to produce rapidly enzymes that hydrolyse wall constituents are important determinants of malt extract values.

$(1 \rightarrow 3, 1 \rightarrow 4)$-$\beta$-glucan and $(1 \rightarrow 3, 1 \rightarrow 4)$-$\beta$-glucanase levels

Barley $(1 \rightarrow 3, 1 \rightarrow 4)$-$\beta$-glucans are generally regarded as undesirable components in the malting and brewing processes. Firstly, high levels of $(1 \rightarrow 3, 1 \rightarrow 4)$-$\beta$-glucan in malt reflect incomplete cell wall degradation and an attendant diminution of storage polymer mobilization, and are therefore associated with lowered malt extract values. A second undesirable property is related to their propensity to form aqueous solutions of high viscosity. Inadequate degradation of $(1 \rightarrow 3, 1 \rightarrow 4)$-$\beta$-glucans during malting may produce viscous extracts that lead to filtration difficulties in the brewery. Furthermore, undegraded $(1 \rightarrow 3, 1 \rightarrow 4)$-$\beta$-glucan in beer may precipitate at elevated ethanol concentrations or at low temperatures, thus contributing to the formation of certain hazes or precipitates in beer (Woodward & Fincher, 1983; Bamforth, 1985).

High (1 → 3, 1 → 4)-β-glucan levels in malt can be attributed to a high initial (1 → 3, 1 → 4)-β-glucan content of the barley sample, to an inability of the grain to synthesize sufficient (1 → 3, 1 → 4)-β-glucanase for the enzymic depolymerization of the polysaccharide during malting and mashing, to the addition of unmalted barley or other cereals as a mash adjunct, or to a combination of these. Levels of (1 → 3, 1 → 4)-β-glucan, which range from 2 to 10% by weight of the grain but usually fall between 3 and 6% (Fincher & Stone, 1986), are determined by both genetic factors and environmental conditions during grain ripening. Genetic factors appear to be more important than environmental factors as determinants of (1 → 3, 1 → 4)-β-glucan content (Henry, 1986; Loi *et al.*, 1988). Genetic analysis using random inbred lines produced by doubled haploidy and single-seed descent indicates that (1 → 3, 1 → 4)-β-glucan content is controlled by a simple additive genetic system of three to five effective factors (Powell *et al.*, 1989).

Total (1 → 3, 1 → 4)-β-glucan in ungerminated barley grain is not highly correlated with malt extract (Henry, 1986; Loi *et al.*, 1988). However, the residual (1 → 3, 1 → 4)-β-glucan in malted barley is highly correlated, in a negative sense, with malt extract (Bourne *et al.*, 1982; Henry, 1986; Loi *et al.*, 1988) and this residual polysaccharide reflects a combination of the initial (1 → 3, 1 → 4)-β-glucan levels in the barley and, more importantly, the capacity of the grain to produce high levels of (1 → 3, 1 → 4)-β-glucanase during malting (Loi *et al.*, 1988; Table 20.3). The (1 → 3, 1 → 4)-β-glucanase potential of barley cultivars is also dependent on both genotype and environment, although environmental conditions during grain maturation appear to be particularly important in the development of the enzymes (Loi *et al.*, 1988). Monoclonal antibodies specific for barley (1 → 3, 1 → 4)-β-glucanases have been used in enzyme-linked immunosorbent assays (ELISA) that may be useful for the quantitation of (1 → 3, 1 → 4)-β-glucanase levels in large numbers of barley lines generated in breeding programmes (Høj *et al.*, 1990). Furthermore, mutant barleys with altered (1 → 3, 1 → 4)-β-glucan content (Aastrup, 1983; Molina-Cano *et al.*, 1989) or (1 → 3, 1 → 4)-β-glucanase potential will be useful in future studies on the effects of these components on malting quality and may be valuable in breeding programmes.

It is worth re-emphasizing that the (1 → 3, 1 → 4)-β-glucanases in germinated grain, or green malt, are inactivated at elevated temperatures (Woodward & Fincher, 1982b). More than 60% of (1 → 3, 1 → 4)-β-glucanase activity in green malt is destroyed during commercial kilning processes and the remaining activity, which is attributable entirely to isoenzyme EII, is quickly lost during laboratory-scale mashing experiments at 55 or 65°C. Significant activity is retained at 45°C in these simulated mashing trials (Loi *et al.*, 1987). It is clear, therefore, that while levels of (1 → 3, 1 → 4)-β-glucanases that develop during

Table 20.3. Linear correlations (r) between barley quality parameters.

	Total β-glucan	Malt β-glucan	% Loss β-glucan	β-glucanase	Malt extract
Total β-glucan	1.00	0.18	0.22	0.27	0.15
Malt β-glucan	–	1.00	−0.92*	−0.61	−0.74*
% loss β-glucan	–	–	1.00	0.72*	0.81*
β-glucanase	–	–	–	1.00	0.85*
Malt extract	–	–	–	–	1.00

*Significant to $P < 0.01$ (Stuart *et al.*, 1988).

germination are correlated with malting quality, it may also be advantageous to manipulate kilning protocols to retain as much enzyme activity as possible in the dry malt, particularly when initial mashing temperatures are sufficiently low to allow further action of the enzyme on residual (1 → 3, 1 → 4)-β-glucan.

Barley (1 → 3, 1 → 4)-β-glucans also pose problems in the stockfeed industry. The elevated viscosity of the gut contents of chickens fed barley diets impairs the diffusion of digestive enzymes and the mixing of gut contents, resulting in slower growth rates and in sticky fecal droppings that make hygienic handling of eggs and carcases difficult (Burnett, 1966; White *et al.*, 1983; Fincher & Stone, 1986). Conversely, cereal (1 → 3, 1 → 4)-β-glucans contribute to the salutary effects of dietary fibre in the human diet, where they have been implicated as potential modulators of blood glucose, blood cholesterol, hormone responses, colonic cancer, and micronutrient availability (Klopfenstein, 1988; Bengtsson *et al.*, 1990).

Arabinoxylan and xylanase levels

In contrast to the (1 → 3, 1 → 4)-β-glucans and (1 → 3, 1 → 4)-β-glucanases, little attention has been afforded to the effects of arabinoxylan levels or the relative importance of enzymes that catalyse arabinoxylan hydrolysis on barley quality. Because arabinoxylans are major constituents of the starchy endosperm cell wall and because they exhibit solution properties similar to the (1 → 3, 1 → 4)-β-glucans, they are likely to be important contributors to malting and brewing characteristics that are generally attributed to the (1 → 3, 1 → 4)-β-glucans. Arabinoxylan contents of 4.4–7.8% by weight of barley grain have been reported, and the values are influenced not only by genotype, but also by environmental conditions (Henry, 1986). The initial grain arabinoxylan content is poorly correlated with malt extract, but much of the arabinoxylan is associated with the husk and other outer layers of the grain (Fincher & Stone, 1986), and analysis of changes in the starchy endosperm component of arabinoxylans during malting is therefore difficult (Henry, 1988).

Application of gene technology

The availability of the gene encoding (1 → 3, 1 → 4)-β-glucanase isoenzyme EI and near full-length cDNAs for isoenzymes EI and EII (Slakeski *et al.*, 1990) presents an opportunity to accelerate or enhance (1 → 3, 1 → 4)-β-glucanase development in germinated grain through gene technology. Increased enzyme activity might be achieved by several means, for example by splicing more efficient promoters onto the gene, by altering the existing promoter to enhance expression levels, by the use of translational enhancers, or by increasing the copy number of the genes. While these options are theoretically plausible, there are practical constraints that must first be overcome. Detailed information on the nucleotide sequence elements responsible for transcription of the (1 → 3, 1 → 4)-β-glucanase genes is not available, nor do we have a thorough understanding of the mechanisms of gene regulation, either at the molecular or the cellular level. Potential *trans*-acting protein factors have not been characterized. The isolation of the isoenzyme EI gene will enable these investigations to proceed, but the gene for isoenzyme EII has not been isolated and the relative importance of each gene in malting quality is still not known. A more fundamental shortcoming that currently limits the genetic engineering approach is our inability to transform barley by stable integration of foreign DNA into the genome, and to regenerate fertile plants from single transformed cells. However, recent successes in the regeneration of transformed rice (Datta *et al.*, 1990) and wheat (Vasil *et al.*, 1990), together with the regeneration of barley plants from embryogenic cultures (Luhrs & Lorz, 1987) are encouraging, and it appears likely that stable transformation of barley will soon be possible on a routine basis. Manipulated genes could then be re-introduced into the barley genome for evaluation of their effects on phenotypic characteristics. This technology may be further enhanced through the development of methods for targeting the introduced DNA to regions of the genome that will permit appropriate expression.

Another strategy that might be applied to increase (1 → 3, 1 → 4)-β-glucanase levels in malt is to engineer increased thermostability into the enzymes themselves. In this way the loss of activity suffered during kilning and mashing might be alleviated, allowing the enzymes more time to depolymerize residual (1 → 3, 1 → 4)-β-glucan during mashing. Although the basis of protein stability is still being defined and the constraints associated with barley transformation remain to be overcome, this approach may ultimately prove more effective than promoter manipulations in attempts to increase (1 → 3, 1 → 4)-β-glucanase activity in malt.

Yet another strategy to overcome the difficulties caused by residual (1 → 3, 1 → 4)-β-glucans in malt extracts involves tackling the problem during the fermentation stage of the brewing process. There has been considerable interest in inserting (1 → 3, 1 → 4)-β-glucanase genes into brewing yeasts, in the expectation that low-level, constitutive expression would lead to the secretion of active enzyme and the depolymerization of residual (1 → 3, 1 → 4)-β-glucan during fermentation (Hinchliffe, 1988). A barley (1 → 3, 1 → 4)-β-glucanase cDNA (Fincher *et al.*, 1986) fused with a mouse α-amylase signal peptide is expressed and secreted from yeast under the direction of the yeast alcohol dehydrogenase I gene promoter (Jackson *et al.*, 1986). Similarly, a (1 → 3, 1 → 4)-β-glucanase gene from *Bacillus subtilis* (Murphy *et al.*, 1984) is expressed at a low level in yeast (Hinchliffe & Box, 1984). However, the pH optimum of the bacterial enzyme is 6.0–6.5, compared with a value of 4.7 for the barley enzyme. Because the pH of malt extracts decreases from values of 5.2–5.7 to 4.0–4.5 during the fermentation phase of the brewing process, engineering of yeast with the barley (1 → 3, 1 → 4)-β-glucanase gene appears to be more appropriate (Fincher, 1987; Hinchliffe, 1988), although the pH optimum of the bacterial enzyme might also be engineered.

Role of (1 → 3)-β-glucanases

In addition to the (1 → 3, 1 → 4)-β-glucanases, high levels of (1 → 3)-β-glucan 3-glucanohydrolase (EC 3.2.1.39) are detected in extracts of germinated barley grain (Manners & Marshall, 1969; Manners & Wilson, 1974). Activity is detected in the embryo and scutellum of ungerminated grain and rises markedly in the aleurone during germination (Ballance *et al.*, 1976). The enzyme is also secreted by isolated aleurone layers in response to gibberellic acid (Taiz & Jones, 1973).

The (1 → 3)-β-glucanases exhibit a substrate specificity that is related to, yet quite distinct from, the (1 → 3, 1 → 4)-β-glucanases. They hydrolyse (1 → 3)-β-glucosyl linkages in (1 → 3)-β-glucans in an essentially random fashion. The enzymes usually require a short segment of unbranched, adjacent (1 → 3)-β-glucosyl residues for hydrolysis. Their action is generally measured reductometrically, using soluble (1 → 3)-β-glucans such as laminarin as substrates. Purified barley (1 → 3)-β-glucanases have no action on (1 → 3, 1 → 4)-β-glucans that contain no contiguous (1 → 3)-β linkages (Woodward *et al.*, 1983b; Høj *et al.*, 1988, 1989).

Enzymic properties

Three (1 → 3)-β-glucanases purified from extracts of germinated barley have been designated isoenzyme GI (Høj *et al.*, 1988), isoenzyme GII (Ballance & Svendsen, 1988; Høj *et al.*, 1989) and isoenzyme GIII (P.B. Høj, J. Wang & G.B. Fincher, unpublished data), in accordance with their increasing isoelectric points. No attempt has been made to dissect the various tissues of the germinated grain before purification, so the precise location of each enzyme cannot be defined. The three purified isoenzymes are basic proteins with apparent M_r

values of 32 000 on SDS polyacrylamide gels, and are almost identical in size to the barley (1 → 3, 1 → 4)-β-glucanases.

The (1 → 3)-β-glucanase isoenzyme GI has a pH optimum of 5.6 (Høj *et al.*, 1988) and all three exhibit action patterns typical of polysaccharide endohydrolases, releasing laminaribiose and laminaritriose as major products from laminarin.

Gene structure and evolution

Southern analyses of genomic DNA probed with a near full-length cDNA for (1 → 3)-β-glucanase isoenzyme GII (Høj *et al.*, 1989) suggest that the barley (1 → 3)-β-glucanases are encoded by a family of approximately six genes. The exact number of genes is difficult to define because they show considerable sequence diversity and at the lower stringency required to detect all members of the family the probe begins to bind to (1 → 3, 1 → 4)-β-glucanase and possibly other gene sequences (P. Xu, J. Wang & G.B. Fincher, unpublished data). In addition to the cDNA encoding (1 → 3)-β-glucanase isoenzyme GII (Høj *et al.*, 1989), a gene for isoenzyme GIII (P. Xu, J. Wang & G.B. Fincher, unpublished data) has been isolated and characterized. Another gene has also been isolated and has been tentatively designated isoenzyme GIV, although the corresponding enzyme has not yet been purified (P. Xu, J. Wang & G.B. Fincher, unpublished data). In both genes, a single intron splits a codon in a putative signal peptide, in a position that is almost identical to the position of intron insertion in the (1 → 3, 1 → 4)-β-glucanase isoenzyme EI gene (Slakeski *et al.*, 1990). Further, the intron marks a transition point in codon usage in the two (1 → 3)-β-glucanase genes; in the 5′ exon a balanced codon usage is observed, while in the protein-coding region of the larger 3′ exon there is a strong bias towards the use of G and C in the wobble base position of codons (P. Xu, J. Wang & G.B. Fincher, unpublished data). A similar transition in codon usage can be detected in the isoenzyme GII cDNA (Høj *et al.*, 1989), suggesting that an intron might also be inserted in the corresponding gene at the same position.

Nucleotide sequence analysis of the two genes and the cDNA has allowed the complete primary structures of three (1 → 3)-β-glucanase isoenzymes to be deduced. The three enzymes are similar in length, although the sequence deduced from the putative isoenzyme GIV gene has an extra 19 amino acids at the COOH-terminus. Only one *N*-glycosylation site is apparent in the isoenzyme GIV gene sequence, and this is located in the COOH-terminal extension. Properties of the (1 → 3)-β-glucanase isoenzymes are shown in Table 20.4.

Within the barley (1 → 3)-β-glucanase isoenzymes GII, GIII and GIV, many homo-

Table 20.4. Properties of barley (1 → 3)-β-glucanase isoenzymes.

Property	Isoenzyme GI	Isoenzyme GII	Isoenzyme GIII	Isoenzyme GIV
Molecular weight	32 000	32 000	32 000	34 000
Isoelectric point	8.5	10.0	10.1	11.5
Amino acids	306(?)	306	305	327
Glycosylation sites	1	0	5	1
Transcription sites	Young roots Young leaves	Aleurone	Young roots Young leaves	Aleurone

From Høj *et al.* (1988, 1989) and P. Xu, J. Wang & G.B. Fincher (unpublished data).

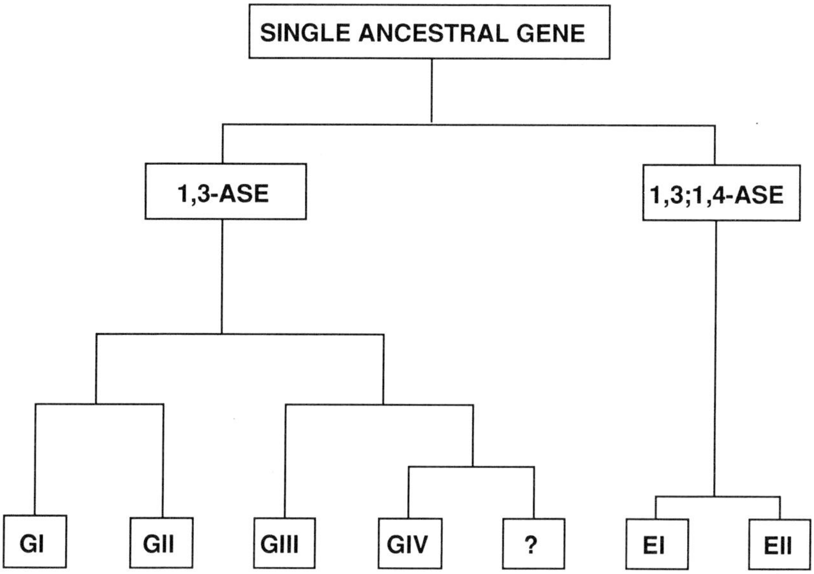

Fig. 20.4. Possible evolutionary relationships of the barley (1→3,1→4)-β-glucanase and (1→3)-β-glucanase gene families.

logous regions of up to 12 identical amino acids are interspersed with variable regions. The overall positional identity is 53% at the amino acid level. Individual (1 → 3)-β-glucanase isoenzymes exhibit 50–60% positional identity with the amino acid sequences of barley (1 → 3, 1 → 4)-β-glucanases (Høj *et al.*, 1989; P. Xu, J. Wang & G.B. Fincher, unpublished data).

The similarities in amino acid sequences, enzyme length, intron position and patterns of codon usage in the genes strongly suggest that the barley (1 → 3)-β-glucanase gene family shares a common evolutionary origin with the (1 → 3, 1 → 4)-β-glucanase gene family (Fig. 20.4). The (1 → 3)-β-glucanase family is now larger and its individual genes show more sequence divergence than the (1 → 3, 1 → 4)-β-glucanase genes.

Expression sites

Barley (1 → 3)-β-glucanase genes are subject to tissue-specific regulation. Northern analyses show that isoenzyme GII is expressed predominantly in the aleurone of germinated barley (P. Xu, J. Wang & G.B. Fincher, unpublished data), and presumably corresponds to the enzyme that is secreted from isolated aleurone layers under the influence of gibberellic acid (Taiz & Jones, 1973). No expression could be detected in scutella, coleoptiles, roots, leaves or stems. Isoenzyme GIV is also expressed predominantly in the aleurone (P. Xu, J. Wang & G.B. Fincher, unpublished data). The genes for isoenzyme GI and GIII are transcribed in young roots and young leaves.

Functional significance

The functional role of the individual (1 → 3)-β-glucanase isoenzymes in germinated barley grain or in other tissues is not yet defined. Endogenous (1 → 3)-β-glucan is restricted to small callosic deposits that are scattered throughout the starchy endosperm (MacGregor *et al.*, 1989) but the low abundance of these deposits does not seem to warrant the relatively high levels of enzyme produced. The enzymes function in plant physiological processes, such as pollen formation, the removal of dormancy or wound callose, and senescence (Fincher & Stone, 1981), but these processes are not occurring in the tissues so far examined. The (1 → 3)-β-glucanases can, however, hydrolyse the (1 → 3)- and (1 → 3, 1 → 6)-β-glucans that are major cell wall constituents of many fungi (Wessels & Sietsma, 1981), which suggests that the enzymes could participate in a general, non-specific protection of the germinating grain against pathogen invasion, at a time when high concentrations of sugars, oligosaccharides, amino acids, peptides and other nutrients render the grain particularly vulnerable to microbial infection (Fincher, 1989; Høj *et al.*, 1989). There is considerable interest in this possibility in many other plants, where (1 → 3)-β-glucanases are important members of the soluble 'pathogenesis-related' (PR) proteins that are synthesized when plant tissues are challenged with viruses, viroids, fungi or bacteria (Boller, 1987; Dixon & Lamb, 1990). The (1 → 3)-β-glucanases may combine with other hydrolases to degrade hyphal cell walls directly, thereby causing lysis of hyphal tips (Mauch *et al.*, 1988), but are perhaps more likely to function by partially hydrolysing the (1 → 3, 1 → 6)-β-glucans of fungal walls to release oligosaccharides that elicit further reactions in defence of the plant against microbial colonization (Boller, 1987; Dixon & Lamb, 1990).

It should be noted that isoenzymes GI, GII and GIII have been purified from barley grain germinated under rigorously aseptic conditions and RNA used in Northern analyses was also isolated from plants grown under sterile conditions (P. Xu, J. Wang & G.B. Fincher, unpublished data). The pre-existence of the enzymes in tissues particularly susceptible to microbial penetration would ensure an instant or at least very rapid response to microbial attack and would enhance the plant's ability to resist infection. The presence of other components of plant defence systems in sterile, germinated barley grains, including chitinases, thionins, inhibitors of exogenous hydrolytic enzymes and protein synthesis inhibitors, lends weight to the argument that (1 → 3)-β-glucanases represent important components of a complex defence mechanism aimed at countering pathogen invasion (Fincher, 1989). In preliminary experiments with mature barley leaves infected with a pathogenic fungus, very high levels of the isoenzyme GII mRNA are specifically expressed; this mRNA is not detected in uninfected leaves (P. Xu & G.B. Fincher, unpublished data).

As discussed above, the barley (1 → 3)-β-glucanase and (1 → 3, 1 → 4)-β-glucanase gene families probably share a common evolutionary pathway, and the divergent substrate specificities of the two classes of enzyme may have evolved following duplication of a single ancestral gene (Høj *et al.*, 1988; Fig. 20.4). This possibility raises the question as to whether the (1 → 3, 1 → 4)-β-glucanase activity preceded the (1 → 3)-β-glucanase activity, on the basis of its central importance in endosperm mobilization (Høj *et al.*, 1988), or vice versa, on the basis of the grain's need to resist pathogen attack. Perhaps the putative ancestral enzyme could hydrolyse either (1 → 3)-β-glucans or (1 → 3, 1 → 4)-β-glucans and was therefore able to perform both functions. Interestingly, there is an analogous 'modern day' enzyme, the (1 → 3)/(1 → 3, 1 → 4)-β-glucan 3(4)-glucanohydrolase (EC 3.2.1.6) from *Rhizopus arrhizus*, that can hydrolyse both (1 → 3)- and (1 → 3, 1 → 4)-β-glucans (Anderson & Stone, 1975). Divergence of substrate specificity from such an ancestral enzyme would permit independent regulation of genes encoding enzymes for endosperm wall hydrolysis and for fungal wall hydrolysis (Høj *et al.*, 1988).

The functional need for multiple isoforms of the (1 → 3)-β-glucanases also remains a matter for speculation. Multiple genes would obviously permit independent control of individual gene expression in different tissues. Multiple genes would also permit individual isoenzymes to be targeted to different subcellular or extracellular sites. The characteristic signal peptide and short COOH-terminal extension of (1 → 3)-β-glucanase isoenzyme GIV are suggestive of a vacuolar location in barley leaves and roots, while isoenzyme GII is probably secreted into the starchy endosperm from aleurone cells during germination (P. Xu, J. Wang & G.B. Fincher; unpublished data). Finally, the presence of multiple genes may reflect the evolution of (1 → 3)-β-glucanases with slightly different or refined substrate specificities, in particular with regard to their ability to hydrolyse (1 → 3)-β-glucans as the degree of main chain substitution or branching increases. This possibility can be tested experimentally, and if particular isoenzymes do show preferences for branched or substituted (1 → 3)-β-glucans, one might speculate that multiple isoforms of the enzymes permit the hydrolysis of a diverse and continually evolving population of fungal cell wall β-glucans, from a broad spectrum of potential pathogens.

Acknowledgements

I am indebted to Danny Doan, Peter Høj, Nada Slakeski, Peilin Xu and Jun Wang for their enthusiastic pursuit of barley β-glucanase enzymes, cDNAs and genes, and to Dr Naoki Sakurai for providing unpublished data on the composition of cell walls from barley leaves. The support of the Australian Research Council and the Barley Research Council of Australia is gratefully acknowledged. I wish to thank Dr David Baulcombe, Sainsbury Laboratory, Norwich, UK, for his support and assistance.

References

Aastrup, S. (1983) Selection and characterization of low β-glucan mutants in barley. *Carlsberg Research Communications* 48, 307–16.

Anderson, M.A. & Stone, B.A. (1975) A new substrate for investigating the specificity of β-glucan hydrolases. *FEBS Letters* 52, 202–7.

Andrewartha, B., Phillips, D.R. & Stone, B.A. (1979) Solution properties of wheat-flour arabinoxylans and enzymically modified arabinoxylans. *Carbohydrate Research* 77, 191–204.

Bacic, A. & Stone, B.A. (1981a) Isolation and ultrastructure of aleurone cell walls from wheat and barley. *Australian Journal of Plant Physiology* 8, 453–74.

Bacic, A. & Stone, B.A. (1981b) Chemistry and organization of aleurone cell wall components from wheat and barley. *Australian Journal of Plant Physiology* 8, 475–95.

Ballance, G.M. & Manners, D.J. (1978) Structural analysis and enzymic solubilization of barley endosperm cell walls. *Carbohydrate Research* 61, 107–18.

Ballance, G.M. & Svendsen, I. (1988) Purification and amino acid sequence determination of an endo, 1,3-β-glucanase from barley. *Carlsberg Research Communications* 53, 411–19.

Ballance, G.M., Meredith, W.O.S. & Laberge, D.E. (1976) Distribution and development of endo-β-glucanase activities in barley tissues during germination. *Canadian Journal of Plant Science* 56, 459–66.

Bamforth, C.W. (1985) Biochemical approaches to beer quality. *Journal of the Institute of Brewing* 91, 154–60.

Beckmann, J.S. & Soller, M. (1986) Restriction fragment length polymorphisms in plant genetic improvement. *Oxford Surveys in Plant Molecular Cell Biology* 3, 196–250.

Benjavongkulchai, E. & Spencer, M.S. (1986) Purification and characterization of barley-aleurone xylanase. *Planta* 169, 415–19.

Benjavongkulchai, E. & Spencer, M.S. (1989) Barley aleurone xylanase: its biosynthesis and possible role. *Canadian Journal of Botany* 67, 297–302.

Bengtsson, S., Aman, P., Graham, H., New, C.W. & Newman, R.K. (1990) Chemical studies on mixed-linked β-glucans in hull-less barley cultivars giving different hypocholesterolaemic responses in chickens. *Journal of the Science of Food and Agriculture* 52, 435–45.

Boller, T. (1987) Hydrolytic enzymes in plant disease resistance. In: Kosuge, T. & Nester, E.W. (eds), *Plant–Microbe Interactions: Molecular and Genetic Perspectives*. Macmillan, New York, Vol. 2, pp. 385–413.

Bourne, D.T., Powlesland, T. & Wheeler, R.E. (1982) The relationship between total β-glucan of malt and malt quality. *Journal of the Institute of Brewing* 88, 371–5.

Brinkmann, H., Martinez, P., Quigley, F., Martin, W. & Cerff, R. (1987) Endosymbiotic origin and codon bias of the nuclear gene for chloroplast glyceraldehyde-3-phosphate dehydrogenase from maize. *Journal of Molecular Evolution* 26, 320–8.

Buliga, G.S., Brant, D.A. & Fincher, G.B. (1986) The sequence statistics and solution configuration of barley (1 → 3, 1 → 4)-β-D-glucan. *Carbohydrate Research* 157, 139–56.

Burnett, G.S. (1966) Studies of viscosity as the probable factor involved in the improvement of certain barleys for chickens by enzyme supplementation. *British Poultry Science* 7, 55–75.

Campbell, W.H. & Gowri, G. (1990) Codon usage in higher plants, green algae, and cyanobacteria. *Plant Physiology* 92, 1–11.

Carpita, N., Sabularse, D., Montezinos, D. & Delmer, D. (1979) Determination of the pore size of cell walls of living plant cells. *Science* 205, 1144–7.

Cassab, G.I. & Varner, J.E. (1988) Cell wall proteins. *Annual Review of Plant Physiology and Plant Molecular Biology* 39, 321–53.

Dashek, W.V. & Chrispeels, M.J. (1977) Gibberellic-acid-induced synthesis and release of cell-wall degrading endoxylanase by isolated aleurone layers of barley. *Planta* 134, 251–6.

Datta, S.K., Peterhans, A., Datta, K. & Potrykus, I. (1990) Genetically engineered fertile indica-rice recovered from protoplasts. *Biotechnology* 8, 736–40.

Deikman, J. & Jones, R.L. (1986) Regulation of the accumulation of mRNA for α-amylase isoenzymes in barley aleurone. *Plant Physiology* 80, 672–5.

Dixon, R.A. & Lamb, C.J. (1990) Molecular communication in interactions between plants and microbial pathogens. *Annual Review of Plant Physiology and Plant Molecular Biology* 41, 339–67.

Fincher, G.B. (1975) Morphology and chemical composition of barley endosperm cell walls. *Journal of the Institute of Brewing* 81, 116–22.

Fincher, G.B. (1976) Ferulic acid in barley cell walls: a fluorescence study. *Journal of the Institute of Brewing* 82, 347–9.

Fincher, G.B. (1987) Molecular biology of barley (1 → 3, 1 → 4)-β-glucanases. In: Mares, D.J. (ed.), *Proceedings of the 4th International Symposium on Pre-Harvest Sprouting in Cereals*. Westview Press, Boulder, CO, USA, pp. 304–13.

Fincher, G.B. (1989) Molecular and cellular biology associated with endosperm mobilization in germinating cereal grains. *Annual Review of Plant Physiology and Plant Molecular Biology* 40, 305–46.

Fincher, G.B. & Stone, B.A. (1981) Metabolism of noncellulosic polysaccharides. In: Tanner, W. & Loewus, F.A. (eds), *Encyclopedia of Plant Physiology, New Series*, Vol. 13B. Springer-Verlag, Berlin, pp. 68–132.

Fincher, G.B. & Stone, B.A. (1986) Cell walls and their components in cereal grain technology. *Advances in Cereal Science and Technology* 8, 207–95.

Fincher, G.B., Lock, P.A., Morgan, M.M., Lingelbach, K., Wettenhall, R.E.H., Mercer, J.F.B., Brandt, A. & Thomsen, K.K. (1986) Primary structure of the (1 → 3, 1 → 4)-β-glucan 4-glucanohydrolase from barley aleurone. *Proceedings of the National Academy of Sciences (USA)* 83, 2081–5.

Forrest, I.S. & Wainwright, T. (1977) The mode of binding of β-glucans and pentosans in barley endosperm cell walls. *Journal of the Institute of Brewing* 83, 279–86.

Fry, S.C. (1986) Cross-linking of matrix polymers in the growing cell walls of angiosperms. *Annual Review of Plant Physiology* 37, 165–86.

Gilbert, W. (1985) Genes-in-pieces revisited. *Science* 228, 823–4.

Gubler, F., Ashford, A.E., Bacic, A., Blakeney, A.B. & Stone, B.A. (1985) Release of ferulic acid esters from barley aleurone. II. Characterization of the feruloyl compounds released in response to GA_3. *Australian Journal of Plant Physiology* 12, 307–17.

Gubler, F., Ashford, A.E. & Jacobsen, J. (1987) The release of α-amylase through gibberellin-treated barley aleurone cell walls. *Planta* 172, 155–61.

Henry, R.J. (1986) Genetic and environmental variation in the pentosan and β-glucan contents of barley, and their relation to malting quality. *Journal of Cereal Science* 4, 269–77.

Henry, R.J. (1987) Pentosan and (1 → 3, 1 → 4)-β-glucan concentrations in endosperm and whole grain of wheat, barley, oats and rye. *Journal of Cereal Science* 6, 253–8.

Henry, R.J. (1988) Changes in β-glucan and other carbohydrate components of barley during malting. *Journal of the Science of Food and Agriculture* 42, 333–41.

Henry, R.J. (1990) Quantitative analysis of barley (1 → 3), (1 → 4)-β-glucanase isoenzymes by high performance liquid chromatography. *Journal of Cereal Science* 12, 187–92.

Hinchliffe, E. (1988) The genetic improvement of brewing yeast. *Transactions of the Biochemical Society* 16, 1077–9.

Hinchliffe, E. & Box, W.G. (1984) Expression of the cloned endo-1,3-1,4-β-glucanase gene of *Bacillus subtilis* in *Saccharomyces cerevisiae*. *Current Genetics* 8, 471–5.

Høj, P.B., Slade, A.M., Wettenhall, R.E.H. & Fincher, G.B. (1988) Isolation and characterization of a (1 → 3)-β-glucan endohydrolase from germinating barley (*Hordeum vulgare*): amino-acid sequence similarity with barley (1 → 3, 1 → 4)-β-glucanases. *FEBS Letters* 230, 67–71.

Høj, P.B., Hartman, D.J., Morrice, N.A., Doan, D.N.P. & Fincher, G.B. (1989) Purification of (1 → 3)-β-glucan endohydrolase isoenzyme II from germinated barley and determination of its primary structure from a cDNA clone. *Plant Molecular Biology* 13, 31–42.

Høj, P.B., Hoogenraad, N.J., Hartman, D.J., Yannakena, H. & Fincher, G.B. (1990) Identification of individual (1 → 3, 1 → 4)-β-D-glucanase isoenzymes in extracts of germinated barley using specific monoclonal antibodies. *Journal of Cereal Science* 11, 261–8.

Hoson, T. & Nevins, D.J. (1989) β-D-glucan antibodies inhibit auxin-induced cell elongation and changes in the cell wall of *Zea* coleoptile segments. *Plant Physiology* 90, 1353–8.

Hoy, J.L., Macauley, B.J. & Fincher, G.B. (1981) Cellulases of plant and microbial origin in germinating barley. *Journal of the Institute of Brewing* 87, 77–80.

Iiyama, K., Lam, T.B.T. & Stone, B.A. (1990) Phenolic acid bridges between polysaccharides and lignin in wheat internodes. *Phytochemistry* 29, 733–7.

Islam, A. & Shepherd, K.W. (1981) Wheat–barley addition lines: their use in genetic and evolutionary studies of barley. *Proceedings of the 4th International Barley Genetics Symposium, Edinburgh*, pp. 729–39.

Jackson, E.A., Ballance, G.M. & Thomsen, K.K. (1986) Construction of a yeast vector directing the synthesis and release of barley (1 → 3, 1 → 4)-β-glucanase. *Carlsberg Research Communications* 51, 445–58.

Jacobsen, J.V. & Chandler, P.M. (1987) Gibberellin and abscisic acid in germinating cereals. In: Davies, P.J. (ed.), *Plant Hormones and Their Role in Plant Growth and Development*. Martinus Nijhoff, Dordrecht, pp. 164–93.

Jeffree, C.E., Dale, J.E. & Fry, S.C. (1986) The genesis of intercellular spaces in developing leaves of *Phaseolus vulgaris* L. *Protoplasma* 132, 90–8.

Kieliszewski, M.J., Leykam, J.F. & Lamport, D.T.A. (1990) Structure of the threonine-rich extensin from *Zea mays*. *Plant Physiology* 92, 316–26.

Klopfenstein, C.F. (1988) The role of cereal β-glucans in nutrition and health. *Cereal Foods World* 33, 865–9.

Knox, C.A.P., Sonthayanon, B., Chandra, G.R. & Muthukrishnan, S. (1987) Structure and organization of two divergent α-amylase genes from barley. *Plant Molecular Biology* 9, 3–17.

Knox, J.P., Linstead, P.J., King, J., Cooper, C. & Roberts K. (1990) Pectin esterification is spatially regulated both within cell walls and between developing tissues of root apices. *Planta* 181, 512–21.

Koehler, S.M. & Ho, T-H.D. (1990) Hormonal regulation, processing, and secretion of cysteine proteinases in barley aleurone layers. *Plant Cell* 2, 769–83.

Kokubo, A., Kuraishi, S. & Sakurai, N. (1989) Culm strength of barley. *Plant Physiology* 91, 876–82.

Loi, L., Barton, P.A. & Fincher, G.B. (1987) Survival of barley (1 → 3, 1 → 4)-β-glucanase isoenzymes during kilning and mashing. *Journal of Cereal Science* 5, 45–50.

Loi, L., Ahluwalia, B. & Fincher, G.B. (1988) Chromosomal location of genes encoding barley (1 → 3, 1 → 4)-β-glucan 4-glucanohydrolases. *Plant Physiology* 87, 300–2.

Luhrs, R. & Lorz, H. (1987) Plant regeneration in vitro from embryogenic cultures of spring- and winter-type barley (*Hordeum vulgare* L.) varieties. *Theoretical and Applied Genetics* 75, 16–25.

MacGregor, A.W., Ballance, G.M. & Dushnicky, L. (1989) Fluorescence microscopy studies on (1,3)-β-D-glucan in barley endosperm. *Food Microstructure* 8, 235–44.

Manners, D.J. & Marshall, J.J. (1969) Studies on carbohydrate metabolizing enzymes. XXII. The β-glucanase system of malted barley. *Journal of the Institute of Brewing* 75, 550–61.

Manners, D.J. & Wilson, G. (1974) Purification and properties of an endo-β-D-glucanase from malted barley. *Carbohydrate Research* 37, 9–22.

Mauch, F., Hadwiger, L.A. & Boller, T. (1988)

Antifungal hydrolases in pea tissue. I. Purification and characterization of two chitinases and two β-1,3-glucanases differentially regulated during development and in response to fungal infection. *Plant Physiology* 87, 325–33.

McFadden, G.I., Ahluwalia, B., Clarke, A.E. & Fincher, G.B. (1988) Expression sites and developmental regulation of genes encoding (1 → 3, 1 → 4)-β-glucanases in germinated barley. *Planta* 173, 500–8.

McNeil, M., Albersheim, P., Taiz, L. & Jones, R.L. (1975) The structure of plant cell walls. VII. Barley aleurone cells. *Plant Physiology* 55, 64–8.

Molina-Cano, J.L., de Togores, F.R., Royo, C. & Perez, A. (1989) Fast germinating low β-glucan mutants induced in barley with improved malting quality and yield. *Theoretical and Applied Genetics* 78, 748–54.

Morrall, P. & Briggs, D.E. (1978) Changes in cell wall polysaccharides of germinating barley grains. *Phytochemistry* 17, 1495–502.

Mundy, J. & Fincher, G.B. (1986) Effects of gibberellic acid and abscisic acid on levels of translatable mRNA for (1 → 3, 1 → 4)-β-D-glucanase in barley aleurone. *FEBS Letters* 198, 349–52.

Mundy, J., Brandt, A. & Fincher, G.B. (1985) Messenger RNAs from the scutellum and aleurone of germinating barley encode (1 → 3, 1 → 4)-β-glucanase, α-amylase and carboxypeptidase. *Plant Physiology* 79, 867–71.

Murphy, N., McConnell, D.J. & Cantwell, B.A. (1984) The DNA sequence of the gene and genetic control sites for the excreted *B. subtilis* enzyme β-glucanase. *Nucleic Acids Research* 12, 5355–67.

Nieuwdorp, P.J. & Buys, M.C. (1964) Electron microscopic structure of the epithelial cells of the scutellum of barley. II. Cytology of the cells during germination. *Acta Botanica Neerlandica* 13, 559–65.

Podrazky, V. (1964) Some characteristics of cereal gums. *Chemistry and Industry (London)*, pp. 712–13.

Powell, W., Caligari, P.D.S., Swanston, J.S. & Jinks, J.L. (1989) Genetical investigations into beta-glucan content in barley. *Theoretical and Applied Genetics* 71, 461–6.

Preece, I.A. & Hoggan, J. (1957) Carbohydrate modification during malting. *European Brewing Convention: Proceedings of 6th Congress, Copenhagen*, pp. 72–83.

Preece, I.A. & MacDougall, M. (1958) Enzymic degradation of cereal hemicelluloses. II. Pattern of pentosan degradation. *Journal of the Institute of Brewing* 64, 489–500.

Roberts, K. (1990) Structures at the plant cell surface. *Current Opinion in Cell Biology* 2, 920–8.

Sakurai, N. & Masuda, Y. (1978) Auxin-induced changes in barley coleoptile cell wall composition. *Plant and Cell Physiology* 19(7), 1217–23.

Selvig, A., Aarnes, H. & Lie, S. (1986) Cell wall degradation in endosperm of barley during germination. *Journal of the Institute of Brewing* 92, 185–87.

Slade, A.M., Høj, P.B., Morrice, N.A. & Fincher, G.B. (1989) Purification and characterization of three (1 → 4)-β-D-xylan endohydrolases from germinated barley. *European Journal of Biochemistry* 185, 533–9.

Slakeski, N., Baulcombe, D.C., Devos, K.M., Ahluwalia, B., Doan, D.N.P. & Fincher, G.B. (1990) Structure and tissue-specific regulation of genes encoding barley (1 → 3, 1 → 4)-β-glucan endohydrolases. *Molecular and General Genetics* 224, 437–49.

Smart, M.G. & O'Brien, T.P (1979a) Observations on the scutellum. II. Histochemistry and autofluorescence of the cell wall in mature grain and during germination of wheat, barley, oats and ryegrass. *Australian Journal of Botany* 27, 403–11.

Smart, M.G. & O'Brien, T.P. (1979b) Observations on the scutellum. III. Ferulic acid as a component of the cell wall in wheat and barley. *Australian Journal of Plant Physiology* 6, 485–91.

Staudte, R.G., Woodward, J.R., Fincher, G.B. & Stone, B.A. (1983) Water-soluble (1 → 3, 1 → 4)-β-D-glucans from barley (*Hordeum vulgare*) endosperm. III. Distribution of cellotriosyl and cellotetraosyl residues. *Carbohydrate Polymers* 3, 299–312.

Stuart, I.M. & Fincher, G.B. (1983) Immunological determination of (1 → 3), (1 → 4)-β-D-glucan endohydrolase development in germinating barley (*Hordeum vulgare*). *FEBS Letters* 155, 201–4.

Stuart, I.M., Loi, L. & Fincher, G.B. (1986) Development of (1 → 3, 1 → 4)-β-D-glucan endohydrolase isoenzymes in isolated scutella and aleurone layers of barley (*Hordeum vulgare*). *Plant Physiology* 80, 310–14.

Stuart, I.M., Loi, L. & Fincher, G.B. (1987) Immunological comparison of (1 → 3, 1 → 4)-β-glucan endohydrolases in germinating cereals. *Journal of Cereal Science* 6, 45–52.

Stuart, I.M., Loi, L. & Fincher, G.B. (1988) Varietal and environmental variations in (1 → 3, 1 → 4)-β-glucan levels and (1 → 3, 1 → 4)-β-glucanase potential in barley: relationships to malting quality. *Journal of Cereal Science* 7, 61–71.

Taiz, L. & Honigman, W.A. (1976) Production of cell wall hydrolyzing enzymes by barley aleurone layers in response to gibberellic acid. *Plant Physiology* 58, 380–6.

Taiz, L. & Jones, R.L. (1973) Plasmodesmata and an associated cell wall component in barley aleurone tissue. *American Journal of Botany* 60, 67–75.

Vasil, V., Redway, F. & Vasil, I.K. (1990) Regeneration of plants of embryogenic suspension culture protoplasts of wheat (*Triticum aestivum* L.). *Biotechnology* 8, 429–34.

Wessels, J.G.H. & Sietsma, J.H. (1981) Fungal cell walls: a survey. In: Tanner, W. & Loewus, F.A. (eds), *Encyclopedia of Plant Physiology, New Series*, Vol. 13B. Springer-Verlag, Berlin, pp. 352–94.

White, W.B., Bird, H.R., Sunde, M.L. & Marlett, J.A. (1983) Viscosity of β-D-glucan as a factor in the enzymatic improvement of barley for chicks. *Poultry Science* 62, 853–62.

Whittier, R.F., Dean, D.A. & Rogers, J.C. (1987) Nucleotide sequence analysis of alpha-amylase and thiol protease genes that are hormonally regulated in barley aleurone cells. *Nucleic Acids Research* 6, 2515–35.

Woodward, J.R. & Fincher, G.B. (1982a) Purification and chemical properties of two (1 → 3), (1 → 4)-β-glucan endohydrolases from germinating barley. *European Journal of Biochemistry* 121, 663–9.

Woodward, J.R. & Fincher, G.B. (1982b) Substrate specificities and kinetic properties of two (1 → 3), (1 → 4)-β-D-glucan endohydrolases from germinating barley (*Hordeum vulgare*). *Carbohydrate Research* 106, 111–22.

Woodward, J.R. & Fincher, G.B. (1983) Water-soluble barley β-glucans. Fine structure, solution behaviour and organization in the cell wall. *Brewers Digest* 58, 28–32.

Woodward, J.R., Morgan, F.J. & Fincher, G.B. (1982) Amino acid sequence homology in two (1 → 3), (1 → 4)-β-glucan endohydrolases from germinating barley (*Hordeum vulgare*) endosperm. *FEBS Letters* 138, 198–200.

Woodward, J.R., Fincher, G.B. & Stone, B.A. (1983a) Water-soluble (1 → 3), (1 → 4)-β-D-glucans from barley (*Hordeum vulgare*) endosperm. II. Fine structure. *Carbohydrate Polymers* 3, 207–25.

Woodward, J.R., Phillips, D.R. & Fincher, G.B. (1983b) Water-soluble (1 → 3), (1 → 4)-β-D-glucans from barley (*Hordeum vulgare*) endosperm. I. Physicochemical properties. *Carbohydrate Polymers* 3, 143–56.

Woodward, J.R., Phillips, D.R. & Fincher, G.B. (1988) Water-soluble (1 → 3), (1 → 4)-β-D-glucans from barley (*Hordeum vulgare*) endosperm. IV. Comparison of 40°C and 65°C soluble fractions. *Carbohydrate Polymers* 8, 85–97.

V

Pathogen Resistance

21

Sources and Genetics of Resistance to Fungal Pathogens

J. Helms Jørgensen

Fungal diseases on barley result from interactions between the host plant and the pathogen. The two organisms and their interactions are affected by environmental factors during their life cycles. This has led to the concept of 'the disease pyramid' (Browning *et al.*, 1977) illustrating that disease is constituted by four elements: host, pathogen, environment and time. The emphasis will be on the genetics of the barley host in the present context, but it should be kept in mind that this represents only one element of the four.

Barley may suffer from many diseases. At least 30 fungal diseases are listed in various books and compendia (cf. Matre, 1982). Their relative importance is affected by the local 'disease pressure', i.e. the level of resistance in the host, the prevalence, aggressiveness and virulence of the pathogen, and the environmental factors which promote or restrain the disease. Therefore, different diseases differ in importance from one ecogeographical region to another. In this chapter, emphasis is placed on barley diseases, primarily powdery mildew and other diseases prevalent in northwest Europe, and general principles and considerations, rather than on a detailed account of the sources and genetics of resistance to all fungal diseases of barley. Only recent, selected references are given, primarily those pertinent to sources and genetics of resistance that may provide references to earlier literature.

Susceptibility

Barley is usually considered susceptible to the barley diseases unless resistance breeding has been conciously carried out. This is, however, an oversimplification. The definition of resistance used here is 'any level of incompatability relative to the most compatible interaction known' (Gabriel & Rolfe, 1990). The evolution of barley and its development as a crop plant have occurred simultaneously with the evolution of pathogens. In some cases, the pathogens have become obligate. It is therefore sensible to assume that host and pathogen have evolved mutually balanced levels of resistance and agressiveness/virulence, respectively. In addition, barley cultivation during recent centuries and early breeding aimed at saving seeds of healthy and high-yielding plants, i.e. barley was exposed to selection against disease susceptibility. Our northwest European barley germplasm therefore has an appreciable level of resistance to several of the prevalent diseases, such that several pathogens are currently considered merely as potential dangers. A discussion of resistance to fungal pathogens has therefore to consider the inherent and sometimes very appreciable levels of resistance in the local barley germplasm.

There are cases where a 'new' fungal disease has been introduced into a geographical region previously devoid of the pathogen. A recent example is the occurrence in 1975 of barley yellow rust in Colombia, South America, from where the disease has spread southwards throughout most of the continent (Dubin & Stubbs, 1986). More recently it has moved northwards to Mexico, and is at present considered a potential threat to barley-growing areas in the USA and Canada (Anonymous, 1989). The damage of such 'new' diseases may be particularly severe since the host germplasm has not co-evolved with the pathogen, and in consequence does not possess any appreciable level of inherent resistance.

Plant protection

Plant protection of barley crops comprises three measures (cf. Jenkyn & Plumb, 1981). The first is good agricultural husbandry: for example, crop rotation to prevent build-up of soil-borne diseases such as take-all caused by *Gaeumannomyces graminis* (Sacc.) v. Arx & Olivier; careful autumn soil preparation whereby all crop residues are mixed into the soil and none left on the surface. In the case of barley net blotch [*Drechslera teres* (Sacc.) Shoem.], the main source of primary inoculum is infested crop residues from a preceding barley crop.

The second plant protection measure, resistance breeding, is currently considered to be a highly profitable measure for society (Priestley & Bayles, 1988; Jørgensen, 1989), in that the cost of resistance breeding constitutes on average only about 10% of the value of the yield increase obtained by growing (more or less) resistant varieties. Cultivation of resistant varieties not only diminishes crop losses and fungicide spraying costs, but also provides more stable yields. The shift in emphasis from application of fungicides to other plant protection measures in Europe has increased the demand for intensified resistance breeding based on present and new sources of resistance in order to provide a more diversified and durable genetic resistance.

The third plant protection measure is application of fungicides, either as a seed dressing to prevent seed-borne diseases or as a spray to inhibit the development of epidemic diseases in the growing crop. Seed-dressing, mainly by mercury, has been used extensively for many years, but recent environmental concern has led to the abolition of mercury-containing seed-dressing compounds. They have been replaced by other compounds and intensified resistance breeding, for example to barley leaf stripe (*Drechslera graminea* (Rabenh. ex Schlecht.) Shoem.). Concern for the environment has also triggered legislation in some countries in Europe, aiming at a substantial reduction in the use of pesticides in the 1980s and 1990s. Recent studies in the UK (Cook & Jenkins, 1988) and Denmark (Jørgensen, 1989) illustrate that a considerable portion of the gain from fungicide spraying is counteracted by the costs of the chemicals and their application.

Sources of resistance

Resistance to fungal pathogens of barley can be found in a variety of sources, ranging from resistance 'floating around' in commercial varieties to that present in wild species of *Hordeum* (cf. Brown *et al.*, 1989). The introduction of resistance by gene technology may occur in the future.

Commercial varieties

Modern varieties of barley possess an appreciable level of resistance to several diseases. This resistance mainly represents partial resistance accumulated during evolution and cultivation. In the case of powdery mildew (*Erysiphe graminis* DC. ex Mérat *hordei* Marchal), a considerable level of partial resistance is present in the northwest European barley germplasm, compared with the extreme susceptibility of barley lines/varieties from northern Scandinavia or from more remote areas where the disease has not been prevalent for centuries. The greater part of the resistance accumulated over centuries still appears to be present in many modern varieties (Knudsen *et al.*, 1987). The use of race-specific resistance genes in barley breeding during the last few decades has relaxed the selection pressure necessary to maintain this level of partial resistance. Some examples suggest the loss of partial resistance, because it has become 'hidden' behind race-specific resistance genes and therefore not exposed to selection, i.e. a 'vertifolia' effect (cf. Munk *et al.*, in press).

Net blotch [*Pyrenophora teres* (Sacc.) Shoem.] is not usually prevalent on spring barley in northern Europe. However, severe epidemics in the early 1980s disclosed large varietal differences in resistance, from close to that known in recognized but primitive sources of resistance, to extreme susceptibility (Skou & Haahr, 1987a).

The abolition of mercury compounds for seed-dressing encouraged extensive screening of old and new European barleys for resistance to the seed-borne disease barley leaf stripe [*Drechslera graminea* (Rabenh. ex Schlecht.) Shoem.] (Skou & Haahr, 1987b). Subsequent studies of a cross of a highly resistant European variety and an extremely susceptible line revealed a single resistance gene (Haahr *et al.*, 1989) which turned out to be race-specific (Knudsen, 1986). In addition, the progeny that did not possess this gene exhibited a wide range in partial resistance originating from the European parent. The same range in partial resistance was seen in the screening of varieties.

In the above examples barley breeders in Europe benefit from a fortunate situation in which resistances or genes are 'floating around' in commercial varieties, and thus probably also in their own breeding material. Consequently, testing for susceptibility to pathogens such as net blotch and leaf stripe is advisable to eliminate breeding lines that are insufficiently resistant.

Commercial varieties have also constituted the main 'source' of resistance in breeding for powdery mildew resistance in Europe. Only about ten unique resistances have been introduced, mainly by individual scientists, but these resistances have been used extensively as resistance 'sources' by practically all European barley breeders (Brown & Jørgensen, in press). More recently, these resistances have been pyramidized to produce multigene-resistant varieties (e.g. Jensen *et al.*, in press).

The resistance of cultivated barleys to economically important diseases has been, and is currently being, monitored in variety testing trials, and by the Agricultural Advisory Services during cultivation of the varieties. Commercial varieties can provide new sources of resistance only to diseases previously neglected or considered unimportant, for example because they have previously been controlled by chemical treatment (mainly seed-borne diseases) or agricultural practices.

Primitive barleys

Resistances or resistance genes to pathogens such as barley leaf rust and powdery mildew have been identified in many more or less primitive lines or populations of barley (*Hordeum vulgare* L.) (e.g. Sharp, 1985). Some of them have been transferred by backcrossing to advanced European germplasm. In fact, all the powdery mildew resistances used commercially except one, the 'Spontaneum' resistance, are derived from primitive *H. vulgare*, mainly from West Asia, Ethiopia and North Africa. In many cases one or more additional resistances or genes have been introduced, several of which are more or less tightly linked to the 'main' gene. Eighteen different powdery mildew resistances and genes are recognized in this way (Anonymous, in press) in European spring and winter barley varieties. In addition, several more have been recognized very recently (Jensen *et al.*, in press) or are designated 'unknown'. Another reason for the discrepancy between the about ten unique 'sources' and the 18 resistances is that the 'main' gene may have been introduced more than once, for example *Mla* 7 has been introduced from four different sources accompanied by four apparently different 'additional' resistance genes (Jensen *et al.*, in press).

Barley in gene banks

Landraces and other primitive forms of cultivated barley (*Hordeum vulgare* subsp. *vulgare*), mainly from gene banks, are being screened for sources of disease resistance (Jørgensen, 1988b; van Leur *et al.*, 1989). This has been carried out continously at the gene bank in Gatersleben, Germany, and reported in 'Kulturpflanze' since 1955 (e.g. Walther & Lehmann, 1980). It is to be expected that these efforts will disclose many potential sources of resistance in the vast holdings of lines, landraces, and populations of *H. vulgare* in gene banks, several of which already have records of the disease reactions of much of their germplasm. Barley breeders can access these data through their local gene bank, e.g. by formulating a search profile for the material desired. The European gene banks are interconnected by the European Barley Data Base (Knüpffer, 1988) which will be part of the planned Global Barley Germplasm Network (IBPGR, 1989).

Hordeum species other than *H. vulgare* subsp. *vulgare*

The wild barley *Hordeum vulgare* subsp. *spontaneum* grows abundantly in the eastern Mediterranean (see Chapter 2, this volume). It shares genome with cultivated barley, and the two are completely interfertile. Today they are combined in a single species (*Hordeum vulgare* L.) with two subspecies, the cultivated form subsp. *vulgare* and the wild form subsp. *spontaneum* (Bothmer & Jacobsen, 1985).

Extensive tests of wild barley for reaction to powdery mildew (e.g. Lange & Balkema-Boomstra, 1988), snow mould (*Typhula incarnata* Lasch ex Fr. and *T. ishikariensis* Imai) (Bothmer *et al.*, 1987), spot blotch [*Bipolaris sorokiniana* (Sacc.) Shoem.], net blotch [*Pyrenophora teres* (Sacc.) Shoem.] (Lehmann & Bothmer, 1988) and leaf rust (*Puccinia hordei* Otth) (Moseman *et al.*, 1990) have revealed extensive genetic variation in race-specific as well as in partial resistance (see Chapter 2 this volume). However, as far as I am aware, only one resistance to a fungal disease present in commercial varieties has been derived from the wild barley; this is the two closely linked powdery mildew resistance genes *Mla6* and *Mla14* that were transferred to German spring and winter barley varieties nearly 50 years ago. The interest in sources of resistance from wild barley has been intensified in the last decade, and several new powdery mildew resistance genes are currently being incorporated into European germplasm, for example the six genes designated *Mla16* to *Mla21* (Jahoor & Fischbeck, 1987).

Some accessions of the wild species *Hordeum bulbosum* L. produce an acceptable frequency of partially fertile hybrids with *H. vulgare*. However, I am aware of only one case where resistances from *H. bulbosum* (in this case one or more resistances to powdery mildew) have been transferred from *H. bulbosum* to *H. vulgare* (R. Pickering, pers. comm.).

The other species of *Hordeum* contain a wealth of resistances, not only to fungal diseases, but also to stresses such as salinity and drought (see Chapter 1 this volume). It is, however, difficult to produce viable hybrids with *H. vulgare* and virtually all the hybrids are sterile. The chromosomes of the two species combined in the hybrids do not pair, and recombination does not take place. Even callus culture and irradiation have failed to induce 'recombination' between chromosomes in the species hybrids (Jørgensen & Andersen, 1989). Therefore, the resistances of most *Hordeum* species are inaccessible at present as sources of resistance for breeding cultivated barley.

Mutation

In the early phase of the 'mutation era', there were great expectations that induced mutations would provide new sources of disease resistance, which could be used to improve highly bred genotypes. The results have, however, been disappointing in that apparently no new sources of resistance have been induced in cereals (Jørgensen, 1991). This may be because the usually dominant gene-for-gene resistance genes each have a unique and highly specific function, whereas most mutations disrupt gene functions leading to recessive genes with non-functional products. The production of a highly specific gene function by induced mutation is very unlikely.

One outstanding exception is the *mlo* powdery mildew resistance genes in barley. Extensive studies on mutagen-induced Mlo resistance have paved the way for the present extensive use of Mlo resistance in barley breeding and production in Europe (Jørgensen, 1991; Andersen, in press). The Mlo resistance used is, however, derived from a spontaneously occurring *mlo* gene present in Ethiopian landrace barleys.

In future, induced mutations in relation to disease resistance will probably be used mainly as a tool for mutational dissection of resistance. Subsequent studies of such mutants may reveal the detailed operation of genes involved in disease reactions (see Jørgensen, 1991 and Chapters 9 and 24, this volume).

Gene technology

Molecular genetic techniques have been developed, mainly in the course of studies on the molecular genetics of micro-organisms. They are being adapted to more complex organisms including crop plants. Three aspects are particularly promising in relation to sources of resistance to fungal diseases of barley. The first is the possibility of overcoming sterility barriers between plant species enabling the transfer of genetic material, for example transferring disease resistance between plant species to produce disease-resistant transgenic plants. However, recent results with wheat and powdery mildew (Tosa & Sakai, 1990) have shown that related host species and pathogens (*formae speciales*) share resistance and virulence genes, respectively (see section on 'Function of race-specific genes' below). The second is the possibility of 'constructing' desired modifications of the genetic material in the laboratory, for example to 'construct' new resistance genes and/or to combine genetic material encoding disease resistance from different sources and transfer it *in vitro* to the recipient barley plant. The third possibility is to expose cell cultures, preferably haploid, to pathogens or their toxins so that resistance screening of millions of individuals may be carried out on the laboratory bench. However, much more basic and applied research has to be done before the potential of biotechnology for providing new sources of resistance to fungal diseases in barley can be assessed.

A considerable breeding investment will be required before such new sources of resistance can be used commercially in a highly bred crop plant. Experience gained with induced mutants shows that mutant genes have to be manipulated in relation to the overall genotype to counteract undesirable pleiotropic effects of the mutant gene or the 'imbalance' between the effect of the mutant gene and those of the rest of the genotype (see Jørgensen, 1991). The same general pattern is also experienced with polypoids produced by humans. New genetic variation may be produced, some of which is desirable and some undesirable. Against this background it is expected that the genetic variation produced by introducing foreign genetic material also must be moulded by conventional plant breeding, at least as much (if not more) than in the case of induced mutant genes.

The main impact of biotechnology on disease resistance in barley will probably be the gaining of a detailed understanding of the genetic and biochemical events involved in host resistance (see Chapters 22 and 23, this volume). Providing new sources of resistance will probably be of less importance.

The preceding discussion may leave the reader with the impression that there is a wealth of sources of resistance to fungal diseases of barley available to plant breeders. In some cases this is true, e.g. for powdery mildew, leaf rust and leaf stripe. In other cases, resistance is rare or even unknown. Resistance to the soil-borne disease take-all [*Gaeumannomyces graminis* (Sacc.) v. Arx and Olivier var. *tritici* Walker] has been desired for many years (Scott, 1981), but extensive screenings of *H. vulgare*, of other *Hordeum* species, and other related species have revealed no promising sources of resistance. None of the accessions tested has exceeded substantially the modest level of resistance expessed by rye. The absence of resistance to take-all may be an extreme and outstanding case, but it illustrates that resistance should not be taken for granted.

Genetics of resistance

Only four kinds of resistance are discussed here: race-specific, partial, Mlo, and passive. They are all accessible to manipulation in barley breeding. 'Adult' or 'mature plant' resistance is assumed to be covered in the section on partial resistance because one is genetically indistinctly separated from the other, though some authors suggest that they may be different (Smit & Parlevliet, 1990). 'Induced' resistance, i.e. the reduced susceptibility induced by inoculation with saprophytes and pathogens to subsequent inoculation with virulent pathogen isolates (e.g. Thordal-Christensen & Smedegaard-Petersen, 1988; Christiansen & Smedegaard, 1990), is not known to be controlled by genes that can be manipulated in plant breeding. 'Non-host' resistance from closely related species is briefly described, but the exploitation of resistance from more distantly related species awaits further progress in gene technology.

The terms used are those in common use, but with the following qualifications. In the gene-for-gene system the 'resistance' trait in the host is conferred by a gene in a resistance locus with (at least) two allelic states termed genes for 'resistance' and 'susceptibility'. Correspondingly, the 'virulence' trait in the pathogen is conferred by a gene in a virulence locus with the two allelic states termed genes for 'virulence' and 'avirulence'. In case of 'partial' resistance the term 'partial' is used to describe quantitatively reduced levels of susceptibility in the host and, correspondingly, the term 'agressiveness' is used to describe (quantitative) grading in the ability of the pathogen to cause disease.

Race-specific resistance

Many disease resistance breeding programmes in barley are based on one or few race-specific resistance genes (Table 21.1), all of which are known or assumed to operate in the gene-for-gene system and most of which confer a resistance response resulting in hypersensitive cell death. The genes have the advantage of being easy to recognize, by conferring drastic effects on the pathogen, and they are relatively easy to discriminate from one another. The disadvantages are that highly mobile pathogen populations can eliminate their effect by acquiring the matching virulence genes, and that they may 'hide' genetic variation in partial resistance inherent in the original host population, which may then be diluted or lost. Race-specific resistance genes usually have a poor durability. The T-gene for resistance to stem rust (*Puccinia graminis* Pers. f.sp. *tritici*) in barley (Table 21.1) has, however, remained effective, even though widely used in North America for more than 60 years (Roelfs, 1985). Table 21.1 lists all the loci reported to harbour genes conferring resistance to fungal pathogens in barley.

Table 21.1. List of gene loci affecting fungal disease reaction in barley and their chromosomal location (from Jørgensen, in press).

Disease	Locus	Synonym(s)	Chromosome
Leaf rust[1]	Rph1	Pa, Pa2 C	2
(*Puccinia hordei*)	Rph2	Pa2	
	Rph3	Pa3 A	1 or 2
	Rph4	Pa4 D	5
	Rph5	Pa5 B	3
	Rph6	Pa6, Pax X	
	Rph7	Pa7, Pay	3
	Rph8	Pa8	
	Rph9	Pa9	
	Rph10		3
	Rph11		6
Stem rust	Rpg1	T	1
(*Puccinia graminis*)	Rpg2	T2	
Stripe rust	rps1	yr1	
(*Puccinia striiformis*)	rps2	yr2	
	rps3	yr3	
	Rps4	Yr4	5
Loose smut	Run1	Un	1
(*Ustilago nuda*)	Run2	Un2	
	Run3	Un3	
	Run4	Un4	
	Run5	Un5	
	Run6	Un6, Ts	
	run7	un7	
Covered smut	Ruh1	Uh	
(*Ustilago hordei*)	Ruh2	Uh2	
	ruh3	uh3	
	ruh4	uh4	
		Un A[2]	
Semiloose smut	Ung		
(*Ustilago nigra*)			
Powdery mildew[3]	Reg1	Mla, Mlb, Mlgo, Mlnn,	
(*Erysiphe graminis* f.sp. *hordei*)		Mlm, Ml(1063), JMl_{sn}	5
	Reg2	Mlg, JMlg	4
	Reg3	Mlh, JMl_{h}	6
	Reg4	Mlk, JMl_{k}	5
	Reg5	Mlp, JMl_{p}	
	reg6	mlo	4
	Reg7	Mlra, Ml(41/145)	5
	Mlat	JMl_{r12}	5
	Ml(CP)		4
	mld		
	Mln		5
	Mlnn	JMl_{nn}	5
	N182		4
Net blotch	Rpt1	Pt	3
(*Pyrenophora teres*)	Rpt2	Pt2	5
	Rpt3	Pt3	2
	Pt,,d		

Table 21.1. *cont.*

Disease	Locus	Synonym(s)	Chromosome
Barley stripe	Rhg1	Hg	
(*Pyrenophora graminea*)	Rhg2	Hg2	
	Rhg3	Hg3	
Spot blotch	hl		2
(*Cochliobolus sativus*)	hl2		5
	hl3		7
	hl4		
Septoria leaf blotch	Rsp1	Sep1	
(*Leptosphaeria avenaria* f.sp. *triticea*)	Rsp2	Sep2	
	Rsp3	Sep3	
Scald	Rrs1	Rh, Rha	3
(*Rhynchosporium secalis*)	Rrs2	Rh2	
	Rrs3	Rh3	3
	Rrs4	Rh4	3
	Rrs5	Rh5	
	rrs6	rh6	4
	rrs7	rh7	
	rrs8	rh8	
	Rrs9	Rh9	4
	Rrs10	Rh10	
	rrs11	rh11	
Scab (head blight)	fb	sc	
(*Fusarium* spp.)			
Blast	Rpo	PHR-1	
(*Pericularia oryzae*)			

[1] Clifford (1985) and Feuerstein *et al.* (1990).
[2] Khokhlova (1987).
[3] Synonyms in *Pm* and *Er* series omitted.

There are 13 loci with genes conferring resistance to powdery mildew (Table 21.1), whereas the number of genes/alleles described is close to 92 (see Jørgensen, 1987). There are two main reasons for this discrepancy in numbers: one is that *Mla* locus on barley chromosome 5 is a complex locus in which many different alleles are named. This locus appears to harbour multigene families representing nature's own pyramidized resistances. Another reason for the large number of genes relative to that of loci is that many resistance genes have not been genetically mapped. Somewhat similar discrepancies may be observed for loci and genes for resistance to other diseases.

Identification of resistance genes

Each resistance gene was originally identified in a particular barley line by the reaction to a particular pathogen isolate, i.e. by a matching avirulence gene. In gene-for-gene systems each resistance gene (allele) is identified by its corresponding avirulence gene (allele) and vice versa. The scientific literature stores the data published and, at least sometimes, the barley lines with newly identified resistance genes are stored in gene banks. To maintain an unequivocal correspondence between the data (in the libraries) and the gene (in the gene banks) the published data should also comprise the gene bank name and the accession number of the barley line in which the newly described gene has been identified. Editors of scientific journals should consider this requirement to be a 'must' for publications on newly described genes.

It is equally necessary to maintain the pathogen isolates (i.e. the newly detected avirulence gene), because the avirulence gene is the tool for unequivocal identification of the resistance gene. This may be done by depositing the isolates in a microbial culture collection or a rust culture collection. With powdery mildew, long-term storage can be achieved only by producing cleistothecia from crosses between isolates known to possess the actual avirulence gene.

Several formerly described resistance genes as well as most pathogen isolates used to identify many resistance genes are no longer available. Therefore, it is virtually impossible to keep track of all the resistance genes described in the literature.

Susceptibility genes

In the absence of a (known) resistance gene (allele), we assume the presence of an alternative allele, usually termed the susceptibility gene (allele). Another possibility is that the susceptibility allele may be a 'defeated' resistance gene. Experiences with species such as barley (Jørgensen, 1987), flax and lettuce show that crop plant varieties susceptible in one geographical region can express resistance in another, and that resistance expressed in different geographical regions may be conferred by different resistance genes. This discloses that several (if not most) susceptibility genes are ordinary resistance genes (Jørgensen, 1990). The decisive factor is whether the frequency of the corresponding virulence gene is fixed at 100% in the pathogen population in one region and not in the population in another region.

Function of race-specific genes

The mechanisms by which genetically controlled resistance influences the colonization of fungal pathogens have been studied extensively (see, for example, Crute *et al.*, 1985), but the detailed functions of genes involved in resistance to fungal pathogens in barley are unknown. They will be dealt with only briefly here because the subject is described in detail elsewhere in this volume. The consensus is that two groups of genes are involved: the race-specific genes, i.e. those responsible for recognizing the avirulent pathogen isolates, and the response genes, i.e. those responsible for the defence reactions of the plant. The race-specific genes appear to comprise two mutationally independent functions (Jørgensen, 1988a). This is illustrated in Fig. 21.1. One of the functions is the (qualitative) recognition of the product of the pathogen's avirulence gene; the other is the (quantitative) activation and regulation of the disease response genes involved in the biochemical and physical events leading to hypersensitive cell death. Working models of specific recognition in plant–microbe interactions have recently been reviewed by Gabriel & Rolfe (1990).

Recent research concerning wheat and powdery mildew (Tosa & Sakai, 1990) indicates that the *forma specialis*-genus specificity may follow the gene-for-gene system, i.e. that resistance (genes) in one host species to an inappropriate *forma specialis* is essentially cultivar (gene-for-gene) resistance, not non-host resistance (see also Niks, 1988). The implication is that race-specific powdery mildew resistance genes transferred to barley from a related species (wheat or rye) being attacked by other *formae speciales* of *Erysiphe graminis* will be as durable or non-durable as the resistance genes inherent in barley.

This is supported by experience from wheat, where powdery mildew and rust resistance genes from the related species rye and wheatgrass, respectively, have been overcome by the wheat pathogens. Race-specific resistance genes may thus function across host species which are so closely related that gene transfer is possible by conventional means, and the corresponding virulence genes may function across *formae speciales* of pathogens. An exception to this

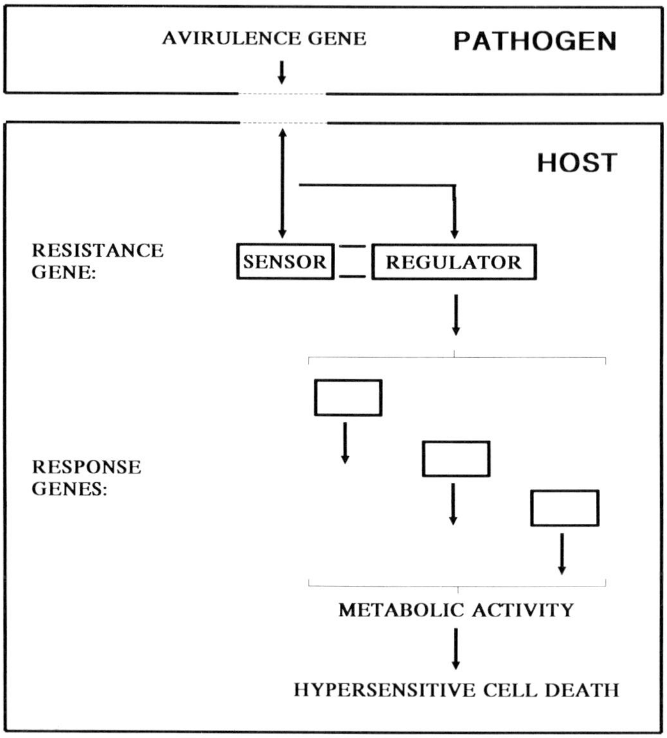

Fig. 21.1. Model for genetic components in race-specific powdery mildew resistance in barley. The resistance gene has two functions/elements. The sensor recognizes (qualitatively) the pathogen's avirulence gene (product). Recognition activates the regulator that regulates (quantitatively) the disease response genes. The resulting metabolic activity leads to hypersensitive cell death (from Jørgensen, 1988a).

pattern is some stem rust (*Puccinia graminis* Pers. f.sp. *tritici*) resistances transferred from related species, which have provided resistance to all races (see Gabriel & Rolfe, 1990). The efficiency of these resistances may, however, be due to several or many genes on the (usually large) alien chromosome segment transferred. These resistances may even be more like non-host resistance. The subject of 'non-host' resistance has recently been discussed by Niks (1988).

Mlo resistance

The Mlo resistance is a relatively new kind of resistance and is, as yet, known only from barley. It was originally described in a mutagen-induced powdery mildew-resistant barley mutant in 1942 and has been induced repeatedly since then. About 1970 it was also discovered as a spontaneous gene in barley landraces from Ethiopia. It is unique in that: (i) Mlo resistance does not conform to the gene-for-gene system; (ii) *mlo* genes originating from different mutational events map as non-complementing recessive alleles in one locus, *mlo* on barley chromosome 4; (iii) all alleles confer the same resistance reaction, although with small quantitative differences (Hentrich, 1979); and (iv) it is effective against all isolates of the pathogen (cf. Andersen, in press).

The phenotype of the resistance reaction is also unique. Leaves of Mlo resistant barley inoculated with powdery mildew spores remain resistant with no visible sign of infection except for occasional small infection type 4 (compatible) mildew colonies. These colonies originate predominantly from successful infections in the subsidiary cells next to the stomata on the barley epidermis (Andersen, in press). These cells do not have the ability to prevent infection. The frequency of occasional mildew colonies on the leaves depends on several factors, most of which are genetically determined.

One non-genetic factor is density of inoculum; a higher density resulting in a higher frequency of colony formation. A second factor is the genotype of the powdery mildew isolate. All isolates collected in nature cause the formation of about the same low frequency of colonies. Only an isolate selected for high aggressiveness in the laboratory may cause many more colonies, about 10% of those occurring in a compatible interaction, i.e. by a virulent isolate on a susceptible barley (Andersen, in press). A third factor is the particular *mlo* allele. Studies in Germany suggest that *mlo* alleles of different origin may differ in the quantitative expression of the resistance, and in pleiotropic effects in the form of more or less severe necrotic leaf spotting (Hentrich, 1979). Fourthly, the general gene background may strongly affect the degree of expression of Mlo resistance (in terms of the number of colonies that may vary from close to zero to close to compatibility) and the severity of necrotic leaf spotting (from virtually zero to very severe). A fifth factor affecting the frequency of mildew colonies on Mlo resistant barley is the presence of other powdery mildew resistance genes. Where the barley line has an infection type 0 resistance gene (and the isolate the matching avirulence gene), the Mlo resistant barley will not show any mildew colonies at all. If, however, an infection type 2 resistance gene is present, the occasional colonies will also be of infection type 2.

The genetically most important implications of these experiences is that: (i) *mlo* alleles of different origin constitute only one source of resistance; (ii) the Mlo resistance is effective world-wide; (iii) the undesirable pleiotropic effects of the *mlo* genes can be modified by the overall gene background and be virtually eliminated; and (iv) Mlo resistance can be freely combined with both race-specific resistance, and partial resistance.

Function of *mlo* resistance genes

When Mlo resistant barley leaf tissue is inoculated with powdery mildew, the host epidermal cells develop cell wall appositions more rapidly and of a greater size than non-Mlo barley. These papillae prevent the pathogen from penetrating (Skou *et al.*, 1984; Bayles *et al.*, 1990).

One of the most common responses of plant cells to invasions of fungi or other agents is the formation of localized cell wall appositions (papillae) at the encounter site (see Crute *et al.*, 1985). However, they also occur in compatible host–pathogen interactions, indicating that papillae formation is a very general phenomenon; it is most likely a wound-healing mechanism in higher plants (Skou *et al.*, 1984). The papillae consist of a variety of chemical elements and compounds (see Bayles *et al.*, 1990). This probably requires the function of a number of genes, which in turn may be regulated by one or more genes. Against this background, it is tempting to assume that the *mlo* locus is harbouring a wild-type gene with regulatory function for papillae formation, and that an *mlo* mutation has more or less efficiently inactivated this gene. This hypothesis has two significant implications. One is that it is hard to imagine a single-step mutation (to virulence) in powdery mildew that is sufficient to overcome Mlo resistance completely; one can more likely imagine a gradual accumulation of genes affecting the amount and efficiency of cell wall-degrading enzymes (to increased aggressiveness). This suggests that Mlo resistance will be durable. The second implication is derived from the fact that Mlo resistance is an enhancement of a mechanism inherent in many plant species, including wheat and oats, suggesting that Mlo resistance may also occur or be induced by mutation in these species (Jørgensen, 1991).

The Mlo resistance mechanism is apparently confined to the epidermal cells of the barley plant and is triggered by any physical damage, whether from a germinating mildew conidium, a thin needle, or the stylet of an aphid. It does not affect the development of many other barley diseases, such as the rusts (*Puccinia* sp.) which enter the stomata and encounter the mesophyll cells, and scald [*Rhynchosporium secalis*

(Oudem.) J.J. Davis] and net blotch [*Pyrenophora teres* (Sacc.) Shoem.] which kill the epidermal cells by toxins before and during penetration. Mlo resistant barley varieties grown in Europe apparently vary their degree of resistance to other diseases as much as non-Mlo resistant barley varieties.

Partial resistance

The inheritance of partial resistance in barley is not clear. The number of genes involved in partial resistance to barley leaf rust has been estimated as 'several' (see Parlevliet, 1988). The actual figure estimated may vary considerably depending on the components of partial resistance assessed, the genetic model chosen for analysing the data, and the assumptions built into the models. The only safe conclusion is that partial resistance is governed by a number of mostly recessive genes acting mainly additively, and affecting several traits of the host plant. Some studies suggest that the plant traits are genetically correlated. These observations are, however, made mainly on series of varieties spanning a broad range of partial resistance, which may bias the results. Varieties with a high level of partial resistance may possess several genes affecting each of the traits, whereas varieties with a low level of partial resistance may possess few genes affecting each of the traits. The problem can be solved only by using segregating populations, preferably in the form of randomly chosen homozygous lines, for example chromosome doubled haploid lines from crosses.

Assessment of partial resistance

All barley genotypes can be classified with respect to partial resistance. Some barleys, usually exotic ones, are extremely susceptible in Europe, and possess virtually no partial resistance, whereas most (usually old) European varieties have a high level.

The phenotypic expression of grades of partial resistance may be the amount of leaf area becoming diseased following uniform inoculation. In the field, the disease assessment can be repeated during the growing season (Fig. 21.2) and the area under the disease progress curve will be an accumulated estimate of the level of partial resistance (see Knudsen *et al*., 1987). This way of determining partial resistance is laborious and costly, and can be accomplished only by comparing groups of barleys without effective race-specific resistance genes, or groups possessing the same race-specific resistance gene (or genes).

A more tempting way to estimate partial resistance is to inoculate uniformly seedlings or young plants in a controlled environment and assess one or more of the components of partial resistance. This may be done by estimating one or more of the following traits: (i) the infection frequency, i.e. the percen-

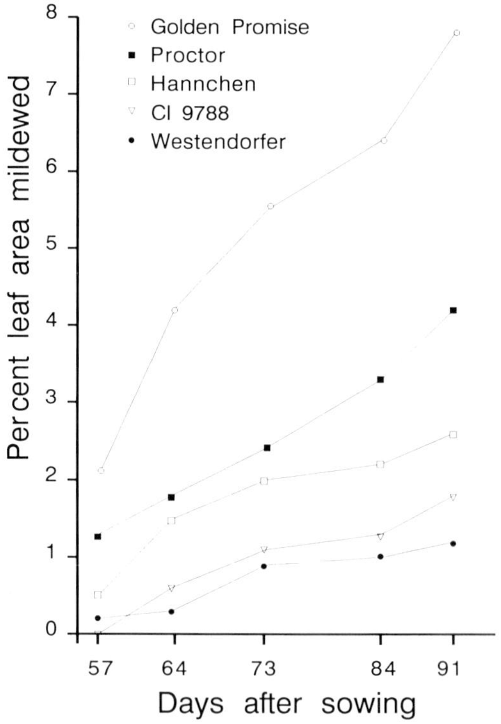

Fig. 21.2. Partial resistance of five spring barley lines without effective race-specific powdery mildew resistance genes. Assessed in field plots on five dates as percentage of the area of the upper four leaves covered with mildew (from Knudsen *et al*., 1987).

tage of spores applied that develop into sporulating colonies; (ii) the latent period, i.e. the time from inoculation to emergence of visible or sporulating colonies; (iii) the number of spores produced per colony or unit leaf area; or (iv) the size of the sporulating colony (or pustule). These parameters relate primarily to the development of the fungus. The corresponding three components of the host plant are resistance to (i) infection, (ii) colonization, and (iii) reproduction. This has been attempted repeatedly. The overall impression is that a significant portion of the variation in partial resistance expressed in the field can be accounted for by these methods. Therefore, it may be advisable for barley breeders to screen their breeding material systematically by one of these methods and discard the more susceptible portion. This will retain a constant selection pressure on the breeding material resulting in a progressively increased level of partial resistance. In the long term it may raise the level of partial resistance so much that most fungicide treatments may become superfluous. Farmers will then have a constant but economically tolerable level of disease. The state of no disease is hardly a realistic goal in disease-prone areas. Fleming & Person (1982) have discussed this subject from a theoretical viewpoint, and Simmonds (1988) from a practical one.

Passive resistance

A very attractive method to obtain effective and durable resistance to pathogens is to alter the morphology or anatomy of the plant so that the pathogen is prevented from developing; another is to alter the timing of the life cycle of the plant relative to that of the epidemic development of disease.

One example of the former method is resistance of barley to loose smut [*Ustilago nuda* (Jensen) Rostrup f.sp. *hordei* Schaffnit]. The smut spores are spread by wind from diseased heads of barley; they then enter passively the open flowers and infect the stigma and ovary wall of healthy plants. Several race-specific smut resistance genes are known in barley (see Table 21.1) but they are prone to be overcome by virulent races of the pathogen. Barley is, however, a naturally self-pollinating species which does not require opening of the flowers. Therefore, breeding for closed flowering has been one goal in barley breeding in Europe. The effect is a passive but impenetrable barrier that the smut pathogen can never overcome. An additional advantage is that closed flowering of barley also prevents infection with ergot [*Claviceps purpurea* (Fr. ex Fr.) Tul.]

An example of the latter method to obtain passive resistance is the early maturing barleys that start senescence before the peak of the epidemic. As far as I am aware, earliness has not been a target in barley breeding with the primary goal of escaping disase. The situation has rather been the reverse, in that disease resistance is unnecessary in early barleys. For example, winter barley in Denmark develops rapidly in the spring and early summer and 'escapes' the later developing powdery mildew epidemic. A somewhat similar effect may be attained by breeding barley for higher levels of partial resistance, possibly combined with strategies for using race-specific resistance in agriculture, e.g. by using variety mixtures, diversification schemes, and gene deployment. These measures aim at delaying the epidemic development of the diseases. But the principle is the same as in the case of early crops: separation of the timing of crop development from that of the disease.

A further example contributes to the relative resistance of rye to take-all [*Gaeumannomyces graminis* (Sacc.) v. Arx & Olivier var. *tritici* Walker]. Rye is moderately susceptible to take-all, but it also has the ability to regenerate more new roots than barley and wheat (see Scott, 1981). In this way rye suffers less from take-all since the increased numbers of roots, though infected, are able to maintain a good supply of water and mineral nutrients to the plant.

The genetics of passive resistance mechanisms in barley vary from one case to another. Earliness is governed mainly by a few major genes that are easy to manipulate in breeding, closed flowering is more complex, while root regeneration ability is probably very complex.

The genetics of resistance to fungal diseases are extremely varied. In each host–pathogen combination there are usually two or more kinds of resistance, mainly combinations of race-specific, partial and passive resistances. Each of these may be governed by one or more major genes and a considerable number of minor genes. The expression of the genes is also affected by genetic modifiers and suppressors and by the overall genotype of the host plant. The above grouping of resistances is, therefore, much more rigid than it is in nature.

Another general impression of the state of knowledge on the genetics of resistance to fungal pathogens is that most data that have been gained derive from relatively simple analyses of the phenotypic expressions of genes affecting more or less drastically the phenotypic development of the disease. It is primarily within the last decade that the underlying physiological, biochemical and biophysical mechanisms of resistance to fungal diseases have been studied (see elsewhere in this volume).

Conclusion

Fungal diseases of barley have been studied extensively, and much effort has been invested over many decades in breeding for disease resistance. The disease resistance increments attained by plant breeding have been additions to the levels of resistance present in the actual barley germplasm that has been exposed through evolution and cultivation to selection for healthy plants and high-quality seeds, i.e. selection against disease susceptibility.

The sources of resistance available today are varied and are mainly confined to those present in commercial cultivars, primitive forms of cultived barley and wild barley. The holdings of gene banks, mainly of cultivated barleys, should be more systematically screened for disease reaction, and the results made available. Induced mutations, formerly considered a potential source of genetic variability, are today thought to be an unlikely source of disease resistance, particularly of the gene-for-gene type. The potential of gene technology to provide new and durable resistance remains to be realized. We anticipate that disease resistance in transgenic plants will need to be balanced in relation to the overall genotype of the plant to eliminate undesirable pleiotropic effects of the foreign genetic material.

The genetics of resistance is immensely varied and related to the four different kinds of resistances (race-specific, partial, Mlo, and passive), each of which may be manipulated independently or combined in plant breeding. It is anticipated that Mlo resistance may not be confined to barley, but may also be found in other cereal species. It is foreseen that an effort to accumulate gradually higher levels of partial resistance, backed up by other resistances and appropriate strategies for diversification of the host plant population, will provide agriculture with barley varieties with durable resistance at a level that is economically tolerable to the farmer, and sufficiently high to make fungicides superfluous except for unusual 'hot-spots'.

Genetic resistance to barley diseases is, however, only one of the components that are decisive for disease severity. The genetics of the pathogens and environmental factors are equally important. Manipulation of the genetics of the host plant may shift the timing of crop development in relation to the timing of the disease epidemic to the extent that the actual disease may become unimportant.

Irrespective of the sources of resistance and their genetic control, a thorough understanding of the underlying physiological, biochemical and biophysical processes is required. The contributions from the 'new' disciplines – molecular biology and biotechnology – to disease resistance and pathogen virulence are eagerly anticipated by plant pathologists, geneticists and breeders. Added to the experiences from using 'conventional' techniques, the totality of knowledge gained will give us a better foundation on which to plan and execute efficient disease resistance breeding in the future.

References

Andersen, L. Mlo aggressiveness in European barley powdery mildew. In: Jørgensen, J.H. (ed.), *Integrated Control of Cereal Mildew: Virulence Patterns and Their Change* (in press).

Anonymous (1989) Barley stripe rust. *Barley Newsletter* 33, 42–3.

Anonymous. Designation of barley powdery mildew resistance and virulence in Europe. In: Jørgensen, J.H. (ed.), *Integrated Control of Cereal Mildews: Virulence Patterns and Their Change* (in press).

Bayles, C.J., Ghemawat, M.S. & Aist, J.R. (1990) Inhibition by 2-deoxy-D-glucose of callose formation, papilla deposition, and resistance to powdery mildew in an *ml-o* barley mutant. *Physiological and Molecular Plant Pathology* 36, 63–72.

Bothmer, R.v. & Jacobsen, N. (1985) Origin, taxonomy, and related species. In: Rasmusson, D.C. (ed.), *Barley*, Agronomy Vol. 26, Amer. Soc. Agron. Inc., Madison, WI, USA, pp. 19–56.

Bothmer, R.v., Jørgensen, R.B. & Linde-Laursen, I. (1987) Natural variation, phylogeny and genetic resources in *Hordeum*. In: Yasuda, S. & Konishi, T. (eds), *Barley Genetics V*, Okayama, Japan, pp. 23–33.

Brown, A.H.D., Frankel, O.H., Marshall, D.R. & Williams, J.T. (eds) (1989) *The Use of Plant Genetic Resources*. Cambridge University Press, Cambridge, UK, 382 pp.

Brown, J.K.M. & Jørgensen, J.H. A catalogue of mildew resistance genes in European barley varieties. In: Jørgensen, J.H. (ed.), *Integrated Control of Cereal Mildews: Virulence Patterns and Their Change* (in press).

Browning, J.A., Simons, M.D. & Torres, E. (1977) Managing host genes: epidemiologic and genetic concepts. In: Horsfall, J.G. & Cowling, E.B. (eds), *Plant Disease – An Advanced Treatise*, Vol. I, *How Disease Is Managed*. Academic Press, New York, USA, pp. 191–212.

Christiansen, S.K. & Smedegaard, V. (1990) Microscopic studies of the interaction between barley and the saprophytic fungus, *Cladosporium macrocarpum*. *Journal of Phytopathology* 128, 209–19.

Clifford, B.C. (1985) Barley leaf rust. In: Roelfs, A.P. & Bushnell, W.R. (eds), *The Cereal Rusts*, Vol. II, *Diseases, Distribution, Epidemiology, and Control*. Academic Press, Orlando, FL, USA, pp. 173–205.

Cook, R.J. & Jenkins, J.E.E. (1988) The contribution and value of pesticides to disease control in cereals. In: Clifford, B.C. & Lester, E. (eds), *Control of Plant Diseases: Costs and Benefits*. Blackwell Scientific, Oxford, UK, pp. 3–12.

Crute, I.R., De Wit, P.J.G.M. & Wade, M. (1985) Mechanisms by which genetically controlled resistance and virulence influence host colonization by fungal and bacterial parasites. In: Fraser, R.S.S. (ed.), *Mechanisms of Resistance to Plant Diseases*. Martinus Nijhoff/Dr W. Junk, Dordrecht, pp. 197–309.

Dubin, H.J. & Stubbs, R.W. (1986) Epidemic spread of barley stripe rust in South America. *Plant Disease* 70, 141–4.

Feuerstein, U., Brown, A.H.D. & Burdon, J.J. (1990) Linkage of rust resistance genes from wild barley (*Hordeum spontaneum*) with isozyme markers. *Plant Breeding* 104, 318–24.

Fleming, R.A. & Person, C.O. (1982) Consequences of polygenic determination of resistance and aggressiveness in nonspecific host:parasite relationships. *Canadian Journal of Plant Pathology* 4, 89–96.

Gabriel, D.W. & Rolfe, B.G. (1990) Working models of specific recognition in plant–microbe interactions. *Annual Review of Phytopathology* 28, 365–91.

Haahr, V., Skou, J.P. & Jensen, H.P. (1989) Inheritance of resistance to barley leaf stripe (*Drechslera graminea*) (Abstract) *Vorträge für Pflanzenzüchtung* 15, no. 3–15.

Hentrich, W. (1979) Allelwirkung und Pleiotropie mehltauresistenter Mutanten des ml-o Locus der Gerste. *Archiv für Züchtungsforschung* 9, 283–91.

IBPGR (1989) *International Crop Network Series 1. Report of an International Barley Working Session*. International Board for Plant Genetic Resources, Rome, 24 pp.

Jahoor, A. & Fischbeck, G. (1987) Genetical studies of resistance of powdery mildew in barley lines derived from *Hordeum spontaneum* collected from Israel. *Plant Breeding* 99, 265–73.

Jenkyn, J.F. & Plumb, R.T. (eds) (1981)

Strategies for the Control of Cereal Disease. Blackwell Scientific, Oxford, UK, 219 pp.

Jensen, H.P., Christensen, E. & Jørgensen, J.H. Powdery mildew resistance genes in 127 new Northwest European spring barley varieties. *Plant Breeding* (in press).

Jørgensen, J.H. (1987) Three kinds of powdery mildew resistance in barley. In: Yasuda, S. & Konishi, T. (eds), *Barley Genetics V*, Okayama, Japan, pp. 583–92.

Jørgensen, J.H. (1988a) Genetic analysis of barley mutants with modifications of powdery mildew resistance gene *Ml-a12*. *Genome* 30, 129–32.

Jørgensen, J.H. (1988b) Screening of *Hordeum vulgare* for powdery mildew resistance. (Abstract) *Nordisk Jordbrugsforskning* 70, 529.

Jørgensen, J.H. (1989) Resistente bygsorter, sortsblandinger og/eller fungicider? *6. Danske Planteværnskonference. Sygdomme og skadedyr 1989*, 159–71 (in Danish, English summary).

Jørgensen, J.H. (1990) Effect of 'unnecessary' powdery mildew resistance genes on agronomic properties of spring barley. *Norsk Landbruksforskning* Suppl. 9, 125–30.

Jørgensen, J.H. (1991) Experiences from mutation studies on cereal disease resistance. In: *Plant Mutation Breeding for Crop Improvement*. Vol. 2. IAEA, Vienna, Austria, pp. 81–91.

Jørgensen, J.H. Coordinator's report: disease and pest resistance genes. *Barley Genetics Newsletter* (in press).

Jørgensen, R.B. & Andersen, B. (1989) Karyotype analysis of regenerated plants from callus cultures of interspecific hybrids of cultivated barley (*Hordeum vulgare* L.). *Theoretical and Applied Genetics* 77, 343–51.

Khokhlova, A.P. (1987) Genetic analysis of resistance to loose smut in Ethiopian barleys. From: *Review of Plant Pathology* 68, p. 325, no. 2631.

Knudsen, J.Chr.N. (1986) Resistance to barley leaf stripe. *Zeitschrift für Pflanzenzüchtung* 96, 161–8.

Knudsen, J.Chr.N., Dalsgaard, H.-H. & Jørgensen, J.H. (1987) Partial resistance to barley powdery mildew. In: Yasuda, S. & Konishi, T. (eds), *Barley Genetics V*, Okayama, Japan, pp. 645–50.

Knüpffer, H. (1988) The European barley data base of the European cooperative programme for the conservation and exchange of crop genetic resources (ECP/GR). *Plant Genetic Resources Newsletter* 75/76, 17–20.

Lange, W. & Balkema-Boomstra, A.G. (1988) The use of wild species in breeding barley and wheat, with special reference to the progenitors of the cultivated species. In: Jorna, M.L. & Slootmaker, L.A.J. (comp.), *Cereal Breeding Related to Integrated Cereal Production*. Pudoc, Wageningen, The Netherlands, pp. 157–62.

Lehmann, L. & Bothmer, R.v. (1988) *Hordeum spontaneum* and land races as a gene resource for barley breeding. In: Jorna, M.L. & Slootmaker, L.A.J. (comp.), *Cereal Breeding Related to Integrated Cereal Production*. Pudoc, Wageningen, The Netherlands, pp. 190–4.

Matre, D.E. (ed.) (1982) *Compendium of Barley Diseases*. American Phytopathological Society Press, St Paul, Minnesota, USA, 73 pp.

Moseman, J.G., Nevo, E. & El-Morshidy, M.A. (1990) Reactions of *Hordeum spontaneum* to infection with two cultures of *Puccinia hordei* from Israel and United States. *Euphytica* 49, 169–75.

Munk, L., Jensen, H.P. & Jørgensen, J.H. Virulence and severity of barley powdery mildew in Denmark 1974–1989. In: Jørgensen, J.H. (ed.), *Integrated Control of Cereal Mildew: Virulence Patterns and Their Change* (in press).

Niks, R.E. (1988) Nonhost plant species as donors for resistance to pathogens with narrow host range. II. Concepts and evidence on the genetic basis of nonhost resistance. *Euphytica* 37, 89–99.

Parlevliet, J.E. (1988) Strategies for the utilization of partial resistance for the control of cereal rusts. In: *Breeding Strategies for Resistance to the Rusts of Wheat*, Mexico, D.F., CIMMYT 1988, pp. 48–62.

Priestley, R.H. & Bayles, R.A. (1988) The contribution and value of resistant cultivars to disease control in cereals. In: Clifford, B.C. & Lester, E. (eds) *Control of Plant Diseases: Costs and Benefits*. Blackwell Scientific, Oxford, UK, pp. 54–65.

Roelfs, A.P. (1985) Wheat and rye stem rust. In: Roelfs, A.P. & Bushnell, W.R. (eds) *The Cereal Rusts*, Vol. II, *Diseases, Distribution, Epidemiology, and Control*. Academic, Orlando, FL, USA, pp. 4–37.

Scott, P.R. (1981) Variation in host susceptibility. In: Asher, M.J.C. & Shipton, P.J. (eds) *Biology and Control of Take-all*. Academic, London, UK, pp. 219–36.

Sharp, E.L. (1985) Breeding for pest resistance. In: Rasmusson, D.C. (ed.), *Barley*, Agronomy Vol. 26, American Society of Agronomy Inc., Madison, WI, USA, pp. 313–33.
Simmonds, N.W. (1988) Synthesis: the strategy of rust resistance breeding. In: *Breeding Strategies for Resistance to the Rusts of Wheat*, Mexico, D.F., CIMMYT 1988, pp. 119–36.
Skou, J.P. & Haahr, V. (1987a) Screening for and inheritance of resistance to barley leaf stripe (*Drechslera graminea*). Risø Report No. 554, Risø National Laboratory, Denmark, 96 pp.
Skou, J.P. & Haahr, V. (1987b) Field screening for resistance to barley net blotch. *Annals of Applied Biology* 111, 617–27.
Skou, J.P., Jørgensen, J.H. & Lilholt, U. (1984) Comparative studies on callose formation in powdery mildew compatible and incompatible barley. *Phytopathologische Zeitschrift* 109, 147–68.
Smit, G. & Parlevliet, J.E. (1990) Mature plant resistance of barley to barley leaf rust, another type of resistance. *Euphytica* 50, 159–62.
Thordal-Christensen, H. & Smedegaard-Petersen, V. (1988) Comparison of resistance-inducing abilities of virulent and avirulent races of *Erysiphe graminis* f.sp. *hordei* and a race of *Erysiphe graminis* f.sp. *tritici* in barley. *Plant Pathology* 37, 20–7.
Tosa, Y. & Sakai, K. (1990) The genetics of resistance of hexaploid wheat to the wheatgrass powdery mildew fungus. *Genome* 33, 225–30.
van Leur, J.A.G., Ceccarelli, S. & Grando, S. (1989) Diversity of disease resistance in barley landraces from Syria and Jordan. *Plant Breeding* 103, 324–35.
Walther, U. & Lehmann, Chr.O. (1980) Resistenzeigenschaften im Gersten- und Weizensortiment Gatersleben. 24. Prüfung von Sommer- und Wintergersten auf ihr Verhalten gegenüber Zwergrost (*Puccinia hordei* Otth). *Kulturpflanze* 28, 227–38.

22

Biochemical and Molecular Analyses of the Response of Barley to Infection by Powdery Mildew

Tomas Bryngelsson & David B. Collinge

Pathogen attack is probably the most important stress factor plants are exposed to. The ability to stop an invasion depends on the efficiency of preformed barriers and induced responses. The constitutive defences, such as the cuticle, the cell wall, and compounds toxic to micro-organisms are generally effective as barriers to invasion. However, plants can also respond defensively to colonization by pathogenic micro-organisms. These active responses encompass a complex of strategies to minimize the impact of the invading micro-organism. The response may be very complex, but the responses to different pathogens, both in different plant species and within a single species, are surprisingly similar. There is evidence that resistance against the many different micro-organisms a plant might encounter is provided not by defences tailored to each organism but rather by a very flexible detection system which triggers a defence complex of broad effectiveness (Collinge & Slusarenko, 1987; Bowles, 1990).

There have been many reports describing physiological and cytological changes in barley which follow infection by *Erysiphe graminis* DC. ex Mérat (Kita *et al.*, 1980; Smedegaard-Petersen & Tolstrup, 1985). However, knowledge of the biochemical and molecular processes underlying both these changes and induced resistance is still very limited (Smedegaard-Petersen *et al.*, 1991). The application of existing techniques of molecular biology and biochemistry is currently giving a more detailed picture of the defence response: genes and gene products are being identified, and their expression analysed in response to different pathogens.

This chapter reviews current knowledge of resistance responses to pathogens, with emphasis on powdery mildew. To avoid duplication within this book, discussion of results from the laboratories of García-Olmedo, Scott and Somerville is largely excluded.

Characteristics of the interaction between barley and mildew

Barley powdery mildew has particular advantages for studies of the interaction between host and parasite. These include distinct morphological stages of differentiation in primary infection, high infection efficiencies and reasonably synchronous development of the parasite population during early stages of the interaction. The obligate parasitism does not seriously hinder studies of the plant response, but is clearly a major handicap for studies of the parasite independent of the host (Ellingboe, 1978).

Erysiphe graminis goes through several distinct morphological stages in the process of infection of host plants (Figs 22.1 and 22.2). On the host epidermis, the infection starts by the formation of a short non-appressorial germ tube by the germinating conidia. Shortly afterwards, an appressorial germ tube emerges which forms an appressorium below which a penetration peg develops. Once the infection peg has crossed

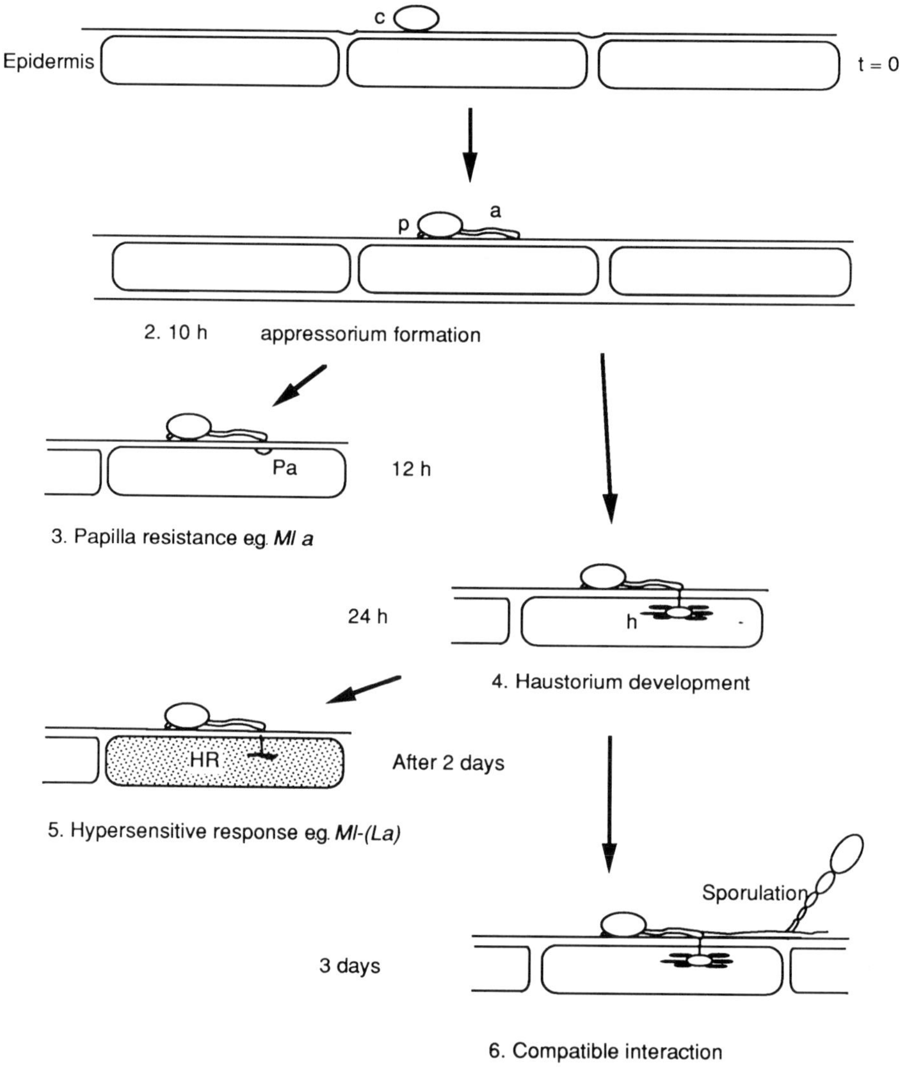

Fig. 22.1. The development of an *Egh* conidium (c) on a barley leaf illustrating different interaction types. 2. primary (p) and appressorial (a) germ tubes. 3. Papillae resistance: papilla formation (Pa). 4. Haustorium formation (h). 5. Hypersensitive response (HR). 6. Sporulation in a compatible interaction.

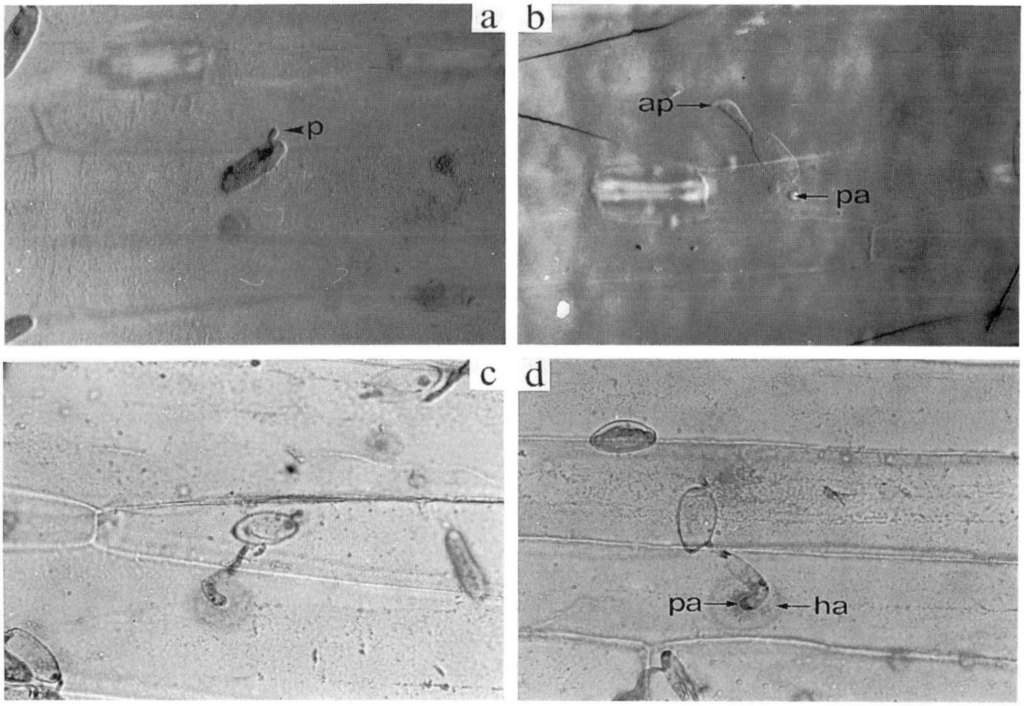

Fig. 22.2. Germination and development of *Egh* spores on barley leaves showing: (a) 3 h, primary germ tube; (b) 9 h, papilla development to primary germ tube; (c) 12 h, papilla development to appressorial germ tube; (d) 24 h, papilla and haloes associated with penetration attempts. (Photographs by Per L. Gregersen.)

the cell wall, it enters the epidermal cell lumen where it expands and begins to form the primary haustorium. The fungus is nourished via the haustorium, and hyphae start to grow on the leaf surface, from which new haustoria and hyphae are formed (Jenkyn & Bainbridge, 1978).

The resistance reaction in the barley / mildew system

Race-specific resistance of barley to mildew is manifested at two distinct developmental stages, each associated with specific cytological reactions in the host. In many cases, resistance is exhibited at the penetration stage (refered to here as papilla resistance). This type of resistance is associated with race-specific resistance conferred by, for example, certain *Ml-a* genes, *Ml-c, Ml-g, Ml-h*, with non-specific resistance conferred by the *ml-o* gene, and with non-host resistance to *E. graminis* f. sp. *tritici* Em. Marchal (Thordal-Christensen & Smedegaard-Petersen, 1988a) and *Erysiphe pisi* DC. ex St.-Anans (Kunoh *et al.*, 1985). The hyphae can nevertheless penetrate the papillae in 1–2% of cases in certain interactions, but development is invariably terminated subsequently. A hypersensitive response (HR) is exclusively associated with race-specific resistance controlled by other resistance genes, such as *Ml-at, Ml-k, Ml-La, Ml-p* (see Kølster *et al.*, 1986).

The rapidity of hypersensitive cell death of challenged cells is variable and dependent on the host. Indeed, host cell death affecting 20–40% of infecting propagules has been reported in susceptible barley lines as well as in lines exhibiting partial resistance (Asher & Thomas, 1983).

The initial stages of the host reaction to mildew in any penetration attempt are accompanied by the formation of cytoplasmic aggregates. These aggregates deposit internal epidermal cell wall appositions, papillae, which are surrounded by haloes. Penetration pegs of *E. graminis* often fail to pierce host papillae even in compatible interactions, and the epidermal cells will return to an apparently normal state. This is always the case in papilla resistance. The macromolecular constituents of papillae, such as phenolic compounds and callose, are usually considered to be the reason for resistance. In other situations, the germinating conidia continue to attack the host cells, and as a result the epidermal cells will collapse (hypersensitive reaction). HR is accompanied by rapid cell death and a decreased rate of colonization in the vicinity of the lesion. In some host–pathogen interactions, HR is also accompanied by the production of phytoalexins in surrounding cells. Whether this is also the case for barley is unclear. There is only one report of phytoalexin-like substances in barley (Oku *et al.*, 1975), although no phytoalexin has been isolated. Nevertheless, the key enzyme of the phenylpropanoid pathway, PAL (phenylalanine ammonia-lyase), is apparently induced in response to infection with *E. graminis* (Shiraishi *et al.*, 1989). Phenolic compounds which autofluoresce have also been detected in papillae (Mayama & Shishiyama, 1978; Aist & Israel, 1986).

Induced resistance and non-host resistance

Induction of resistance in plants is well documented, and can be defined as a defence based on physical and chemical barriers activated in the host plant as response to a previous infection/attempted infection, or to elicitor-like substances. It acts against subsequent infection attempts by otherwise pathogenic organisms. The consequences of induced resistance on mildew-infected barley leaves is lower penetration efficiency, smaller colonies and less sporulation. Thus, induced resistance resembles the genetically determined race-non-specific resistance (Stenzel *et al.*, 1985). Induced resistance is largely independent of race-specificity (Thordal-Christensen & Smedegaard-Petersen, 1988b).

Induced resistance in barley can be localized (Ouchi *et al.*, 1974, 1976; Cho & Smedegaard-Petersen, 1986) or systemic (Hwang & Heitefuss, 1982). Locally induced resistance appears to be much more effective than systemically induced resistance. It is also likely that different mechanisms underlie these two types of resistance. Locally induced resistance is apparently confined to single cells where the pathogen attempts to penetrate (Ouchi *et al.*, 1974, 1976), and, indeed, there is evidence for localization within regions of single cells (Woolacott & Archer, 1984). Locally induced resistance appears to involve the activation of the papilla response since the fluorescent papillae are larger in induced cells (Thordal-Christensen & Smedegaard-Petersen, 1988a).

Non-host pathogens and avirulent races of a pathogen are usually stopped by the preformed defence of the plant, or by an early induction of the active response. Johnson *et al.* (1982) inoculated 15 monocots with *E. graminis* f. sp. *hordei* Em. Marchal (*Egh*) and analysed the defence response. The defences were expressed on or in the following host structures: (i) the leaf surface, which was often inhospitable to germination and differentiation of infection structures; (ii) the cuticle and cell wall, which the fungi frequently failed to penetrate; (iii) the papillae, which were frequently induced but rarely penetrated; and (iv) the first cell invaded, which had a high probability of collapsing hypersensitively in the few instances in which the fungus successfully entered a cell. Similar types of response were found by Gustafsson & Claesson (1988) when populations from 24 wild *Hordeum* species were screened for resistance against four isolates of *Egh*.

The non-host *E. graminis* f. sp. *tritici* *(Egt)* can induce a relatively high level of resistance in barley against subsequent penetration attempts by a virulent race of *Egh* (Thordal-Christensen & Smedegaard-Petersen, 1988b). In addition, saprophytic fungi are capable of inducing resistance in barley (Smedegaard-Petersen & Tolstrup, 1985). Thus, *Cladosporium macrocarpum* can induce a resistance response in barley, although not as rapidly and efficiently as the non-host *Egt* (Gregersen & Smedegaard, 1989). The presence of *C. macrocarpum* elicited marked reactions in epidermal cells such as aggregation of the cell content and development of papilla-like structures (Christiansen & Smedegaard, 1990).

An important feature of the induced resistance, and the resistance to non-pathogenic or saprophytic fungi, is that it probably is a general defence response. Sahashi *et al.* (1989) inoculated germ-free barley plants with *Egh* or *Erysiphe polygoni* DC. ex St.-Anans, and found an increased susceptibility in comparison with those not grown under germ-free conditions. Furthermore, Smedegaard-Petersen & Tolstrup (1985) found that fungicide treatment which suppressed the saprophytic flora at the leaf surface resulted in increased yield, thus indicating that saprophytes extracted a metabolic cost from the host plant. These results suggest that, in plants grown in nature, a general response is more or less constantly activated by the interaction of the plant with surface microbes.

Elicitors of defence responses

The rapid induction of defence responses in plants is generally believed to be the result of the interaction between elicitors and putative receptors of the host plant. The elicitor molecules can be various types of molecules, for example, glucans, proteins, glycoproteins, fatty acids and cell wall fragments (De Wit, 1986; Dixon & Lamb, 1990), and can have an origin in the parasite (exogenous elicitors) or in the host plant (endogenous elicitors). Very few elicitor-active substances isolated exhibit race-specificity; many do not even show species-specificity (Dixon & Lamb, 1990). Indeed, many abiotic factors such as heavy metals, detergents, polyanions and physical damage can act as elicitors.

Race-specific elicitors have been predicted to represent direct, or indirect, products of avirulence genes (Keen, 1986). However, only two reports provide evidence for the isolation of race-specific elicitors from pathogenic fungi (Tepper & Anderson, 1986; De Wit *et al.*, 1989). The concept that a receptor for a race-specific elicitor represents the product of the complementary host resistance gene is highly attractive, and generally accepted, but remains to be proven.

The existence of elicitors from powdery mildew has not yet been described, although Kristensen (1989), in preliminary experiments, reported elicitor-active substances in germinating mildew spores. Yamada *et al.* (1990) demonstrated the presence of endogenous elicitors in barley seeds capable of inducing local resistance. These compounds also induce systemic resistance in seedlings, which indicates a role of the endogenous elicitors in the seed for juvenile resistance against pathogen attack.

Pathogen stress causes changes in protein synthesis

There have been numerous reports on physiological and biochemical changes in barley tissues following infection with powdery mildew. Early findings demonstrated a pronounced increase in the respiration rate (Shaw & Samborski, 1957) and a decline in the rate of photosynthesis in mildewed tissues (Scott & Smillie, 1963; Edwards & Allen, 1966), probably due to a reduction of the amount of ribulose bisphosphate carboxylase/oxygenase (Walters & Ayres, 1984). Alterations in enzyme activities involved in major metabolic pathways, for example, phosphoenolpyruvate carboxylase, 3-phosphoglycerate kinase and glyceraldehyde-3-phosphate dehydrogenase was also reported (Walters & Ayres, 1984).

Sako & Stahmann (1972) investigated the isozyme patterns of 14 enzymes in near-isogenic barley lines inoculated with *Egh*. Several new bands appeared in the compatible reaction as well as quantitative alterations of pre-existing bands. A hypersensitive line (*Ml-k*) showed changes after inoculation similar to those of the susceptible line. Systemic infection of barley with barley stripe mosaic virus, and local infection with powdery mildew, also induce quantitative changes in the protein pattern and reductions in the level of rRNA and chlorophyll (White & Brakke, 1983; Brakke *et al.*, 1988).

These, and many other investigations, have conclusively demonstrated major alterations in metabolism when barley plants have been inoculated with virulent or avirulent races of *E. graminis*. It has, however, not been possible until recently to define specific genes and gene products involved in the resistance reaction of barley to powdery mildew and other pathogens.

Induction of biosynthetic pathways

The production of secondary metabolites as a response to pathogens has been extensively studied (Dixon, 1986; Collinge & Slusarenko, 1987). The metabolic pathway which has received most attention in relation to plant defences is the phenylpropanoid pathway, involved in the synthesis of phenolic compounds, such as lignin and many phytoalexins.

The initial reaction in this pathway is the conversion of phenylalanine to cinnamic acid catalysed by phenylalanine ammonia-lyase (PAL). This enzyme, for which several isozymes are known, is constitutively expressed in various tissues and induced by different stress stimuli such as wounding, ultraviolet radiation, elicitors and pathogen attack. Other important enzymes in this pathway induced by various pathogens and elicitors are chalcone synthase and chalcone isomerase, cinnamic acid 4-hydroxylase and 4-coumarate: CoA ligase (Dixon, 1986; Collinge & Slusarenko, 1987; Hahlbrock & Scheel, 1989; Dixon & Lamb, 1990).

Studies of the involvement of phenylpropanoid metabolism in resistance responses have mostly involved dicotyledonous plant species, but lignification and enhancement of PAL and tyrosine ammonia-lyase activity have also been reported in graminaceous plants infected with various pathogens (Ride, 1975; Yamamoto *et al.*, 1977; Mayama *et al.*, 1981). In wheat inoculated with *Egt*, Green *et al.* (1975) found a significant increase in PAL activity only 4 h after inoculation and lignin synthesis after 92 h.

There have been few reports on pathogen activation of the phenylpropanoid pathway in barley. However, Shiraishi *et al.* (1989) found that PAL activity increased 2 h after infection with *Egh*, reaching a maximum at 4 h. A tentative phytoalexin in barley has been proposed (Oku *et al.*, 1975), and various phenolic compounds are apparently associated with papilla development.

Cell wall modifications

The extent to which cell walls act as a natural barrier to invasion by micro-organisms is a function of composition. Consequently, lignified secondary walls are likely to be more effective against penetration than non-lignified primary walls due to the physical and chemical resistance of lignin (Lewis & Yamamoto, 1990). Cuticular surfaces and suberized cell walls present additional barriers to microbial penetration. However, most fungal pathogens can penetrate cell walls either by wall-degrading enzymes, physical penetration, or a combination of both (Bacic *et al.*, 1988).

A number of possible changes can occur in the cell wall in response to infection: thickening, lignification, suberization, synthesis of hydroxyproline-rich glycoproteins, phenolic compounds, localized wall modifications (papillae and haloes) and silica deposition. In most cases, the chemical changes in host walls are limited to cells adjoining or near those invaded by the pathogen. Among these cell wall modifications, papillae are perhaps the most studied in barley responding to mildew infection. Callose, considered to be the principal component of papillae, is synthesized by the Ca^{2+}-dependent enzyme callose synthase (β-1,3-glucan synthase) and added to the inside of the cell wall just outside the plasmalemma (Kauss & Jeblick, 1986). Callose synthase, which is located in the plasma membrane, has been purified from *Brassica oleracea*, and probably consists of five polypeptides. The purified enzyme is activated by Ca^{2+} and spermine and requires a short oligosaccharide as primer (Fredrikson *et al.*, 1991). Papillae also contain polyphenolic compounds, perhaps including lignin, as well as cellulose and silica (reviewed in Aist, 1983; Heitefuss & Ebrahim-Nesbat, 1986).

Resistance to powdery mildew in barley controlled by the recessive *ml-o* mutant gene is expressed at the early penetration stage of the fungus (Jørgensen & Mortensen, 1977), and there is evidence that the genetic lesion affects papilla formation. Thus, Skou (1982) observed, using the histochemical stain lacmoid (resorcin blue), which is believed to be specific for callose, that papillae formed in ten *ml-o* mutants were 1.4–1.9 times larger in diameter than in the respective mother varieties. Plasma membrane fractions prepared from epidermal strips of *ml-o* and *Ml-o* barley lines were assayed for β-1,3-glucan synthase activity. *In vitro* activity was apparently unaffected in both genotypes following infection (L.H. Pedersen, pers. comm.). Thus, there is currently no evidence to suggest that the lesion lies in the structural genes encoding this enzyme. Coleoptiles carrying the *ml-o* gene incubated in Ca^{2+}-rich medium had a high degree of resistance to *Egt* when compared with coleoptiles incubated at a low Ca^{2+} concentration. Treatment with calcium ion chelators also inhibited resistance (Gold *et al.*, 1986). This inhibition could be overcome by adding more calcium (Bayles & Aist, 1987). In contrast, Takamatsu *et al.* (1978) noted that the addition of calcium ions increased the incidence of haustoria formation in a compatible interaction, and decreased papilla formation. The inhibitor of callose synthesis, 2-deoxy-D-glucose, causes a decrease in papillae frequency and an increase in penetration efficiency (Bayles *et al.*, 1990). The presence of phenolic compounds in the papillae also seems to be of importance. Aist *et al.* (1988) demonstrated that papillae in a resistant *ml-o* isoline contain a UV-absorbing component apparently absent from the corresponding susceptible isoline.

Hydroxyproline-rich glycoproteins (HRGPs)

An important cell wall modification as response to infection is the synthesis of various proteins of unusual composition, including hydroxyproline-rich glycoproteins (HRGPs). HRGPs are characteristic of dicotyledon primary walls but are also found at low levels in graminaceous species, although considerable variability exists between species with respect to their hydroxyproline richness and abundance in the cell wall (Bacic *et al.*, 1988). HRGPs occur as rod-shaped molecules coated with covalently bound sugar residues, and they are insolubilized through cross-linking by wall-associated peroxidases (Cooper & Varner, 1984). HRGPs accumulate in the cell wall of dicotyledonous plants in response to wounding and pathogen stress indicating a role in plant defence; there have been numerous reports in support of this (reviewed in Bacic *et al.*, 1988; Bostock & Stermer, 1989; Showalter & Varner, 1989).

Hydroxyproline is relatively abundant in cell walls of maize. HRGP gene sequences from several dicots have been shown to cross-hybridize to barley, maize, wheat, oat and rice DNA or RNA, which indicates a general occurrence in Poaceae (Showalter & Varner, 1989; Munk-Kruse & Meldgaard-Madsen, 1990). The role of these HRGPs in plant defence in the cereals is largely unknown, and there is currently little evidence for this (Clarke *et al.*, 1983).

Thionins are a family of low-molecular-weight, sulphur-containing cell wall proteins which are widely spread in the plant kingdom (García-Olmedo *et al.*, 1989). They have bactericidal and fungicidal properties and are induced in barley inoculated with *Egt* (Bohlmann *et al.*, 1988). Their possible role in resistance is described in more detail in Chapter 16 of this volume.

Peroxidase

Peroxidase reactions primarily involve H_2O_2 and a diverse array of hydrogen donors (van Huystee, 1987). A wide range of compounds present in plant tissues known to act as substrate for plant peroxidase include hydroxyphenolic compounds, suberin monomers and proteins (Cooper & Varner, 1984; Lewis & Yamamoto, 1990). The function of peroxidase *in vivo* is unclear, but polymerization of lignin monomers is generally considered the most important, at least in healthy tissues. Such reinforced walls are believed to be more resistant to penetration by fungal pathogens in particular. Given the chemical diversity of these substrates, it is not surprising that a number of peroxidase isozymes are recognized in many plant tissues. Highly acidic and basic, as well as neutral forms are widely represented, but are relatively homogeneous in size (30 000–35 000 M_r), and are all glycosylated haemoproteins. Such isozymes are generally defined using guaiacol (*o*-methoxyphenol) as the hydrogen donor. Several isozymes have been shown to possess phenoloxidase (diphenol oxidase, laccase and monophenol oxidase) activities in addition to peroxidase activity (De Biasi & Badiani, 1990).

Some peroxidase proteins, and the corresponding genes, have been sequenced from both monocot and dicot species (Mazza & Welinder, 1980; Lagrimini *et al.*, 1987; Roberts *et al.*, 1988; Schweizer *et al.*, 1989; Rasmussen *et al.*, 1991a, b; Thordal-Christensen, Brandt, Cho, Gregersen, Rasmussen, Smedegaard-Petersen & Collinge, unpublished). Similarity comparisons show considerable sequence diversity, from 37 to 56% amino acid identity, except for three horseradish isozymes, which exhibit up to 94% sequence identity to each other at the amino acid level (Fujiyami *et al.*, 1988). This seems low in comparison with other protein families, but this diversity presumably reflects the wide range of specialized functions *in vivo*.

Peroxidase enzyme activity increases following infection of both dicot and monocot plant species with pathogens (Hislop & Stahmann, 1971; Arora & Bajaj, 1985). The differential induction of peroxidase isozymes in barley has been investigated following inoculation with *Egh*. There is evidence that the major forms induced differ in extracellular washings (Kerby & Somerville, 1989) and total leaf extracts (Hislop & Stahmann, 1971), although these studies are not directly comparable due to differences in both electrophoretic and peroxidase staining techniques. An extracellular location is consistent with a role in defence: the three possible functions concerning peroxidases most often quoted in this context are lignification, suberization and polymerization of hydroxyproline-rich glycoproteins, all cell wall depositions. Furthermore, there is evidence for hydrophobic N-terminal extensions of the cloned horseradish, potato and tobacco peroxidases, thought to represent signals for secretion of the mature enzyme (Lagrimini *et al.*, 1987; Fujiyami *et al.*, 1988; Roberts *et al.*, 1988).

Increases in both peroxidase and polyphenol oxidase activities are associated with induced resistance to *Rhizoctonia solani* in mung bean (Arora & Bajaj, 1985): specific peroxidase isozymes were induced by ethephon and the pathogen. Six peroxidase isozymes were induced in cucumber hypocotyls infected with *Cladosporium cucumerinum*, only one of which was constitutive. The same isozymes were also induced by wounding but not by an oligogalacturonide elicitor. In contrast, lignification was induced by the heptagalacturonide, but not wounding (Robertsen, 1986; Robertsen & Svalheim, 1990; Svalheim & Robertsen, 1990). This implies that the peroxidases are not responsible for lignification in this system.

A peroxidase gene (WIR 3) was identified (on the basis of sequence similarity) among six cDNA clones induced in wheat as a response to *Egh*. Expression of this gene correlated with the development of induced resistance (Schweizer *et al.*, 1989). A similar study in barley inoculated with *Egt* also revealed a peroxidase gene sequence (pBT6-3: Thordal-Christensen, Brandt, Cho, Gregersen, Smedegaard-Petersen &

Collinge, unpublished). Hybridization experiments involving cDNA clones for these two peroxidases, and cDNA clones for a leaf peroxidase gene and a seed-specific peroxidase from barley (Rasmussen *et al.*, 1991 and in press) indicate that WIR 3 and pBT6–3 are more related in sequence to each other than to the other barley genes (Thordal-Christensen, Brandt, Cho, Gregersen, Rasmussen, Smedegaard-Petersen & Collinge, unpublished).

A peroxidase cDNA isolated from potato tissue represents an anionic isozyme believed to be responsible for suberin polymerization (Roberts *et al.*, 1988). Induction of corresponding mRNA occurs as a response to wounding of tuber tissue and tomato fruit. Use of the potato cDNA clone as a hybridization probe demonstrated wound induction of peroxidase mRNA earlier in resistant than in susceptible, near-isogenic tomato lines differing in their resistance to *Verticillium* spp. Elevated mRNA levels were detected within minutes of treatment of suspension cells derived from resistant plants by an elicitor prepared from *Verticillium* culture filtrate, but were scarcely detectable in cell cultures derived from susceptible tissues (Mohan & Kolattukudy, 1990). A vascular coating of suberin was deposited in infected resistant but not in susceptible plants, and this was associated with trapped spores (Street *et al.*, 1986). *In situ* hybridization demonstrated that mRNA for the peroxidase was confined to associated vascular bundles (P.E. Kolattukudy, pers. comm.). Tobacco plants transformed with glucuronidase (GUS) under the regulation of the promotor of the peroxidase expressed β-GUS in wounded regions. In addition, transformation of tobacco plants with the peroxidase gene under the control of the S35 promotor of tobacco mosaic virus apparently developed normally (Kolattukudy *et al.*, 1991). So, although previous studies indicated a role for peroxidase in several species through correlation with defence responses, these studies will provide the opportunity for obtaining the necessary genetic proof. Furthermore, this type of study provides the tools for determining the biochemical role(s) for different peroxidases in the plant.

PR proteins and hydrolytic enzymes

Pathogenesis-related (PR) proteins are induced in the host plant in response to pathogen infection or related stress (Antoniw *et al.*, 1980). They were first discovered in 1970 in tobacco reacting hypersensitively to tobacco mosaic virus infection, and considerable information has accumulated since then concerning their biochemical properties, induction and relation to resistance. When hypersensitive response (HR) is induced by microbial pathogens, a set of defence genes are activated in the host, and large amounts of soluble, low-molecular-weight proteins are transported into the intercellular space. The presence of these PR proteins is closely correlated with HR and with induced resistance. Biochemically, PR proteins are either acidic or basic, although most of them exist as both acidic and basic isoforms. Until 1987, no physiological function was known for PR proteins in spite of their characteristic properties and the large amounts that accumulated during active defence (for review see Carr & Klessig, 1989; Bol *et al.*, 1990).

During the past four years, there has been a dramatic increase in the understanding of the structure, synthesis, location and, to some extent, the function of PR proteins. In tobacco, the best-studied species, more than 20 PR proteins have been defined. The PR 1 proteins (PR 1a, 1b and 1c) were the subject of much of the earlier work but their biochemical function is still unknown. An important group of PR proteins are the chitinases and β-1, 3-glucanases, which have been shown to inhibit the growth of a number of different fungi (Schlumbaum *et al.*, 1986; Mauch *et al.*, 1988) and to have a lytic action on bacteria (Boller *et al.*, 1983). Another group of PR proteins exhibits high homology to thaumatin, an intensely sweet-tasting protein found in the arils of the African

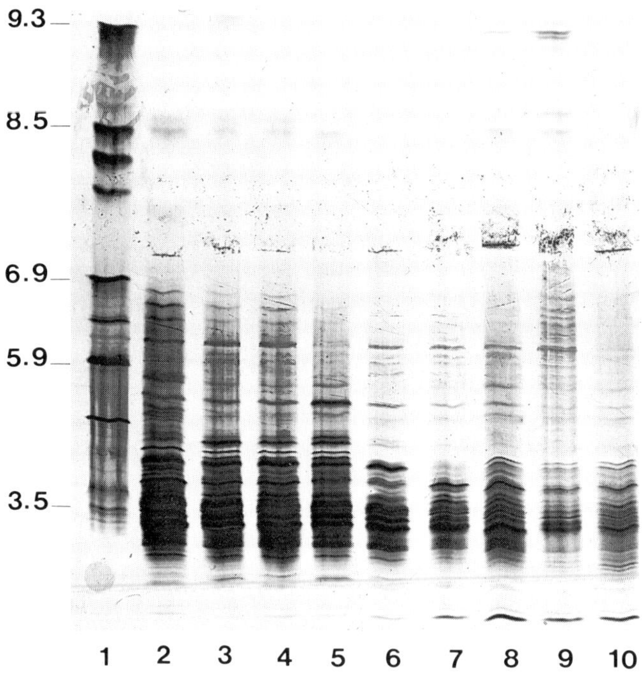

Fig. 22.3. Induction of pathogenesis-related proteins in barley inoculated with an incompatible race of *Egh*. Soluble proteins were extracted and analysed by isoelectric focusing. (1) pI marker, (2) 0 h after inoculation, (3) 6 h, (4) 9 h, (5) 15 h, (6) 24 h, (7) 3 days, (8) 6 days, (9) 9 days, (10) 12 days after inoculation.

shrub *Thaumatococcus daniellii*. Other extracellular proteins induced in response to pathogens, such as the proposed glycine-rich cell wall protein in tobacco (van Kan *et al.*, 1988) and extracellular peroxidases could also be considered as PR proteins. However, the function of the remaining PR proteins is still unknown (Bol *et al.*, 1990).

The first indication for the presence of PR proteins in monocots came when White *et al.* (1987) demonstrated the presence of PR 1-type proteins in mildew-infected barley with immunochemical techniques. A year later, PR proteins, including four chitinases, were found in maize infected with brome mosaic virus (Nasser *et al.*, 1988) and in barley infected with *Egh*. In barley, some ten new proteins were visualized with SDS-PAGE and isoelectric focusing in five varieties carrying the *Ml-La* gene in different genetic backgrounds. These proteins were only induced during HR (Fig. 22.3). The new proteins have low molecular weights and either very low or very high isoelectric points. Thus, they resemble PR proteins found previously in dicotyledonous species (Bryngelsson *et al.*, 1988).

Several of the barley PR proteins have been purified to homogeneity (Fig. 22.4), and one of the acidic proteins has also been characterized in more detail. Hv-1 is a highly acidic (pI 3.4), monomeric protein with an apparent molecular weight of 19 kDa. It is probably not a glycoprotein, as no binding was detected to ^{125}I-labelled concanavalin A or peanut agglutinin in a solid-phase radio-immunoassay. When the N-terminal sequence was determined and compared with known protein sequences, a striking homology was found to the N-termini of six other plant proteins (Fig. 22.5). Alignment of the amino acids revealed 48% homology to osmotin and NP24, proteins which are synthesized in response to salinity stress in tobacco and tomato, respectively. The same level of sequence identity was apparent with a maize trypsin/α-amylase bifunctional inhibitor, MAI, and with thaumatin. The PR protein pot-C from potato (Pierpoint *et al.*, 1990)

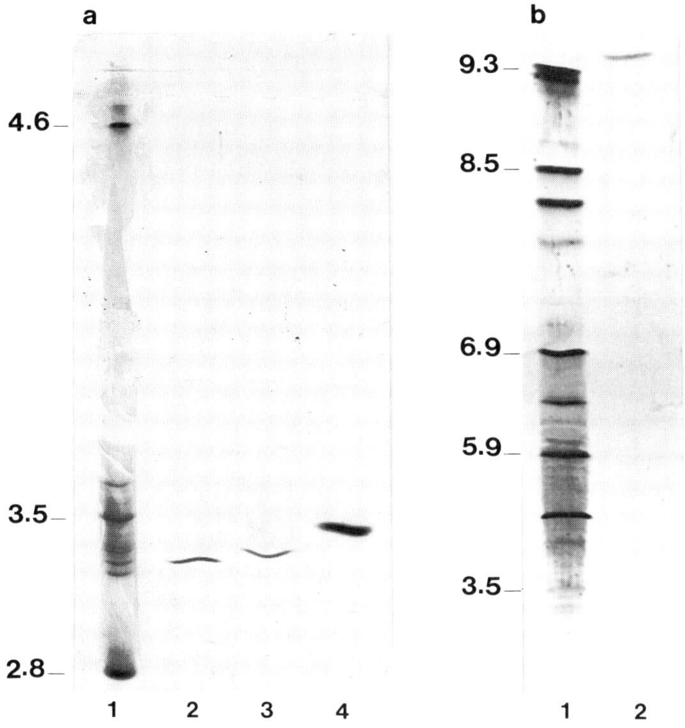

Fig. 22.4. Purified pathogenesis-related proteins from barley analysed by isoelectric focusing in (a) a pH 2.5–5 gel: (1) pI marker, (2) Hv-1, (3) Hv-2, (4) Hv-3 and (b) a pH 3–10 gel: (1) pI marker, (2) Hv-4.

	Sequence (positions 1, 10, 20, 30)
Hv-1	A T F N I K N N C G S T I W P A G I P V - - - - - G G G F E L X S
PR-5	A T F D I V N K C T Y T V W A A A S P - - - - - - G G G R R L N S
pot-C	A T F D I T N R C T Y P V W A A A S P - - - - - - G G G R R L D S
Osm	A T I E V R N N C P Y T V W A A S T P I - - - - - G G G R R L D R
NP24	A T I E V R N N C P Y T V W A A S T P I - - - - - G G G R R L N R
MAI	A V F T V V N Q C P F T V W A A S V P V - - - - - G G G R Q L N R
Tha	A T F E I V N R C S Y T V W A A A S K G D A A L D A G G R Q L N S

Fig. 22.5. Alignment of the N-terminal amino acid sequences of the thaumatin-like proteins. Boxed residues are homologous with the Hv-1 protein sequence from barley (Bryngelsson & Gréen, 1989). Numbering is based on the thaumatin (Tha) sequence from *Thaumatococcus daniellii* (Edens *et al.*, 1982). PR-5, PR (pathogenesis-related) protein from tobacco (Cornelissen *et al.*, 1986); pot-C, PR protein from potato (Pierpoint *et al.*, 1990); Osm, osmotin, a salt-induced protein from tobacco (Singh *et al.*, 1987); NP 24, a salt-induced protein from tomato (King *et al.*, 1986); MAI, a trypsin/α-amylase inhibitor from maize (Richardson *et al.*, 1987).

shows 52% similarity to Hv-1, while even higher homology (58%) was discovered with the N-terminus of the PR-5 protein of tobacco (Bryngelsson & Gréen, 1989). At present, it is difficult to envisage a functional relationship among the thaumatin-like (TL) proteins. However, their occurrence in both monocot and dicot species, and their structural homology suggest a common unique tertiary structure. The TL proteins have a tightly coiled structure stabilized by disulphide bonds, and have been classified as a separate family within a larger protein superfamily (Pierpoint & Shewry, 1987).

We have tried to isolate cDNA clones for the Hv-1 protein using the heterologous probe PROB 12 (Cornelissen *et al.*, 1986) which encodes the thaumatin-like PR-5 protein in tobacco. A λgt10 cDNA library, consisting of about 100 000 clones was constructed from barley, cv. Alva, *Ml-La*, exhibiting a hypersensitive reaction to the mildew isolate MK 81–10. No signal, however, was detected using PROB 12 for plaque hybridizations or genomic Southern blots of 'Alva' DNA. Therefore, an oligonucleotide mixture based on part of the N-terminal amino acid sequence was used as a probe. This gave a strong signal in plaque hybridizations and we have now isolated some 20 clones from the library (Bryngelsson *et al.*, unpublished).

The barley PR proteins Hv-2 and Hv-3 are highly acidic proteins (Fig. 22.4) with molecular weights of 33 and 36 kDa, respectively. Hv-4 is a basic protein (pI > 10) with an apparent molecular weight of 15 kDa. All three proteins are blocked at the N-terminus and so far we have only obtained sequence information from Hv-4 (Bryngelsson, Green & Ek, unpublished). This protein yielded eight polypeptides when cleaved with Lys-C, of which two have been sequenced (13 + 17 amino acids). Sequence comparisons revealed high homology (ca. 55%) to the acidic PR 1-type proteins of tobacco (Cornelissen *et al.*, 1987) and the basic counterpart p14 in tomato (Lucas *et al.*, 1985). The highest level of sequence identity, 70%, was found with the PROB 51 cDNA clone from tobacco which codes for a suspected basic isoform of PR 1 (Cornelissen *et al.*, 1987).

Chitinase activity in barley increased two- and four-fold by 7 days after inoculation with *Egh* in compatible and incompatible interactions, respectively. However, this increase occurred several days after manifestation of resistance in the interactions studied, and therefore does not appear to be associated with *Ml-a* resistance. Of six chitin-binding proteins extracted from these leaves, one was an endochitinase of M_r 30 000, pI 9.1, which showed amino acid sequence similarity with several published chitinases. A minor part of the chitinase activity was associated with intercellular washing fluids (Kragh *et al.*, 1990). Several polypeptides present in intercellular washing fluids of *Ml-La* but not *Ml-a* resistant plants exhibiting resistance against avirulent *Egh*, cross-reacted with antiserum raised against chitinase and β-1,3-glucanase from sugarbeet and PR-1 from tobacco (Andersen, 1989). Furthermore, antiserum raised against the maize PR proteins m3 and m4 (acidic β-1,3-glucanases) cross-reacted with barley PR proteins in immunoblotting experiments. The antiserum (Nasser *et al.*, 1990) reacted with PR proteins from barley showing HR to mildew infection, but not to acid-soluble proteins from the uninfected control (Bryngelsson *et al.*, unpublished). However, neither chitinase sequences cloned from bean, sugarbeet and tobacco, nor β-1, 3-glucanase from tobacco showed cross-hybridization with barley (Munk-Kruse & Meldgaard-Madsen, 1990). Nevertheless, Scott *et al.* (1990) identified a β-1, 3-glucanase among six sequences induced by mildew infection.

The shotgun approach

The studies just described primarily represent the results of the direct approach to studying resistance responses. This is, however, intrinsically conservative as the only responses studied are those previously known, or for which cytological clues are available. An alternative molecular strategy to this targeted approach has been dubbed the shotgun approach (Collinge & Slusarenko, 1987). This allows the identification of the genes expressed as a response to pathogens without setting preconditions as to the nature of any response identified. Initial studies used two-dimensional electrophoresis and fluorography to analyse radioactively labelled *in vitro* translation products prepared using mRNA extracted at different times following inoculation. Thus the appearance of a number of new mRNA species 48 h after inoculation with *Egh* correlated with the development of the HR which exhibit papilla resistance (Davidson *et al.*, 1987). A different set of polypeptides was detected as early as 3 h after inoculation in barley leaves inoculated heavily with *Egt* (Gregersen *et al.*, 1990).

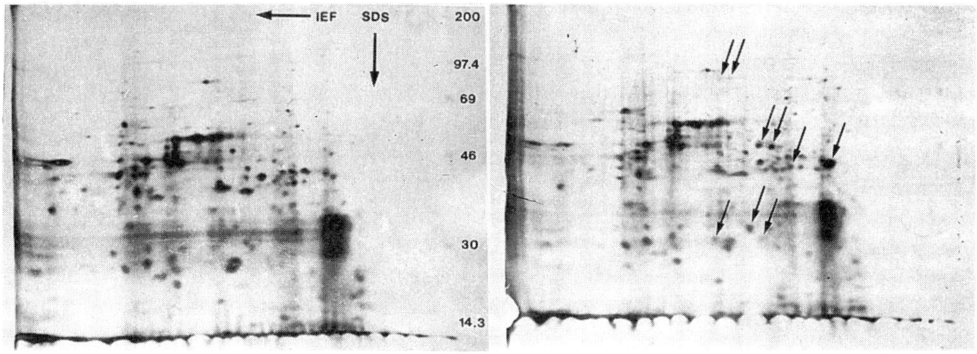

Fig. 22.6. Analysis of *in vitro* translation products by two-dimensional electrophoresis and fluorography. Polypeptide species increasing following inoculation are indicated with arrows. mRNA was isolated from leaves 8 h following inoculation with *Egh*. Total RNA was translated in the presence of [^{35}S]methionine using rabbit reticulocyte lysate (see Gregersen *et al.*, 1990 for experimental details).

Similar results were obtained using *Egh* in interactions with *Ml-a* resistant plants, which exhibit papilla resistance (see Fig. 22.6, Thordal-Christensen, Mikkelsen & Gregersen, pers. comm.).

This technique gives an overview of the complexity and timing of the changes in gene expression, but does not provide information as to the physiological function of the gene products concerned. Furthermore, the translation efficiency varies among different RNA species, as do the relative amounts of label incorporated. [^{35}S]methionine is often used: the amounts present in certain proteins (notably chitinase) can be very low. Arguably the most important contributions of this type of study concern the timing and complexity of the response. The results facilitate the preparation of cDNA libraries which can be screened for resistance response genes: that is, genes expressed as a response to challenge with a pathogen.

Differential or subtractive hybridization techniques allow the selection of clones which represent genes whose transcription is induced by challenge with a pathogen. cDNA libraries prepared against mildew-inoculated barley leaves have been screened using these techniques in two separate studies (Davidson *et al.*, 1987; Thordal-Christensen, Brandt, Cho, Gregersen, Smedegaard-Petersen & Collinge, unpublished). Eight cDNA clones, representing six different genes, were identified using differential hybridization from a cDNA bank of 3500 clones prepared from mRNA extracted 48 h after inoculation with an avirulent isolate of *Egh* from plants carrying the *Ml-p* resistance gene (Davidson *et al.*, 1987). Six cDNA clones, representing four genes, were isolated by subtractive hybridization from approximately 1000 cDNA clones in the second study. The cDNA bank was prepared using RNA isolated 6 h after inoculation with *Egt*, at which point papilla development was being manifested. None of the six HR-related cDNA clones cross-reacted with any of the *early* resistance-related sequences (Thordal-Christensen, Brandt, Cho, Gregersen, Smedegaard-Petersen & Collinge, unpublished). This is not surprising given the dissimilarity in the patterns of gene expression detected using two-dimensional electrophoresis and fluorography to analyse radioactively labelled *in vitro* translation products prepared from RNA isolated in the two studies (Davidson *et al.*, 1987; Gregersen *et al.*, 1990). Studies of the expression and identity of the gene sequences associated with *Ml-p* resistance by K.J. Scott and co-workers are described in Chapter 23, this volume.

Studies concerning the expression of the *Egt*-induced genes (Thordal-Christensen,

Brandt, Cho, Gregersen, Smedegaard-Petersen & Collinge, unpublished) indicate that the timing of induction of transcription of these genes is co-ordinated among all four (Fig. 22.7 and unpublished results), but the levels of expression differ markedly among the four genes. All four genes are induced both by *Egt* and in compatible and incompatible interactions with *Egh*, associated with *Ml-a* resistance. The apparent two-phase induction of the signals, which is pronounced especially with pBT6-3 and pBT6-6 as probes, is reproducible. The timing of the two peaks coincides with the appearance of papillae associated with the primary and appressorial germ tubes (Figs 22.1 and 22.2). This may imply a role of the corresponding gene products

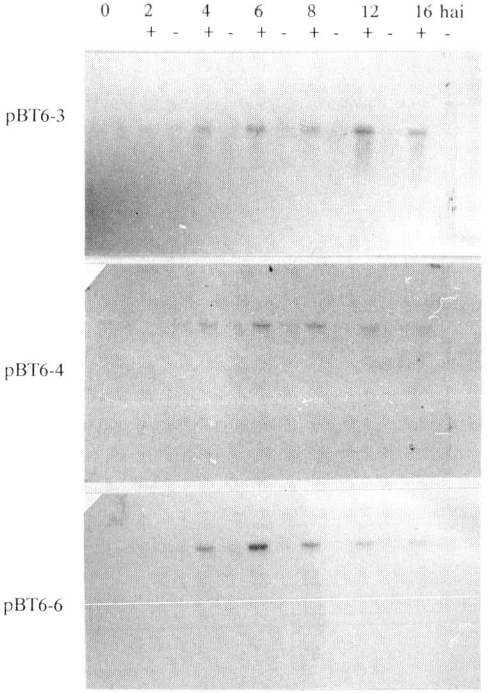

Fig. 22.7. Northern blot analyses using pBT6-3, 4, 6 to study alteration in gene expression in plants exhibiting papilla resistance to *Egh*. RNA was isolated from inoculated (+) and uninoculated (–) primary leaves over 16 h after inoculation. pBT6-3, peroxidase; pBT6-4, sucrose synthase; pBT6-6, GRP 94 homologue. (Technical details will be published elsewhere.)

in papilla resistance. All six cDNA clones have been sequenced and contain single open reading frames (Brandt, Thordal-Christensen, Cho, Winther-Larsen, Smedegaard-Petersen & Collinge, unpublished). The sequences have been used to search DNA and protein data banks, which indicated a biochemical function for three of the four genes. One of these is a peroxidase, described earlier.

Three clones (pBT6-1, pBT6-2 and pBT6-6) represent the same gene. The sequence, 1394 bp in pBT6-1 containing an open reading frame (ORF) corresponding to 384 amino acids, shows remarkable similarity to a group of stress-induced proteins (HSP 90) with apparent molecular weights in the range 80 000–94 000. The sequences are among the most conserved known, especially among eukaryotes (Lindquist & Craig, 1988). The barley sequences show greatest similarity (46% of the ORF in pBT6-1) to a member of the family known as GRP 94 (glucose-regulated protein) or endoplasmin (Sorger & Pelham, 1987; Chang *et al.*, 1989). The expression of these genes, previously characterized only in animals, is not induced by heat-shock treatment, but by stimuli such as glucose starvation of cultured cell lines. This group of proteins also differs from the HSP 90 proteins in their cellular localization: the HSP 90 proteins appear to be located in the cytoplasm whereas GRP 94 proteins appear to be located in the endoplasmic reticulum (Lindquist & Craig, 1988). The C-terminus of the GRP 94 proteins (including pBT6-1) is lysine–asparagine–glutamine–leucine (KDEL; Pelham, 1989), in contrast to the HSP 90 proteins (GEVD; Lindquist & Craig, 1988). KDEL has been shown to be responsible for retention of proteins in the endoplasmic reticulum, preventing their export (Herman *et al.*, 1990; Vaux *et al.*, 1990). It is considered, although unsubstantiated, that these proteins function as molecular chaparones/chaparonins (Hardesty & Kramer, 1989; Pelham, 1989): proteins which are responsible for ensuring the correct assembly of multicomponent proteins from their subunits. It is suggested that the changes in cellular environment caused by stresses such as heat shock and glucose starvation necessitate the induction of chaparonins with altered activity. pBT6-6 recognizes a transcript present in uninoculated tissues (Fig. 22.7), but that transcription is enhanced during papilla development. Since protein export occurs via the endoplasmic reticulum, we suggest that the corresponding protein is involved in controlling or assisting the secretion of proteins involved in papilla formation through the plasmalemma. This might also explain the relatively high levels of transcript required.

The ORF of clone pBT6-4 shows very high similarity (73–93% amino acid identity) to sucrose synthase from maize, potato and wheat (Salanoubat & Belliard, 1987; Maraña *et al.*, 1988a; Bhave *et al.*, 1990). Two wheat gene sequences are available, one of which is apparently expressed constitutively, the other, like the maize gene, primarily in developing seed. The derived amino acid sequence of pBT6-4 shows higher identity to the seed-specific maize and wheat sequences (91–3% amino acid similarity), than to the constitutive sequences. The two wheat sequences show very high similarity to each other and both cross-hybridize to the same sequences in barley (Maraña *et al.*, 1988b), so it is likely that two transcripts are involved in the Northern blots (Fig. 22.7).

Sucrose synthase catalyses both the production of sucrose from fructose and UDP-glucose, and its dissociation into these components. Two hypotheses on the transcription of a sucrose synthase gene during papilla resistance are suggested. The induction of resistance is associated with increased respiratory activity in the host (Smedegaard-Petersen & Tolstrup, 1985). Sucrose synthase activity is associated with enhanced metabolism but its role is controversial (Sung *et al.*, 1988). Thus the increased sucrose synthase gene transcription rates might reflect a need to provide metabolites in order to support this enhanced respiratory activity. The enzyme callose synthase uses UDP-glucose as its substrate; sucrose synthase could therefore have a role in providing this substrate for callose deposition.

Not all studies of this type are equally successful, as until recently, the databases were too small to allow a reasonable chance of success in identifying similar sequences (Fristensky *et al.*, 1988; Jutidamrongphan *et al.*, 1989). Caution must be observed in interpreting apparent identifications from sequence similarities: whereas 93% identity between two full-length nucleotide sequences is beyond dispute, formal proof comes only by association of the amino acid sequence of a polypeptide with known enzyme activity. However, immunoprecipitation of hybrid-arrested *in vitro* translation products, or fusion proteins derived from a cloned sequence, provides sufficient evidence to convince the majority of molecular biologists, if not all biochemists. Indeed, this touches the central problem of resistance response studies: most of the evidence is corroborative, little provides rigorous proof. Even the demonstration of the association of transcripts by *in situ* hybridization or of the derived proteins by immunocytological techniques with reacting cells is not *proof* of a role of these products in resistance. It is necessary to perform genetic experiments, either by site-directed mutagenesis – currently unrealistic in graminaceous plants – or by testing resistance of transgenic plants containing either antisense constructs or constitutively expressed transcripts. The latter has yielded mixed results in plant–pathogen studies (see e.g. Bol *et al.*, 1990).

Conclusions and future prospects

We have summarized the knowledge of resistance responses in plants with special reference to the interaction between barley and powdery mildew. It is clear that, whereas much is now known about the complexity of resistance responses, this knowledge is still superficial. There are, nevertheless, a few threads linking information obtained from cytological, physiological, biochemical, molecular and genetic studies of resistance. Thus, although phenolic compounds have been detected in papillae, their nature and hence means of production remain unknown. Both locally and systemically induced resistance are induced, but their mode of action is undefined. Perhaps the most enigmatic problem is the nature of race-specific resistance. In the barley/mildew interaction, this would appear to be manifested at two stages of development in different specific interactions: correlating with the development of either papillae or HR. Whereas the occurrence of PR proteins in barley appears to be restricted to HR, no papilla-specific resistance response genes have been identified to date whose expression differs in plants exhibiting resistance at the papilla stage. It is to be hoped that the accumulation of further data on the nature of the biochemical and molecular changes associated with race-specificity will lead to elucidation of the underlying mechanisms. *In situ* hybridization and immunocytological techniques are providing information as to the likely role of individual components in resistance. However, a formal proof that any single component is essential to resistance awaits the development of a practical method for stable transformation and regeneration of barley plants.

Acknowledgements

We are most grateful to the following for permission to quote unpublished data, and for commenting on drafts of this manuscript: Jakob Brandt, Baik Ho Cho, Per L. Gregersen, Lone Rossen, V. Smedegaard and Hans Thordal-Christensen, Copenhagen; Per Kjellbom, Lund; Karsten M. Kragh, Lars H. Pedersen and Søren K. Rasmussen, Risø; Mats Gustafsson, Svalöv. We also wish to thank L. Berglund, Århus; C.J. Lamb, La Jolla; J. Bol and H. Hoge, Leiden; J. Varner, St Louis; K.J. Scott, St Lucia for providing cDNA clones and J.D. Mikkelsen, Copenhagen and G. Burkard, Strasbourg for providing antibodies used in studies described. Unpublished work from Copenhagen is supported by The Center for Plant Biotechnology, Copenhagen, The Danish Agricultural and Veterinary Science Research Council (SJVF) and The Royal Veterinary and Agricultural University Research Council. Baik Ho Cho is supported by the Yonam Foundation in the Lucky & Kumsung Group, South Korea. Collaboration between Svalöv and Copenhagen is supported by the Nordic Council of Ministers.

References

Aist, J.R. (1983) Structural responses as resistance mechanisms. In: Bailey, J.A. & Deverall, B.J. (eds), *The Dynamics of Host Defense*. Academic Press, Sydney, pp. 33–70.

Aist, J.R. & Israel, H.W. (1986) Autofluorescent and ultraviolet-absorbing components in cell walls and papillae of barley coleoptiles and their relationship to disease. *Canadian Journal of Botany* 64, 266–72.

Aist, J.R., Gold, R.E., Bayles, C.J., Morrison, G.H., Chandra, S. & Israel, H.W. (1988) Evidence that molecular components of papillae may be involved in ml-o resistance to barley powdery mildew. *Physiological and Molecular Plant Pathology* 33, 17–32.

Andersen, J.B. (1989) Studies on PR-proteins of barley after inoculation with barley powdery mildew. PhD thesis, The Royal Veterinary and Agricultural University, Copenhagen.

Antoniw, J.F., Ritter, C.E., Pierpoint, W.S. & van Loon, L.C. (1980) Comparison of three pathogenesis-related proteins from plants of two cultivars of tobacco infected with TMV. *Journal of General Virology* 47, 79–87.

Arora, Y.K. & Bajaj, K.L. (1985) Peroxidase and polyphenol oxidase associated with induced resistance of mung bean to *Rhizoctonia solani* Kuhn. *Phytopathologische Zeitschrift* 114, 325–31.

Asher, M.J. & Thomas, C.E. (1983) The expression of partial resistance to *Erysiphe graminis* in spring barley. *Plant Pathology* 32, 79–89.

Bacic, A., Harris, P.J. & Stone, B.A. (1988) Structure and function of cell walls. In: Preiss, J. (ed.), *The Biochemistry of Plants*, vol. 14. Academic Press, San Diego, pp. 297–371.

Bayles, C.J. & Aist, J.R. (1987) Apparent calcium mediation of resistance of an ml-o barley mutant to powdery mildew. *Physiological and Molecular Plant Pathology* 30, 337–45.

Bayles, C.J., Ghemawat, M.S. & Aist, J.R. (1990) Inhibition by 2-deoxy-D-glucose of callose formation, papilla deposition, and resistance to powdery mildew in an *ml-o* barley mutant. *Physiological and Molecular Plant Pathology* 36, 63–72.

Bhave, M.R., Lawrence, S., Bareton, C. & Hannah, L.C. (1990) Identification and molecular characterization of *Shrunken*-2 cDNA clones of maize. *Plant Cell* 2, 581–8.

Bohlmann, H., Clausen, S., Behnke, S., Giese, H., Hiller, C., Reimann-Philipp, U., Schrader, G., Barkholt, V. & Apel, K. (1988) Leaf-specific thionins of barley – a novel class of cell wall proteins toxic to plant-pathogenic fungi and possibly involved in the defence mechanism of plants. *EMBO Journal* 7, 1559–65.

Bol, J.F., Lindhorst, H.J. & Cornelissen, B.J. (1990) Plant pathogenesis-related proteins induced by virus infection. *Annual Review of Phytopathology* 28, 113–38.

Boller, T., Gehri, A., Mauch, F. & Vögeli, U. (1983) Chitinase in bean leaves: induction by ethylene, purification, properties, and possible function. *Planta* 157, 22–31.

Bostock, R.M. & Stermer, B.A. (1989) Perspectives on wound healing in resistance to pathogens. *Annual Review of Phytopathology* 27, 343–71.

Bowles, D.J. (1990) Defense-related proteins in higher plants. *Annual Review of Biochemistry* 59, 873–907.

Brakke, M.J., White, J.L., Samson, R.G. & Joshi, J. (1988) Chlorophyll, chloroplast ribosomal RNA, and DNA are reduced by barley stripe mosaic virus systemic infection. *Phytopathology* 78, 570–4.

Bryngelsson, T. & Gréen, B. (1989) Characterization of a pathogenesis-related, thaumatin-like protein isolated from barley challenged with an incompatible race of mildew. *Physiological and Molecular Plant Pathology* 35, 45–52.

Bryngelsson, T., Gustafsson, M., Ramos Leal, M. & Bartonek, E. (1988) Induction of pathogenesis-related proteins in barley during the resistance reaction to mildew. *Journal of Phytopathology* 123, 193–8.

Carr, J.P. & Klessig, D.F. (1989) The pathogenesis-related proteins of plants. In: Setlow, J.K. (ed.), *Genetic Engineering. Principles and Methods*, Vol. 11. Plenum Press, New York, pp. 65–109.

Chang, S.C., Erwin, A.E. & Lee, A.S. (1989) Glucose-regulated protein (GRP 94 and GRP 78) genes share common regulatory domains and are coordinately regulated by common *trans*-acting factors. *Molecular and Cellular Biology* 9, 2153–62.

Cho, B.H. & Smedegaard-Petersen, V. (1986) Induction of resistance to *Erysiphe graminis* f. sp. *hordei* in near-isogenic barley lines. *Phytopathology* 76, 301–5.

Christiansen, S.K. & Smedegaard, V. (1990) Microscopic studies of the interaction between barley and the saprophytic fungus *Cladosporium macrocarpum*. *Journal of Phytopathology* 128, 209–19.

Clarke, J.A., Lisker, N., Ellingboe, A.H. & Lamport, D.T. (1983) Cell wall composition of near isogenic wheat lines carrying genes for resistance to *Erysiphe graminis* f. sp. *tritici*. *Plant Science Letters* 30, 339–46.

Collinge, D.B. & Slusarenko, A.J. (1987) Plant gene expression in response to pathogens. *Plant Molecular Biology* 9, 389–410.

Cooper, J.B. & Varner, J.E. (1984) Cross-linking of soluble extensin in isolated cell walls. *Plant Physiology* 76, 414–17.

Cornelissen, B.J., Hooft van Huijsduijnen, R.A. & Bol, J.F. (1986) A tobacco mosaic virus-induced tobacco protein is homologous to the sweet-tasting protein thaumatin. *Nature* 321, 531–2.

Cornelissen, B.J., Horowitz, J., van Kan, J.A., Goldberg, R.B. & Bol, J.F. (1987) Structure of tobacco genes encoding pathogenesis-related proteins from the PR-1 group. *Nucleic Acids Research* 15, 6799–811.

Davidson, A.D., Manners, J.M., Simpson, R.S. & Scott, K.J. (1987) cDNA cloning of mRNAs induced in resistant barley during infection by *Erysiphe graminis* f. sp *hordei*. *Plant Molecular Biology* 8, 77–85.

De Biasi, M.G. & Badiani, M. (1990) The phenoloxidase-like activity of purified peroxidase from *Triticum aestivum* L. seedling leaves. *Plant Science* 67, 29–37.

De Wit, P.J. (1986) Elicitation of active resistance mechanisms. In: Bailey, J.A. (ed.), *Biology and Molecular Biology of Plant–Pathogen Interactions*. Springer Verlag, Berlin, pp. 149–69.

De Wit, P.J., van den Ackerveken, G.F., Joosten, M.H. & van Kan, J.A. (1989) Apoplastic proteins involved in communication between tomato and the fungal pathogen *Cladosporium fulvum*. In: Lugtenberg, B.J. (ed.), *Signal Molecules in Plants and Plant–Microbe Interactions*. Springer Verlag, Berlin, pp. 273–80.

Dixon, R.A. (1986) The phytoalexin response: elicitation, signalling and the control of host gene expression. *Biological Reviews* 61, 239–91.

Dixon, R.A. & Lamb, C.J. (1990) Molecular communication in interactions between plants and microbial pathogens. *Annual Review of Plant Physiology and Plant Molecular Biology* 41, 339–67.

Edens, L., Heslinga, L., Klok, R., Ledeboer, A.M., Maat, J., Toonen, M.Y., Visser, C. & Verrips, C.T. (1982) Cloning of a cDNA encoding the sweet-tasting plant protein thaumatin and its expression in *Escherichia coli*. *Gene* 18, 1–12.

Edwards, H.H. & Allen, P.J. (1966) Distribution of the products of photosynthesis between powdery mildew and barley. *Plant Physiology* 41, 683–8.

Ellingboe, A.H. (1978) A genetic analysis of host–parasite interactions. In: Spencer, D.M. (ed.), *The Powdery Mildews*. Academic Press, London, pp. 159–81.

Fredrikson, K., Kjellbom, P. & Larsson, C. (1991) Isolation and polypeptide composition of 1,3-β-glucan synthase from plasma membranes of *Brassica oleracea*. *Physiologia Plantarum* 81, 289–94.

Fristensky, B., Horowitz, D. & Hadwiger, L.A. (1988) cDNA sequences for pea disease resistance response genes. *Plant Molecular Biology* 11, 713–15.

Fujiyami, K., Takemura, H., Shibiyama, S., Kobayashi, K., Choi, J.-K., Shinmyo, A., Takano, M., Yamada, Y. & Okada, H. (1988) Structure of the horseradish peroxidase isozyme C genes. *European Journal of Biochemistry* 173, 681–7.

García-Olmedo, F., Rodriguez-Palenzuela, P., Hernández-Lucas, C., Ponz, F., Maraña, C., Carmona, M.J., Lopez-Fando, J., Fernandez, J.A. & Carbonero, P. (1989) The thionins: a protein family that includes purothionins, viscotoxins and crambins. *Oxford Surveys of Plant Molecular & Cell Biology* 6, 31–60.

Gold, R.E., Aist, J.R., Hazen, B.E., Stolzenburg, M.C., Marshall, M.R. & Israel, H.W. (1986) Effects of calcium nitrate and chlortetracycline on papilla formation, *ml-o* resistance and susceptibility of barley to powdery mildew. *Physiological and Molecular Plant Pathology* 29, 115–29.

Green, N.E., Hadwiger, L.A. & Graham, S.O. (1975) Phenylalanine ammonia-lyase, tyrosine ammonia-lyase, and lignin in wheat inoculated with *Erysiphe graminis* f. sp. *tritici*. *Phytopathology* 65, 1071–4.

Gregersen, P.L. & Smedegaard, V. (1989) Induction of resistance in barley against *Erysiphe graminis* f. sp. *hordei* after preinoculation with the saprophytic fungus, *Cladosporium macrocarpum*. *Journal of Phytopathology* 124, 128–36.

Gregersen, P.L., Collinge, D.B. & Smedegaard-Petersen, V. (1990) Early induction of new mRNAs accompanies the resistance reaction of barley to the wheat pathogen *Erysiphe graminis* f. sp. *tritici*. *Physiological and Molecular Plant Pathology* 36, 471–81.

Gustafsson, M. & Claesson, L. (1988) Resistance to powdery mildew in wild species of barley. *Hereditas* 108, 231–7.

Hahlbrock, K.N. & Scheel, D. (1989) Physiology and molecular biology of phenylpropanoid metabolism. *Annual Review of Plant Physiology and Plant Molecular Biology* 40, 347–69.

Hardesty, B. & Kramer, G. (1989) The 90 000 dalton heat shock protein, a lot of smoke but no function as yet. *Biochemistry and Cell Biology* 67, 749–50.

Heitefuss, R. & Ebrahim-Nesbat, F. (1986) Ultrastructural and histochemical studies on mildew of barley (*Erysiphe graminis* DC. f. sp. *hordei* Marchal). III. Ultrastructure of different types of papillae in susceptible and adult plant resistant leaves. *Journal of Phytopathology* 116, 358–73.

Herman, E.M., Tague, B.W., Hoffman, L.M., Kjemtrup, S.E. & Chrispeels, M.J. (1990) Retention of phytohemagglutinin with carboxyterminal tetrapeptide KDEL in the nuclear envelope and the endoplasmic reticulum. *Planta* 182, 305–12.

Hislop, E.C. & Stahmann, M.A. (1971) Peroxidase and ethylene production by barley leaves infected with *Erysiphe graminis* f. sp. *hordei*. *Physiological Plant Pathology* 1, 297–312.

Hwang, B.K. & Heitefuss, R. (1982) Induced resistance of spring barley to *Erysiphe graminis* f. sp. *hordei*. *Phytopathologische Zeitschrift* 103, 41–7.

Jenkyn, J.F. & Bainbridge, A. (1978) Biology and pathology of cereal powdery mildews. In: Spencer, D.M. (ed.), *The Powdery Mildews*. Academic Press, London, pp. 283–321.

Johnson, L.E., Bushnell, W.R. & Zeyen, R.J.

(1982) Defense patterns in nonhost higher plant species against two powdery mildew fungi. I. Monocotyledonous species. *Canadian Journal of Botany* 60, 1068–83.

Jørgensen, J.H. & Mortensen, K. (1977) Primary infection by *Erysiphe graminis* f. sp. *hordei* of barley mutants with resistance genes in the *ml-o* locus. *Phytopathology* 67, 678–85.

Jutidamrongphan, W., Mackinnon, G., Manners, J.M. & Scott, K.J. (1989) Sequence of a near-full length clone for a mRNA of barley induced by fungal infection. *Nucleic Acids Research* 17, 9478.

Kauss, H. & Jeblick, W. (1986) Synergistic activation of 1,3-β-D-glucan synthase by Ca^{2+} and polyamines. *Plant Science* 43, 103–7.

Keen, N.T. (1986) Pathogenic strategies of fungi. In: Lugtenberg, B. (ed.), *Recognition in Microbe–Plant Symbiotic and Pathogenic Interactions*. Springer Verlag, Berlin, pp. 171–88.

Kerby, K. & Somerville, S. (1989) Enhancement of specific intercellular peroxidases following inoculation of barley with *Erysiphe graminis* f. sp. *hordei*. *Physiological and Molecular Plant Pathology* 35, 323–37.

King, G.J., Hussey Jr, C.E. & Turner, V.A. (1986) A protein induced by NaCl in suspension cultures of *Nicotiana tabacum* accumulates in whole plant roots. *Plant Molecular Biology* 7, 441–9.

Kita, N., Toyoda, H. & Shishiyama, J. (1980) Histological reactions of papilla and cytoplasmic aggregate in epidermal cells of barley leaves infected by *Erysiphe graminis hordei*. *Annals of the Phytopathology Society, Japan* 46, 263–5.

Kolattukudy, P.E., Podila, G.K., Sherf, B.A., Bajar, M.A. & Mohan, R. (1991) Mutual triggering of gene expression in plant–fungus interactions. In: Hennecke, H. & Verma, D.P. (eds), *Advances in Molecular Genetics of Plant–Microbe Interactions*. Kluwer Academic, Dordrecht, pp. 242–9.

Kølster, P., Munk, L., Stølen, O. & Løhde, J. (1986) Near-isogenic barley lines with genes for resistance to powdery mildew. *Crop Science* 26, 903–7.

Kragh, K.M., Jacobsen, S. & Mikkelsen, J.D. (1990) Induction, purification and characterization of barley leaf chitinase. *Plant Science* 71, 55–68.

Kristensen, H.J. (1989) Induktion af meldugresistens og -modtagelighed i byg. Master's thesis, The Royal Veterinary and Agricultural University, Copenhagen.

Kunoh, H., Kuno, K. & Ishizaki, H. (1985) Cytological studies of the early stages of powdery mildew in barley and wheat. XI. Autofluorescence and haloes at penetration sites of appressoria of *Erysiphe graminis hordei* and *Erysiphe pisi* on barley coleoptiles. *Canadian Journal of Botany* 63, 1535–9.

Lagrimini, L.M., Burkhart, W., Moyer, M. & Rothstein, S. (1987) Molecular cloning of complementary DNA encoding the lignin-forming peroxidase from tobacco: molecular analysis and tissue-specific expression. *Proceedings of the National Academy of Sciences (USA)* 84, 7542–6.

Lewis, N.G. & Yamamoto, E. (1990) Lignin: occurrence, biogenesis and biodegradation. *Annual Review of Plant Physiology and Plant Molecular Biology* 41, 455–96.

Lindquist, S. & Craig, E.A. (1988) The heat shock proteins. *Annual Review of Genetics* 22, 631–77.

Lucas, J., Camacho Henriquez, A., Lottspeich, F., Henschen, A. & Sänger, H.L. (1985) Amino acid sequence of the pathogenesis-related leaf protein p14 from viroid-infected tomato reveals a new type of structurally unfamiliar proteins. *EMBO Journal* 4, 2745–9.

Maraña, C., García-Olmedo, F. & Carbonero, P. (1988a) Linked sucrose synthase genes in group-7 chromosomes in hexaploid wheat (*Triticum aestivum*). *Gene* 63, 253–60.

Maraña, C., García-Olmedo, F. & Carbonero, P. (1988b) Equivalent locations of sucrose synthase genes in chromosomes 7D of wheat, 7Ag of *Agropyron elongatum*, and 7H of barley. FEBS *Letters* 234, 417–20.

Mauch, F., Mauch-Mani, B. & Boller, T. (1988) Antifungal hydrolases in pea tissue. *Plant Physiology* 88, 936–42.

Mayama, S. & Shishiyama, J. (1978) Localized accumulation of fluorescent and u.v.-absorbing compounds at penetration sites in barley leaves infected with *Erysiphe graminis hordei*. *Physiological Plant Pathology* 13, 347–54.

Mayama, S., Tani, T., Matsuura, Y., Ueno, T. & Fukami, H. (1981) The production of phytoalexins by oat in response to crown rust, *Puccinia coronata* f. sp. *avenae*. *Physiological Plant Pathology* 19, 217–26.

Mazza, G. & Welinder, K.G. (1980) Covalent structure of turnip peroxidase 7. Cyanogen bromide fragments, complete structure and comparison to horse radish peroxidase. *European Journal of Biochemistry* 108, 481–9.

Mohan, R. & Kolattukudy, P.E. (1990) Differential activation of expression of a suberization-associated anionic peroxidase gene in near-isogenic resistant and susceptible tomato lines by elicitors of *Verticillium albo-atrum*. *Plant Physiology* 92, 276–80.

Munk-Kruse, A.-G. & Meldgaard-Madsen, K.M. (1990) Undersøgelse af resistensresponsgener hos ært (*Pisum sativum*) i interaktion med *Ascochyta pisi* Lib. ved hjælp af krydshybridisering med klonede hydroxyprolin-rige glycoprotein-og kitinase-sekvenser. Master's Thesis. The Royal Veterinary and Agricultural University, Copenhagen.

Nasser, W., de Tapia, M., Kauffmann, S., Montasser-Kouhsari, S. & Burkard, G. (1988) Identification of maize pathogenesis-related proteins. Four maize PR proteins are chitinases. *Plant Molecular Biology* 11, 529–38.

Nasser, W., de Tapia, M. & Burkard, G. (1990) Maize pathogenesis-related proteins: characterization and cellular distribution of 1,3-β-glucanases and chitinases induced by brome mosaic virus infection or mercuric chloride treatment. *Physiological and Molecular Plant Pathology* 36, 1–14.

Oku, H., Ouchi, S., Shiraishi, T., Komoto, Y. & Oki, K. (1975) Phytoalexin activity in barley powdery mildew. *Annals of the Phytopathological Society of Japan* 41, 185–91.

Ouchi, S., Oku, H., Hibino, C. & Akiyama, I. (1974) Induction of accessibility to a nonpathogen by preliminary inoculation with a pathogen. *Phytopathologische Zeitschrift* 79, 142–54.

Ouchi, S., Oku, H. & Hibino, C. (1976) Localization of induced resistance and susceptibility in barley leaves inoculated with the powdery mildew fungus. *Phytopathology* 66, 901–5.

Pelham, H.R. (1989) Control of protein exit from the endoplasmic reticulum. *Annual Review of Cell Biology* 5, 1–23.

Pierpoint, W.S. & Shewry, P.R. (1987) Amino acid homologies suggest a function for pathogenesis-related proteins. *Oxford Surveys of Plant Molecular & Cell Biology* 4, 337–42.

Pierpoint, W.S., Jackson, P.J. & Evans, R.M. (1990) The presence of a thaumatin-like protein, a chitinase and a glucanase among the pathogenesis-related proteins of potato (*Solanum tuberosum*). *Physiological and Molecular Plant Pathology* 36, 325–38.

Rasmussen, S.K., Welinder, K.G. & Heigaard, J. (1991) cDNA cloning, characterization and expression of an endosperm specific barley peroxidase. *Plant Molecular Biology* 16, 317–27.

Rasmussen, S.K., Johansson, A., Rasmussen, H.N. & Thelade, B. Molecular analysis and cloning of barley peroxidase genes. In: Lobarzewski, J., Greppin, H., Gaspar, T. & Penel, C. (eds), *Biochemical Molecular and Physiological Aspects of Plant Peroxidases*, University of Geneva, Geneva (in press).

Richardson M., Valdes-Rodriguez, S. & Blanco-Labra, A. (1987) A possible function for thaumatin and a TMV-induced protein suggested by homology to a maize inhibitor. *Nature* 327, 432–4.

Ride, J.P. (1975) Lignification in wounded wheat leaves in response to fungi and its possible rôle in resistance. *Physiological Plant Pathology* 5, 125–34.

Roberts, E., Kutchan, T. & Kolattukudy, P.E. (1988) Cloning and sequencing of a cDNA for a highly anionic peroxidase from tomato and the induction of its mRNA in suberizing potato tubers and tomato fruits. *Plant Molecular Biology* 11, 15–26.

Robertsen, B. (1986) Elicitors of the production of lignin-like compounds in cucumber hypocotyls. *Physiological and Molecular Plant Pathology* 28, 137–48.

Robertsen, B. & Svalheim, Ø. (1990) The nature of lignin-like compounds in cucumber hypocotyls. *Physiologia Plantarum* 79, 512–18.

Sahashi, N., Tsuji, H. & Shishiyama, J. (1989) Barley plants grown under germ-free conditions have increased susceptibility to two powdery mildew fungi. *Physiological and Molecular Plant Pathology* 34, 163–70.

Sako, N. & Stahmann, M.A. (1972) Multiple molecular forms of enzymes in barley leaves infected with *E. graminis* f. sp. *hordei*. *Physiological Plant Pathology* 2, 217–26.

Salanoubat, M. & Belliard, G. (1987) Molecular cloning and sequencing of sucrose synthase cDNA from potato (*Solanum tuberosum* L.): preliminary characterization of sucrose synthase mRNA distribution. *Gene* 60, 47–56.

Schlumbaum, A., Mauch, F., Vögeli, U. & Boller, T. (1986) Plant chitinases are potent inhibitors of fungal growth. *Nature* 324, 365–7.

Schweizer, P., Hunziker, W. & Mösinger, E. (1989) cDNA cloning, *in vitro* transcription and partial sequence analysis of mRNAs from winter wheat (*Triticum aestivum* L.) with induced resistance to *Erysiphe graminis*

f. sp. *tritici*. *Plant Molecular Biology* 12, 643–54.

Scott, K.J. & Smillie, R.M. (1963) Possible relationship between photosynthesis and the rise in respiration in diseased leaves. *Nature* 197, 1319–20.

Scott, K.J., Davidson, A.D., Jutidamrongphan, W., Mackinnon, G. & Manners, J.M. (1990) The activation of genes of wheat and barley by fungal pathogens. *Australian Journal of Plant Physiology* 17, 229–38.

Shaw, M. & Samborski, D.J. (1957) The physiology of host–parasite relations. III. The pattern of respiration in rusted and mildewed cereal leaves. *Canadian Journal of Botany* 35, 389–407.

Shiraishi, T., Yamaoka, N. & Kunoh, H. (1989) Association between increased phenylalanine ammonia lyase activity and cinnamic acid synthesis and the induction of temporary inaccessibility caused by *Erysiphe graminis* primary germ tube penetration of the barley leaf. *Physiological and Molecular Plant Pathology* 34, 75–83.

Showalter, A.M. & Varner, J.E. (1989) Plant hydroxyproline-rich glycoproteins. In: Marcus, A. (ed.), *The Biochemistry of Plants*. Vol. 15. Academic Press, San Diego, pp. 485–520.

Singh, N.K., Bracker, C.A., Hasegawa, P.M., Handa, A.K., Buckel, S., Hermodson, M.A., Pfankoch, E., Regnier, F.E. & Bressan, R.A. (1987) Characterization of plant osmotin. A thaumatin-like protein associated with osmotic adaptation in plant cells. *Plant Physiology* 85, 529–36.

Skou, J.P. (1982) Callose formation responsible for the powdery mildew resistance in barley with genes in the *ml-o* locus. *Phytopathologische Zeitschrift* 104, 90–5.

Smedegaard-Petersen, V. & Tolstrup, K. (1985) The limiting effect of disease resistance on yield. *Annual Review of Phytopathology* 23, 475–90.

Smedegaard-Petersen, V., Brandt, J., Cho, B.H., Collinge, D.B., Gregersen, P.L. & Thordal-Christensen, H. (1991) Molecular and energetic aspects of induced resistance in barley. In: Riesselman, J. (ed.), *Proceedings of the International Symposium on Biotic Stress of Barley in Arid and Semiarid Environments*, pp. 123–34.

Sorger, P.K. & Pelham, H.R. (1987) Glucose-regulated protein grp94 is related to heat shock protein hsp90. *Journal of Molecular Biology* 194, 341–4.

Stenzel, K., Steiner, U. & Schönbeck, F. (1985) Effect of induced resistance on the efficiency of powdery mildew haustoria in wheat and barley. *Physiological Plant Pathology* 27, 357–67.

Street, P.F., Robb, J. & Ellis, B.E. (1986) Secretion of vascular coating components by xylem parenchyma cells of tomatoes infected with *Verticillium albo-atrum*. *Protoplasma* 132, 1–11.

Sung, S.-J.S., Xu, D.-P., Galloway, C.M. & Black, C.C. (1988) A reassessment of glycolysis and gluconeogenesis in higher plants. *Physiologia Plantarum* 72, 650–54.

Svalheim, Ø. & Robertsen, B. (1990) Induction of peroxidases in cucumber hypocotyls by wounding and fungal infection. *Physiologia Plantarum* 78, 261–7.

Takamatsu, S., Ishizaki, H. & Kunoh, H. (1978) Cytological studies of early stages of powdery mildew in barley and wheat. V. Effects of calcium on the infection of coleoptiles of barley by *Erysiphe graminis hordei*. *Canadian Journal of Botany* 56, 2544–9.

Tepper, C.S. & Anderson, A.J. (1986) Two cultivars of bean display a differential response to extracellular components from *Colletotrichum lindemuthianum*. *Physiological and Molecular Plant Pathology* 29, 411–20.

Thordal-Christensen, H & Smedegaard-Petersen, V. (1988a) Correlation between induced resistance and host fluorescence in barley inoculated with *Erysiphe graminis*. *Journal of Phytopathology* 123, 34–46.

Thordal-Christensen, H. & Smedegaard-Petersen, V. (1988b) Comparison of resistance inducing abilities of virulent and avirulent races of *E. graminis* f. sp. *hordei* and a race of *E. graminis* f. sp. *tritici* in barley. *Plant Pathology* 37, 20–7.

van Huystee, R.B. (1987) Some molecular aspects of plant peroxidase biosynthetic studies. *Annual Review of Plant Physiology* 38, 205–19.

van Kan, J.A., Cornelissen, B.J. & Bol, J.F. (1988) A virus-inducible tobacco gene encoding a glycine-rich protein shares putative regulatory elements with the ribulose bisphosphate carboxylase small subunit gene. *Molecular Plant–Microbe Interactions* 1, 107–12.

Vaux, D., Tooze, J. & Fuller, S. (1990) Identification by anti-idiotype antibodies of an intercellular membrane protein that recognizes a mammalian endoplasmic reticulum retention signal. *Nature* 345, 495–502.

Walters, D.R. & Ayres, P.G. (1984) Ribulose bisphosphate carboxylase protein and enzymes of CO_2 assimilation in barley infected by powdery mildew (*Erysiphe graminis hordei*). *Phytopathologische Zeitschrift* 109, 208–18.

White, J.L. & Brakke, M.K. (1983) Protein changes in wheat infected with wheat streak mosaic virus and barley infected with barley stripe mosaic virus. *Physiological Plant Pathology* 22, 87–100.

White, R.F., Rybicki, E.P., von Wechmar, M.B., Dekker, J.L. & Antoniw, J.F. (1987) Detection of PR 1-type proteins in *Amaranthaceae, Chenopodiaceae, Graminae and Solanaceae* by immunoelectroblotting. *Journal of General Virology* 68, 2043–8.

Woolacott, B. & Archer, S.A. (1984) The influence of the primary germ tube on infection of barley by *Erysiphe graminis* f. sp. *hordei*. *Plant Pathology* 33, 225–31.

Yamada, T., Hiramoto, T., Tobimatsu, R., Shiraishi, T. & Oku, H. (1990) Elicitor-like substances present in barley and wheat seeds. *Journal of Phytopathology* 128, 89–98.

Yamamoto, H., Hokin, H., Tani, T. & Kadota, G. (1977) Phenylalanine ammonia-lyase in relation to the crown rust resistance of oat leaves. *Phytopathologische Zeitschrift* 90, 203–11.

Note in Proof

Two peroxidase sequences, which are induced by powdery mildew, have been isolated from wheat (Hertig *et al.*, 1991; Rebmann *et al.*, 1991), one corresponding to the WIR3 cDNA. Sequence analysis demonstrates that these show over 90% amino acid sequence identity to pBH6-601, a full length cDNA clone corresponding to pBT6-3 (Thordal-Christensen, Brandt, Cho, Gregersen, Rasmussen, Smedegaard-Petersen & Collinge, unpublished).

Hertig, C., Rebmann, G., Bull, J., Mauch, F. & Dudler, R. (1991) Sequence and tissue-specific expression of a putative peroxidase gene from wheat (*Triticum aestivum* L.). *Plant Molecular Biology* 16, 171–4.

Rebmann, G., Hertig, C., Bull, J., Mauch, F. & Dudler, R. (1991) Cloning and sequencing of cDNAs encoding a pathogen-induced putative peroxidase of wheat (*Triticum aestivum* L.). *Plant Molecular Biology* 16, 329–31.

23

The Molecular Analysis of Barley Resistance to Powdery Mildew

K.J. Scott

Infection of barley by the powdery mildew fungus *Erysiphe graminis* f.sp. *hordei* has provided a useful model for examining host–pathogen specificity and the molecular basis of disease resistance. This system has four main advantages: (i) ease of handling where optimal growth conditions for both host and infecting fungus are clearly defined; (ii) the fungus is an ectoparasite (Day & Scott, 1973) (Fig. 23.1) and hence it is possible to remove most of the fungal hyphae from the host; (iii) detailed information is available on the development of the pathogen; and (iv) a number of near-isogenic lines differing in resistance to infection by races of the powdery mildew fungus by single identified genes are available. The main disadvantage of this system is that *Erysiphe graminis* is an obligate parasite and it is not possible to culture it independently from the host.

This chapter examines the differences between pairs of resistant and susceptible near-isogenic lines of powdery mildew-infected barley and corresponding non-infected controls, in terms of: (i) early biochemical and cytological changes; and (ii) patterns of protein and mRNA synthesis. The production of cDNA clones from infection-related mRNAs (IR-mRNAs), and the potential uses of these clones in elucidating the molecular basis of disease resistance in barley are discussed.

Near-isogenic barley lines: early biochemical and cytological changes in response to powdery mildew infection

On the basis of extreme host specificity exhibited by the powdery mildew fungus, a series of near-isogenic resistant and susceptible barley lines has been developed by Moseman (1972). These lines contain one of several dominant genes specifying resistance to the powdery mildew fungus, or the corresponding recessive allele, in a common genetic background. They have been used to demonstrate that fungal development as determined by individual resistance genes can differ considerably (Masri & Ellingboe, 1966; Ellingboe, 1972; Hsu &

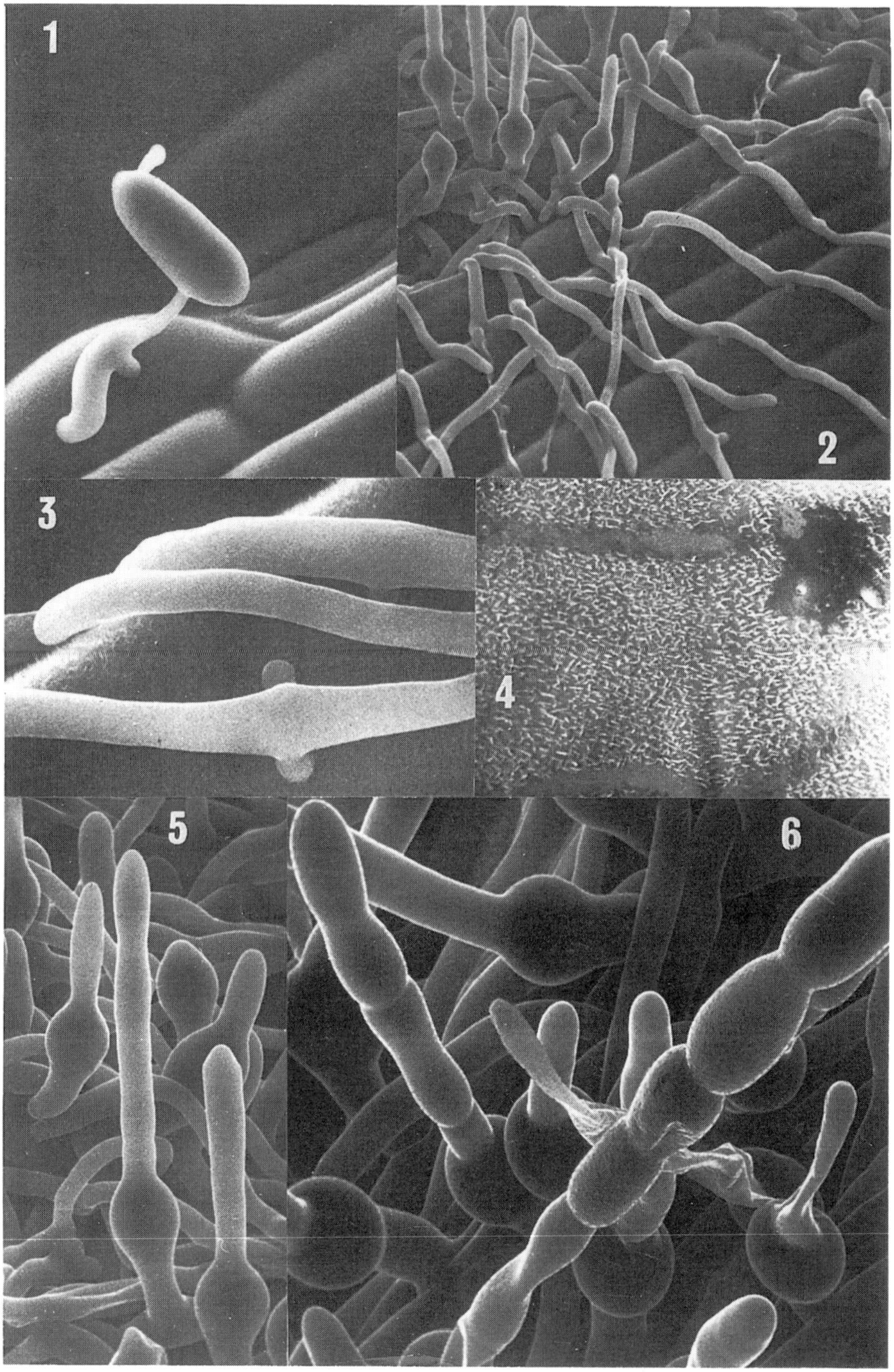

Fig. 23.1. (1) Germinated conidium at 24 h showing appressorium and hyphal development. Magnification ×720. (2) Lesion margin at 4 days showing penetration pegs and conidiophore initials. Magnification ×430. (3) Ramifying hyphae at 7 days with paired appressoria. Magnification ×1800. (4) Cuticle exposed by removing superficial mycelium to show hyphal tracks in wax crystals, position of appressorium and hole made by penetration peg. Magnification ×1800. (5) Early stages of conidiophore development. Magnification ×855. (6) Immature conidia at early (right) and late (left) four-celled stage. Note damage caused by electron beam. Magnification ×1260. After Day & Scott, 1973.

Ellingboe, 1972; Yang *et al.*, 1972). For example, the Algerian gene, at the *Ml-a* locus, conditions an early and highly effective resistance response with very little development of the pathogen occurring after one day (Johnson *et al.*, 1979; Bushnell, 1981). The Kwan (*Ml-k*) and Psaknon (*Ml-p*) genes condition a weaker response and some development of the pathogen continues for several days (Masri & Ellingboe, 1966; Hsu & Ellingboe, 1972; Yang *et al.*, 1972).

The cytological and biochemical responses of barley to infection with the powdery mildew fungus have been closely followed during the period 0–48 h following inoculation. Initial host responses occur within 1–2 h after inoculation with both a compatible and incompatible race of the fungus (Cho & Smedegaard-Petersen, 1986). Inoculation of leaves with an incompatible race of the fungus results in protection of those leaves within 1–2 h from infection by compatible races of the fungus (Cho & Smedegaard-Petersen, 1986). In the period 2–14 h following inoculation an accumulation of fluorescent compounds occurs together with cytoplasmic aggregation and papillae formation in the region of the primary penetration peg (McKeen & Bhattacharya, 1970; Lin & Edwards, 1974; Bushnell & Bergquist, 1975; Kunoh & Ishikahi, 1976; Mayama & Shishiyama, 1976, 1978; Aist & Israel, 1977; Kunoh *et al.*, 1977, 1978, 1982; Johnson *et al.*, 1979; Zeyen & Bushnell, 1979; Koga *et al.*, 1980; Kita *et al.*, 1981). Deposition of minerals, such as silicon, is observed 12 h after inoculation (Kunoh & Ishizaki, 1976).

In the period 14–48 h following inoculation, certain responses occur primarily in resistant lines which are associated with the disruption of haustorium formation. These responses are the formation of a collar, composed of amorphous cell components, around the neck of the haustorium and a general cytoplasmic aggregation and host cell fluorescence in the epidermal cells in response to the establishment of fungal haustoria (Sargent & Gay, 1977; Johnson *et al.*, 1979; Kita *et al.*, 1981). Bushnell & Bergquist (1975) and Bushnell (1981) observed that in a resistant line, carrying the *Ml-a* gene, the cytoplasmic streaming stops 16–36 h after inoculation and this is followed by cell collapse in single infected host epidermal cells.

Biochemical host responses occur 8–24 h after inoculation with the increased production of phytoalexin-like compounds (Oku *et al.*, 1975). This was found in resistant but not susceptible lines. An enhanced synthesis of IR-mRNAs was also observed from 12 h onwards in lines resistant to the powdery mildew fungus (Davidson *et al.*, 1987). The onset of this enhancement is determined by the resistant gene present and will be discussed in some detail later. There is an increased production of ethylene 16–30 h after inoculation in the resistant line cvMl-a compared with its near-isogenic susceptible partner cvml-a (Hislop & Stahmann, 1971). However, this was not observed in the resistant lines cvMl-k and cvMl-p suggesting it may not be a general response in resistant lines. These authors found an increased production of peroxidase in both compatible and incompatible interactions. Hsu & Ellingboe (1972) found increased rates of ^{35}S transfer from the host to fungus, at 18–36 h after inoculation, correlated with the degree of infection.

Certain changes have been described in susceptible barley lines but not in the corresponding near-isogenic line carrying the *Ml-a* gene for resistance, at 24 h after inoculation. A decrease has been observed in both polyribosomes (Dyer & Scott, 1972), and *in vivo* protein synthesis in the chloroplast but not in the cytoplasm (Manners & Scott, 1983, 1984). This specific degradation of chloroplast polysomes in susceptible but not resistant leaves appears to be crucial in the establishment of a compatible interaction and may be triggered off by the induction of specific host ribonucleases (Scott, 1972). At 24–48 h after inoculation there is a decrease in the translational activity of mRNA isolated from infected susceptible leaves compared with non-infected controls (Manners & Scott, 1985).

Near-isogenic barley lines: patterns of protein synthesis associated with resistance

In vivo protein synthesis

The near-isogenic lines afford a system which allows the onset of resistance to be correlated with the induction of biochemical responses. Manners *et al.* (1985) have used this system to investigate protein synthesis associated with the expression of resistance. Pairs of susceptible and resistant near-isogenic lines of barley which differed in the *Ml-a, Ml-k* and *Ml-p* genes for resistance to the powdery mildew fungus, were inoculated with this pathogen. Patterns of protein synthesis associated with primary infection were mapped using pulse-labelling with [^{35}S] methionine and two-dimensional electrophoresis to separate the radioactive proteins. The results showed enhanced synthesis of five and eight polypeptides at 25 and 30 h, respectively, after inoculation of barley carrying the *Ml-a* gene. The enhanced synthesis of these same polypeptides together with 11 additional polypeptides was observed at 48 and 72 h after inoculation of barley carrying either the *Ml-p* or *Ml-k* genes. These enhancements correlated with the onset of resistance suggesting a possible role for the polypeptides in the resistance reaction. They were termed resistance-related, or infection-related (IR), proteins and it was suggested that resistance genes have a regulatory role in determining the synthesis of resistance-related proteins during the infection process (Manners *et al.*, 1985).

In vitro protein synthesis

In an attempt to understand how the barley resistance loci may exert a regulatory effect on expression of other host genes, Davidson *et al.* (1987) investigated the possibility that host mRNA species are induced in leaves of barley carrying the *Ml-p* resistance gene during infection by the powdery mildew fungus. Near-isogenic lines of barley, which differ in the *Ml-p* gene for resistance, were inoculated with powdery mildew and the *in vitro* translation products of mRNA populations compared by two-dimensional gel electrophoresis and fluorography. Considerable alterations in the translatable polyadenylated (poly(A^+)) RNA populations of barley were detected as early as 12 h after inoculation. These major changes in host gene expression precede, or at least coincide with, penetration of the epidermal cells by the fungus (Bushnell, 1981). However, at this time similar post-inoculation changes were observed in the poly(A^+) RNA populations of both cvMl-a and cvml-a (Davidson *et al.*, 1987), suggesting that these observed changes are unlikely to play a role in resistance. Experiments carried out 24 and 48 h after inoculation revealed a number of mRNAs showing increased translational activity in resistant lines carrying either the *Ml-a* or *Ml-p* genes when compared with the corresponding infected susceptible line and the non-infected control (Davidson *et al.*, 1987).

These results are consistent with models which propose that resistance is determined by alteration in the host gene expression following interaction of the resistance gene product to a race-specific ligand of the pathogen (Keen, 1982). Further, it appears that separate genes for resistance can share a regulatory role in enhancing the accumulation of specific host mRNAs and that this regulatory role occurs shortly after primary penetration of the pathogen.

The identification of infection-related cDNA clones

The molecular cloning of complementary DNA (cDNA) copies of mRNAs associated with disease resistance in plants has led to a better understanding of the changes in host gene expression during plant–pathogen interactions (Dixon, 1986; Ebel, 1986). Using cloned cDNAs it has been possible to quantitate accurately specific changes in mRNA induction during resistance interactions (Bell *et al.*, 1984; Schmelzer *et al.*,

Table 23.1. Characteristics of cDNA clones.

Clone no.	Insert size	mRNA size	Maximal ratios of mRNA induction associated with infection and resistance*	
			I-cvMl-p/C-cvMl-p	I-cvMl-p/I-cvml-p
pRP1	820	1350	8.6	1.5
pRP2	414	1610	15.8	4.6
pRP3	357	2430	3.0	2.8
pRP4	280	1130	4.3	1.5
pRP5	270	1320	7.8	2.9
pRP6	160	1190	11.3	5.7

*Data from either 1 or 2 days after inoculation, I, infected; C, control.
After Davidson *et al.*, 1987.

1984; Fristensky *et al.*, 1985; Schowalter *et al.*, 1985; Daniels *et al.*, 1987), to investigate the chromosomal location, regulation and organization of the genes corresponding to the cDNA clones (Chen & Varner, 1985; Douglas *et al.*, 1987), and to identify previously unidentified proteins which have roles in host–parasite interactions (Cornelissen *et al.*, 1986; Richardson *et al.*, 1987).

Davidson *et al.* (1987) prepared a cDNA library of infection-related mRNAs isolated from leaves of cvMl-p 2 days after inoculation. The mRNA population selected encoded *in vitro* translation products ranging in M_r from 25 to 50 kDa. The cDNAs were synthesized from these selected poly(A^+) RNA fractions (Wickens *et al.*, 1978) and the resulting blunt-ended double-stranded cDNA was size-fractionated to remove cDNA smaller than 200 base pairs. For cloning, the cDNA was inserted into a plasmid by oligo(dC) tailing and annealing to *Pst*I linearized oligo(dG)-tailed pBR 322 (Maniatis *et al.*, 1982). The reannealed plasmids containing the cDNA inserts were transformed (Hanahan, 1983) into *Escherichia coli* strain DH1 yielding 9000 recombinant colonies. From this cDNA library 3500 colonies were replica-plated and screened by hybridization to [^{32}P]cDNA prepared from control or infected leaves of cvMl-p 2 days after inoculation.

Six cDNA clones showing greater hybridization to cDNA prepared from infected leaves were selected (Table 23.1). These six cDNA clones hybridized to DNA isolated from barley leaves but not to DNA from conidia of the fungus. In Northern blot analysis of RNA from infected leaves, the six cDNA clones each hybridized to mRNA of different species. Translation products of three of the cDNA clones correspond to infection-related translation products identified on two-dimensional fluorogrammes.

The induction of host infection-related mRNAs

Davidson *et al.* (1988) used the cDNA clones to study the kinetics of host mRNA induction in near-isogenic lines of barley. The host mRNA species corresponding to the cDNA clones were induced before 24 h after inoculation during the primary penetration process. In addition, the mRNA corresponding to four of the cDNA clones increased to greater amounts in cvMl-p than in the near-isogenic susceptible cvml-p over a 2-day period following inoculation. It was suggested that the *Ml-p* gene has a regulatory role in host gene expression, resulting in enhanced expression of several host mRNA species following infection by the powdery mildew fungus, and that these mRNAs have a role in determining resistance.

An example of differential induction for some IR-mRNAs in barley conditioned by the *Ml-p* gene is shown in Fig. 23.2. The analysis of these events has led to the

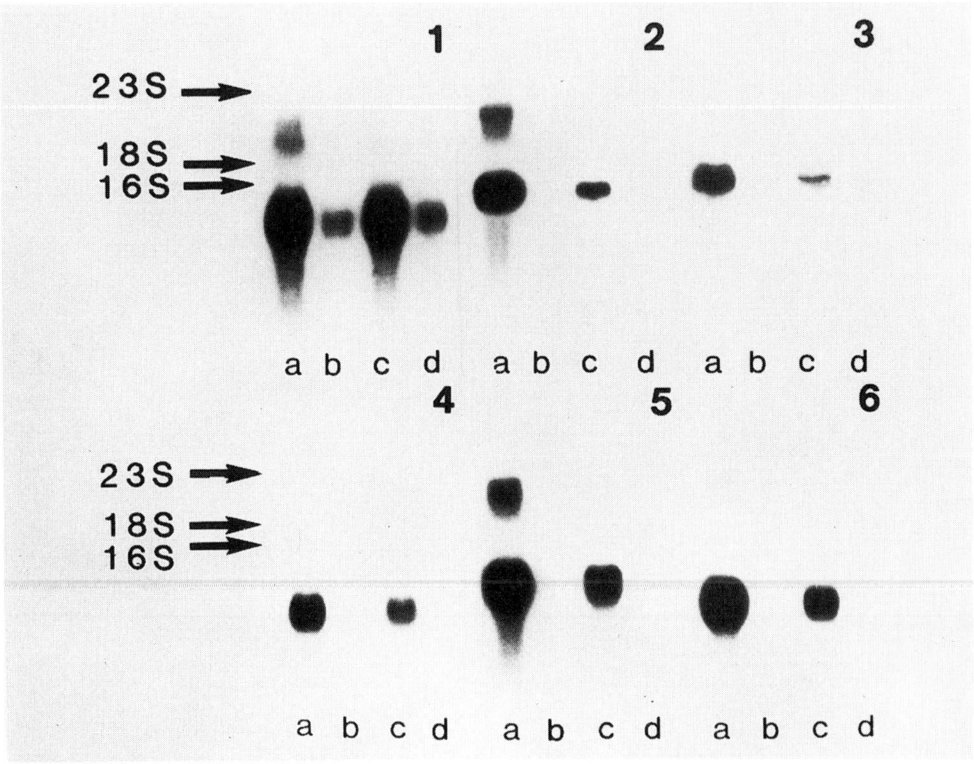

Fig. 23.2. Effect of the *Ml-p* gene on the steady state amounts of IR-mRNAs of barley during powdery mildew infection. Northern blots of total RNA (15 μg per lane) were probed with inserts from six cDNA clones (pRP1–6) corresponding to distinct IR-mRNAs (panels 1–6). RNA was obtained from control (lanes b and d) and inoculated (lanes a and c) leaves of near-isogenic resistant (lanes a and b) and susceptible (lanes c and d) barleys differing in the *Ml-p* gene, at 48 h after inoculation with the powdery mildew fungus. After Scott *et al.*, 1990.

hypothesis outlined in Fig. 23.3. It is suggested that a product of an avirulence gene in the pathogen interacts with a regulatory gene, or gene product, in the host which results in the expression of structural genes. These genes mediate the resistance reaction via their products.

This model is consistent with results obtained in studies of host–parasite interactions in dicotyledonous plants. The induction of specific mRNAs following infection with facultative fungal pathogens has been unequivocally demonstrated in several legumes (Dixon, 1986; Collinge & Slusarenko, 1987). The mRNAs induced in legumes include those coding for enzymes involved in the synthesis of phytoalexins, lignin, the cell wall glycoprotein 'extensin', chitinases and other unidentified infection-related proteins (Collinge & Slusarenko, 1987). The induction of infection-related mRNAs in legumes is associated with both non-host (Fritensky *et al.*, 1985) and race-line specific (Bell *et al.*, 1984; Cramer *et al.*, 1985; Showalter *et al.*, 1985), resistance to fungal pathogens. These host responses can also be elicited by wounding, ethylene, fungal cell wall extracts and abiotic stresses (Fritensky *et al.*, 1985; Dixon, 1986; Collinge & Slusarenko, 1987). Thus in legumes, infection-related mRNAs are thought to be part of a general host defence response which can be directly stimulated by fungal elicitors (Dixon, 1986) and

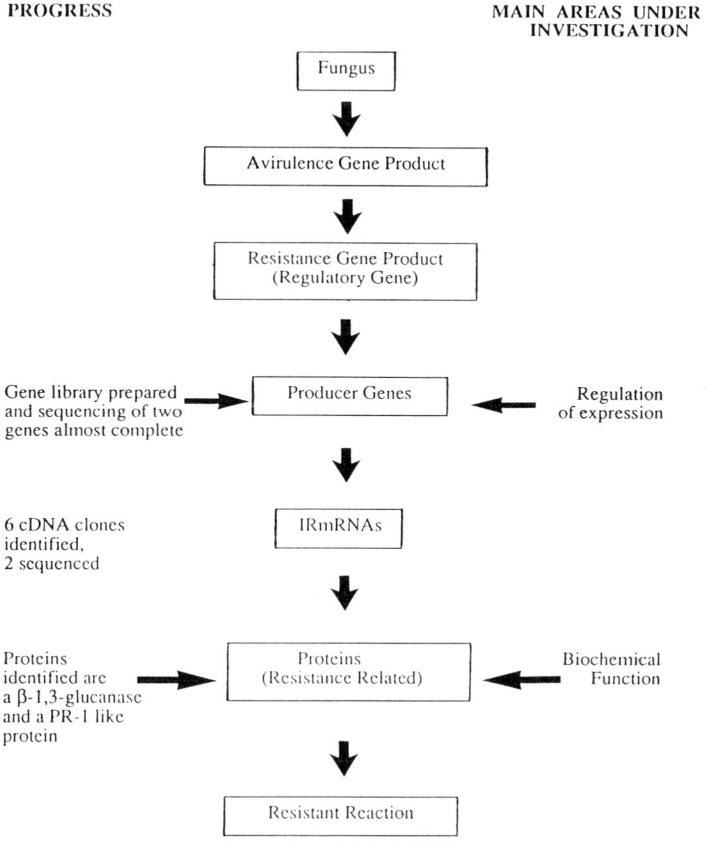

Fig. 23.3. Proposed mechanisms of disease resistance in barley to the powdery mildew fungus.

indirectly by either wounding or the release of endogenous elicitors by pathogens during infection (Darvill & Albersheim, 1984; Dixon, 1986). The activation of genes transcribing these mRNAs is extremely rapid and is tightly co-ordinated. It appears that co-ordinated transcriptional activation of defence genes may be a common and basic factor in active host resistance in dicotyledonous plants (Carr & Klessig, 1989).

In contrast to the IR-mRNAs and proteins of dicotyledonous plants, the IR-mRNAs of cereals are less well defined. For example, it was not known whether the induction of IR-mRNAs of barley is specific to infection by the obligate powdery mildew pathogen or more generally associated with infection by other types of pathogens. Scott *et al.* (1990) examined the effect of a range of pathogens on the induction of six cloned IR-mRNAs in barley and wheat. Barley was inoculated with the compatible pathogens chlorostriate mosaic virus, barley stripe virus, *Pseudomonas syringae, Pyrenophora teres, Bipolaris sorokiniana, E. graminis* f.sp. *hordei* and the incompatible pathogens *E. graminis* f.sp *tritici, Puccinia graminis* f.sp. *tritici* and *Py. tritici-repentis*. Wheat was inoculated with the compatible pathogens *B. sorokiniana, Pu. graminis* f.sp. *tritici* and *E. graminis* f.sp. *tritici* and the incompatible pathogen *E. graminis* f.sp. *hordei*. Induction of the IR-mRNAs, previously identified in barley infection by the powdery mildew fungus, was observed after inoculation with necrotrophic fungi but not following bacterial or viral infection. Moreover, the results demonstrated that infection by the pathogens *Py. tritici-repentis* and *B. sorokiniana* resulted in extremely rapidly induction of all the IR-mRNAs, indicating that the IR-mRNAs are not specifically involved in host responses to biotrophic obligate fungal parasites.

The results of these studies (Scott *et al.*, 1990) indicated that the mRNA response of barley to powdery mildew infection is a general response to fungal infection. The model proposed in Fig. 23.3 is probably an over-simplification of the factors regulating the expression of IR genes in specific host–pathogen interactions. Multiple non-specific elicitors and suppressors may also be involved as fungal signals in addition to avirulence gene products. Cereals may resist fungal pathogens by the early activation of a series of defence genes leading to the production of IR-mRNAs, with resistance being determined by the timing and extent of induction of these genes during infection. The IR-mRNA response in barley appears to resemble the induction of plant defence genes described in detail in dicotyledonous plants (Dixon, 1986; Collinge & Slusarenko, 1987).

The sequencing of cDNA clones

From the above, it appears that the induction of mRNAs in barley during fungal infection resembles that of dicotyledonous plants where a number of host defence genes are known to be activated (Collinge & Slusarenko, 1987; Ebel & Grisebach, 1988; Templeton & Lamb, 1988). In the dicotyledonous plants the identity of many of the induced host IR-mRNAs has been established, showing that they encode proteins that are either antifungal (for example chitinases and β-1,3-glucanase), or are involved in well-characterized cellular defence processes such as phytoalexin synthesis and lignification. In barley the functions of the proteins encoded by the six IR-mRNAs have been investigated by sequencing the cDNAs. Sequence analysis of pRP-1 has failed to reveal any homology with previously published sequences and the role of the protein encoded by this gene is presently unknown (Jutidamrongphan *et al.*, 1989).

However, the proteins encoded by two other (pRP-2 and pRP-6) of the six previously cloned IR-mRNAs (Davidson *et al.*, 1987, 1988) have been identified (Jutidamrongphan *et al.*, 1991; Wolfe *et al.*, in preparation) as a β-1,3-glucanase and a pathogenesis-related protein of the Family I group (Carr & Klessig, 1989) respectively.

A comparison of the nucleotide and deduced amino acid sequences of the cDNA clone of pRP-2 showed that it matched almost exactly with part of the deduced amino acid sequence of β-1,3-glucanase in tobacco (Shinshi *et al.*, 1988) and a β-1,3/1,4-glucanase from barley (Fincher *et al.*, 1986). The amino acid sequence was identical to that reported for barley β-1,3-glucanase by Ballance & Svendsen (1988), and the nucleotide sequence matched almost perfectly with that of a cDNA clone, coding for β-1,3-glucanase, isolated from barley scutellum (Hoj *et al.*, 1989).

Estimations of the amount of this IR-mRNA in near-isogenic resistant and susceptible barleys during infection with powdery mildew fungus (Jutidamrongphan *et al.*, 1991) showed an early induction correlated with resistance conditioned by the *Ml-a*, *Ml-p* and *Ml-k* genes; this indicates that the regulation of genes encoding for β-1,3-glucanase may be important in determining resistance to this pathogen. Western blots of intercellular proteins isolated from different resistant and susceptible strains of barley inoculated with powdery mildew fungus showed the greatest induction of β-1,3-glucanase occurred in the two resistant lines. This was confirmed by assays of β-1,3-glucanase activity in resistant and near-isogenic susceptible lines (Jutidamrongphan *et al.*, 1991).

The induction of β-1,3-glucanase was also demonstrated following infection with the necrotrophic pathogen *Bipolaris sorokiniana*, not only in barley, but also in wheat, rice and sorghum (Jutidamrongphan *et al.*, 1991). The identification of β-1,3-glucanase as a cereal defence response in cereals may closely resemble that previously demonstrated in the dicotyledons (Mauch *et al.*, 1984; Boller, 1985; Kauffman *et al.*, 1987; Kombrink *et al.*, 1988). However, cDNA clones for phenylalanine ammonia lyase, chalone synthetase (Krenzaler *et al.*, 1983; Kuhn *et al.*, 1984) and extensin (Chen & Varner, 1985) did not hybridize to mRNA populations in powdery mildew-infected barley (Davidson *et al.*, 1987).

The appearance of β-1,3-glucanase in the intercellular fluid of tobacco leaves (Kauffman *et al.*, 1987) and potato leaves (Kombrink *et al.*, 1988) has been associated with chitinases and other 'pathogenesis-related' proteins. Similarly, in barley, chitinase and a protein serologically related to tobacco PR-1 have been identified (Andersen, 1989; Wolfe *et al.*, in preparation). The induction of both chitinase and β-1,3-glucanase has been demonstrated in pea tissue as fungal inhibitors (Mauch *et al.*, 1988). In addition, Bohlmann *et al.* (1988) have demonstrated that another class of antifungal proteins, the leaf-specific thionins, are induced in barley during infection by the powdery mildew fungus. It seems that barley has a multicomponent response to fungal infection and further investigation of the IR-mRNAs may elucidate the functions of other components in this response. Ultimately, investigation of the regulation of these genes should permit manipulation of expression of these genes to assess their roles in resistance to fungal pathogens.

Results based on sequence analysis and Western blotting indicate that pPR-6 isolated by Davidson *et al.* (1987) corresponds to proteins in Family 1 of the pathogenesis-related protein group (Wolfe *et al.*, in preparation). The function of this group of proteins is presently unknown. Pathogenesis-related proteins are discussed in more detail in Chapter 22 of this volume.

Summary and conclusions arising from IR-mRNAs studies

The amounts of IR-mRNAs are always greater in powdery mildew-infected barley than in non-inoculated controls. Furthermore, it has been shown that the IR-mRNAs are also induced in barley in response to infection with other fungi, both obligate parasites and necrotrophs. This indicates that the IR-mRNAs are associated with a general response of barley which may be induced independently of the race-line specific resistance response (Ellingboe, 1972; Hsu & Ellingboe, 1972; Yang *et al.*, 1972).

The enhanced accumulation of the IR-mRNAs in resistant lines generally corresponds to the period when incompatibility occurs. In cvMl-a, when compared with cvml-a, the enhanced accumulation of the IR-mRNAs occurs 12–48 h after inoculation when incompatibility is being determined (Ellingboe & Slesinski, 1971; Johnson *et al.*, 1979). In cvMl-p and cvMl-k this enhanced accumulation occurs during later periods following inoculation, corresponding to the time of the determination of incompatability in these plants (Ellingboe & Slesinski, 1971; Day, 1974). This differential induction of the resistance genes suggests that their activities are pleiotropic affecting the expression of several structural genes.

Although the IR-mRNAs accumulate in both susceptible and resistant lines, the enhanced accumulation of the IR-mRNAs in the resistant lines may account for resistance. Similarly, in other systems, accumulation of mRNAs involved in disease resistance and compounds inhibitory to fungal pathogens occur in both compatible and incompatible interactions (Bell *et al.*, 1984; Hahn *et al.*, 1985; Showalter *et al.*, 1985; Mehdy & Lamb, 1987), but the earlier induction and the enhanced accumulation of these compounds may account for resistance.

Future directions

Genomic studies

As discussed, a number of proteins and mRNAs have been identified in barley which are induced by fungal infection (Manners *et al.*, 1985; Davidson *et al.*, 1987, 1988; Bohlmann *et al.*, 1988). On the basis of the onset and magnitude of induction and their antifungal activity, it is likely that these mRNAs and the proteins they encode play an important role in resistance. Indeed, it has been shown by sequence homology and Western blotting that one of these mRNAs encodes β-1,3-glucanase, an enzyme with known antifungal activity (Mauch *et al.*, 1988). Another encodes a protein classified in Family I of the pathogenesis-related proteins which are thought to function in systemic acquired resistance against a range of plant pathogens (Carr & Klessig, 1989). These results indicate that barley has a multicomponent inducible defence system for resistance to infection by fungal pathogens and this defence system is regulated at the level of gene expression. A possible strategy for enhancing barley resistance to fungal infection is to engineer a high level of expression of these barley genes. In this connection it has been shown that the infection-related genes have homologous counterparts in wheat, rice and sorghum (Scott *et al.*, 1990). Therefore, the over-expression of these genes may constitute a general approach to engineering resistance mechanisms to infection by fungal phytopathogens in all cereals.

Induction of eukaryotic genes is determined by the interaction of proteins (transcription factors) with specific sequence motifs in the promoter region of genes. To identify the promoter regions involved in gene regulation, it is necessary to carry out a functional analysis of the putative promoter region, linked to an easily assayed reporter gene, until the particular sequence which is responsible for induction is identified. To use such constructs in functional analysis it is necessary to develop techniques for the transient and stable expression of foreign genes either in barley or a closely related cereal.

If the ultimate goal is to manipulate the expression of IR genes in barley there are two identifiable steps which need to be taken.

1. Sequencing of infection-related genes. Graham *et al.* (1990) have described procedures for the routine construction of genomic libraries in cereals and these are applicable to barley. Genomic libraries of barley DNA have been probed with cDNAs of IR-mRNAs and several clones for the IR-genes have been purified. Once the sequence of these genes has been determined, sequences 5′ of the initiation site for transcription can be fused to reporter genes. Similarly sequences located 3′ to the coding region can be identified and fused to a reporter gene carrying a constitutive promoter. These chimeric genes can be used to determine whether or not either the 5′ or 3′ regions contain elements which are induced by fungal pathogens.
2. The development of transient and stable gene expression in barley. (see also Chapter 22, this volume). Since *Agrobacterium tumefaciens* cannot be used to transfer genes into barley, direct delivery systems for DNA are being developed. The three main systems currently under investigation are the use of protoplasts, microprojectiles and microinjection. All these systems have technical problems to overcome. Although the use of protoplasts has resulted in the regeneration of fertile transformed rice plants (Shimamoto *et al.*, 1989), there are still problems to be overcome before routine regeneration of competent barley protoplasts can be achieved.

The use of projectiles for gene implantation into plant cells was pioneered by Sanford *et al.* (1987). In the original guns, explosives were used to accelerate heavy metal particles coated with DNA for transport into plant cells. The method has the advantage of ease of handling, one shot can lead to multiple hits, genes coated on the particles resume biological activity in the cell, and a range of different materials can be used as targets. The system is proving particularly useful for testing gene expression in transient systems. However, the procedure does have have inherent problems for although there are many particle guns in use by many research workers world-wide, with the exception of maize (Fromm *et al.*, 1990), no transgenic cereals have yet been produced. The two chief problems with the explosive gun are the very low frequency of transformation and the damage caused to the cells on impact, often leading to a cessation of cell division.

Gas guns are being developed which may overcome the problems associated with explosive microprojectile procedures. The gas gun provides a well-controlled method for delivering a uniform cloud of high-velocity microprojectiles over a large area.

Another approach to direct DNA transfer is microinjection. Microcapillaries are used to deliver DNA into defined cells such that the injected cells survive and proliferate. So far no transgenic cereals have been produced using this procedure.

cDNA clones of IR-mRNAs as tools for studying the molecular basis of disease resistance

Full-length cDNA clones may be subcloned into expression vectors and used in *in vitro* systems to synthesize polypeptides corresponding to the cDNA clones. Antisera may be obtained against the expressed polypeptides and then used to detect immunologically the *in vivo* synthesized proteins in cells of barley. Information on the intracellular localization of the IR-mRNAs, or their corresponding protein products, would contribute significantly to elucidating the role of these proteins in resistance. To be actively involved in resistance these proteins would need to be localized in the vicinity of the invading pathogen. Recently two techniques have shown promise for the localization of defence reactions in plants. O'Connell *et al.* (1986) successfully used immunogold labelling of fungal antisera to locate fungal antigens within the cells of *Phaseolus vulgaris* infected with *Colletotrichum lindemuthianum*. In another study (Cornish *et al.*, 1987) a cDNA clone, believed to be the S2-allele of *Nicotiana alata*, was used for *in situ* hybridization experiments to localize the corresponding mRNA in tobacco leaves. If such techniques are applicable to barley it may be possible to:

1. Identify the cells involved in the defence reactions of barley to the powdery mildew fungus. This will establish whether only the epidermal cells are involved, or whether endogenous or fungal elicitors cause the mRNAs to accumulate in other cells such as the mesophyll;
2. Perform precise quantitative experiments on the accumulation of IR-mRNAs *in situ* to differentiate between a very localized accumulation and a widespread induction of mRNAs, or compounds, inhibitory to the fungus. This difference may be significant in determining incompatibility and therefore would give more information regarding the importance of these mRNAs in the early stages of the resistance response and the possible regulation of IR-mRNAs by resistance loci;
3. Determine whether the timing and location of the *in situ* accumulation of the mRNAs, or the proteins they encode, correspond with the cytological changes in the host tissue which occur in response to fungal infection. This would indicate a further physiological function for the IR-mRNAs.

The role of the IR-mRNAs in resistance could be examined further if transformation methods could be developed for barley. For example the coding regions of the IR-mRNAs could be linked to a high-level promoter, thereby elevating the levels of the protein products in a normally susceptible host, and the effect of this on susceptibility could be assessed. Alternatively, it may be possible to produce transgenic barley carrying antisense IR-mRNA (Ecker & Davis, 1986) and examine whether this impairs the resistance response.

Although the identification of resistance genes is important, this knowledge is of little value without an understanding of their role in the plant's overall metabolism. Establishing the identity and the role for the cloned IR-mRNAs together with understanding the regulation of genes encoding these mRNAs of barley may characterize both the biochemical and genetic basis for disease resistance in this important cereal. This information may also be useful in other plant systems which are more amenable to genetic manipulation.

References

Aist, J.R. & Israel, H.W. (1977) Papilla formation: timing and significance during penetration of barley coleoptiles by *Erysiphe graminis* f.sp. *hordei*. *Phytopathology* 67, 455–61.

Andersen, J.B. (1989) Studies on PR-proteins of barley inoculated with barley powdery mildew. Unpublished PhD thesis, The Royal Veterinary and Agricultural College, Copenhagen, Denmark.

Ballance, G.M. & Svendsen, I. (1988) Purification and amino acid sequence determination of an endo-β-glucanase from barley. *Carlsberg Research Communications* 53, 411–19.

Bell, J.N., Dixon, R.A., Bailey, J.A., Rowell, P.M. & Lamb, C.J. (1984) Differential induction of chalone synthetase mRNA activity at the onset of phytoalexin accumulation in compatible and incompatible plant–pathogen interactions. *Proceedings of the National Academy of Sciences (USA)* 81, 3384–8.

Bohlmann, H., Clausen, S., Behnke, S., Giese, H., Hiller, C., Reinmann-Phillipp, U., Schrader, G., Barkholt, V. & Apel, K. (1988) Leaf-specific thionins of barley – a novel class of cell wall proteins toxic to plant-pathogenic fungi and possibly involved in the defence mechanism of plants. *EMBO Journal* 7, 1559–65.

Boller, T. (1985) Induction of hydrolases as a defence reaction against pathogens. In: Key, J.L. & Kosuge, T. (eds), *Cellular and Molecular Biology of Plant Stress*. Alan R. Liss, New York, pp. 247–62.

Bushnell, W.R. (1981) Incompatibility conditioned by the Mla gene in powdery mildew of barley: the halt in cytoplasmic streaming. *Phytopathology* 71, 1062–6.

Bushnell, W.R. & Bergquist, S.E. (1975) Aggregation of host cytoplasm and the formation of papillae and haustoria in powdery mildew of barley. *Phytopathology* 65, 310–18.

Carr, J.P. & Klessig, D.F. (1989) The pathogenesis-related proteins of plants. In: Setlow, J. (ed.), *Genetic Engineering Principles and Methods*, Vol. 11. Plenum, New York, pp. 65–100.

Chen, J. & Varner, J.E. (1985) An extracellular matrix protein in plants: characterization of a genomic clone for carrot extensin. *EMBO Journal* 4, 2145–51.

Cho, B.H. & Smedegaard-Petersen, V. (1986) Induction of resistance to *Erysiphe graminis* f.sp. *hordei* in near-isogenic barley lines. *Phytopathology* 76, 301–5.

Collinge, D.B. & Slusarenko, A.J. (1987) Plant gene expression in response to pathogens. *Plant Molecular Biology* 9, 389–410.

Cornelissen, B.J.C., Hooft van Huijsduijnen, R.A.M., Van Loon, L.C. & Bol, J.F. (1986) Molecular characterization of messenger RNAs for 'pathogenesis-related' proteins 1a, 1b and 1c, induced by TMV infection of tobacco. *EMBO Journal* 5, 37–40.

Cornish, E.C., Pettit, J.M., Bonig, I. & Clarke, A.E. (1987) Developmentally controlled expression of a gene associated with self-incompatibility in *Nicotiana alata*. *Nature* 326, 99–102.

Cramer, C.L., Bell, J.N., Ryder, T.B., Bailey, J.A., Schuch, W., Bolwell, G.P., Robbins, M.P., Dixon, R.A. & Lamb, C.J. (1985) Co-ordinated synthesis of phytoalexin biosynthetic enzymes in biologically-stressed cells of bean (*Phaseolus vulgaris L.*). *EMBO Journal* 4, 285–9.

Daniels, C.H., Fristensky, B., Wagoner, W. & Hadwiger, L.A. (1987) Pea genes associated with non-host disease resistance to *Fusarium* are also active in race-specific disease resistance to *Pseudomonas*. *Plant Molecular Biology* 8, 309–16.

Darvill, A.G. & Albersheim, P. (1984) Phytoalexins and their elicitors – a defence against microbial infection in plants. *Annual Review of Plant Physiology* 35, 243–75.

Davidson, A.D., Manners, J.M., Simpson, R.S. & Scott, K.J. (1987) cDNA cloning of mRNAs induced in resistant barley during infection by *Erysiphe graminis* f.sp. *hordei*. *Plant Molecular Biology* 8, 77–85.

Davidson, A.D., Manners, J.M., Simpson, R.S. & Scott, K.J. (1988) Altered host gene expression in near-isogenic barley conditioned by different genes for resistance during infection by *Erysiphe graminis* f.sp. *hordei*.

Physiological and Molecular Plant Pathology 32, 127–39.

Day, P.R. (1974) *Genetics of Host–Parasite Interaction.* Freeman, San Francisco, CA, USA.

Day, P.R. & Scott, K.J. (1973) Scanning electron microscopy of fresh material of *Erysiphe graminis* f.sp. *hordei. Physiological Plant Pathology* 3, 433–5.

Dixon, R.A. (1986) The phytoalexin response: elicitation, signalling and control of host gene expression. *Biological Reviews* 61, 239–91.

Douglas, C., Hoffman, H. Schulze, W. & Hahlbrock, K. (1987) Structure and elicitor or u.v. light-stimulated expression of two 4-coumarate: CoA ligase genes in parsley. *EMBO Journal* 6, 1189–95.

Dyer, T.A. & Scott, K.J. (1972) Decrease in chloroplast content of barley leaves infected with powdery mildew. *Nature* 263, 237–8.

Ebel, J. (1986) Phytoalexin synthesis: the biochemical analysis of the induction process. *Annual Review of Phytopathology* 24, 235–64.

Ebel, J. & Grisebach, H. (1988) Defence strategies of soybean against the fungus *Phytophthora megasperma* f.sp. *glycinea*: a molecular analysis. *Trends in Biochemical Sciences* 13, 23–7.

Ecker, J.R. & Davis, R.W. (1986) Inhibition of gene expression in plant cells by expression of antisense RNA. *Proceedings of the National Academy of Sciences (USA)* 83, 5372–6.

Ellingboe, A.H. (1972) Genetics and physiology of primary infection by *Erysiphe graminis. Phytopathology* 62, 401–6.

Ellingboe, A.H. & Slesinski, R.S. (1971) Genetic control of mildew development. In: Nilan, R.A. (ed.), *Barley Genetics II.* Washington State University Press, USA, pp. 472–4.

Fincher, G.B., Lock, P.A., Morgan, M.M., Lingelbach, K., Wattenhall, R.E.H., Mercer, J.F.B., Brandt, A. & Thomsen, K.K. (1986) Primary structure of the (1–3, 1–4)-β-D-glucan 4-glucohydrolase from barley aleurone. *Proceedings of the National Academy of Sciences (USA)* 83, 2081–5.

Fritensky, B., Riggleman, R.C., Wagoner, W. & Hadwiger, L.A. (1985) Gene expression in susceptible and disease resistant interactions of peas induced with *Fusarium solani* pathogens and chitosan. *Physiological Plant Pathology* 27, 15–28.

Fromm, M.N., Morrish, F., Armstrong, C., Williams, R., Thomas, J. & Klein, T.M. (1990) Inheritance and expression of chimeric genes in the progeny of transgenic maize plants. *Biotechnology* 8, 833–8.

Graham, M.W., Doherty, J.P. & Woodcock, D.M. (1990) Efficient construction of plant genomic libraries requires the use of mcr host strains and packaging mixes. *Plant Molecular Biology Reporter* 8, 18–27.

Hahn, M.G., Bonhoff, A. & Grisebach, H. (1985) Quantitative localization of the phytoalexin glyceollin I in relation to fungal hyphae in soybean roots infected with *Phytophthora megasperma* f.sp. *glycinea. Plant Physiology* 77, 591–601.

Hanahan, D. (1983) Studies on transformation of *Escherichia coli* with plasmids. *Journal of Molecular Biology* 166, 557–80.

Hislop, E.C. & Stahmann, M.A. (1971) Peroxidase and ethylene production by barley leaves infected with *Erysiphe graminis* f.sp. *hordei. Physiological Plant Pathology* 1, 297–312.

Hoj, O.B., Hartman, D.J., Morrice, N.A., Doan, D.N.P. & Fincher, G.B. (1989) Purification of (1–3)-β-glucan endohydrolase isoenzyme II from germinating barley and determination of its primary structure from a cDNA clone. *Plant Molecular Biology* 13, 31–42.

Hsu, S.C. & Ellingboe, A.H. (1972) Elongation of secondary hyphae and transfer of ^{35}S from barley to *Erysiphe graminis* f.sp. *hordei* during primary infection. *Phytopathology* 62, 876–82.

Johnson, L.E.B., Bushnell, W.R. & Zeyen, R.J. (1979) Binary pathways of analysis of primary infection and host response in populations of powdery mildew fungi. *Canadian Journal of Botany* 57, 497–511.

Jutidamrongphan, W., Mackinnon, G., Manners, J.M. & Scott, K.J. (1989) Sequence of a near-full length cDNA clone for a mRNA of barley induced by fungal infection. *Nucleic Acids Research* 17, 9478.

Jutidamrongphan, W., Anderson, J.B., Mackinnon, G., Manners, J.M., Simpson, R.S. & Scott, K.J. (1991) Induction of β-1,3-glucanase in barley in response to infection by the powdery mildew fungus. *Molecular Plant–Microbe Interactions* 4, 234–8.

Kauffmann, S., Legrand, M., Geoffroy, P. & Fritig, G. (1987) Biological function of pathogenesis-related proteins: four PR proteins of tobacco have 1,3-β-glucanase activity. *EMBO Journal* 6, 3209–12.

Keen, N.T. (1982) Specific recognition in gene-for-gene host–parasite systems. *Advances in Plant Pathology* 1, 35–82.

Kita, N.T., Toyoda, H. & Shishiyama, J. (1981) Chronological analysis of cytological responses in powdery-mildewed barley leaves. *Canadian Journal of Botany* 59, 1761–8.

Koga, H., Mayama, S. & Shishiyama, J. (1980) Correlation between the deposition of fluorescent compounds in papillae and resistance in barley against *Erysiphe graminis hordei*. *Canadian Journal of Botany* 58, 536–41.

Kombrink, E., Schoroder, M. & Halhbrock, K. (1988) Several 'pathogenesis-related' proteins in potato are 1,3-β-glucanases and chitinases. *Proceedings of the National Academy of Sciences (USA)* 85, 782–6.

Kreuzaler, F., Ragg, H., Fautz, E., Kuhn, D.N. & Hahlbrock, K. (1983) UV-induction of chalcone synthase mRNA in cell suspension cultures of *Petroselinum hortense*. *Proceedings of the National Academy of Sciences (USA)* 80, 2591–3.

Kuhn, D.N., Chappell, J., Boudet, A. & Hahlbrock, K. (1984) Induction of phenylalanine ammonia-lyase and 4-coumarate:CoA ligase mRNAs in cultured plant cells by UV light or fungal elicitor. *Proceedings of the National Academy of Sciences (USA)* 81, 1102–6.

Kunoh, H. & Ishizaki, H. (1976) Accumulation of chemical elements around the penetration sites of *Erysiphe graminis hordei* on barley leaf epidermis: (III) micro manipulation and X-ray microanalysis of silicon. *Physiological Plant Pathology* 8, 91–6.

Kunoh, H., Ishizaki, H. & Nakaya, K. (1977) Cytological studies of early stages of powdery mildew in barley and wheat leaves: (II) significance of the primary germ tube of *Erysiphe graminis* on barley leaves. *Physiological Plant Pathology* 10, 191–9.

Kunoh, H., Tsuzuki, T. & Ishizaki, H. (1978) Cytological studies of early stages of powdery mildew in barley and wheat. IV. Direct ingress from superficial primary germ tubes and appressoria of *Erysiphe graminis hordei* on barley leaves. *Physiological Plant Pathology* 13, 327–33.

Kunoh, H., Yamamori, K. & Ishizaki, H. (1982) Cytological studies of early stages of powdery mildew in barley and wheat. VIII. Autofluorescence at penetration sites of *Erysiphe graminis hordei* on living barley coleoptiles. *Physiological Plant Pathology* 21, 373–9.

Lin, M.R. & Edwards, H.H. (1974) Primary penetration process in powdery mildewed barley related to host cell age, cell type and occurrence of basic staining material. *New Phytology* 73, 131–7.

Maniatis, T., Fritsch, E.F. & Sambrook, J. (1982) *Molecular Cloning: A Laboratory Manual*. Cold Spring Harbor Laboratory Press, NY.

Manners, J.M. & Scott, K.J. (1983) Translational activity of polysomes of barley leaves during infection by *Erysiphe graminis* f.sp. *hordei*. *Phytopathology* 73, 1386–92.

Manners, J.M. & Scott, K.J. (1984) The effect of infection by *Erysiphe graminis* f.sp. *hordei* on protein synthesis in vivo in leaves of barley. *Plant Cell Physiology* 25, 1307–11.

Manners, J.M. & Scott, K.J. (1985) Reduced translatable messenger RNA activities in leaves of barley infected with *Erysiphe graminis* f.sp. *hordei*. *Physiological Plant Pathology* 26, 297–308.

Manners, J.M., Davidson, A.D. & Scott, K.J. (1985) Patterns of post-infectional protein synthesis in barley carrying different genes for resistance to the powdery mildew fungus. *Plant Molecular Biology* 8, 77–85.

Masri, S.S. & Ellingboe, A.H. (1966) Primary infection of wheat and barley by *Erysiphe graminis*. *Phytopathology* 56, 389–95.

Mauch, F., Hadwiger, L.A. & Boller, T. (1984) Ethylene: Symptom, not signal for the induction of chitinase and β-1,3-glucanase in pea pods by pathogens and elicitors. *Plant Physiology* 76, 607–11.

Mauch, F., Mauch-Mani, B. & Boller, T. (1988) Antifungal hydrolases in pea tissue. II. Inhibition of fungal growth by combinations of chitinase and β-1,3-glucanase. *Plant Physiology* 88, 936–42.

Mayama, S. & Shishiyama, J. (1976) Histological observation of cellular responses of barley leaves to powdery mildew infection by UV-fluorescence microscopy. *Annals of the Phytopathological Society of Japan* 42, 618–20.

Mayama, S. & Shishiyama, J. (1978) Localized accumulation of fluorescent and UV absorbing compounds at penetration sites in barley leaves infected with *Erysiphe graminis hordei*. *Physiological Plant Pathology* 13, 347–54.

McKeen, W.E. & Bhattacharya, P.K. (1970) Limitation of infection by *Erysiphe graminis* f.sp. *hordei* culture CR3 by the Algerian gene Mla in barley. *Canadian Journal of Botany* 48, 1109–13.

Mehdy, M.C. & Lamb, C.J. (1987) Chalcone isomerase cDNA cloning and mRNA induction by fungal elicitor, wounding and infection. *EMBO Journal* 6, 1527–33.

Moseman, J.G. (1972) Isogenic barley lines for reaction to *Erysiphe graminis* f.sp. *hordei*. *Crop Science* 12, 681–2.

O'Connell, R.J., Bailey, J.A., Vose, I.R. & Lamb, C.J. (1986) Immunogold labelling of fungal antigens in cells of *Phaseolus vulgaris* infected by *Colletotrichum lindemuthianum*. *Physiological and Molecular Plant Pathology* 28, 99–105.

Oku, H., Ouchi, S., Shiraishi, T., Komoto, Y. & Oki, K. (1975) Phytoalexin production in barley powdery mildew. *Annals of the Phytopathological Society of Japan* 41, 185–191.

Richardson, M., Valdes-Rodriguez, S. & Blanco-Labra, A. (1987) A possible function for thaumatin and a TMV induced protein suggested by homology to a maize inhibitor. *Nature* 327, 432–4.

Sanford, J.C., Klein, T.M., Wolf, E.P. & Allen, N. (1987) Delivery of substances into cells and tissues using a particle bombardment process. *Particulate Science and Technology* 5, 27–37.

Sargent, C. & Gay, J.L. (1977) Barley epidermal apoplast structure and modification by powdery mildew contact. *Physiological Plant Pathology* 11, 195–205.

Schmelzer, E., Borner, H., Ebel, J. & Hahlbrock, K. (1984) Phytoalexin synthesis in soybean (*Glycine max*). Similar time courses of mRNA induction in hypocotyls infected with a fungal pathogen and in cell cultures treated with fungal elicitor. *FEBS Letters* 172, 59–63.

Scott, K.J. (1972) Obligate parasitism by phytopathogenic fungi. *Biological Reviews* 47, 537–72.

Scott, K.J., Davidson, A.D., Jutidamrongphan, W., Mackinnon, G. & Manners, J.M. (1990) The activation of genes of wheat and barley by fungal phytopathogens. *Australian Journal of Plant Physiology* 17, 229–38.

Shimamoto, K., Terada, R., Izawa, T. & Fujimoto, H. (1989) Fertile rice plants regeneration from transformed protoplasts. *Nature* 338, 274–6.

Shinshi, H., Wenzler, H., Neuhaus, J.M., Felix, G., Hofsteenge, J. & Meins, F. (1988) Evidence for N- and C-terminal processing of a plant defense-related enzyme: primary structure of tobacco prepro-β-1,3-glucanase. *Proceedings of the National Academy of Sciences (USA)* 85, 5541–5.

Showalter, A.M., Bell, J.N., Cramer, C.L., Bailey, J.A., Varner, J.E. & Lamb, C.J. (1985) Accumulation of hydroxyproline-rich glycoprotein mRNAs in response to fungal elicitor and infection. *Proceedings of the National Academy of Sciences (USA)* 82, 6551–5.

Templeton, M.D. & Lamb, C.J. (1988) Elicitors and defence gene activation. *Plant, Cell and Environment* 11, 395–401.

Wickens, M.P., Buell, G.N. & Schmike, R.T. (1978) Synthesis of double stranded DNA complementary to lysozyme, ovomucoid and ovalbumin mRNAs. *Journal of Biological Chemistry* 253, 2483–95.

Yang, S.L., Moseman, J.G. & Ellingboe, A.H. (1972) The formation of elongating secondary hyphae of *Erysiphe graminis* and the segregation of M1 genes in barley. *Phytopathology* 62, 1219–23.

Zeyen, R.J. & Bushnell, W.R. (1979) Papilla response of barley epidermal cells caused by *Erysiphe graminis*: rate and method of deposition determined by microcinematography and transmission electron microscopy. *Canadian Journal of Botany* 57, 898–913.

24

Strategies for Cloning Disease Resistance Genes

Shauna Somerville

Fungal diseases are a major limitation to crop yields world-wide. The primary means available for controlling them include the use of genetically resistant cultivars, fungicides, and cultural methods that do not favour growth of the pathogen. Although time-consuming to develop, genetically resistant cultivars have been important in disease management strategies. De-licensing of established fungicides, the rising occurrence of fungicide tolerant and resistant pathogens, and the increasing difficulty of developing new fungicides have contributed to the reduced usefulness of fungicides. Genetically resistant cultivars will become increasingly important.

The *in vitro* manipulation of disease resistance genes promises alternative sources of resistance for crop improvement programmes. Despite extensive cytological and, in some cases, genetic studies of host–pathogen interactions, virtually nothing is known about host resistance genes or their products. As a result, efforts to identify disease resistance genes or their polypeptide products have met with little success. One notable exception is the target site for the T-toxin produced by *Cochliobolus heterostrophus*, the causal agent of Southern corn leaf blight. Maize genotypes lacking the T-cytoplasm, and the *T-urf13* gene, are largely resistant to Southern corn leaf blight (Levings, 1990). Those plants with T-cytoplasm and the unusual *T-urf13* gene produce a 13 000 Da polypeptide that binds T-toxin and are susceptible to Southern corn leaf blight (Dewey *et al.*, 1988; Braun *et al.*, 1990).

This chapter looks at the cloning of the *Ml-a* mildew resistance locus of barley to illustrate some of the strategies currently available for cloning host resistance genes. Ellis *et al.* (1988) provide additional discussion of some of the techniques outlined here.

Disease resistance

It has been proposed that disease resistance can be best understood if host defences are grouped into non-host resistance and host resistance mechanisms (Heath, 1980, 1981, 1985). Non-host defences are those that are active against a broad range of potential pathogens that a plant may encounter. For example, barley plants do not succumb to diseases caused by pathogens that normally cause diseases on plant species other than barley. Successful pathogens have acquired mechanisms for defeating non-host resistance mechanisms. It is debatable whether such organisms must also acquire traits that permit them to colonize a specific host and cause disease (Heath, 1985). The thigmotrophic detection of the stomatal lip by the bean rust pathogen, *Uromyces appendiculatus* (Hoch *et al.*, 1987), or the release of mucilage by conidia for attachment to rice leaves by the rice blast pathogen, *Magnaportha grisea* (Hamer *et al.*, 1988), are examples of traits that enhance pathogenicity even though they do not act by defeating active host defences. In response to the selection pressure applied by organisms that have acquired the ability to cause disease, plants have evolved host resistance mechanisms that operate against specific pathogens. In addition, highly race-specific host resistance has been described for a number of diseases, especially for those caused by obligate pathogens such as the mildews and rusts. These more specific forms of host resistance have been exploited by plant breeders in the development of genetically resistant cultivars.

The gene-for-gene hypothesis was developed by Flor (1955, 1956, 1971) to describe the highly specific race–cultivar interactions in the flax rust disease (Flor, 1946, 1947). This hypothesis states that incompatible relationships develop when a host carrying a dominant resistance allele interacts with a pathogen having a complementary dominant avirulence allele (Table 24.1; Flor, 1955, 1971; Ellingboe, 1981; Crute & Norwood, 1986). Other combinations involving a recessive allele for susceptibility in the host or a recessive allele for virulence in the pathogen lead to the establishment of a compatible interaction and disease. A common mechanistic interpretation of the gene-for-gene hypothesis is that the host resistance gene encodes a receptor that intercepts a signal molecule encoded by the pathogen avirulence gene (Ellingboe, 1979). This recognition event triggers a cascade of host defensive responses leading to the expression of resistance.

Table 24.1. Disease reaction phenotype for host-pathogen interactions according to the gene-for-gene hypothesis.

Pathogen genotype	Host genotype	
	R–	rr
A–	Incompatible	Compatible
aa	Compatible	Compatible

Incompatible phenotype: resistant host and avirulent pathogen. Compatible phenotype: susceptible host and virulent pathogen.

The powdery mildew disease of barley, caused by the obligate fungal pathogen *Erysiphe graminis* f. sp. *hordei* is thought to conform to the gene-for-gene hypothesis and the *Ml-a* locus is considered typical of resistance genes exemplified by this model (Moseman, 1966). A number of races of *E. g. hordei* exist (Moseman, 1956). Each race exhibits a characteristic spectrum of avirulence and virulence patterns when tested on a range of barley genotypes (Moseman, 1968; Wiberg, 1974; Torp *et al.*, 1978). Subsequent genetic analyses of field isolates of *E. g. hordei* demonstrated that several virulence loci exist, each one effective with a different host resistance allele (Moseman, 1959, 1963; Moseman & Schaller, 1960, 1962). Most avirulence loci have two allelic forms. A rough genetic map of *E. g. hordei* with three linkage groups and a limited number of avirulence loci has been published (Jørgensen, 1988b). More recently, a map with seven linkage groups based on 31 restriction fragment length polymorphism (RFLP) loci and five avirulence loci was constructed for *E. g. hordei* (Christiansen & Giese, 1990).

Genetic analyses in barley have led to the identification and mapping of about ten mildew disease reaction loci (Jensen & Jørgensen, 1975a, b; Giese *et al.*, 1981; Søgaard & Jørgensen, 1982). With the exception of the *ml-o* locus, alleles conferring resistance at these mildew loci are generally dominant or semidominant. As is common to other gene-for-gene diseases, one mildew disease reaction locus is highly polymorphic. More than 30 resistance alleles have been mapped to the *Ml-a* locus (Jørgensen & Moseman, 1971; Moseman & Jørgensen, 1973; Giese, 1981). To explain this relatively high degree of polymorphism, the *Ml-a* locus has been referred to either as a complex locus (Wiberg, 1974) or as a cluster of closely linked genes (Giese, 1981). Two kinds of evidence are given to support the latter claim. In studies of the fine structure of the *Ml-a* locus, putative intragenic recombinants arose at a relatively high frequency in one experiment (0.083%; Jørgensen & Moseman, 1972; Giese, 1981). For comparison, the genetic distance between allelic markers within the *cer-cqu* locus, controlling wax biosynthesis, was estimated at 0.012 cM (von Wettstein-Knowles & Søgaard, 1980). Furthermore, the *Hor1* and *Hor2* loci, which bracket the *Ml-a* locus, consist of multiple genes (Bunce *et al.*, 1986). The *Hor2* locus extends over approximately 360 kilobase pairs (kbp) and may consist of up to 13 genes (Sørensen, 1989). The genetic mechanisms giving rise to the complex, multigenic *Hor1* and *Hor2* loci may also act at the *Ml-a* locus (Giese, 1981). Ultimately, a molecular description of *Ml-a* locus is needed if the mechanism for generating new resistance alleles at this locus is to be understood.

During the infection sequence of the powdery mildew disease, a range of cytological responses have been described (Ellingboe, 1972). A number of these responses occur in both resistant and susceptible hosts and, thus, are not directly related to race-specific resistance. Papillae formation (Koga *et al.*, 1978; Wright & Heale, 1988), the deposition of autofluorescent compounds (Mayama & Shishiyama, 1978; Koga *et al.*, 1980), and the alteration of cytoplasmic aggregates (Bushnell & Zeyen, 1976; Koga *et al.*, 1983) are examples of cytological responses that have been observed. A number of enzyme activities increase following inoculation of barley with *E. g. hordei* including phenylalanine ammonia lyase (Shiraishi *et al.*, 1989) and extracellular peroxidases (Kerby & Somerville, 1989). In addition, mRNA pools for a small number of anonymous genes have been shown to increase following inoculation with *E. g. hordei* (Manners *et al.*, 1985; Davidson *et al.*, 1987, 1988).

Cytological events that have been correlated with the expression of race-specific resistance include the deposition of papillae (Johnson *et al.*, 1979), a halt in cytoplasmic streaming at about 13 h post-inoculation (Bushnell, 1981), the deposition of fluorescent compounds in the host cytoplasm by 15 h post-inoculation (Kita *et al.*, 1981) and the hypersensitive necrosis response (Johnson *et al.*, 1979; Wright and Heale, 1988; Koga *et al.*, 1990). Although papillae are produced in both resistant and susceptible host plants, it has been suggested that in resistant barley plants, papillae are formed earlier or are structurally modified and, thus, more difficult for infection pegs to breach (Aist, 1976; Israel *et al.*, 1980). The biochemical components underlying these cytological events are largely unknown. Thus, if clones for a resistance locus such as *Ml-a* were isolated, they would provide invaluable tools for understanding the biochemical basis for resistance as well as the genetic basis of the highly polymorphic nature of these loci.

This chapter concentrates on the cloning of a classic race-specific resistance gene because such genes and their products are central determinants of resistance (Flor, 1955, 1971; Ellingboe, 1979, 1981). However, as our knowledge base of the biochemical components of mildew infection expands, alternative strategies for genetically engineering disease resistance may suggest themselves.

Strategies

Cloning by 'complementation'

A dominant allele can be identified by its ability to restore a dominant phenotype in a recipient carrying a recessive allele of the gene. Thus, in principle, a dominant disease resistance allele can be cloned by identifying a DNA fragment that confers resistance when introduced into a susceptible host plant. This method is only effective if the dominant and recessive phenotypes are dramatically different so that the change in phenotype of the transformed recipient can be easily and reliably identified in individual plants. Furthermore, a transformation protocol in which thousands of independent transformants can be readily generated is required. A major advantage of this method is that no knowledge of the gene is required other than its dominance relationship. No assumptions are made about the expression of the resistance gene or the gene product.

In the field of plant pathology, cloning based on 'complementation' has been used extensively to recover avirulence genes from bacterial pathogens. A notable example is the cloning of *avr6* from *Pseudomonas syringae* pv. *glycinea* by Staskawicz *et al.* (1984). Race 6 of *P. s. glycinea* is avirulent on the soybean cultivars Peking and Harosoy. Race 5 is virulent. Staskawicz and co-workers prepared a genomic cosmid library from race 6, and transformed the library into race 5 to generate a large population of transformed race 5 colonies each carrying a different segment of the race 6 genome. Roughly 680 race 5 transformants were individually tested for their ability to cause disease on Peking and Harosoy. One race 5 transformant that became avirulent on Harosoy and Peking was recovered. The race 6 DNA was recovered from the transformant and sequenced (Napoli & Staskawicz, 1987). By this means the first avirulence gene was cloned from a plant pathogen. Cloning by 'complementation' is also being evaluated for cloning avirulence genes from fungal plant pathogens (Valent, 1990).

Cloning genes from higher eukaryotes by complementation has found limited application due both to the large genome sizes and to the lack of highly efficient transformation protocols for most eukaryotes. In a demonstration experiment in tobacco, a herbicide resistance gene was recovered from a cosmid library (average insert ca. 20 kbp) prepared from a chlorsulphuron-resistant mutant of *Arabidopsis thaliana* (Olszewski *et al.*, 1988). It was estimated that 20 000 tobacco transformants would have to be screened following transformation with the *Arabidopsis* cosmid library to ensure that a specific sequence from the library was recovered. Given a genome size of 5.5×10^6 kbp and an average insert size of 20 kbp, a minimum of 824 000 transformants would need to be scored to be 95% sure of recovering a clone for a barley gene of interest (Clarke & Carbon, 1976). This example illustrates the difficulty of applying this method to organisms with large genomes. Because no transformation method for barley has been reported in the literature, the discussion of this method is moot. Even when a transformation method does become available, cloning by 'complementation' will remain unattractive due to the very large number of transformants that must be regenerated. Thus, this method is likely to find use only for organisms with relatively small genomes, like flax (Jones *et al.*, 1985) and *Arabidopsis* (Olszewski *et al.*, 1988).

A cDNA clone for the *Ml-a* mildew disease reaction gene might occur more frequently in a cDNA library prepared from epidermal mRNA, than the *Ml-a* gene occurs in the barley genome, such that cloning by 'complementation' could be applied to cDNA clones. Since the growth of *E. g. hordei* is limited to epidermal tissue, isolating epidermal mRNA should increase the relative abundance of *Ml-a* mRNA species over that found in leaf mRNA. Even with a relative abundance of 0.01%, more than 30 000 transformants would be needed. The relative abundance of *Ml-a* mRNA is not known; they could be very rare or moderately abundant.

Cloning by 'complementation' may be useful in screening for host receptors for pathogen-produced toxins. If resistance to such toxins can be screened for in transformed calluses, rooty tumours produced by *Agrobacterium rhizogenes*, yeast, or *Escherichia coli*, then it would be possible to generate and screen large numbers of transformants. As a hypothetical example, the target site for the T-toxin, the T-URF13 polypeptide, confers sensitivity to T-toxin on *E. coli* when the *T-urf13* gene is expressed in this prokaryote (Braun *et al.*, 1990). Thus, one could envisage preparing a library from susceptible maize plants, introducing the library into *E. coli* and then screening the transformants for those that became sensitive to T-toxin. Since little is known about the mode of action of most pathogen-produced toxins, it is not clear whether this approach is applicable to cloning any of the other host genes that encode toxin receptors (Scheffer & Livingston, 1984).

Cloning by transposon tagging or T-DNA tagging

Transposon tagging, like cloning by 'complementation', requires no prior knowledge of the biochemical properties of the resistance gene or its product. If a mutation at a disease reaction locus dramatically alters the phenotype, then such a gene can be cloned by transposon tagging. A recent example is the tagging with *Ds* of the maize gene *Knotted*, which controls leaf morphology (Hake *et al.*, 1989). A gene that has been mutated by transposon insertion can be cloned by recovering the transposable element and flanking genomic sequences from the mutant line once the mutation and the transposable element have been shown to co-segregate. These flanking sequences can then be used to identify the wild-type allele of the gene by routine molecular genetic methods. For *Arabidopsis*, tagging with T-DNA has also proved successful for cloning genes (Feldman *et al.*, 1989; Yanofsky *et al.*, 1990).

Attempts were made to clone the *Rp1* locus of maize, controlling resistance to the rust pathogen, *Puccinia sorghi*, by transposon tagging with *Mu* (Bennetzen *et al.*, 1988a) and *Ac/Ds* (Pryor, 1987). Unfortunately, some alleles at the *Rp1* locus revert to susceptibility at a rate that overlaps the rate of transposon-induced mutagenesis (Bennetzen *et al.*, 1988a). Therefore, it is difficult to distinguish transposon-induced from spontaneous mutants and no success in cloning *Rp1* by transposon tagging has yet been reported.

Although genetic instabilities have been reported for barley (Wise & Ellingboe, 1985) and transposon-like elements have been found in cloned barley genes (Whittier *et al.*, 1987), there is at present no transposable element available for use in barley. As an alternative to an endogenous element, a heterologous transposon can be introduced into barley once a transformation system becomes available. The maize transposon system *Ac/Ds* has been demonstrated to excise from sites of insertion in a number of plant species (Van Sluys *et al.*, 1987; Jones *et al.*, 1990; Yoder, 1990).

Because no active endogenous barley transposable element is known and since it is not possible to introduce a heterologous element into barley, discussion of transposon tagging in barley is theoretical at present. However, since these limitations may be transient, it is useful to discuss some of the features of transposon tagging. Successful application of this strategy requires that the rate of transposon-induced mutagenesis is significantly higher than the rate of spontaneous mutation (Bennetzen *et al.*, 1988a). Also, the target gene must not be essential to the welfare of the host. The temperature-sensitive alleles described for some rust resistance genes in wheat suggest that resistance genes are not essential and that, in the absence of the pathogen, the plant can operate normally at non-permissive temperatures (Skipp & Samborski, 1974; Gousseau *et al.*, 1985). Also, it is essential that the mutated gene exhibit a phenotype distinct from the wild-type parent to facilitate the screening of large numbers of individuals for rare mutants with the altered disease resistance phenotype. As an example, 530 000 kernels from a maize line with an active *En/Spm* element were screened to recover three transposon-induced *opaque-2* mutants (Schmidt *et al.*, 1987). These properties can be evaluated in experiments using conventional chemical or ionizing radiation mutagenesis. Jørgensen (1976, 1983) has reported the recovery of *ml-o* mutants in mutagenesis experiments in which powdery mildew resistant barley lines were selected. More recently, this research group has reported the recovery of mutants of the *Ml-a*12 allele with increased susceptibility to the powdery mildew pathogen (Torp & Jørgensen, 1986; Jørgensen, 1987, 1988a). These experiments suggest that transposon-induced mutants at the *Ml-a* or *ml-o* loci could be recovered if a highly efficient method of transposon tagging does become available for use in barley.

Cloning by chromosome walking

For this method, the map location of the resistance gene must be known. Much of this information has been compiled for barley powdery mildew resistance loci by plant breeders and geneticists. Roughly ten mildew resistance genes have been mapped and several more have been defined but not mapped in barley (Søgaard & Jørgensen, 1982). In addition, an RFLP map of barley is being developed (Blake *et al.*, 1989; Chapter 6, this volume). Thus, the basic materials are available for chromosome walking to powdery mildew disease reaction loci.

Chromosome walking consists of mapping a disease resistance locus relative to RFLP markers in F_2 individuals, F_3 families, test-cross or similar segregating populations. Ideally, two flanking RFLP markers, each within a few hundred kilobases of the disease resistance locus, are identified. These RFLPs serve as starting points for isolating overlapping clones that cover the region between the two RFLPs. The individual clones can then be mapped to identify those clones spanning the disease reaction locus. Eventually, clones thought likely to contain the resistance allele are introduced into a susceptible host, so that those clones that restore the dominant resistance phenotype can be identified.

A major concern in chromosome walking projects is the relationship between genetic distance, based on recombination data, and physical distance. In maize, 43.3 kbp/map unit has been reported for the alcohol dehydrogenase gene (Freeling & Bennett, 1985). For *Arabidopsis* 140 kbp/cM has been estimated from the total amount of DNA per haploid genome and the total genetic map size (Chang *et al.*, 1988). In barley, a similar calculation gives 1500–3000 kbp/cM (Bunce *et al.*, 1986). The genetic distance between the *Hor1* and *Hor2* loci, which flank the *Ml-a* locus, is estimated to be 7.3 cM, roughly 11 000–22 000 kbp. Such distances are greater than any that have been covered by chromosome walking to date. Thus for this method to be applied to cloning the *Ml-a* locus, RFLP markers within about 0.1–0.3 cM of *Ml-a* must be identified so that the total distance to be covered is less than about 1000 kbp. To map RFLP markers to within 0.2 cM would require scoring roughly 750 F_3 families, assuming that all genotypic classes can be scored (95% probability level; Allard, 1956). The average density of RFLP maps, which are being developed to facilitate breeding for quantitative traits in crop species, is roughly one marker per 10–20 cM, a density that is insufficient for chromosome walking (Paterson *et al.*, 1988).

Markers for developing an RFLP map can include RFLPs based on clones from a genomic library (Helentjaris *et al.*, 1986; Young *et al.*, 1987), cDNA clones (Laundry *et al.*, 1987; Burr *et al.*, 1988), or RAPD (Random Amplified Polymorphic DNA) markers (Williams *et al.*, 1990). RAPD markers are bands created in polymerase chain reactions (PCRs) directed by random decameric oligonucleotides. By empirical observation from a range of species, addition of one primer, 9–10 nucleotides in length, will yield 4–8 bands from a template of genomic DNA in a PCR mix. The presence and absence of specific bands can be compared in two genotypes in a fashion analogous to traditional RFLP markers. The significant advantages of the use of RAPD markers are that PCR reactions are simple to perform relative to the Southern blots used to score traditional RFLPs and no radioactive reagents are required. Thus, many RAPD oligomers can be quickly evaluated and then used to score large segregating populations.

Three sets of isolines exist in barley for the powdery mildew resistance genes. Paired lines in the Manchuria collection are related by 16–18 generations of crossing and are estimated to be 99% homologous (Moseman, 1972). Those of the Pallas and Siri isolines are thought to be roughly 96% homologous (Kølster *et al.*, 1986; Kølster & Stølen, 1986). However, lines in the Pallas set may in practice be more homologous than 96% since they were selected as having an introgressed region less than the distance between the two *Hor* loci (Kølster *et al.*, 1986). Thus, if one selected for RFLP or RAPD markers that exhibited polymorphisms between highly isogenic lines then these markers would most likely be linked to the mildew resistance locus (Young *et al.*, 1988). *In situ* hybridization techniques have been developed to an extent that it may be possible to place RFLP clones on specific chromoscomes as a means of confirming that the RFLP clones are located on chromosome 5, the site of the *Ml-a* locus (Clark & Karp, 1989; Fan *et al.*, 1990). It may also be possible to use the trisomics (Tsuchiya, 1987; Young *et al.*, 1987) or wheat–barley addition lines (Islam *et al.*, 1981) to confirm that clones hybridize to the chromosome of interest. Then, a detailed genetic map of the region surrounding the mildew locus can be developed with the selected RFLP and RAPD markers.

Jumping libraries have been used in mammalian species to provide DNA markers and such libraries may also contribute to chromosome walking experiments in barley (Collins *et al.*, 1987). A jumping library is one in which genomic DNA, which is partially digested with *Mbo*I or a similar enzyme, is selected for fragments of roughly 100 kbp and then ligated at low concentration in the presence of a vector containing a selectable marker. These large plasmids are digested with a restriction endonuclease like *Eco*RI and permitted to self-ligate to create a set of smaller plasmids in which DNA sites separated by about 100 kbp are placed adjacent to one another on the plasmid. The DNA site about 100 kbp distant from a start site (e.g. an RFLP site) can be isolated from the jumping clone, and used in plaque hybridizations to identify clones containing DNA sequences 100 kbp removed along the genome. Thus, it is possible to jump in 100 kbp jumps from one site in the genome to the next. When coupled with genetic mapping of RFLPs based on the *Mbo*I/*Eco*RI fragments or with physical mapping, a dense map of the genomic region of interest can be produced.

The clones used for chromosome walking may be any one of several types. Yeast artificial chromosomes (YAC) libraries (Burke *et al.*, 1987; Coulson *et al.*, 1988) and, more recently, P1 libraries (Sternberg, 1990) and F plasmid-based libraries (Leonardo and Sedivy, 1990) were developed for covering long distances in eukaryotic genomes. For an organism like *Arabidopsis*, a complete YAC library, made up of 150 kbp YAC clones, consists of about 1400–2000 clones (Ward & Jen, 1990; Grill & Somerville, 1991). For barley, such a library would consist of about 110 000 clones. Thus, it may not be practical to maintain and screen a totally representative barley YAC library. However, in conjunction with physical mapping of specific regions of the genome (Chu *et al.*, 1986; Dunham *et al.*, 1987; Sørensen, 1989), partial YAC libraries may find use in barley. Cosmid libraries and λ libraries may also be prepared for chromosome walking but, because of the large distances that are likely to be covered, such libraries will probably not play a role in chromosome walking in barley.

With the anticipated improvements in physical mapping and in chromosome walking methods that result from the Human Genome Project, chromosome walking may yet prove to be a viable method for recovering resistance genes from barley.

Plus/minus screening methods

Enrichment methods can be used to identify clones or polypeptides that are unique to the resistant host. These methods depend on the existence of some difference between the resistant and susceptible genotypes that can be exploited. For barley powdery mildew, several sets of isolines are available for such experiments. These lines have the advantage that the likelihood of recovering spurious differences between the two genotypes is reduced.

The application of this group of methods is hampered by lack of knowledge of the resistance gene and its mRNA and polypeptide products. For example, it is not known whether transcription of the *Ml-a* gene is induced following inoculation of barley with an avirulent race of *E. g. hordei*. If the *Ml-a* gene is constitutively expressed, then the plus/minus screens can be carried out on healthy uninoculated tissues, thereby avoiding the recovery of infection-related gene products.

In the discussion that follows, the assumption is made that the resistance allele is actively transcribed to produce a polypeptide with an active function in expression of resistance. Genetic arguments support this assumption. Generally the dominant allele is thought to provide the active gene product and the recessive allele to encode an inactive gene product. A mutant allele, which encodes an inactive, non-functional enzyme activity, commonly behaves as a recessive allele. However, there are examples in the genetic literature of dominant mutant alleles. The second argument comes from the analysis of temperature-sensitive rust resistance alleles in wheat. Wheat lines carrying the *Sr6* rust resistance allele are resistant to the rust pathogen at the lower, permissive temperature and are susceptible at the higher, non-permissive temperature (Skipp & Samborski, 1974). Such temperature-sensitivity is often associated with the inactivation of a protein gene product at the non-permissive temperature, suggesting that *Sr6* rust resistance depends on an active gene product. However, it is theoretically possible that the resistance allele is not expressed but rather that the susceptibility allele is expressed. Although this assumption is less favoured, there is no compelling biochemical evidence to discount it. Thus, in the methods described next it may be useful to conduct two sets of experiments in parallel, screening for susceptibility and selection of resistance gene products.

The plus/minus experiments can be grouped into three types based on whether the starting materials are DNA, mRNA, or protein. These methods are considered below in greater detail.

Genomic subtraction

Recently, methods for recovering DNA sequences that are unique to one genotype and absent in a second genotype have been reported (Straus & Ausubel, 1990; Wieland *et al.*, 1990). Genomic subtraction relies on tagging DNA from a deletion mutant with biotin. Following denaturation and reassociation of a mixture of DNA from the mutant and wild-type, streptavidin is added. The protein–DNA complexes are removed from solution either by extraction with phenol (Sive & St John, 1988) or by binding to a solid, biotin-containing matrix (Straus & Ausubel, 1990). The process is repeated several times to enrich for sequences present only in the wild-type. These sequences can then be amplified by PCR and cloned into a vector.

An essential component of the method is the availability of defined deletion mutants covering the disease reaction gene. The common criteria for distinguishing a deletion mutation from a point mutation are that a deletion mutation reverts at a much lower frequency than a point mutation; that a deletion mutant does not recombine with two or more point mutations; or that loss of a segment of the genome can be detected by cytogenetic analysis. The last method is suitable only for relatively large deletions. The first two criteria are not routinely met because it is often too demanding of time and space to collect such data in plants.

A recent variation on this method has been reported (Yokota & Oishi, 1990). DNA from two genotypes (e.g. a wild-type and a deletion mutant) is digested with a restriction endonuclease and then the DNA from one genotype (e.g. the deletion mutant) is treated with phosphatase. The DNA from the two genotypes is mixed, separated on an agarose gel and, then, subjected to denaturation and reassociation in the gel. The DNA is eluted from the gel and cloned. Because it has not been treated with phosphatase, only the DNA from the wild-type will ligate into the vector. Hybrids of wild-type and mutant DNA or of mutant DNA alone should not be recovered. In a reconstruction experiment, Yokota & Oishi (1990) demonstrated that the method was not effective for recovering single-copy genes from the human genome. If the hybridization conditions can be optimized, then this method may be applicable to cloning resistance genes from barley isolines. The method is only dependent on an RFLP for the gene of interest. Thus, the isolines could be used directly in these experiments. Since the isolines may differ by as much as 1% of the genome (equivalent to about 55 000 kbp), it is possible that several clones from the introgressed region will be recovered.

Subtractive hybridization

Screens for cDNA clones unique to one tissue type or to a specific set of environmental conditions are powerful methods for identifying changes in gene expression. A variety of methods have been developed for identifying and recovering mRNA species that are present in one sample and absent in a second sample. Newer methods promise increased sensitivity and the ability to recover very rare mRNA species. For example, Travis & Sutcliffe (1988) report the recovery of a cDNA for an mRNA that is 0.001% of the monkey brain cortex mRNA pool. This cDNA was recovered by subtractive hybridization in which phenol emulsions were used to enhance the rapidity of the reassociation of hybrids between rare cDNA species.

A number of variations on this method can be applied to recovering a cDNA for a powdery mildew resistance gene. If the gene is constitutively expressed, then plus/minus screens between the resistant line and a susceptible mutant should yield resistance-specific cDNA clones. In the absence of clearly defined null mutants, the resistant and susceptible isolines can be used if the susceptible line either does not produce a transcript for the mildew disease reaction gene or does not produce a transcript that can hybridize to the resistance transcript. As with genomic subtraction, the number of cDNA clones recovered will depend on the number of transcribed genes in the introgressed region that exhibit significant differences between the resistant and susceptible isolines.

If the resistance gene is induced following inoculation with an avirulent race of *E. g. hordei*, then plus/minus screens between cDNA populations from inoculated resistant and susceptible barley isolines can be performed. cDNA clones for mRNA species that are induced in the resistant but not in the susceptible isoline and clones of mRNA species that decrease in relative abundance in the susceptible but not the resistant isoline following inoculation with *E. g. hordei* will be recovered. An alternative screen would be to isolate cDNA clones present in inoculated resistant tissue but missing from healthy resistant tissue. Among the cDNA clones isolated by this method, both infection-related cDNA clones and clones for the disease reaction resistance allele may be recovered.

An appropriate time following inoculation would need to be chosen. Since near-synchronous cultures of *E. g. hordei* can be established, one should choose a time during the development of the first haustorium (i.e. during the first 24 h post-inoculation). Haustoria are laid down at about 24 h intervals. For times >24 h post-inoculation, both infection-related and resistance-specific cDNA may be recovered. Even if the resistance gene is transiently expressed, it is likely to be re-induced with each attempt to establish a haustorium. Therefore, at later times both early and late responses are likely to overlap. The underlying assumption of this approach is that the expression of the resistance gene is an early event in the incompatible interaction. Cytological features distinguishing resistance from susceptibility occur at about 12–15 h post-inoculation. Presumably transcription of the resistance gene must precede these events. Therefore, a time before about 12 h post-inoculation should be suitable for harvesting the mRNA that will provide the 'inoculated' cDNA pool.

The tissue to be sampled should also be considered. Since the powdery mildew disease occurs on all aerial plant parts, any plant tissue other than root tissue would be appropriate. In addition, the infection is localized in the epidermis. Therefore, enriching for epidermal tissue may potentially enrich for mildew-related mRNA species. The coleoptile may be a useful starting tissue for such experiments because this organ is highly enriched in epidermal tissue and devoid of mesophyll tissue that dominates leaves (Bushnell *et al.*, 1967).

A number of infection-related changes in mRNA species occur in barley leaves following inoculation with *E. g. hordei*, as measured by *in vitro* translation or in Northern blots (Higgins *et al.*, 1985; Manners *et al.*, 1985; Davidson *et al.*, 1988). Up to about 24 h post-inoculation, six infection-related genes were induced in both the resistant line, with the *Ml-p* resistance allele, and the susceptible isoline. Some of these genes were similarly induced in other barley genotypes and other plant species (Davidson *et al.*, 1988; Jutidamrongphan *et al.*, 1989; Scott *et al.*, 1990). Thus, the infection-related cDNA clones are unlikely to encode the *Ml-p* gene. These experiments suggest that the *Ml-p* gene is not dramatically induced following inoculation, that this gene is only transiently induced, or that the *Ml-p* transcript is rare and, therefore, difficult to recover.

Polypeptide polymorphisms

Two methods can be adopted here: analysis of polypeptide differences between isolines by two-dimensional polyacrylamide gel electrophoresis (2D-PAGE); and enriching for resistance-specific polypeptides using antibodies. As in screens for resistance-specific cDNA clones, considerations as to whether to compare healthy or inoculated tissue, what time to sample, and what tissue to examine pertain to the search for a polypeptide product of a mildew resistance allele. An additional consideration is the method of preparing tissue extracts. It may be difficult to design a sample preparation method that ensures the recovery of the *Ml-a* gene product, since it is not known where this polypeptide is localized. Also, if the resistance-specific domain of the polypeptide is readily lost due to proteolysis during sample preparation, then it will be difficult to identify the disease reaction gene product. A variety of methods exists for identifying clones encoding a specific polypeptide once the correct polypeptide has been identified (Suggs *et al.*, 1981; Young & Davis, 1983).

Comparisons of polypeptide patterns by 2D-PAGE were used to search for the *Pm* mildew resistance gene product in congenic wheat lines (Gabriel & Ellingboe, 1982). However, no candidates for the *Pm* gene product were identified and the method was concluded not to be sufficiently sensitive to detect subtle differences between polypeptides from congenic lines. Methods for preparing samples, of running 2D-PAGE and of image analysis have improved since 1982. Thus, comparisons of additional isoline pairs among the three sets of congenic lines or of mutants at the *Ml-a* locus may yield a polypeptide polymorphism that is correlated with a mildew disease reaction locus following high-resolution 2D-PAGE analysis.

Antibodies can be highly sensitive and selective screening reagents. Therefore, it is possible that immunological methods can be used to identify a resistance gene product as an antigenic difference between congenic lines. To be effective, highly congenic lines are required so that antigenic differences are limited to the disease reaction gene product. Polyclonal antisera raised against tissue extracts from resistant plants will induce antibodies reactive with antigens unique to the resistant genotype and antibodies to antigens that are shared with the susceptible genotype. Repeated adsorption of such antisera with tissue extracts from the susceptible host line should yield a pool of antibodies that react only with resistance-specific epitopes. The resistance-specific antibodies can then be used to identify antigens unique to the resistant isoline. As with all immunological methods, this approach will fail if the immunogenicity of the resistance gene product is low (Harlow & Lane, 1988).

Unlike polyclonal antibodies, which consist of a diverse collection of antibodies with varying specificities, a given monoclonal antibody will react with only one epitope (Kohler, 1985). Thus, monoclonal antibodies offer the potential for identifying very subtle antigenic differences between congenic lines. A second advantage of using monoclonal antibodies is that very crude preparations can be used as immunogens and for screening, provided that a sufficiently discriminating and informative screen can be developed. For example, panels of monoclonal antibodies could be screened for those antibodies that react with tissue from the resistant host but not with tissue from a susceptible, near-isogenic line (Somerville & Sherwood, 1985). Methods designed to suppress the immune response to epitopes shared by two samples may facilitate the recovery of resistance-specific monoclonal antibodies (Barclay & Smith, 1986). The major disadvantages of the use of monoclonal antibodies are the high cost and the labour generally required to identify and establish hybridoma lines producing antibodies with the desired specificity and characteristics.

Cloning disease resistance genes in a plant species with sophisticated molecular genetic tools

A variety of molecular genetic tools that have been adapted for use with *Arabidopsis*, maize, and tomato are not currently available in barley. For example, several transposable elements have been described in maize. Efficient transformation methods are available and the *Ac* transposable element functions in tomato (Yoder *et al.*, 1988; Yoder, 1990) and *Arabidopsis* (Van Sluys *et al.*, 1987; Valvekens *et al.*, 1988; Schmidt & Willmitzer, 1989). Also, T-DNA tagging has been successfully used to clone genes from *Arabidopsis* (Feldman *et al.*, 1989; Yanofsky *et al.*, 1990). Detailed RFLP maps exist for all three organisms (Helentjaris *et al.*, 1986; Chang *et al.*, 1988; Paterson *et al.*, 1988; Nam *et al.*, 1989). In addition, several YAC libraries of *Arabidopsis* have been generated (Ward & Jen 1990; Grill & Somerville, 1991). Because of its small genome and limited amount of repetitive DNA, chromosome walking is likely to progress more quickly in *Arabidopsis* than in other plant species (Pruitt & Meyerowitz, 1986; Meyerowitz, 1987).

It may be more economical of time and effort to clone powdery mildew resistance genes from *Arabidopsis*. Once cloned, these genes could be used as heterologous probes to identify the corresponding resistance genes from barley. The map position of the barley genes identified by the *Arabidopsis* clone could be used to determine which of the mildew genes had been cloned (e.g. *Ml-a, Ml-g, Ml-p*). This approach assumes, however, that the same resistance mechanisms against *Erysiphe* spp. are used in *Arabidopsis* and barley such that the mildew resistance genes in these two plant species will be sufficiently similar to hybridize.

A limitation of this approach is that no powdery mildew resistance genes have been described in *Arabidopsis*. Koch & Slusarenko (1990) have reported that *Arabidopsis* is partially susceptible to *Erysiphe cruciferarum*, although no macroscopic symptoms were visible on inoculated plants. A problem with working with a pathogen such as *Erysiphe*, which is capable of forming highly race–cultivar specific interactions, is that it may be difficult to find isolates that are well adapted for growth on *Arabidopsis*. Perhaps, if sufficient numbers of *Arabidopsis* ecotypes and *E. cruciferarum* isolates are tested, suitable genetic variability will be identified.

Using pathogen avirulence gene products to identify host resistance gene products

This approach is based on the common mechanistic models of the gene-for-gene hypothesis that suggest the resistance gene product and the avirulence gene product interact directly in a lock-and-key fashion (Ellingboe, 1979). Avirulence genes may be recovered from *E. g. hordei* before a host resistance gene can be cloned. Christiansen & Giese (1990) have developed an RFLP map of *E. g. hordei*, which can be used for cloning avirulence genes by chromosome walking. An affinity column based on the avirulence gene product could potentially be used to recover host proteins that bind the avirulence gene product. As a control, the susceptible isoline should lack this binding activity. Alternatively, anti-idiotype antibodies might be prepared to the anti-avirulence polypeptide antibodies to generate an antiserum that would cross-react with the host receptor for the avirulence gene product (Sege & Peterson, 1978).

A number of avirulence genes have been cloned from bacterial plant pathogens. However, these avirulence genes appear to encode cytoplasmic proteins, suggesting that the avirulence gene product does not interact directly with the host resistance gene product (Napoli & Staskawicz, 1987; Staskawicz *et al.*, 1987; Tamaki *et al.*, 1988). For example, the *avrD* gene from *Pseudomonas syringae* pv. *tomato* encodes a protein that directs the synthesis of a low-molecular-weight compound (Keen *et al.*, 1990). It is the low-molecular-weight compound that appears to interact with the host. Although no avirulence genes have been cloned from a fungal plant pathogen, these results with bacterial avirulence gene products suggest some caution should be exercised in applying this approach to identifying host resistance gene products.

Criteria for identifying clones for disease resistance genes

Transformation of a susceptible barley line to resistance or vice versa would be the ideal method for testing the cDNA or genomic clones isolated by any of the methods described above. *E. g. hordei* does not appear to infect callus tissue, therefore, transformed barley seedlings would be required for testing putative clones encoding resistance genes. At present no suitable transformation method exists for barley. However, transformation techniques have been described for maize (Rhodes *et al.*, 1988) and rice (Shimamoto *et al.*, 1989; Terada & Shimamoto, 1990), and the Biolistic Particle Delivery System (Klein *et al.*, 1987) appears to be particularly promising for transforming monocots (Fromm *et al.*, 1990; Gordon-Kamm *et al.*, 1990). It seems likely that a barley transformation method will be developed in the near future.

An intermediate stage in the development of maize transformation protocols was the production of chimeric plants with sectors of transformed somatic tissue. Assuming that barley transformation methods develop in a similar fashion, chimeric transformants can be exploited for testing any putative resistance clones that are recovered. Work with maize suggests that resistance to the rust pathogen *Puccinia sorghi* acts in a cell autonomous manner (Bennetzen *et al.*, 1988b). Thus, it should be possible to identify resistant sectors in a background of susceptible tissue and, then, to screen for a marker of transformation, such as β-glucuronidase, to demonstrate the coincidence of transformation and resistance phenotype.

In the absence of a transformation system for barley, the way to determine whether the recovered clones encode the *Ml-a* locus is to evaluate RFLPs among the various barley genotypes and to map the clones relative to the *Ml-a* locus. Three collections of isolines representing 13 different resistance alleles at the *Ml-a* locus have been generated (Moseman, 1972; Kølster *et al.*, 1986; Kølster & Stølen, 1987), and collections of partially susceptible barley *Ml-a* mutants have been produced (Torp & Jørgensen, 1986). Those clones exhibiting distinct RFLP patterns among the various barley genotypes can be mapped relative to the *Ml-a* locus in a large segregating population. Co-segregation of an RFLP based on a cDNA or genomic clone and the disease reaction phenotype would provide strong support that the clone encodes the powdery mildew disease reaction gene.

Conclusion

The cloning of traditional race-specific resistance genes has been hampered by a lack of suitable techniques. Methods for genome analysis that are likely to result from the Human Genome Project should be directly applicable to cloning disease resistance genes from barley. At present, chromosome walking or plus/minus screens appear most promising because they make use of the rich genetic resources available in barley. Current inability to test any putative resistance clones by transformation into susceptible host plants limits the ability to evaluate clones critically.

Lack of biochemical information about race-specific resistance genes has hampered understanding of host–pathogen interactions. Once clones and antibody probes for race-specific resistance genes are available, then the potentially difficult task of determining the biochemical function of the resistance gene product can begin. The primary amino acid sequence, deduced from cDNA clones, may suggest a function for the resistance gene product based on comparisons to known sequences in the various data banks. Since the resistance gene product is often depicted as a receptor it may share partial sequence similarity with receptors that transduce signals across the plasma membrane (e.g. the various G-protein complexes; Bourne *et al.*, 1990). Models of host–pathogen interactions suggest that the resistance gene product will be localized in the plasma membrane and that expression of the resistance gene will be constitutive. With the cDNA and antibody probes these assumptions can be tested. In addition, the antibody can be used to purify the resistance gene product in its native form so that the physical properties of the protein, including post-translational modifications, can be determined. It is hoped that from such analyses of the resistance gene and its polypeptide product some insight into the biochemical function of the gene product will result.

Sequence comparisons among the various alleles may suggest models for the generation of allelic diversity at the *Ml-a* locus. In addition, the relatedness of the various powdery mildew resistance loci can be addressed once clones for the *Ml-a* locus are available. A detailed knowledge of the structure of race-specific resistance genes will be required if new, effective sources of resistance are to be created by *in vitro* manipulation.

Acknowledgements

I am grateful for the financial support of the National Institute of Health (GM40084), the Michigan Agriculture Experiment Station and the US Department of Energy (DE-AC02-76-ER01338).

References

Aist, J.R. (1976) Papillae and related wound plugs of plant cells. *Annual Review of Phytopathology* 14, 145–63.

Allard, R.W. (1956) Formulas and tables to facilitate the calculation of recombination values in heredity. *Hilgardia* 24, 235–78.

Barclay, S.L. & Smith, A.M. (1986) Rapid isolation of monoclonal antibodies specific for cell surface differentiation antigens. *Proceedings of the National Academy of Sciences (USA)* 83, 4336–40.

Bennetzen, J.L., Qin, M.-M., Ingels, S. & Ellingboe, A.H. (1988a) Allele-specific and *Mutator*-associated instability at the *Rp1* disease-resistance locus of maize. *Nature* 332, 369–70.

Bennetzen, J.L., Blevins, W.E. & Ellingboe, A.H. (1988b) Cell-autonomous recognition of the rust pathogen determines *Rp1*-specified resistance in maize. *Science* 241, 208–10.

Blake, T., Shin, J.S. & Sanchez, M. (1989)

Development of the barley RFLP map. In: Helentjaris, T. & Burr, B. (eds), *Development and Application of Molecular Markers to Problems in Plant Genetics.* Current Communications in Molecular Biology. Cold Spring Harbor Laboratory, Cold Spring Harbor, NY, USA, pp. 139–43.

Bourne, H.R., Sanders, D.A. & McCormick, F. (1990) The GTPase superfamily: a conserved switch for diverse cell function. *Nature* 348, 125–32.

Braun, C.J., Siedow, J.N. & Levings, C.S. (1990) Fungal toxins bind to the URF13 protein in maize mitochondria and *Escherichia coli*. *Plant Cell* 2, 153–61.

Bunce, N.A.C., Forde, B.G., Kreis, M. & Shewry, P.R. (1986) DNA restriction fragment length polymorphism at hordein loci: application to identifying and fingerprinting barley cultivars. *Seed Science and Technology* 14, 419–29.

Burke, D.T., Carle, G.F. & Olson, M.V. (1987) Cloning of large segments of exogenous DNA into yeast by means of artificial chromosome vectors. *Science* 236, 806–12.

Burr, B., Burr, F.A., Thompson, K.H., Albertsen, M.C. & Stuber, C.W. (1988) Gene mapping with recombinant inbreds in maize. *Genetics* 118, 519–26.

Bushnell, W.R. (1981) Incompatibility conditioned by the *Mla* gene in powdery mildew of barley: the halt in cytoplasmic streaming. *Phytopathology* 71, 1062–6.

Bushnell, W.R. & Zeyen, R.J. (1976) Light and electron microscope studies of cytoplasmic aggregates formed in barley cells in response to *Erysiphe graminis*. *Canadian Journal of Botany* 54, 1647–55.

Bushnell, W.R., Dueck, J. & Rowell, J.B. (1967) Living haustoria and hyphae of *Erysiphe graminis* f. sp. *hordei* with intact and partly dissected host cells of *Hordeum vulgare*. *Canadian Journal of Botany* 45, 1719–32.

Chang, C., Bowman, J., DeJohn, A., Lander, E. & Meyerowitz, E. (1988) Restriction fragment length polymorphism linkage map for *Arabidopsis thaliana*. *Proceedings of the National Academy of Sciences (USA)* 86, 6856–60.

Christiansen, S.K. & Giese, H. (1990) Genetic analysis of the obligate parasitic barley powdery mildew fungus based on RFLP and virulence loci. *Theoretical and Applied Genetics* 79, 706–12.

Chu, G., Vollrath, D. & Davis, R.W. (1986) Separation of large DNA molecules by contour-clamped homogeneous electric fields. *Science* 234, 1582–5.

Clark, M. & Karp, A. (1989) Physical mapping of the B-hordein loci on barley chromosome 5 by *in situ* hybridization. *Genome* 32, 925–9.

Clarke, L. & Carbon, J. (1976) A colony bank containing synthetic *ColE1* hybrid plasmids representative of the entire *E. coli* genome. *Cell* 9, 91–9.

Collins, F.S., Drumm, M.L., Cole, J.L., Locwood, W.K., Vande Worde, G.F. & Iannuzzi, M.C. (1987) Construction of a general human chromosome jumping library, with application to cystic fibrosis. *Science* 235, 1046–9.

Coulson, A., Waterson, R., Kiff, J., Sulston, J. & Kohara, Y. (1988) Genome linking with yeast artificial chromosomes. *Nature* 335, 184–6.

Crute, I.R. & Norwood, J.M. (1986) Gene-dosage effects on the relationship between *Bremia lactucae* (downy mildew) and *Lactuca sativa* (lettuce): the relevance to a mechanistic understanding of host–parasite specificity. *Physiological and Molecular Plant Pathology* 29, 133–45.

Davidson, A.D., Manners, J.M., Simpson, R.S. & Scott. K.J. (1987) cDNA cloning of mRNAs induced in resistant barley during infection by *Erysiphe graminis* f. sp. *hordei*. *Plant Molecular Biology* 8, 77–85.

Davidson, A.D., Manners, J.M., Simpson, R.S. & Scott, K.J. (1988) Altered host gene expression in near-isogenic barley conditioned by different genes for resistance during infection by *Erysiphe graminis* f. sp. *hordei*. *Physiological and Molecular Plant Pathology* 32, 127–39.

Dewey, R.E., Siedow, J.N., Timothy, D.H. & Levings, C.S. (1988) A 13-kilodalton maize mitochondrial protein in *E. coli* confers sensitivity to *Bipolaris maydis* toxin. *Science* 239, 293–5.

Dunham, I., Sargent, C.A., Trowsdale, J. & Campbell, R.D. (1987) Molecular mapping of the human major histocompatibility complex by pulsed-field gel electrophoresis. *Proceedings of the National Academy of Sciences (USA)* 84, 7237–41.

Ellingboe, A.H. (1972) Genetics and physiology of primary infection by *Erysiphe graminis*. *Phytopathology* 62, 401–6.

Ellingboe, A.H. (1979) Inheritance of specificity: the gene-for-gene hypothesis. In: Daly, J.M. & Uritani, I. (eds), *Recognition and Spe-*

cificity in Plant Host–Parasite Interactions. Japan Scientific Society Press and University Park Press, Baltimore, MD, USA, pp. 3–17.

Ellingboe, A.H. (1981) Changing concepts in host–pathogen genetics. *Annual Review of Phytopathology* 19, 125–43.

Ellis, J.G., Lawrence, G.J., Peacock, W.J. & Pryor, A.J. (1988) Approaches to cloning plant genes conferring resistance to fungal pathogens. *Annual Review of Phytopathology* 26, 245–63.

Fan, Y.-S., Davis, L.M. & Shows, T.B. (1990) Mapping small DNA sequences by fluorescene *in situ* hybridization directly on banded metaphase chromosomes. *Proceedings of the National Academy of Sciences (USA)* 87, 6223–7.

Feldman, K.A., Marks, M.D., Christianson, M.L. & Quatrano, R.S. (1989) A dwarf mutant of *Arabidopsis* generated by T-DNA insertion mutagenesis. *Science* 243, 1351–4.

Flor, H.H. (1946) Genetics of pathogenicity in *Melampsora lini*. *Journal of Agricultural Research* 73, 335–57.

Flor, H.H. (1947) Inheritance of reaction to rust in flax. *Journal of Agricultural Research* 74, 241–62.

Flor, H.H. (1955) Host–parasite interaction in flax rust – its genetics and other implications. *Phytopathology* 45, 680–5.

Flor, H.H. (1956) The complementary genic systems in flax and flax rust. *Advances in Genetics* 8, 29–54.

Flor, H.H. (1971) Current status of the gene-for-gene concept. *Annual Review of Phytopathology* 9, 275–96.

Freeling, M. & Bennett, D.C. (1985) Maize *Adh1*. *Annual Review of Genetics* 19, 297–323.

Fromm, M.E., Morrish, F., Armstrong, C., Williams, R., Thomas, J. & Klein, T.M. (1990) Inheritance and expression of chimeric genes in the progeny of transgenic maize plants. *Biotechnology* 8, 833–9.

Gabriel, D.W. & Ellingboe, A.H. (1982) High resolution two-dimensional electrophoresis of protein from congenic wheat lines differing by single resistance genes. *Physiological Plant Pathology* 20, 349–57.

Giese, H. (1981) Powdery mildew resistance genes in the *Ml-a* and *Ml-k* regions on barley chromosome 5. *Hereditas* 95, 51–62.

Giese, H., Jørgensen, J.H., Jensen, H.P. & Jensen, J. (1981) Linkage relationships of ten powdery mildew resistance genes on barley chromosome 5. *Hereditas* 95, 43–50.

Gordon-Kamm, W.J. *et al.* (14 additional authors) (1990) Transformation of maize cells and regeneration of fertile transgenic plants. *Plant Cell* 2, 603–18.

Gousseau, H.D.M., Deverall, B.J. & McIntosh, R.A. (1985) Temperature-sensitivity of the expression of resistance to *Puccinia graminis* conferred by the *Sr15*, *Sr9b* and *Sr14* genes in wheat. *Physiological Plant Pathology* 27, 335–43.

Grill, E. & Somerville, C. (1991) Construction and characterization of a yeast artificial chromosome library of *Arabidopsis*. *Molecular and General Genetics* 226, 484–90.

Hake, S., Vollbrecht, E. & Freeling, M. (1989) Cloning *Knotted*, the dominant morphological mutant in maize using *Ds2* as a transposon tag. *EMBO Journal* 8, 15–22.

Hamer, J.E., Howard, R.J., Chumley, F.G. & Valent, B. (1988) A mechanism for surface attachment in spores of a plant pathogenic fungus. *Science* 239, 288–90.

Harlow, E. & Lane, D. (1988) *Antibodies, A Laboratory Manual*. Cold Spring Harbor Laboratory, Cold Spring Harbor, NY, USA.

Heath, M.C. (1980) Reactions of nonsuscepts to fungal pathogens. *Annual Review of Phytopathology* 18, 211–36.

Heath, M.C. (1981) A generalized concept of host–parasite specificity. *Phytopathology* 71, 1121–3.

Heath, M.C. (1985) Implications of nonhost resistance for understanding host–parasite interactions. In: Groth, J.V. & Bushnell, W.R. (eds), *Genetic Basis of Biochemical Mechanisms of Plant Disease*. American Phytopathological Society Press, St Paul, MN, USA, pp. 25–42.

Helentjaris, T., Slocum, M., Wright, S., Schaefer, A. & Nienhuis, J. (1986) Construction of genetic linkage maps in maize and tomato using restriction fragment length polymorphisms. *Theoretical and Applied Genetics* 72, 761–9.

Higgins, C.M., Manners, J.M. & Scott, K.J. (1985) Decrease in three messenger RNA species coding for chloroplast proteins in leaves of barley infected with *Erysiphe graminis* f. sp. *hordei*. *Plant Physiology* 78, 891–4.

Hoch, H.C., Staples, R.C., Whitehead, B., Comeau, J. & Wolf, E.D. (1987) Signaling for growth orientation and cell differentiation by surface topography in *Uromyces*. *Science* 235, 1659–62.

Islam, A.K.M.R., Shepherd, K.W. & Sparrow, D.H.B. (1981) Isolation and characterization of euplasmic wheat–barley chromosome addition lines. *Heredity* 46, 161–74.

Israel, H.W., Wilson, R.G., Aist, J.R. & Kunoh, H. (1980) Cell wall appositions and plant disease resistance: acoustic microscopy of papillae that block fungal ingress. *Proceedings of the National Academy of Sciences (USA)* 77, 2046–9.

Jensen, J. & Jørgensen, J.H. (1975a) The barley chromosome 5 linkage map. I. Literature survey and map estimation procedure. *Hereditas* 80, 5–16.

Jensen, J. & Jørgensen, J.H. (1975b) The barley chromosome 5 linkage map. II. Extension of the map with four loci. *Hereditas* 80, 17–26.

Johnson, L.E.B., Bushnell, W.R. & Zeyen, R.J. (1979) Binary pathways for analysis of primary infection and host response in populations of powdery mildew fungi. *Canadian Journal of Botany* 57, 497–511.

Jones, D.A., Zahn, X.-C. & Kerr, A. (1985) To clone a gene from flax (*Linum usitatissimum*) for resistance to flax rust (*Melampsora lini*). In: Sussex, I., Ellingboe, A., Crouch, M. & Malmberg, R. (eds), *Plant Cell/Cell Interactions*. Current Communications in Molecular Biology. Cold Spring Harbor Laboratory, Cold Spring Harbor, NY, USA, pp. 115–19.

Jones, J.D.G., Carland, F., Lim, E., Ralston, E. & Dooner, H.K. (1990) Preferential transposition of the maize element *Activator* to linked chromosomal locations in tobacco. *Plant Cell* 2, 701–7.

Jutidamrongphan, W., Mackinnon, G., Manners, J.M. & Scott, K.J. (1989) Sequence of a near-full length cDNA clone for a mRNA of barley induced by fungal infection. *Nucleic Acids Research* 17, 9478.

Jørgensen, J.H. (1976) Identification of powdery mildew resistant barley mutants and their allelic relationship. In: Gaul, H. (ed.), *Barley Genetics III*. Karl Thiemig, Munich, Germany, pp. 446–55.

Jørgensen, J.H. (1983) Experience and conclusions from the work at Risø on induced mutations for powdery mildew resistance in barley. In: *Induced Mutations for Disease Resistance in Crop Plants II*. International Atomic Energy Agency, Vienna, Austria, pp. 73–87.

Jørgensen, J.H. (1987) Genetic dissection of a powdery mildew resistance gene in barley. *Barley Genetics Newsletter* 17, 99–100.

Jørgensen, J.H. (1988a) Genetic analysis of barley mutants with modifications of powdery mildew resistance gene *Ml-a12*. *Genome* 30, 129–32.

Jørgensen, J.H. (1988b) *Erysiphe graminis*, powdery mildew of cereals and grasses. *Advances in Plant Pathology* 6, 137–57.

Jørgensen, J.H. & Moseman, J.G. (1971) Two new genes at the *Ml-a* locus in barley for resistance to *Erysiphe graminis* f. sp. *hordei*. *Zeitschrift für Pflanzenzüchtung* 66, 67–75.

Jørgensen, J.H. & Moseman, J.G. (1972) Recombination at the *Ml-a* locus in barley conditioning resistance to *Erysiphe graminis* f. sp. *hordei*. *Canadian Journal of Genetics and Cytology* 14, 43–8.

Keen, N.T., Tamaki, S., Kobayashi, D., Gerhold, D., Stayton, M., Shen, H., Gold, S., Lorang, J., Thordal-Christensen, H., Dahlbeck, D. & Staskawicz, B. (1990) Bacteria expressing avirulence gene D produce a specific elicitor of the soybean hypersensitive reaction. *Molecular Plant–Microbe Interactions* 3, 112–21.

Kerby, K. & Somerville, S. (1989) Enhancement of specific intercellular peroxidases following inoculation of barley with *Erysiphe graminis* f. sp. *hordei*. *Physiological and Molecular Plant Pathology* 35, 323–37.

Kita, N., Toyoda, H. & Shishiyama, J. (1981) Chronological analysis of cytological responses in powdery-mildewed barley leaves. *Canadian Journal of Botany* 59, 1761–8.

Klein, T.M., Wolf, E.D., Wu, R. & Sanford, J.C (1987) High-velocity microprojectiles for delivering nucleic acids into living cells. *Nature* 327, 70–3.

Koch, E. & Slusarenko, A.J. (1990) Fungal pathogens of *Arabidopsis thaliana* (L.) Heyhn. *Botanica Helvetica* 100, 257–68.

Koga, H., Mayama, S. & Shishiyama, J. (1978) Microscopic specification of compatible and incompatible interactions in barley leaves inoculated with *Erysiphe graminis hordei*. *Annals of the Phytopathology Society of Japan* 44, 111–19.

Koga, H., Mayama, S. & Shishiyama, J. (1980) Correlation between the deposition of fluorescent compounds in papillae and resistance in barley against *Erysiphe graminis hordei*. *Canadian Journal of Botany* 58, 536–41.

Koga, H., Toyoda, H., Mayama, S., Shishiyama, J. & Hiura, U. (1983) Characteristic cellular responses as expression of genes for resistance to *Erysiphe graminis* f. sp. *hordei* in barley. *Phytopathology* 73, 907–10.

Koga, H., Bushnell, W.R. & Zeyen R.J. (1990)

Specificity of cell type and timing of events associated with papilla formation and the hypersensitive reaction in leaves of *Hordeum vulgare* attacked by *Erysiphe graminis* f. sp. *hordei*. *Canadian Journal of Botany* 68, 2344–52.

Kohler, G. (1985) Derivation and diversification of monoclonal antibodies. *EMBO Journal* 4, 1359–65.

Kølster, P. & Stølen, O. (1987) Barley isolines with genes for resistance to *Erysiphe graminis* f. sp. *hordei* in the recurrent parent 'Siri'. *Plant Breeding* 98, 79–82.

Kølster, P., Munk, L., Stølen, O. & Løhde, J. (1986) Near-isogenic barley lines with genes for resistance to powdery mildew. *Crop Science* 26, 903–7.

Landry, B.S., Kesseli, R., Leung, H. & Michelmore, R.W. (1987) Comparison of restriction endonucleases and sources of probes for their efficiency in detecting restriction fragment length polymorphisms in lettuce (*Lactuca sativa* L.) *Theoretical and Applied Genetics* 74, 646–53.

Leonardo, E.D. & Sedivy, J.M. (1990) A new vector for cloning large eukaryotic DNA segments in *Escherichia coli*. *Biotechnology* 8, 841–4.

Levings, C.S. (1990) The Texas cytoplasm of maize: cytoplasmic male sterility and disease susceptibility. *Science* 250, 942–7.

Manners, J.M., Davidson, A.D. & Scott, K.J. (1985) Patterns of post-infectional protein synthesis in barley carrying different genes for resistance to the powdery mildew fungus. *Plant Molecular Biology* 4, 275–83.

Mayama, S. & Shishiyama, J. (1978) Localized accumulation of fluorescent and u.v.-absorbing compounds at penetration sites in barley leaves infected with *Erysiphe graminis hordei*. *Physiological Plant Pathology* 13, 347–54.

Meyerowitz, E. (1987) *Arabidopsis thaliana*. *Annual Review of Genetics* 21, 93–111.

Moseman, J.G. (1956) Physiological races of *Erysiphe graminis* f. sp. *hordei* in North America. *Phytopathology* 46, 318–22.

Moseman, J.G. (1959) Host–pathogen interactions of the genes for resistance in *Hordeum vulgare* and for pathogenicity in *Erysiphe graminis* f. sp. *hordei*. *Phytopathology* 49, 469–72.

Moseman, J.G. (1963) Relationship of genes conditioning pathogenicity of *Erysiphe graminis* f. sp. *hordei* on barley. *Phytopathology* 53, 1326–30.

Moseman, J.G. (1966) Genetics of powdery mildews. *Annual Review of Phytopathology* 4, 269–90.

Moseman, J.G. (1968) Reactions of barley to *Erysiphe graminis* f. sp. *hordei* from North America, England, Ireland, and Japan. *Plant Disease Reporter* 52, 463–7.

Moseman, J.G. (1972) Isogenic barley lines for reaction to *Erysiphe graminis* f. sp. *hordei*. *Crop Science* 12, 681–2.

Moseman, J.G. & Jørgensen, J.G. (1973) Differentiation of resistance genes at the *Ml-a* locus in six pairs of isogenic barley lines. *Euphytica* 22, 189–96.

Moseman, J.G. & Schaller, C.W. (1960) Genetics of the allelic series at the *Ml-a* locus in barley and cultures of *Erysiphe graminis* f. sp. *hordei* that differentiate these alleles. *Phytopathology* 50, 736–41.

Moseman, J.G. & Schaller, C.W. (1962) Interaction of genes conditioning reaction of Algerian barley and pathogenicity of *Erysiphe graminis* f. sp. *hordei*. *Phytopathology* 52, 529–33.

Nam, H.-G., Giraudat, J., den Boer, B., Moonan, F., Loos, W.D.B., Hauge, B.M. & Goodman, H.M. (1989) Restriction fragment length polymorphism linkage map of *Arabidopsis thaliana*. *Plant Cell* 1, 699–705.

Napoli, C. & Staskawicz, B. (1987) Molecular characterization and nucleic acid sequence of an avirulence gene from race 6 of *Pseudomonas syringae* pv. *glycinea*. *Journal of Bacteriology* 169, 572–8.

Olszewski, N.E., Martin, F.B. & Ausubel, F.M. (1988) Specialized binary vector for plant transformation: expression of the *Arabidopsis thaliana* AHAS gene in *Nicotiana tabacum*. *Nucleic Acids Research* 16, 10765–82.

Paterson, A.H., Lander, E.S., Hewitt, J.D., Peterson, S., Lincoln, S.E. & Tanksley, S.D. (1988) Resolution of quantitative traits into Mendelian factors by using a complete RFLP linkage map. *Nature* 335, 721–6.

Pruitt, R.E. & Meyerowitz, E.M. (1986) Characterization of the genome of *Arabidopsis thaliana*. *Journal of Molecular Biology* 187, 169–83.

Pryor, A. (1987) The origin and structure of fungal disease resistance genes in plants. *Trends in Genetics* 3, 157–61.

Rhodes, C.A., Pierce, D.A., Mettler, I.J., Mascarenhas, D. & Detmer, J.J. (1988) Genetically transformed maize plants from protoplasts. *Science* 240, 204–7.

Scheffer, R.P. & Livingston, R.S. (1984) Host-selective toxins and their role in plant diseases. *Science* 223, 17–21.

Schmidt, R. & Willmitzer, L. (1989) The maize autonomous element *Activator (Ac)* shows a minimal germinal excision frequency of 0.2–0.5% in transgenic *Arabidopsis thaliana* plants. *Molecular and General Genetics* 220, 17–24.

Schmidt, R., Burr, F.A. & Burr, B. (1987) Transposon tagging and molecular analysis of the maize regulatory locus *opaque-2*. *Science* 238, 960–3.

Scott, K.J., Davidson, A.D., Jutidamrongphan, W., Mackinnon, G. & Manners, J.M. (1990) The activation of genes of wheat and barley by fungal phytopathogens. *Australian Journal of Plant Physiology* 17, 229–38.

Sege, K. & Peterson, P.A. (1978) Use of anti-idiotypic antibodies as cell-surface receptor probes. *Proceedings of the National Academy of Sciences (USA)* 75, 2443–7.

Shimamoto, K., Terada, R., Izawa, T. & Fujimoto, H. (1989) Fertile transgenic rice plants regenerated from transformed protoplasts. *Nature* 338, 274–6.

Shiraishi, T., Yamaoka, N. & Kunoh, H. (1989) Association between increased phenylalanine ammonia-lyase activity and cinnamic acid synthesis and the induction of temporary inaccessibility caused by *Erysiphe graminis* primary germ tube penetration of the barley leaf. *Physiological and Molecular Plant Pathology*, 34, 75–83.

Sive, H.L. & St John, T. (1988) A simple subtractive hybridization technique employing photoactivatable biotin and phenol extraction. *Nucleic Acids Research* 16, 10937.

Skipp, R.A. & Samborski, D.J. (1974) The effect of the *Sr6* gene for host resistance on histological events during the development of stem rust in near-isogenic wheat lines. *Canadian Journal of Botany* 52, 1107–15.

Somerville, S.C. & Sherwood, J. (1985) Strategies for identification of disease reaction loci. In: Freeling, M. (ed.), *Plant Genetics*. Alan R. Liss, New York, USA, pp. 661–9.

Staskawicz, B., Dahlbeck, D. & Keen, N. (1984) Cloned avirulence gene of *Pseudomonas syringae* pv. *glycinea* determines race-specific incompatibility on *Glycine max* (L.) Merr. *Proceedings of the National Academy of Sciences (USA)* 81, 6024–8.

Staskawicz, B.J., Dahlbeck, D., Keen, N. & Napoli, C. (1987) Molecular characterization of cloned avirulence genes from race 0 and race 1 of *Pseudomonas syringae* pv. *glycinea*. *Journal of Bacteriology* 169, 5789–94.

Sternberg, N. (1990) Bacteriophage P1 cloning system for the isolation, amplification, and recovery of DNA fragments as large as 100 kilobase pairs. *Proceedings of the National Academy of Sciences (USA)* 87, 103–7.

Straus, D. & Ausubel, F. (1990) Genomic subtraction for cloning DNA corresponding to deletion mutations. *Proceedings of the National Academy of Sciences (USA)* 87, 1889–93.

Suggs, S.V., Wallace, R.B., Hirose, T., Kawashima, E.H. & Itakura, K. (1981) Use of synthetic oligonucleotides as hybridization probes: isolation of cloned cDNA sequences for human β2-microglobin. *Proceedings of the National Academy of Sciences (USA)* 78, 6613–17.

Søgaard, B. & Jørgensen, J.H. (1982) Genes for reaction to *Erysiphe graminis hordei* (powdery mildew). *Barley Genetics Newsletter* 12, 161–8.

Sørensen, M.B. (1989) Mapping of the *Hor2* locus in barley by pulsed field gel electrophoresis. *Carlsberg Research Communications* 54, 109–20.

Tamaki, S., Dahlbeck, D., Staskawicz, B. & Keen, N.T. (1988) Characterization and expression of two avirulence genes cloned from *Pseudomonas syringae* pv. *glycinea*. *Journal of Bacteriology* 170, 4846–54.

Terada, R. & Shimamoto, K. (1990) Expression of CaMV35S-GUS gene in transgenic rice plants. *Molecular and General Genetics* 220, 389–92.

Torp, J. & Jørgensen, J.H. (1986) Modification of barley powdery mildew resistance gene *Ml-a12* by induced mutation. *Canadian Journal of Genetics and Cytology* 28, 725–31.

Torp, J., Jensen, H.P. & Jørgensen, J.H. (1978) Powdery mildew resistance genes in 106 Northwest European spring barley varieties. *Kgl. Vet.-og Landbohojsk. Aarsskr.* pp. 75–102.

Travis, G.H. & Sutcliffe, J.G. (1988) Phenol emulsion-enhanced DNA-driven subtractive cDNA cloning: isolation of low-abundance monkey cortex-specific mRNA's. *Proceedings of the National Academy of Sciences (USA)* 85, 1696–700.

Tsuchiya, T. (1987) Gene analysis and linkage studies in barley. In: Yasuda, S. & Konishi, T. (eds), *Barley Genetics V*. Sanyo Press, Okayama, Japan, pp. 175–87.

Valent, B. (1990) Rice blast as a model system

for plant pathology. *Phytopathology* 80, 33–6.

Valvekens, D., van Montagu, M. & van Lijsebettens, M. (1988) *Agrobacterium tumefaciens*-mediated transformation of *Arabidopsis thaliana* root explant by using kanamycin selection. *Proceedings of the National Academy of Sciences (USA)* 85, 5536–40.

Van Sluys, M.A., Tempé, J. & Federoff, N. (1987) Studies on the introduction and mobility of the maize *Activator* element in *Arabidopsis thaliana* and *Daucus carota. EMBO Journal* 6, 3881–9.

von Wettstein-Knowles, P. & Søgaard, B. (1980) The *cer-cqu* region in barley: gene cluster or multifunctional gene. *Carlsberg Research Communications* 45, 125–41.

Ward, E.R. & Jen, G.C. (1990) Isolation of single-copy-sequence clones from a yeast artificial chromosome library of randomly-sheared *Arabidopsis thaliana. Plant Molecular Biology* 14, 561–8.

Whittier, R.F., Dean, D.A. & Rogers, J.C. (1987) Nucleotide sequence analysis of *alpha*-amylase and thiol protease genes that are hormonally regulated in barley aleurone cells. *Nucleic Acids Research* 15, 2515–35.

Wilberg, A. (1974) Genetical studies of spontaneous sources of resistance to powdery mildew in barley. *Hereditas* 77, 89–148.

Wieland, I., Bolger, G., Asouline, G. & Wigler, M. (1990) A method for difference cloning: gene amplification following subtractive hybridization. *Proceedings of the National Academy of Sciences (USA)* 87, 2720–4.

Williams, J.G.K., Kubelik, A.R., Livak, K.J., Rafalski, J.A. & Tingey, S.V. (1990) DNA polymorphisms amplified by arbitrary primers are useful as genetic markers. *Nucleic Acids Research* 18, 6531–5.

Wise, R.P. & Ellingboe, A.H. (1985) Fine structure and instability of the *Ml-a* locus in barley. *Genetics* 111, 113–30.

Wright, A.J. & Heale, J.B. (1988) Host responses to fungal penetration in *Erysiphe graminis* f. sp. *hordei* infections in barley. *Plant Pathology* 37, 131–40.

Yanofsky, M.F., Ma, H., Bowman, J.L., Drews, G.N., Feldman, K.A. & Meyerowitz, E.M. (1990) The protein encoded by the *Arabidopsis* homeotic gene *agamous* resembles transcription factors. *Nature* 346, 35–9.

Yoder, J.I. (1990) Rapid proliferation of the maize transposable element *Activator* in transgenic tomato. *Plant Cell* 2, 723–30.

Yoder, J.I., Palys, J., Alpert, K. & Lassner, M. (1988) *Ac* transposition in transgenic tomato plants. *Molecular and General Genetics* 213, 291–6.

Yokota, H. & Oishi, M. (1990) Differential cloning of genomic DNA: cloning of DNA with an altered primary structure in in-gel competitive reassociation. *Proceedings of the National Academy of Sciences (USA)* 87, 6398–402.

Young, N.D., Miller, J.C. & Tanksley, S.D. (1987) Rapid chromosomal assignment of multiple genomic clones in tomato using primary trisomics. *Nucleci Acids Research* 15, 9339–48.

Young, N.D., Zamir, D., Ganal, M.W. & Tanksley, S.D. (1988) Use of isogenic lines and simultaneous probing to identify DNA markers tightly linked to the *Tm-2a* gene in tomato. *Genetics* 120, 579–85.

Young, R.A. & Davis, R.W. (1983) Efficient isolation of genes by using antibody probes. *Proceedings of the National Academy of Sciences (USA)* 80, 1194.

VI

Biotechnology

25
Haploid Production: Approaches and Use in Plant Breeding

R.A. Pickering & P. Devaux

Haploids in plant breeding and related research have been used increasingly in recent years in many agricultural and horticultural crop plants. Their applications and development have been the subject of several recent reviews (e.g. Choo *et al.*, 1985; Dunwell, 1985; Kasha *et al.*, 1989) and conferences (Kasha, 1974; Davies & Hopwood, 1980; Maluszynski, 1989) and they are now attracting interest in the field of molecular biology. In the Gramineae, success rates have lagged behind species more amenable for tissue culture, but results are now becoming sufficiently promising to warrant investment in large breeding programmes. In barley, several cultivars have been released as doubled haploids, and there is no reason to doubt that many more will be bred using the techniques outlined in the following pages.

In vitro gynogenesis (ovary culture)

The possibility of producing barley haploid plants by unpollinated ovary culture was first reported by San Noeum (1976), and studies on factors which influence success rates have been undertaken (Yean-San, 1987). The most important of these are:

- the developmental stage of the embryo sac, which is optimal when almost mature. This corresponds to microspores having two nuclei in young pollen grains.
- cold pretreatment (5–14 days at 3°C) applied to barley spikes before ovary collection and plating.
- the culture medium composition such as a high concentration of sucrose (80–120 g l^{-1}), the presence of 2,4-D (2,4-dichlorophenoxyacetic acid) and the use of a combination of growth substances (Wang & Kuang, 1981).
- dorsal orientation of the ovaries on to the medium.

One striking phenomenon which occurs just a few days after inoculation, is that a high proportion of ovaries enlarge (Tang & Xie, 1987; Kasha, pers. comm.; Devaux, unpublished data). Unfortunately the inefficiency of this technique, usually less than one plantlet for 100 ovaries plated (Yean-San, 1987), makes its use limited in breeding and insufficient data exist to draw conclusions about genotypic influence on success rates. However, ovaries or ovules may be suitable for transforming barley by microinjection of foreign DNA (Steinbiss *et al.*, 1985).

Pseudogamy (haploid initiator gene)

This system enables haploid plants to be directly produced by pseudogamy using a haploid initiator gene *hap* (Hagberg & Hagberg, 1980). Although normal development of the endosperm takes place after the polar nuclei fuse with one of the sperm, the other sperm fails to enter the egg which subsequently develops into a haploid embryo (Mogensen, 1982). The numbers of haploids produced in progeny from plants homozygous for the gene (*hap/hap*) ranged from 1 to 40% depending on genotype, and 0.1 to 10% in progeny from the heterozygote (Hagberg & Hagberg, 1987). However, some of the fertile diploid plants were spontaneously doubled haploids, and these occurred three times as frequently as normal haploids. To identify them, a strategy has been devised using marker genes, such as recessive mildew resistance alleles *mlo* (Hagberg & Hagberg, 1987).

The main advantage over other haploid systems is that an *in vitro* culture phase is not required, but the number of crosses needed is greater than with the 'bulbosum' technique, and limits its practical applications.

In vitro androgenesis (anther culture)

Research on barley anther culture was initiated by Clapham (1971) whose techniques and media have been used extensively since, but several important modifications (Table 25.1) have now enabled anther culture to be used in part of a breeding programmes. The following survey outlines the stages of the method and describes the developments which have contributed most to improving success rates, a sample of which is presented in Table 25.2.

In vivo anther development

Following meiosis and tetrad release the isolated microspore is enclosed in a thin membrane and the nucleus becomes centrally situated (Gaul *et al.*, 1976; Sunderland & Dunwell, 1977). A thick wall gradually forms with a single pore and the nucleus then migrates to the cell periphery during vacuolation. Before first pollen grain mitosis the nucleus increases in size, often being located opposite the pore at this stage. First pollen grain mitosis results in the formation of a large vegetative cell and a small genera-

Table 25.1. Recent progress in anther culture of barley.

Source of progress	Reference
Ficoll *vs* agar	Kao, 1981
Cold pretreatment of spikes	Xu *et al.*, 1981 Dunwell *et al.*, 1987
Agarose *vs* agar	Lyne *et al.*, 1986
Anther orientation on the medium	Shannon *et al.*, 1985 Hunter, 1985
Melibiose *vs* sucrose in starch medium	Sorvari & Schieder, 1987
Low [NH_4NO_3] + Glutamine *vs* High [NH_4NO_3]	Olsen, 1987
Maltose *vs* sucrose	Hunter, 1987
Pretreatment of anthers on mannitol	Roberts-Oehlschlager & Dunwell, 1990

Table 25.2. Regeneration rates of green plants from 100 cultured anthers, presented in chronological order, with examples of cultivar used.

Cultivar or line	Medium*	Maximum no. of green plants per 100 cultured anthers	Specific purpose of experiment	Reference
'Sabarlis'/'Akka' etc.	mod. LS	3.0	Media and albinism study	Clapham, 1973
'Amsel'	mod. LS	1.0	Genotype and cytological study	Malepszy & Grunewaldt, 1974
'Dissa'	mod. LS	8.0	Genotype influence	Foroughi-Wehr *et al.*, 1976
'Dissa'	mod. LS	1.5	Environmental influence	Foroughi-Wehr & Mix, 1979
'Dissa'	mod. LS	2.7	Genotype influence and heritability of response	Foroughi-Wehr *et al.*, 1982
Cvs + Hybrids	Kao, 1981	10.0	Ficoll study	Kao & Horn, 1982
'Igri'	mod. LS	2.4	Production of virus-resistant barley	Foroughi-Wehr & Friedt, 1984
'Sabarlis'	mod. LS	28.9	Anther orientation	Hunter, 1985
'Sabarlis'	mod. LS	4.2	Media and gelling agent comparisons	Lyne *et al.*, 1986
'Dissa'	mod. LS	2.0	Gelling agent comparisons	Sorvari, 1986
'Sabarlis' (S) 'Dissa' (D) S × (D × S)	mod. MS	0.6 0.3 1.0	Genetic study of response	Dunwell *et al.*, 1987
'Igri' 'Sabarlis'	mod. LS	590 122	Carbon source comparisons	Hunter, 1987
'Igri'	mod. LS	464	Media modifications	Olsen, 1987
'Arra' 'Dissa'	N6	43.3 34.8	Osmoticum study in medium	Sorvari & Schieder, 1987
–	Kao, 1981	40	Media development	Kao, 1988
'Doublet' 'Igri' 'Dissa' 'Sabarlis'	mod. MS	18.3 11.1 5.6 2.8	Anther orientation	Powell *et al.*, 1988
Hybrid	mod. BAC1	14.4	Media development	Szarejko & Kasha, 1988
'Tweed'	mod. LS (Hunter, 1987)	190	Carbon source comparison	Finnie *et al.*, 1989
'Igri'	mod. LS	18.6	Genotype and environmental effects on response	Knudsen *et al.*, 1989

Table 25.2. *cont.*

Cultivar or line	Medium*	Maximum no. of green plants per 100 cultured anthers	Specific purpose of experiment	Reference
'Igri'	mod. LS (Hunter, 1987; Kao, 1981; Sorvari, 1986)	10.4	Media comparisons	Kuhlmann & Foroughi-Wehr, 1989
Hybrid	mod. LS (Hunter, 1987)	11.3	Media comparisons	Luckett (pers. comm.)
'Igri'	mod. LS (Hunter, 1987)	69.4	Media development	Devaux, 1990
'Igri'	mod. LS (Hunter, 1987)	30.6	Anther pretreatment; inoculation density; anther orientation	Roberts-Oehlschlager & Dunwell, 1990
Hybrid	mod. LS (Olsen, 1987)	65	Spike pretreatment; environmental influence	Hou & Ullrich, 1990

*BAC1 = Marsolais & Kasha, 1985
LS = Linsmaier & Skoog, 1965
MS = Murashige & Skoog, 1962
N6 = Chu, 1978
mod. = modified

tive cell, the latter cut off by a wall. Starch formation begins before second pollen grain mitosis. To obtain the best results from anther culture, it is generally agreed that anthers with microspores at the mid-uninucleate stage (stage 4 – Gaul *et al.*, 1976; stage 2 – Sunderland, 1974) are most responsive when the nucleus is in G1 interphase (Sunderland, 1974; Sunderland & Huang, 1985).

Anomalous development

The previous description applies to most developing microspores. However, Sunderland (1974) observed pollen dimorphism in barley in which 15–20% non-staining grains were present amongst those which had developed normally. Dale (1975) subsequently observed that the proportions of non-staining pollen grains and pollen calluses were comparable. The frequency of these grains was also greater in the proximal anther region, where calluses were most often formed. Sunderland *et al.* (1979) and Sunderland & Evans (1980) reported further on these anomalies, calling the routes of development the A, B and C pathways, after those previously described in *Nicotiana tabacum* (Sunderland, 1974). The routes differ after first pollen grain mitosis in the presence or absence of cell wall formation and whether the products of division resemble generative and vegetative cells (nuclei) or two morphologically similar nuclei. Cell fusion may or may not occur resulting in haploid, diploid, polyploid or mixoploid multicellular structures. The ability to switch pollen development from the normal gametophytic into the sporophytic pathways just outlined can be induced by an environmental stress applied as a treatment to spikes gathered at the appropriate developmental stage.

Spike pretreatment

The proportions of responsive pollen grains have been increased by pretreating spikes, with microspores at the mid-uninucleate stage, at various temperatures before anther culture. Huang & Sunderland (1982) obtained a positive anther response (as measured by callus production) by pretreating excised spikes kept in darkness in sealed Petri dishes at various temperatures. At 25°C the optimum duration of pretreatment was 2–4 days, but progressively greater responses were achieved by reducing the temperature to 4°C and increasing the treatment time to 28 days. Highest rates of green plant regeneration were obtained at 4°C for 35 days. Spike pretreatment has subsequently been used extensively to improve success of anther culture (Hunter, 1985; Shannon *et al.*, 1985; Dunwell *et al.*, 1987; Olsen, 1987). However, the technique has not been universally adopted (Kao, 1988) possibly because of differences in the environmental conditions under which donor plants are grown (Kasha, 1989).

The development of anthers from pretreated spikes differs from those *in vivo* in that the tapetal cells, and the fluid released from them into the anther cavity, disappear prematurely (Sunderland *et al.*, 1984). This disruption to the synchrony of microspore and tapetal development is thought to influence the switch from the gametophytic to the sporophytic phase (Sunderland & Huang, 1985).

Donor plant growth conditions

Pollen dimorphism is also influenced by the environment in which donor plants are raised. Sunderland (1978) observed that the proportion of pollen grains which undergo cell division after culture (E-type or embryogenic) was increased from around 3 to 20% when donor plants were grown at 10°C compared with 15°C; this could partly explain the increase in success using anthers from plants grown below 15°C. For example, Foroughi-Wehr & Mix (1979) using controlled environments compared several temperatures and reported that highest callus and green plant production was obtained from plants grown at 12/5°C, with progressively poorer results at higher temperatures. Lyne *et al.* (1986) also identified 12°C as being more suitable for green plant production than 15°C. Spikes harvested from the field and glasshouse at temperatures lower than those encountered at the height of summer were also more amenable to callus and green plant production (Clapham, 1973; Foroughi-Wehr *et al.*, 1976; Kuhlmann & Foroughi-Wehr, 1989).

Although 12°C is now frequently quoted as the optimum temperature for donor plant growth (Hunter, 1985; Olsen, 1987) a preference for 16–20°C, without a spike cold pretreatment, has been expressed (Kasha, 1989) which may be more appropriate for barley cultivars adapted to the Canadian environment. Operating costs may also preclude the use of controlled environment chambers and many research workers still raise their plants in glasshouses at varying temperatures (Devaux, 1987; Dunwell *et al.*, 1987; Lee & Chen, 1987; Powell *et al.*, 1988), although this is likely to result in seasonal fluctuations in success.

Genotypic influence

Variation in anther response among cultivars has been widely reported (Wilson *et al.*, 1977; Foroughi-Wehr & Friedt, 1984; Sorvari & Schieder, 1987; Powell, 1988). The first report illustrating a marked genotypic effect on callusing and plant production was that of Malepszy & Grunewaldt (1974) who found 'Amsel' the most responsive of eight cultivars tested, but 'Dissa' appeared to be even more suitable for anther culture in a later comparison involving 19 cultivars including 'Amsel' (Foroughi-Wehr *et al.*, 1976). The high callus and green plant production of 'Dissa' was inherited partly from 'Amsel', one of its parents (Friedt & Foroughi-Wehr, 1981), although these two characteristics are thought to be controlled by independent genetic systems (Foroughi-Wehr *et al.*, 1982; Huang *et al.*, 1984). It has since been shown that ethylene plays a major role in embryogenesis, and as cultivars differ in their capacity to produce ethylene, this may explain some of the differences in anther response (Cho & Kasha, 1989). Although a cytoplasmic effect on green plant production has been implicated (Foroughi-Wehr *et al.*, 1982) this has not been confirmed (Foroughi-Wehr & Friedt, 1984; Dunwell *et al.*, 1987).

Many other investigations into anther culture methodology have involved single cultivars. Clapham (1971) in the earliest report on barley anther culture, carried out his research on 'Sabarlis', a spring cultivar which has since been used extensively by groups elsewhere (e.g. Huang & Sunderland, 1982; Hunter, 1985; Lyne *et al.*, 1986). The winter cultivar 'Igri' was subsequently found to be particularly responsive (Foroughi-Wehr & Friedt, 1984) and is now widely recommended for research purposes (Olsen, 1987; Hunter *et al.*, 1989; Devaux, 1990).

Spike and anther culture

There have been two reports of culturing intact barley spikes in liquid media. Wilson (1977) either stood excized tillers of 'Dissa' in water for 1–2 days before culture, or transferred freshly harvested spikes, with microspores at first pollen grain mitosis to a shaken liquid medium based on Linsmaier & Skoog (1965). Inconsistent callus formation occurred and of the 40 plants regenerated, only two were green. Datta (1987) pretreated spikes with early uninucleate microspores at 5°C for 12 days before culture in a medium containing Ficoll 400 at 100 g l^{-1}. Embryogenic callus was observed which may have been induced partly because of the addition of Ficoll to the medium, and 16 green plants were regenerated. All other published reports describe the culture of individual anthers. These are removed from spikes after surface sterilization of the enclosing last leaf sheath, or the spike itself, with ethanol or a dilute solution of sodium or calcium hypochlorite.

The orientation of anthers on solid media has been shown to be important for subsequent callus production. This was enhanced when anthers were placed with only one of the two loculi in contact with the medium (Hunter, 1985; Shannon *et al.*, 1985) and embryogenic callus formation was only observed in this situation, from which increased green plant production occurred (Hunter, 1985). Powell *et al.* (1988) using several cultivars, confirmed that anther orientation had a significant effect on green plant production, but not on the proportions of responding anthers over all cultivars. They attributed this result to possible differences in growth regulators in the medium although IAA (indole-3-acetic acid) and BAP (6-benzylaminopurine) were both included at the same concentrations as those used by Hunter (1985). A possible explanation for the influence of anther orientation on response is that the microspores in the upper loculus escape the inhibitory effects of sucrose (Wei *et al.*, 1986) in the medium, or its breakdown products (Hunter *et al.*, 1989).

Media

Carbon source and gelling agent

The use of maltose as carbon source is perhaps the most significant recent development for improving success rates in barley. Green plant production is greatly increased compared with sucrose (Hunter, 1987; Finnie *et al.*, 1989) probably because maltose is hydrolysed to glucose more slowly than sucrose thus preventing inhibitory concentrations of glucose from accumulating (Roberts-Oehlschlager *et al.*, 1990). Before this research, sucrose at 6–12% was the most frequently used carbon source in media and had been shown to raise the medium osmolality with increases in concentration. This resulted in the promotion of callus (Kao, 1981) and embryoid formation, in contrast to plant regeneration via organogenesis from calluses on media with low osmolality (Sorvari & Schieder, 1987). In the same report, sucrose was replaced with melibiose as osmoticum in combination with barley starch. As a result, significant improvements in overall response and green plant production were obtained compared with the control medium containing sucrose.

Sorvari (1986) observed that embryo formation and green plant production could be increased by using starches from various sources in place of agar, and Lyne *et al.* (1986) obtained an improved anther response using sea-plaque agarose instead of agar. Other types of agarose may be as effective (Luckett, pers. comm.) and Ziauddin *et al.* (1988a) failed to confirm the superiority of sea-plaque agarose over agar in their BAC1 medium. Nevertheless it is known that agar has deleterious effects on embryo development (Kohlenbach & Wernicke, 1978) and agarose, a purer compound, may be preferable in some media. Kao (1981) omitted agar from the callus-inducing medium but added Ficoll 400 to support the anthers in an otherwise liquid medium. Anther response was significantly improved over media without Ficoll 400, and embryo formation was observed on calluses 2–3 mm in diameter, a feature not previously recorded on calluses derived from barley microspores. It appears, therefore, that to achieve embryo formation, which is preferable to organogenesis for plant regeneration (Lyne *et al.*, 1986), it is necessary to stabilize media osmolality by replacing sucrose with maltose, and to substitute starch (Kuhlmann & Foroughi-Wehr, 1989) or Ficoll 400 (Kao, 1981) for agar. To reduce the cost of media components, Kasha (1989) has obtained comparable results to Ficoll 400 with dextran and soluble starch both of which are cheaper than Ficoll 400, the cost of which may be prohibitive for routine haploid production.

Plant growth regulators

Auxins and cytokinins are incorporated into media to induce callusing from anthers, and their concentrations are reduced to promote plant regeneration after callus subculture. The most frequently used are 2,4-D (0.5–2 mg l^{-1}) and kinetin (0.25–0.5 mg l^{-1}) or IAA (indole-3-acetic acid) and BAP (6-benzylaminopurine) (both at 1 mg l^{-1}). NAA (L-naphthaleneacetic acid) (2 mg l^{-1}) with BAP (1 mg l^{-1}) has also been adopted (Devaux, 1987).

Media development

In almost 20 years of research, many media have been formulated, some of which will be outlined here. Clapham (1971) used the medium of White (1943) with 12% sucrose, 1 mg l^{-1} NAA, coconut milk and glutamine, and obtained multicellular structures and plantlets from anthers cultured at the uninucleate stage. Clapham (1973) subsequently reported that calluses were induced on a Linsmaier & Skoog (1965) medium modified by reducing the NH_4NO_3 concentration and adding IAA and BAP (both at 1 mg l^{-1}) and sucrose at 12%. Plants were regenerated on a similar medium with 3% sucrose. Foroughi-Wehr *et al.* (1976) further modified Clapham's medium by omitting $CoCl_2.6H_2O$, changing the iron source and reducing the sucrose concentration to 6%. Calluses were subcultured to this basal medium with reduced IAA, BAP (0.4 mg l^{-1}) and sucrose (3%). This medium has subsequently been used extensively but other groups have developed their own. Sunderland and co-workers at the John Innes Institute, UK, have employed media based on one devised for wheat which includes potato extract (Chuang *et al.*, 1978) and the N6 medium of Chu (1978) developed originally for rice anther culture. This has reduced ammonium ions compared with that of Miller (1963) on which it is based. Kao (1981) in Canada published details of a relatively complex medium for use with Ficoll 400. Because of a lack of success with the above media, Marsolais & Kasha (1985) at Guelph University, Canada, developed a defined medium – BAC1 – which also incorporated Ficoll 400. Both Kao (1988) and the Guelph group (Szarejko & Kasha, 1988) have since modified their media by the addition of $KHCO_3$ in attempts to stabilize the pH, which tended to drop during culture from 5.5 to 4.5 (Kao, 1981).

The important role of inositol in anther response has been reported by Xu & Sunderland (1981) who observed a 20-fold increase in callus production over that obtained from the Chuang and N6 media without inositol, although maximum effect was only gained at an anther density of 40 per ml compared with 20 per ml of medium.

It is not possible to elaborate greatly on all the changes in media composition, and only the major ones have been described here. However, there have been several reports comparing the relative efficiencies of media. Kuhlmann & Foroughi-Wehr (1989) reported that for four different hybrids and 'Igri' a liquid medium based on the formulation of Foroughi-Wehr *et al.* (1976) produced better results than the N6 (Chu, 1978). Both media were modified by substituting maltose for sucrose, and Ficoll 400 plus barley starch for agar. Lyne *et al.* (1986) also obtained best overall response with the medium of Foroughi-Wehr *et al.* (1976) compared with those of Chu (1978), Chuang *et al.* (1978) and Kao (1981). However, the relative efficiencies of the media varied for the different stages during culture. Huang *et al.* (1984) did not find it possible to identify which of the two media – N6 (Chu, 1978) and that of Chuang *et al.* (1978) – was superior because of genotype and media interactions.

Although Wilson *et al.* (1977) obtained an increase in the proportions of green plants after filter sterilizing their medium, autoclaving has frequently been used for sterilization (Foroughi-Wehr *et al.*, 1976; Chuang *et al.*, 1978; Xu *et al.*, 1981; Sorvari, 1986; Lee & Chen, 1987). Other researchers either failed to report the sterilization method or autoclaved all nutrients except a few thermolabile components (Clapham, 1973; Xu & Sunderland, 1981; Shannon *et al.*, 1985; Devaux, 1987). Complete filter-sterilization of all components, except gelling agent, is preferred by several other groups (Kao, 1981; Marsolais & Kasha, 1985; Olsen, 1987; Hunter *et al.*, 1989).

Media preconditioning and inoculation density

To maximize callus production, Xu *et al.* (1981) advocated using conditioned media. This was effected by culturing anthers with microspores at the binucleate to starch formation stages for 7 days before their removal and culture of uninucleate anthers in the normal way for callus induction. Percentage anther response could not be assessed accurately, but it was clear that callus formation was greatly increased after conditioning, especially when anthers for callus induction were plated at densities of 20 compared with 10 per ml. A conditioning factor present in anthers is thought to be responsible for this effect, and its effect was synergistic with inositol in the medium. Preconditioning of media in this way reduced the need for high anther inoculation densities which, without media preconditioning, produced maximum response and calluses per anther at 60–120 anthers per ml (Xu & Sunderland, 1982). This positive effect of high inoculation density was not recorded when anthers were initially cultured on a medium containing 3.2% mannitol for 4 days at 25°C before re-culturing on a medium with maltose and incubation at 25°C (Roberts-Oehlschlager & Dunwell, 1990). Anther inoculation densities in other published reports range from 2–4 per ml (Dunwell *et al.*, 1987; Lee & Chen, 1987; Olsen, 1987; Powell, 1988) to 10–20 per ml (Kao, 1981, 1988; Marsolais & Kasha, 1985; Kuhlmann & Foroughi-Wehr, 1989) to 40–70 per ml (Shannon *et al.*, 1985).

Incubation

This is usually carried out at between 22 and 28°C in darkness, and embryos and/or calluses become visible around 21 days after culture. A reduced temperature of 4°C (Jing & Hu, 1981) or 15°C (Kao, 1981) for the first 3 or 5 days, respectively, may enhance callus formation, and even an increased temperature of 28°C for up to 3 days followed by 25°C improved the response of 'Bruce' although temperatures greater than 28°C were deleterious for all cultivars tested (Ziauddin *et al.*, 1988a).

Regeneration

To promote plant regeneration, calluses greater than 1 mm in diameter are usually subcultured to a solid medium, for example B5 (Gamborg *et al.*, 1968), or, more frequently, one based on that described by Foroughi-Wehr *et al.* (1976), with 2–3% sucrose and either a reduced concentration of growth hormones or a change from 2,4-D plus kinetin to BAP plus IAA (Huang & Sunderland, 1982). Kao (1988) however, actually increased the sucrose level from 4.25 to 6% and used polyester floats on a liquid medium to support calluses. Calluses and embryos are incubated for 5–7 days in darkness until the start of plant regeneration. Further development takes place under lights, the daylength ranging from 10 h to continuous light (12–70 μE m^{-2} s^{-1}), at 20–26°C. Green plantlets are then either subcultured again to a medium without hormones (Lyne *et al.*, 1986; Olsen, 1987) or IAA only (Devaux, 1987), or transferred to potting compost.

Progeny from anther culture

Albinos

Albinism has been and remains a problem amongst regenerants, and is known to be strongly influenced by genotype (Dunwell *et al.*, 1987; Powell, 1988; Knudsen *et al.*, 1989). Clapham (1973) obtained 80–100% albino plants and these proportions could not be reduced by changes in media composition or culture conditions. Fewer albinos were, however, obtained from late-season plants (Clapham, 1973) an observation confirmed by other researchers (Foroughi-Wehr *et al.*, 1976; Foroughi-Wehr & Mix, 1979; Kuhlmann & Foroughi-Wehr, 1989). This may be attributed to a temperature effect, as Lyne *et al.* (1986) obtained greater proportions of green plants by growing donor plants at 12°C compared with 15°C. Significant improvements in green plant production have also been achieved as follows:

1. Spike pretreatment at 4°C (Huang & Sunderland, 1982);
2. Filter sterilization of media (Wilson, 1977);
3. Sea-plaque agarose (Lyne *et al.*, 1986) or starch (Sorvari, 1986) as gelling agents in place of agar;
4. Melibiose as osmoticum at 80–120 g l^{-1} in sucrose-free medium (Sorvari & Schieder, 1987);
5. Substitution of maltose for sucrose in the medium (Hunter 1987; Hunter *et al.*, 1989). By way of confirmation, Devaux (1990) subsequently reported that 58, 72, and 79% of total plants regenerated were green on media containing sucrose, maltose or maltose plus glutamine respectively.

There is still speculation about the causes of albinism in regenerants. Clapham (1973) observed proplastids and chloroplasts of almost normal appearance in barley albinos but Day & Ellis (1985) reported deletions in plastid DNA.

Ploidy level

Diploidization can occur during the first cell divisions after culture following cell fusion and endomitosis (Wilson *et al.*, 1978; Sunderland *et al.*, 1979; Sunderland & Evans, 1980; Chen *et al.*, 1984; Lee & Chen, 1987) resulting in callus with haploid, diploid and polyploid cells (Malepszy & Grunewaldt, 1974; Grunewaldt & Malepszy, 1975; Wilson, 1977) and successive subculturing of calluses leads to a gradual reduction in haploid cell numbers (Grunewaldt & Malepszy, 1975). Regenerated plants, both albino (Grunewaldt & Malepszy, 1975; Kao, 1981) and green (Mix *et al.*, 1978; Foroughi-Wehr *et al.*, 1982; Devaux, 1988) can be haploid, diploid, tetraploid, aneuploid or mixoploid. Spontaneous chromosome doubling results in fertility levels ranging from just under 50% (Powell, 1988) to over 90% (Huang & Sunderland, 1982; Lyne *et al.*, 1986; Olsen, 1987). This may eliminate the need for colchicine treatment to induce chromosome doubling. However, should the breeder consider it necessary to determine the ploidy level of regenerants, several methods can be used: chromosome counting (Foroughi-Wehr *et al.*, 1976); phenotypic identification (Mix *et al.*, 1978); stomatal guard cell length (Tometorp, 1939); and flow cytometry (Brown, 1989). Flow cytometry may be the most promising as it is rapid, reliable and only requires small amounts of tissue. Some profiles of haploid, doubled haploid and tetraploid microspore-derived barley plants (MDP) analysed by flow cytometry are illustrated in Fig. 25.1.

As well as variation in ploidy level, chromosomal abnormalities have been reported

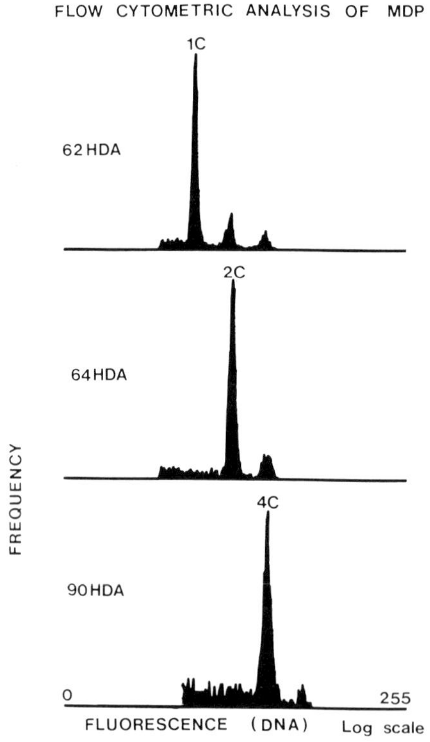

Fig. 25.1. Flow cytometric profiles of haploid (62HDA), diploid (64HDA) and tetraploid (90HDA) plants derived from anther culture (MDP). Analyses made at CNRS, Gif (France) in collaboration with S. Brown.

in mixoploid regenerants such as chromosome breakage, ring chromosomes, fragments and bridges (Mix *et al.*, 1978) which may have resulted from translocations and inversions. Powell *et al.* (1986) also recorded that three out of 13 diploid regenerants had reduced chiasma frequency, and another was heterozygous for a translocation.

Changes in ploidy levels, recessive mutants (including albinos) and a maternally inherited mutant have also been observed in regenerants from calluses derived from haploid explants after crossing *H. vulgare* and *H. bulbosum* (Saalbach & Koblitz, 1977; Jensen, 1981; Pickering, 1989). It seems, therefore, that genetic and cytoplasmic changes do take place during callus induction regardless of explant source although regeneration via embryogenesis may reduce induced variation.

In addition to cytological aberrations, differences between regenerants and control barley cultivars have been reported for several quantitatively inherited traits (San Noeum & Ahmadi, 1982; Powell *et al.*, 1984). However, pre-existent heterogeneity within the parental cultivar may sometimes explain this variation (Snape *et al.*, 1988). Deviations from expected segregation ratios have also been recorded for simply inherited characters such as disease resistance (Foroughi-Wehr & Friedt, 1984; Friedt *et al.*, 1986) and row type (Foroughi-Wehr & Friedt, 1984; Kao, 1988) in regenerants derived from hybrid material. For example, an excess of six-rowed types over two-rowed has been obtained and, as the proportions of each could be modified by changes to the concentration of zeatin riboside in the medium, specific media for two-rowed versus six-rowed barley were developed (Kao, 1988).

Microspore culture

This technique, relying as it does on isolated microspore culture, is possibly more amenable than anther culture for research into transformation, mutations and *in vitro* selection. It eliminates the effect of diploid anther tissue and competitive effects within the anther. Further advantages over anther culture are the potential regeneration of larger numbers of haploid plants without the interference of somatic callus derived from the anther wall although this problem has not arisen in barley anther culture. Despite these advantages less research has been carried out on microspore compared with anther culture.

Spike pretreatment

Spikes, with anthers containing uninucleate microspores, may be cold pretreated in a similar way to those used for anther culture (Huang & Sunderland, 1982). Olsen (pers. comm.) found 28 days at 4°C to be suitable, although Sunderland & Xu (1982) preferred 14 days at 7°C to obtain rapid anther dehiscence and microspore liberation. Datta & Wenzel (1988) on the other hand, obtained differential responses of spring and winter cultivars at 5°C for 8 and 12 days respectively. Köhler & Wenzel (1985) obtained greatest success by omitting this step altogether, using instead media preconditioned for 7 days with 10 ovaries per ml after the method of Xu *et al.* (1981). Sunderland & Xu (1982) also found a beneficial effect of preconditioning of media, most pronounced when microspores were liberated from anthers inoculated at a density of 80 per ml compared with 20 per ml. The positive effects on response of: (i) spike cold pretreatment; (ii) media preconditioned with anthers or ovaries; or (iii) inositol or mannitol in the initial culture medium (Sunderland & Xu, 1982; Wei *et al.*, 1986) led Ziauddin *et al.* (1989) to suggest that there may be a conditioning factor present in the anther for successful induction to occur.

Microspore isolation

There are two methods for isolating microspores, namely shed pollen and mechanical separation. Sunderland & Xu (1982) preferred the former for its simplicity and avoidance of damage to the pollen. In their method, anthers were cultured on drops of preconditioned medium with 1–2 g l^{-1} inositol (Sunderland & Xu, 1982; Köhler & Wenzel, 1985; Datta & Wenzel, 1988). Dehiscence took place 24 h later when the anthers could be removed after about half (1500) of the total number of microspores per anther had been liberated (Sunderland & Xu, 1982). Inoculation densities greater than 20 anthers per ml increased productivity (Sunderland & Xu, 1982; Köhler & Wenzel, 1985). Ziauddin *et al.* (1990) have also used the shed pollen technique by culturing anthers of 'Igri' on 0.3 M mannitol for 3 days at 25–28°C. The liberated microspores were then separated by centrifuging in mannitol, water or medium depending on cultivar (Ziauddin *et al.*, 1989), and recultured on fresh ovary-conditioned medium based on those of Olsen (1987) and Hunter (1987). Ficoll 400 was found to be inhibitory (Ziauddin *et al.*, 1988b) and may result in the collapse of microspores (Thörn, pers. comm.). A faster response and embryo formation (18 days) was obtained using this method compared with mechanical separation (3–4 weeks) (Ziauddin *et al.*, 1989).

Sunderland & Xu (1982) also observed greater productivity from shed pollen compared with isolated pollen released mechanically by gentle tapping. In addition to this latter method of mechanical separation (which has been used to great effect by Hunter, 1987), Wei *et al.* (1986) released microspores, by means of an internal pestle or injector, into a 0.3 M mannitol solution before filtering, rinsing, centrifuging and inoculating on to medium (Xu & Sunderland, 1981). Other methods of mechanical separation have been compared (Olsen, pers. comm.), namely maceration by Teflon rod (Nitsch, 1974) and microblending (Swanson *et al.*, 1987). Microblending released more microspores and had a gentler action as assessed by frequencies of large viable microspores (20% vs 10%). A cold spike pretreatment may assist the mechanical separation of microspores because tapetal degeneration enables them to be squeezed out more easily (Sunderland *et al.*, 1984).

Media

For callus induction, liquid or semiliquid media of similar composition to those used for anther culture have been found to be satisfactory, for example, Foroughi-Wehr *et al.* (1976), Chuang *et al.* (1978) and Chu (1978). *m*-Inositol at 1 or 2 g l^{-1} was found to be beneficial and reduced the need for media preconditioning (Sunderland & Xu, 1982; Köhler & Wenzel, 1985). More recently, microspore response and/or embryo formation have been improved by the addition of phenyl acetic acid to the medium in place of more commonly used growth hormones (Kasha, pers. comm.).

Sucrose at 6–9% used to be the usual carbon source in media, but as with anther culture, when maltose was substituted for sucrose a greatly increased response was obtained (Hunter, 1987).

After a period of incubation which may include 1–3 days initially at 4–5°C (Thörn, pers. comm.), calluses 1 mm or more in diameter can be subcultured on to media with reduced auxin concentration and finally to solid media for plant regeneration (e.g. Köhler & Wenzel, 1985; Wei *et al.*, 1986). A simpler method has been adopted by R.B. Jørgensen (pers. comm.) in which microspores are isolated for initial culture in droplets of media. After plating out on a solid medium (Hunter, 1987) embryo development and plant regeneration occur without subculture.

Response

Köhler & Wenzel (1985) obtained 1.2 plants per 100 cultured anthers of 'Dissa', but 93% were albino. Datta & Wenzel (1988) improved success rates with this cultivar to 14 green plants per 100 cultured anthers, by modifying the medium (e.g. increased *m*-inositol; addition of Ficoll 400) and cold-pretreating the spikes. Wei *et al.* (1986) had previously obtained up to 60% plant regeneration from inoculated calluses of 'Sabarlis', and 37.5% of regenerants were green. Hunter (1987), Olsen (pers. comm.) and Ziauddin *et al.* (1990), all using 'Igri', regenerated 608, 94 and 292 green plants per 100 anthers cultured respectively and Thörn (pers. comm.) has obtained 330 green plants per 100 anthers with the spring cultivar 'Alva'.

If these regeneration rates can be consistently repeated with a range of cultivars, microspore culture could become a more appropriate technique than anther culture for haploid production in a breeding programme and in research.

Interspecific hybridization

Selective chromosome elimination has been reported in crosses within the Triticeae, frequently resulting in haploids of one of the parental species (e.g. Bothmer *et al.*, 1983). Barley dihaploids and/or haploids have been produced from crosses of *Hordeum vulgare* L. × *Psathyrostachys fragilis* (Boiss.) Nevski (Bothmer *et al.*, 1984) and *H. vulgare* × *Secale cereale* L. (Fedak, 1977). However, the numbers of plants obtained were very low, and the following description of barley haploid production will be concerned only with *H. vulgare* × *H. bulbosum* crosses. As a measure of success rates, the numbers of green haploid plants per 100 pollinated florets are presented in Table 25.3.

Hordeum bulbosum L. is a perennial outcrossing species found in the Mediterranean region, and is thought to be closely related to *H. vulgare* (Bothmer *et al.*, 1983). It occurs as two cytotypes, tetraploid ($2n = 4x = 28$) and diploid ($2n = 2x = 14$), and has been crossed frequently with barley in attempts to transfer desirable characters from the wild into the cultivated species.

Successful crosses between *Hordeum vulgare* L. and *H. bulbosum* L. were first performed by Kuckuck (1934), who obtained one sterile triploid hybrid from *H. vulgare* ($2n = 2x = 14$) × *H. bulbosum* ($2n = 4x = 28$). In later studies Konzak *et al.*

Table 25.3. Regeneration rates of green haploid plants per 100 pollinated florets of *H. vulgare* × *H. bulbosum*, presented in chronological order.

Cultivar	Maximum no. of green haploid plants per 100 pollinated florets	Reference
F_1 hybrid	46.7	Jensen, 1976
'York'	26.8	Kasha *et al.*, 1978
F_1 hybrid	ca. 18–20	Pickering, 1980b
Hybrid	12.6	Simpson *et al.*, 1980
Hybrid	16	Simpson & Snape, 1981
'Sabarlis'	28.9	Pickering, 1983a
F_1 hybrid	15.4	Huang *et al.*, 1984
F_1 hybrid	ca. 11.5	Devaux, 1987
F_1 hybrid	41.6	Chen & Hayes, 1989
'Schuyler'	47.8	Hayes & Chen, 1989
F_1 hybrid	13.2	Devaux *et al.*, 1990
Hybrid	35.0*	Guerrero, pers. comm.
Hybrid	ca. 15* (7.5–10 average)	Jestin, pers. comm.
Hybrid	10 (average)	Shugar, pers. comm.
Hybrid	ca. 21	Torp, pers. comm.

*Doubled haploids.

(1951) and Morrison *et al.* (1959) amongst others also produced hybrids. Kao & Kasha (1969) obtained progeny resembling *H. vulgare* from similar hybridizations and proposed that the *H. bulbosum* chromosomes had been eliminated from the developing embryo after fertilization. This was confirmed by Kasha & Kao (1970) after hybridizations of *H. vulgare* ($2n = 2x = 14$) with *H. bulbosum* ($2n = 2x = 14$) resulted in the production of haploid *H. vulgare* plants. These had been regenerated after rescuing embryos to a nutrient medium. Haploid formation is known to depend on genetic factors (Ho & Kasha, 1975), genome ratios (Subrahmanyam & Kasha, 1973; Subrahmanyam, 1980) and temperature after fertilization (Pickering, 1985). Subrahmanyam (1982) proposed a hierarchical system in *Hordeum* based on species dominance to explain differences between crosses whereby chromosomes of the non-dominant species are eliminated. Several haploids and polyhaploids obtained after interspecific hybridization have been meiotically analysed to obtain data about pairing regulatory mechanisms and genome homologies. For example, Subrahmanyam (1978) and Bothmer & Subrahmanyam (1988) have presented evidence from which it was concluded that the polypoid species examined were alloploid with some genome homology.

The potential value of interspecific hybridization between *H. vulgare* and *H. bulbosum* in breeding new cultivars faster than conventional methods was predicted by Kasha & Kao (1970). To use this interspecific cross for successful doubled haploid production, the technique has been developed further by making improvements at each stage of the process.

Plant growth conditions

Diploid *H. bulbosum* plants are vernalized at 4°C for 6–8 weeks (8 h daylength) before repotting and transfer to a suitable environment for use as pollinator with *H. vulgare*. The environmental requirements for crossing, although important, are probably not as stringent as those for anther culture, and a heated glasshouse maintained at 14–24°C is satisfactory. Continuous temperatures lower than 18°C during the first 5 days of embryo development can, however, result in higher proportions of hybrids, which retain both sets of parental chromosomes (Pickering, 1985).

To make use of field-grown material for crossing, Kasha *et al.* (1978) recommended placing detached tillers into nutrient solution + GA_3 (gibberellic acid) and maintaining them in a controlled environment. Although inconsistent results are frequently obtained with this technique (Pickering, 1980a), Chen & Hayes (1989) and Hayes & Chen (1989) reported regenerating 30.6 and 13.5 haploid plants per 100 pollinated florets (respectively). Unlike anther culture, where large spike to spike variation in response occurs (Dunwell *et al.*, 1987) no significant differences in success rates were obtained between pollinated spikes from the main shoot and first and second tillers (Pickering, 1983a).

Hybridization

Emasculation of barley florets is carried out conventionally, and freshly collected *H. bulbosum* pollen is applied to receptive stigmas 2–3 days later. Some researchers prefer to use pollen from individual clones of *H. bulbosum* (Guerrero, pers. comm.) whereas others use mixtures of pollen collected from selected clones (Devaux & Jean, 1985; Chen & Hayes, 1989; Jestin, pers. comm.). A possible disadvantage of pollen mixtures is that one of the components may preferentially effect fertilization (Pickering, 1984). It may therefore be more effective to pollinate individually, with the two best *H. bulbosum* clones available, on opposite sides of the same spike, but this is a decision each breeder must make individually.

Seed setting

Seed sets on most barley cultivars pollinated with *H. bulbosum* are over 80%, but after crossing certain cultivars with *H. bulbosum*, reduced seed setting occurs. This was recorded on 'Vada' and found to be controlled by a single dominant gene located on barley chromosome 7 (Pickering, 1983b). By preparing sections of pollinated ovaries of 'Vada' × *H. bulbosum*, it was observed that pollen tubes successfully penetrated the stigma branches, but were subsequently arrested in the stylar transmitting tract as a result of inflation and bursting of tube tips (Pickering, 1981). Similar manifestations of incompatibility have been observed in wheat × rye (e.g. Jalani & Moss, 1980) and wheat × *H. bulbosum* (Snape *et al.*, 1980; Sitch & Snape, 1987), crossabilities of which are governed by two dominant genes on wheat chromosomes 5A and 5B (Riley & Chapman, 1967). As this chromosome is homoeologous with barley chromosome 7 (Islam & Shepherd, 1981) these three incompatibility systems, as well as crossability within the Triticeae generally, may be closely related (Thomas *et al.* 1981; Fedak & Jui, 1982; Pickering, 1984).

Other cultivars possibly unrelated to 'Vada' have also exhibited incompatibility in the form of reduced seed setting, for example 'Firlbeck III' (Pickering, 1980b), 'Agneta' (Bjørnstad, 1986) and 'Apex' (Devaux *et al.*, 1990). Unlike 'Vada', the incompatibility in 'Agneta' is strongly influenced by environment and appears not to be dominantly inherited (Bjørnstad, 1986). Winter cultivars of barley, especially six-rowed types (e.g. 'Vogelsanger Gold', Thörn pers. comm.), tend to be more strongly incompatible with *H. bulbosum* (Simpson *et al.*, 1980; Chen & Hayes, 1989) which partly explains the reduced overall success rates with winter compared with spring barleys (Torp, pers. comm.; Devaux, 1986).

The problem has only been overcome to some extent by screening and selecting amongst different *H. bulbosum* accessions (Pickering, 1980b). Changes in technique and post-pollination sprays of growth hormones such as IAA + GA_3 did not achieve any marked improvements, although an application of 200 g l^{-1} proline at the time of emasculation tended to increase seed setting (Navolotsky, pers. comm.).

Seed quality

The major advance in improving longevity of developing seeds in intergeneric and interspecific hybrids in the Gramineae was the application of GA_3 to florets after fertilization (Larter & Chaubey, 1965). Without this plant growth hormone developing seeds degenerate before embryo rescue. GA_3 (75 mg l^{-1}) is therefore routinely applied as a spray to florets 1–2 days after pollination to enhance seed development and embryo size (Kasha *et al.*, 1978; Adamski, 1979). Parental genotype of the pollinator does not appear greatly to affect seed quality but there are differences between barley cultivars in this respect (Pickering, 1983a; Bjørnstad, 1986). Chen & Hayes (1989) in comparisons with the detached tiller technique of Kasha *et al.* (1978) have obtained improvements in the numbers of green seeds by culturing individual pollinated florets before embryo excision. This had previously been used effectively in wheat × maize crosses to prolong seed development (Laurie & Bennett, 1988). Although the procedure creates an additional step in the scheme, the advantages of greater success rates may offset the extra labour required.

Embryo differentiation

In order to regenerate haploid plants, embryos are cultured at an optimum length of 1–1.5 mm (Adamski, 1979; Pickering, 1980a). This is reached 8.5 to 28 days after pollination at temperatures ranging from 25.5 to 10°C respectively. Apart from small globular embryos obtained from *H. vulgare* × *H. bulbosum* Cb 2951 (selected from P.I.240163 by Pickering), there are three other classes of embryo:

1. Well-differentiated haploid;
2. Undifferentiated, often cuneiform, haploid;
3. Well-differentiated hybrid.

The frequencies of the various types of embryo are influenced by male and female genotype (Jensen, 1976; Pickering, 1983a) and season (Pickering, 1980a; Devaux & Jean, 1985). For example, type 3 embryos (hybrids) are more frequently encountered during winter months at suboptimal temperatures (Pickering, 1985). To improve success rates *H. bulbosum* genotypes have been selected to produce higher proportions of 'type 1' embryos (Pickering & Rennie, 1990).

For embryo culture, the medium most favoured is that based on B5 (Gamborg *et al.*, 1968) modified by omitting 2,4-D and adding sucrose and agar at 20 g l^{-1} and 7 g l^{-1} respectively (Kasha & Kao, 1970). Incubation is generally carried out in darkness at 20–22°C, and once the coleoptile is about 1 cm in length, the culture vials are put under lights (16 h daylength) at a similar temperature. These conditions are not critical, and cultured embryos, plantlets, and even pollinated spikes, can be maintained at 2–4°C to slow down growth and smooth out work schedules.

Plantlets at the two-leaf stage are removed directly to potting compost, and the weaker ones covered with clear plastic beakers to maintain high humidity. Around 50–60% of those embryos classified as type 1 will regenerate viable haploid plantlets after embryo culture compared with only 10–15% of type 2 embryos (Kasha *et al.*, 1978; Pickering, 1980a).

Chromosome doubling with colchicine

Despite reports of mutations having been induced by colchicine treatment in barley (Gilbert & Patterson, 1965) we have not observed their occurrence in our breeding programmes. Finch & Bennett (1979) did, however, record a slight reduction in chiasma frequency in colchicine-treated barley seedlings compared with a 'once-grown' doubled haploid control, but the differences were not considered important. Although Snape *et al.* (1988) detected 'gametoclonal variation' in 'bulbosum-derived' wheat doubled haploids, which may have been attributable to cochicine treatment, no such variation was reported in barley doubled haploids using this technique. Thus, unlike anther culture, gametoclonal variation has not proved to be a problem and its effects can be disregarded in breeding and research. Furthermore, segregation ratios amongst doubled haploid progenies derived from early generation material are generally as expected (for example, Simpson & Snape, 1979; Doll *et al.*, 1989), and therefore random gamete selection takes place for simply inherited characters.

Regarding colchicine treatment, all hybrid plants can usually be identified morphologically and discarded at tillering (Jensen, 1976). Haploid plants are removed from compost when they have produced three or four tillers; they are then washed and their roots trimmed. Incisions can be made at each of the stem bases before immersion to a depth of 5 cm in an aerated 0.05% aqueous solution of colchicine + 2% dimethyl sulphoxide (Jensen, 1976; Thiebaut *et al.*, 1979; Pickering, 1980a). The plants are treated for 5 h at 25–30°C then washed for 5 min in tap water and replanted into compost. Chromosome doubling rates of >80% are commonly achieved, although there are significant differences in the proportions and fertility of doubled haploid plants obtained (Adamski, 1979; Pickering, 1983a; Devaux, 1989a).

Comparisons between anther culture, 'bulbosum' technique and conventional breeding programmes

The 'bulbosum' method has been shown to be more efficient for haploid production than *in vitro* androgenesis (Huang *et al.*, 1984) but in a similar comparison, Friedt *et al.* (1987) reported greater success with anther culture for winter barley hybrids. Using seven F_1 winter barley hybrids, Devaux (1987) found that the response and the time required to produce doubled haploids from the two techniques were similar, but it was proposed to use both systems since a highly significant genotype × technique interaction was obtained. Furthermore, as the two methods demand dissimilar environmental conditions for donor plant growth, they can be carried out at different seasons to obtain maximum success.

The time taken to process a spike is about 30 min from emasculation or spike excision to transfer of green plants to soil with both techniques (Devaux, 1987) and in each case, 7–9 months are needed from the time of sowing donor spring barley plants to harvesting doubled haploid progeny (Choo *et al.*, 1985). On this basis, both methods offer similar returns of doubled haploids for the amount of labour expended (cf. Tables 25.2 and 25.3) although it must be remembered that recent improvements in anther culture success have been obtained with a limited range of hybrids and responsive cultivars (e.g. 'Igri').

A scheme for rapidly obtaining spring barley doubled haploids from an F_1 hybrid is presented in Table 25.4. Only one year is required to enter doubled haploids in field nurseries from the time of sowing F_1 seeds. By using off-season nurseries for multiplication, the subsequent testing process can be accelerated even further. Time-savings are greater when the system is used with winter barley, because every selfed generation in a conventional breeding programme requires vernalization to induce flowering.

A refinement to the breeding programme is the use of recurrent selection whereby superior doubled haploids from a diallel are used as parents in another round of crossing before further doubled haploid production (Choo *et al.*, 1979; Patel *et al.*, 1985). By accumulating desirable genes, gradual improvements in the performance of the progeny should be achieved more efficiently following the use of haploids to obtain increased additive genetic variance (Kasha & Reinbergs, 1981).

Because of the very rapid production of populations of homozygous lines which have similar yield potential, and in some cases means and variances, to those produced conventionally, the 'bulbosum' technique has been compared favourably with single seed descent (SSD) pedigree and bulk breeding methods (Park *et al.*, 1976; Song *et al.*, 1978; Choo *et al.*, 1985). The doubled

Table 25.4. Current scheme for obtaining spring barley doubled haploid lines from two parents A and B.

Year 0	15 Feb.	Sowing of the two parents A & B (glasshouse)
	15 April	Crossing A × B
	5 May	Sowing of F_1 grains AB
	15–20 June	Anther culture and/or crossing F_1 plants with *H. bulbosum*
	20–30 July	Transfer of gametophyte-derived plants into soil (glasshouse)
	10 August	Colchicine treatment of haploid plants
	Oct.–Dec.	Harvest of grains from doubled haploids
Year 1	March	Sowing of doubled haploids in field nurseries, for first multiplication and selection

haploids have also shown similar stability compared with control cultivars over a range of environments (Reinbergs *et al.*, 1978). At Florimond Desprez, comparisons have also been made between breeding methods for several agronomic traits. Grain yields of F_5 pedigree-bred and 'bulbosum'-derived doubled haploids were assessed 4 years after making the initial cross. Mean yield was higher and standard deviation lower for the pedigree lines compared with the doubled haploids (60.13 ± 4.98 dt ha^{-1} and 56.88 ± 6.75 dt ha^{-1} respectively). This probably reflects the lack of selection pressure on the doubled haploids derived from an F_1 hybrid.

Regarding anther culture derived lines, Friedt & Foroughi-Wehr (1983) obtained no differences in mean yield between doubled haploid lines and their mid-parent values, but there was a shift in the frequency distribution towards lower yields in the doubled haploids. In a later study, however, identical means, variances and frequency distributions were obtained for yield from *H. bulbosum*-derived and anther culture-derived doubled haploids (Friedt *et al.*, 1987). In another investigation, mean values of anther culture-derived doubled haploid lines for yield and its components, were inferior to those lines obtained by the 'bulbosum' technique and pedigree and SSD systems (Rossnagel *et al.*, 1987). Selection of superior lines among segregating populations during pedigree breeding would account for their improved attributes as the anther culture-derived lines were obtained directly from F_1s. Despite their poorer overall performance, it was possible to identify lines from anther culture equal in performance to those selected conventionally (Rossnagel *et al.*, 1987). A summary of the main points to consider when comparing the two techniques is presented in Table 25.5, but in the future, greater improvements in success will more likely come from anther or isolated microspore culture than the 'bulbosum' technique. This forecast is based on the significant increases recently obtained in plant numbers regenerated from single anthers by changes in media and cultural conditions.

Optimum F generation for doubled haploid production

Although doubled haploid production from small numbers of F_1 hybrids offers great savings in time over conventional breeding programmes, to increase efficiency and break linkages it may be more effective to produce fewer doubled haploids from many more agronomically selected F_2 or F_3 plants (Simpson & Snape, 1981). One result of linkage disequilibrium present in hybrid material is that differences in means (if epistasis is present) and variances can be detected between doubled haploid populations derived from the F_1 and F_2 generations (Snape & Simpson, 1981). However, except for 'Lester', all of the doubled haploid

Table 25.5. A summary of the differences in requirements and results between the 'bulbosum' technique and anther culture (for spring barley).

Requirements and results	'Bulbosum' technique	Anther culture
No. of donor plants needed	*H. bulbosum* *H. vulgare*	*H. vulgare*
Labour requirements	——— similar ———	
Environmental requirements:		
vernalization chamber	✓	
heated glasshouse	✓	✓
growth room		(✓)
Callus stage		✓
Regenerants:		
albinos	−	+
spontaneously chromosome doubled	−	+
gametoclonal variation	−	+
random gamete selection	✓	Some deviations from expected
Success rates	Becoming comparable, but anther culture more genotype-dependent	

cultivars currrently registered or recommended have been obtained using the 'bulbosum' technique with F_1 hybrids (Table 25.6), or selected F_1s from first backcross ('Pipkin'; Habgood, pers. comm.). This may reflect the more usual practice in breeding programmes of using F_1s as starting material (Foroughi-Wehr *et al.*, 1982; Foroughi-Wehr & Friedt, 1984; Kao, 1988; Chen & Hayes, 1989; Devaux, 1989a; Kuhlmann & Foroughi-Wehr, 1989). However, Luckett (pers. comm.) prefers F_2s for anther culture having obtained a better *in vitro* response compared with F_1s and with the 'bulbosum' technique, selected F_2s and/or F_3s have also been used (Pickering,

Table 25.6. Doubled haploid barley cultivars currently available.

Name	Institution/country	Reference
'Mingo' (S)	Ciba-Geigy, Canada	Ho & Jones, 1980
'Rodeo' (S)	Ciba-Geigy, Canada	Campbell *et al.*, 1984
'Doublet' (S)	WPBS, UK	Jones *et al.*, 1985
'Gwylan' (S)	WPBS, UK*; DSIR, NZ**	Coles, 1986
'Pipkin' (W)	WPBS, UK	Jones *et al.*, 1986
'Waveney' (W)	ICI Seeds, UK	NIAB, 1988
'Craig' (S)	Thompson, Canada	Shugar, 1989
'Winthrop' (S)	Thompson, Canada	Shugar (pers. comm.)
'Michka' (S)	Desprez, France	Devaux, 1989b
'Odessky 115' (S)	Institute of Plant Breeding, Odessa, USSR	Navolotsky (pers. comm.)
'Lester' (S)†	Thompson, Canada	Shugar (pers. comm.)
'Lombard' (S)†	Desprez, France*; Italy**	Devaux, 1989b

S, spring; W, winter cultivar.
* Country where the doubled haploid has been produced.
** Country where the doubled haploid has been selected and cultivated.
† In course of registration.

1980a; Simpson & Snape, 1981). Doubled haploid production from later generation material (e.g. F_4, F_5) is a means to achieve homozygosity to satisfy the requirements of Plant Variety Rights (Griffin, pers. comm.; Jestin, pers. comm.).

Applications of haploids

The use of haploids in breeding new cultivars has been described briefly, and some examples of other applications in breeding techniques will now be outlined.

Identification of superior hybrid combinations

Plant breeders need to predict the potential value of crosses in breeding programmes at the earliest opportunity to discard inferior combinations. As few as 20 doubled haploids produced from early generation hybrids were required for this, after assessing yield in hill plots (Reinbergs *et al.*, 1976). More doubled haploids could then be produced from superior hybrid combinations. However, Bjørnstad & Aastveit (1990) pointed out that larger numbers of doubled haploids would be needed to obtain this result in cases where there are negative pleiotropic effects on the mean of a character.

Linkage analysis

Doubled haploids have been valuable for detecting linkage associated with quantitatively inherited characters, and can also be used for calculating recombination values between linked genes. The production of haploids from a heterozygous F_1 (assuming random gamete selection), reflects the genotype of the egg ('bulbosum' technique) or sperm (anther culture) and is equivalent to backcrossing the F_1 to the double recessive. Any significant deviation from expected ratios is therefore a result of linkage. The method has been successfully used to determine linkage and recombination values between genes on chromosome 7 (Thomas *et al.*, 1984) and chromosome 5 (Shewry *et al.*, 1980). In the latter case, the test involved analysing seed storage proteins so homozygous triploid endosperm tissue present in the doubled haploids was advantageous.

Pleiotropic effects of specified genes

The production of doubled haploids from heterozygous F_1s enables the researcher to study the effects of particular genes on quantitatively inherited characters. The procedure entails separating the doubled haploids into two groups according to which allele is present, and studying the effect on the quantitative characters under investigation. Any significant differences between the means of the two groups can then be considered to be a pleiotropic effect of the major gene. Snape & Simpson (1981) reported a deleterious effect of a dwarfing gene on several characters. Bjørnstad (1987), Bjørnstad & Aastveit (1990) and Kjaer *et al.* (1990) also recorded pleiotropic effects on several agronomic characters of the gene on chromosome 7 controlling grain lysine content and the *mlo* powdery mildew resistance genes on chromosome 4.

Restriction fragment length polymorphisms (RFLPs)

Many genes governing morphological characters and isozymes have been mapped on the barley genome (Tsuchiya, 1986; Søgaard & von Wettstein-Knowles, 1987), and multiple marker stocks have been produced for accurately locating and assigning genes to particular chromosomes (Franckowiak, 1987). The location of polygenes controlling quantitative characters such as yield, has not been easily carried out but could be aided by extracting doubled haploids from crosses between parents with contrasting marker genes (Choo, 1983). There is now, however, increasing interest in constructing a genome map based on molecular markers by RFLP analysis. In the same way that doubled haploids can assist the research into conventional polygene mapping, homozygotes should therefore prove invaluable as an aid in linking polygenes to the molecular markers. Such a programme is being developed in barley in the United States (Hayes, pers. comm.) using the 'bulbosum' technique partly to avoid any variants which may arise with anther culture. The programme is only in its early stages and few data are available as yet.

Genetic engineering and mutations

These topics are discussed elsewhere in this volume, but haploid tissues could be used in biotechnology programmes. For example, after mutagenic treatment, recessive mutations are likely to be expressed in haploid regenerants without being masked by the dominant allele, and homozygotes obtained after chromosome doubling. Gaj (pers. comm.) has produced up to 21.7% mutants in the MII generation after treating haploid embryos (from *H. vulgare* × *H. bulbosum* crosses) with 1 mM *N*-methyl-*N*-nitrosourea for 3 h. As well as chlorophyll deficiencies most of the other mutants which arose were associated with isozyme variations but two semidwarfs were also identified from the same treatment. Immature haploid embryos have also been successfully cultured on media to induce callus formation and then regenerate plants after subculture of the calluses to auxin-free media. Regeneration was reported to be higher than from diploid embryo-derived calluses and the technique was subsequently used to assess the frequency of somaclonal variation, but unfortunately, very few useful variants were obtained (Pickering, 1989).

Calluses or embryoids derived either from haploid embryos or anther/microspore culture may provide a source of tissue for transformation research without the disadvantages associated with diploid cultures. For example, embryoids from anther culture have been microinjected with two plasmid constructs and the progeny screened for the gene encoding neomycine phosphotransferase (Olsen, pers. comm.). Sixteen plants which appeared to have increased resistance to kanamycin are being analysed further. Protoplasts derived from similar sources have also been successfully transformed using polyethylene glycol treatment and the reporter gene encoding chloramphenicol acetyl transferase (Olsen, pers. comm.).

Pathogen and *in vitro* screening

In addition to assisting with linkage studies described earlier, haploid production has been used effectively to regenerate recombinants rapidly from F_1 hybrids derived from crosses between cultivars resistant and susceptible to barley yellow mosaic virus (Foroughi-Wehr & Friedt, 1984). *In vitro* screening using haploid embryos or microspore callus has, however, rarely been reported, although diploid embryos have proved to be suitable explants for selecting biochemical mutants (Bright *et al.*, 1979) and plants with increased tolerance to herbicides (Gifford *et al.*, 1986). Anthers from a cross between salt tolerant and sensitive cultivars have been cultured on liquid media containing various concentrations of Na_2SO_4. All (six) green plants regenerated from the 'high salt' medium were more tolerant than the sensitive parent in subsequent germination tests (Ye *et al.*, 1987).

Regarding plant pathogens, the results of direct inoculation of powdery mildew (*Erysiphe graminis* f. sp. *hordei*) on to microspore calluses of barley have been unsuccessful (Franzone *et al.*, 1982). However, as it is known that certain plant pathogens produce phytotoxins which interfere with the host metabolism and are capable of reproducing some or all of the disease symptoms associated with the pathogen [e.g. *Drechslera* spp. (Shotwell & Ellis, 1976); *Rhynchosporium secalis* (Auriol *et al.*, 1978)] a more successful approach has been the incorporation of the toxin in tissue culture media before *in vitro* selection. Regeneration of green plants tolerant to *Drechslera* spp. has been carried out in maize (Gengenbach *et al.*, 1977), barley and wheat (Chawla & Wenzel, 1987), but the technique was not successful in identifying any resistance to rhynchosporoside, a toxin produced by *R. secalis* (Branchard, 1982).

Although there is only a limited number of barley pathogens for which these screening techniques would be suitable for haploid explants, the savings in labour, materials and space would be considerable.

Conclusions

In this chapter we have outlined the history of doubled haploid barley production over a 20-year period. The techniques of anther culture and interspecific hybridization are now used extensively for cultivar production, genetic analysis and molecular biological research. Improvements in success rates should continue to be achieved in the 1990s and provide us with a greater understanding of the mechanisms involved in albinism, chromosome elimination and incompatibility systems. Furthermore, the regeneration of green barley plants from protoplasts (e.g. Jähne *et al.*, 1990) makes it clear that haploid cells and explants will increasingly be sought after for genetic manipulation experiments such as biolistics and microinjection. We can therefore look forward to an exciting decade in which haploid barley will play a key role in maintaining the momentum of agricultural research and development.

Acknowledgements

We wish to thank the many research workers who supplied us with data for use in this chapter. We are also grateful to Dr Eva Thörn (Svalöf AB) for her helpful comments on the text.

References

Adamski, T. (1979) The obtaining of auto-diploid barley lines using haploids from the cross *Hordeum vulgare* L. × *Hordeum bulbosum* L. *Genetica Polonica* 20, 31–42.

Auriol, P., Strobel, G., Beltran, J.P. & Gray, G. (1978) Rhynchosporoside, a host-selective toxin produced by *Rhynchosporium secalis*, the causal agent of scald disease of barley. *Proceedings of the National Academy of Sciences (USA)* 75, 4339–43.

Bjørnstad, Å. (1986) Partial incompatibility between Scandinavian six-rowed barleys (*Hordeum vulgare* L.) and *Hordeum bulbosum* L., and its genetical basis. *Hereditas* 104, 171–91.

Bjørnstad, Å. (1987) The role of linkage and pleiotropy in doubled haploid lines from a barley cross. *Hereditas* 106, 155–68.

Bjørnstad, Å. & Aastveit, K. (1990) Pleiotropic effects of the *ml-o* mildew resistance gene in barley in different genetical backgrounds. *Euphytica* 46, 217–26.

Bothmer, R. von & Subrahmanyam, N.C. (1988) Assessment of chromosome associations in haploids and their parental accessions in *Hordeum*. *Genome* 30, 204–10.

Bothmer, R. von, Flink, J., Jacobsen, N., Kotimäki, M. & Landström, T. (1983) Interspecific hybridization with cultivated barley (*Hordeum vulgare* L.). *Hereditas* 99, 219–44.

Bothmer, R. von, Jacobsen, N., Jørgensen, R.B. & Linde-Laursen, I. (1984) Haploid barley from the intergeneric cross *Hordeum vulgare* × *Psathyrostachys fragilis*. *Euphytica* 33, 363–7.

Branchard, M. (1982) *In vitro* culture of barley: a method to study *Rhynchosporium* scald disease and select plants resistant to the toxin, Rhynchosporoside. In: Earle, E.D. & Demarly, Y. (eds), *Variability in Plants Regenerated from Tissue Culture*. Praeger, New York, USA, pp. 343–50.

Bright, S.W.J., Norbury, P.B. & Miflin, B.J. (1979) Isolation of a recessive barley mutant resistant to *S*-(2-aminoethyl) L-cysteine. *Theoretical and Applied Genetics* 55, 1–4.

Brown, S. (1989) Sondes fluorescentes, analyse et tri de cellules par cytofluorométrie. *Le Selectionneur Français* 40, 21–30.

Campbell, K.W., Brown, R.I. & Ho, K.M. (1984) Rodeo barley. *Canadian Journal of Plant Science* 64, 203–5.

Chawla, H.S. & Wenzel, G. (1987) *In vitro* selection of barley and wheat for resistance against *Helminthosporium sativum*. *Theoretical and Applied Genetics* 74, 841–5.

Chen, F.Q. & Hayes, P.M. (1989) A comparison of *Hordeum bulbosum*-mediated haploid production efficiency in barley using *in vitro* floret and tiller culture. *Theoretical and Applied Genetics* 77, 701–4.

Chen, C.C., Kasha, K.J. & Marsolais, A. (1984) Segmentation patterns and mechanisms of genome multiplication in cultured microspores of barley. *Canadian Journal of Genetics and Cytology* 26, 475–83.

Cho, U.H. & Kasha, K.J. (1989) Ethylene production and embryogenesis from anther cultures of barley (*Hordeum vulgare*). *Plant Cell Reports* 8, 415–17.

Choo, T.M. (1983) Doubled haploids for locating polygenes. *Canadian Journal of Genetics and Cytology* 25, 425–9.

Choo, T.M., Christie, B.R. & Reinbergs, E. (1979) Doubled haploids for estimating genetic variances and a scheme for population improvement in self-pollinating crops. *Theoretical and Applied Genetics* 54, 267–71.

Choo, T.M., Reinbergs, E. & Kasha, K.J. (1985) Use of haploids in breeding barley. *Plant Breeding Reviews* 3, 219–52.

Chu, C.C. (1978) The N_6 medium and its applications to anther culture of cereal crops. In: *Proceedings of Symposium on Plant Tissue Culture*. Science Press, Beijing, China, pp. 43–50.

Chuang, C.C., Ouyang, T.W., Chia, H., Chou, S.M. & Ching, C.K. (1978) A set of potato media for wheat anther culture. In: *Proceedings of Symposium of Plant Tissue Culture*. Science Press, Beijing, China, pp. 51–6.

Clapham, D. (1971) *In vitro* development of callus from the pollen of *Lolium* and *Hordeum*. *Zeitschrift für Pflanzenzüchtung* 65, 285–92.

Clapham, D. (1973) Haploid *Hordeum* plants from anthers *in vitro*. *Zeitschrift für Pflanzenzüchtung* 69, 142–55.

Coles, G.D. (1986) 'Gwylan' barley (*Hordeum vulgare* L.). *New Zealand Journal of Experimental Agriculture* 14, 101–3.

Dale, P.J. (1975) Pollen dimorphism and anther culture in barley. *Planta* 127, 213–20.

Datta, S.K. (1987) Plant regeneration by pollen embryogenesis from cultured whole spikes of barley (*Hordeum vulgare*). *Theoretical and Applied Genetics* 74, 121–4.

Datta, S.K. & Wenzel, G. (1988) Single microspore derived embryogenesis and plant formation in barley (*Hordeum vulgare* L.). *Archiv für Züchtungsforschung, Berlin* 18, 125–31.

Davies, D.R. & Hopwood, D.A. (eds) (1980) *Proceedings of the Fourth John Innes Symposium the Plant Genome and Second International Haploid Conference*. The John Innes Charity, Norwich, England.

Day, A. & Ellis, T.H.N. (1985) Deleted forms of plastid DNA in albino plants from cereal anther culture. *Current Genetics* 9, 671–8.

Devaux, P. (1986) Yield of haploid production through the bulbosum method in a winter barley breeding programme. *Cereal Research Communications* 14, 273–9.

Devaux, P. (1987) Comparison of anther culture and *Hordeum bulbosum* method for the production of doubled haploids in winter barley. I. Production of green plants. *Plant Breeding* 98, 215–19.

Devaux, P. (1988) Comparison of anther culture and *Hordeum bulbosum* method for the production of doubled haploids in winter barley. II. Variation of chromosome number and seed set in the progeny. *Plant Breeding* 100, 181–7.

Devaux, P. (1989a) Variations in the proportions of fertile colchicine-treated haploid plants derived from winter barley hybrids. *Plant Breeding* 103, 247–50.

Devaux, P. (1989b) Haplodiploidisation chez l'orge. In: *Biotechnology for Major Crops*. Salon International des Techniques et Energies du Futur, Toulouse, France, pp. 12–18.

Devaux, P. (1990) Individual and combined effects of glutamine and maltose on anther culture response of the winter barley cultivar 'Igri'. *Bulletin de la Société Botanique de France, Actualitiés Botaniques* 137, 142–3.

Devaux, P. & Jean, R. (1985) Influence du mois et de la technique de castration dans la réussite de l'haploid diploidisation de l'orge par la méthode *bulbosum*. *Agronomie* 5, 795–800.

Devaux, P., Adamski, T. & Surma, M. (1990) Studies on low crossabilities encountered with the *Hordeum bulbosum* method for haploid production of barley, *Hordeum vulgare* L. *Plant Breeding* 104, 305–11.

Doll, H., Haahr, V. & Søgaard, B. (1989) Relationship between vernalization requirement and winter hardiness in doubled haploids of barley. *Euphytica* 42, 209–13.

Dunwell, J.M. (1985) Anther and ovary culture. In: Bright, S.W.J. & Jones, M.J.K. (eds), *Cereal Tissue and Cell Culture*. Martinus Nijhoff/Dr W. Junk, Dordrecht, The Netherlands, pp. 1–44.

Dunwell, J.M., Francis, R.J. & Powell, W. (1987) Anther culture of *Hordeum vulgare* L.: a genetic study of microspore callus production and differentiation. *Theoretical and Applied Genetics* 74, 60–4.

Fedak, G. (1977) Haploids from barley × rye

crosses. *Canadian Journal of Genetics and Cytology* 19, 15–19.

Fedak, G. & Jui, P.Y. (1982) Chromosomes of Chinese Spring wheat carrying genes for crossability with Betzes barley. *Canadian Journal of Genetics and Cytology* 24, 227–33.

Finch, R.A. & Bennett, M.D. (1979) Meiotic stability in control and newly colchicine-induced dihaploid barley. *Canadian Journal of Genetics and Cytology* 21, 33–5.

Finnie, S.J., Powell, W. & Dyer, A.F. (1989) The effect of carbohydrate composition and concentration on anther culture response in barley (*Hordeum vulgare* L.). *Plant Breeding* 103, 110–18.

Foroughi-Wehr, B. & Friedt, W. (1984) Rapid production of recombinant barley yellow mosaic virus resistant *Hordeum vulgare* lines by anther culture. *Theoretical and Applied Genetics* 67, 377–82.

Foroughi-Wehr, B. & Mix, G. (1979) *In vitro* response of *Hordeum vulgare* L. anthers cultured from plants grown under different environments. *Environmental and Experimental Botany* 19, 303–9.

Foroughi-Wehr, B., Mix, G., Gaul, H. & Wilson, H.M. (1976) Plant production from cultured anthers of *Hordeum vulgare* L. *Zeitschrift für Pflanzenzüchtung* 77, 198–204.

Foroughi-Wehr, B., Friedt, W. & Wenzel, G. (1982) On the genetic improvement of androgenetic haploid formation in *Hordeum vulgare* L. *Theoretical and Applied Genetics* 62, 233–9.

Franckowiak, J.D. (1987) Multiple recessive marker stocks for mapping male-sterile genes in barley. In: Yasuda, S. & Konishi, T. (eds), *Barley Genetics V* (Proceedings of the 5th International Barley Genetics Symposium). Maruzen, Okayama, Japan, pp. 209–12.

Franzone, P.J., Foroughi-Wehr, B., Fischbeck, G. & Friedt, W. (1982) Reaction of microspore callus, androgenetic albino plantlets and roots of barley to *Erysiphe graminis* f.sp. *hordei*. *Phytopathologische Zeitschrift* 105, 170–4.

Friedt, W. & Foroughi-Wehr, B. (1981) Anther culture of barley (*Hordeum vulgare* L.): plant regeneration and agronomic performance of homozygous diploid progenies. In: Asher, M.J.C., Ellis, R.P., Hayter, A.M. & Whitehouse, R.N.H. (eds), *Barley Genetics IV* (Proceedings of the 4th International Barley Genetics Symposium). Edinburgh University Press, Edinburgh, UK, pp. 690–8.

Friedt, W. & Foroughi-Wehr, B. (1983) Field performance of androgenetic doubled haploid spring barley from F_1 hybrids. *Zeitschrift für Pflanzenzüchtung* 90, 177–84.

Friedt, W., Breun, J., Züchner, S. & Foroughi-Wehr, B. (1986) Comparative value of androgenetic doubled haploid and conventionally selected spring barley lines. *Plant Breeding* 97, 56–63.

Friedt, W., Foroughi-Wehr, B. & Snape, J.W. (1987) The significance of biotechnology for the evolution of barley breeding methods. In: Yasuda, S. & Konishi, T. (eds), *Barley Genetics V* (Proceedings of the 5th International Barley Genetics Symposium). Maruzen, Okayama, Japan, pp. 903–13.

Gamborg, O.L., Miller, R.A. & Ojima, K. (1968) Nutrient requirements of suspension cultures of soybean root cells. *Experimental Cell Research* 50, 151–8.

Gaul, H., Mix, G., Foroughi-Wehr, B. & Okamoto, M. (1976) Pollen grain development of *Hordeum vulgare*. *Zeitschrift für Pflanzenzüchtung* 76, 77–80.

Gengenbach, B.G., Green, C.E. & Donovan, C.M. (1977) Inheritance of selected pathotoxin resistance in maize plants regenerated from cell cultures. *Proceedings of the National Academy of Sciences (USA)* 74, 5113–17.

Gifford, S.C., Collin, H.A. & Putwain, P.D. (1986) Enhancement of asulam resistance in barley. *Weed Research* 26, 389–96.

Gilbert, S.K. & Patterson, F.L. (1965) Colchicine-induced mutants in Decatur barley. *Crop Science* 5, 44–7.

Grunewaldt, J. & Malepszy, S. (1975) Observations on anther callus from *Hordeum vulgare* L. In: Gaul, H. (ed.), *Barley Genetics III* (Proceedings of the 3rd International Barley Genetics Symposium). Verlag Karl Thiemig, Munich, Germany, pp. 367–73.

Hagberg, A. & Hagberg, G. (1980) High frequency of spontaneous haploids in the progeny of an induced mutation in barley. *Hereditas* 93, 341–3.

Hagberg, A. & Hagberg, G. (1987) Production of spontaneously doubled haploids in barley using a breeding system with marker genes and the 'hap'-gene. *Biologisches Zentralblatt* 106, 53–8.

Hayes, P.M. & Chen, F.Q. (1989) Genotypic variation for *Hordeum bulbosum* L.-mediated haploid production in winter and facultative barley. *Crop Science* 29, 1184–8.

Ho, K.M. & Jones, G.E. (1980) Mingo barley. *Canadian Journal of Plant Science* 60, 279–80.

Ho, K.M. & Kasha, K.J. (1975) Genetic control of chromosome elimination during haploid formation in barley. *Genetics* 81, 263–75.

Hou, L. & Ullrich, S.E. (1990) Effect of genotype and donor plant environment on anther culture of barley. *14th American Barley Research Workshop*, 13–17 May 1990, Sacramento, USA (abstract).

Huang, B. & Sunderland, N. (1982) Temperature-stress pretreatment in barley anther culture. *Annals of Botany* 49, 77–88.

Huang, B., Dunwell, J.M., Powell, W., Hayter, A.M. & Wood, W. (1984) The relative efficiency of microspore culture and chromosome elimination as methods of haploid production in *Hordeum vulgare* L. *Zeitschrift für Pflanzenzüchtung* 92, 22–9.

Hunter, C.P. (1985) The effect of anther orientation on the production of microspore-derived embryoids and plants of *Hordeum vulgare* cv. Sabarlis. *Plant Cell Reports* 4, 267–8.

Hunter, C.P. (1987) Plant generation method. *European Patent Application*, No. 87200773.7.

Hunter, C.P., Loose, R.W., Clerk, S.P. & Lyne, R.L. (1989) Maltose – the preferred carbon source for barley anther culture. In: *International Congress by EUCARPIA, Genetic Manipulation in Plant Breeding – Biotechnology for the Breeder*.

Islam, A.K.M.R. & Shepherd, K.W. (1981) Wheat-barley addition lines: their use in genetic and evolutionary studies of barley. In: Asher, M.J.C., Ellis, R.P., Hayter, A.M. & Whitehouse, R.N.H. (eds), *Barley Genetics IV* (Proceedings of the 4th Barley Genetics Symposium). Edinburgh University Press, Edinburgh, UK, pp. 729–39.

Jähne, A., Lazzeri, P.A., Jäger-Gussen, M. & Lörz, H. (1990) Anther-culture-derived suspension cultures of barley (*Hordeum vulgare*); establishment, plant regeneration and protoplast culture. In: *VIIth International Congress on Plant Tissue and Cell Culture*, Amsterdam 24–29 June 1990 (abstract).

Jalani, B.S. & Moss, J.P. (1980) The site of action of the crossability genes (Kr_1,Kr_2) between *Triticum* and *Secale*. I. Pollen germination, pollen tube growth, and number of pollen tubes. *Euphytica* 29, 571–9.

Jensen, C.J. (1976) Barley monoploids and doubled monoploids: techniques and experience. In: Gaul, H. (ed.), *Barley Genetics III* (Proceedings of the 3rd International Barley Genetics Symposium). Verlag Karl Thiemig, Munich, Germany, pp. 316–45.

Jensen, C.J. (1981) Regenerating monoploid barley plants from cells, tissues and organs from various origins. In: Asher, M.J.C., Ellis, R.P., Hayter, A.M. & Whitehouse, R.N.H. (eds), *Barley Genetics IV* (Proceedings of the 4th International Barley Genetics Symposium). Edinburgh University Press, Edinburgh, UK, pp. 677–85.

Jing, J. & Hu, H. (1981) Anther and immature inflorescence culture in *Hordeum vulgare*. In: Asher, M.J.C., Ellis, R.P., Hayter, A.M. & Whitehouse, R.N.H. (eds), *Barley Genetics IV* (Proceedings of the 4th International Barley Genetics Symposium). Edinburgh University Press, Edinburgh, UK, pp. 945–8.

Jones, D.L., Jones, D.M., Clifford, B.C., Clothier, R.B., Cook, R., Griffiths, T.E.R., Jones, E.W.C. & Jones, I.T. (1985) Cereal breeding, barley. In: *Annual Report of the Welsh Plant Breeding Station for 1984*, pp. 65–6.

Jones, D.M., Clothier, R.B. & Griffiths, T.E.R. (1986) Cereal breeding, barley. In: *Annual Report of the Welsh Plant Breeding Station for 1985*, pp. 74–5.

Kao, K.N. (1981) Plant formation from barley anther cultures with Ficoll media. *Zeitschrift für Pflanzenphysiologie* 103, 437–43.

Kao, K.N. (1988) *In vitro* culture in barley breeding. In: *Semi-Dwarf Cereal Mutants and their Use in Cross-Breeding III* (Proceedings of the Final Research Coordination Meeting on Evaluation of Semi-Dwarf Cereal Mutants for Cross-Breeding). International Atomic Energy Agency, Vienna, Austria, pp. 183–91.

Kao, K.N. & Horn, D.C. (1982) A method for induction of pollen plants in barley. In: Fujiwara, A. (ed.), *Plant Tissue Culture* (Proceedings of the 5th International Congress of Plant Tissue and Cell Culture). Japanese Association of Plant Tissue Culture, Tokyo, Japan, pp. 529–30.

Kao, K.N. & Kasha, K.J. (1969) Haploidy from interspecific crosses with tetraploid barley. In: Nilan, R.A. (ed.), *Barley Genetics II* (Proceedings of the 2nd International Barley Genetics Symposium). Washington State University Press, Pullman, USA, pp. 82–8.

Kasha, K.J. (ed.) (1974) *Haploids in Higher Plants: Advances and Potential* (Proceedings of the 1st International Symposium). University of Guelph, Guelph, Canada.

Kasha, K.J. (1989) Production of haploids in cereals. In: Maluszynski, M. (ed.), *Current Options for Cereal Improvement: Doubled Haploids, Mutants and Heterosis*. (Proceedings of the First FAO/IAEA Research Co-ordination Meeting on 'Use of Induced Mutations in Connection with Haploids and Heterosis in Cereals'. Kluwer, Dordrecht, The Netherlands, pp. 71–80.

Kasha, K.J. & Kao, K.N. (1970) High frequency haploid production in barley (*Hordeum vulgare* L.). *Nature* 225, 874–6.

Kasha, K.J. & Reinbergs, E. (1981) Recent developments in the production and utilization of haploids in barley. In: Asher, M.J.C., Ellis, R.P., Hayter, A.M. & Whitehouse, R.N.H. (eds), *Barley Genetics IV* (Proceedings of the 4th International Barley Genetics Symposium). Edinburgh University Press, Edinburgh, UK, pp. 655–65.

Kasha, K.J., Subrahmanyam, N.C. & Ali, A. (1978) Effect of gibberellic acid treatment, and nutrient supply through detached tillers, upon haploid frequency in barley. *Theoretical and Applied Genetics* 51, 169–75.

Kasha, K.J., Ziauddin, A. & Cho, U.H. (1989) Haploids in cereal improvement: anther and microspore culture. In: *XIXth Stadler Genetics Symposium*, pp. 213–36.

Kjaer, B., Jensen, H.P., Jensen, J. & Jørgensen, J.H. (1990) Associations between three *ml-o* powdery mildew resistance genes and agronomic traits in barley. *Euphytica* 46, 185–93.

Knudsen, S., Due, I.K. & Andersen, S.B. (1989) Components of response in barley anther culture. *Plant Breeding* 103, 241–6.

Kohlenbach, H.W. & Wernicke, W. (1978) Investigations on the inhibitory effect of agar and the function of active carbon in anther culture. *Zeitschrift für Pflanzenphysiologie* 86, 463–72.

Köhler, F. & Wenzel, G. (1985) Regeneration of isolated barley microspores in conditioned media and trials to characterize the responsible factor. *Journal of Plant Physiology* 121, 181–91.

Konzak, C.F., Randolph, L.F. & Jensen, N.F. (1951) Embryo culture of barley species hybrids. Cytological studies of *Hordeum sativum* × *Hordeum bulbosum*. *Journal of Heredity* 42, 124–34.

Kuckuck, H. (1934) Artkreuzungen bei Gerste. *Der Züchter* 6, 270–3.

Kuhlmann, U. & Foroughi-Wehr, B. (1989) Production of doubled haploid lines in frequencies sufficient for barley breeding programs. *Plant Cell Reports* 8, 78–81.

Larter, E. & Chaubey, C. (1965) Use of exogenous growth substances in promoting pollen tube growth and fertilization in barley–rye crosses. *Canadian Journal of Genetics and Cytology* 7, 511–18.

Laurie, D.A. & Bennett, M.D. (1988) The production of haploid wheat plants from wheat × maize crosses. *Theoretical and Applied Genetics* 76, 393–7.

Lee, F.M. & Chen, C.C. (1987) Nuclear fusion in cultured microspores of barley. *Plant Cell Reports* 6, 191–3.

Linsmaier, E.M. & Skoog, F. (1965) Organic growth factor requirements of tobacco tissue cultures. *Physiologia Plantarum* 18, 100–27.

Lyne, R.L., Bennett, R.I. & Hunter, C.P. (1986) Embryoid and plant production from cultured barley anthers. In: Withers, L.A. & Alderson, P.G. (eds), *Plant Tissue Culture and its Agricultural Applications, University of Nottingham*. Butterworths, Guildford, UK, pp. 405–11.

Malepszy, S. & Grunewaldt, J. (1974) Ein Beitrag zur Erzeugung von Haploiden bei *Hordeum vulgare* L. *Zeitschrift für Pflanzenzüchtung* 72, 206–11.

Maluszynski, M. (ed.) (1989) *Current Options for Cereal Improvement: Doubled Haploids, Mutants and Heterosis* (Proceedings of the First FAO/IAEA Research Co-ordination Meeting on 'Use of Induced Mutations in Connection with Haploids and Heterosis in Cereals'). Kluwer, Dordrecht, The Netherlands.

Marsolais, A.A. & Kasha, K.J. (1985) Callus induction from barley microspores. The role of sucrose and auxin in a barley anther culture medium. *Canadian Journal of Botany* 63, 2209–12.

Miller, C.O. (1963) Kinetin and kinetin-like compounds. In: Linskens, H.F. & Tracey, M.V. (eds), *Moderne Methoden in der Pflanzenanalyse*, Vol. VI. Springer, Berlin, Germany, pp. 194–202.

Mix, G., Wilson, H.M. & Foroughi-Wehr, B. (1978) The cytological status of plants of *Hordeum vulgare* L. regenerated from microspore callus. *Zeitschrift für Pflanzenzüchtung* 80, 89–99.

Mogensen, H.L. (1982) Double fertilization in barley and the cytological explanation for haploid embryo formation, embryoless caryopses, and ovule abortion. *Carlsberg Research Communications* 47, 313–54.

Morrison, J.W., Hannah, A.E., Loiselle, R. & Symko, S. (1959) Cytogenetic studies in genus *Hordeum* II. Interspecific and intergeneric crosses. *Canadian Journal of Plant Science* 39, 375–83.

Murashige, T. & Skoog, F. (1962) A revised medium for rapid growth and bio-assays with tobacco tissue cultures. *Physiologia Plantarum* 15, 473–97.

NIAB (1988) *Recommended Varieties of Cereals, Farmers Leaflet Number 8*. National Institute of Agricultural Botany, Cambridge, UK.

Nitsch, C. (1974) Pollen culture – a new technique for mass production of haploid and homozygous plants. In: Kasha, K.J. (ed.), *Haploids in Higher Plants: Advances and Potential* (Proceedings of the 1st International Symposium). University of Guelph, Guelph, Canada, pp. 123–35.

Olsen, F.L. (1987) Induction of microspore embryogenesis in cultured anthers of *Hordeum vulgare*. The effects of ammonium nitrate, glutamine and asparagine as nitrogen sources. *Carlsberg Research Communications* 52, 393–404.

Park, S.J., Walsh, E.J., Reinbergs, E., Song, L.S.P. & Kasha, K.J. (1976) Field performance of doubled haploid barley lines in comparison with lines developed by the pedigree and single seed descent methods. *Canadian Journal of Plant Science* 56, 467–74.

Patel, J.D., Reinbergs, E. & Fejer, S.O. (1985) Recurrent selection in doubled-haploid populations of barley (*Hordeum vulgare* L.). *Canadian Journal of Genetics and Cytology* 27, 172–7.

Pickering, R.A. (1980a) Use of the doubled haploid technique in barley breeding at the Welsh Plant Breeding Station. In: *Annual Report of the Welsh Plant Breeding Station for 1979*, pp. 208–26.

Pickering, R.A. (1980b) Attempts to overcome partial incompatibility between *Hordeum vulgare* L. and *H. bulbosum* L. *Euphytica* 29, 369–77.

Pickering, R.A. (1981) Pollen tube–stylodium interaction in *Hordeum vulgare* L. × *H. bulbosum* L. In: Asher, M.J.C., Ellis, R.P., Hayter, A.M. & Whitehouse, R.N.H. (eds), *Barley Genetics IV* (Proceedings of the 4th International Barley Genetics Symposium). Edinburgh University Press, Edinburgh, UK, pp. 666–76.

Pickering, R.A. (1983a) The influence of genotype on doubled haploid barley production. *Euphytica* 32, 863–76.

Pickering, R.A. (1983b) The location of a gene for incompatibility between *Hordeum vulgare* L. and *H. bulbosum* L. *Heredity* 51, 455–9.

Pickering, R.A. (1984) Crossability relationships between certain species in the Hordeae. *Barley Genetics Newsletter* 14, 14–17.

Pickering, R.A. (1985) Partial control of chromosome elimination by temperature in immature embryos of *Hordeum vulgare* L. × *H. bulbosum* L. *Euphytica* 34, 869–74.

Pickering, R.A. (1989) Plant regeneration and variants from calli derived from immature embryos of diploid barley (*Hordeum vulgare* L.) and *H. vulgare* L. × *H. bulbosum* L. crosses. *Theoretical and Applied Genetics* 78, 105–12.

Pickering, R.A. & Rennie, W.F. (1990) The evaluation of superior *Hordeum bulbosum* L. genotypes for use in a doubled haploid barley breeding programme. *Euphytica* 45, 251–5.

Powell, W. (1988) Diallel analysis of barley anther culture response. *Genome* 30, 152–7.

Powell, W., Hayter, A.M., Wood, W., Dunwell, J.M. & Huang, B. (1984) Variation in the agronomic characters of microspore-derived plants of *Hordeum vulgare* cv. Sabarlis. *Heredity* 52, 19–23.

Powell, W., Borrino, E.M., Allison, M.J., Griffiths, D.W., Asher, M.J.C. & Dunwell, J.M. (1986) Genetical analysis of microspore derived plants of barley (*Hordeum vulgare*). *Theoretical and Applied Genetics* 72, 619–26.

Powell, W., Borrino, E.M. & Goodall, V. (1988) The effect of anther orientation on microspore-derived plant production in barley (*Hordeum vulgare* L.). *Euphytica* 38, 159–63.

Reinbergs, E., Park, S.J. & Song, L.S.P. (1976) Early identification of superior barley crosses by the doubled haploid technique. *Zeitschrift für Pflanzenzüchtung* 76, 215–24.

Reinbergs, E., Song, L.S.P., Choo, T.M. & Kasha, K.J. (1978) Yield stability of doubled haploid lines of barley. *Canadian Journal of Plant Science* 58, 929–33.

Riley, R. & Chapman, V. (1967) The inheritance in wheat of crossability with rye. *Genetical Research* 9, 259–67.

Roberts-Oehlschlager, S.L. & Dunwell, J.M. (1990) Barley anther culture: pretreatment on mannitol stimulates production of microspore derived embryos. *Plant Cell, Tissue and Organ Culture*, 20, 235–40.

Roberts-Oehlschlager, S.L., Dunwell, J.W. & Foulks, R. (1990) Changes in the sugar content of barley anthers during culture on different carbohydrates. *Plant Cell, Tissue and Organ Culture* 22, 77–85.

Rossnagel, B.G., Sariah, M.A. & Kao, K.N. (1987) Field evaluation of anther culture derived breeding materials compared to materials developed by the pedigree and single seed descent methods in barley (*Hordeum vulgare* L.). In: Yasuda, S. & Konishi, T. (eds), *Barley Genetics V* (Proceedings of the 5th International Barley Genetics Symposium). Maruzen, Okayama, Japan, pp. 997–1004.

Saalbach, G. & Koblitz, H. (1977) Karyological instabilities in callus cultures from haploid barley plants. *Biochemie und Physiologie der Pflanzen* 171, 469–73.

San Noeum, L.H. (1976) Haploïdes d'*Hordeum vulgare* L. par culture *in vitro* d'ovaires non fécondés. *Annales de l'Amélioration des Plantes* 26, 751–4.

San Noeum, L.H. & Ahmadi, N. (1982) Variability of doubled haploids from *in vitro* androgenesis and gynogenesis in *Hordeum vulgare* L. In: Earle, E.D. & Demarly, Y. (eds), *Variability in Plants Regenerated from Tissue Culture*. Praeger, New York, USA, pp. 273–83.

Shannon, P.R.M., Nicholson, A.E., Dunwell, J.M. & Davies, D.R. (1985) Effect of anther orientation on microspore-callus production in barley (*Hordeum vulgare* L.). *Plant Cell, Tissue and Organ Culture* 4, 271–80.

Shewry, P.R., Faulks, A.J., Pickering, R.A., Jones, I.T., Finch, R.A. & Miflin, B.J. (1980) The genetic analysis of barley storage proteins. *Heredity* 44, 383–9.

Shotwell, O.L. & Ellis, J.J. (1976) *Helminthosporium, Drechslera*, and *Bipolaris* toxins. *Advances in Chemistry* 149, 318–43.

Shugar, L.P. (1989) Commercialization of the bulbosum method in barley (*Hordeum vulgare* L.). In: Maluszynski, M. (ed.), *Current Options for Cereal Improvement: Doubled Haploids, Mutants and Heterosis* (Proceedings of the 1st FAO/IAEA Research Co-ordination Meeting on 'Use of Induced Mutations in Connection with Haploids and Heterosis in Cereals'). Kluwer, Dordrecht, The Netherlands, pp. 161–70.

Simpson, E. & Snape, J.W. (1979) Differences between erect and prostrate genotypes of spring barley. *Barley Genetics Newsletter* 9, 97–8.

Simpson, E. & Snape, J.W. (1981) The use of doubled haploids in a winter barley programme. In: Asher, M.J.C., Ellis, R.P., Hayter, A.M. & Whitehouse, R.N.H. (eds), *Barley Genetics IV* (Proceedings of the 4th International Barley Genetics Symposium). Edinburgh University Press, Edinburgh, UK, pp. 716–20.

Simpson, E., Snape, J.W. & Finch, R.A. (1980) Variation between *Hordeum bulbosum* genotypes in their ability to produce haploids of barley, *Hordeum vulgare*. *Zeitschrift für Pflanzenzüchtung* 85, 205–11.

Sitch, L.A. & Snape, J.W. (1987) Factors affecting haploid production in wheat using the *Hordeum bulbosum* system. I. Genotypic and environmental effects on pollen grain germination, pollen tube growth and the frequency of fertilization. *Euphytica* 35, 483–96.

Snape, J.W. & Simpson, E. (1981) Uses of doubled haploid lines for genetical analysis in barley. In: Asher, M.J.C., Ellis, R.P., Hayter, A.M. & Whitehouse, R.N.H. (eds), *Barley Genetics IV* (Proceedings of the 4th International Barley Genetics Symposium). Edinburgh University Press, Edinburgh, UK, pp. 704–9.

Snape, J.W., Bennett, M.D. & Simpson, E. (1980) Post-pollination events in crosses of hexaploid wheat with tetraploid *Hordeum bulbosum*. *Zeitschrift für Pflanzenzüchtung* 85, 200–4.

Snape, J.W., Sitch, L.A., Simpson, E. & Parker, B.B. (1988) Tests for the presence of gametoclonal variation in barley and wheat doubled haploids produced using the *Hordeum bulbosum* system. *Theoretical and Applied Genetics* 75, 509–13.

Song, L.S.P., Park, S.J., Reinbergs, E., Choo, T.M. & Kasha, K.J. (1978) Doubled haploid vs the bulk plot method for production of homozygous lines in barley. *Zeitschrift für Pflanzenzüchtung* 81, 271–80.

Sorvari, S. (1986) The effect of starch gelatinized nutrient media in barley anther cultures. *Annales Agriculturae Fenniae* 25, 127–33.

Sorvari, S. & Schieder, O. (1987) Influence of sucrose and melibiose on barley anther cultures in starch media. *Plant Breeding* 99, 164–71.

Steinbiss, H.H., Stabel, P., Töpfer, R., Hirtz, R.D. & Schell, J. (1985) Transformation of plant cells by microinjection of DNA. In: Chapman, G.P., Mantell, S.H. & Daniels, R.W. (eds), *The Experimental Manipulation of Ovule Tissues*. Longman, New York, USA, pp. 64–75.

Subrahmanyam, N.C. (1978) Meiosis in polyhaploid *Hordeum*: hemizygous ineffective control of diploid-like behaviour in a hexaploid? *Chromosoma* 66, 185–92.

Subrahmanyam, N.C. (1980) Haploidy from *Hordeum* interspecific crosses. 3. Trihaploids of *H. arizonicum* and *H. lechleri*. *Theoretical and Applied Genetics* 56, 257–63.

Subrahmanyam, N.C. (1982) Species dominance in chromosome elimination in barley hybrids. *Current Science* 51, 28–31.

Subrahmanyam, N.C. & Kasha, K.J. (1973) Selective chromosomal elimination during haploid formation in barley following interspecific hybridization. *Chromosoma* 42, 111–25.

Sunderland, N. (1974) Anther culture as a means of haploid induction. In: Kasha, K.J. (ed.), *Haploids in Higher Plants: Advances and Potential* (Proceedings of the 1st International Symposium). University of Guelph, Guelph, Canada, pp. 91–122.

Sunderland, N. (1978) Strategies in the improvement of yields in anther culture. In: *Proceedings of Symposium on Plant Tissue Culture*. Science Press, Beijing, China, pp. 65–86.

Sunderland, N. & Dunwell, J.M. (1977) Anther and pollen culture. In: Street, H.E. (ed.), *Plant Tissue and Cell Culture*, 2nd edn. Blackwell, Oxford, pp. 223–65.

Sunderland, N. & Evans, L.J. (1980) Multicellular pollen formation in cultured barley anthers. *Journal of Experimental Botany* 31, 501–14.

Sunderland, N. & Huang, B. (1985) Barley anther culture – the switch of programme and albinism. *Hereditas* supplement 3, 27–40.

Sunderland, N. & Xu, Z.H. (1982) Shed pollen culture in *Hordeum vulgare*. *Journal of Experimental Botany* 33, 1086–95.

Sunderland, N., Roberts, M., Evans, L.J. & Wildon, D.C. (1979) Multicellular pollen formation in cultured barley anthers. *Journal of Experimental Botany* 30, 1133–44.

Sunderland, N., Huang, B. & Hills, J. (1984) Disposition of pollen *in situ* and its relevance to anther/pollen culture. *Journal of Experimental Botany* 35, 521–30.

Swanson, E.B., Coumans, M.P., Wu, S.C., Barsby, T.L. & Beversdorf, W.D. (1987) Efficient isolation of microspores and the production of microspore-derived embryos from *Brassica napus*. *Plant Cell Reports* 6, 94–7.

Szarejko, I. & Kasha, K.J. (1988) Media developments for barley anther culture. In: *University of Guelph, Department of Crop Science, Annual Report 1988*, pp. 119–21.

Søgaard, B. & von Wettstein-Knowles, P. (1987) Barley: genes and chromosomes. *Carlsberg Research Communications* 52, 123–96.

Tang, Z.K. & Xie, J.X. (1987) Investigations of *in vitro* induced parthenogenesis in barley (*Hordeum vulgare* L.). In: Yasuda, S. & Konishi, T. (eds), *Barley Genetics V* (Proceedings of the 5th International Barley Genetics Symposium). Maruzen, Okayama, Japan, pp. 1005–7.

Thiebaut, J., Kasha, K.J. & Tsai, A. (1979) Influence of plant development stage, temperature, and plant hormones on chromosome doubling of barley haploids using colchicine. *Canadian Journal of Botany* 57, 480–3.

Thomas, J.B., Kaltsikes, P.J. & Anderson, R.G. (1981) Relation between wheat–rye crossability and seed set of common wheat after pollination with other species in the Hordeae. *Euphytica* 30, 121–7.

Thomas, W.T.B., Powell, W. & Wood, W. (1984) The chromosomal location of the dwarfing gene present in the spring barley variety Golden Promise. *Heredity* 53, 177–83.

Tometorp, G. (1939) Cytological studies on haploid *Hordeum distichum*. *Hereditas* 25, 241–54.

Tsuchiya, T. (1986) List of genetic stocks with BGS numbers. *Barley Genetics Newsletter* 16, 81–121.

Wang, C.C. & Kuang, B.J. (1981) Induction of haploid plants from the female gametophyte of *Hordeum vulgare* L. *Acta Botanica Sinica* 23, 329–30.

Wei, Z.M., Kyo M. & Harada, H. (1986) Callus formation and plant regeneration through direct culture of isolated pollen of *Hordeum vulgare* cv. 'Sabarlis'. *Theoretical and Applied Genetics* 72, 252–5.

White, P.R. (1943) *A Handbook of Plant Tissue Culture*. Jacques Cattell, Lancaster, Pennsylvania, USA.

Wilson, H.M. (1977) Culture of whole barley spikes stimulates high frequencies of pollen calluses in individual anthers. *Plant Science Letters* 9, 233–8.

Wilson, H.M., Foroughi-Wehr, B. & Mix,

G. (1977) Haploids in *Hordeum vulgare* through anther culture: potential as starting material for genetic manipulation *in vitro*. In: Sanchez-Monge, E. & García-Olmedo, F. (eds), *Interspecific Hybridization in Plant Breeding* (Proceedings of the 8th Eucarpia Congress). Universidad Politecnica de Madrid, Madrid, Spain, pp. 243–52.

Wilson, H.M., Mix, G. & Foroughi-Wehr, B. (1978) Early microspore divisions and subsequent formation of microspore calluses at high frequency in anthers of *Hordeum vulgare* L. *Journal of Experimental Botany* 29, 227–38.

Xu, Z.H. & Sunderland, N. (1981) Glutamine, inositol and conditioning factor in the production of barley pollen callus *in vitro*. *Plant Science Letters* 23, 161–8.

Xu, Z.H. & Sunderland, N. (1982) Inoculation density in the culture of barley anthers. *Scientia Sinica B* 9, 961–8.

Xu, Z.H., Huang, B. & Sunderland, N. (1981) Culture of barley anthers in conditioned media. *Journal of Experimental Botany* 32, 767–78.

Ye, J.M., Kao, K.N., Harvey, B.L. & Rossnagel, B.G. (1987) Screening salt-tolerant barley genotypes via F_1 anther culture in salt stress media. *Theoretical and Applied Genetics* 74, 426–9.

Yean-San, L.H. (1987) 'Gynogenèse *in vitro*. Variabilité des haploïdes doublés issus d'androgenèse, de gynogenèse *in vitro* et de croisement interspécifique chez *Hordeum vulgare* L.' Thèse Doctorat d'Etat Université Paris-sud, France.

Ziauddin, A., Kasha, K.J. & Szarejko, I. (1988a) Evaluation of factors in barley anther culture. In: *University of Guelph, Department of Crop Science, Annual Report 1988*, p. 118.

Ziauddin, A., Simion, E., Glover, K. & Kasha, K.J. (1988b) Microspore culture in cereals – barley and wheat. In: *University of Guelph, Department of Crop Science, Annual Report 1988*, pp. 122–3.

Ziauddin, A., Simion, E. & Kasha, K.J. (1989) Microspore culture in barley. In: *University of Guelph, Department of Crop Science, Annual Report 1989*, pp. 76–8.

Ziauddin, A., Simion, E. & Kasha, K.J. (1990) Improved plant regeneration from shed microspore culture in barley (*Hordeum vulgare* L.) cv. Igri. *Plant Cell Reports* 9, 69–72.

26

Regeneration, Stability and Transformation of Barley

Angela Karp & Paul A. Lazzeri

Genetic manipulation has advanced considerably over the past decade and engineered plants, carrying newly introduced genes, are being produced in major crops, including gramineaceous species such as maize and rice. This technology has far-reaching potential, enabling direct modification of the plant genome for crop improvement without dependency on applied chemicals. Specific breeding aims include engineering resistance against pathogens and pests, improving tolerance to environmental stresses such as drought and salinity, and improving quality by modifying key proteins and metabolic processes.

For this technology to become routinely applicable to the breeding of barley (*Hordeum vulgare*), techniques for the efficient introduction of new genes (transformation) and their stable integration in the barley genome need to be established. Most transformation approaches use tissue culture systems of plant regeneration from cultured cells. The development of suitable regeneration systems is therefore a prerequisite for application of the genetic manipulation approach. Although these regeneration systems are asexual, they can be associated with genetic instability or somaclonal variation. This could cause loss of morphogenetic potential, reduced fertility and unwanted variations in regenerated plants. The stability of newly introduced genes may also be affected by the *in vitro* culture process.

In this chapter we review the advances made in the establishment of efficient regeneration and transformation systems in barley and assess the importance of somaclonal variation in these systems. Their potential application for improvement of the barley crop is then discussed. Emphasis is placed on regeneration from immature embryos and inflorescences and on advances in cell suspension and protoplast culture. Methods of anther and microspore culture and their applications are described in detail in Chapter 25.

Plant regeneration systems in barley

Plant regeneration can be achieved from cultured cells and tissues by organogenesis or somatic embryogenesis. In the former, new plants are derived from adventitious shoot meristems, whilst in the latter, distinct embryos are formed. Although large numbers of plants can be derived from organogenic cultures, somatic embryogenesis is the preferred mode of regeneration because it parallels normal plant development from zygotic embryos (Haccius, 1978; Vasil & Vasil, 1982). Embryogenic cultures have also become very important as a source of morphogenetic cell suspensions for isolation and culture of protoplasts.

Gramineaceous species have, in general, proved difficult to regenerate and barley has been ranked among the less responsive. Nevertheless, regeneration has been achieved from a range of cultured tissues. In addition to anther and microspore culture (described in Chapter 25), these include regeneration from cultured apical meristems, leaves, mature embryos and seeds and from immature embryos, inflorescences and ovarian tissue.

Apical meristems, leaves, mature embryos and seeds

The use of cultured apical meristems for production of morphogenetic callus was among the early approaches that yielded successful plant regeneration in barley. Cheng & Smith (1975) dissected apical meristems from one-week-old seedlings grown on media lacking growth regulators. Sizeable callus was produced within 3–5 weeks by culturing the apices on modified Murashige & Skoog (1962) medium (MS) containing 1.75 mg l^{-1} indole-3-acetic acid (IAA), 3.3 mg l^{-1} 2,4-dichlorophenoxyacetic acid (2,4-D) and 0.3 mg l^{-1} 6-(3-methyl-2-butenyl-amino) purine (2iP). Regeneration by organogenesis was achieved by transfer of calluses to MS medium without growth regulators. More recently, Weigel & Hughes (1985) have shown that regeneration by somatic embryogenesis can also be obtained in callus derived from apical meristem explants.

Attempts to regenerate plants from cultured leaf pieces have not been so successful. For example, Saalbach & Koblitz (1978) found that callus could be produced by culturing the basal portion of leaf segments isolated from barley seedlings on MS medium containing 10 mg l^{-1} 2,4-D, even when segments 3–5 mm above the apical meristem were chosen; however, shoot formation from the calluses was only rarely observed. Jelaska *et al.* (1984) also reported limited success with cultured leaf segments from germinated seedlings. However, good callus induction and regeneration was obtained for cultured mesocotyl segments in which the scutellum was retained and placed side-down on MS medium supplemented with high concentrations (2.6–10 mg l^{-1}) of 2,4-D or 2,4,5-trichlorophenoxyacetic acid (2,4,5-T). Plants were regenerated from the mesocotyl explants by organogenesis on MS medium with 1.5 mg l^{-1} 2,3,5-triiodobenzoic acid (TIBA).

Successful plant regeneration has also been obtained from cultured mature embryos. For example, Lupotto (1984) placed excised embryos dissected from mature seed scutellum side down on agar medium containing the growth regulators of Cheng & Smith (1975) with 5 mg l^{-1} 2,4-D. Callus proliferated within 10 days from the embryo axis and, after 4 months, plant regeneration was achieved by organogenesis.

Immature ovaries, embryos and inflorescences

Where callus production and regeneration from cereal explants of different ages have been compared, immature tissues give the most favourable responses. Orton (1979) assessed a range of different explants including 3-day-old embryos and endosperms, immature ovaries, stem and leaf sections, root meristems and anthers and found that immature ovaries gave the most vigorous and easily regenerated callus. Five different organogenic callus types were distinguishable based on their tissue consistency, pigmentation, growth rate and regeneration. Dale & Deambrogio (1979) also noted that the production and quality of callus cultures depended on explant source and found that the best results were obtained with callus derived from culture of immature embryos.

In more recent years, the immature embryo has become a popular choice of explant for regeneration in cereals in general, and barley has been no exception. Many studies have indicated the suitability of this tissue source not only for the production of morphogenetic callus but also for establishment of somatic embryogenesis in *H. vulgare* and related species (Breiman, 1985; Goldstein & Kronstrad, 1986; Breiman *et al.*, 1987a; Karp *et al.*, 1987; Jørgensen *et al.*, 1986; Lührs & Lörz, 1987).

A number of factors have been shown to be important for efficient regeneration but the general procedure can be outlined as follows. Immature embryos, 1–2 mm long (see later) are dissected from dehusked grain which has been surface-sterilized in 70% ethanol and 1% sodium hypochlorite and placed on MS medium containing 1–2 mg l^{-1} 2,4-D (Fig. 26.1a). Coconut milk (10% v/v) may also be added to the medium. Within a few days the embryos can be seen to swell and, 10–15 days later, callus proliferation follows from the scutellum (Fig. 26.1b). Cultures can be maintained

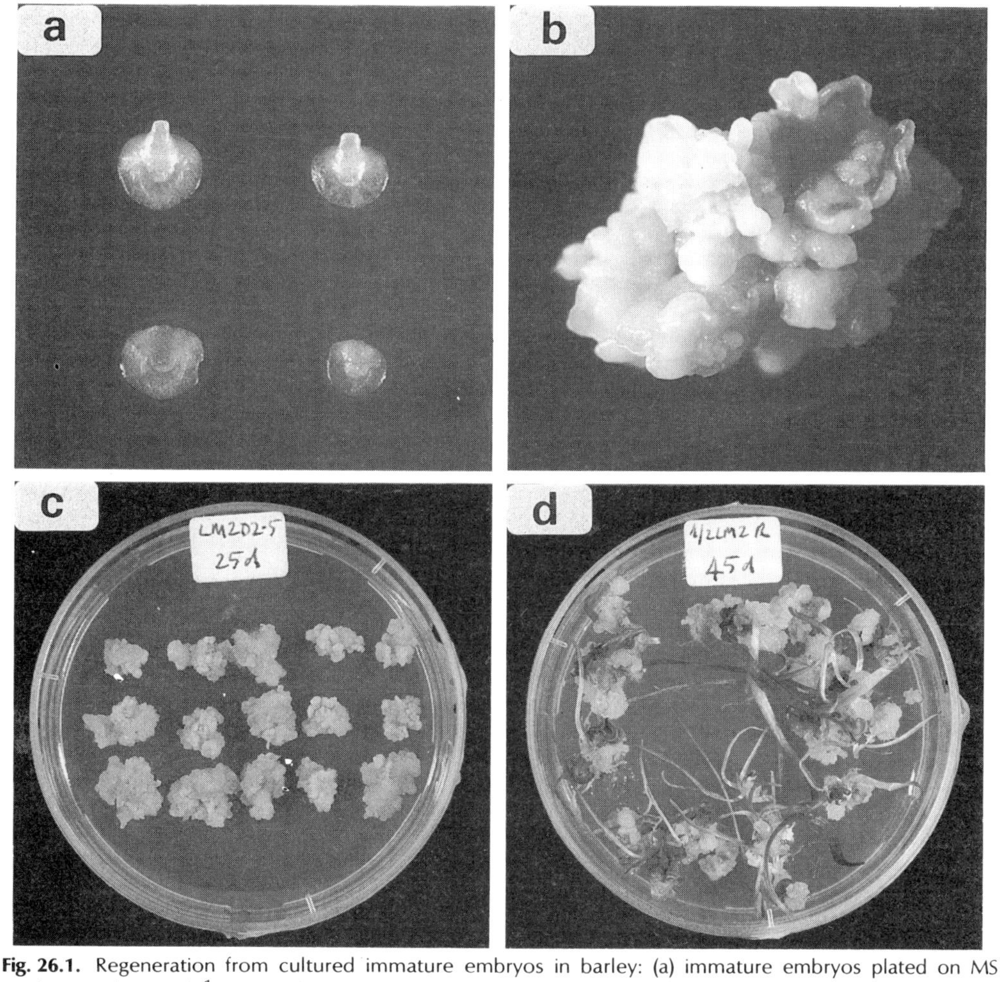

Fig. 26.1. Regeneration from cultured immature embryos in barley: (a) immature embryos plated on MS medium with 1 mg l^{-1} 2,4-D , (b) immature embryo-derived embryogenic callus, (c) callus cultures from immature embryos, (d) shoot formation on callus culture.

on the same medium (Fig. 26.1c) until shoot formation (Fig. 26.1d) after which the shoots are removed and transferred to MS without 2,4-D for rooting. Once roots have become established, plantlets can be transplanted into pots containing vermiculite or soil and transferred to the greenhouse.

Immature inflorescences can be used as explants in procedures similar to those for immature embryos. Inflorescences are selected for culture at around the stage when the fifth leaf has emerged (see below). Once the outer leaves are removed, the inflorescences are surface sterilized in 70% ethanol, cut into 1–2 mm pieces and cultured on MS medium supplemented with 1–2 mg l^{-1} 2,4-D. Callus formation occurs from the cut ends of the inflorescence segments after 8–15 days and is followed by shoot formation. Shoots can be rooted and grown into whole plants as just described.

Amongst the factors of importance for regeneration from cultured immature embryos and inflorescences are explant size, medium composition and genotype. Orientation of the immature embryo on the medium does not appear to be critical. Although Dale & Deambrogio (1979) reported that 48% more callus was obtained when the embryo was placed with the scutellum in contact with the medium, both orientations of scutellum exposed (Thomas & Scott, 1985; Karp *et al.*, 1987) and scutellum in contact (Dale & Deambrogio, 1979; Hanzel *et al.*, 1985) have been used successfully.

The size and age of the explant are important factors. Thomas & Scott (1985) achieved regeneration by somatic embryogenesis in barley from culture of both immature embryos and inflorescences. For immature embryos, sizes between 1 and 2 mm were found to be optimal. Embryos that were too small (<0.8 mm) did not survive well, whilst precocious germination was a problem with embryos >2 mm long. Other studies have indicated similar results (Dale & Deambrogio, 1979; Breiman, 1985; Goldstein & Kronstad, 1986; Karp *et al.*, 1987). For immature inflorescences, Thomas & Scott (1985) found that a size range of 1.5–3.5 mm gave good regeneration.

The quality and efficiency of callus produced is dependent on media composition. The most suitable basal medium for callus induction is MS (Hanzel *et al.*, 1985; Thomas & Scott, 1985; Breiman, 1985; Karp *et al.*, 1987) but B5 medium (Gamborg *et al.*, 1968) has also been used (Dale & Deambrogio, 1979; Breiman, 1985; Goldstein & Kronstad, 1986). From a number of studies where the effects of different growth regulators have been tested for both callus induction and shoot formation (Dale & Deambrogio, 1979; Hanzel *et al.*, 1985; Goldstein & Kronstad, 1986), 1 mg l^{-1} 2,4-D has proved suitable for both stages of regeneration, but shoot formation is generally improved when cytokinins and/or coconut milk are added to the medium (Breiman, 1985; Lührs & Lörz, 1987).

Genotype has also been recognized as an important factor in barley regeneration (Dale & Deambrogio, 1979). Goldstein & Kronstad (1986) tested 20 genetically diverse genotypes of spring barley for culture characteristics and regenerative capacity. The genotypes varied in their ability to develop vigorously growing callus and in their capacity for shoot production and regeneration. Out of the 19 genotypes tested, only 50% gave at least one regenerated plant per immature embryo culture. The greatest mean regeneration value was found in one cultivar ('Multum') which gave an average of two shoots per explant. Similar results were obtained by Hanzel *et al.* (1985). Of 91 barley genotypes tested for their ability to form callus, only 45 gave some callus growth. Interactions between genotype and media were highly significant, as were genotypic differences observed in morphogenetic response. Of the regenerated plants, 70% were derived from only two out of the 45 responsive genotypes. Although strong genotypic effects are still a general characteristic of morphogenetic response in barley, more recent studies have shown that, with improved techniques, some of these genotype variations can be overcome.

Cell suspensions and protoplast culture

In common with other graminaceous species, the establishment of regenerable cell suspensions in barley has proved difficult. Embryogenic callus is most often used to initiate cell suspensions in grasses (Vasil, 1987; Lazzeri *et al.*, 1990) but embryogenic barley callus is usually compact, comparatively slow growing and does not adapt well to growth in liquid culture. Suspension initiation is easier with friable callus which disperses readily in liquid, but such callus is generally only produced after extended periods of culture during which the regenerative capacity of the callus declines, and the frequency of variant cells increases (see section on Somaclonal variation below). Despite these difficulties, barley suspensions have been established and the regeneration of plants from suspension cells recently reported.

To date, most workers have used callus derived from immature embryos for suspension initiation. Kott & Kasha (1984) and Seguin-Swartz *et al.* (1984) used callus from haploid embryos (produced by the *bulbosum* method) to initiate suspensions which after several months were found to be mixoploid, although they still contained some (>10%) haploid cells (Seguin-Swartz *et al.*, 1984). Embryoids were produced when cell-aggregates were homogenized and plated on solid medium, but only albino plantlets could be regenerated (Kott & Kasha, 1984). Lührs & Lörz (1988) also used embryo-derived callus to establish suspensions of four different barley genotypes. Morphogenic suspensions were obtained from the genotypes 'Dissa' and 'Igri', but again only albino plants could be regenerated. Subsequently, Lazzeri & Lörz (1990) used a system of recurrent selection on solid medium to select for a granular callus phenotype allowing reproducible suspension initiation. Several embryogenic cell lines of cv. Dissa were obtained, and green plants regenerated from the lines were transferred to greenhouse conditions, but the plants were sterile. Mesocotyl-derived callus has also been used for suspension initiation (Müller *et al.*, 1989), but the resulting cell lines were non-morphogenic.

As it is clear that the regenerative capacity of graminaceous tissue cultures, in particular cell suspensions, is age-dependent, alternative methods for suspension establishment avoiding an extended callus phase, have been sought. Two possibilities that have been explored are the culture of immature embryos directly in liquid medium and the use of microspore-derived cultures as source material. The first approach has been used by Yan *et al.* (in press) to produce suspensions from which regenerable protoplasts could be isolated. The total culture period for suspension establishment was around 5 months: about half the time usually needed for establishment via embryogenic callus.

The potential of embryogenic microspore-derived cultures for suspension production has been demonstrated by Jähne *et al.* (in press). Embryogenic cell lines could be reproducibly produced by transferring one-month-old microspore callus to liquid medium, followed by selection for fine cell aggregates. With the cultivar 'Igri', more than 150 green plants were regenerated from 2- to 6-month-old suspensions, of which 50% were morphologically normal and fertile. The utility of microspore cultures for suspension initiation has been confirmed by Lührs and Nielsen (in press), who reported higher establishment frequencies in comparison with callus cultures.

Barley protoplasts, isolated from various tissues (e.g. mesophyll, roots, aleurone), have been widely used in physiological and biochemical investigations, but, like other grass protoplasts derived from differentiated cells, they have limited capacity for development in culture (Jones, 1985) and there have been few studies on regeneration, cell fusion or protoplast-mediated transformation. Several workers have attempted to induce divisions in barley mesophyll protoplasts (Mesencev *et al.*, 1976; Chin & Scott, 1979), but although these protoplasts may regenerate cell walls and survive for several days, sustained cell division has not been achieved. To obtain grass protoplasts capable of division it is usually necessary to isolate them from a tissue itself in active division (Jones, 1985; Lörz *et al.*, 1988). The most often used source is suspension cultures (Fig. 26.2a), but it is also possible to use fast-growing callus cultures. Koblitz (1976) isolated protoplasts from friable barley callus and these protoplasts could divide and give rise to cell lines, but no mention of morphogenesis was made. Subsequently, several groups have reported the isolation of division-competent protoplasts (Fig. 26.2b, c) from cell suspensions (Ishii *et al.*, 1986; Lührs & Lörz, 1988; Eigel & Koop, 1989; Müller *et al.*, 1989; Lazzeri & Lörz, 1990; Jähne *et al.*, 1991; Yan *et al.*, in press). Such suspensions are typically homogeneous, containing small cytoplasmic cell aggregates, and are relatively fast-growing (Fig. 26.2a).

The regenerative capacity of protoplast-derived cultures is usually similar to that of the donor suspensions. Lührs & Lörz (1988) first regenerated albino plantlets from suspension protoplasts of the cultivar 'Dissa', while Lazzeri & Lörz (1990) obtained green plantlets, from cell suspension protoplasts of the same cultivar (Fig. 26.2d). These plantlets did not survive transfer to soil (Lazzeri, unpublished). Using protoplasts derived from younger suspensions, however, Yan *et al.* (in press) and Jähne *et al.* (1991) were able to regenerate plants which sur-

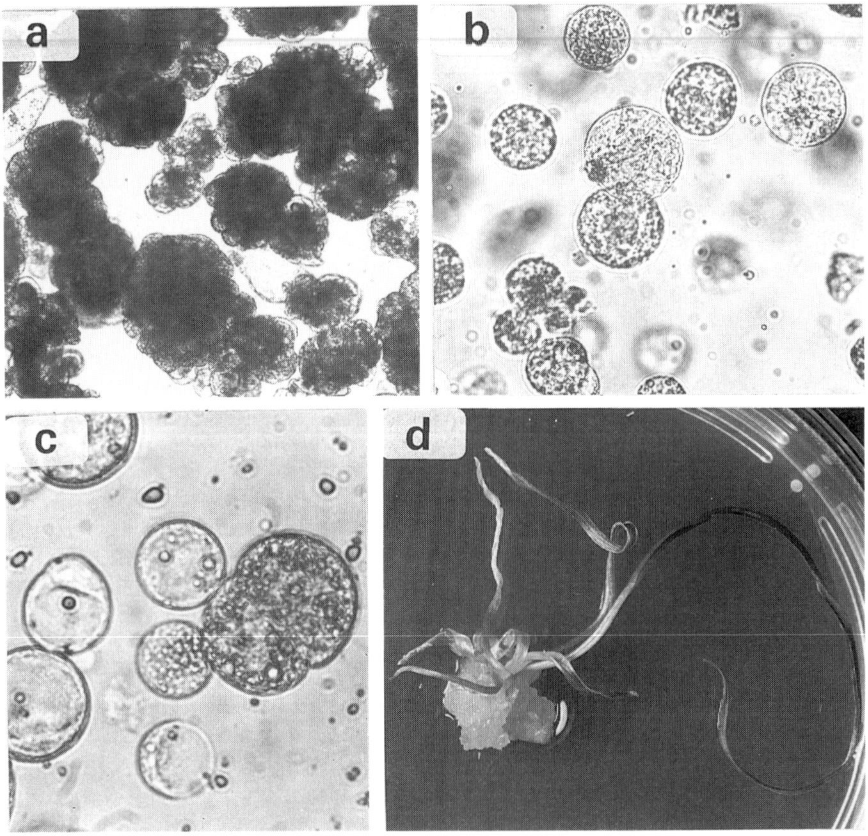

Fig. 26.2. Cell suspension and protoplast culture in barley: (a) cell aggregates of an embryogenic barley cell suspension (cv. Dissa), (b) cell suspension protoplasts directly after embedding in agarose, (c) first protoplast division (5 days after plating), (d) plantlet regenerated from protoplast-derived callus.

vived and exhibited normal morphology. Several plants from the latter study have now flowered, and set seed (A. Jähne, pers. comm.).

These independent regeneration experiments now make the transformation of barley via DNA uptake into protoplasts feasible (see later). For barley, the major obstacle to the development of efficient protoplast systems remains the difficulty of establishing suitable embryogenic suspensions which provide good yields of viable protoplasts. The 'quality' of the suspension is by far the most important factor in deciding protoplast response; when healthy protoplasts can be obtained, relatively simple culture procedures allow plating efficiencies in excess of 5%. This is more than adequate for manipulations such as protoplast fusion or transformation. In this context, the recent development of the use of microspore cultures for more efficient suspension production (see above) is an important advance.

Somaclonal variation in barley tissue culture

The growth of plant cells in culture and their subsequent regeneration into whole plants is subject to uncontrolled variability or 'somaclonal variation' (Larkin & Scowcroft, 1981). Passage through disorganized growth, such as a callus or cell suspension phase, appears to be a critical feature in terms of instability. Only meristem-tip culture can be considered free of somaclonal variation as all other tissue culture systems can involve varying degrees of callus formation (Karp, 1991). The regeneration systems described above for barley all involve disorganized cell growth. Before discussing their use in transformation, it is therefore pertinent to consider the extent to which uncontrolled variations occur in barley tissue culture.

Studies on the heritability of changes arising from plant tissue culture have shown that somaclonal variation encompasses three types of variation (Karp, 1991). Heritable somaclonal variation arises from changes in the genome, ranging from gross alterations in chromosome number or structure to qualitative and quantitative changes in DNA sequences. Heritable changes in gene expression can also occur, but these may be unstable or reversible. Non-heritable or 'epigenetic' variation may also arise as transient phenotypic modifications which are not transmissable through sexual reproduction. All classes of change could be present in regenerated barley plants but most information available relates to heritable variation.

Chromosome variation

That changes in the barley genome do occur in tissue culture is evident from cytological studies of callus and cell suspensions. Aneuploidy, polyploidy and structural rearrangements have been reported in cultured cells of *H. vulgare* and related species (Orton, 1980; Papes *et al.*, 1983; Lupotto, 1984; Singh, 1986; Gaponenko *et al.*, 1988; Lührs & Lörz, 1988). Time-course studies have shown that these chromosome aberrations appear after several (5–6) days in culture and increase in frequency as the culture phase progresses. This indicates that they arise from errors in *in vitro* cell division and not from pre-existing changes in the barley tissues (Singh, 1986; Gaponenko *et al.*, 1988).

The frequency of chromosome changes is dependent on the type and rate of callus growth. Two callus types were distinguished by Orton (1980) in a study of cultured immature ovarian tissues of *H. vulgare, H. jubatum* and their interspecific hybrid. Cells of type A callus were associated by an intracellular matrix and retained some tissue organization. Mitoses in this callus type were predominantly diploid with slow accumulation of aneuploid cells over time in culture. In contrast, a greater array of chromosome variation was observed in cells of type E callus which showed no intracellular association or tissue organization.

In regeneration from callus of mature embryo explants, Lupotto (1984) also distinguished two callus types. The primary callus was soft, watery, friable and pale green and contained considerable variation in chromosome number after 3 months in culture: 20% of cells were aneuploid and over 10% were tetraploid. Conversely, white, compact, nodular type B callus which arose on the surface of callus A consisted of predominantly diploid cells (85.3%) and some tetraploid cells (10.7%). It was noted that regeneration only occurred from callus B and therefore that morphogenetic potential appeared to be related to cytological stability.

The relationship between morphogenetic potential and chromosome instability in barley tissue cultures has been further studied by Singh (1986) who compared chromosome variation in ten morphogenetic (regenerable) and seven non-morphogenetic (non-regenerable) calluses derived from immature embryos. In non-morphogenetic callus, which was friable and yellow-brown, the mitotic index was low and chromosomes were examined in only two out of the seven calluses. The frequency of normal diploid cells was also low (12–33%) and many of the aneuploid cells carried chromosome structural rearrangements. In contrast, in morphogenetic callus, which was compact and dark green, the expected diploid number of 14 chromosomes predominated, and haploid, triploid, tetraploid and octaploid cells were only present at low frequencies. The extent of instability observed varied (0–30% cells) and calluses with high numbers of aberrant cells were not as morphogenic as calluses which were largely diploid.

It would appear from these studies that morphogenetic capacity in barley tissue culture is related to the organization of callus growth and the associated level of chromosomal instability. Loss of regenerative capacity after prolonged time in culture (Jelaska *et al.*, 1984; Goldstein & Kronstad, 1986; Gaponenko *et al.*, 1988) has been attributed to accumulation of chromosomal changes. However, a recent study suggests that the relationship between chromosomal instability and morphogenetic potential may not be as simple as originally supposed. Ziauddin & Kasha (1990) initiated callus from immature embryos on B5 medium (Gamborg *et al.*, 1968). All the plants regenerated had the expected diploid chromosome number although considerable chromosomal instability was present in some calluses. Cultures which showed the best embryogenic potential had the least chromosome variation but no plants were regenerated from any of the calluses after 7 months in culture, even though some remained predominantly diploid. The general relationship between chromosome instability and morphogenetic response did not appear to hold after prolonged periods of culture and the sudden loss of morphogenetic capacity could not be related to the level of chromosome variation.

Ziaddun & Kasha (1990) did find evidence of the influence of growth regulators on chromosomal instability. Five different concentrations of 2,4-D (0.1, 1.0, 2.0, 3.0 and 4.0 mg l^{-1}) and four different concentrations of naphthalene acetic acid (NAA) (0.1, 1.0, 5.0 and 10.0 mg l^{-1}) were tested as supplements to the B5 medium. Prolonged exposure to high levels of 2,4-D (2.0 mg l^{-1} or more) significantly increased the proportion of polyploid cells and, after 12 months in culture, 11 and 13 times as many polyploid cells were observed in treatments with 3.0 mg l^{-1} and 4.0 mg l^{-1} 2,4-D, respectively. This was not attributable to an effect of culture age as the levels of polyploid cells did not increase significantly over time in cultures on media containing concentrations of 0.1 and 1.0 mg l^{-1}. A study of mitotic aberrations revealed an increase in the incidence of abnormalities in the spindle apparatus with increased levels and exposure time to 2,4-D. The effects of NAA were not as dramatic and ploidy levels increased only gradually in all concentrations tested. However, more structural changes were observed on NAA. The frequency of aneuploid cells did not appear to be influenced by growth regulator type, concentration or period of exposure.

It would appear from these studies that although gross changes in the barley genome, in the form of numerical and structural chromosome variation, do occur in

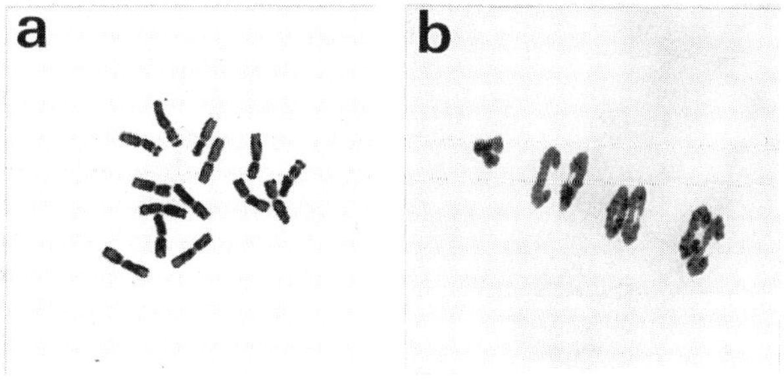

Fig. 26.3. Chromosome stability in regenerated barley plants: (a) normal karyotype in root-tip cell ($2n = 2x = 14$), (b) normal meiosis at metaphase I. (From Karp *et al.*, 1987.)

culture, strong diplontic selection operates and regeneration is largely confined to cultured cells which have retained their normal diploid complement. This is supported by findings that cytological abberrations are normally rare in regenerated barley plants (Fig. 26.3) (Gaponenko *et al.*, 1988; Lupotto, 1984; Singh, 1986; Karp *et al.*, 1987) unlike other diploid cereals, such as rye, where selective constraints appear to be more relaxed and numerical and structural chromosome changes have been observed in regenerated plants (Linacero & Vazques, 1986; Bebeli *et al.*, 1990). The occurrence of strong diplontic selection in barley tissue culture is in accordance with more general observations of the low tolerance of barley to chromosomal aberrations (Tsuchiya, 1960; Singh & Tsuchiya, 1977).

There are regeneration systems in barley, however, where chromosome variation plays an important role. Spontaneous chromosome doubling during the culture phase can remove the requirement for colchicine treatment in the production of doubled haploids by anther and microspore culture. Frequencies of doubling vary from under 50% (Powell *et al.*, 1988) to over 90% (Lyne *et al.*, 1986) (see Chapter 25).

Culture-induced chromosome doubling can also be advantageous for production of amphidiploids in interspecific hybrids which fail to respond to colchicine application. Fedak & Grainger (1986) induced callus on cultured immature inflorescences of a *Triticum crassum* × *H. vulgare* hybrid and obtained some mixoploidy in regenerated plants. Similar results have been obtained for two *H. vulgare* × *T. aestivum* hybrids (Chu *et al.*, 1984). The potential application of this approach for introgression has also been explored in nine different interspecific combinations of *H. vulgare* with wild barley species (Jorgensen & Andersen, 1989). Numerical and structural chromosome changes, including aneuploidy, amphidiploidy, deletions, inversions and extra heterochromatic segments, were detected in 82 regenerated plants, but there was little evidence of translocation heterozygozity.

Morphological variation

The rarity of cytological abnormalities among regenerated barley plants provides an explanation for the numerous reports of general morphological uniformity in plants derived from barley tissue cultures (Cheng & Smith, 1975; Lupotto, 1984; Karp *et al.*, 1987). However, albino plants, have been reported to occur at low frequencies (Dale & Deambrogio, 1979; Breiman, 1985; Thomas & Scott, 1985; Goldstein & Kronstad, 1986) and are a particular problem in anther culture, as discussed in Chapter 25.

Field trials of plants regenerated from cultured immature embryos have likewise revealed little variation in morphological characters (Luckett *et al.*, 1989). Pickering (1989) screened some 200 fertile progeny lines of regenerated plants in which selfed *H. vulgare* (VV), haploid (V) or hybrid (HV) immature embryos from crosses between *H. vulgare* and *H. bulbosum* were used as explants. Plants were assessed for a range of agronomic traits with the aim of selecting useful variants. The vast majority of plants was identical to their respective parents. Around 12% of regenerants from V cultures had undergone chromosome doubling but tetraploids were not present among VV regenerated plants and were found only at low frequencies (1.3%) in VB regenerants. A number of chlorophyll anomalies were also observed, including white longitudinal leaf stripe, segregation of green and albino plants, segregation of albino and striped plants and purple or yellow leaf coloration. Apart from the occasional tetraploids and albinos, three variants were identified. One showed maternal inheritance for yellow leaf striping and another carried a single recessive mutation which resulted in a phenotype resembling a *chlorina* mutant. The third variant, the only one of potential use, carried a recessive single gene mutation which conferred increased resistance to scald (*Rhynchosporium secalis*).

Variation in morphological traits has also been assessed in the field for doubled haploid (DH) plants derived from barley anther culture. The agronomic performances of 74 DH lines have been compared with 16 selfed seed-derived lines in cultivar Sabarlis. Significant variation was observed in the DH lines for seven out of the nine characters assessed (days to awn emergence, plant height, neck length, number of fertile tillers, number of grains on the main tiller, grain yield on the main tiller and thousand grain weight). Overall, the DH plants tended to be later to awn emergence, taller, with longer necks and had fewer tillers than the selfed seed-derived lines (Powell *et al.*, 1984).

In a later study, F_1 hybrid material (Golden Promise × Mazurka) was used and agronomic performance compared in DH lines from anther culture with those from the *H. bulbosum* chromosome elimination technique. Significant variation was induced by anther culture and, in general, lower mean values were obtained for the characters scored compared with DH plants from the *H. bulbosum* technique (Powell *et al.*, 1986a). A genetical analysis revealed significant deviations in expected 1:1 segregation ratios for three out of five genes analysed in the anther culture-derived DH lines and for only one gene in the *H. bulbosum* lines. Mean chiasma frequencies were found to be similar to the original hybrid in the *H. bulbosum* lines, whilst a reduced chiasma frequency was present in four out of ten DH lines produced by anther culture (Powell *et al.*, 1986b).

Further evidence of segregation distortion has been found in a more recent study of populations of anther culture-derived lines from two pairs of reciprocal hybrids. The segregations of ten marker genes located on five of the seven barley chromosomes were followed using biochemical, molecular and morphological markers. Four of the ten markers showed significant deviations from the expected Mendelian segregation ratios. The distorted ratios were associated with loci on chromosomes 4 and 6 and were in favour of the responsive parent (Blenheim) used in the anther culture studies (Thompson *et al.*, in press).

In the anther culture systems referred to, the regenerants are often described as 'microspore-derived' plants to clarify that they originate from microspores of the anther rather than from tissues such as the anther wall. Such systems should not be confused with direct regeneration from microspores which are first isolated from the anther mechanically or as shed pollen (see Chapter 25). In the latter regeneration systems, there is less information on 'gametoclonal variation', although there have been reports of albinism.

Molecular and biochemical variation

The low frequency of gross genomic changes in barley plants regenerated from tissue cultures does not preclude the possibility that variations occur at a finer level, as detectable by molecular and biochemical studies. Data on changes at this level are limited in barley. In one study, over 40 barley plants regenerated from cultured immature embryos were analysed cytologically and their seed progenies assayed for the isoenzymes esterase and glutamate-oxaloacetate transaminase, for ribosomal DNA (rDNA) spacer length polymorphism and for hordein patterns on sodium dodecylsulphate polyacrylamide gel electrophoresis (SDS-PAGE). Only minimal variation was revealed by these studies. One regenerated plant had aberrant meiosis. No variation was observed in the two isoenzymes studied, or in rDNA spacer lengths, but one selfed progeny plant from the regenerant with abnormal meiosis had a variant pattern for B-hordein on SDS-PAGE (Fig. 26.4). Molecular analysis of this variant using a 620 base pair genomic clone encoding the non-repetitive carboxyterminal domain of a B-hordein polypeptide did not reveal differences at the molecular level. Variation undetected by the probe may have been present, or, alternatively, the variant pattern may have resulted from mutation in a regulatory gene (Karp *et al.*, 1987). In a similar study, progenies of *H. spontaneum* regenerated from immature embryo culture were analysed for variation in cytology, rDNA spacer organization, hordein pattern on SDS-PAGE and mitochondrial DNA organization. One progeny plant showed an altered hordein pattern on SDS-PAGE and loss of 1.8 and 2.5 kb Taq 1 rDNA spacer fragments. No variation was detected in chromosome number or structure or in mitochondrial DNA organization (Breiman *et al.*, 1987a). Molecular changes have also been described in plants derived from barley anther culture where albino plants were found to contain large deletions in the chloroplast genome (Day & Ellis, 1985).

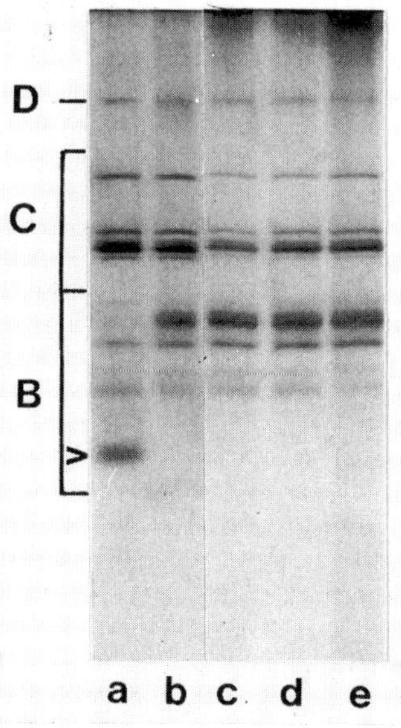

Fig. 26.4. Somaclonal variation in barley seed storage proteins: SDS-PAGE of hordein fractions from half seeds of barley regenerants (a–b) and controls (c–e) showing variant pattern in the B hordeins of a single seed (track a). (From Karp *et al.*, 1987.)

These studies indicate that somaclonal variation does occur in barley, but even changes detectable at the molecular level do not appear to arise as frequently as reported for other cereal regenerants such as wheat (Breiman *et al.*, 1987b), triticale (Brettel *et al.*, 1986a), maize (Brettel *et al.*, 1986b) and rice (Brown *et al.*, 1990; Müller *et al.*, 1990). This would suggest either that the barley genome is relatively stable in culture or that strong selective constraints operate to minimize the variation observed in plants regenerated from cultured barley cells. More detailed molecular studies are required before any further conclusions can be drawn.

Barley transformation

Since the first demonstrations of the transfer and expression of foreign genes in plants in the early 1980s, the area of crop plant transformation has developed rapidly so that many of the major crop species have now been transformed and the use of genetic engineering for crop improvement can be seriously considered (Gasser & Fraley, 1989; Vasil, 1990).

For most dicotyledonous crops, the transformation systems developed are based on the use of *Agrobacterium tumefaciens* as a vector, but the Gramineae are not natural hosts for *Agrobacterium* and alternative transformation techniques must be found if the important cereals are to be genetically engineered (Göbel & Lörz, 1988). A number of different transformation methods have been tested in cereals (Potrykus, 1990), but, to date, only two (protoplast transfor-

Table 26.1. Transformation methods tested in barley.

Method	Results
Macroinjection of DNA into developing tillers or ovaries	Equivocal evidence for gene transfer
Pollen transformation with DNA during tube growth/fertilization	Equivocal evidence for gene transfer
Pollen transformation by *Agrobacterium* during tube growth/fertilization	Equivocal evidence for gene transfer
DNA uptake by imbibing embryos	Transient gene expression
Electrophoresis of DNA into seeds	Equivocal evidence for gene transfer
Transformation of tissues by *Agrobacterium*	Transfer of viral DNA; virus replication
Microinjection of DNA into cells of embryos or tissues	Transient gene expression
Particle bombardment of cells or tissues	Transient gene expression
Electrically induced DNA uptake into protoplasts	Transient gene expression
Chemically induced (PEG) DNA uptake into protoplasts	Stable transformation

mation and particle bombardment) have clearly yielded transgenic plants and much work is still required before transformation systems are generally applicable (Lazzeri & Lörz, 1988).

In the following sections, we will discuss the various experiments which have been made on gene transfer to barley (summarized in Table 26.1) and present the current status and future prospects for barley transformation. We begin with methods using whole plants and progress through methods using organs and tissues to those operating at the cellular level.

Macroinjection of DNA into whole plants

Among early experiments attempting plant transformation were those of Soyfer and colleagues (Soyfer, 1980), who injected six-row *waxy* mutant barley plants (with abnormal starch synthesis) with DNA isolated from two-row wild-type plants. Injections were made into developing seeds and the resulting plants were screened in the field. Among plants from injected seeds, a few had normal starch and two-row morphology. The expression of these characteristics in the progeny did not, however, follow normal Mendelian segregation.

Such experiments, using phenotypic markers, are in danger of erroneous results from pollen and seed contaminants, and it is difficult to assess their validity. Interest in macroinjection was renewed, however, following the report of successful transformation of rye via the injection of plasmid DNA into developing floral tillers (De la Pena *et al.*, 1987) and several independent laboratories have attempted barley transformation by this method. Mendel *et al.* (1990) injected more than 2000 barley plants with plasmids containing the neomycin-phosphotransferase II (NPT) gene, allowing kanamycin selection. From 28 300 selected seeds, 446 (1.6%) were antibiotic resistant, although 0.6% of control seeds were also resistant. When tested for NPT activity, 10% of these resistant seedlings had low-level activity. Among the progeny from the selfed resistant seedlings, very few plants had detectable NPT activity, although they appeared to show segregation on kanamycin. DNA isolated from putative transformants was analysed by polymerase chain reaction (PCR) amplification and Southern blotting. With the PCR technique, sequences unique to the plasmids were detected. However, hybridization patterns observed on Southern blotting analysis were difficult to interpret, with independent transformants appearing to have identical integration patterns. Similar tiller macroinjection experiments have been carried out by Rogers & Hennessey (J.C. Hennessey, pers. comm.), who used the taumatin gene as a marker. From 1536 seed from treated plants, 51 apparently expressed the gene and 12 were 'positive' on Southern hybridization, but, again, the hybridization patterns and inheritance data obtained were abnormal. In contrast with the results of Mendel *et al.* (1990) and Rogers & Hennessey, Lörz found no evidence for transformants among 1500 seedlings derived from barley plants injected with a plasmid containing the NPT gene, although 12 plants had been selected as being resistant (H. Lörz, pers. comm.).

Pollen tube transformation

The gametes and developing zygote are prime targets for gene transfer, because transformation of these targets would give rise to transformed germline cells, which would ensure transmission of the integrated genes to the progeny. The female gametic cells of plants are generally few in number and not readily accessible. Most attention has, therefore, been given to the transformation of pollen.

In grasses, De Wet *et al.* (1985), Ohta (1986) and Luo & Wu (1988) have reported transformation by application of DNA solutions, either to pollen or to the style, at the time of fertilization. Several groups have also tested this approach in barley, but the only published study is that of Mendel *et al.* (1990). These workers applied DNA plasmids containing the NPT and β-glucuronidase (GUS) genes to the stigmas of more than 1000 plants, from which 11 200 seeds were recovered. After kanamycin selection, 305 putative transformants remained; of these, 101 were tested for NPT and 25 found to have low activity. Using PCR, NPT- and GUS-promotor-specific sequences were detected, but in Southern analysis, identical hybridization patterns were seen in all transformants, with the same size fragments as in the macroinjected plants (see earlier) again appearing. NPT expression, weak in primary transformants, was not transmitted to the F_2 generation.

A variation on the pollen/DNA transformation methods described here has been the use of *Agrobacterium* as a vector, again to attempt pollen tube or zygote transformation. Brettschneider *et al.* (1990) inoculated barley florets at anthesis with acetosyringone-activated *Agrobacterium* stains harbouring plasmids containing the NPT and GUS genes. Seeds were selected for kanamycin resistance, or screened for NPT activity. From about 1000 seeds tested, 5 putative transformants were identified and further analysed by Southern blotting. Two candidates contained fragments hybridizing to the NPT and GUS genes and showed no hybridization when probed with the *Agrobacterium vir* region, ruling out bacterial contamination. However, when progeny from these putative primary transformants were analysed, no hybridization to the NPT or GUS gene probes was detected (Langridge, Brettschneider, Lazzeri & Lörz, manuscript in preparation).

In conclusion, from the present data, macroinjection and pollen transformation, whether by naked DNA or by *Agrobacterium*, cannot be said to be viable methods for barley transformation, although there is evidence for the transfer and integration of foreign DNA.

DNA uptake by embryos and seeds

Plant cell transformation by the incubation of tissue in a DNA solution is an attractive idea and, not surprisingly, there have been numerous experiments to test this method, some of which are professed to have been successful (reviewed by Kleinhofs & Behki, 1977). Early experiments were difficult to interpret as evidence for transformation was usually based on proving the physical presence of introduced DNA. With the advent of heterologous scorable and selectable marker genes and molecular biological techniques, it has become possible to detect transformation events with greater accuracy and thus to reassess the potential of DNA uptake procedures.

In barley, Ledoux & Huart (1969) made a series of experiments on the uptake of exogenous bacterial DNA by germinating seedlings and reported integration of input DNA, but these results were not universally accepted (Kleinhofs & Behki, 1977). Twenty years later, Töpfer *et al.* (1989) showed that exogenous DNA can be taken up by imbibing cereal embryos, leading to transient gene expression. In this technique, dry embryos of wheat, barley or other cereals are mechanically isolated, washed in organic solvent and then incubated in a DNA solution. DNA uptake occurs during imbibition and gene expression is measured after 3 days of culture at 24°C. There is presumably some disruption of cell walls during embryo isolation and hydration, allowing penetration of the DNA across the plasmalemma, which is thought to be unstable during the hydration period. Using this method Töpfer *et al.* (1989) reported expression of the NPT gene with several different promotors and also used autonomously replicating wheat dwarf virus (WDV) constructs to amplify the transient expression signal. Plants have been grown from DNA-treated embryos but, to date, there is no evidence for stable transformation by this method (R. Töpfer, pers. comm.).

Another method for transient gene expression in barley seeds, that of 'electrophoretic transfection' has recently been proposed by Ahokas (1989). DNA was electroporated into imbibed barley seeds supported between microelectrodes. The movement of labelled DNA along cell walls was followed by autoradiography, while evidence for gene expression was based on a histochemical assay for expression of the GUS gene. As the tissues of barley embryos have considerable endogenous GUS activity in this assay, these experiments need to be repeated with alternative marker genes before the validity of the method can be judged.

Even if such DNA uptake procedures can be demonstrated to function and to result in stably transformed cells, the difficulty remains of targeting the germ line cells of the shoot meristem for transformation. This is likely to prove problematic as these cells are few in number and relatively inaccessible.

Transformation of tissues by *Agrobacterium*

For most dicot species the most efficient transformation method is the use of *Agrobacterium* as a vector (Gasser & Fraley, 1989). Many attempts have been made to apply these methods to cereal species, but the Gramineae are not readily infected by *Agrobacterium* and until recently most results were negative. In 1987 Grimsley *et al.*, demonstrated that viral DNA molecules capable of replication and spread in host tissues could be transferred to maize cells by the T-DNA of *Agrobacterium*. This system allows the detection of extremely low-frequency transformation events, due to the amplifying effect of the virus. This so-called 'agroinfection' technique has since been utilized in other cereals, including barley (Boulton *et al.*, 1989; Creissen *et al.*, 1990).

Boulton *et al.* (1989) used various *A. tumefaciens* and *A. rhizogenes* strains with dimeric maize streak virus (MSV) molecules cloned into their Ti or Ri plasmids for injection into the stems of maize, barley and several other Gramineae. Two of the 25 barley plants injected with two different *A. rhizogenes* strains developed MSV disease symptoms, indicating *Agrobacterium*-mediated transfer of the virus to plant cells. Subsequently, Creissen *et al.* (1990) used four of the *A. rhizogenes* strains tested by Boulton *et al.* (1989) but with wheat dwarf virus dimers instead of MSV dimers cloned into their Ri plasmids. The four strains were first tested for barley seedling inoculations and the two most infectious strains then used in co-cultivation experiments with microspore-derived embryos (cv. Igri).

Co-cultivation was performed by applying bacterial suspension to embryos collected on filter paper, followed by overnight incubation at 25°C. Replicative WDV forms were detected one week after cultivation. Co-cultivation using *Agrobacterium* strains lacking T-DNA, or incubation of embryos with plasmid DNA containing WDV dimers did not result in WDV infection, giving evidence that transformation was T-DNA-mediated. Although these experiments show that *Agrobacterium* can deliver DNA to barley cells, there is as yet no evidence for integration of the T-DNA. Also, the virus molecules are non-integrative and are excluded from the meristems. For agroinfection to become a method for stable transformation these problems of integration and germline targeting need to be solved.

Microinjection of DNA

Plant transformation by microinjection into single cells (Crossway *et al.*, 1986) is an aesthetically appealing concept: it should remove the need for selectable marker genes to isolate the transformed cells and could provide opportunities for following the expression and regulation of inserted genes during the development of a plant from a very few cells. In practice, the technical difficulties mean that few plant cells are amenable to such manipulations. A major requisite for the use of microinjection is that single cells or small cell clusters can be efficiently cultured and regenerated; such microculture requires very specialized conditions by comparison with mass culture (Schweiger *et al.*, 1987). Furthermore, the microinjection itself must be very precise, the injected cells must survive at high frequency and there must be high frequency integration of introduced DNA. Two potential targets for microinjection are zygotic proembryos and microspore-derived proembryos. In barley the recent advances in anther and isolated microspore culture mean that the latter material can now be produced in large numbers (see Chapter 25). So far microinjection experiments with barley proembryos have yielded putative primary transformants (Potrykus, 1990; G. Spangenberg, pers. comm.).

An alternative approach has recently been reported by Toyoda *et al.* (1990), who microinjected plasmid DNA into barley coleoptile cells. Transient expression of the GUS gene could be detected 6 h after injection, and peaked at 24 h. A technique for plant regeneration from leaf tissue of *Tritordeum* (the fertile barley × wheat amphiploid) has now been developed (Barcelo *et al.*, 1991) and there may be the possibility of combining these two methods to produce a transformation system.

Particle bombardment

Particle bombardment is a very recent technique for plant transformation. The principle is that very small (1–5 μm), dense particles (usually tungsten or gold) are coated with DNA and then accelerated into cells, which become transformed by the input DNA. The original particle gun (Klein *et al.*, 1987) used explosive discharge as the accelerating force, but subsequently electrostatic discharge (McCabe *et al.*, 1988), compressed air- and gas-powered guns have been developed.

As the particles can penetrate most cell walls this method is theoretically independent of tissue and species. Practically, however, there are several constraints on the use of the technique; cells must remain viable after receiving particles; the DNA must remain viable after delivery and must then be integrated; it must be possible to select the few transformed cells from the mass of non-transformed cells, and the transformed cells must be able to regenerate. In consequence, particle bombardment has proved to be an efficient method for achieving transient gene expression, but the recovery of stable, transformed cells is much less efficient.

There are three reports of the use of the particle gun for barley transformation (Kartha *et al.*, 1989; Mendel *et al.*, 1989; Creissen *et al.*, 1990). Mendel *et al.* used non-morphogenetic cell suspensions as target tissue and bombarded the cells with tungsten particles coated with plasmids containing the GUS and NPT genes under the control of two different promotors. Both of the marker genes could be expressed in bombarded cells and it was possible to co-transform cells with a mixture of two plasmids. Both promotors tested, the cauliflower mosaic virus (CaMV) 35S promotor and the *Agrobacterium* $T_R1'2'$ promotor, functioned in the barley cells. In the experiments of Kartha *et al.* (1989) cell cultures and immature embryos were bombarded with a construct containing the chloramphenicol acetyltransferase (CAT) gene, driven by the CaMV 35S promotor. Transient CAT expression was detected 20 hours after bombardment and the authors suggested that CAT expression was dependent on the recipient cells being in active division. Creissen *et al.* (1990) used particle bombardment to transfer wheat dwarf virus (WDV) constructs to microspore-derived embryos, in a comparison of this method of delivery with *Agrobacterium*-mediated transformation (see earlier). WDV replication was detected in bombarded embryos, and at higher frequency than in the agro-infected cultures.

In these reports no mention is made of selection for stably transformed cells; for this to be realistic, large numbers of transient expression signals must be routinely obtained, as the frequency of integrative events is usually at least two orders of magnitude less than that of transient expression. In addition, antibiotic selection of barley tissues or even large cell aggregates is difficult, making the rescue of transformed cells inefficient. In this context, isolated microspore cultures are a good target tissue for particle bombardment as they are more easily selected and can show high frequencies of embryogenesis. Young isolated microspore cultures of barley (a few days post-initiation) have been bombarded with a plasmid containing the GUS gene driven by the CaMV 35S promotor. Dark blue cells indicative of high levels of GUS expression have been recovered, suggesting that this may be a promising approach (R. Brettschneider & A. Jähne, pers. comm.).

Electroporation of cells and tissues

Electroporation is now a standard technique for protoplast transformation (see later), and a number of attempts have also been made to transform intact cells by this method. Until recently, the results have been negative or inconclusive (Potrykus, 1990), but there is now evidence that cell electroporation may be possible. Dekeyser *et al.* (1990) have developed a protocol for the electroporation of rice leaf tissues, involving nuclease elimination, prolonged incubation with DNA and long pulse lengths. Transient expression of the GUS gene and of both light-and tissue-specific constructs has been demonstrated, as has the application of the system to other Gramineae, including barley. Whether this method will prove suitable for stable transformation is not yet known.

Protoplast transformation

Direct gene transfer (Paszkowski *et al.*, 1984), the uptake and expression of exogenous DNA by protoplasts, is now a routine transformation method for many plant species, including a number of the Gramineae (Lazzeri *et al.*, 1990). The two methods most often used to induce DNA uptake are treatment with polyethylene glycol (PEG) (Paszkowski *et al.*, 1984) and electroporation (Fromm *et al.*, 1985). Both methods induce transient pores in the protoplast membranes, allowing DNA to enter from the surrounding solution. Each method has its own advantages and disadvantages and the choice of procedure depends on the type of protoplasts used and the intended application.

Barley mesophyll protoplasts were used in early studies on DNA uptake in a series of experiments by Hughes *et al.* (1979). These workers showed that uptake was stimulated by di- and polycations and that most of the

DNA taken up was subsequently associated with protoplast nuclei. The plasmid DNA used in these experiments did not, however, contain a marker gene and gene expression could not be investigated.

The first report of transformation of barley protoplasts was that of Junker *et al.* (1987). The protoplasts were isolated from a suspension of cv. Golden Promise and transformed with a plasmid containing the NPT gene via PEG-mediated gene transfer. Transient expression of the NPT gene was maximal 2 days after transformation. PEG-mediated transient gene expression has also been reported in protoplasts from a *Hordeum bulbosum* cell suspension (Baratz & Breiman, in press). In this study, the GUS gene was used as a marker to analyse various factors affecting expression levels.

Barley protoplasts have also been transformed by electroporation. Teeri *et al.* (1989) electroporated mesophyll protoplasts with a plasmid construct containing the pea ribulose 1,5-bisphosphate carboxylase (RuBisCo) small subunit transit peptide fused to the NPT gene, or with a similar construct lacking the transit peptide. Both plasmids gave transient NPT activity, but only with the transit peptide could activity be detected in chloroplasts, demonstrating that targeting had occurred.

Hantke (1990) used protoplasts from a cell suspension of cv. Dissa to study factors affecting transient gene expression after electroporation. Electrical parameters were optimized and the influence of the electroporation solution and the amount and form of input DNA was investigated.

As well as mesophyll and suspension protoplasts, barley aleurone protoplasts have also been used to study transient gene expression (Lee *et al.*, 1989). These authors developed an aleurone protoplast isolation and transformation system, with the aim of analysing tissue-specific gene expression. In the published work, protoplasts were transformed by PEG treatment and CAT was used as the marker gene, being driven by the CaMV 35S promotor. Levels of CAT expression were seen to be improved by the presence of an intron from the maize alcohol dehydrogenase (*adh1*) gene in the transforming plasmid. A similar effect of the *adh1* intron on transient gene expression in barley was also seen by Kartha *et al.* (1989) and Hantke (1990).

There is at present a single report of stable transformation of barley by DNA uptake into protoplasts (Lazzeri *et al.*, 1991). In these experiments protoplasts were isolated from a cell suspension of cv. Dissa and were transformed by using three different PEG-mediated DNA-uptake protocols. The plasmid used for transformation contained both the NPT and GUS genes, each under the control of CaMV 35S promotors. Transformed microcalluses were selected in media containing G418 sulphate (Fig. 26.5). NPT activity was detected in all antibiotic-resistant cell lines, however, not all these NPT-positive cell lines showed GUS activity, despite both genes being present on the same plasmid. Southern analysis confirmed the presence of the NPT gene in all G418-resistant lines, but some lines appeared to lack the GUS gene, or not to contain a full-sized copy. For these experiments the authors calculated transformation frequencies of between 1 and 10 in 10^5, based on the number of resistant clones recovered as a fraction of the dividing protoplast population.

In summary, a number of different methods for barley transformation have been tested, of which several have given some positive results, but to date a clearly applicable technique has not been developed. Whole-plant transformation methods are attractive in that they avoid *in vitro* culture and the attendant problems of regeneration and variation, but the evidence for transformation by these methods is still unclear. Transformation by DNA uptake into cells, with or without electrophoresis/electroporation, is attractive because of the relative simplicity of the methods, but at present does not seem suitable for stable transformation. Micro-injection requires specialized equipment and culture systems, and as integration usually occurs at low frequency among cells receiving DNA (perhaps 10^{-4} on average, 10^{-2} at best) it

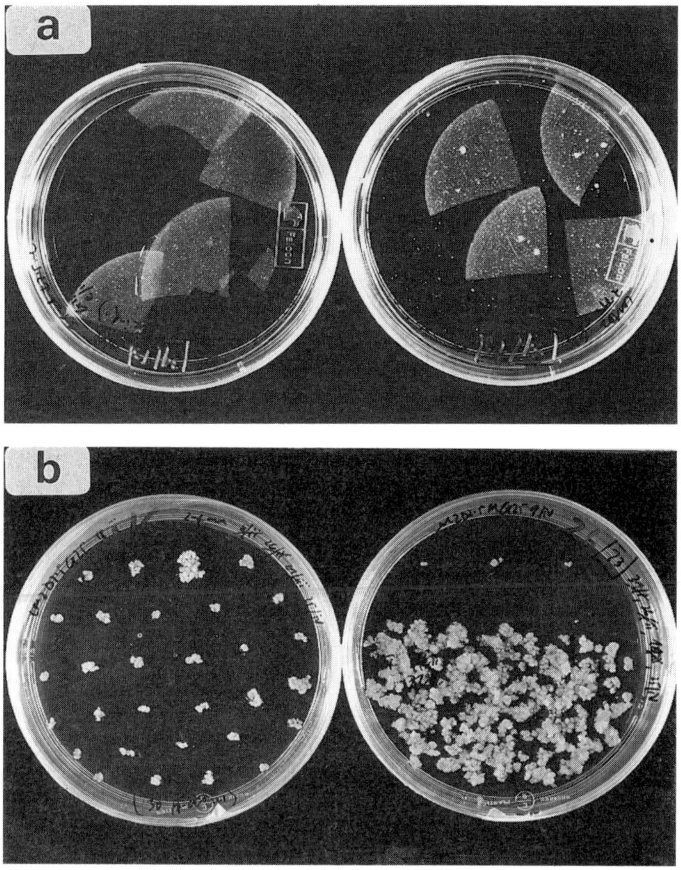

Fig. 26.5. Transformation in barley. (a) Selection of transformed cells by culture of protoplast-derived calluses (embedded in agarose segments) in medium containing an antibiotic (geneticin sulphate): (left) control culture, (right) culture from transformed protoplasts, with antibiotic-resistant calluses visible. (b) Selection of antibiotic-resistant calluses on medium containing geneticin sulphate (left); control calluses (right), three non-resistant calluses (above) and callus from a single transformant (below).

must be possible to recover large numbers of micro-injected individuals for this to be a feasible method. *Agrobacterium* appears to be able to transfer viral DNA to barley cells, but does not direct integration; whether this problem can be overcome remains to be seen. The two techniques which today appear to hold most potential for barley transformation are particle bombardment and DNA uptake by protoplasts. Both are proven methods for transient gene expression and the latter also for stable transformation. For both techniques the present limitation is the provision of the appropriate proliferating embryogenic cultures as protoplast source or target tissue. With the current progress in barley cell culture these limitations should soon be overcome.

Conclusions

Genetic engineering has many attractions for improvement of barley but, to date, this approach has been hindered by the lack of a suitable transformation method. The last few years have seen important progress in the general area of cereal transformation.

Among the various methods available, particle bombardment and DNA uptake by protoplasts appear to hold most promise. However, both techniques are limited by the lack of suitable embryogenic cultures for use as a source of viable morphogenetic protoplasts or regenerable target tissue. Since the most successful transformation methods rely on *in vitro* culture, an understanding of the extent and causes of somaclonal variation is of increasing importance. The present data indicate that plants regenerated from barley cell culture are less variable than those of other cereal species, suggesting that barley is either relatively stable in culture or that stringent selection results in the regeneration of only normal shoots. More extensive analyses at the molecular level are still needed. A further question still to be addressed is the stability of introduced genes in barley transformants. If barley is to be improved by the direct transfer of agronomically important genes, these must be stably integrated, correctly expressed and transmitted to the progeny in normal Mendelian fashion. With the recent production of the first few barley transformants such questions can now be addressed.

References

Ahokas, H. (1989) Transfection of germinating barley seed electrophoretically with exogenous DNA. *Theoretical and Applied Genetics* 77, 469–72.

Baratz, S. & Breiman, A. Parameters affecting transient expression of β-glucuronidase in protoplasts derived from *Hordeum bulbosum* cell suspensions (submitted).

Barcelo, P., Lazzeri, P.A., Martin, A. & Lörz, H. (1991) Totipotency of cereal leaf cells: patterns of proliferation and regeneration capability *in vitro* of the inflorescence sheath leaves of barley, wheat and tritordeum. *Plant Science* 177, 243–51.

Bebeli, P., Karp, A. & Kaltsikes, P.J. (1990) Somaclonal variation from cultured immature embryos of rye differing in heterochromatic content. *Genome* 33, 177–83.

Boulton, M.L., Bucholz, W.G., Marks, M.S., Markham, P.G. & Davies, J.W. (1989) Specificity of *Agrobacterium*-mediated delivery of maize streak virus to members of the Gramineae. *Plant Molecular Biology* 12, 31–40.

Breiman, A. (1985) Plant regeneration from *Hordeum spontaneum* and *Hordeum bulbosum* immature embryo-derived calli. *Plant Cell Reports* 4, 70–3.

Breiman, A., Rotem, D., Karp, A. & Shaskin, H. (1987a) Heritable somaclonal variation in wild barley (*Hordeum spontaneum*). *Theoretical and Applied Genetics* 74, 104–12.

Breiman, A., Felsenburg, T. & Galun, E. (1987b) Nor loci analysis in progenies of plants regenerated from scutellar callus of breadwheat. *Theoretical and Applied Genetics* 73, 827–31.

Brettschneider, R., Lazzeri, P.A., Langridge, P., Hartke, S., Gill, R. & Lörz, H. (1990) Cereal transformation with *Agrobacterium tumefaciens* via the pollen tube pathway. *Abstracts of the VIIth International Congress on Plant Tissue and Cell Culture*, Amsterdam, The Netherlands, June 1990. Abstract A2-22, p. 49.

Brettel, R.I.S., Pallotta, M.A., Gustafson, J.P. & Appels, R. (1986a) Variation at the Nor loci in triticale derived from tissue culture. *Theoretical and Applied Genetics* 71, 637–43.

Brettel, R.I.S., Dennis, E.S., Scowcroft, W.R. & Peacock, W.J. (1986b) Molecular analysis of a somaclonal mutant of maize alcohol dehyrogenase. *Molecular and General Genetics* 202, 235–9.

Brown, P.T.H., Kyozuka, J., Sukekiyo, Y., Kimura, Y., Shimamoto, K. & Lörz, H. (1990) Molecular changes in protoplast-derived rice plants. *Molecular and General Genetics* 223, 324–8.

Cheng, T.Y. & Smith, H.H. (1975) Organogenesis from callus culture of *Hordeum vulgare*. *Planta* 123, 307–10.

Chin, J.C. & Scott, K.J. (1979) A large scale isolation procedure for cereal mesophyll protoplasts. *Annals of Botany* 43, 23–32.

Chu, C.C., Sun, C.S., Chen, X., Zhang, W.X. & Du, Z.H. (1984) Somatic embryogenesis and plant regeneration in callus from inflorescences of *Hordeum vulgare* × *Triticum aestivum* hybrids. *Theoretical and Applied Genetics* 68, 375–83.

Creissen, G., Smith, C., Franic, R., Reynolds, H. & Mullineaux, P. (1990) *Agrobacterium*- and microprojectile-mediated viral DNA delivery into barley microspore-derived cultures. *Plant Cell Reports* 8, 680–3.

Crossway, A., Hauptli, H., Houck, C.M., Irvine, J.M., Oakes, J.V. & Perani, L.A. (1986)

Micromanipulation techniques in plant biology. *BioTechniques* 4, 320–34.

Dale, P.J. & Deambrogio, E. (1979) A comparison of callus induction and plant regeneration from different explants of *Hordeum vulgare*. *Zeitschrift für Pflanzenzüchtung* 94, 65–77.

Day, A. & Ellis, T.H.N. (1985) Deleted forms of plastid DNA in albino plants from cereal anther culture. *Current Genetics* 9, 671–9.

Dekeyser, R.A., Claes, B., De Ryke, R.M.U., Habets, M.E., Van Montagu, M.C. & Caplan, A.B. (1990) Transient gene expression in intact and organised rice tissues. *Plant Cell* 2, 591–602.

De la Pena, A., Lörz, H. & Schell, J. (1987) Transgenic plants obtained by injecting DNA into young floral tillers. *Nature* 325, 274–6.

DeWet, J.M.J., Bergquist, R.R., Harlan, J.R., Brink, D.E., Cohan, C.E., Newell, C.A. & DeWet, A.E. (1985) Exogenous DNA transfer in maize (*Zea mays*) using DNA-treated pollen. In: Chapman, G.P., Mantell, S.H. & Daniels, W. (eds), *Experimental Manipulation of Ovule Tissue*. Longman, London, pp. 197–209.

Eigel, L. & Koop, H.-U. (1989) Nurse culture of individual cells: regeneration of colonies from single protoplasts of *Nicotiana tabacum*, *Brassica napus* and *Hordeum vulgare*. *Journal of Plant Physiology* 134, 577–81.

Fedak, G. & Grainger, J. (1986) Chromosome instability in somaclones of a *Triticum crassum* × *Hordeum vulgare* hybrid. *Canadian Journal of Genetics and Cytology* 28, 618–23.

Fromm, M., Taylor, L.P. & Walbot, V. (1985) Expression of genes transferred into monocot and dicot plant cells by electroporation. *Proceedings of the National Academy of Sciences (USA)* 82, 5824–8.

Gamborg, O.L., Miller, R.A. & Ojima, K. (1968) Nutrient requirements of suspension cultures of soybean root cells. *Experimental Cell Research* 50, 148–51.

Gaponenko, A.K., Petrova, T.F., Iskakov, A. & Sozinov, A.A. (1988) Cytogenetics of *in vitro* cultured somatic cells and regenerated plants of barley (*Hordeum vulgare* L.). *Theoretical and Applied Genetics* 75, 905–11.

Gasser, C.S. & Fraley, R.T. (1989) Genetically engineering plants for crop improvement. *Science* 244, 1293–9.

Göbel, E. & Lörz, H. (1988) Genetic manipulation of cereals. *Oxford Surveys of Molecular and Cell Biology* 5, 1–22.

Goldstein, C.S. & Kronstrad, W.E. (1986) Tissue culture and plant regeneration from immature embryos of barley *Hordeum vulgare*. *Theoretical and Applied Genetics* 71, 631–6.

Grimsley, N.H., Hohn, T., Davies, J.W. & Hohn, B. (1987) *Agrobacterium*-mediated delivery of infectious maize streak virus into maize plants. *Nature* 325, 177–9.

Haccius, B. (1978) Question of unicellular origin of nonzygotic embryos in callus cultures. *Phytomorphology* 28, 74–81.

Hantke, S. (1990) Studien zur Elektroporation von Gerste-Protoplasten aus Suspensionkultur. *Diplomarbeit, Fach Biologie, Universität zu Köln*. Universität zu Köln, 118 pp.

Hanzel, J.J., Miller J.P., Brinkman, M.A. & Fendos, E. (1985) Genotype and media effects on callus formation and regeneration in barley. *Crop Science* 25, 27–31.

Hughes, B.G., White, F.G. & Smith, M.A. (1979) Fate of bacterial plasmid DNA during uptake by barley and tobacco protoplasts: protection by poly-L-ornithine. *Plant Science Letters* 14, 303–10.

Ishii, C., Okamura, M. & Matsuno, T. (1986) Protoplast isolation and plant regeneration from barley (*Hordeum vulgare*) immature embryo callus. In: Yasuda, S. & Konishi, T. (eds), *Barley Genetics V* (Proceedings of the 5th international Barley Genetics Symposium). Maruzen, Okayama, Japan, October 6–11, 1986.

Jähne, A., Lazzeri, P.A. & Lörz, H. (1991) Regeneration of fertile plants from protoplasts derived from embryogenic cell suspensions of barley (*Hordeum vulgare* L.) *Plant Cell Reports* 10, 1–6.

Jähne, A., Lazzeri, P.A., Jäger-Gussen, M. & Lörz, H. Plant regeneration from embryogenic suspensions derived from anther culture of barley. *Theoretical and Applied Genetics* (in press).

Jelaska, S., Rengel, Z. & Cesar, V. (1984) Plant regeneration from mesocotyl callus of *Hordeum vulgare* L. *Plant Cell Reports* 3, 125–9.

Jones, M.G.K. (1985) Cereal protoplasts. In: Bright, S.W.J. & Jones, M.G.K. (eds), *Cereal Tissue and Cell Culture*. Martinus Nijhoff/W. Junk, Dordrecht, pp. 204–30.

Jorgensen R.B. & Andersen, B. (1989) Karyotypic analysis of regenerated plants from callus cultures of interspecific hybrids of cultivated barley (*Hordeum vulgare* L.). *Theoretical and Applied Genetics* 77, 343–51.

Jorgensen, R.B., Jensen, C.J., Andersen, B. &

von Bothmer, R. (1986) High capacity of plant regeneration from callus of interspecific hybrids with cultivated barley (*Hordeum vulgare* L.) *Plant Cell, Tissue and Organ Culture* 6, 199–207.

Junker, B., Zimny, J., Lührs, R. & Lörz, H. (1987) Transient expression of chimaeric genes in dividing and non-dividing cereal protoplasts after PEG-induced DNA uptake. *Plant Cell Reports* 6, 329–32.

Karp, A. (1991) On the current understanding of somaclonal variation. In: Miflin, B.J. (ed.), *Oxford Surveys of Plant Molecular and Cell Biology*, Vol. 7. Oxford University Press, Oxford, pp. 1–58.

Karp, A., Steele, S.H., Breiman, A., Shewry, P.R., Parmar, S. & Jones, M.G.K. (1987) Minimal variation in barley plants regenerated from cultured immature embryos. *Genome* 29, 405–12.

Kartha, K.K., Chibbar, R.N., Georges, F., Leung, N., Caswell, K., Kendall, E. & Qureshi, J. (1989) Transient expression of chloramphenicol acetyltransferase (CAT) gene in barley cell cultures and immature embryos through microprojectile bombardment. *Plant Cell Reports* 8, 429–32.

Klein, T.M., Wolf, E.D., Wu, R. & Sanford, J.C. (1987) High-velocity microprojectiles for delivering nucleic acids into living cells. *Nature* 327, 70–3.

Kleinhofs, A. & Behki, R. (1977) Prospects for plant genome modification by nonconventional methods. *Annual Review of Genetics* 11, 79–101.

Koblitz, H. (1976) Isolation and cultivation of protoplasts from callus cultures of barley. *Biochemie und Physiologie der Pflanzen* 170, 287–93.

Kott, L.S. & Kasha, K.J. (1984) Initiation and morphological development of somatic embryoids from barley cell cultures. *Canadian Journal of Botany* 62, 1245–9.

Larkin, P.J. & Scowcroft, W.R. (1981) Somaclonal variation – a novel source of variability from cell cultures for plant improvement. *Theoretical and Applied Genetics* 60, 197–214.

Lazzeri, P.A. & Lörz, H. (1988) *In vitro* genetic manipulation of cereals and grasses. *Advances in Cell Culture* 6, 291–325.

Lazzeri, P.A. & Lörz, H. (1990) Regenerable suspension and protoplast cultures of barley and stable transformation via DNA uptake into protoplasts. In: Lycett, G.W. & Grierson, D. (eds), *Genetic Engineering of Crop Plants*. Butterworths, London, UK, pp. 231–7.

Lazzeri, P.A., Kollmorgen, J. & Lörz, H. (1990) *In vitro* technology. In: Chapman, G.P. (ed.), *Reproductive Versatility in the Grasses*. Cambridge University Press, Cambridge, UK, pp. 182–219.

Lazzeri, P.A., Brettschneider, R., Lührs, R. & Lörz, H. (1991) Stable transformation of barley via PEG-induced direct DNA uptake into protoplasts. *Theoretical and Applied Genetics* 81, 437–44.

Ledoux, L. & Huart, R, (1969) Fate of exogenous bacterial deoxyribonucleic acids in barley seedlings. *Journal of Molecular Biology* 43, 243–62.

Lee, B., Murdoch, K., Topping, J., Kreis, M. & Jones, M.G.K. (1989) Transient gene expression in aleurone protoplasts isolated from developing caryopses of barley and wheat. *Plant Molecular Biology* 13, 21–39.

Linacero, R. & Vazquez, A.M. (1986) Somaclonal variation in plants regenerated from embryo calluses in rye *Secale cereale* L. In: Horn, W., Jensen, C.J., Odenbach, W. & Schneider, O. (eds), *Genetic Manipulation in Plant Breeding*. Walter de Gruyter, Berlin, Germany, pp. 479–82.

Lörz, H., Gobel, E. & Brown, P.T.H. (1988) Advances in tissue culture and progress towards genetic transformation of cereals. *Plant Breeding* 100, 1–25.

Luckett, D.J., Rose, D. & Knights, E. (1989) Paucity of somaclonal variation from immature embryo culture of barley. *Australian Journal of Agricultural Research* 40, 1155–9.

Lührs, R. & Lörz, H. (1987) Plant regeneration *in vitro* from embryogenic cultures of spring- and winter-type barley (*Hordeum vulgare* L.). *Theoretical and Applied Genetics* 75, 16–25.

Lührs, R. & Lörz, H. (1988) Initiation of morphogenic cell suspension and protoplast cultures of barley (*Hordeum vulgare* L.). *Planta* 175, 71–81.

Lührs, R. & Nielsen, K. Microspore cultures as donor tissue for the initiation of embryogenic cell suspensions in barley. *Plant Cell Reports* (submitted).

Luo, Z. & Wu, R. (1988) A simple method for transformation of rice via the pollen tube pathway. *Plant Molecular Biology Reporter* 6, 165–74.

Lupotto, E. (1984) Callus induction and plant regeneration from barley mature embryos. *Annals of Botany* 54, 523–9.

Lyne, R.L., Bennett, R.I. & Hunter, C.P. (1986) Embryoid and plant production from cultured barley anthers. In: Withers, L.A. & Alderson, P.J. (eds), *Plant Tissue Culture and its Agricultural Applications*, Butterworths, Guildford, UK, pp. 405–11.

McCabe, D.E., Swain, W.F., Martinell, B.J. & Christou, P. (1988) Stable transformation of soybean (*Glycine max*) by particle acceleration. *Bio/Technology* 6, 923–6.

Mendel, R.R., Müller, B., Schulze, J., Kolesnikov, V. & Zelenin, A. (1989) Delivery of foreign DNA to intact barley cells by high-velocity microprojectiles. *Theoretical and Applied Genetics* 78, 31–4.

Mendel, R.R., Clauss, E., Hellmund, R., Schulze, J., Steinbiss, H.H. & Tewes, A. (1990) Gene transfer to barley. *Proceedings VIIth International Congress on Plant Cell and Tissue Culture*, Amsterdam, June 1990. Kluwer Press, Netherlands, pp. 73–8.

Mesencev, A.V., Butenko, R.G. & Rodionova, N.A. (1976) Obtention of isolated protoplasts from mesophyll of perennial crop plants and barley. *Fiziologia Rastenii* 23, 508–12.

Müller, B., Schulze, J. & Wegner, U. (1989) Establishment of barley cell suspension cultures of mesocotyl origin suitable for isolation of dividing protoplasts. *Biochemie und Physiologie der Pflanzen* 185, 123–30.

Müller, E., Brown, P.T.H., Hartke, S. & Lörz, H. (1990) DNA variation in tissue culture-derived rice plants. *Theoretical and Applied Genetics* 80, 673–9.

Murashige, T. & Skoog, F. (1962) A revised medium for rapid growth and bio-assays with tobacco tissue cultures. *Physiologia Plantarum* 15, 473–97.

Otha, Y. (1986) High-efficiency genetic transformation of maize by a mixture of pollen and exogenous DNA. *Proceedings of the National Academy of Sciences (USA)* 83, 715–19.

Orton, T.J. (1979) A quantitative analysis of growth and regeneration from tissue cultures of *Hordeum vulgare*, *H. jubatum* and their interspecific hybrid. *Environmental and Experimental Botany* 19, 319–35.

Orton, T.J. (1980) Chromosomal variability in tissue cultures and regenerated plants of *Hordeum*. *Theoretical and Applied Genetics* 56, 101–12.

Papes, D., Garaj-Vrhovac, V., Jelaska, S. & Kolevska-Pletikapic, B. (1983) Chromosome behaviour in cultured cell populations of higher plants. In: Brandham, P.E. & Bennett, M.E. (eds), *Kew Chromosome Conference II*. George Allen & Unwin, London, UK, pp. 155–63.

Paszkowski, J., Shillito, R.D., Saul, M.W., Mandak, V., Hohn, B. & Potrykus, I. (1984) Direct gene transfer to plants. *EMBO Journal* 3, 2717–22.

Pickering, R.A. (1989) Plant regeneration and variants from calli derived from immature embryos of diploid barley (*Hordeum vulgare* L.) and *H. vulgare* L. × *H. bulbosum* L. crosses. *Theoretical and Applied Genetics* 78, 105–12.

Potrykus, I. (1990) Gene transfer into cereals: an assessment. *Bio/Technology* 8, 535–42.

Powell, W., Hayter, A.M., Wood, W., Dunwell, J.M. & Huang, B. (1984) Variation in the agronomic characters of microspore-derived plants of *Hordeum vulgare* cv. Sabarlis. *Heredity* 52, 19–23.

Powell, W., Caligari, P.D.S. & Dunwell, J.M. (1986a) Field performance of lines derived from haploid and diploid tissues of *Hordeum vulgare*. *Theoretical and Applied Genetics* 72, 458–65.

Powell, W., Borrino, E.M., Allison, M.J., Griffiths, D.W., Asher, M.J.C. & Dunwell, J.M. (1986b) Genetical analysis of microspore-derived plants of barley (*Hordeum vulgare*). *Theoretical and Applied Genetics* 72, 19–26.

Powell, W., Borrino, E.M. & Goodall, V. (1988) The effect of anther orientation on microspore-derived plant production in barley (*Hordeum vulgare* L.). *Euphytica* 38, 159–63.

Saalbach, G. & Koblitz, H. (1978) Attempts to initiate callus formation from barley leaves. *Plant Science Letters* 13, 165–9.

Schweiger, H.G., Dirk, J., Koop, H.U., Kranz, E., Neuhaus, G., Spangenberg, G. & Wolff, D. (1987) Individual selection, culture and manipulation of higher plant cells. *Theoretical and Applied Genetics* 73, 769–83.

Seguin-Swartz, G., Kott, L. & Kasha, K.J. (1984) Development of haploid cell lines from immature barley, *Hordeum vulgare* embryos. *Plant Cell Reports* 3, 95–7.

Singh, R.J. (1986) Chromosomal variation in immature embryo derived calluses of barley (*Hordeum vulgare* L.). *Theoretical and Applied Genetics* 72, 710–16.

Singh, R.J. & Tsuchiya, T. (1977) Morphology, fertility and transmission in seven monotelotrisomics of barley. *Zeitschrift für Pflanzenzüchtung* 78, 327–40.

Soyfer, V.N. (1980) Hereditary variability of plants under the action of exogenous DNA. *Theoretical and Applied Genetics* 58, 225–35.

Teeri, T.H., Patel, G.H., Aspgren, A. & Kauppinen, V. (1989) Chloroplast targeting of neomycin phosphotransferase II with a pea transit peptide in electroporated barley mesophyll protoplasts. *Plant Cell Reports* 8, 187–90.

Thomas, M.R. & Scott, K.J. (1985) Plant regeneration by somatic embryogenesis from callus initiated from immature embryos and immature inflorescences of *Hordeum vulgare*. *Journal of Plant Physiology* 12, 159–69.

Thompson, D.M., Chalmers, K., Waugh, R., Forster, B.P., Thomas, W.T.B., Caligari, P.D.S. & Powell, W. The inheritance of biochemical, molecular and morphological markers in microspore-derived plants of barley *Hordeum vulgare* L. *Theoretical and Applied Genetics* (in press).

Töpfer, R., Gronenborn, B., Schell, J. & Steinbiss, H.H. (1989) Uptake and transient expression of chimaeric genes in seed-derived embryos. *Plant Cell* 1, 133–9.

Toyoda, H., Yamaga, T., Matsuda, Y. & Ouchi, S. (1990) Transient expression of the β-glucuronidase gene introduced into barley coleoptile cells by microinjecton. *Plant Cell Reports* 9, 299–302.

Tuschiya, T. (1960) Cytogenetic studies of trisomics in barley. *Japanese Journal of Botany* 17, 177–213.

Vasil, I.K. (1987) Developing cell and tissue culture systems for the improvement of cereal and grass crops. *Journal of Plant Physiology* 128, 193–218.

Vasil, I.K. (1990) The realities and challenges of plant biotechnology. *Bio/Technology* 8, 296–301.

Vasil, V. & Vasil, I.K. (1982) The ontogeny of somatic embryos of *Pennisetum americanum* (L) K. Schum I in cultured immature embryos. *Botanical Gazette* 143, 454–65.

Weigel, R.C. Jr & Hughes, K.W. (1985) Long term regeneration by somatic embryogenesis in barley (*Hordeum vulgare* L.) tissue cultures derived from apical meristem explants. *Plant Cell, Tissue and Organ Culture* 5, 151–62.

Yan, Q., Zhang, X., Shi, J. & Li, J. Green plant regeneration from protoplasts of barley (*Hordeum vulgare* L.). *Kexue Tonqbao* (in press).

Ziauddin, A. & Kasha, K.J. (1990) Long-term callus cultures of diploid barley (*Hordeum vulgare* L.). II. Effects of auxins on chromosomal status of cultures and regeneration of plants. *Euphytica* 48, 279–86.

27

The Case of High-lysine Barley Breeding

L. Munck

A provisional, primary selection hypothesis

> Only by working out a theory of breeding will the investigator be able to achieve real control over the organism – the ultimate aim of modern biology
>
> N.I. Vavilov

In trying to review the achievements of the scientific community in launching a nutritious barley with a drastically improved amino acid balance, one may wonder from where did the idea originally come? I do not have a simpler answer than that implicit in this paper. The case of high-lysine barley is in itself an excellent example of the close but rather unpredictable bonds between science and society which the puritans of classical science tended to overlook. Let us, therefore, provisionally accept that at the beginning of the 1960s, information about amino acids, plants, animal and human nutrition and social human conditions came together and created, rightly or wrongly, an awareness of a 'protein crisis' on a global level, directed especially towards the poor developing countries with their rapidly expanding populations (Waterlow, 1970). Cereals in particular, with wheat, maize and rice supplying more than half of the food resource, were considered nutritionally deficient, due to a low content of essential amino acids such as lysine, threonine, methionine, cysteine, tryptophan and valine, if not supplemented with proteins from pulses or animals. The conviction of 'the market' to the paradigm of 'the protein crisis' was indeed serious as indicated by FAO/WHO (1965) . . . 'Only 93% of the world protein needs are met' . . . The scientific community responded in a searching way rather vacillatingly to these demands (and opportunities) from 'the market'.

Few of us, if any, were able to forecast at the beginning of the 1960s, the existence of simply inherited Mendelian recessive genes in cereals which could drastically change the protein and amino acid patterns in cereals to approach the nutritional quality of milk protein or meat. In 1963, I joined a research group in Svalöv, Sweden, to tackle the quality concept along the production chain by working on the primary concept of 'Improving the nutritional value of cereals' (Munck, 1972) in which the protein issue was just a part. Since 1973, I have been working in Denmark on elucidating the structure and quality control in production chains employing the essential human approach of 'Man as selector' (Munck, 1991), e.g. with respect to the utilization of the renewable plant resources (Bjørn Petersen & Munck, in press; Munck, in press a) and the production of beer (Munck, in press b). The case of high-lysine barley is a good example of a 'local' effort to address an apparently global problem such as 'the protein crisis' through science. The headings in this chapter feature the stages of the selection cycle, which experimental man uses according to the model (Munck, 1991) to achieve and judge information.

Development of screening methods

At the end of last century, E. Fischer laid the foundations for the basic understanding of proteins and amino acids. Using the laborious hydrolysis and crystallization methods of classical organic chemistry, Osborne discovered the amino acid composition of protein fractions from seeds, extracted with solvents such as water, salt solution, hot ethanol and diluted sodium hydroxide (see review by Osborne, 1928). He then found that there were differences in the amino acid compositions of protein fractions of seeds, both within and between different species. By selecting extreme fractions such as the alcohol-soluble zein from maize and the water/salt-soluble edestin from hemp, Osborne, in co-operation with the nutritionist Mendel (Osborne & Mendel, 1914), formulated critical feed rations to laboratory rats. This showed that, on a maize diet, lysine was the first and tryptophan the second limiting amino acid for growth. In 1935, Rose and his co-workers following essentially the same methodology defined the last of the eight amino acids essential for mammals – threonine (Rose *et al.*, 1935). In the 1940s, paper and ion-exchange chromatography introduced through Moore & Stein revolutionized amino acid analysis by reducing the time required from a matter of months to about a day. This was still, however, too time-consuming for large-scale screening. During the 1950s a library of increasingly reliable amino acid composition data of foods and feeds from animals and plants, their tissues under different physiological conditions and individual purified proteins was established. Thus, the fundamental role of a proper balance between the essential amino acids in foods for optimum growth and health became gradually acknowledged by agriculture, the industry and the public.

The new analytical methods began to be used systematically to reveal 'the laws of nature'. In 1928, Bishop had demonstrated 'the nitrogen regulation principle in barley', where the ethanol-soluble hordein fraction increased with increasing protein content at the expense of water and salt-soluble proteins, with slightly different rates in different varieties (Bishop, 1928). In 1956, Postel indicated the effect of this law on the level of lysine in barley; the percentage of lysine in protein showed a negative regression with increased protein levels in the seed, because the concentrations of hordeins were much lower in lysine than the salt-soluble proteins were (Postel, 1956).

In 1962, when studying the effect of heat treatment on the nutritional quality of barley proteins, with emphasis on lysine which is labile, I sent a few barley samples, which had been shown by mice tests to have a low lysine availability, to Roger Mossberg. He analysed them using the dye-binding technique with diazo-dye acilane orange previously used by Udy in 1954 for protein determination in cereals (Munck, 1966). We found a perfect correlation between mice growth, lysine content and dye-binding on these more or less heat-damaged barley samples, indicating that the acilane orange dye bound to lysine and the other basic amino acids, histidine and arginine. The empirical use of different constants by Udy for different cereals in measuring protein by dye-binding were found to be proportional to their total content of basic amino acids. The dye-binding method was further perfected by Mossberg (1969) as a screening method for lysine using Kjeldahl protein analysis as a reference to compensate for differences in lysine due to fluctuation in protein contents. This meant that about

50–100 lysine analyses could be performed per person per day. It is fascinating to realize that the organic chemists who at the end of the 19th century developed the diazo-dye acilane orange for textile dying could well have launched dye-binding 90 years ago as a screening analysis for lysine if they had been aware of its application (Munck, 1983).

At the end of the 1960s and 1970s the International Atomic Energy Commission in Vienna organized a series of useful symposia stimulating the development of a wide range of screening analyses for lysine and their use to search for high-lysine mutants (see reviews by Kaul, 1973 and Georgi *et al.*, 1979). Further development revealed that there were many more scientists available for making new analyses and studying basic biochemistry than for applying them in long-term efforts in screening programmes and breeding. Since then, a range of convenient and specific screening methods using immunological and molecular biological techniques have been developed, targeting a specific protein or a specific nucleic acid sequence at the RNA or DNA level. At the Carlsberg Research Laboratory, immunological and immunofluorescence methods have been used (Vaag & Munck, 1987) to visualize and quantify the tissue-specific synthesis of a protein (enzyme) such as α-amylase (Gibbons, 1980) and in large-scale screening with chymotrypsin inhibitor-2, for example (Rasmussen, 1985, 1988). We have also applied sensitive ELISA immunological screening techniques, to malting and brewing to follow the α-amylase subtilisin inhibitor (ASI) from barley (Munck *et al.*, 1984). It is clear that the techniques now routinely used in molecular and biological laboratories could also be applied to screening at the breeding level after some modification, if a feasible concept is provided.

As will be discussed later, the selection for genes conferring improved nutritional quality of cereal protein using the lysine trait results in a range of other diverse changes in the plant seed phenotype, some of which can be easily seen. Thus Nelson (see reviews 1969, 1979), working with starch mutants in maize, found that the mutant seeds were often opaque in translucent light because grain filling was incomplete, leading to air spaces. He realized that the same phenotype could be produced by genes influencing protein synthesis and sent a few opaque mutants for amino acid analysis in E.T. Mertz's laboratory, which, in 1964, resulted in the discovery of the first high-lysine mutants *opaque*-2 and *floury*-2 (see review by Mertz, 1976). Another indirect morphological feature enabling the screening for certain types of high-lysine genes is the presence of a large embryo/scutellum. Other secondary characteristics of high-lysine genes such as high amounts of free amino acids can be screened through the ninhydrin reaction, and the absence of prolamins can be screened through the prolamin turbidity test (Rhodes, 1975).

Selection

The identification of the first high-lysine mutant *opaque*-2 in maize and the subsequent range of high-lysine mutants (Nelson, 1969) by screening for morphological characters is an illustrative example of how an uncomplicated approach may reveal mechanisms which are simply inherited but have extremely complex physiological implications. The identification, in Svalöv, Sweden, of the Hiproly (*lys1*) genotype from the world barley collection, Beltsville, USA, in 1967, is the first example of a high-lysine mutant selected by a chemical screening analysis for lysine, namely the dye-binding method (Hagberg & Karlsson, 1969; Munck *et al.*, 1970). In Fig. 27.1, the dye-binding method in combination with Kjeldahl protein analysis is used to demonstrate how normal and high-lysine plants from a cross are segregated for the *lys1* gene. In Fig. 27.2, the same material is analysed for lysine as a check. It is obvious that the gene offsets 'the natural law' previously cited, where, in normal barley, lysine, as a percentage of total protein, is negatively correlated with protein. In the segregating population containing the *lys1* gene, this correlation is now elevated to a level about 30% higher and the negative slope is less pronounced. The

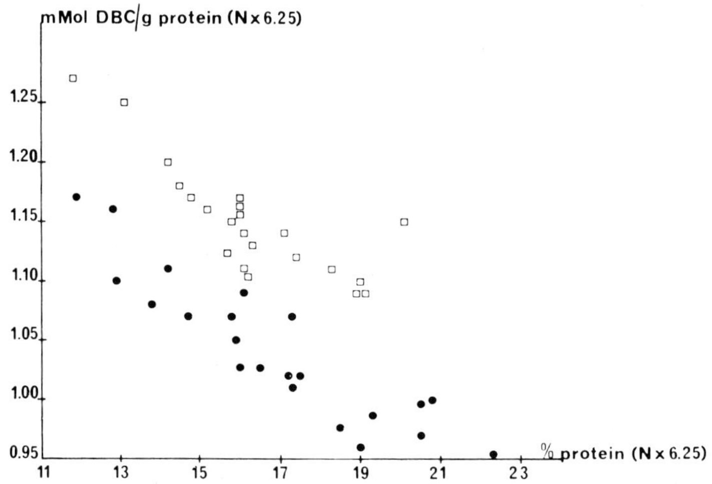

Fig. 27.1. Segregation for dye-binding (mM/g protein) at different crude protein levels in F_2 plants from Hiproly (*lys1*) × normal lysine barley. High DBC (dye-binding capacity) plants are indicated with squares and normal DBC plants with dots (Munck, 1972).

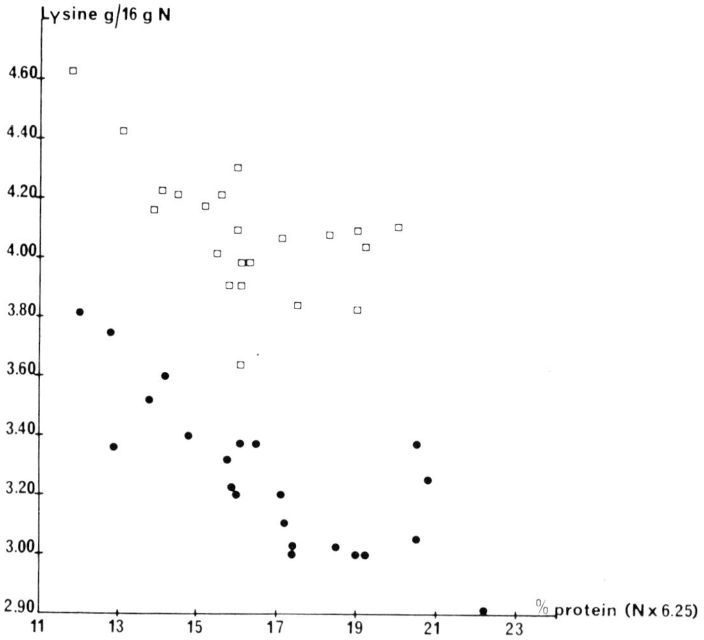

Fig. 27.2. Segregation for lysine (g/16 g N) at different crude protein levels in F_2 from the material in Fig. 27.1 (Munck, 1972).

dye-binding screening analysis has been successfully used by Doll and co-workers at Risø, Denmark since 1969 (see reviews by Doll, 1976, 1983 and 1984). About 17 000 lines from Carlsberg II and Bomi barley backgrounds were treated with radiation and chemicals and screened by the dye-binding method. Other high-lysine barley mutants have been reported by Bansal (1977), Di Fonzo & Stanca (1977) and Oram & Doll (1981).

In sorghum, J. Axtell and his group at the Purdue University (see review by Axtell *et al.*, 1981) utilized the opaque character in screening the world sorghum collection and found two high-lysine lines originating from Ethiopia with slightly dented seeds. A mutagen-treated population of 23 000 M_3 heads gave one opaque high-lysine sorghum line with reasonable seed quality. Mutagen-treated naked barley were screened for opaqueness at the Carlsberg Research Laboratory and a range of mutants with high dye-binding but with poor seed quality was found (unpublished results). Doll (1983) reported the use of single seed electrophoresis of 3000 M_3 seeds giving one high-lysine mutant. In 1983, mutagen-treated materials were screened for the β-glucan trait of importance in malting barley and three low β-glucan mutants were found, one of which, M-1460, was also surprisingly high in lysine (Aastrup, 1983).

In general, it can be concluded that high-lysine mutants can be isolated readily by widely differing methods, showing a surprisingly high mutation rate compared with other characters.

The following discussion concentrates on the high-lysine barley genes which have

Table 27.1. Amino acid composition of a Hiproly line, mutant 1508, mutant 7 and mutant 56 expressed as related to normal barley, var. Tellus, and Carlsberg II.

	Normal barley var. Tellus = 100				Carlsberg II = 100
	Tellus[1] g/16g N	*lys1*[2] line	*lys3a*[1]	Mutant 7[1]	Mutant 56[3]
Lys	3.16	127	148	115	121
His	2.11	111	123	113	103
Arg	4.76	111	139	108	107
Asp[4]	5.30	120	170	113	115
Tre	3.11	113	121	119	97
Ser	4.17	106	106	108	75
Glu[4]	26.35	92	61	92	87
Pro	11.29	90	54	98	90
Gly	3.72	113	139	123	117
Ala	3.85	120	130	111	108
Cys	2.17	88	97	107	-
Val	4.76	113	105	104	98
Met	1.63	119	109	105	97
Ileu	3.46	110	92	97	93
Leu	6.90	111	93	105	97
Tyr	3.20	105	103	108	98
Phe	5.06	108	75	90	106
Protein N × 6.25	12.63	11.60	13.69	12.19	13.10

[1] Tallberg & Eggum (1981)
[2] Triumph × (Varunda × (Hiproly × Mona[5])), Tallberg & Eggum (1986)
[3] Bang-Olsen (1991), pers. comm.
[4] Asp and Glu include their amide forms Asn and Gln

been studied in most detail, namely *lys1* (from Hiproly), *lys3* (with 4 alleles of Risø mutants 1508, 18 and 19 and Carlsberg mutant 1460), mutant 7 from the Bomi background, and *hor2* ca (from Risø mutant 56) from the Carlsberg II background. The high-lysine barley genes are shown in Table 27.1 with regard to amino acid composition. *lys3* is the most radical mutant with regard to lysine increase followed by *lys1*, mutant 56 and mutant 7. Changes in other amino acids generally follow the same pattern with increases in aspartic acid, threonine, arginine, glycine and alanine whilst proline and glutamic acid + glutamine are significantly reduced. The different genes (mutants) differ individually in their effects on specific amino acids. Note, for example, that *lys1* shows a significant decrease in cysteine while the increases in methionine and isoleucine and leucine are especially high compared with the other genotypes.

Testing and characterization

Gross composition, agronomy and basic genetics

Hiproly and *lys1*

It is difficult to select a mutant such as the Hiproly high-lysine barley line and try to define its advantages and drawbacks, agronomically, genetically, biochemically and nutritionally with limited time and material. Firstly, the environmental stability of the high dye-binding high-lysine trait in relation to the crude protein content was established for Hiproly (CI 3947) grown at various localities and in the phytotron and was compared with its putatively isogenic sister-line (CI 4362) as well as with commercial barley varieties (Munck *et al.*, 1970; Munck, 1972). The two Ethiopian cultivars were not adapted to Scandinavian growing conditions and the yield of Hiproly was lower than that of CI 4362 – and only about 30% of that of commercial barley varieties, due to smaller shrivelled seeds. However, high-lysine plants from crosses between Hiproly and normal varieties were only 10% lower in yield compared with the normal lysine segregants from the same cross. The high-lysine seeds were plump although 15% smaller in size (Munck, 1972). The high-lysine gene was preliminarily called *lys1* (Munck, 1972) and was found to be recessive and located near to the character s (short rachilla hairs) on the long arm of chromosome 7 (Karlsson, 1972). The *lys1* gene was associated with a hard endosperm which, however, could be modified into a soft high-lysine type found at a low frequency in crosses.

Interestingly, in a pedigree selection test (Munck, 1972) the soft endosperm high-lysine plants did show a significant negative correlation ($r = -0.9$) between lysine g/16 g N and Kjeldahl protein while the same relation was insignificant ($r = 0.0$) for the hard high-lysine control lines. The lysine content of the protein of Hiproly in relation to its sister-line was increased by 35%, with protein content as high as 16% (Munck, 1972). In the pedigree material, the difference in lysine g/16 g N between the high-lysine lines and the normal ones was reduced to 23% (Munck, 1972) at a normal protein level of about 11–12%. The Svalöf breeding programme involved extensive back-crosses to high yielding, large seeded lines (Persson & Karlsson, 1977). The best cross with the Kristina variety yielded 96–99% relative to the control variety (Ingrid) while the increase in lysine g/16 g N ranged from 16 to 25%. Genetic effects modifying the expression of the *lys1* gene were also observed (see Table 27.2) and apparent in segregation ratios different from 1:3 in two crosses out of four (Munck *et al.*, 1970). Thus, in using the *lys1* trait in breeding, frequent testing with the DBC-test checked with lysine analyses is necessary due to the influence of the genetic background together with the breeding for yield performance. In order to really exploit the *lys1* gene nutritionally, about 2% higher protein level in the seed compared to normal barley is required (Persson & Karlsson, 1977). These criteria are, however, difficult to combine with high yield.

At an early stage it was found that the water-soluble (albumin) fraction which was increased in lysine, also displayed at least

Table 27.2. Amino acid compositions (g/16g N) of barley lines in F_2, F_4 and F_5 (Hiproly × normal barley) (Munck, 1972).

Amino acids	F_2 plants in pots outdoors 1969	Selected line from F_3		Controls	
		F_4 plants in field 1970	F_5 plots in field 1971	1970	1971
Normal segregant 69/21257				Normal CI 4362	
Lys	3.39	3.55	–	2.95	3.10
Asp	6.57	6.44	–	4.87	5.56
Glu	26.35	24.12	–	25.22	25.54
Protein (N × 6.25)	14.8	13.9	–	17.20	14.30
High-lysine modified segregant 69/21301					
Lys	3.91	3.92 ± 0.12 (*n* = 7)	3.97		
Asp	6.64	6.84 ± 0.19	6.26		
Glu	24.80	23.02 ± 0.72	22.66		
Protein (N × 6.25)	16.2	13.4 ± 0.7	13.2		
High-lysine segregant 69/21901				Hiproly CI 3947	
Lys	4.22	4.51 ± 0.02 (*n* = 2)	4.30	3.98	4.19
Asp	7.18	8.04 ± 0.05	7.01	6.33	6.80
Glu	24.72	20.54 ± 0.20	21.60	21.69	22.73
Protein (N × 6.25)	14.5	12.1 + 1.2	13.0	17.7	16.4

four protein bands of increased strength in polyacrylamide electrophoresis (Munck, 1972). These proteins were further defined and studied by Jørgen Hejgaard and his colleagues at the Technical University of Denmark, Lyngby, (Hejgaard & Boisen, 1980) as well as by Ib Jonassen (Jonassen & Svendsen, 1982), John Mundy (Mundy *et al.*, 1983) and Robert Leah (Leah & Mundy, 1989) in our group at the Carlsberg Research Laboratory. It was obvious that the *lys1* gene increased the proportion of the hydrophilic proteins in the barley seed and also changed the extractability of proteins in the fractionation procedure. For example, one band found in the ethanol-soluble prolamine fraction in the sister-line migrated to the water-soluble fraction in Hiproly (Munck, 1972). In Hiproly, cysteine was lower and methionine higher in all fractions except for the ethanol-soluble fraction. In general, ratios such as dye-binding and lysine versus Kjeldahl protein, methionine versus cysteine, lysine versus arginine and aspartic acid versus glutamic acid were useful to define the *lys1* biochemical phenotype. Prolamins seemed only slightly reduced in *lys1* barley seeds compared with normal although the synthesis of prolamins during seed development was significantly delayed in Hiproly compared with the sister-line (Munck, 1972). The *lys1* gene of Hiproly, the first high-lysine gene isolated outside maize, was therefore different from the *opaque*-2 gene of maize which displayed a major depression of the synthesis of the prolamin proteins (Nelson, 1979).

Contemplating the problems encountered in breeding for yield in maize, sorghum and barley where practically all high-lysine mutants display retarded synthesis of both prolamins and seed carbohydrates (starch), Nelson (1979) cites a hypothesis advanced by Tsai *et al.* (1978) that maximum yield is attained only in genotypes that synthesize maximum amounts of prolamin. Nelson also advances a hypothesis that in high-lysine mutants such as *opaque*-2 in maize, the primary lesion affects prolamin synthesis and the enhanced supply of nitrogen made available by this partial block effectively derepresses (stimulates) the synthesis of some of the albumin and globulin proteins. In parallel, Nelson (1979) however, also launched a pragmatic approach to high-lysine plant breeding: 'It is obvious that considerable progress can be made by selecting for genetic backgrounds that interact harmoniously with the high-lysine genes'.

The following shows that, in the discussion on the feasibility of high-lysine barley breeding, these two hypotheses are those of major concern. The first negatively formulated centring on facts obtained through scientific analysis of present achievements; the second positively formulated based mainly on options regarding the pragmatic approach of selection and breeding and a trust in a great adaptability of the plant genome.

Mutant 1508 and *lys3*

Doll and his group started to screen mutagen-treated materials with the dye-binding/protein methods in 1969 and found a wide range of mutants. Three of them will be discussed here. The most well known is the 1508 mutant induced with ethylenimine in Bomi barley (Ingversen *et al.*, 1973), displaying a lysine increase in the barley protein of no less than 44%, approaching the nutritional value of milk protein, and associated increases in several other essential amino acids. The 1508 mutant is highly variable in yield compared to Bomi, ranging in different years from a relative figure of 65 to 93% (see Table 27.7 below). It is about 10% smaller in seed size and the seed is shrivelled. The gene *lys3a*, causing the high-lysine trait, is recessive and situated in the centromere region of chromosome 7 (Jensen, 1979). Two other alleles, mutant 18 (*lys3b*) and mutant 19 (*lys3c*), were found in Bomi material treated with sodium azide. A fourth allelic mutant, mutant 1460, was identified in a sodium azide-treated material of the high β-glucan variety Minerva by analysing for low β-glucan content (namely, viscosity of acid barley extracts) by Aastrup from our research group in 1983. The 1508 mutant seems softer and gives less resistance in milling than the mother variety Bomi (Allison *et al.*, 1979). Like the *opaque*-2 mutant it has an opaque phenotype when compared with the mother variety in transmitted polarized light and, in contrast to *lys1*, it has a 50% increased level of free amino acids in the seed. The prolamin fraction is reduced in mutant 1508 by 66% while the albumin/globulin fraction is increased correspondingly and the glutelin fraction is constant. The Risø group has studied transzygotic lines derived from chromosome-doubled haploids made from F_1 material with both *lys1* and *lys3a* genes. In the *lys1* material (Balasaraswathi *et al.*, 1984), the *lys1* lines had 16% lower grain yield and 13% lower seed weight than the corresponding normal lines. Hordeins were reduced from 49.3% in the normal to 43.0% in the high-lysine lines. In similar material displaying the *lys3a* gene (Doll & Køie, 1978) grain yield was in the high-lysine condition reduced by 20%. More than 90% of this reduction was caused by lower carbohydrate production. The percentage of total DMSO-extractable carbohydrates including starch was on average 54.9 in grains from normal lines and 45.7 in high-lysine lines. Lysine production ($g\,m^{-2}$), however, was increased by 26% while hordein, as a percentage of total protein, decreased from 46.5% in the normal to 13.7% in the *lys3a* plants. Similarly, under Australian conditions (Oram & Doll, 1981) a *lys1* line was 11% lower in DMSO-extractable carbohydrates of the whole ground seed than Bomi, while mutant 1508 was 19% lower. Thus, the available data on practically all high-lysine barley mutants now pointed in favour of the theory advanced by Tsai *et al.* (1978), showing a relationship between carbohydrate yield and the synthesis of prolamins as expressed in the reviews by Doll (1983, 1984) and Tallberg (1982), favouring the *lys1* Hiproly model despite the limited breeding success with this gene. The idea was to increase the production of lysine-rich endosperm proteins with a minimum of interference with prolamin production, to minimize the putatively negative effect of a decrease of the latter on carbohydrate (starch) production and yield.

Much less effort seems to have been directed towards the alternative positively formulated selection hypothesis to evaluate high-lysine barley namely, trying to breed for an improved gene background for the high-lysine gene for yield. Jenkins *et al.* (1978), however, initially had positive experience with regard to yield of mutant 1508 segregants in trials in England and Oram & Doll (1981) in trials in Australia showed that some of their crosses with the *lys1* and *lys3a* genes displayed improved yield and carbohydrate content. In *opaque*-2 maize, Vasal *et al.* (1979) demonstrated that high-lysine levels with improved yield could be obtained by crossing. An estimate of a strategy for breeding of high yielding high-lysine cereals was summarized by Munck, 1978.

1. Find a new genetic background for existing major genes to modify the deleterious part of the pleiotropic complex.
2. Search for new major mutants with better agronomic characteristics using mass screening methods, e.g. the opaqueness of seed or the hordein turbidity method.
3. Identify and combine in the same genotype genes with minor effect on amino acid composition but which through an additive effect might produce a major increase in lysine without substantially negative influence on yield.

When starting up the breeding programme at Carlsberg in 1973 on the first two points, the first priority was to find a favourable genetic background for the Risø 1508 mutant. The reason for selecting this mutant was its superior amino acid composition and the fact that it showed practically no genetical or environmental modification in its expression, thus differing from the *lys1* gene which could be modified genetically as previously discussed (see Table 27.2). The breeding effort, including the work of Kirsten Bang-Olsen and Bodil Stilling, with the *lys3* genes is now described.

Mutant 7 and its double recessives with *lys3a* and *lys1*

In breeding for new alternatives to reach a better yield in high-lysine barley, Tallberg has made an innovative approach to the strategy point 3 described earlier in combining two genes with moderate to low effect, namely the *lys1* gene with the gene from the mutant 7 (Tallberg, 1981), the later induced by thermal neutrons in Bomi, the same background as mutant 1508 (Doll, 1976). Mutant 7 displays a low (15%) increase in lysine g/16 g N (Table 27.3). It is unique because it is the only high-lysine barley mutant that does not have any reduction in seed size and starch content compared to the mother line (Køie & Doll, 1979). The yield of mutant 7 is, however, decreased by 10% mainly due to a reduced seed number compared with Bomi. The mutant also has a normal level of free sugars differing from mutant 1508. In contrast, it resembles mutant 1508 in showing a decrease in prolamins, although only 20%, and a corresponding increase in non-prolamin proteins. It has not been possible yet to place the mutant 7 gene on the barley gene map because of difficulties in screening (H. Doll, pers. comm.). Both the Hiproly *lys1* gene and the mutant 7 gene are characterized by the fact that a significant part of the improvement in lysine content comes from proteins in the glutelin fraction (Tallberg, 1981, 1982).

While the double recessive between *lys1* (Hiproly) and *lys3a* (mutant 1508) is agronomically poor despite a favourable amino acid composition (Table 27.3) – about 5.4% lysine g/16 g N – the double recessive between *lys1* and the high-lysine gene of mutant 7 is agriculturally much more promising (Tallberg, 1981; Tallberg & Eggum, 1981, 1986) with a lysine increase of about 35%. It also displays completely plump seeds, although with a 20% reduc-

Table 27.3. Amino acid composition of double mutants *lys1*/mutant 7 and *lys1*/*lys3a* compared with parent lines and expressed relative to normal barley (see also Table 27.1).

	Normal barley var. Tellus = 100[1]				
	lys1[2]	*lys1*/ mutant 7	Mutant 7	*lys1*/ *lys3a*	*lys3*
Lys	127	135	115	165	148
His	111	109	113	122	123
Arg	111	111	108	163	139
Asp	120	128	113	167	170
Tre	113	119	119	119	121
Ser	106	109	108	96	106
Glu	92	81	92	52	61
Pro	90	77	89	44	54
Gly	113	120	123	137	139
Ala	120	120	111	138	130
Cys	88	84	107	81	97
Val	113	113	104	108	105
Met	119	127	105	110	109
Ileu	110	104	97	95	92
Leu	111	105	105	91	93
Tyr	105	102	108	94	103
Phe	108	93	90	75	75

[1] Tallberg & Eggum (1981)
[2] Triumph × (Varunda × (Hiproly × Mona[5])), Tallberg & Eggum (1986)

tion in seed size. The *lys1*/mutant 7 double recessive has a reduction in available carbohydrates of only 5% and the hordein content is unchanged compared with mutant 7. The increased glutelin fraction is the major contributor to the increased lysine content in the *lys1*/mutant 7 double recessive in spite of the fact that it has a higher lysine content in the water-soluble protein fraction than even the Hiproly backcross line (Tallberg, 1981). As seen in Table 27.3, several of the amino acids of the two double recessives display transgressive expression over their parents, positive for lysine, aspartic acid (only for *lys1*/mutant 7), alanine and methionine and negative for glutamic acid, proline and cysteine. Note the low cysteine values in all *lys1* genotypes.

Thus the *lys1*/mutant 7 double recessive opportunity as investigated by Tallberg gave the first indication in barley that it could be worthwhile to combine genes with minor effects to get enhanced expression of the quality trait, minimizing the negative effects on seed carbohydrate synthesis and yield while keeping the prolamins at a high level. The different *lys1*/mutant 7 lines varied from 74 to 92% in yield compared to a normal barley (Tallberg & Eggum, 1981), which indicated a base for breeding efforts, but this possibility has not been followed up further.

Mutant 56 or *hor2ca*

Another theoretically interesting recessive mutant is 56 induced in Carlsberg II barley by X-rays (Doll, 1976). Seed size is reduced by 10%, starch by 8% and prolamins by 23%, while lysine g/16 g N is increased by 24% compared with Carlsberg II (Køie & Doll, 1979). The changes in amino acid pattern of mutant 56 in comparison with normal barley (see Table 27.1) show similarities to those of mutant 7 and *lys1*. The low content of the prolamin fraction in mutant

56 is due to a mutation at or near the *Hor2* locus coding for hordein-2 (B hordein) polypeptides in chromosome 5, making the mutant deficient in these polypeptides. The mutation site is, therefore, called *hor2ca* (Doll, 1980). This is probably the first case where a high-lysine mutant is apparently proved to be identical with a structural locus for a protein (hordein). The other high-lysine genes investigated are of putative regulatory character and situated on chromosomes other than the structural genes for the proteins, the synthesis of which they depress or enhance. However, mutant 56 is not a clear-cut hordein-2 (B hordein) deficient mutant because the same water-soluble proteins which are overexpressed in the *lys1* gene are also increased in mutant 56. In mutant 56 there is also a translocation between chromosomes 2 and 5. Although no linkage has been detected between the site for *hor2ca* mutation and the translocation, this structural rearrangement makes an evaluation of the effects of the *hor2ca* mutant more complex.

Nutritional quality

Feeding trials with laboratory animals, rats and mice, (Munck *et al.*, 1970; Munck, 1972; Doll *et al.*, 1974) confirmed the improved nutritional balance of the high-lysine barley selections, Hiproly and mutant 1508. The biological value (BV) or nitrogen retention in rats was clearly improved, especially for the mutant 1508 while Hiproly

Table 27.4. Relative yield, 1000 kernel weight, chemical composition and nutritional quality in rat trials (TD, BV and NPU) of normal barley (Tellus) and *lys1*, *lys3a*, mutant 7 and double recessives *lys1* (mutant 7 and *lys3a*). For the corresponding amino acid composition, see Tables 27.1 and 27.3.

	Normal[1,2] barley (Tellus)	*lys1*[2]	*lys3a*/mutant 1508[1]	Mutant 7[1]	*lys1*/mutant 7[1] $n = 7$	*lys1*/ *lys3a*[1]
Relative yield	100	94.0	81.0	90.0	74–92	71.0
1000 KW(g)	50.5[1] 48.3[2]	 42.5	45.5	52.0	32.0–38.5	32.0
Protein N × 6.25 (% DM)	12.6[1] 10.3[2]	 11.6	13.7	12.2	11.9–14.1	14.7
Fat[3] (% DM)	3.8[1] 3.7[2]	 3.7	5.8	3.7	3.2–4.2	5.7
Carbohydrates[4] (% DM)	71.6[1] 69.4[2]	 71.4	60.2	70.7	65.7–69.5	59.4
Lysine g/16 g N	3.16[1] 3.56[2]	 4.52	4.67	3.63	4.1–4.28	5.22
TD True Digestibility (N%)	91.0[1] 87.3[2]	 89.4	86.7	90.2	88.9–90.3	88.4
BV (Biological Value %)	74.4[1] 79.8[2]	 85.2	91.8	85.5	85.5–88.7	89.8
NPU (Net Protein Utilization, %)	67.8[1] 69.7[2]	 76.1	79.6	77.1	77.1–79.5	79.4

[1]Tallberg & Eggum (1981)
[2]Tallberg & Eggum (1986)
[3]Fat determined after hydrolysis
[4]Available carbohydrates determined as starch and sugar

was intermediate between this extreme and normal barley. The increased level of water-soluble high-lysine proteins in Hiproly barley was reflected in improved protein and lysine digestibility in rat nitrogen balance studies. Mutant 1508 compared with the mother variety Bomi was, however, two absolute per cent lower in protein digestibility. In Table 27.4, data from rat nitrogen balance trials of selected high-lysine barleys are given together with their chemical compositions (Tallberg & Eggum, 1981, 1986). As expected, the genotype with the highest lysine content, *lys3a* (mutant 1508, + 17.8 BV units) and its double recessive with *lys1* (+ 15.4 BV units) reached the highest BVs compared with the normal lysine control (var Tellus), while *lys1* (a Hiproly cross + 5.4 BV units) is intermediate despite a relatively high lysine content. Mutant 7 and the double recessive *lys1*/mutant 7 lines have surprisingly high responses in BV (+ 10.9 units and + 10.9–14.3 units respectively) compared with their relatively low lysine contents (+ 15–35% compared with + 48% for *lys3a*). The standard deviation in these experiments was on average 0.98 BV units and the correlation between the lysine content and BV was r = + 0.85. As pointed out by Munck (1972), reviewing experiments comparing the same barley diets supplemented with gluten, lysine and methionine fed to rats, mice and pigs, it is evident that young rats and mice immediately after weaning are much less sensitive to dietary depression of lysine than piglets at a similar developmental stage. This is also demonstrated by comparing the direct responses of *lys3a* to normal barley in BV/nitrogen retention trials with rats in Table 27.4 (Tallberg & Eggum, 1981) (123%) and with young pigs in Table 27.10 (Mortensen *et al.*, 1988) (159%). This pig trial will be discussed in more detail later.

It is doubtful if lysine is still limiting for rats and mice in, for example, feeding on a *lys3a* barley diet. Since the genetical increase in lysine is combined with an increase in other important amino acids such as threonine, methionine and valine which could be limiting in a *lys3a* barley diet for example, the biological value may still be indirectly positively correlated with the lysine content. The drastic effect on pig growth and feed efficiency of feeding the mutant 1508 barley compared with normal barley without protein additive is demonstrated in Table 27.5. Growth is increased by 243% and is approaching the growth rate of a complete feed mixture, including protein additives at a much higher level of protein and lysine in the feed. Feeding trials with pigs (Newman *et al.*, 1977; Thomke & Widströmer, 1975; Mortensen *et al.*, 1988) further confirmed the validity of the lysine increase for growth and nitrogen metabolism when feeding Hiproly and mutant 1508 barley. However, more concern began to be centred around a deficiency in energy

Table 27.5. Feed trial with pigs (Munck *et al.* 1985).

	Without protein additive[1]		With protein additive[2]
	Normal feed[3] barley	1508[3] barley	Complete feed Barley + soybean meal
Protein	10.5	11.5	15.0
Lysine, g/16 g N	3.7	5.2	5.0
Lysine, g/kg	3.8	6.0	7.5
Feed to obtain 90 kg live weight (kg)	538	209	206
Daily growth (g)	261	635	703

[1]*Ad libitum* feeding
[2]Restrictive feeding according to Danish norms
[3]Dehulled barley

digestibility which was registered for both the original shrivelled *lys1* genotype Hiproly and for the 1508 mutant compared to their isogenic counterparts as reviewed by Johnsson *et al.* (1978) and Bach Knudsen & Munck (1981). This deficiency was caused by the decrease in easily metabolized carbohydrates (starch + sugar), tentatively connected as previously discussed with the reduction in hordeins, and was especially pronounced in lines containing the *lys3a* gene from mutant 1508 (starch + sugar in normal barley 71.6%, *lys3a* 60.2%) as demonstrated in Table 27.4. Obviously the increase in acid hydrolysable fat in this mutant, from 3.7 to 5.8%, was not sufficient to compensate for the loss in potentially available energy. It was decided to investigate the digestibility in rat trials of the different botanical components of the mutant 1508 barley in more detail and compare it with its parental variety Bomi (Bach Knudsen, 1982). Digestibility for energy and protein is quite comparable between rats and pigs. A fraction from the central endosperm was isolated by decortication, milling and sieving and compared with the whole seed (Table 27.6). It is seen that the higher contents of fat, ash and hemicellulose in the whole seeds of mutant 1508 when compared with Bomi are further exaggerated when comparing on the central endosperm level. Surprisingly, the 1508 mutant contains more β-glucans in the central endosperm than Bomi, while the situation is reversed in whole seeds. The decrease of 7% in starch + sugar in whole seeds of the 1508 mutant is unchanged when comparing the central endosperm fractions. Thus, these data not only reveal great differences in metabolizable carbohydrates between Bomi and mutant 1508, but indicate a composition change in the cell wall carbohydrates and the seed structure. In the rat trials, the nitrogen true digestibility of mutant 1508 was 97% of that of Bomi for whole seeds and decreased to 93% in the central endosperm fraction, while the relative digestible energy of mutant 1508 vs. Bomi fell only from 98 to 97% from the whole grain to the central endosperm fraction. The difference in protein availability was thus increased comparing the endosperm fractions with the whole seed, while the digestible energy difference was more constant. It is seen that the relative response

Table 27.6. Chemical composition (% DM) and nutritional quality in rat trials (TD, BV, DE) of mutant 1508 expressed as relative to mother variety Bomi = 100 (Bach Knudsen, 1982).

	Whole grain		Central endosperm fraction	
	Bomi	Mutant 1508 (Bomi = 100)	Bomi	Mutant 1508 (Bomi = 100)
Protein (N × 6.25) %	11.0	**109**	8.4	**131**
Fat (%)	3.08	**165**	1.28	**268**
Ash (%)	2.21	**109**	0.87	**157**
Starch + sugar (%)	67.3	**93**	87.3	**93**
Cellulose + lignin (%)	6.82	**101**	–	–
β-glucans (%)	3.75	**65**	0.83	**181**
Hemicellulose (%)	9.56	**116**	2.13	**150**
Lysine %, g/16 g N	3.50	**146**	3.00	**153**
True Digestibility (TD) (%)	89	**97**	100	**93**
Biological Value (BV) (%)	75	**116**	64	**136**
Digestible energy (DE) (%)	80	**98**	96	**97**

in biological value between Bomi and mutant 1508 is much higher than the relative lysine increase at the endosperm level because the barleys are compared at a lower intake of lysine, further supporting the arguments put forward before questioning lysine as the first limiting amino acid in high-lysine (*lys3*) barley for rats.

As protein prices on the market were comparatively cheap compared with carbohydrates, we became increasingly concerned about the energy availability problem of high-lysine barley as expressed in our review at the 4th Barley Genetics Congress in 1981: 'Thus the advantage of mutant 1508 with regard to amino acid composition is cancelled out by its low availability of energy' (Bach Knudsen & Munck, 1981).

Formulating and testing an alternative secondary selection hypothesis in breeding for an improved genetical background for yield in *lys3* barley

The serious drawbacks of high-lysine barley genes were thus acknowledged at the end of the 1970s, and there was a conviction that the decrease in hordein content led to a low seed starch content, resulting in low energy availability in feeding, inadequate seed filling and small seeds, thus reducing yield. Confidence in high-lysine barley breeding was non-existent and most breeders abandoned their efforts. We were also near giving up but then the yields of some of our *lys3a* lines took a leap forward. In 1974/75 we had started with crosses of malting varieties with high starch and low protein content, e.g. Nordal, and experimented on how to improve screening efficiency above that of dye-binding by trying the ninhydrin and the hordein turbidity methods. After a few years, we began to realize (see review by Bang-Olsen *et al.*, 1987) that the trained eye could distinguish the *lys3a* kernel phenotype directly even without dehusking because the large scutellum (germ) changed the 'profile' of the grain. Our strategy from 1979 was then to screen exclusively for plump seeds having the large scutellum *lys3* character. Being able to grow one crop in Denmark in the summer and one in New Zealand during the winter, we crossed first with large seeded plump varieties such as Nordal, Cilla, Minerva, Elbo and Gerkra, selected for the plump recombinants and crossed back a maximum of twice to the high-yielding variety. In a pedigree type of breeding programme (Bang-Olsen *et al.*, 1991) spikes checked in F_3 were sown in rows in F_4, in small plots in F_5 and in large 10 m^2 plots in F_6. The first large-scale test with four replicates was performed in F_7.

In another generation of crosses, we crossed the best lines of the above mentioned crosses with varieties such as Triumph, Nery and Mandolin. About 50 spikes were selected for homogeneity and carefully checked for large scutellum in each generation for each pedigree line. There were large differences in combining ability of different barley varieties with the gene. Thus, about five successive crosses were made to high-yielding lines with a maximum of two to the same variety. Our breeding goals (Bang-Olsen *et al.*, 1991) were:

1. maintain the high *lys3a* protein quality;
2. improve yield and yield stability above the national control (variety Zita until 1983, and thereafter a mix of the four most high-yielding barleys);
3. improve seed quality/size and starch content;
4. improve straw length/elasticity avoiding seed losses at harvest;
5. improve disease resistance, e.g. against mildew.

The best line in our first generation of *lys3a* barley recombinants was Ca 700202 (Piggy) involving a single cross with the starch-rich malting barley Triumph, a short-statured barley with high tillering capacity, to the mutant 1508 × Nordal^2 intermediate. Its yield in numerous trials in the 1980s was found not to differ significantly from the reference variety Zita (Table 27.7). It was much more stable in yield under field condi-

Table 27.7. Yield stability of the stable high-lysine variety Ca 700202 in relation to the original mutant R 1508 and the top-yielding malting barley Triumph at Danish locations 1983–1985 (variety of comparison, Zita = 100; Bang-Olsen *et al.*, 1987).

% of Zita	1983			1984			1985			Average
	H	A	S	H	A	S	H	A	S	
Triumph	105	101	100	115	113	108	111	114	108	108
R 1508	93	79	93	82	93	88	90	92	92	87
Ca 700202 (Piggy)	97	96	103	98	95	99	96	100	97	98

H = Hyldagergård, Sealand; A = Allindemaglegaard, Sealand; S = Sejet, East Jutland.

tions than the original Risø mutant but was smaller in seed size even though seed quality and starch content were improved and the shrunkenness of the typical 1508 mutant reduced. The starch content (Fig. 27.3) approached the same level as in feed barleys such as Zita, Mona and Nery but lagged behind the very high starch content of the malting barley Triumph. However, our improved *lys3a* line Piggy was still more than 10% lower yielding than the new national standard after 1983. We therefore concluded at the 5th International Barley Genetics Symposium in 1986:

> We have fulfilled our original goal to obtain lines which are as high-yielding as the former standard Zita but still, we are only half-way in catching the ever changing national barley standard of comparison. Of primary importance, however, is the fact that we have proved that it is possible to obtain lines with stable yield and reasonable seed quality, and that our rate of improvement has been no slower than the pace of normal barley breeding with regard to yield
>
> (Bang-Olsen *et al.*, 1987).

In 1985/86, we performed a careful feed trial on two protein levels with restrictive and *ad libitum* feeding in co-operation with the National Institute of Animal Science (Mortensen *et al.*, 1988). We confirmed the findings (see Fig. 27.4) that the

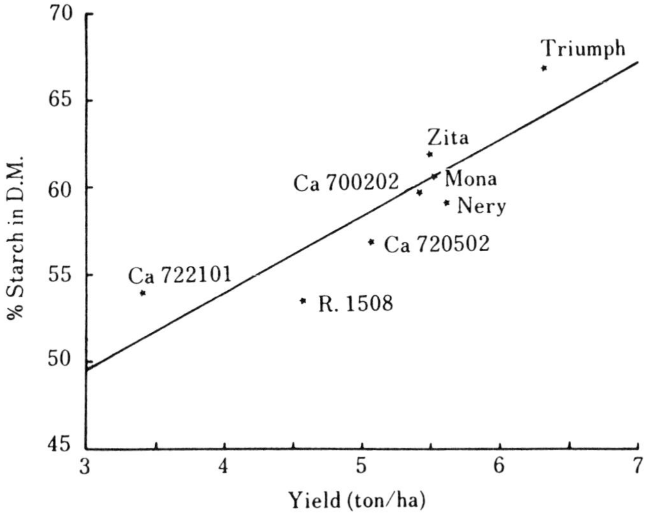

Fig. 27.3. Correlation between yield and starch content (% DM). Hyldagergård, Denmark, 1984, mean of four repetitions (Bang-Olsen *et al.*, 1987).

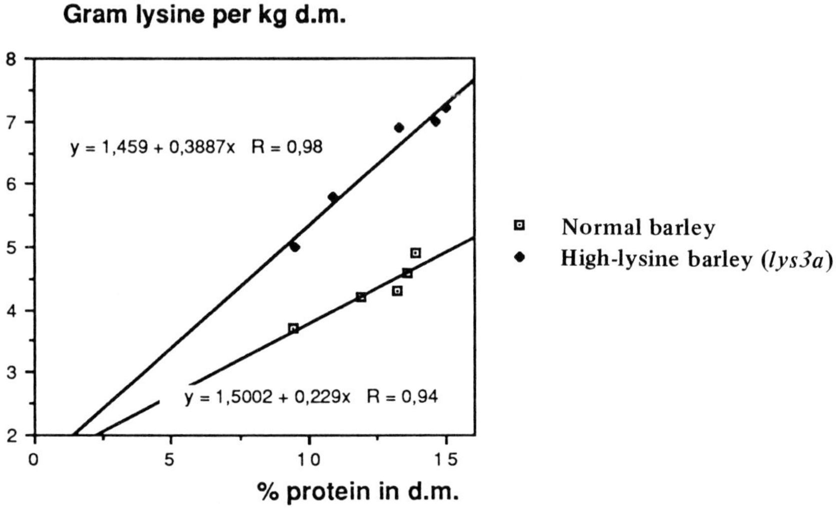

Fig. 27.4. Increase of lysine (g DM) with increase in protein (N × 6.25) content (% DM) for normal and high-lysine (*lys3*) barley (Bang-Olsen *et al.*, 1991).

advantage in lysine of *lys3* barley above that of normal barley increases appreciably with increasing protein content, and that a protein level above 11% should be aimed for either by breeding or by late nitrogen fertilizer application. By late application of urea we reached protein levels in the feed variety Lami and Piggy of 13.2 and 14.0% with lysine g/16 g N of 3.64 and 5.07, respectively (Table 27.8). Readily available carbohydrate was reduced by 6.0 absolute per cent in Piggy while fat was increased by 2.2 absolute per cent. The pig trials (Table 27.9) were arranged with seven individually fed animals on each diet arranged so that soybean meal was added to the diets to reach the same level of intake of digestible lysine according to the Danish restrictive feeding norm (lots 1 and 3). This feeding regime aims at optimizing the final carcass composition at 90 kg live slaughter weight for bacon for the English market with a maximum of meat and a minimum of fat at the expense of daily growth and prolonged feeding time. A comparison was made to *ad libitum* feeding (lots 2, 4 and 5) which is common in e.g. Germany and the USA to rear heavy pigs at 110–120 kg optimizing speed of

Table 27.8. Chemical composition (% DM) of normal high-lysine barley lines used in pig feed trials (in Tables 27.9 and 27.10).

	Protein	Fat*	Carbohydrates**	Amino acids, g/16 g N		
				Lysine	Threonine	Methionine + cysteine
Lami	13.2	3.3	62.7	3.64	3.41	3.85
Ca 700202 (*lys3a*) Piggy	14.0	5.5	57.7	5.07	4.00	3.92
Soybean meal	46.4	4.0	–	6.27	3.94	2.98

* After acid hydrolysis
** Starch + sugar

Table 27.9. Feed trial with pigs 20–90 kg with normal and high-lysine barley (Mortensen *et al.*, 1988).

	1	2	3	4	5
	Lami		Piggy *lys3a*		
Barley feeding regime	Restricted	*ad libitum*	Restricted	*ad libitum*	*ad libitum*
% Soybean meal					
- Before 50 kg	23.0	23.0	16.5	16.5	16.5
- After 50 kg	17.0	17.0	10.0	10.0	0
Soybean meal, (kg)	40.9	47.3	25.9	29.8	14.3
Barley (86% DM) (kg)	162.9	192.1	183.7	211.7	225.8
Feed units consumed	210	247	212	244	241
Daily growth (g)	684[b]	810[a]	686[b]	797[a]	788[a]
Number of feeding days to 90 kg weight	104[b]	88[a]	104[b]	89[a]	91[a]
Live weight	91.1	91.9	90.4	91.1	90.1
Carcass weight	67.3[ab]	70.1[a]	66.2[b]	70.0[a]	68.8[ab]
% Meat	53.7[a]	50.7[c]	53.3[ab]	51.6[bc]	50.3[c]

Results marked with different letters are significantly different ($P < 0.05$)
Chemical composition of the diets is described in Table 27.8

production and emphasizing a large meaty ham where excess fat can be trimmed away. The pigs all had soybean protein meal at a higher concentration at the beginning of the feeding period (20–50 kg) than in the following (50–90 kg) period. To test if Piggy alone given *ad libitum* could fulfil the protein and lysine needs in the 50–90 kg period when lysine needs are reduced, an experiment (lot 5) was added omitting totally the addition of soybean meal during this period. It is evident (Table 27.9) that in restrictive feeding (lots 1 vs 3), 15 kg soybean (− 37%) per pig could be substituted by Piggy barley leading to a consumption of 20.8 kg more barley (+ 13%) without significant changes in daily growth, feed efficiency and carcass quality.

The *ad libitum* feeding in general leads to more rapid growth (+ 12%) and a shorter feeding period (− 14%) to reach a live weight of 90 kg. Feed units consumed in *ad libitum* feeding increased, however, by 16% over restrictive feeding from 20–90 kg, while slaughtered carcass weight increased by 5.5% and the meat in the carcass decreased by 4.8% (relative percentages).

When feeding soybean meal as additive from 20–90 kg, the Piggy *ad libitum* pig (4) consumed 17.5 kg less soybean meal (− 40%) than the Lami pig (2) but ate 19.6 kg more barley. There was no significant difference in the other production parameters between the Piggy and Lami pigs in the *ad libitum* comparison (2 vs 4). In the other *ad libitum* Piggy comparison (5), omitting soybean meal between 50 and 90 kg, there was 33 kg less soybean meal (− 70%) and 33.7 more barley (+ 17.5%) consumed compared to Lami (2), with no significant change in daily growth, number of feeding days or percentage of meat in carcass compared with the other *ad libitum* regimes fed soybean meal in the 50–90 kg period (2 and 4). The carcass weight (5) was, however, significantly reduced by − 1.9% compared with Lami (2) but not with Piggy (4).

The difference in energy digestibility percentage between Lami and Piggy investigated in a special digestibility trial (Table 27.10) with pure barleys (+ minerals) was 2% as expected, but there was no difference in metabolizable energy (MJ/kg) which is the relevant production parameter.

It is evident that the production goals set

Table 27.10. Nitrogen and energy metabolism of normal and high-lysine barley for pigs (Mortensen *et al.*, 1988).

	N Digestibility %	Nitrogen retention	Energy digestibility %	Metabolizable energy MJ/kg DM
Lami	72	22	78	14.1
Piggy (Ca 700202)	69	35	76	14.0

Chemical composition of the diets is described in Table 27.8

and the resulting feeding regimes heavily influenced to what extent soybean meal could be substituted with high-lysine barley (at 14% protein) with a reduction of soybean meal varying from 37 to 70%. If the pigs were slaughtered at 110–120 kg weight instead of 90 kg in the *ad libitum* feeding regime, the percentage reduction of soybean meal when feeding the high-lysine barley would have been even higher. The reduction in available carbohydrates in *lys3* barley seems to be of minor importance in the production trials with pigs.

In order to elucidate the feasibility of high-lysine barley (*lys1* and *lys3*) breeding, a wide range of normal and high-lysine lines derived from them, 10 of each (in high-lysine barley one *lys1* and nine *lys3*), was subjected to yield trials in four localities of varying soil types in Denmark, with four replicates over 2 years (Haastrup Pedersen, 1987). The harvest was analysed for 1000 kernel weight, starch, non-starch carbohydrates (NSC) and hemicelluloses (by difference), β-glucan, crude fibre and protein (Table 27.11 and 27.12). The effect of the year in this trial on the variance (Table 27.11) of most parameters is generally less than that of locality and genotype. Grain yield variance is to a high degree influenced by locality (72.2% for normal barley and 60.6% for high-lysine barley). The genotype effect is much higher for the high-lysine (25.2%) than for the normal barleys (8.6%). A similar tendency is seen for the variance of 1000 kernel weight. The variance of starch is equally divided between the effects of locality (32.9/33.5) and genotype (36.3/31.4) with no major differences due to barley type. There is a similar pattern for NSC, however, with a greater effect of genotype in high-lysine barley (48.8%), stressing the variation in this cell wall character which is so typical of the *lys3* trait

Table 27.11. Contribution of year, 1000 kernel weight and locality to normal and high-lysine [*lys3* (n = 9) and *lys1* (n = 1)] barleys, variation in yield and chemical constituents; contribution in % of total variation (Haastrup Pedersen, 1987).

	Year		Locality		Genotype	
	Normal	High-lysine	Normal	High-lysine	Normal	High-lysine
Grain yield	6.9	0.1	72.2	60.6	8.6	25.2
1000 KW	19.8	4.0	53.4	52.4	13.3	21.9
Starch	3.3	6.8	32.9	33.5	36.3	31.4
NSC	1.8	2.8	38.2	30.6	36.5	48.8
β-Glucan	9.2	0.0	24.5	5.8	41.5	71.7
Crude fibre	26.1	13.6	5.6	12.0	48.5	54.1
Protein	0.5	1.2	61.1	59.8	4.5	6.4

NSC = non-starch carbohydrates calculated by difference
Years n = 2; Localities n = 4; Genotypes n = 2 × 10

Table 27.12. Grain yield, 1000 kernel weight, carbohydrate and protein content of 10 normal and 10 high-lysine barleys grown in 2 years and at 4 localities (Haastrup Pedersen, 1987).

Barley type	Normal $n = 80$ mean	range	High-lysine $n = 80$ mean	range
Grain yield (Hkg/ha)	51.5	39.7–65.4	44.9	23.3–59.5
1000 KW	39.4	31.4–47.5	34.2	27.6–43.7
Starch (%)	60.5	51.2–68.2	54.7	44.1–62.2
NSC (%)	23.4	16.1–28.9	27.0	21.0–32.4
β-Glucan	4.4	2.9–6.3	3.3	2.1–5.7
Crude fibre (%)	4.5	3.5–5.9	5.2	4.1–6.4
Protein (%)	11.7	8.7–17.5	12.7	9.3–18.2

NSC = non-starch carbohydrates calculated by difference

as previously discussed. This tendency is even more pronounced for the β-glucan character which is also of cell wall origin and where there is a predominant genotypic variation in both classes of barleys (41.5/71.7). The crude fibre also belonging to the cell wall complex, displays a major genotypic variance (48.5/54.1), but this is particularly dependent on the year. Finally, the variation of the protein trait is mostly confined to locality with no significant difference between normal and high-lysine barley.

In Table 27.12, the mean values and ranges of the parameters discussed in Table 27.11 are summarized. The high-lysine lines are on average 13% lower in yield, 9.6% in starch, and 25% in β-glucan, but have, on the other hand, 15.4% more NSC, 15.6% crude fibre and 8.5% protein. The ranges of variation are, however, considerable in both barley categories, which overlap. Given the genotype variance (Table 27.11) there should be positive breeding opportunities for important characters such as seed size, yield and starch in high-lysine barley breeding as indicated earlier.

In 1988, we achieved a further boost in yield in *lys3* barley, which reached, and even exceeded, the present demanding national control of a mixture of the four best varieties (Bang-Olsen *et al.*, in press). Involving crosses between Risø mutant 1508, Cilla, Mona, and Mandolin on one side, and Allis and Carla on the other, five high-yielding lines have been obtained from which we have data in 2–3 years from trials in Sealand, Denmark. Two of these lines, Ca 429202 and Ca 533601 are shown in Table 27.13 with regard to yield trials and in Table 27.14 with regard to chemical composition. The high-lysine (*lys3*) phenotypes were confirmed in amino acid analyses. In our *lys3* breeding material there was a clear genetic variability in seed size and quality as well as in starch and protein content. While our best yielding *lys3* lines were reduced in protein by about 2% (absolute %) compared to mutant 1508, the highest yielding *lys3* lines were far from being the best with regard to seed size and quality (Table 27.14). Likewise, in the experiment by Haastrup Pedersen (1987), the very high-yielding normal variety Triumph, displaying the highest starch content, and its related *lys3a* line Ca 700202 (Piggy), which was the best performing high-lysine barley line, were not the best in kernel weight. Accordingly, another *lys3* line, Ca 720502, was on the same level as several normal barley varieties in seed size but significantly lower in yield and starch content than Ca 700202. The year × locality interaction for 1000 kernel weight (not shown in Table 27.11) was 15.1% in high-lysine and 4.5% in normal barleys. Locality × genotype interaction was 4.3% and 6.4% and year × genotype interaction was 2.3 and 2.7%, respectively. The only *lys1* line, Ca 633525, in the experiment was the best in starch

Table 27.13. Agronomical characters of high-lysine *lys3a* breeding lines, mutant 1508 (*lys3a*) and Bomi in relation to national barley reference (Bang-Olsen *et al*., 1991).

		Yield 1988*	Yield 1989*	Yield 1990 I	Yield 1990 II*	1000 KW g 1988	Mildew score 1990
Reference yield[a]		(58)	(79)	(59)	(67)	(51)	
Relative yield (HKg ha^{-1})		100	100	100	100		
Variety	**Pedigree**						
Bomi	Bonus × Minerva	101	102	91	98	52	6
R-1508	Mutant in Bomi	89	83	81	84	46	6
Ca 429202	Ca 040223 × Carla	101	93	96	103	42	2
Ca 533601	Ca 040302 × Alis	–	103	105	110	43	3

* Sprayed against mildew
[a] Mixture of four varieties

Table 27.14 Seed quality and chemical analyses of high-yielding *lys3a* lines compared to controls (harvest 1989; Bang-Oslen *et al*., 1991).

	Bomi	Mutant 1508	Ca 429202	Ca 533601
Kernel size (% > 2.5 mm)	84	64	96	63
1000 KW (g)	48	42	34	32
Starch (% DM)	59	54	56	55
Fat (% DM*)	2.3	3.0	2.9	3.0
Protein (% DM)	11.4	12.9	10.8	11.1
Lysine (g/16 g N)	3.6	5.2	5.6	5.6
g lys/kg DM	4.1	6.7	6.1	6.2

* Ether extractable without acid hydrolysis

content (average 59%) among the high-lysine lines but only fourth best in yield and fifth in kernel weight. The correlation between kernel weight and yield in the total material was as high as r = + 0.79, overshadowing the correlation deviates, Triumph and Ca 700202. However, the best yielding high-lysine line, Ca 700202 (*lys3a*), was also superior in this category with regard to starch content with a mean of 57% compared to 57.5% for the normal feed barley Lami but lagging behind the superb malting barley Triumph (63.9%), confirming the data given in Fig. 27.3 from Bang-Olsen *et al*. (1987).

It can be concluded that there is a striking flexibility in our *lys3* barley breeding material to compensate for a small seed by an increased number of seeds produced per unit area. In the short term, this way to obtain high yields is more successful than improving seed size and quality, including starch content, although positive variation exists for these characters which is positively related to the yield character although not in a simple fashion. A similar capability of compensating for a small seed size by high seed production is seen in the *lys1*/mutant 7 material of Tallberg (see Table 27.4) where the reduction in yield compared with the reference variety Tellus is much less than the reduction in seed size. It is interesting to note that in high-yielding normal barley varieties, Torp (1979) found that the correlation between yield of high-lysine proteins per hectare and total seed yields was steeper than the correlation between yield of hordeins and total seed yields. In material of less adapted varieties, there was no such difference.

Although our high-yielding *lys3a* lines should be tested in a range of environments to define their adaptability, we can conclude (Bang-Olsen *et al.*, 1991) that it is possible in a breeding programme to achieve yields which are competitive with those of the best normal barleys from high-lysine crossbred lines with the *lys3* gene. There exists in the *lys3* crossbred material a variation for yield and seed quality, smooth, non-shrunken seeds and starch content, as wide as in normal barley, which responds to breeding as fast as with the normal material. Thus, using the pleiotropic large scutellum trait as a marker makes it possible to integrate high-lysine barley breeding in a normal barley breeding programme without much extra effort.

A provisional primary evaluation of the results of selection

Coming back to the quotation from N.I. Vavilov (published in 1951) in the introduction, we can conclude that a relevant primary hypothesis is the key to breeding success but if it gets too elaborate and reductionistic it tends to retard development rather than stimulate it. In the limited sense, the search for high-lysine mutants and cultivars in cereals was such a successful hypothesis because it resulted rapidly in a wide range of selections with changed amino acid composition. However, problems were encountered when trying to formulate a new secondary selection hypothesis of how to utilize these mutants in further breeding work. Much more energy was spent on static analytical research on existing mutants and high-lysine derivatives to explore the depressingly negative correlations between the high-lysine trait and yield than on a pragmatic dynamic breeding approach in the spirit of Borlaug (1968), where nature was given a chance to show its plasticity by producing new unique recombinants which could be selected from a great number of crosses with simple and effective screening methods.

Thus, scientists at the appearance of the first new high-lysine genes were more inclined to search for their own solutions to the problem by finding new mutants than to investigate the already existing mutants which nature had offered them. Biochemists (see review by Shewry *et al.*, 1981) used their experience of bacteria and suggested strategies such as overproduction of free lysine through blocking synthesis by mutation or by changing the codons of the structural genes for storage proteins from glutamine and proline to lysine (von Wettstein, 1983). An increase in the nutritious germ and aleurone tissues at the expense of the endosperm containing the low-lysine storage proteins was also suggested. The high-lysine concept, however, initiated and stimulated a wide range of molecular and biochemical work through the 1970s and 1980s which is fundamental to our present level of understanding of plant protein synthesis (see reviews by von Wettstein, 1983 and Shewry & Miflin, 1985). The conclusions from this research are, however, in most cases not sufficiently applicable to be included in plant breeding programmes. The plant breeder manipulates the genes as hidden principles (the genotype) by selection of visible traits (the phenotype) from the barley plant in the field, trying to reproduce them the next year. In doing this systematically by classifying a great number of plants with different combinations of traits, he or she can state, without using sophisticated methods, both the number of chromosome pairs in barley and the relative positions of different genes for specific traits on the chromosomes, the sequence of which can be defined by the molecular geneticist. For the latter, the DNA sequence is the visible (the phenotype) and the barley plant the hidden (the genotype) principle. Today, the plant breeder and the molecular scientist still stand quite far from each other in their evaluation of the same object, e.g. the barley plant (seed). Not only are they separated by the dynamic process of seed development (Olsen *et al.*, 1984), including tissue differentiation (Bosnes *et al.*, 1987) with the regulation and production of its components (Mundy *et al.*, 1986) they are also divided by their different conceptual approaches in synthesis versus analysis and in holism versus reductionism (Munck, 1991).

Thus, when the breeder looks upon a high-lysine gene such as *lys3a* as a simply inherited gene located on chromosome 7 with some drawbacks which could be corrected by breeding, the molecular scientist in his approach to control the plant in accordance with the previous citation from Vavilov (1951), is entrapped in all the several hundreds of pleiotropic effects caused by the gene and faces a very difficult task in telling what is primary and what is secondary. Mertz (1976) summarized this situation as the high-lysine gene syndrome, thus implying almost a pathological situation. We can here see a multitude of factors fenced in a cage of co-variance – a black hole which is swallowing up all the efforts from scientists without rendering any information on the sequence of events hidden to them. Until recently, the efforts to open up the pleiotropic cage have been in vain, both with methods of classical genetics and statistics (Olsen, 1978) and with those of plant physiology and biochemistry, reviewed by Shewry & Miflin (1985). However, recently the hypothesis of Nelson (1979) on the regulation of zein synthesis as the major lesion in *opaque*-2 maize, which shows a striking resemblance to the *lys3* gene in barley, has been verified by Schmidt *et al.* (1990) finding that the *opaque*-2 gene codes for a regulatory protein which binds to the promotor sequences upstream of the structural gene for zein. This is an inspiring beginning but we can still not judge if the regulatory protein encoded by the *opaque*-2 gene is unique for zein only, so that all the other pleiotropic effects are just secondary side-effects of this major lesion. The indication that the production of plant hormones (Skriver & Mundy, 1990) is affected e.g. in high-lysine *lys1* and *lys3a* barley (Mounla *et al.*, 1980) indicates that the case could be much more complex. It may now be possible to define and clone the *lys3* barley gene by using the *opaque*-2 gene as a probe.

However, one other lasting effect of the research on high-lysine cereals has been a deeper understanding of the functions of different proteins, some of which are controlled by the high-lysine genes either directly or indirectly. Thus, the concept from the time of the classical cereal chemists dividing the seed proteins into metabolic/catabolic, structural and storage proteins has been further advanced with the concept of antimicrobial storage proteins like protease inhibitors, protein synthesis inhibitors and enzymes which can stop fungal growth and even degrade fungal cell walls such as chitinase and (1–3)-β-glucanase (Leah *et al.*, 1991). The principle of one protein/one specific function has likewise been modified by finding a multi-purpose protein (ASI) which is an inhibitor of both a barley α-amylase and bacterial subtilisin (Mundy *et al.*, 1983; Leah & Mundy, 1989).

It is evident that the importance of reduced contents of storage proteins and starch as well as small seed size in relation to obtaining a high yield in the breeding process was overestimated while testing the high-lysine mutants and recombinants. The original Risø 1508 mutant is more shrivelled than the selected version used in our crosses, indicating that background mutations in the first crosses may have contributed to the reduced yield of the *lys3a* recombinants. After 3–5 crosses with high-yielding plump varieties, kernel quality was additionally improved along with yield indicating that it is necessary in mutation breeding to utilize intense recombination to sort out eventual deleterious background mutations.

Our final breeding results indicate that there is no apparent negative effect of the *lys3* gene on the photosynthetic capacity and the production of photosynthesis products in our most high-yielding lines. This confirms the observation by Salomonsson *et al.* (1983) that there is no shortage of the precursors to the high-molecular grain constituents during seed development in the 1508 mutant compared with Bomi. Thus, in our breeding programme, we have compensated the small seed size of the *lys3a* genotype with a high tillering capacity and many seeds per spike, so that its potential production capability could be effectively utilized. It is also clear that lines with a consistent, improved seed quality without shrivelling and with a relatively high level of starch could be obtained from our crosses, but are not the most high-yielding. In future breeding, the combination of improved seed quality and high yield should be obtainable. The seed size character seems more environmentally influenced than starch content and it has been difficult to obtain high-yielding lines with larger seeds than the original 1508 mutant which is one of the most large-seeded *lys3* cultivars. We expect that it will be difficult but possible in future breeding to combine large seed size and yield but this now seems less critical because of the proven flexibility of our breeding material to compensate for the small seed size under Danish conditions. It is interesting to note that the protein content of our improved high-yielding high-lysine lines is about 2 absolute per cent lower than the original 1508 mutant and on the same level as normal (Bomi) barley, indicating that the new *lys3* lines have a higher content of carbohydrates and a more plump seed. However, increasing protein content up to 14% under Danish conditions, either by breeding or by late fertilizer application, tend to reduce yield, partly because the energy cost for the plant to produce protein is double that of starch. Thus a target of 11–12% protein in high yielding *lys3* barley seems to be a realistic compromise.

The reduced content of easily available carbohydrates in our most high-yielding *lys3a* lines resulted in a decreased digestibility of energy of about 2% but the net energy content was unchanged or only slightly affected as demonstrated in rat (Munck *et al.*, 1985) and pig (Mortensen *et al.*, 1988) trials, partly because of the higher fat content of *lys3* barley and partly due to other effects which will be discussed later.

A surprising result, which we had not expected at the start of our programme, was that by the use of high-lysine barley or by amino acid supplementation, it is possible to lower the total protein level of the slaughter pig feed, diminishing nitrogen excretion by 10–20% without a change in production efficiency (B.O. Eggum, pers. comm.). In densely populated countries like Denmark and Holland with yearly pig populations 2–4 times that of the human population, this could imply a significant reduction in nitrogen pollution and the possibility to launch the high-lysine barley as an ecologically adopted product.

For optimum feed value, the protein content of the *lys3* variety should be increased above 11% with a late fertilizer application which should result in a better nitrogen utilization of the plant and thus less pollution. The reduction of the protein level in the feed by utilizing *lys3* barley has a positive effect by relieving the urea load on the kidney (of importance especially in lactating sows). It also exerts a positive effect on reducing the energy needs for catabolism, partly compensating for the lower content of available carbohydrates. Optimizing the feeding of high-lysine barley thus stimulates new thinking in pig feeding and nutrition. At present, low fat content and high meat percentage for bacon are of highest priority and have long been so in the Danish pig industry. This requires restrictive feeding and a relatively high feed efficiency at the expense of a longer feeding period and less productivity per m^2 in the piggery. The rising demand for meaty hams and fore-ends now focuses on the production of animals with a final weight of 110–120 kg rather than 90 kg, speeding up growth rate with less restrictive feeding which as discussed will enable a substantial increase in the use of high-lysine *lys3* barley. The needs of the final period from 50 kg onwards could be met by the *lys3* barley at 14% protein content with minerals and vitamins alone. This new trend in *ad libitum* pig feeding calls for a reduced energy content in the pig feed in order to avoid a fatty carcass, which again may relax the needs to breed for an improved content of starch in *lys3* barley.

Barley is an established crop for the production of beer, porridge, soups and bread. The latter uses are now almost abandoned but barley is still grown for direct human consumption in several developing countries. By dehulling and grinding *lys3* barley, a complete gruel food for children from 1–2 years old can be obtained after adding just 1% barley malt (to reduce viscosity and increase caloric density in cooking) plus minerals and vitamins (Pedersen *et al.*, 1989).

We can now envisage the possibility of a high-yielding high-lysine *lys3a* variety which is competitive with the highest yielding barleys and with clear nutritional benefits. One would then expect that such a barley would sell itself to the market and the industry. This is, however, not the case. The feed industry is not used to having a continuous control of the feedstuff raw materials it is using. Instead it generally goes for table values of the different feed ingredients correcting for differences between years and suppliers by just a few analytical checks. The industry is also not used to separating different qualities of barley to optimize its use. Thus, as discussed by Munck (1972)

> When the beneficial effects of the new high-lysine barley variety have been established in practice, the problem remains as to how this variety is to be distinguished and separated from other barley varieties and how the increase in the economically limiting factor is to be analysed for feed optimization. The introduction of high-lysine barleys without matching, appropriate screening techniques would probably result in an increased variability of lysine in complete feeds and would thus in turn even result in a wastage of high quality protein.

This is because the pig breeder must know the quality of the feed to adopt the optimal amounts of it in his feeding regime for the best compromise between growth velocity and meat quality.

Obviously, the feed industry market was not asked when the protein issue was formulated during the 1960s and the idea of high-lysine barley breeding was introduced.

In northern Europe, barley has traditionally been the preferred cereal for feeding pigs and other animals. This fact is lucidly expressed in the concept of the Scandinavian feed unit system for pigs equal to 1 kg of barley as discussed by Bach Knudsen & Munck (1981) for evaluating empirically the energy values of different feeds. However, as indicated by the trial in Table 27.12, the variation among commercial barley varieties in starch (which is the major contributor to metabolizable energy) amounts from 51.2 to 68.2% which is rather high for a 'standard'. While the malting barley is checked for extract (starch) in the official trials by the maltsters and brewers, there is no such evaluation for either starch (metabolizable energy) or protein (amount of essential amino acids) in the breeding and marketing of feed barley (Haastrup Pedersen, 1987). Such an evaluation will only develop when the quality control in the feed industry has been upgraded. There are now rapid non-destructive online control methods such as near infrared (NIR) analysis which could already be used today. Even though there are rapid chemical methods to evaluate lysine, e.g. with dye-binding (Munck, 1983), a high lysine (*lys3*) variety could be distinguished after some training by kernel phenotype. Lysine $g\,kg^{-1}$ could then, according to Fig. 27.4, be calculated if the protein content was estimated, e.g. with NIR analysis. However, this requires separate storage of high-lysine barley which together with the analysis input increases costs. The message back to the barley grower could either be by a payment system for quality or by a certification of high quality feed barley varieties with morphological markers through the national seed control. Before such a system is introduced, a high-yielding high-lysine barley could be used by pig producers who grow their own barley and mix their own feed. This is now how *opaque*-2 maize is marketed in the US.

Traditionally in western Europe, cereals like barley are protected relative to world market prices while protein additives like soybean meal are imported without tariff at a low world market price. Thus, the internal price (e.g. in the EC) of energy (starch) is artificially higher than the price of protein. This makes the marketing of high-lysine barley a tough undertaking. We have, in an economic model (Bach Knudsen & Munck, 1981), described how the price of soybean relative to barley and the energy contents are decisive for the competitiveness of high-lysine barley. With the present cheap soybean meal prices, the energy content plays a greater role than the lysine level. Even if the drawbacks of the lower starch content of *lys3a* barley in practical feeding tests have been overestimated and the positive effects of the improved amino acid content on the metabolism of the animal and on the environmental aspect have been underestimated, further improvements in the starch content of the high-yielding *lys3* lines are desirable for competitiveness. Because of the easily visible germ character, *lys3* barley can now be bred in conventional barley breeding trials without lysine analysis. Hopefully, high-lysine barley breeding will now be facilitated and undramatically identified as just one of the valid breeding goals. Protein is an inevitable component of barley; by far the largest part of barley produced is used for feed. Why not breed this protein to a better nutritional quality if yield and net energy content are not compromised and if the feeding of the barley could give other advantages, e.g. on environment and animal health? The fact that a similar effect could be obtained by supplementation with lysine and other amino acids made industrially should not prevent further high-lysine barley breeding. The plant breeding effort needed is not going to be a major investment but rather a continuous, consistent development along with other important traits in breeding aiming at quality combined with yield, utilizing the classical tools of breeding within a normal barley plant breeding programme.

Thus, the primary breeding hypothesis launching a high-lysine barley was found to involve unexpected options and drawbacks at practically all levels in society, in plant husbandry, animal production and utilization in food and feed industry, and in science as well as in the environment. The complexity of the implications of high-lysine breeding as an example of the two-way adaptation process between science and society seems to surprise most of the participants in this research, including the financiers. In future developments, we have to seriously improve our methodology of evaluation in the primary area for conceptual consistency, combining analysis with synthesis in a true interdisciplinary effort aiming at a much improved applicability in our efforts (for further discussion, see Munck, 1991). Facing the claims of biotechnology today, the case of high-lysine barley gives food for thought.

Acknowledgements

This paper represents the conclusion (hopefully not the final one) of working with high-lysine barley research from 1963 at the Plant Breeding Institute in Svalöv, Sweden, and from 1973–1991 at the Carlsberg Research Laboratory, Copenhagen, Denmark, together with a wide range of co-workers and friends both outside and inside these institutes in a true interdisciplinary effort. My sincere thanks to all these people, especially K. Bang-Olsen (plant breeding), B. Stilling (analyses), L. Haastrup Pedersen (agronomy), J. Mundy, R. Leah and H. Tommerup (biochemistry) at the Carlsberg Research Laboratory, and to the two institutes for their keen and generous support to make high-lysine barley and its many spin-offs a collective learning experience which I have tried to make a case for in this paper. Mrs Kirsten Kirkegaard's professional assistance in developing this manuscript is greatly acknowledged.

References

Aastrup, S. (1983) Selection and characterization of low β-glucan mutants from barley. *Carlsberg Research Communications* 48, 307–16.

Allison, M.J., Borzacki, R., Cowe, I.A. & McHale, R. (1979) Variation in barley collection for endosperm attributes that relate to malting quality. *Journal of the Institute of Brewing* 85, 86–8.

Axtell, J.D., Gebisa Ejeta & Munck, L. (1981) Sorghum nutritional quality – progress and prospects. In: *Sorghum in the Eighties*. International Crops Research Institute for the Semi-Arid Tropics, Patancheru, A.P., India, pp. 589–604.

Bach Knudsen, K.E. (1982) The nutritive value of botanically defined mill fractions of barley. 1. The protein and energy value of husk and endosperm of Bomi and the high-lysine variety M-1508. *Zeitschrift für Tierphysiologie, Tierernähring und Futtermittelkunde* 48(1–2), 90–104.

Bach Knudsen, K.E. & Munck, L. (1981) The feasibility of breeding barley for feed quality exemplified by the nutritional analysis of the botanical components of Bomi and the 1508 high lysine mutant. In: *Barley Genetics IV* (Proceedings of the 4th International Barley Genetics Symposium). University of Edinburgh, Edinburgh, UK, pp. 320–9.

Balasaraswathi, R., Køie, B. & Doll, H. (1984) The concentration and yield of hordein and some lysine-rich proteins as influenced by the *lys* gene of Hiproly barley. *Hereditas* 100, 225–31.

Bang-Olsen, K., Stilling, B. & Munck, L. (1987) Breeding for yield in high-lysine barley. In: Yasuda, S. & Konishi, T. (eds), *Barley Genetics V* (Proceedings of the 5th International Barley Genetics Symposium). Maruzen, Okayama, Japan, pp. 865–70.

Bang-Olsen, K., Stilling, B. & Munck, L. (1991) The feasibility of high-lysine barley breeding – a summary. In: Munck, L. (ed.), *Barley Genetics VI* (Proceedings of the 6th International Barley Genetics Symposium Helsingberg, Sweden). Munksgaard International, Copenhagen, Denmark, pp. 433–8.

Bansal, H.C., Srivastata, N.K., Eggum, B.O. & Metha, L.S. (1977) Nutritional evaluation of high protein genotypes of barley. *Journal of the Sciences of Food and Agriculture* 28, 157–60.

Bishop, L.R. (1928) Barley proteins. 1. The composition of quantitative estimation of barley proteins. *Journal of the Institute of Brewing* 34, 101–8.

Bjørn Petersen, P. & Munck, L. Whole crop utilization of barley, including new potential uses. In: MacGregor, A.W. & Bhatty, R.S. (eds), *AACC Monograph on Barley, Barley: Chemistry and Technology*. American Association of Cereal Chemists, St Paul, MN, USA (in press).

Borlaug, N.E. (1968) Wheat breeding and its impact on world food supply. In: *Proceedings of the 3rd International Wheat Genetics Symposium*. Butterworths, Canberra, Australia, pp. 1–36.

Bosnes, M., Harris, E., Aigeltinger, L. & Olsen, O.-A. (1987) Morphology and ultrastructure of 11 barley shrunken endosperm mutants. *Theoretical and Applied Genetics* 74, 177–87.

Di Fonzo, N. & Stanca, A.M. (1977) EMS derived barley mutants with increased lysine content. *Genetica Agraria* 31, 401–9.

Doll, H. (1976) Genetic studies of high-lysine mutants. In: *Barley Genetics III*. Verlag Karl Thiemig, München, Germany, pp. 542–6.

Doll, H. (1980) A nearly non-functional mutant allele of the storage protein locus *Hor2* in barley. *Hereditas* 93, 217–22.

Doll, H. (1983) Barley seed proteins and possibilities for their improvement. In: Gottschalk, W. & Müller, H.P. (eds), *Seed Proteins: Biochemistry, Genetics, Nutritive Value*. Martinus Nijhoff Publishers, The Hague, The Netherlands, pp. 207–23.

Doll, H. (1984) Nutritional aspects of cereal proteins and approaches to overcome their deficiences. *Philosophical Transactions of the Royal Society of London B* 304, 373–80.

Doll, H. & Køie, B. (1978) Influence of the high lysine gene from barley mutant 1508 on grain, carbohydrate and protein yield. In: *Seed Protein Improvement by Nuclear Techniques*. IAEA, Vienna, Austria, pp. 107–14.

Doll, H., Køie, B. & Eggum, B.O. (1974) Induced high-lysine mutants of barley. *Radiation Botany* 14, 73–80.

FAO/WHO (1965) Protein requirements. Report of a joint FAO/WHO Expert group. *Report Series No. 37*, WHO; Technical report series No. 301, FAO. Nutrition meetings, FAO, Rome, Italy.

Georgi, B., Niemann, E.G., Brock, R.D. & Axmann, G. (1979) Comparison of analytical techniques for seed protein and amino acid analysis: An interlaboratory comparison. In: *Seed Protein Improvement in Cereals and Grain Legumes*. International Atomic Energy Agency, Vienna, Austria, STI/PUB/496, pp. 311–39.

Gibbons, G.C. (1980) On the sequential determination of α-amylase transport and cell wall breakdown in germinating seeds of *Hordeum vulgare*. *Carlsberg Research Communications* 45, 177–84.

Haastrup Pedersen, L. (1987) Development of screening methods for evaluation of starch structure and synthesis in barley. PhD Thesis, Danish Academy of Technical Sciences, Lyngby, Denmark.

Hagberg, A. & Karlsson, K.E. (1969) Breeding for high protein and quality in barley. In: *New Approaches to Breeding for Improved Plant Protein*. IAEA, Vienna, Austria, STI/PUB/212.

Hejgaard, J. & Boisen, S. (1980) High-lysine proteins in Hiproly barley breeding: identification, nutritional significance and new screening methods. *Hereditas* 93, 311–20.

Ingversen, J., Køie, B. & Doll, H. (1973) Induced seed protein content of barley endosperm. *Experientia* 29, 1151–2.

Jenkins, G.J., Rhodes, A.P., Gill, A.A. & Hansson, P.R. (1978) The effect of irrigation and nitrogen supply on the yield and quality of protein in high lysine barleys. *Journal of the Sciences of Food and Agriculture* 30, 647–52.

Jensen, J. (1979) Location of a high-lysine gene and the DDT-resistance gene on barley chromosome 7. *Euphytica* 24, 47–56.

Johnsson, I.L., Carpenter, K.J., Hurrel, R.F., Miller, E.L. & Rhodes, A.P. (1978) Some studies of the nutritive value of high lysine barleys. *Journal of the Sciences of Food and Agriculture* 29, 127–35.

Jonassen, I. & Svendsen, I. (1982) Identification of the reactive sites in two homologous serine proteinase inhibitors isolated from barley. *Carlsberg Research Communication* 47, 199–203.

Karlsson, K.E. (1972) Linkage studies on a gene for high lysine content in Hiproly barley. *Barley Genetics Newsletter* 2, 34.

Kaul, A.K. (1973) Mutation breeding and crop protein improvement. In: *Nuclear Technique for Seed Protein Improvement*. IAEA, Vienna, Austria, STI/PUB/310, pp. 1–106.

Køie, B. & Doll, H. (1979) Protein and carbohydrate components in the Risø high-lysine barley mutants. In: *Seed Protein Improvement in Cereals and Grain Legumes*. IAEA, Vienna, Austria, pp. 205–15.

Leah, R. & Mundy, J. (1989) The bifunctional α-amylase/subtilisin inhibitor of barley: Nucleotide sequence and patterns of seed specific expression. *Plant Molecular Biology* 12, 673–82.

Leah, R., Tommerup, H., Svendsen, I. & Mundy, J. (1991) Biochemical and molecular characterization of three barley seed proteins with antifungal properties. *Journal of Biological Chemistry* 266, 1564–73.

Mertz, E.T. (1976) Case histories of existing models. In: *Genetic Improvement of Seed Proteins*. National Academy of Sciences, Washington, DC, USA, pp. 57–70.

Mortensen, H.P., Madsen, A., Hall, D.D., Munck, L., Bang-Olsen, K. & Stilling, B. (1988) Protein og aminosyrer til slagtesvin. 4. Lysinrig byg. *Beretning fra Statens Husdyrbrugsforsøg nr. 640* (in Danish).

Mossberg, R. (1969) Evaluation of protein quality and quantity by dye-binding capacity: a tool in plant breeding. In: *New Approaches to Breeding for Improved Plant Protein*. IAEA/FAO,STI/PUB 212, Vienna, Austria, pp. 151–61.

Mounla, M.A.K.H., Bangerth, F. & Stoy, V. (1980) Gibbereline-like substances and indolyl type auxins in developing grains of normal and high lysine genotypes of barley. *Physiologia Plantarum* 48, 568–73.

Munck, L. (1966) The problem of quality concerning cereal for feed with special respect to the effect of heat at different moisture contents on amino acid availability. *Journal of the Swedish Seed Association* 76, 194–212.

Munck, L. (1972) Improvement of nutritional value in cereal. *Hereditas* 72, 1–128.

Munck, L. (1978) The genetical basis for protein improvement in cereals. In: Vartanian, M.E. (ed.), *Proceedings of the XIVth Congress of Genetics*. MIR Publishers Moscow, USSR (1980), pp. 249–59.

Munck, L. (1983) Optimization of lysine composition in plant breeding programmes and in feed technology by the dye-binding analysis. In: Lásztity, R. & Hidvégi, M. (eds), *Symposium on Amino Acid Composition and Biological Value of Cereal Proteins*, International Association for Cereal Science and Technology, Vienna, Austria, pp. 325–36.

Munck, L. Man as selector – A Darwinian boomerang striking through natural selection.

In: Hansen, J.Aa. (ed.), *Environmental Concerns – A Disciplinary Exercise*. Elsevier, Amsterdam, The Netherlands, pp. 211–27.

Munck, L. On the utilization of the renewable plant resources. In: Hayward, M.D., Bosemark, N.O. & Romagosa, I. (eds), *Plant Breeding: Principles and Prospects*, Chapman and Hall, London, UK (in press a).

Munck, L. Quality criteria in the production chain from malting barley to beer – Some views on the logistic problems with a few examples (dormancy, vitality and β-glucans). *Ferment* (in press b).

Munck, L., Karlsson, K.E., Hagberg, A. & Eggum, B.O. (1970) Gene for improved nutritional value in barley seed protein. *Science* 168, 985–7.

Munck, L., Mundy, J. & Vaag, P. (1984) Characterization of enzyme inhibitors in barley and their tentative role in malting and brewing. *ASBC Journal* (1985) 43, 35–8.

Munck, L., Bang-Olsen, K. & Stilling, B. (1985) Genome adjustment by breeding to balance yield defects in high-lysine mutants of barley. In: Horn, W., Jensen, C.J., Odenbach, W. & Schieder, O. (eds), *Genetic Manipulation in Plant Breeding*. Walter de Gruyter, Berlin, Germany, pp. 49–60.

Mundy, J., Svendsen, I. & Hejgaard, J. (1983) Barley α-amylase/subtilisin inhibitor. I. Isolation and characterization. *Carlsberg Research Communications* 48, 81–90.

Mundy, J., Hejgaard, J., Hansen, A., Hallgren, L., Jørgensen, K.G. & Munck, L. (1986) Differential synthesis *in vitro* of barley aleurone and starchy endosperm proteins. *Plant Physiology* 81, 630–6.

Nelson, O.E. (1969) Genetic modification of protein quality in plants. *Advances in Agronomy* 21, 171–94.

Nelson, O.E. (1979) Inheritance of amino acid contents in cereals. In: *Seed Protein Improvement in Cereals and Grain Legumes*. IAEA, Vienna, Austria, pp. 79–86.

Newman, C.W., Eslik, R.F., Moss, B.R. & El-Negoumy, A.M. (1977) Hiproly barley as a source of protein and amino acids for growing-finishing pigs. *Nutrition Report International* 15(4), 383–8.

Olsen, O.-A. (1978) Diallel analysis of high lysine barley. *Hereditas* 88, 151–63.

Olsen, O.-A., Krekling, T., Nissen, Ø. & Ayaz, M. (1984) An ultrastructural approach to the problem of barley high-lysine gene action. In: *Cereal Protein Improvement*. IAEA, Vienna, Austria, pp. 241–57.

Oram, R.N. & Doll, H. (1981) Yield improvement in high lysine barley. *Australian Journal of Agricultural Research* 32, 425–34.

Osborne, T.B. (1928) *The Vegetable Proteins*. 2nd edn Longman, Green and Co., London, UK.

Osborne, T.B. & Mendel, L.B. (1914) Amino acids in nutrition and growth. *Journal of Biology and Biochemistry* 17, 325–49.

Pedersen, B., Hansen, M., Munck, L. & Eggum, B.O. (1989) Weaning foods with improved energy and nutrient density prepared from germinated cereals. 2. Nutritional evaluation of gruels based on barley. *Food and Nutrition Bulletin* 11, 46–52.

Persson, G. & Karlsson, K.E. (1977) Progress in breeding for improved nutritive value in barley. *Cereal Research Communications* 5(2), 169–78.

Postel, W. (1956) Der Einfluss genetischer und ökologische Faktoren auf den Eiweisshaushalt von Sommergasten Caryopsen unter besonderer Berücksichtigens der exogenen Aminosäuren. *Züchter* 26, 211–39.

Rasmussen, U. (1985) Immunological screening for specific protein content in barley seeds. *Carlsberg Research Communications* 50, 83–93.

Rasmussen, U. (1988) The application of immunofluorescence for detecting specific proteins in barley seeds. In: Linskens, H.F. & Jackson, J.F. (eds), *Modern Methods of Plant Analysis* New Series 7. Springer Verlag, Heidelberg, Germany, pp. 35–55.

Rhodes, A.P. (1975) A comparison of two rapid screening methods for the selection of high-lysine barley. *Journal of the Sciences of Food and Agriculture* 26, 1703–10.

Rose, W.C., Meyer, C.E. & McCoy, R.H. (1935) Feeding experiments with mixtures of highly purified amino acids. VIII. Isolation and identification of a new highly essential amino acid. *Journal of Biological Chemistry* 112, 283–93.

Salomonsson, A.-C., Bertholdsson, N.-O., Stoy, V., Theander, O. & Åman, P. (1983) Carbohydrate composition of and ^{14}C incorporation into developing grains of normal and high-lysine genotypes of barley. *Swedish Journal of Agricultural Research* 13, 75–89.

Schmidt, R.J., Burr, F.A., Aukerman, M.J. & Burr, B. (1990) Maize regulatory gene opaque-2 encodes a protein with a 'leucine-zipper' motif that binds to Zein DNA. *Proceedings of the National Academy of Sciences (USA)* 87, 46–50.

Shewry, P.R. & Miflin, B.J. (1985) Seed storage proteins of economically important cereals. In: Pomeranz, Y. (ed.), *Advances in Cereal Science and Technology*, Vol. VII, American Association of Cereal Chemists, St Paul, MN, USA, pp. 1–83.

Shewry, P.R., Miflin, B.J., Forde, B.G. & Bright, S.W.J. (1981) Conventional and novel approaches to the improvement of the nutritional quality of cereal and legume seeds. *Science Progress Oxford*, pp. 575–600.

Skriver, K. & Mundy, J. (1990) Gene expression in response to abscisic acid and osmotic stress. *Plant Cell* 2, 503–12.

Tallberg, A. (1981) Protein and lysine content in high-lysine, a double-recessive of barley. II. Combinations between mutant 7 and a Hiproly back cross. *Hereditas* 94, 261–8.

Tallberg, A. (1982) Characterization of high-lysine barley genotypes. *Hereditas* 96, 229–45.

Tallberg, A. & Eggum, B.O. (1981) The nutritional value of high lysine barley genotypes. *Qualitas Plantarum, Plant Foods for Human Nutrition*.

Tallberg, A. & Eggum, B.O. (1986) Grain yields and nutritional qualities of some high-lysine barley hybrids. *Journal of Cereal Science* 4, 345–52.

Thomke, S. & Widströmer, B. (1975) Nutritional evaluation of high lysine barley fed to pigs. In: Friedman, M. (ed.), *Protein, Nutritional Quality of Foods and Feeds*. Marcel Dekker, New York, USA, Part 2, pp. 79–100.

Torp, J. (1979) Relations between production of starch and percentage quality and yield of protein in barley. *Journal of Agricultural and Crop Science* 148, 367–77.

Tsai, C.Y., Huber, D.M. & Warren, H.L. (1978) Relationship of the kernel sink for N to maize productivity. *Crop Science* 18, 399–404.

Udy, D.C. (1954) Dye-binding capacities of wheat flour protein fractions. *Cereal Chemistry* 31, 389–95.

Vaag, P. & Munck, L. (1987) Immunochemical methods in cereal research and technology. *Cereal Chemistry* 64(2), 59–72.

Vasal, S.K., Villegas, E. & Bauer, R. (1979) Present status of breeding quality protein in maize. In *Seed Protein, Improvement in Cereal and Grain Legumes II*, Proceedings of the IAEA/FAO Symposium. Neuherberg, IAEA, Vienna, Austria.

Vavilov, N.I. (1951) The origin, variation, immunity and breeding of cultivated plants. Waltham, MA, USA, *Chronica Botanica* 13, 1–366.

Waterlow, J.C. (1970) Human protein requirements and malnutrition. In: Bender, A.E., Kilberg, R., Löfqvist, B. & Munck, L. (eds), *Evaluation of Novel Protein Products*. Pergamon Press, London, UK, pp. 15–22.

von Wettstein, D. (1983) Genetic engineering in the adaptation of plants to evolving human needs. *Experientia* 39(7), 687–804.

Index

abscisic acid (ABA) 271, 380, 388
  drought resistance 280
  responsive genes 280
addition lines
  wheat–barley 101, 122, 142, 144, 148
  wheat–*H. chilense* 14
Agrobacterium rhizogenes 573
Agrobacterium tumefaciens 559, 561
Agrobacterium transformation 562
agroinfection 562
albino plants 553, 554, 557
alcohol dehydrogenase 104, 282
aleurone cell walls 355
  layer 273, 299, 319, 370, 376, 381–8, 405, 418
  protoplasts 565
allergens 342
aminoethylcysteine (AEC) 194
aminotransferases
  glutamate: glyoxylate 185
  serine: glyoxylate 185, 200
aminopeptidases 379
α-amylase 105, 293, 297, 310, 374, 381, 406
  genus 273
  isoenzymes 273
  secretion 274
  types 341
α-amylase inhibitors 312, 336–42, 345–7
α-amylase/trypsin inhibitors 336–42, 345–7, 575
  activity *in vitro* 341
  genes 337
  gene expression 340
  role in defence 344
  sequences 338–9
  structure 336
β-amylase 122, 312, 321, 376
  gene expression 328
aneuploids 101, 555, 556
anther culture (androgenesis) 119, 520–37, 558
  albino progeny 527
  anonomous development 522
  donor plant conditions 523
  genotype effect 523
  incubation 526
  inoculation density 526
  in vitro development 520
  media-carbon source 524
  media-development 525
  media-gelling agent 525
  media-PGRs 525
  media-preconditioning 526
  ploidy of progeny 527
  regeneration 527
  spike/anther culture 524
  spike pretreatment 523

anthocyanin biosynthesis 122
apical meristems 550
arabinoxylans 354, 415, 417
  degradation 421
  role in malting 428
  role in nutrition 356
  structure 354
  varietal variation 355
aspartate kinane 194, 195
aspartic acid metabolism 193
autonomously replicating sequences 138
azetidine-2-carboxylic acid 199

barley yellow dwarf virus (BYDV) resistance 100, 110, 111
base composition 133
basic proteins 104
betaine 281
biochemical mutants 181
breeding with *lys3a* 586–93
  agronomy and quality 587
  objectives 586

C banding 124, 126, 136
cadmium binding proteins 283
calose synthase 482
catalase 183
carboxy peptidases 379
Calvin cycle enzymes 106
cell cycle 267
cell division 295, 296
cell extension 267
cell suspension protoplasts 553–4
cell suspensions 553–5
cell turgor 267
cell walls
  composition 414
  degradation 417–22
  metabolism 269, 413–37
  role in malting 426–9
cellulose 354
  nutritional implications 356
chitinase 470
chloroplasts
  enzymes 154
  genetic map 81
  genome 155
  mutants 153
  proteins 154
chloroplast biosynthetic pathways 167
  amino acids 169
  betaine 169
  chlorophyll 168
  fatty acids 167
  isoprenoids 168
  lipids 167
  nitrogen assimilation 169
  starch 169
chromosome maps
  chromosome 1 74
  chromosome 2 75
  chromosome 3 76
  chromosome 4 77
  chromosome 5 78
  chromosome 6 79
  chromosome 7 80
  RFLP 82
chromosome size 126
  variation 555–6
chymotrypsin inhibitors 321
  gene expression 328
clones
  cDNAs 85
  genes 88
  repetitive sequences 89
coconut milk 551
co-cultivation 562
complementation of mutants 499
CpG islands 116
cultivar selection 21
cystathionine synthase 104

debranching enzyme 377
development 231–60
  homoeotic genes 258
  molecular analysis 255–60
  mutants 247–55
  proto-oncogenes 259
development mutants 251
  addition of phytomers 254
  organ modification 254
  reduction of phytomers 254
  transitional 252
dialect 1 109, 141
dihydrodipicolinate synthase (DHPS) 194
dipeptidases 379
disease resistance genes 497
  chromosome walking 502
  cloning by complementation 499

disease resistance genes *Contd.*
  gene-for-gene hypothesis 498
  genomic subtraction 504
  identification of genes 508
  identification of products 508
  polypeptide polymorphisms 506
  plus/minus screening 503
  subtractive hybridization 505
  transposon tagging 501
distribution 19
DNA uptake 561
domestication 19
doubled haploids 119
drought response 280–1
  abscisic acid 80
  betaine 281
  polyamines 281
  proline 281

ear type 20
electrophoretic transfection 561
electroporation 564
embryo 319, 370, 381
embryonic sac 136, 143
endoplasmic reticulum 376, 389
endopolyploidy 136, 143
endosperm 136, 319, 370, 419
enzymes 373
epigenetic variation 555
esterases 27, 104, 558
ethylene 338

ferredoxin 191
flow cytometry 128
fungal pathogens 345, 442
fungal pathogen resistance 441–55
  definition 441
  genetic engineering 446
  mutation-induced 445
  sources 442
  wild *Hordeum* spp. 444
fungal pathogen resistance genes 445–55
  functions 448, 451
  partial resistance 452
  passive resistance 453
  race-specific resistance 446
  susceptibility genes 449
  Mlo resistance 450

G418 sulphate 565
gametoclonal variation 558
gene expression 215, 219, 327, 328, 344, 349, 424
genetic diversity 21
genetic engineering 35, 446, 549
genetic resources 21
genetic maps 74–82, 115, 120
genetic mapping 104
genome organization 133–48
genome size 124, 133
germination 371
gibberellins 101, 269–71, 372, 381, 383, 403–11
  action in endosperms 405
  α-amylase synthesis 406
  dwarf mutants 265, 407
  effects on growth 403
  GA-responsive mutants 403
  slender mutant 407
globoids 299
β-glucans
  effects on poultry 356
  effects on cholesterol 356
  mutants 577
  nutritional implications 356
  structure 354
  synthesis 309
β-(1–3) glucans 415–17
  and malting 426
β-(1–3) glucanases 379, 381, 429–33, 470, 488
  biological role 431
  enzymic properties 429
  gene evolution 430
  gene expression 431
  gene structure 430
β-(1–3, 1–4) glucans 299
  degradation 419–21
β-(1–3, 1–4) glucanases 379, 382, 389–91, 419–27, 488
  gene regulation in coleoptiles 426
  gene regulation in grain 424
  gene regulation in leaves 425
  gene regulation in roots 426
  gene structure 423
  genetics 422
  role in cell wall digestion 419–21
  role in malting 427
α-glucosidase (maltase) 377
β-glucosidases 379

β-glucuronidase (GUS) 561–4
glutamine synthetase 189
glutamate oxaloacetate transaminase 558
glutamate synthase 190, 200
glycine decarboxylase 185
grain composition 302–10, 369–71, 385–6
  hordein 307, 320, 357, 559
  lipids 310
  non-starch polysaccharides 309
  proteins 307, 357–8
  starch 303, 352
  sugars 352
grain development
  aleurone 299
  chalaza 292
  crease 300
  embryo 299
  endosperm 295
  fertilization 292
  inflorescence 291
  nucellus 295
  pericarp 292
  testa 294
grain respiration 310
grain structure 300

H element 139, 145
haploid initiator gene 520
haploid production 519–47
  androgenesis 520–30
  applications 535–9
  comparison of methods 531–4
  gynogenesis 519–47
  H. bulbosum method 531–4
  pseudogamy 520
haploidy, applications 535–9
  genetic engineering 538
  identification of superior hybrids 537
  linkage analysis 537
  pleiotropic genes 537
  RFLP mapping 537
  resistance to pathogens 538
heat shock response 277
  HSPs 277
  HSP90 472
heavy metal toxicity 283
herbicide resistance 170
heterochromatin 126
high-lysine breeding 573–601
  Hiproly 575
  philosophy 593–601
  Risø mutants 577
  screening 574
  selection 575
high-lysine mutants 320, 325, 359, 578–86
  composition 325, 341, 578–83
  genetics 578–83
  Hiproly 325, 578–80
  nutritional quality 583–6
  Risø ISO 580–3
  Risø 56 583–6
Hiproly 325, 575, 578–80
histogenesis 237–43
  floral 240–3
  internode/axillary bud 238
  leaf primordia 238
  shoot apex 237
homoserine dehydrogenase 194
hordeins 26, 307, 320, 357, 559
  B hordein 126–7, 559
hordein genes 325
  gene expression 327
  loci 126–7, 322–30, 358, 499
  synthesis 308
Hordeum bulbosum 51, 119, 139, 143, 444, 558, 565
  haploid production 531–4
  chromosome doubling 533
  embryo differentiation 533
  hybridization 532
  seed quality 533
  seed setting 532
Hordeum chilense 14, 51
  addition lines in wheat 14
Hordeum species 3–15, 139, 146, 444
  aneuploids 13
  breeding systems 6
  distribution 7
  domestication 8
  ecology 5
  flower structure 5
  haploidy 13
  hybrids 8
  in vitro techniques 12
  life forms 5
  malting quality 12
  resistance genes 9, 11, 14
  salt tolerance 12, 14

Hordeum species *Contd.*
  seed dispersal 6
  seed dormancy 6
  somaclonal variation 12
  taxonomy 5
  winter hardiness 11
Hordeum spontaneum 19–38, 123, 139, 144, 336, 340, 444, 559
  adaptive polymorphisms 27
  agronomic traits 29
  chloroplast DNA 27
  drought tolerance 31
  earliness 31
  enzyme polymorphisms 23
  esterases 27
  genetic diversity 27
  herbicide resistance 37
  hordein polymorphism 26
  photosynthesis 37
  population genetics 23
  protein content 32
  resistance genes 33
  salt tolerance 31, 35
husk 369
hybrids (interspecific) 45–70, 530, 557
  agronomic potential 62
  barley–wheat 45, 100
  Hordeum–Elymus 57
  Hordeum–Secale 54
  Hordeum–Triticum 47
  miscellaneous 60
  trigeneric 53
  wheat–barley 46
hydroxyproline 199
hydroxyproline-rich glycoproteins 465
hydroxypyruvate reductase 187
hypoxia response 282–3

immature embryo culture 550, 556, 558
immature inflorescence culture 552, 557
immature ovary culture 550, 555
improvement 36
indoleacetic acid 388
inhibitors
  α-amylase 312
  α-amylase–trypsin 336–42, 345–7, 575
  chymotrypsin 321, 328
  protease 361
  subtilisin 105
insect α-amylases 341
insect pests 341, 346
in situ hybridization 127, 136, 143, 503
interspecific hybrids 45–70, 530, 557
isoenzymes 104

kanamycin 560
karyotype 136
kaurene 386

landraces 21
leaf segment culture 557
leucine aminopeptidase 104
life cycle 232
linkage analysis 119, 120
lipase 380
lipids
  composition 358
  droplets 295, 299
  nutritional quality 359
  starch 310, 359
  synthesis 310
  tocophenols 359
  tocotrienols 359
  waxy barley 359
lipoxygenase 104
low temperature response 278
  enzyme activities 279
  gene expression 279
  membrane structure 279
  soluble components 279
low temperature stress 278
lysine 194, 358

macroinjection 560
maize streak virus 562
malate dehydrogenase 104
malting 371, 426
maps
  chloroplast 81
  chromosome 1 74
  chromosome 2 75
  chromosome 3 76
  chromosome 4 77
  chromosome 5 78
  chromosome 6 79
  chromosome 7 80
  RFLP 82

markers (genes) in
  chloroplast DNA 83
  chromosome 1 73
  chromosome 2 74
  chromosome 3 74
  chromosome 4 77
  chromosome 5 78
  chromosome 6 79
  chromosome 7 80
mature embryo culture 550, 555
mesocotyl culture 550
mesophyll protoplasts 553, 564
metallothionins 283
metal toxicity stress 283
methylation 133, 141, 145, 146
microinjection 563
microspore culture 119, 528–30, 562, 563, 564, 567
  media 530
  microspore isolation 529
  response 530
  spike pretreatment 529
minerals (grain)
  composition 360
  nutritional quality 360
mitochondrial DNA 559
modification 371
morphology 231–7
  culm 232
  floral apex 233
  floral axis 237
  inflorescence 235
  leaf 234
  root 232

naked barley 20
neomycin phosphotransferase 560–4
nitrate assimilation 220–2
nitrate metabolism 209–29
nitrate reductase 212–19
  genes 215
  mutants 215
nitrate transport kinetics 211
nitrate transporter induction 210
nitrate uptake 210
nitrite reductase 219–20
  genes 219
non-transcribed spacer 143
nucellus 370
nutritional quality 321, 351–62

origins of barley 19, 20
origins of replication 138
organogenesis 549
2-oxoglutarate transport 192

particle bombardment 563
pathogenesis-related (PR) proteins 467–74, 488
pentosans
  structure 354
  nutritional implications 356
pericarp 370
peroxidase 466
phenolics
  composition 361
  nutritional impact 361
phenylalanine ammonia lyase 462, 464
phosphatase 380
phosphoglycollate phosphatase 183
photorespiration 185
photosynthesis 123, 155–66, 293
  ATP synthase 166
  cytochrome *b6-f* complex 159
  enzymes 123
  light harvesting complex 165
  photosystem I 159–65
  Rubisco 166, 182
phytoalexins 462, 483
phytomer concept 243–51
  type 1 phytomers 243
  type 2 phytomers 245
  grass flowers 245
  barley 248
pollen tube transformations 564
polyethylene glycol 564–5
polymerase chain reaction (PCR) 108, 116, 141, 143, 502, 504, 506
polyploidy 555
powdery mildew infection, resistance 443, 459–94
  cell wall modification 464–7
  effects on enzymes 463
  elicitors 463
  hypersensitive response 461
  induced resistance 462
  induction of pathways 464
  infection-related cDNAs 471, 484, 488
  infection-related mRNAs 485, 489
  infection-related proteins 470, 484
  infection process 459

powdery mildew infection, resistance *Contd.*
non-host resistance 462
PR proteins 488
proline 198, 281
protease inhibitors
nutritional implications 361
types 361
proteinases (endopeptidases) 378
protein bodies 298, 322
proteins
albumins 357
globulins 357
glutelins 357
hordeins 320, 357, 559
lysine content 357
nutritional quality 358
protein Z 376
proteolysis 378
protoplasts 553–5, 564–6
transient expressions 328
transformation 564
pulsed field gel electrophoresis (PFGE) 126, 147

quantitative trait loci (QTLs) 116, 118

random amplified polymorphic DNAs (RAPDs) 108, 117, 123, 450
regeneration 549–55
relic DNA 134, 145
repeated sequences 107, 116, 124–6, 133, 138
ribosomal RNA 105
ribosomal RNA genes 140–5, 150
ribulose bisphosphate carboxylase/oxygenase (Rubisco) 166, 182
restriction fragment length polymorphisms (RFLPs) 36, 100, 104, 105, 107, 116, 142, 322, 498, 502, 505, 507, 537
maps 82

S-adenosylmethionine 194
satellite DNA 134
salt stress 281–2
scutellum 319, 387, 389, 419
seed dispersal 20
shrunken endosperm mutants 301
slender (*sl*) mutant 265, 407
starch
degradation 374
deposition 306
digestibility 353
effects on serum 354
granules 293, 296, 322
high-amylose 353
lipids 310, 359
structure 304
synthesis 304
waxy type 353
stature mutants 265
stress responses 378–85
drought 280
heat 277
heavy metals 283
hypoxia 282
low temperature 278
salt 281
somaclonal variation 549, 555–9
somatic embryogenesis 549, 553–5, 564
subtilisin inhibitors 105
sucrose synthase 473

testa 370
taumatin 561
thionins 105, 283, 342–7, 465, 489
activity *in vitro* 345
gene expression 344
genes 342
role in defence 345
types 342
threonine 194
threonine synthase 194
transformation 387, 509, 549, 559–66
transgenic tobacco 327
transient expression 327
transposable elements 128, 501
triosephosphate isomerase 183
trisaccharides 354
Tritordeum 563

wheat–barley addition lines 101, 122, 142, 144, 147, 503
wheat–barley recombinant lines 102
wheat–barley substitution lines 102
wheat–*H. chilense* addition lines 14

wheat dwarf virus 562–4

xylanases 380

yeast artificial chromosomes (YACs) 127, 503, 507